FROM NEURON TO BRAIN

FROM NEURON TO BRAIN

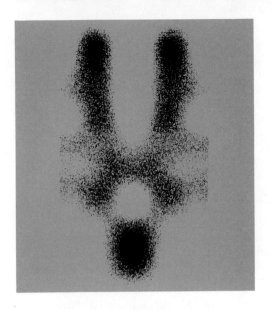

A Cellular and
Molecular Approach
to the Function of
the Nervous System

THIRD EDITION

JOHN G. NICHOLLS
Biocenter
Basel University

A. ROBERT MARTIN
University of Colorado
School of Medicine

BRUCE G. WALLACE
University of Colorado
School of Medicine

Sinauer Associates, Inc. • Publishers
Sunderland, Massachusetts   U.S.A.

Cover design by Laszlo Meszoly. The micrograph of the nicotinic acetylcholine receptor on the title page appears courtesy of Dr. Nigel Unwin.

FROM NEURON TO BRAIN:
A Cellular and Molecular Approach to the Function of the Nervous System

THIRD EDITION

Library of Congress Cataloging-in-Publication Data

Nicholls, John G.
    From neuron to brain : a cellular and molecular approach to the function of the nervous system.—3rd ed. / John G. Nicholls, A. Robert Martin, Bruce G. Wallace.
        p.   cm.
    Rev. ed. of : From neuron to brain / Stephen W. Kuffler, John G. Nicholls, A. Robert Martin. 2nd ed. c1984.
    Includes bibliographical references and index.
    ISBN 0-87893-580-0 (cloth)
    1. Neurophysiology. 2. Brain. 3. Neurons.   I. Martin, A. Robert, 1928–   . II. Wallace, Bruce G., 1947–   . III. Kuffler, Stephen W. From neuron to brain.   IV. Title.
    [DNLM: 1. Nervous System—physiology.   WL 102 N615f]
QP355.2.K83 1992
591.1'88—dc20
DNLM/DLC
for Library of Congress   92-15974
                                                                                      CIP

Printed in U.S.A.

10 9 8 7 6 5 4 3 2 1

This book is dedicated to our friend and colleague, Steve Kuffler.

Stephen W. Kuffler (1913–1980)

In a career that spanned 40 years, Stephen Kuffler made experiments on fundamental problems and laid paths for future research to follow. A feature of his work is the way in which the right problem was tackled using the right preparation. Examples are his studies on denervation, stretch receptors, efferent control, inhibition, GABA and peptides as transmitters, integration in the retina, glial cells, and the analysis of synaptic transmission. What gave papers by Stephen Kuffler a special quality were the clarity, the beautiful figures, and the underlying excitement. Moreover, he himself had done *every* experiment he described. Stephen Kuffler's work exemplified and introduced a multidisciplinary approach to the study of the nervous system. At Harvard he created the first department of neurobiology, in which he brought together people from different disciplines who developed new ways of thinking. Those who knew him remember a unique combination of tolerance, firmness, kindness, and good sense with enduring humor. He was the J. F. Enders University Professor at Harvard, and was associated with the Marine Biological Laboratory at Woods Hole. Among his many honors was his election as a foreign member of the Royal Society.

BRIEF CONTENTS

CONTENTS

PART ONE

Properties of Neurons and Glia

CHAPTER 1   ANALYSIS OF SIGNALS IN THE NERVOUS SYSTEM   1

CHAPTER 2   MEMBRANE CHANNELS AND SIGNALING   27

PART TWO

Communication between Excitable Cells

PART THREE

Development and Regeneration in the Nervous System

CHAPTER 11 NEURONAL DEVELOPMENT AND THE FORMATION OF SYNAPTIC CONNECTIONS 339

CHAPTER 12 DENERVATION AND REGENERATION OF SYNAPTIC CONNECTIONS 388

PART FOUR

Integrative Mechanisms

CHAPTER 13 LEECH AND *APLYSIA*: TWO SIMPLE NERVOUS SYSTEMS 422

CHAPTER 14 TRANSDUCTION AND PROCESSING OF SENSORY SIGNALS 467

PART FIVE

The Visual System

PART SIX

Conclusion

PREFACE TO THE
THIRD EDITION

This edition of *From Neuron to Brain* has been revised extensively. All the chapters have been rewritten, and new chapters on such topics as membrane channels, neuromodulation, sensory transduction, motor systems, and development have been introduced. To keep the book manageable in size, some chapters have been amalgamated or eliminated. Glial cells and the blood–brain barrier are now dealt with in a single chapter; pumps and ion transport, muscle spindles, and integrative mechanisms no longer constitute separate chapters. A major change has been to shift the section on the visual system toward the end. The wealth of unexpected, interesting, and detailed information discovered since the last edition appeared in 1984 has made it impractical for a student unfamiliar with the nervous system to begin by reading even a simplified account of the retina, lateral geniculate nucleus, and visual cortex.

Permeating the book are the exciting new concepts derived from experiments at the molecular level. Techniques such as patch clamp, gene cloning, expression of molecules in oocyte membranes, site-directed mutagenesis, and monoclonal antibodies have provided information that not only deepens but often simplifies the presentation of topics relating to signaling, synaptic transmission, plasticity, and development. At the same time, we have retained accounts of older work that still provide a vital framework for understanding the functions of neurons and integrative systems. For example, the Hodgkin and Huxley experiments on action potentials explain mechanisms underlying important phenomena such as threshold and conduction that are not revealed by patch clamp or molecular analysis of single channels. Similarly, for the study of neuronal architecture and integration, the old technique of Golgi staining is still essential; it has been enriched but not replaced by antibody techniques. Most chapters combine such "classic" material together with "open" avenues for research. As in previous editions, we have chosen not to provide a comprehensive account of modern neurobiology. Rather, we describe in detail those topics that provide a coherent account of how molecular and cellular approaches can be used to study the workings of the brain. We have tried to retain the flavor of the previous editions and to convey the excitement of observing single molecules as they alter their configuration to produce electrical signals, and of correlating those signals with higher functions of the nervous system such as perception. Although Stephen Kuffler's name is no longer on the title page of this edition, he was in our thoughts as we wrote. The preface to the second edition of *From Neuron to Brain* ended thus:

> The pleasure and satisfaction that we might hope to feel in recreating a book that has seemed to fill a need has been diminished by the death of our friend and colleague, Steve Kuffler. We have tried to produce a book he would not have minded keeping his name on.

J. G. Nicholls
A. R. Martin
B. G. Wallace

ACKNOWLEDGMENTS

We are grateful to the numerous colleagues who have encouraged us and influenced our thinking. We particularly thank those who have offered valuable comments on individual chapters of the book: Drs. W. Adams, D. A. Baylor, W. J. Betz, W. K. Chandler, R. Cooper, P. Drapeau, F. Fernandez de Miguel, H.-J. Freund, P. A. Fuchs, O. W. Hill, J. W. Karpen, S. R. Levinson, M. Luskin, K. J. Muller, J. M. Ritchie, W. T. Thach, M. Treherne, D. Weisblat, and W. O. Wickelgren.

We wish to thank our colleagues who kindly provided original illustrations from published and unpublished work. Original plates for this edition were kindly provided by Drs. A. J. Aguayo, W. J. Betz, J. Black, R. Boch, T. Bonhoeffer, H.-J. Freund, A. Grinvald, S. Grumbacher-Reinert, M. B. Hatten, J. E. Heuser, J. Jellies, E. A. Knudsen, M. B. Luskin, U. J. McMahan, K. J. Muller, E. Newman, M. Rayan, D. F. Ready, J. H. Rogers, M. M. Salpeter, S. Schacher, R. Seitz, D. J. Selkoe, S. J. Smith, P. N. T. Unwin, R. D. Vale, R. B. Vallee, F. Valtorta, H. Wässle, W. O. Wickelgren, W. Wisden, and S. Zeki. Plates used from the last edition of the book were provided through the courtesy of Drs. B. Boycott, M. Brightman, J. Dowling, A. Kaneko, S. LeVay, B. Nunn, and S. L. Palay.

We also thank the editors of the *Journal of Physiology*, the *Journal of Neurophysiology*, the *Journal of Neuroscience*, and *Neuron*, from which many of the illustrations were adapted.

J.G.N. wishes to express his appreciation to the Fogarty Center at NIH, Bethesda for help and support during the tenure of a Fogarty Fellowship, with particular thanks to Dr. Jack Schmidt and Ms. Sheila Feldman.

We also wish to thank Ms. Kathy Fernandez and Ms. I. Wittker for their invaluable help with the manuscript.

To Laszlo Meszoly, whose artwork for this and the preceding editions has established the style of the book; to John Woolsey, who has contributed fine new art for this edition; to Joseph Vesely and Janice Holabird, who have handled design and production matters; to our copyeditor, Gretchen Becker; and to Carol Wigg and Andy Sinauer, we owe special thanks for their skill, insight, and taste, and for making the collaboration such a pleasure.

FROM THE PREFACE
TO THE FIRST EDITION

Our aim is to describe how nerve cells go about their business of transmitting signals, how these signals are put together, and how out of this integration higher functions emerge. This book is directed to the reader who is curious about the workings of the nervous system but does not necessarily have a specialized background in biological sciences. We illustrate the main points by selected examples, preferably from work in which we have first-hand experience. This approach introduces an obvious personal bias and certain omissions.

We do not attempt a comprehensive treatment of the nervous system, complete with references and background material. Rather, we prefer a personal and therefore restricted point of view, presenting some of the advances of the past few decades by following the thread of development as it has unraveled in the hands of a relatively small number of workers. A survey of the table of contents reveals that many essential and fascinating fields have been left out: subjects like the cerebellum, the auditory system, eye movements, motor systems, and the corpus callosum, to name a few. Our only excuse is that it seems preferable to provide a coherent picture by selecting a few related topics to illustrate the usefulness of a cellular approach.

Throughout, we describe experiments on single cells or analyses of simple assemblies of neurons in a wide range of species. In several instances the analysis has now reached the molecular level, an advance that enables one to discuss some of the functional properties of nerve and muscle membranes in terms of specific molecules. Fortunately, in the brains of all animals that have been studied there is apparent a uniformity of principles for neurological signaling. Therefore, with luck, examples from a lobster or a leech will have relevance for our own nervous systems. As physiologists we must pursue that luck, because we are convinced that behind each problem that appears extraordinarily complex and insoluble there lies a simplifying principle that will lead to an unraveling of the events. For example, the human brain consists of over 10,000 million cells and many more connections that in their detail appear to defy comprehension. Such complexity is at times mistaken for randomness; yet this is not so, and we can show that the brain is constructed according to a highly ordered design, made up of relatively simple components. To perform all its functions it uses only a few signals and a stereotyped repeating pattern of activity. Therefore, a relatively small sampling of nerve cells can sometimes reveal much of the plan of the organization of connections, as in the visual system.

We also discuss "open-ended business," areas that are developing and whose direction is therefore uncertain. As one might expect, the topics cannot at present be fitted into a neat scheme. We hope, however, that they convey some of the flavor that makes research a series of adventures.

From Neuron to Brain expresses our approach as well as our aims. We work mostly on the machinery that enables neurons to function. Students who become interested in the nervous system almost always tell us that their curiosity stems from a desire to understand perception, consciousness, behavior, or other higher functions of the brain. Knowing of our preoccupation with the workings of isolated nerve cells or simple cell systems, they are frequently surprised that we ourselves started with similar motivations, and they are even more surprised that we have retained those interests. In fact, we believe we are working toward that goal (and in that respect probably do not differ from most of our colleagues and predecessors). Our book aims to substantiate this claim and, we hope, to show that we are pointed in the right direction.

S. W. Kuffler
J. G. Nicholls
Woods Hole
August 1975

THE AUTHORS

JOHN G. NICHOLLS

is Professor of Pharmacology at the Biocenter, Basel University. He was born in London in 1929 and received a medical degree from Charing Cross Hospital and a Ph.D. in Physiology from the Department of Biophysics at University College, London, where he did his research under the direction of Sir Bernard Katz. He has worked at University College, at Oxford, and at Harvard, Yale, and Stanford Medical Schools. In 1988 he was made a Fellow of the Royal Society. With Stephen Kuffler, he analyzed the physiological properties of neuroglial cells and wrote the first edition of this book. He has given courses in Neurobiology at Woods Hole and Cold Spring Harbor, and in universities in Chile, China, India, Israel, Mexico and Venezuela. For some years he has used the nervous system of the leech to study the regeneration of synaptic connections. Recently he has worked on repair of mammalian spinal cord, using the nervous system of the neonatal opossum in culture.

A. ROBERT MARTIN

is Professor and Chairman of the Department of Physiology at the University of Colorado School of Medicine. He was born in Saskatchewan in 1928 and majored in mathematics and physics at the University of Manitoba. He received a Ph.D. in Biophysics in 1955 from University College, London, where he worked on synaptic transmission in a mammalian muscle under the direction of Sir Bernard Katz. From 1955 to 1957 he did postdoctoral research in the laboratory of Herbert Jasper at the Montreal Neurological Institute, studying the behavior of single cells in the motor cortex. He has taught at McGill University, the University of Utah, Yale University, and the University of Colorado Medical Schools, and has been a visiting professor at Monash University, Edinburgh University, and the Australian National University. His research has contributed to the understanding of synaptic transmission, including the mechanisms of transmitter release, electrical coupling at synapses, and properties of postsynaptic ion channels.

BRUCE G. WALLACE

is Associate Professor of Physiology at the University of Colorado School of Medicine. He was born in Plainfield, New Jersey in 1947 and majored in biophysics at Amherst College. He received a Ph.D. in Neurobiology in 1974 from Harvard University, where he worked with Edward Kravitz on transmitter biochemistry. From 1974 to 1977 he did postdoctoral research at Stanford University with John Nicholls, studying the function and regeneration of synapses in the leech nervous system. He has taught at Stanford and the University of Colorado Medical Schools. His research on the molecular mechanisms of synapse formation includes studies done in collaboration with U. J. McMahan that led to the identification of agrin and its role in regulating the differentiation of postsynaptic specializations.

PART ONE

Properties of Neurons and Glia

CHAPTER ONE

INTRODUCTION: ANALYSIS OF SIGNALS IN THE NERVOUS SYSTEM

The brain uses stereotyped electrical signals to process all the information it receives and analyzes. The signals are symbols that do not resemble in any way the external world they represent, and it is therefore an essential task to decode their significance.

The origins of nerve fibers and their destinations within the brain determine the content of the information they transmit. Thus, fibers in the optic nerve carry visual information while signals in another type of sensory nerve—for example, one arising in the skin—convey quite different meaning. Individual neurons can encode complex information and concepts into simple electrical signals; the meaning behind these signals is derived from the specific interconnections of neurons. The signals themselves consist of potential changes produced by electrical currents flowing across the cell membranes. The currents are carried by ions such as sodium, potassium, and chloride. Neurons use only two types of signals: localized potentials and action potentials. The localized, graded potentials can spread only short distances, which are usually limited to 1 or 2 millimeters. They play an essential role at special regions, such as sensory nerve endings (where they are called receptor potentials) or at junctions between cells (where they are called synaptic potentials). The localized potentials enable individual nerve cells to perform their integrative functions and to initiate action potentials. The action potentials are regenerative impulses that are conducted rapidly over long distances without attenuation. These two types of signals are the universal language of nerve cells in all animals that have been studied. Basic principles of signaling and neuronal interconnections are briefly summarized as a preparation for the following chapters. Simple reflexes and processing of signals in the retina provide examples to illustrate how various techniques can be applied to analyze the electrical activity, the neurochemistry, and the functional architecture of neurons.

The central nervous system is an unresting assembly of cells that continually receives information, analyzes and perceives it, and makes decisions. At the same time, the brain can also take the initiative and act upon various sense organs to regulate their performance.

From neuronal signals to perception

To carry out its tasks of determining the many aspects of behavior and of controlling directly or indirectly the rest of the body, the nervous system possesses an immense number of lines of communication provided by the nerve cells (NEURONS). These cells are the fundamental units, or building blocks, of the brain. It is our task to find out the meaning behind their signaling.

That neurobiologists who study single nerve cells should be in a position to discuss higher functions of the brain, such as perception, has been an unexpected development. It has long been realized, of course, that knowledge of the cellular properties of neurons is essential for any detailed study of the brain. Nevertheless, there were no clear indications of how an understanding of membrane properties, signaling, or connections could help to explain intricate psychological phenomena such as depth perception and pattern recognition. It seemed quite possible that the workings of the cerebral cortex would still remain a mystery even when a great deal was known about signaling in individual nerve cells. Such a pessimistic view has now lost much of its force. Although new functions that are not to be found in a single cell do emerge from networks, the modeling of neural circuits becomes a sterile exercise unless it incorporates the known properties of nerve cells.

Fortunately, there are many simplifying features in the nervous system. First, the electrical signals are virtually identical in all nerve cells of the body, whether they carry messages from the periphery about painful stimuli or touch, or simply interconnect various portions of the brain. Second, signals are so similar in different animals that even a sophisticated investigator is unable to tell with certainty whether a photographic record of a nerve impulse is derived from the nerve fiber of whale, mouse, monkey, worm, tarantula, or professor. In this sense, nerve impulses can be considered to be stereotyped units. They are the universal coins for the exchange of communication in all nervous systems that have been investigated. Similarly, at SYNAPSES, the sites at which signals are handed on from one cell to the next, neurons secrete various chemical substances (the TRANSMITTERS) that are often common to different species of animals, and the chemoreceptor molecules in their surface membranes with which the transmitters interact are also highly conserved.

The neurophysiologist has learned to deal reasonably well with the initial stages of sensory signaling that eventually lead to perception, and also with the final stages of output from the nervous system, for example, movement of skeletal muscles; but problems of different scope and magnitude arise in connection with questions concerning the neural basis of perception and the initiation of "willed" movements. The complications introduced by such "higher functions" can no longer be avoided. For example, the functions of the cerebral cortex cannot be considered without reference to consciousness, perception, and volition. Many tools are now available for making a meaningful analysis of cortical mechanisms.

To account for the neural events involved in the perception of touch, we can start by recording signals from a neuron that terminates in the skin. These signals are described in greater detail below; they consist of brief electrical pulses, about 0.1 volt in amplitude, that last for about 0.001 second (1 millisecond). They move along the nerve at a speed up to 180 miles/hr (80 meters/sec). Although the impulses in a cell responsive to touch are virtually identical with those in other nerve cells, the significance and meaning are quite specific for that cell; for example, they convey to the central nervous system that a particular part of the skin has been pressed. A further important generalization, first made by Adrian,[1] is that the frequency of impulse firing in a nerve cell is a measure of the intensity of the stimulus. In the above example, the stronger the pressure applied to the skin, the higher the frequency and the better maintained the firing of the cell. As Hodgkin has written, Adrian "made what in the jargon of today (which Adrian detested) would be called a break-through."[2]

Adrian himself described the circumstances of the experiment:

'That particular day's work, I think, had all the elements that one could wish for. The new apparatus seemed to be misbehaving very badly indeed, and I suddenly found that it was behaving so well that it was opening up an entire new range of data. I'd been bogged down in a series of very unprofitable experiments and here suddenly was the prospect of getting direct evidence instead of indirect, and direct evidence about all sorts of problems which I had set aside as outside the range of the techniques that one could use . . . it didn't involve any particular hard work, or any particular intelligence on my part. It was one of those things which sometimes just happens in a laboratory if you stick apparatus together and see what results you get.'[2]

As Hodgkin says, "The comment that one wants to make about the last sentence is that when most people stick apparatus together and look around they do not make discoveries of the same importance as those of Adrian."

So far, then, there seems little difficulty in interpretation: Information is provided about (1) the modality of the stimulus by the particular type of sensory neuron it influences, (2) its location by the position and the connections of the sensory cell, and (3) its intensity by the frequency of firing. We can now go a step further and discuss simple reflexes involving two or three sequential steps. However, much less is known about the meaning of signals generated by a neuron deep within the brain, a neuron that receives its input from many cells and in turn supplies many others. Before the analysis can be started, a great deal of information is needed. Does the neuron under study handle information derived from the skin, the eye, the ear, or all three? If it is

[1]Adrian, E. D. 1946. *The Physical Background of Perception*. Clarendon Press, Oxford.
[2]Hodgkin, A. L. 1977. *Nature* 269: 543–544.

influenced by the eye, does it in turn regulate the size of the pupil, does it move the eye, or is it involved in perception of form? Or does it perhaps secrete a transmitter or hormone that profoundly influences the emotional state of the animal? A remarkable lesson, emphasized in this book, is that considerable progress can be made in understanding higher functions of the brain by correlating the activity of individual nerve cells and the transmitters they release with complex behavioral or perceptual activities.

Pattern of neuronal connections determines the meaning of electrical signals

At first it may seem surprising that the nervous system uses only stereotyped electrical messages. The signals themselves cannot be endowed with special properties because they are much the same in all nerves. The mechanisms by which signals are generated are also similar, though with interesting variants. The brain deals only with symbols of external events, symbols that do not resemble the real objects any more than the letters D O G, taken together, resemble a spotted Dalmatian. Rather, a particular set of signals must have a precise and special relation to an event.

Theoretically, there is no reason why a great deal of information could not be conveyed by an agreed-upon code made up of different frequencies. In the nervous system, however, the frequency or pattern of discharges will not do on its own as a code, for the following reason: Even though impulses and frequencies are the same in diverse cells responding to light, touch, or sound, the content of information is quite different. The quality or meaning of a signal depends on the origins and destinations of the nerve fibers, that is, on their connections. Various types of sensory modalities (light, sound, touch) are linked to different parts of the brain; even within each modality and in each area of the cortex, specific stimuli (such as lines or rectangles for the visual system) act selectively on specific populations of neurons. This organization is brought about by well-defined connections. Frequency coding is used by the nervous system to convey information only about the intensity of a stimulus. Occasional exceptions to this rule do occur, for example, in vibration sense or in the pathways for deep-pitched sounds, where the frequency of firing may follow the frequency of the source. In addition, maintained firing at different frequencies or bursts of impulse activity can exert profound effects on transmission at synapses or can alter the pattern of connections made by neurons.

It is worth pointing out that the conclusions presented here were expressed in 1868 by the German physicist–biologist Helmholtz. Starting from first principles, long before the facts as we know them were available, he reasoned:[3]

> The nerve fibers have often been compared with telegraphic wires traversing a country, and the comparison is well fitted to illustrate the striking and important peculiarity of their mode of action. In the network of telegraphs we find everywhere the same copper or iron wires carrying the same kind of movement, a stream of electricity, but producing the most different results

[3]Helmholtz, H. 1889. *Popular Scientific Lectures*. Longmans, London.

in the various stations according to the auxiliary apparatus with which they are connected. At one station the effect is the ringing of a bell, at another a signal is moved, at a third a recording instrument is set to work. . . . In short, every one of the hundred different actions which electricity is capable of producing may be called forth by a telegraphic wire laid to whatever spot we please, and it is always the same process in the wire itself which leads to these diverse consequences. . . . All the difference which is seen in the excitation of different nerves depends only upon the difference of the organs to which the nerve is united and to which it transmits the state of excitation.

An understanding of the importance of the pattern of connections is enhanced by looking at physical examples that show how information about events is inherent in connections. A computer, for example, uses stereotyped components and signals, yet performs a variety of tasks. The specialization resides in the design of the wiring. The diversity of connections, not the types of signals, increases the complexity of the tasks that can be undertaken. A further requirement that goes with complex, satisfactory computer performance is an adequate number of components; this condition is also met by the nervous system. The numbers of cells in the cortex are so great (probably more than 20,000 cells/mm^3) that they do not yet present a limitation to speculation. The brain, then, is an instrument, made of 10^{10} to 10^{12} components of rather uniform materials, that uses a few stereotyped signals. What seems so puzzling is how the proper assembly of the parts endows the instrument with the extraordinary properties that reside in the brain. Whereas it takes a brain to wire up a computer, the brain must establish and tune its own connections by itself.

The preceding sections point out some of the difficulties that enter into considerations of conscious perception. These difficulties are much reduced in the visual system, particularly when responses in cortical cells are analyzed. This is illustrated in Chapters 16 and 17, which deal mainly with experiments made by recording from single cells in the visual pathways. Although much information is available about the auditory and other sensory systems in the body, the mammalian visual system has several advantages, owing particularly to the relative technical simplicity of many experiments and their direct relevance to perception. It offers clues for an understanding of the code used by neurons to transmit not just simple information about light and darkness but also sophisticated concepts. For example, reasonable hypotheses can now be formulated in terms of neural signals and organization, relating to the following questions: What neural mechanisms can explain the recognition of shapes, such as light edges or corners of certain dimensions, positioned at one angle rather than another in the visual field? How can triangles or squares be recognized independent of their position on the retina or of their brightness and size? How can we perceive with both eyes one fused image rather than two, even though it is known that each eye really sees a somewhat different part of the world? (You don't get a sensation in the middle of your body when someone touches your two hands . . . usually.) The visual system is different

Neuronal signaling and higher functions

from a photographic plate in that it takes account of contrast or differences rather than of the absolute level of brightness; visual perception ignores information about absolute levels and can detect subtle differences even if the background illumination changes over a range of many orders of magnitude. An apparent paradox pointed out by Helmholtz[4] is

> that white paper in full moonlight is darker than black satin in daylight, but we never find any difficulty in recognizing the paper as white and the satin as black. Every painter represents a white object in shadow by means of grey pigment, and if he has correctly imitated nature, it appears pure white.

The visual system also provides one of the most favorable systems for studying fundamental questions relating to the development and maturation of the nervous system. Are the neural circuits used for perception already present at birth or are they formed as a result of visual experience? What sort of stimuli must be present in the environment to prevent sensory systems from becoming atrophied and useless? The questions are considered in Chapter 18. It is now possible to discuss these problems in cellular terms since so much is known about the hierarchy of connections in the visual system and in other sensory systems. The arrangement of these connections accounts for the almost infinite wealth of information that reaches us as we look at the world around us. This is not to deny that we still remain profoundly ignorant about higher functions such as perception. But it seems reasonable to expect that the processes that lead to perception can be analyzed by applying the same principles that govern the other functions of the nervous system.

At this stage it is convenient to summarize briefly the essential aspects of the electrical signals and the morphology of neurons. The remainder of this chapter provides a basis for the detailed descriptions in the following chapters at the cellular and molecular levels.

BACKGROUND INFORMATION ABOUT ELECTRICAL SIGNALS

Current flow in nerve cells

In order to understand the processes underlying electrical signaling, it is useful to have in mind a picture of the relevant structural components of the nerve fiber that carries the signals. The fiber, or AXON, can be considered as a tube filled with a watery solution of salts (dissociated into positively and negatively charged ions) and proteins, separated from the extracellular solution by a membrane. The solutions are of the same ionic strength but of different ionic composition, and the membrane is relatively, but not totally, impermeable to the ions present on either side. Ions move through specific channels that span the membrane. Electrical and chemical stimuli cause the various channels to open or to close.

The internal fluid, or AXOPLASM, is analogous to a copper wire and

[4]Helmholtz, H. 1962. *Physiological Optics*. P. C. Southall (ed.). Dover, New York.

the membrane to a layer of insulation around a wire, but the two systems are quantitatively quite different. First, the axoplasm is about 10^7 times worse than a metal wire as a conductor of electricity. This is because the density of charge carriers (ions) in the axoplasm is very much less than that of free electrons in a wire; and, in addition, their mobility is less. Second, movement of currents along the axon for any great distance is hampered by the fact that the membrane, although relatively impermeable to ions, is not a perfect insulator. Consequently, any current flowing along the axoplasm is gradually lost to the outside by leakage through ion channels in the membrane. Finally, the fact that nerve fibers are extremely small (usually not exceeding 20 μm in diameter in vertebrates) further limits the amount of current they can carry. Hodgkin has provided a striking illustration of the consequences these limitations have on the spread of electrical signals:[5]

> If an electrical engineer were to look at the nervous system he would see at once that signalling electrical information along the nerve fibers is a formidable problem. In our nerves the diameter of the axis cylinder varies between about 0.1 μ and 20 μ. The inside of the fiber contains ions and is a reasonably good conductor of electricity. However, the fiber is so small that its longitudinal resistance is exceedingly high. A simple calculation shows that in a 1 μ fiber containing axoplasm with a resistivity of 100 ohm·cm, the resistance per unit length is about 10^{10} ohm per cm. This means that the electrical resistance of a meter's length of small nerve fiber is about the same as that of 10^{10} miles of 22-gauge copper wire, the distance being roughly ten times that between the earth and the planet Saturn.

Passively conducted electrical signals, then, are severely reduced, or *attenuated*, over a relatively short length of nerve fiber. In addition, when such signals are brief, their time course may be severely distorted and their amplitude further attenuated by the electrical capacitance of the cell membrane. The properties of membrane resistance, capacitance, and axoplasm resistance will be discussed in detail in Chapter 5.

Many recordings of electrical activity within the central nervous system have been made with extracellular electrodes placed close to a nerve cell. The electrode itself can be either a fine wire insulated almost to its tip or a capillary tube filled with salt solution. Figure 1 shows the arrangement for extracellular recording. This technique supplies information about whether a cell is firing or is quiescent and about whether the rate of firing is increasing or decreasing. With care, the signals from a single neuron can be identified.

Recording techniques for monitoring electrical activity in nerve cells

Intracellular recording is used to obtain information about the processes of excitation and inhibition and the mechanisms that initiate nerve impulses. The tip of a microelectrode is inserted into the cell, with the aid of a micromanipulator (Figure 2). The electrode is a fine glass capillary with a tip 0.1 μm in diameter or smaller and is completely filled with a salt solution such as 4 molar potassium acetate. The micro-

[5]Hodgkin, A. L. 1964. *The Conduction of the Nervous Impulse*. Liverpool University Press, Liverpool.

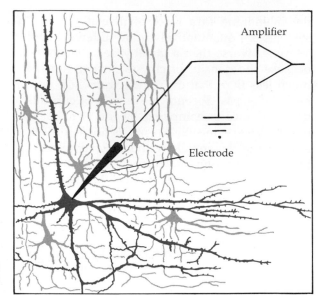

1

EXTRACELLULAR RECORDING with a fine wire electrode. The tip is located close to a nerve cell in the cortex.

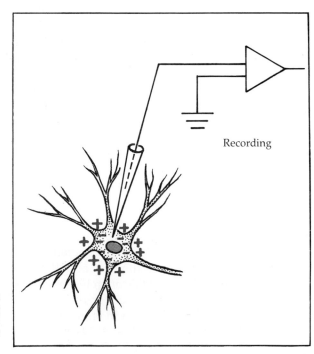

2

INTRACELLULAR RECORDING. The tip of a microelectrode has been inserted into a nerve cell. In a neuron that is at rest, there is a potential difference of about 70 mV; the inside is negative with respect to the outside.

electrode measures the potential difference between the inside and the outside of the cell without causing damage. Intracellular electrodes can also be used for passing electrical currents or for injecting molecules into the cytoplasm.

Measurements of electrical activity can also be made without electrodes, by the use of optical recording. In brief, certain dyes that bind to cell membranes act as sensitive indicators of membrane potential; depolarization causes a change in the absorbance of light, producing signals that can be measured with photodetectors (see Chapter 16). Patch clamp recording, which provides a direct measurement of events occurring at the level of single channels, is described in Chapter 2.

An important simplification is that the nervous system makes use of only two kinds of electrical signals. The first kind are graded, passive, or LOCALIZED POTENTIALS, which are subject to the attenuation and distortion discussed earlier; the second kind are impulses, or ACTION POTENTIALS, which involve active processes and can travel rapidly and without distortion from one end of a nerve to another.

Types of signals

The main characteristic of local potentials is that they can be graded continuously in size. In sensory endings, such local potentials are known as generator potentials or RECEPTOR POTENTIALS. In a sensory nerve ending sensitive to pressure on the skin, for example, the size of the receptor potential increases in relation to the magnitude of the applied pressure. There are many types of such endings or receptors, each responsive to one type of physical stimulus such as the bending of a hair, changes in temperature, changes in angle of a joint, or (in the retina of the eye) light. The job of such receptors is to change, or transduce, the physical stimulus into a receptor potential that then can be processed further by the nerve cell so that information about the stimulus eventually reaches the central nervous system. Other types of local potentials occur at synapses, where they are known as POST-SYNAPTIC POTENTIALS or more simply as SYNAPTIC POTENTIALS.

Synaptic potentials can be EXCITATORY (if they tend to give rise to nerve impulses) or INHIBITORY (if they tend to counteract excitation). (For a summary of nomenclature, see Box 1.) The size of a synaptic potential can also be graded and is a reflection of the number and rate of activity of excitatory or inhibitory presynaptic nerve terminals giving rise to it.

In contrast to local potentials, the action potential is a brief event that travels unattenuated along an axon. Although the action potential appears just about the same in an optic nerve, in an auditory nerve, or in a sensory fiber carrying information from the big toe to the spinal cord, what does vary is its speed of propagation: Speed of propagation is greater in large axons than in smaller ones. Initiation of the nerve impulse and its mechanisms of propagation are essentially the same in neurons from a wide range of invertebrates and vertebrates, as are the mechanisms underlying receptor potentials and postsynaptic potentials.

BOX 1   A NOTE ON NOMENCLATURE

The localized changes in membrane potential that depend on the passive electrical properties of the membrane are all basically similar, but they have been given a variety of names de-pending on the mechanism that generates the potential change, its effect, and the site at which it occurs. Defined here are some commonly used terms.

Designation	Mechanism
ELECTROTONIC POTENTIAL	Brought about by passing current through electrodes
SYNAPTIC POTENTIAL Often abbreviated as epsp and ipsp (excitatory and inhibitory postsynaptic potentials); in muscle the excitatory postsynaptic potential is called an end plate potential (epp)	Brought about by chemical transmitters at synapses
RECEPTOR POTENTIAL	In sensory receptors; brought about by an adequate stimulus

Thus the combination of local potentials and propagated action potentials constitutes the universal language of all known nervous systems.

Properties of local and action potentials

Intracellular recordings of local and action potentials are shown in Figures 3 and 4. When the microelectrode and the indifferent electrode are in the extracellular fluid, no potential difference is recorded; as the microelectrode is pushed through the cell membrane, a sudden jump of about 80 mV in the negative direction is registered on the oscilloscope. This potential difference is called the RESTING POTENTIAL. DE-POLARIZATION is a reduction in the magnitude of the resting potential (toward zero); HYPERPOLARIZATION is an increase in magnitude. In many of the figures in this book, potential is plotted against time, as on an oscilloscope trace, and the potentials represent those at the microelectrode tip, recorded with respect to the indifferent electrode in the extracellular solution. Negativity is shown in the downward direction.

In the experiment illustrated in Figure 3, a second microelectrode is inserted near the first and is used to pass current through the membrane. In addition, yet another electrode is placed in the axon about 1 mm away to record events at some distance from the point of current injection. The illustration shows the effect of passing hyperpolarizing or depolarizing current pulses into the axon. The local potentials produced by the current pulses have the following properties:

1. They are graded. The amplitudes of the voltage changes at both the near and the distant electrodes increase with an increase in the amount of injected current.
2. The duration of the potential changes also varies with the duration of the current pulse, but the potentials rise and fall more slowly because of the capacitance of the membrane (Chapter 5).

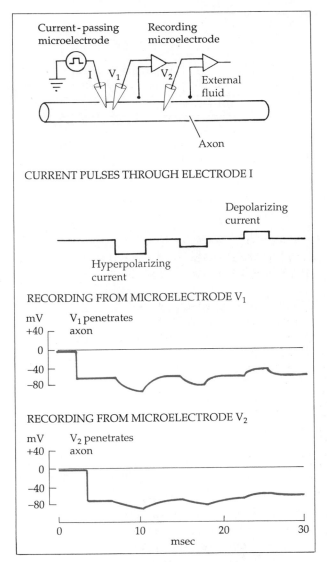

3

INTRACELLULAR RECORDING from a large axon with microelectrodes. One electrode (V_1) is inserted into the axon and records a resting potential of -70 mV (inside negative with respect to the outside). A second electrode (I), next to V_1, is used to pass pulses of current that produce localized graded potentials. The first two are hyperpolarizations and the third is a depolarization. Electrode V_2, about 1 mm away from V_1, also measures a resting potential of -70 mV when it penetrates the axon, but the localized graded potentials are smaller and slower than at V_1, owing to the passive electrical properties of the axon.

3. At the distant electrode, the change in potential is much smaller than at the point of current injection, and it rises and falls more slowly. At greater distances (several millimeters), little or no potential change is observed as a result of the current injection. Localized, graded potentials of this type, which depend on the passive properties of the membrane, are therefore useless for long-range signaling.

Action potentials are initiated by increasing the amount of depolarizing current we inject into the axon. The results of such depolarization are shown in Figure 4. In this part of the experiment, the second voltage-recording electrode (V_2) has been moved a greater distance from the point of current injection than previously—say, 2 cm rather than 2 mm.

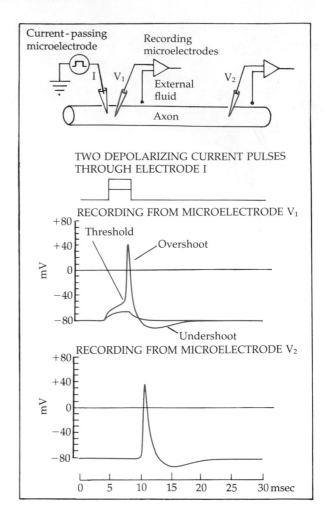

4

ACTION POTENTIALS recorded from a large axon by means of intracellular electrodes V_1 and V_2. The resting potential is −70 mV (inside negative); each trace shows two successive sweeps of the oscilloscope beam. During one sweep, a relatively small depolarizing current pulse is passed through microelectrode I, causing a small localized potential (recorded in V_1) that does not reach threshold. The second current pulse is larger, giving rise to a depolarization that reaches threshold and initiates an action potential that propagates rapidly along the axon and reaches V_2 at a distance of 2 cm. Unlike the graded localized potential, which cannot spread more than 1 to 2 mm and is therefore not recorded at V_2, the action potential is the same all along the axon.

As before, a modest amount of depolarizing current injected into the axon produces a local potential that is recorded by the nearby voltage electrode. Now, however, the distant voltage-recording electrode sees nothing because it is well outside the range of spread of the electrical potential. When we inject more depolarizing current into the axon, a totally new type of event occurs: The membrane depolarizes rapidly so that the inside of the fiber becomes transiently positive and then returns rapidly to near its resting level. This event is the action potential, and it occurs whenever the nerve cell membrane is depolarized to a certain critical level, called the THRESHOLD. Once the threshold is reached, the response is automatic and bears no relation to the form of the original stimulus. In this sense, action potentials are "all-or-none" events.

The reversal of the membrane potential beyond 0 mV is called the OVERSHOOT. In most nerve cells, the action potential is terminated by a

brief hyperpolarization, called the UNDERSHOOT (Figure 4). A most remarkable property of the action potential is that it is propagated along
the axon and arrives at the distant voltage-recording electrode unaltered
in size and form. The properties of the action potential may be summarized as follows:

1. The action potential is a triggered, explosive, all-or-nothing event. It has
   a distinct threshold, and once initiated, its amplitude and duration are
   not determined by the amplitude and duration of the initiating event:
   Larger currents do not give rise to larger action potentials, and currents
   of longer duration do not prolong the action potential.
2. The entire action potential sequence must be completed before another
   action potential can be initiated. After each action potential, there is a
   period of enforced silence (the REFRACTORY PERIOD) during which a second impulse cannot be initiated. If depolarization of the nerve beyond
   threshold outlasts the refractory period, a second action potential may
   be initiated. In many neurons, prolonged depolarization may produce a
   train of action potentials that lasts as long as the depolarization. The
   frequency of the repeated action potentials is limited by the refractory
   period.
3. The action potential is propagated along the axon and does not decline
   with distance. In mammals, the fastest action potentials travel in the
   largest fibers at a rate of about 120 m/sec (430 km/hr, or 270 miles/hr) and
   are therefore capable of conveying information rapidly over a long distance. The mechanism whereby the action potential travels along the
   axon is discussed in detail in Chapter 5 and in Chapter 8 it is shown that
   important changes in action potential shape can occur.

The use made by the nervous system of local potentials and action
potentials can be illustrated in a simple form by the STRETCH REFLEX. A
familiar reflex of this nature is the knee jerk, which is initiated by
tapping the patellar tendon below the knee. (Doctors use this reflex to
test for syphilis and to play for time.) The tap on the tendon stretches
a group of muscles that extend the leg, and, as a result of this stretch,
they undergo a reflex contraction. As only two types of neuron are
involved in the reflex, it provides a simple example of the ways in
which the nervous system performs its tasks. The functional properties
of the sensory cells, the motor cells, and their synaptic connections and
integrative mechanisms are fully described in Chapters 14 and 15.

Signals used in a simple reflex

The structural elements that subserve the stretch reflex are shown
schematically in Figure 5. The first to be involved is a sensory neuron
that has its cell body in a dorsal root ganglion near the spinal cord. In
the periphery its sensory ending is in intimate contact with a specialized
structure called the MUSCLE SPINDLE. Muscle spindles lie within the
mass of the muscle and respond to muscle stretch by producing a
receptor potential in the sensory nerve ending. The sensory neuron
sends a fiber centrally to form synapses on many other cells in the
spinal cord. Among such connections are excitatory synapses on motoneurons supplying the same muscle. These motoneurons, then, are

Neurons involved in a stretch reflex

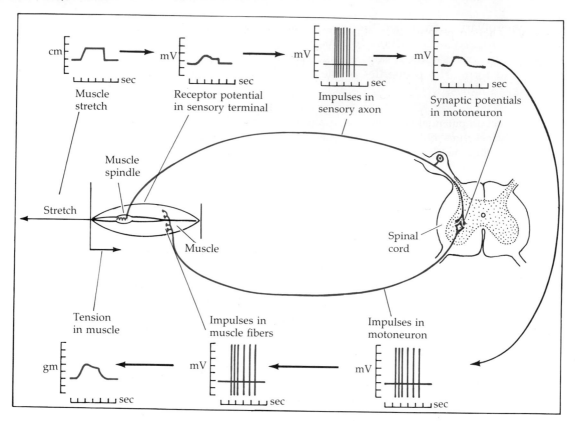

**5   STRUCTURES AND SIGNALS INVOLVED IN THE STRETCH REFLEX. Some
important structural and functional features are omitted for clarity and will be
described in detail in later chapters.**

the second type of neuron involved in the reflex; they complete the
reflex arc by forming excitatory synapses on the muscle fibers. This
reflex connection is another example of the remarkable specificity of
neural connections to be discussed below. During development, the
sensory fiber selects not just any motoneurons for its excitatory reflex
connections but only those supplying the same muscles. This orderly
anatomical arrangement is, of course, a prerequisite for the functioning
of the reflex.

Figure 5 also depicts sequentially the excitatory events occurring
during the reflex. The first event is, of course, the production of a
receptor potential in the sensory ending as a result of the muscle stretch.
The size and duration of the receptor potential reflect the intensity and
duration of the applied stretch (in this case, the magnitude and duration
of the tap to the tendon). The receptor potential itself is confined to the
first millimeter of the sensory terminal but is of sufficient amplitude to
depolarize the adjacent nerve fiber beyond threshold and initiate a series
of action potentials (measured by extracellular electrodes). The fre-

quency of action potentials in the train is related to the amplitude of the receptor potential (and hence to the intensity of the stretch) and limited, as noted previously, by the refractory period. The duration of the train reflects the duration of the stretch. This train of action potentials, then, is the means whereby information about the stretch reaches the central nervous system; intensity is coded by action potential frequency within the train and duration by the length of the train. The action potentials travel from the muscle toward the spinal cord at a velocity of about 120 m/sec and arrive at the synapses formed on the motoneurons by the central terminals of the sensory nerve. Here each impulse releases from each terminal a small quantity of neurotransmitter, which causes a depolarizing synaptic potential in the motoneuron. The duration of each synaptic potential is relatively long, so that the train of excitatory postsynaptic potentials produced by the train of presynaptic action potential results in a prolonged, relatively smooth depolarization of the motoneuron resembling the receptor potential that initiated the sequence (Figure 5). This depolarization of the motoneuron in turn initiates a train of action potentials that travel rapidly out again to the nerve–muscle synapses to produce depolarizing synaptic potentials (end plate potentials) in the muscle fibers. Here again the synaptic depolarizations produce a train of action potentials in the muscle fiber; these cause a muscle contraction. This whole sequence of events (receptor potential → sensory nerve impulses → synaptic potentials in motoneuron → end plate potentials in muscle → muscle impulses → muscle contraction) is rapid, taking less than 50 milliseconds in humans, with a significant fraction of the time occupied by the last step, namely, initiation of the contractile process in the muscle. The reflex is therefore adequate for speedy adjustments in muscle tension, for example, to maintain a desired posture while riding a bicycle or to blast space invaders in a video game.

The sensory fibers subserving the stretch reflex constitute only a very small fraction of the synaptic inputs to a motoneuron. Thousands of other neurons converge on a motoneuron to form many thousands of synaptic connections—some excitatory, some inhibitory—and the stretch reflex can be overruled in many ways. For example, a pin stuck in the toe will cause the knee to bend—a reaction opposite that of the stretch reflex. This is because a painful stimulus to the foot causes contraction of other muscle groups, which produce bending of the knee, and at the same time gives rise to inhibition of the motoneurons responsible for the knee jerk. The process by which the myriad of excitatory and inhibitory influences on a neuron are sorted out was called by Sherrington *the integrating action of neurons.*[6] Integration at the cellular level is simply the way in which action potentials in fibers converging on a neuron become converted into postsynaptic potentials, the sum

How does a neuron take account of different converging influences?

[6]Sherrington, C. S. 1906. *The Integrative Action of the Nervous System*, 1961 Ed. Yale University Press, New Haven.

total of which determines its firing pattern. This firing pattern, then, is the synthesis of all the various inputs. Integration will be discussed in more detail in Chapter 15. The point is that in all the complex activities of the nervous system, only two basic types of signals are used to convey the abstractions of the surrounding world and to implement actions. Integration by the motoneuron, which adds up excitation and inhibition and then fires one or more impulses or remains silent, is strikingly similar to integration by the nervous system as a whole. The cell and the brain both decide whether or not to act on the basis of information received from a wide variety of sources.

Many of these principles we owe to Sherrington, who discovered them through recording tension in skeletal muscle by the stretch reflex before electrical recording from individual nerve cells was possible. The following quotation is still a useful, concise description of different neural signals.[7]

> The nerve nets are patterned networks of threads. The human brain is a vast example, offering immense numbers of determinate paths, and numbers of junctional points. At these latter the travelling signal, so to speak, hesitates and sets up a local gradable state which may have to accumulate before transmitting further, or indeed may there subside and fail. These junctional points are often convergent points for lines from several directions. Arrived there signals convergent from several lines may coalesce and thus reinforce each other's exciting power.
>
> At such points too appears a process which instead of exciting, quells and precludes excitation. This inhibition, like its opposite process, excitation, does not travel. It is evoked, however, by travelling signals not distinguishable from those which call forth excitement. The travelling signals calling up excitement and those calling up inhibition never, however, reach the nodal point by the same path, never have paths in common. [*This is no longer considered completely true; there are exceptions.*]
>
> The two are relatively antagonistic. Each can be neutralized gradually by a dosage of the other. The inhibition may be a temporary stabilization of the membrane at the nodal point, which is potentially a relay station. The inhibitory stabilization produced by a travelling signal is evanescent; a train of signals is required to maintain it. While it lasts, the nodal point is blocked to signals, or only transmits them slowly.
>
> These two opposed processes, excitation and inhibition, cooperate at nodal point after nodal point in the nerve-circuits. Their joint operation at any moment settles what will be the conduction pattern, and so the motor outcome, of the signalling going forward to the brain.

BACKGROUND INFORMATION ABOUT NEURONAL STRUCTURE

For an easier understanding of the material in this book, we present in the following section a few basic facts about the structure of neurons and their interconnections. Some key terms and definitions are sum-

---

[7]Sherrington, C. S. 1933. *The Brain and Its Mechanism.* Cambridge University Press, London.

marized at the end of this chapter. Major structures and pathways of the brain are shown in Appendix C.

The shape of a neuron, as well as information about its position, origin, and destination in the neural network, supplies valuable clues to its function. For example, the arborization of a neuron provides a notion of how many connections a cell can accommodate and to how many sites it sends its own processes.

Shapes and connections of neurons

In practice it is difficult to find out about the configuration of neurons because they are so densely packed. Early anatomists had to tease nervous tissue apart to see individual neurons. Figure 6 shows a spinal motoneuron dissected and drawn by Deiters more than 100 years ago. Staining methods that impregnate all neurons are virtually useless for investigating cell shapes and connections because a structure like the cortex appears as a dark blur of intertwined cells and processes. Many of the pictures in Figures 6 and 8 were made with the Golgi staining method, which has become an essential tool because by some unknown mechanism it stains just a few random neurons out of the whole population. Furthermore, the technique tends to stain individual cells in their entirety.

The illustrations in Figure 6 are based mainly on the work of Ramón y Cajal, done before the turn of the century. Ramón y Cajal was one of the greatest students of the nervous system, selecting samples from a wide range of the animal kingdom with an almost unfailing instinct for the essential.[8] The illustrations show several distinct cell types, some relatively simple, such as the motor neuron, others with a highly complex arborization. Cytology has demonstrated that what appears at first sight to be a staggering array of shapes and processes can be divided into meaningful groupings; thus, cells can be recognized and classified in much the same ways as trees can. Although differences within a group can be considerable, one can distinguish a spinal motor neuron from a pyramidal cell, just as one can tell a birch tree from a palm tree.

Ramón y Cajal, about 1914

In recent years, selectivity of staining has been increased by a number of new techniques that have enabled single cells or cells with common properties to be marked. For example, cells can be injected with a fluorescent dye such as Lucifer Yellow, or with a metal such as cobalt, or with the enzyme horseradish peroxidase. These methods enable the investigator to obtain the entire outline and geometry of cells from which recordings have been made and whose physiological performance are known. Moreover, after horseradish peroxidase injection, the detailed morphology of the cell and its contacts can be seen by electron microscopy. Remarkably good agreement exists between the principal features of cell structure and organization first described by Ramón y Cajal and those revealed by these newer techniques.

A further important technique for labeling cells with distinctive properties is to stain them selectively with antibodies. The transmitters

[8]Ramón y Cajal, S. 1955. *Histologie du Système Nerveux*, Vol. II. C.S.I.C., Madrid.

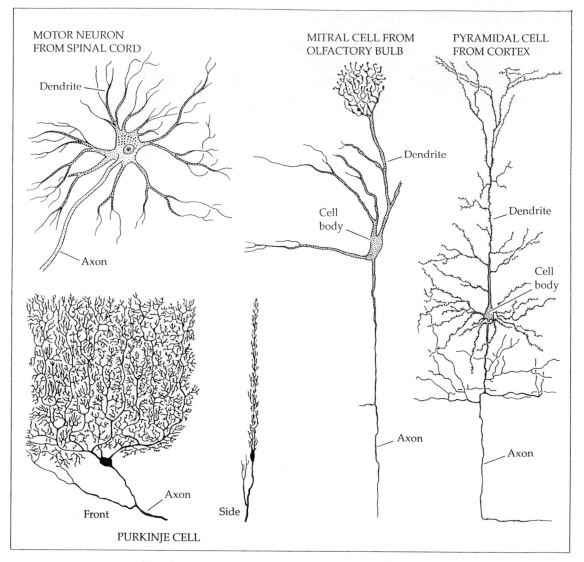

MOTOR NEURON
FROM SPINAL CORD

MITRAL CELL FROM
OLFACTORY BULB

PYRAMIDAL CELL
FROM CORTEX

Dendrite

Axon

Dendrite

Cell
body

Axon

Dendrite

Cell
body

Axon

Front   Side   Axon

PURKINJE CELL

6   **SHAPES AND SIZES OF NEURONS.** The cells have processes, the dendrites, upon which other neurons form synapses. Each cell in turn makes connections with other neurons. The motor neuron, drawn by Deiters in 1869, was dissected from a mammalian spinal cord. The other cells, stained by the Golgi method, were drawn by Ramón y Cajal. The pyramidal cell is from the cortex of a mouse, the mitral cell from the olfactory bulb (a relay station in the pathway concerned with smell) of a rat, and the Purkinje cell from human cerebellum.

within the terminals of a nerve cell as well as synthetic enzymes can often be identified in this way. Antibodies can be used to highlight neurons displaying characteristic membrane-bound or intracellular epitopes. Other techniques can be used to follow changes in gene expression as a neuron develops or acquires new properties.

Methods for tracing the entire course taken by axons from their cell bodies to their final destination are also available. For example, the terminals of axons in the central nervous system or in a muscle take up horseradish peroxidase that has been injected extracellularly in their vicinity. RETROGRADE transport of the enzyme along the axon carries it to the cell body, which can then be stained and identified in another, perhaps distant, region of the central nervous system. A neuron that has axons distributed to two separate regions of the brain can be doubly labeled by retrograde transport. Two markers with different colors can be injected, each into a different area containing the terminals. This technique enables one to distinguish which individual nerve cells are doubly stained and supply both areas. ANTEROGRADE transport from the cell body to the terminals can also be used to define cells and their pathways. Amino acids that are taken up by the cell body, converted into protein, and shipped down the axon have proved particularly valuable. In the visual system, transsynaptic transfer occurs from one cell to the next along the pathway.

Various methods that do not use electrodes are now available for detecting individual cells or populations of neurons that are or have been active. Optical recording methods, with or without fluorescent dyes, can monitor signals in single cells or in large numbers of cells by measuring small changes in the emission or absorbance of light at a particular wavelength. Nuclear magnetic resonance and positron emission tomography use computer-assisted reconstructions to register activity deep within the brain. Those cells in a particular region of the nervous system that respond to a particular stimulus can also be distinguished from those that do not by a biochemical technique using a radioactive analogue of glucose (2-deoxyglucose). In principle, the active cells can take up the radioactive glucose but cannot metabolize it; the radioactivity is then detected in those active cells. Examples of these techniques are shown in Chapters 17 and 18.

At the level of the electron microscope, the complexity of the nervous system seems at first glance insoluble. Yet orderliness becomes apparent once again by following processes through successive sections, by selectively staining cells, and by correlating the appearance with that seen in light micrographs. Figure 7 is a section through the spinal cord and shows several presynaptic nerve terminals apposed to a motoneuron. The synapses, sites at which information is transferred from cell to cell, appear as well-defined structures. The presynaptic terminals contain numerous vesicles close to the cell membrane. Separating the membranes of the two cells is a cleft that is filled with extracellular fluid and is somewhat wider than clefts elsewhere. Dense material is often seen in the cleft and on the two membranes. Many aspects of synaptic transmission are now well enough understood for the morphology to be correlated with the functional behavior observed by recording electrically.

The presynaptic terminals that release transmitter can themselves

Structure of synapses

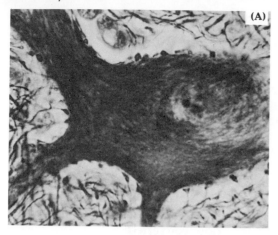

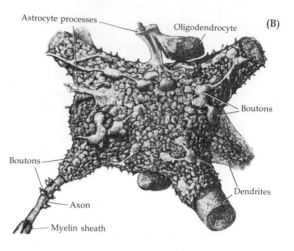

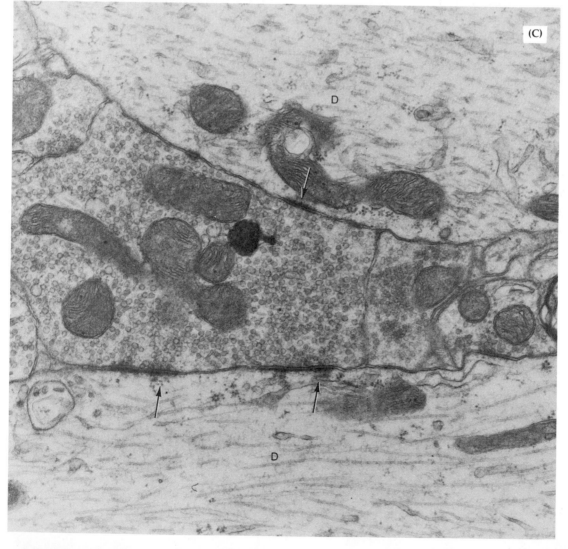

◄ **7** SYNAPSES ON MAMMALIAN MOTONEURONS. (A) Silver-stained sections of a spinal motoneuron. Axons and their terminal synaptic boutons converging on the dendrites and cell body are stained black. Their diameters are several micrometers. (B) Drawing based on an electron microscopic study of a motoneuron cell body. (C) Several nerve terminals are apposed to two dendrites, D, of a motoneuron. Three chemical synapses are marked by arrows. (A unpublished photographs by F. DeCastro; B from Poritsky, 1969; C from Peters, Palay and Webster, 1976.)

be influenced by other nerve endings. Such endings of other fibers can depolarize or hyperpolarize the presynaptic ending and modulate the amount of transmitter it releases. The identification of chemical transmitters is further discussed in Chapters 8 and 9.

The retina stands as one of the best examples of the orderliness of the design of the neuronal connections in the brain. The various types of cells and their arrangement in the retina are shown in Figure 8. Light entering the eye flows through the layers of cells to reach the photoreceptors.[9] The signals, leaving the eye through the optic nerve fibers of ganglion cells, are entirely responsible for vision. The anatomy and signaling are discussed more fully in Chapter 16.

Design of connections as exemplified by the retina

The schematic presentation in Figure 8 gives an idea of the orderly arrangements of the retinal neurons. It is possible to distinguish immediately the photoreceptors, the bipolar cells, and the ganglion cells. These cells are clearly arranged in layers, and the lines of transmission are from input to output, from receptor to ganglion cell. In addition, two other types of cells, the horizontal and amacrine cells, make predominantly lateral connections, linking the through-pathways with each other. Two non-neuronal cell types are also apparent: the Müller cells that are glial satellite cells, and the pigmented epithelial cells.

This oversimplified description, however, omits many essential features of the architecture. First, there is a dramatic reduction in numbers and convergence from receptors to ganglion cells. Thus, 125 million receptors provide input to 1 million ganglion cells, the axons of which make up the optic nerve leading to higher centers. Second, the grouping of cells into these few characteristic types conceals a wealth of variation. There are many distinctive types of ganglion cells, horizontal cells, bipolar cells, and amacrine cells, each with its characteristic morphology and physiological properties. The simple through-processing of signals is dramatically influenced by numerous parallel and feedback interactions. In Chapter 16, it will be shown further that the various cells use a wide variety of chemical transmitter molecules.

In spite of these complexities, it is now possible to correlate structure and function at the cellular level in the retina. The branching patterns of the cells, their interconnections, and the properties of the synapses between them can be used to explain the steps that lead from transduction of light into meaningful electrical signals. Here too are exemplified the roles of local and action potentials: rods, cones, bipolar cells,

[9]Dowling, J. E. 1987. *The Retina*. Harvard University Press, Cambridge, MA.

(A)

(B)

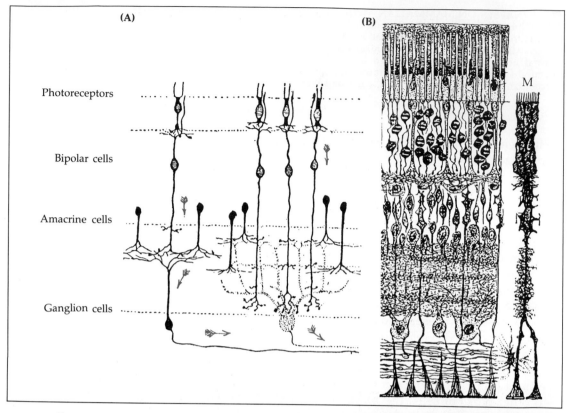

Photoreceptors

Bipolar cells

Amacrine cells

Ganglion cells

M

8 THE RETINA. (A, B) Structure and connections of cells in the mammalian retina drawn by Ramón y Cajal. The photoreceptors—rods and cones—connect to bipolar cells; these in turn connect to ganglion cells, whose axons constitute the optic nerve. Horizontal cells and amacrine cells make connections that are predominantly horizontal. The Müller cell (M) on the right is a satellite glial cell. In Chapter 16 it is shown that this scheme still holds, in general, but that essential new pathways and feedback loops have been discovered since Cajal's time. (C) Photomicrograph of chick retina stained by immunofluorescence for proteins in the retina that bind calcium. With the technique of conofocal microscopy it is possible in this preparation to observe the photoreceptors, bipolar, amacrine, and ganglion cells. (C from Rogers, 1989; photograph courtesy of J. H. Rogers.)

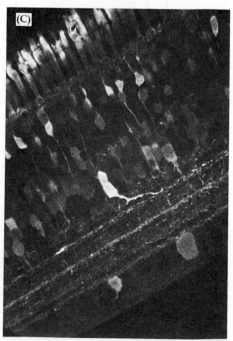

(C)

and horizontal cells transmit their signals entirely by local potentials spreading passively. Only amacrine and ganglion cells are able to generate propagated impulses (Chapter 16).

SUGGESTED READING

Adrian, E. D. 1946. *The Physical Background of Perception*. Clarendon Press, Oxford.

Hodgkin, A. L. 1964. *The Conduction of the Nervous Impulse*. Liverpool University Press, Liverpool.

Kuypers, H. G. J. M. and Ugolini, G. 1990. Viruses as transneuronal tracers. *Trends Neurosci.* 13: 71–75.

Nauta, W. J. H. and Feirtag, M. 1986. *Fundamental Neuroanatomy*. Freeman, New York.

Peters, A., Palay, S. L. and Webster, H. de F. 1991. *The Fine Structure of the Nervous System*, 3rd Ed. Oxford University Press, New York.

Poritsky, R. 1969. Two and three dimensional ultrastructure of boutons and glial cells on the motoneuronal surface in the cat spinal cord. *J. Comp. Neurol.* 135: 423–452.

Ramón y Cajal, S. 1955. *Histologie du Système Nerveux*, Vol. II. C.S.I.C., Madrid.

Sherrington, C. S. 1906. *The Integrative Action of the Nervous System*, 1961 Ed. Yale University Press, New Haven.

BOX 2   REVIEW OF KEY CONCEPTS AND TERMS

Recording techniques

Extracellular recordings are made with fine wire electrodes or glass pipettes placed close to the cell body or the axon of a neuron.

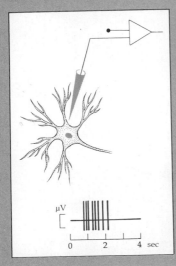

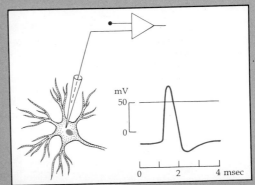

Intracellular recordings are made by inserting the tip of a microelectrode into a nerve cell. The same electrode can be used to pass currents through the membrane or to inject substances into the cytoplasm.

Recordings of activity in single ion channels can be made with a patch clamp. A pipette with a fire-polished tip binds tightly to the cell membrane. As an individual ion channel opens and closes, a small current flows through the pipette. Patch clamp recordings reveal the fine grain and properties of the individual molecular events that make up neuronal signals.

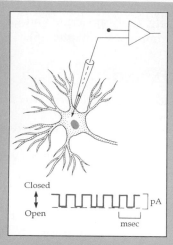

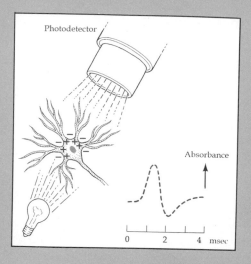

Optical recordings are made by applying to the cell dyes that bind to its membrane. In response to depolarization or hyperpolarization, changes occur in the absorbance or emission of light at a characteristic wavelength. As a cell lights up or darkens, its signals can be measured without electrodes. Specific dyes can also provide optical recordings of changes in levels of intracellular ions such as calcium.

Resting and action potentials

The ACTION POTENTIAL in a nerve fiber is fixed in size. It is a brief, stereotyped electrical event lasting about 1 millisecond. It moves rapidly along the nerve from one end to another. Another term for action potential is IMPULSE.

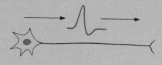

DEPOLARIZATION is a reduction in magnitude of the membrane potential toward 0 mV, the inside becoming less negative with respect to the exterior; HYPERPOLARIZATION is an increase in the magnitude of the potential, the inside becoming more negative. Depolarization to a critical potential level, the THRESHOLD, causes the initiation of an impulse. At the peak of the impulse, the inside of a cell becomes positive with respect to the outside.

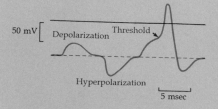

The specialized ending of a sensory nerve that responds to an external physical stimulus (touch, light, heat) is called a SENSORY RECEPTOR.

In a typical sensory nerve, the ADEQUATE STIM-ULUS (for example, stretch) depolarizes and sets up impulses.

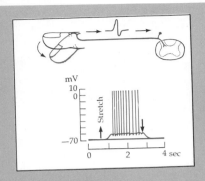

Stronger stimuli produce higher frequencies of impulse firing. For example, a sensory nerve responding to stretch of a muscle fires at a rate proportional to the stretch.

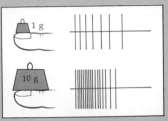

Neuronal connections

The entire arborization of a neuron can be revealed by staining it with a variety of techniques. These include silver staining (Golgi technique), the intracellular injection of a marker (such as horseradish peroxidase or the dye Lucifer Yellow), or labeling it with a monoclonal antibody (which in turn is linked to a visible marker). A typical nerve cell receives inputs from other neurons on its dendrites; from the cell body the axon emerges and later divides into numerous branches.

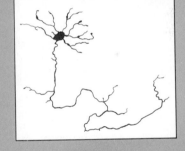

Nerve cells influence each other by (1) EXCITATION, that is, they tend to produce impulses in another cell; and by (2) INHIBITION, that is, they tend to prevent impulses from arising in another cell.

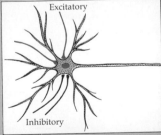

A cell receives many excitatory and inhibitory inputs from other cells (called CONVERGENCE) and in turn supplies many others (DIVERGENCE). The process whereby a cell adds together all the incoming signals that excite and inhibit it is known as INTEGRATION.

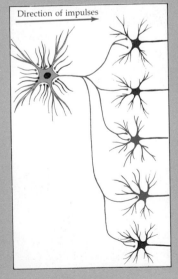

The junctions between nerve cells are called SYN-APSES. These are the sites at which cells transfer signals.

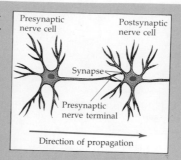

At CHEMICAL SYNAPSES the presynaptic terminal liberates a chemical, the TRANSMITTER, in response to a depolarization. Characteristically, at the site of a chemical synapse the two nerve cell membranes are thickened and are farther apart than elsewhere. Vesicles containing transmitter are aggregated close to the presynaptic membrane. Dense material is often present in the synaptic cleft and under the membrane of the postsynaptic cell. The transmitter binds to specialized chemoreceptor molecules that are situated in the postsynaptic membrane. At certain "electrical" synapses, current spreads from cell to cell through GAP JUNCTIONS.

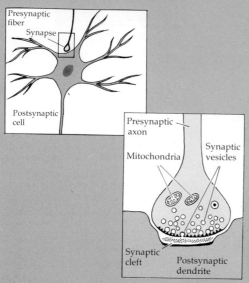

At an EXCITATORY SYNAPSE the transmitter liberated by the presynaptic ending depolarizes the postsynaptic cell, driving its membrane potential toward threshold.

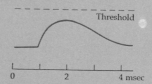

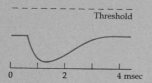

At an INHIBITORY SYNAPSE the transmitter tends to keep the membrane potential of the postsynaptic cell below threshold.

Slow MODULATORY synaptic potentials can increase or decrease the excitability of a neuron for prolonged periods of seconds, minutes, or hours. Such potentials often arise through the intervention of second messengers.

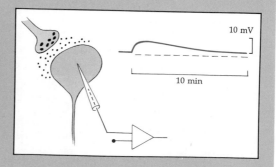

CHAPTER TWO

MEMBRANE CHANNELS AND SIGNALING

Ionic currents across cell membranes move through aqueous pores called channels. Such channels may have relatively little ion selectivity or may be selective for small anions, or for small cations, or for a single ion such as sodium. In general, ion channels fluctuate between open and closed states. The closed state usually predominates in the resting cell membrane, except for channels responsible for the resting membrane potential. When channels are activated, the frequency of openings increases and ionic currents flow through the open channels into or out of the cell. In neurons, some channels are activated by changes in membrane potential (voltage-activated) or by chemical ligands such as neurotransmitters (ligand-activated). Others are activated by chemicals in the cytoplasm (intracellular messengers), and still others by mechanical deformation of the cell membrane. Currents through such channels are responsible for the electrical signals generated in the nervous system.

The technique of patch clamping can be used to observe the activity of single channels in the membrane. Such observations are made by forming a tight electrical seal between the circular tip of a glass pipette and the cell membrane. Currents through individual channels in the enclosed membrane patch can then be recorded. Once a particular type of channel is isolated in a patch, it is possible to measure its kinetic behavior and permeability properties and to relate these to its overall function in the cell. Similar information can also be obtained indirectly, without patch clamping, by observing fluctuations in membrane current (noise) due to random opening and closing of channels.

In addition to their mode of activation and their ionic specificity, channels are characterized experimentally by their kinetic behavior and their conductance properties. Channels, when activated, may open at random intervals with open times distributed around a characteristic mean. Alternatively, channel openings may occur in irregular bursts. A channel may pass ions more readily or less readily, that is, it may have a relatively large or small conductance. It may also fluctuate between two or more open states with different conductances.

Channel conductance to a particular ion depends on ionic permeability, which is an intrinsic property of the channel, and on the ionic concentrations in the cytoplasm and extracellular fluid. Channel current depends on permeability, concentrations, and the driving force for ionic flux through the channel. The driving force may be a concentration gradient across the membrane, a potential difference, or a combination of both.

Modern methods of biochemistry and molecular biology have been used to deduce the molecular structure of membrane channels. For example, the cationic channel activated by the chemical acetylcholine in vertebrate muscle and in electric organs of fish (the nicotinic acetylcholine receptor or nAChR) has been shown to be composed of five polypeptide subunits (two of them identical) arranged in a circular array around a central pore. The amino acid sequence of each polypeptide is known and its arrangement in the membrane postulated. Channels activated by other ligands, such as glycine (the glycine receptor) and γ-aminobutyric acid (the GABA$_A$ receptor) have highly similar molecular structures.

The molecular structure of the voltage-activated sodium channel, responsible for action potential generation, has also been deduced. A number of sodium channel types have been studied, from eel electric organ, mammalian brain, and skeletal muscle. The major active component is a single large polypeptide chain containing four repeating domains in its polypeptide sequence. These domains display amino acid sequences that are highly similar (homologous) to one another. Voltage-activated calcium channels have a similar structure. Voltage-activated potassium channels are composed of subunits whose structures resemble closely the individual domains of the sodium and calcium channels.

The relation between molecular structure and channel function can now be studied by using techniques of molecular biology and electrophysiology. For example, nicotinic acetylcholine receptors from fetal and adult mammalian muscle have slightly different functional properties, and different subunit structures. Injection of messenger RNAs for various combinations of adult and fetal subunits into oocytes results in expression in the oocyte membrane of chimeric channels with properties dependent on subunit composition. In addition, mutant cDNAs can be constructed, with mutations directed at particular parts of a channel structure, for example at an amino acid in a membrane-spanning helix. When messenger RNA derived from the mutants is injected into oocytes, channels are expressed with modified functional properties. With such experiments the molecular mechanisms underlying channel function are now being revealed.

The business of the nervous system is carried out by electrical signals that either act locally or are propagated along nerve fibers. These potential changes occur across the plasma membrane of nerve cells and are mediated by movement of ions back and forth across the membrane.

To have a proper appreciation of how neurons function, we must understand how such ionic movements occur and how they are regulated. Indeed, a good deal of modern research in neurobiology is directed toward that precise goal.

Cell membranes consist of a more or less fluid mosaic of lipid and protein molecules. As shown in Figure 1A, the lipid molecules are arranged in a bilayer about 6 nanometers thick, with their polar, or hydrophilic, heads facing outward and their hydrophobic tails extending to the middle of the layer. Embedded in this lipid bilayer are protein molecules, some of which span the membrane, making contact with both the extracellular fluid and the cytoplasm.

Many substances can move across the membranes of cells from the extracellular bathing solution to the cytoplasm and in the reverse direction. For some small molecules, the ease of penetration into the cells depends upon their oil–water partition coefficient, that is, upon their relative solubility in lipids. Such molecules (for example, alcohols or glycerol) can cross the cell membrane by dissolving in the lipid layer and emerging on the other side. However, many substances involved in the everyday function of the cell are virtually insoluble in lipid and must traverse the membrane in some other way. From the point of view of neuronal function, the most important of these substances are inorganic ions—particularly sodium, potassium, calcium, and chloride. These move across the membrane in two ways: through water-filled pores, or CHANNELS, formed by membrane-spanning proteins, or by binding to CARRIER MOLECULES that transport them across the membrane. Transport by carrier molecules is discussed in Chapter 3. The basic features of a channel protein are illustrated in Figure 1B. The protein forms an aqueous pore through which ions can move. A region of the pore acts as a SELECTIVITY FILTER, regulating ion permeation by virtue of both its size and its molecular structure, for example, the presence of oxygen dipoles in the lining of the pore to substitute for waters of hydration stripped from the penetrating ions.[1] Finally the channel is GATED, that is, it fluctuates between open and closed states, the dominance of one over the other depending upon, for example, the potential across the membrane or the presence of a ligand at its outer end.

The modern concept of channels forming aqueous pores in nerve cell membranes began in the early 1950s, when Hodgkin and Huxley[2] suggested that the fluxes of sodium and potassium ions associated with the action potentials were too large to be attributed to the action of transport, or carrier, molecules. This led to the alternative idea of "holes," or aqueous channels though which ions could move. Over the past two decades, experimental tools have been developed to measure ion fluxes through individual channels, first indirectly by analysis of

[1]Hille, B. 1992. *Ionic Channels of Excitable Membranes*, 2nd Ed. Sinauer, Sunderland, MA. pp. 355–361.
[2]Hodgkin, A. L. and Huxley, A. F. 1952. *J. Physiol.* 117: 500–544.

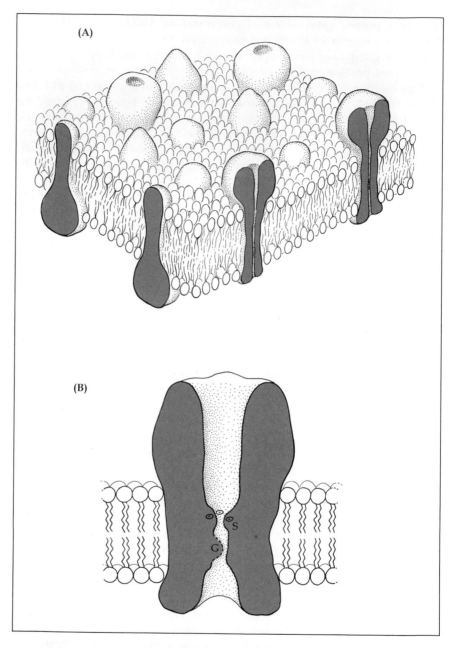

1   CELL MEMBRANE AND ION CHANNEL. (A) A cell membrane is composed of a lipid bilayer embedded with proteins. Some of the proteins traverse the lipid layer and some of these, in turn, form membrane channels. (B) Schematic representation of a membrane channel with a central water-filled pore, selectivity filter (S), and channel gate (G). The selectivity filter restricts ion permeation according to size and ionic charge. The gate opens and closes irregularly, with a probability of opening that may be regulated by membrane potential, the binding of a ligand at the outer end of the channel, or other biophysical or biochemical conditions.

membrane noise,[3,4] and later by direct observation.[5,6] Such fluxes have been shown to exceed 10^6 ions/sec—far greater than can be accounted for by a carrier mechanism.

Membrane channels vary considerably in their selectivity. Some are permeable to cations, some to anions. Cationic channels may be relatively nonspecific or specific, some for sodium, others for potassium, still others for calcium. Anionic channels tend to be less specific but are referred to as "chloride channels" because chloride is the major permeant anion in biological solutions. There are also channels that connect adjacent cells (see the discussion of connexons in Chapter 5) and allow the passage of most inorganic ions and many small organic molecules.

Most ion channels are gated. This means that they fluctuate between closed and open states, often with characteristic open times. For a few channels the open state predominates in the resting membrane; these are mostly potassium and chloride channels responsible for the resting membrane potential. The remainder are predominantly in the closed state, and the probability of an individual channel opening is low. When such channels are activated by an appropriate stimulus, the frequency of openings increases sharply. On the other hand, some channels that open frequently at rest may be inactivated by a stimulus; that is, their frequency of opening may be decreased. An important point to remember is that activation or inactivation of a channel means an increase or decrease in the *probability* of channel openings, not (as the words might imply) the production of a continuously open or closed state of the channel by the stimulus. Some channels are sensitive primarily to changes in membrane potential; they are VOLTAGE-ACTIVATED. In particular, the voltage-activated sodium channel is responsible for the regenerative depolarization that underlies the rising phase of the action potential (Chapter 4). Others, known as LIGAND-ACTIVATED channels, respond to extracellularly-applied chemicals of various kinds. For example, the chemical acetylcholine, released from motor nerve terminals at synapses on skeletal muscle (Chapter 7), causes muscle depolarization by activating cationic channels that allow sodium ions to enter the cell. MESSENGER-ACTIVATED ion channels respond to intracellular molecules (Chapter 8). Finally, STRETCH-ACTIVATED channels respond to mechanical distortion of the cell membrane (Chapter 14). It should be noted that these classifications are not rigid. For example, ligand-activated channels can be voltage-sensitive, and some voltage-activated channels can be modified by intracellular messengers.

Many channels are named for the ions to which they are permeable, for example sodium channels or chloride channels. However, channels

Channel properties

Nomenclature

[3]Katz, B. and Miledi, R. 1972. *J. Physiol.* 224: 665–699.
[4]Anderson, C. R. and Stevens, C. F. 1973. *J. Physiol.* 235: 665–691.
[5]Neher, E., Sakmann, B. and Steinbach, J. H. 1978. *Pflügers Arch.* 375: 219–228.
[6]Hamill, O. P. et al. 1981. *Pflügers Arch.* 391: 85–100.

permeable to a particular ion may fall into several subclassifications, depending upon their mode of activation, kinetic behavior, or voltage-dependent properties. This is particularly true of the bewildering array of potassium channels. These are subdivided into more than 10 types, characterized by their behavior in response to changes in membrane potential as delayed rectifier channels, inward rectifier channels, A channels, and so on; or by other modes of activation, such as calcium-activated or sodium-activated. Other channels are named only for the chemical that activates them. Thus the cationic channels activated by acetylcholine (ACh) are referred to as ACh channels. In the following discussion, we shall define such terms as they occur.

Direct measurement of single-channel currents

One of the important technical achievements of the last two decades was the development of patch clamp recording methods.[5,6] These methods provided direct answers to questions of obvious physiological interest about channels, for example: How much current does a single channel carry? How long does a channel stay open? How do its open and closed times depend on voltage? on the activating molecule? Patch clamp recording methods involve sealing the tip of a small (~1 μm internal diameter) glass pipette to the membrane of a cell. The technique works best with individual cells in tissue culture but is applicable in other circumstances as well, for example on exposed cells in slices of brain.[7,8] Under ideal conditions, with slight suction on the pipette, a seal resistance of greater than 10^9 ohms (hence a "gigaohm" seal) is formed around the rim of the pipette tip between the cell membrane and the glass (Figure 2B). When the pipette is connected to an appropriate amplifier, small currents across the patch of membrane inside the pipette tip can be recorded (Figure 3). Such events consist of rectangular pulses of current, reflecting the opening and closing of single channels. In other words, one observes in real time the activity of *single-protein molecules* in the membrane! In their simplest form the current pulses appear irregularly, with nearly fixed amplitudes and variable durations (Figure 3A). However, in some cases the currents exhibit open states with more than one conductance; in Figure 3B, for example, the open channels often close to a smaller "sub-state" conductance. Similarly, channels may display complicated kinetics. For example, channel openings may occur in bursts (Figure 3C). In summary, patch clamp techniques enable us to measure (1) the amplitudes of currents through single channels, and (2) the kinetic behavior of channels.

Erwin Neher (left) and Bert Sakmann in their laboratory (1985).

Patch clamp methods permit other recording configurations. Having made a seal to form a *cell-attached patch*, we can then pull the patch from the cell (Figure 2C) to form an *inside-out patch*, with the cytoplasmic face of the patch membrane facing the bathing solution. Alternatively, with slight additional suction we can rupture the membrane inside the patch (Figure 2D) to provide access to the cell cytoplasm. In this condition

[7]Gray, R. and Johnston, D. 1985. *J. Neurophysiol.* 54: 134–142.
[8]Edwards, F. A. et al. 1989. *Pflügers Arch.* 414: 600–612.

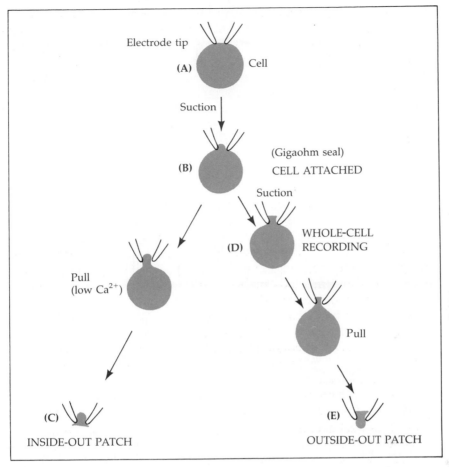

2 **PATCH CLAMP RECORDING CONFIGURATIONS. (A) On contact with the cell membrane, the electrode forms a seal, which is converted to a gigaohm seal (B) by gentle suction. Records may then be made from the patch of membrane within the electrode tip (cell-attached patch), or a cell-free, inside-out patch may be made by pulling (C). Alternatively, the membrane within the electrode tip may be ruptured by further suction to obtain a whole-cell recording (D), or by subsequent pulling to obtain an outside-out patch (E). (After Hamill et al., 1981.)**

currents are recorded from the entire cell (*whole-cell recording*). Finally, we may first obtain a whole-cell recording and then pull the electrode away from the cell to form a thin neck of membrane that separates and seals to form an *outside-out patch* (Figure 2E). As we will see later, each of these configurations has an advantage, depending upon the type of channels we are studying and the kind of information we wish to obtain. For example, if we wish to apply a variety of chemical ligands to the outer membrane of the patch, then an outside-out patch is most convenient.

Patch clamp techniques offer two advantages for studying the behavior of channels. First, the isolation of a small patch of membrane

Channel "noise"

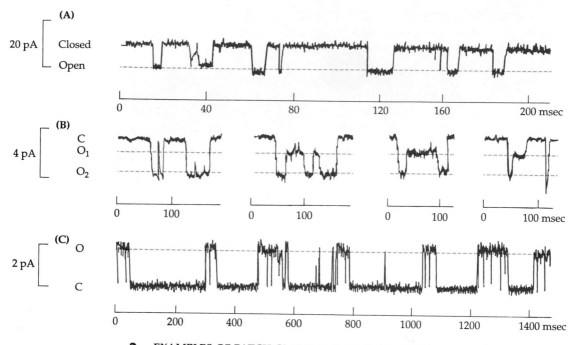

3   EXAMPLES OF PATCH CLAMP RECORDINGS. (A) Glutamate-activated channel currents recorded in a cell-attached patch from locust muscle occur irregularly, with a single amplitude and varied open times. Downward deflections indicate current flowing into the cell. **(B)** Acetylcholine-activated currents from single channels in an outside-out patch from cultured embryonic rat muscle reach an amplitude of about 3 pA (O_1), and relax to a "sub-state" current of about 1.5 pA (O_2). Downward deflections indicate inward current. **(C)** Pulses of outward current through glycine-activated chloride channels in an outside-out patch from cultured mouse spinal cord cells are interrupted by fast closing and reopening transitions to produce bursts. (A after Cull-Candy, Miledi and Parker, 1980; B after Hamill and Sakmann, 1981; C courtesy of A. I. McNiven.)

allows us to observe the activity of only a few channels, rather than the thousands that may be active in an intact cell; second, the very high resistance of the seal enables us to observe extremely small currents. Before these techniques became available, the general properties of some membrane channels were known from experiments in which microelectrodes were used to measure membrane potential or membrane current in whole cells. With intracellular recording it was possible to measure "noise" produced in the membrane when large numbers of channels were activated. An example is activation of channels by acetylcholine at the neuromuscular junction of the frog. The ACh released from the presynaptic nerve terminal opens ligand-gated channels that allow cations to cross the membrane, depolarizing the muscle. It was known for many years that when ACh was applied to the postsynaptic membrane, the resulting depolarization was "noisy"—there were variations about the mean depolarization that were larger than the baseline fluctuations at rest. This increase in noise was due to the random opening and closing of the cationic channels activated by ACh. In other words, application of ACh resulted in the opening of a large number of channels, and this number fluctuated in a random way as ACh molecules bombarded the membrane. In the early 1970s, Katz and Miledi[3] realized that increases in noise contained information about the size and time course of the potential changes produced by individual channel openings, and they applied techniques of noise analysis to obtain such information.

After Katz and Miledi's original observations, Anderson and Stevens[4] used similar techniques to measure fluctuations in inward current through the membrane (rather than fluctuations in membrane potential). The results of one such experiment are shown in Figure 4. Two characteristics of the response to ACh were recorded: (1) the mean current flowing through the end plate membrane (Figure 4A), and (2) the increase in noise, which was recorded at much higher amplification

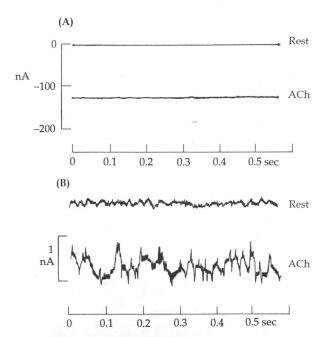

4 CURRENT NOISE PRODUCED BY ACETYLCHOLINE APPLICATION at the frog neuromuscular junction. (A) membrane currents at the end plate. At rest (upper trace) there is no current across the membrane; application of ACh produces about 130 nA of inward current (lower trace). (B) Traces in (A) shown at greater amplification. There is little fluctuation in the baseline at rest; the inward current produced by ACh shows relatively large fluctuations (noise), due to random opening and closing of ACh-activated channels. Analysis of the increased noise (see Box 1) gives the single-channel current and mean open time. (Modified from Anderson and Stevens, 1973.)

(Figure 4B). In (A) we see that application of ACh resulted in an inward current of about 130 nA. The corresponding records in (B) show that at rest the baseline fluctuations were less than 0.1 nA in amplitude, while after application of ACh the peak-to-peak amplitude of the current noise approached 1 nA. Analysis of the additional noise produced by ACh application provides information about the size of the single-channel

BOX 1   NOISE ANALYSIS: AN INDIRECT MEASURE OF CHANNEL CONDUCTANCE AND OPEN TIME

Given experimental observations such as those shown in Figure 4, what information about the underlying ACh-activated channels can we extract from them? First, it might be expected intuitively that large single-channel currents should produce large noise. For example, if an average current of 120 nA were due to activation by ACh of an average of 12 channels, each carrying 10 nA of current, we would expect relatively large fluctuations in the current as the channels opened and closed at random. If, on the other hand, 120 nA of current were due to an average of 1200 open channels, each carrying 0.1 nA, then smaller fluctuations would be expected. This intuitive expectation is borne out theoretically. The increased variance (*Var*), relative to the mean current (*I*), is directly related to the size of the single-channel current (*c*); that is, $Var/I = c$. The theoretical derivation assumes that the channels open and close independently of one another, and that only a small percentage of their total number is activated. If all were activated, there would be no fluctuation; hence the precise relation is $Var/I = c(1 - p)$, where *p* is the fraction activated. Rearranging the equation, we get

$$c = \frac{Var}{I(1 - p)}$$

By applying only a small quantity of ligand, *p* can be kept small enough to be ignored.

To determine the single-channel current, it is necessary only to measure the mean current and its variance. In the example given in Figure 4,

the inward current *I* was 134 nA, and *Var* was 2.5×10^{-19} A^2. Thus the calculated single-channel current is 1.9 pA.

As we might expect, the mean open time of the channels can be obtained from the frequency composition of the fluctuations. Thus if most channels stay open for a relatively long time (for example, for 30 msec), then the fluctuations occur at a relatively low frequency. If, on the other hand, most of the channels remain open for only a brief period (say, 1 msec), then higher-frequency fluctuations occur. In practice, the frequency composition of the extra noise is determined by computer analysis and presented in the form of a SPECTRAL DENSITY DISTRIBUTION. Such a distribution from the experiment illustrated in Figure 4 is shown here. The POWER DENSITY S is plotted as a function of frequency *f*. What is power? In this case it simply means current variance (*VarI*, in A^2). Power *density* is a measure of the amount of variance at each increment of frequency; that is, the amount of variance due to baseline fluctuations between 1 and 2 Hz, the amount due to fluctuations between 2 and 3 Hz, 3 and 4 Hz, and so on. Power density, then, has the units amperes squared per hertz (A^2/Hz). Because Hz = sec^{-1}, this results in the seemingly strange units A^2·sec. If all these incremental bits are added up, their sum is the total variance, *Var*. The exact form expected for the spectral density distribution depends on how one imagines the channels to behave. If the transition from the open to the closed state is a first-

currents, and the mean open time of the channels. The analysis itself (Box 1) is relatively straightforward: Large fluctuations, relative to the mean current, indicate large single-channel currents, and high-frequency fluctuations mean short open times.

Although noise analysis has been largely supplanted by patch clamp methods, the technique is still useful for studying membrane channels

order process, the channel open times will be distributed exponentially with a single time constant, τ. This means that if a large number of channels (N_0) were opened simultaneously, then the number N remaining open at any later time t would decay according to the relation

$$N = N_0 e^{-t/\tau}$$

The time constant τ has the same value as the *mean open time* of the channels. (This is a property of exponential distributions, not an intrinsic property of the channels.) If the channel times are indeed distributed exponentially, it can be shown that the spectral distribution will have the form

$$S = \frac{S(0)}{1 + (2\pi f\tau)^2}$$

This relation is shown by the continuous line. Its shape is determined by the two constants in the equation, $S(0)$ and τ. $S(0)$ is the maximum height of the curve at the low-frequency end. Its value increases with the size of the single-channel currents (remember that large single-channel currents produce large fluctuations). The mean open time τ determines how far the curve extends into the higher-frequency range before it bends over. As we have already guessed intuitively, the shorter the mean open time, the farther the curve extends into the high-frequency range. It can be seen from the equation that S is equal to half of $S(0)$ when $2\pi f\tau = 1$. The frequency at which this occurs is called the CORNER FREQUENCY, f_c (arrow). Thus we can calculate from the corner frequency: $\tau = 1/(2\pi f_c)$.

The spectral density distribution provides three important pieces of information. First, the fact that the experimental points fit reasonably well to the theoretical curve indicates that the channel open times are distributed exponentially with a single time constant. Second, the corner frequency f_c gives us the time constant, which is the same as the mean channel open time. Finally, the area under the distribution provides a measure of the total variance *Var*, which, when divided by the mean current, gives us the single-channel current c.

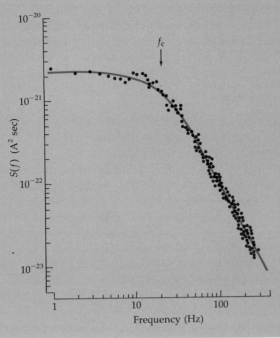

in cells that are not amenable to patch clamping, such as those in the intact central nervous system.[9] In addition, it is a relatively quick method of obtaining information about the properties of a population of channels. It cannot, however, provide detailed information about single-channel behavior, such as the existence of multiple conductance states.

Channel conductance

The kinetic behavior of a channel—that is, the durations of its closed and open states—can be used to make models of the steps involved in channel opening and closing, and the rate constants associated with these steps. The channel current, on the other hand, is a direct measure of how rapidly ions move through a channel. The current depends not only on channel properties but also on the transmembrane potential. Consider, for example, the outside-out membrane patch shown in Figure 5, which contains a single spontaneously active channel that is permeable to potassium. Both the patch electrode solution and the bathing solution contain 150 mM potassium. Potassium ions move in both directions through the open channel but, because the concentrations are equal, there is no net movement in either direction. Thus no current is seen (Figure 5B). Fortunately, the patch clamp recording system has an important feature that has not yet been mentioned: We can apply a voltage to the recording electrode and thus to the outer surface of the membrane patch. When a potential of +20 mV is applied to the electrode (Figure 5C), each channel opening results in a pulse of outward current from pipette to the bathing solution. On the other hand, when the electrode is made negative by 20 mV (Figure 5D), current flow through the open channel is into the pipette. The effect of voltage on the size of the current is plotted in Figure 5E. The relationship is linear: the current (i) through the channel is proportional to the voltage (V) applied to it:

$$i = \gamma V$$

The constant of proportionality, γ, is the channel CONDUCTANCE. For a particular applied voltage, a high-conductance channel will carry a lot of current, a low-conductance channel only a little.

Conductance has the units of siemens (S). In nerve cells, the potential across membrane channels is usually in units of millivolts (1 mV = 10^{-3} V), currents in picoamperes (1 pA = 10^{-12} A), and conductances in picosiemens (1 pS = 10^{-12} S). In Figure 5C, a potential of +20 mV produced a current of about 2.2 pA, so the channel conductance ($g = i/V$) was 2.2 pA/20 mV = 110 pS.

In some channels, the current–voltage relation is not linear; the conductance depends on membrane potential. When the conductance of such a channel is specified, it must be stated at what membrane potential the measurement was made. In addition, it is necessary to state whether the measurement was of CHORD CONDUCTANCE or SLOPE

[9]Gold, M. R. and Martin, A. R. 1983. *J. Physiol.* 342: 99–117.

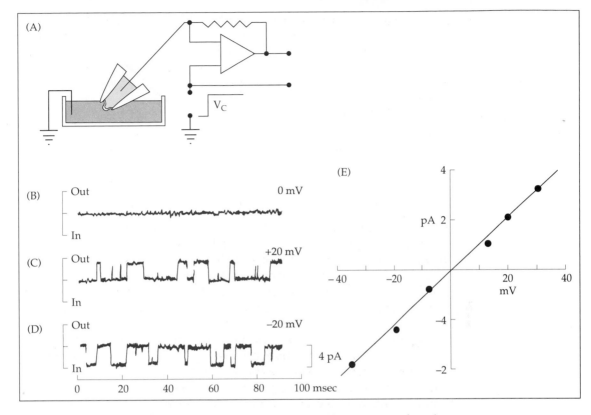

5 **EFFECT OF POTENTIAL ON CURRENTS** though a single, spontaneously active potassium channel in an outside-out patch, with 150 mM potassium in both the electrode and the bathing solution. (A) A sketch of the recording system. The output from the patch clamp amplifier is proportional to the current through the patch. A potential is applied to the electrode, and hence the patch, by applying a command potential (V_c) to the amplifier as shown. Current flowing into the electrode is shown as negative. (B) When no potential is applied to the patch, no channel currents are seen because there is no net flux of potassium through the channels. (C) Application of +20 mV to the electrode results in an outward current of about 2 pA through the channels. (D) A −20-mV potential results in inward channel currents of the same amplitude. (E) Channel currents as a function of applied voltage. Slope of the line is the channel conductance (γ). In this case $\gamma = 110$ pS. (A. R. Martin, unpublished.)

CONDUCTANCE at that potential. Chord conductance is a simple practical measure of how much current flows through the channel for a given applied potential, as illustrated in Figure 6. The point chosen for measurement is +50 mV, and the current at this potential is +0.6 pA; the chord conductance of the channel is 0.6 pA/50 mV = 12 pS (the slope of the solid line). It is plain, however, that an additional increase in the potential applied to the channel (e.g., to +60 mV) will not produce an increase in current corresponding to a 12-pS conductance. In other words, the chord conductance does not describe the condition of the channel at +50 mV, which in this illustration is virtually saturated.

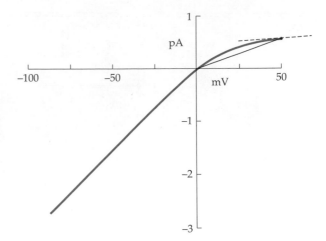

6

NONLINEAR RELATION BETWEEN CURRENT AND VOLTAGE in a channel, from an idealized experiment on an outside-out patch. The channel passes inward (negative) current more readily than outward current. The conductance of the channel at +50 mV can be defined in two ways: the first (*chord conductance*) by dividing the current by the driving potential (i.e., the slope of the solid black line); the second (*slope conductance*) by the slope (*dI/dv*) of the current–voltage relation (dashed line).

Instead, the state of the channel is described more accurately by the slope of the current–voltage relation; that is, by the slope conductance (dashed line).

Equilibrium potential

In the examples of channel current we have just considered, the concentration of potassium ions was the same on both sides of the membrane patch. What happens when we make the concentrations different? In order to examine this question, imagine that we make an outside-out patch and that the potassium concentration in the bath is 3 m*M* (similar to the concentration outside many cells), and in the electrode 90 m*M* (similar to the cytoplasmic concentration in many cells), as indicated in Figure 7. Now when the channel opens there will be a net movement of potassium ions through the channel from the pipette to the bath, even when no potential is applied to the pipette (Figure 7B); potassium ions simply move down their concentration gradient. If we now make the pipette positive, the potential gradient across the membrane will accelerate the outward potassium ion movement, and the channel current will increase (Figure 7C). If, on the other hand, we make the pipette negative, outward movement of potassium will be retarded, and the channel current will decrease (Figure 7D). If the negativity is sufficiently large, potassium ions will flow *inward* across the membrane against their concentration gradient (Figure 7E). If we make a number of such observations and plot channel current against applied voltage, we will get a result like that shown in Figure 7F. Unlike the result when the potassium concentration was the same on both sides of the membrane (Figure 5), the channel current is zero when the potential applied to the pipette is about −85 mV. In this condition the concentration gradient, which would otherwise produce an outward flux of potassium through the channel, is balanced exactly by the electrical potential gradient tending to move the ions in the channel in the opposite direction. The potential needed to achieve this condition is called the POTASSIUM EQUILIBRIUM POTENTIAL, E_K. The relation between

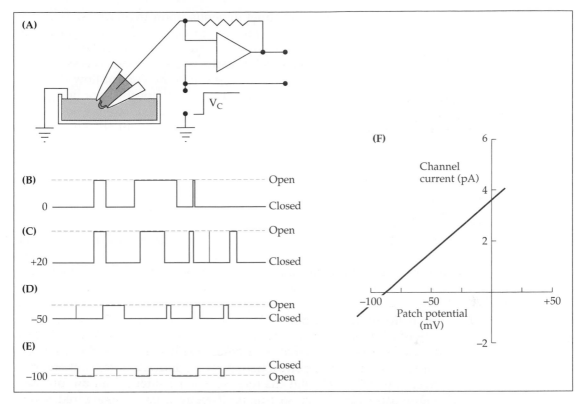

7   REVERSAL POTENTIAL FOR POTASSIUM CURRENTS in a hypothetical experiment using an outside-out patch with the concentration of potassium in the recording pipette ("intracellular") 90 mM and in the bathing solution ("extracellular") 3
mM. (A) Sketch of the recording system. (B) With no potential applied to the pipette,
flux of potassium from the electrode to the bath along its concentration gradient produces
outward channel currents. (C) When a potential of +20 mV is applied to the pipette,
outward currents increase in amplitude. (D) Application of −50 mV to the pipette
reduces outward currents, and (E) at −100 mV currents are reversed. (F) Current–voltage
relation indicates zero current at −85 mV, which is the potassium equilibrium potential
(E_K).

concentration and equilibrium potential for potassium and other ions is
discussed in the next chapter.

Figure 7F illustrates an important practical point about current flow
through channels: The magnitude of the current is not necessarily proportional to the absolute potential across the patch membrane. With no
applied potential, there is an outward current of almost 4 pA and with
−85 mV applied to the pipette, the current is zero. The current is
determined, then, not by the absolute membrane potential, but rather
by the difference between the membrane potential and the potential at
which the current is zero. It is convenient to call this difference the
DRIVING POTENTIAL. Referring again to Figure 7F, we see that when the
membrane potential is zero, the driving potential is +85 mV.

Conductance and permeability

The conductance of a channel depends upon two factors. The first is the ease with which the ions can pass through the channel, a property we call channel PERMEABILITY; the second is the concentration of the ions in the region of the channel. Clearly, if there are no potassium ions in the inside or outside solution, there can be no current flow through a potassium channel, no matter how large its permeability or how great a potential is applied. If only a few potassium ions are present, then for a given permeability and a given driving potential, the channel current will be smaller than when potassium ions are present in abundance. One way to think of these relations is as follows:

$$\text{Open channel} \rightarrow \text{permeability}$$
$$\text{Open channel} + \text{ions} \rightarrow \text{conductance}$$
$$\text{Open channel} + \text{ions} + \text{driving potential} \rightarrow \text{current}$$

The reader may be relieved to know that no general mathematical relation between permeability and conductance can be derived. This is because the relation depends on the manner in which the ions find their way through the channel.

Ion permeation through channels

How do ions actually pass through channels? One obvious way is simply by diffusion through the water-filled pore. This idea formed the basis of many early ideas about ion permeation, but for most (perhaps all) channels, it is not entirely accurate. This is because the channels themselves affect the permeation process. For example, if there is a physical restriction in the channel (selectivity filter), then an ion may require a certain energy to tear itself away from its associated water molecules (water of hydration) in order to squeeze through the channel neck; at some other location the ion may be attracted to or repelled by electrostatic charges lining the channel wall, or may be bound to a site within the channel from which it must escape to continue its journey. Such interactions of the channel with the permeating ion affect the rate of ion flux through the channel. Channel models that deal with ion permeation in this way are called Eyring rate theory models and are characterized by a number of energy barriers and binding sites.[10] For example, a "two-site, three-barrier" model is one in which an ion, to traverse the channel, has to hop over a peripheral energy barrier to a binding site, then over a central barrier to a second site and, finally, over a third barrier out of the channel. In general, such models are much more successful than simple diffusion models for describing ion movements through channels and selectivity between ions of like charge.

Channel structure

What does an actual channel molecule look like? How does it work? Complete and unequivocal answers to these questions are not yet available, but remarkable progress has been made in characterizing a number of ion channels, so that the general principles of channel structure and functional organization are understood in some detail. We will discuss

[10]Johnson, F. H., Eyring, H. and Polissar, M. J. 1954. *The Kinetic Basis of Molecular Biology*. Wiley, New York.

here two such channels: the nicotinic ACh receptor and the voltage-activated sodium channel. These are representative of two classes, or families, of channels—ligand-activated and voltage-activated. Channels within each family closely resemble one another in their biochemical and structural characteristics, but there is little detailed resemblance between families.

The first channel to be studied in detail was the nicotinic ACh receptor (nAChR, or simply AChR). [Note that for ligand-activated channels, the word "receptor" rather than "channel" is used routinely; this is because studies of the molecular structure of such molecules have relied primarily on binding of activating molecules (agonists), antagonists, toxins, and antibodies rather than on the specific channel properties of the protein.] The nicotinic ACh receptor underlies excitation not only at the vertebrate neuromuscular junction, but also at synapses in many vertebrate and invertebrate ganglia, and at the neuroeffector junctions of electric organs of a number of electric fish. The receptor is designated "nicotinic" because the actions of ACh are mimicked by nicotine, and to distinguish it from very different acetylcholine receptors activated by muscarine. Excellent summaries of the original biochemical and physical studies of the AChR in the laboratories of Changeux, Karlin, Raftery, Stroud, and many other investigators, and of the sequencing of the polypeptide subunits by Numa and his colleagues, are found in the 1983 Cold Spring Harbor Symposium.[11] Biochemical isolation and characterization of the AChR was facilitated by the existence of a remarkable density of synapses in electrocyte membranes in the electric organ of the ray *Torpedo*. After solubilization, AChR molecules were distinguished from other membrane proteins by their ability to bind α-bungarotoxin, a neurotoxin that binds to the channel in intact electrocytes and other tissues, such as the vertebrate neuromuscular junction. Similar toxins can be used in an affinity column to isolate the channel. The *Torpedo* AChR was found to comprise four glycoprotein subunits (α, β, γ, and δ), of about 40, 50, 60, and 65 kilodaltons respectively, arranged in a pentameric configuration in which there are two copies of the α subunit. The isolated AChR was shown to retain the major functional properties of native ionic channel when reincorporated into lipid vesicles.[12]

The nicotinic acetylcholine receptor

After identification, the sequence of each of the four subunits was obtained by a combination of biochemical and molecular genetic techniques. For each subunit, the sequences of the first 54 residues on the amino-terminal end were determined biochemically.[13] Knowledge of the partial amino acid sequence then made it possible to clone and sequence the cDNA for each subunit[14–16] (Box 2) and thus to deduce the corre-

[11]*Cold Spring Harbor Symp. Quant. Biol.* 48. 1983. pp.1–146.

[12]Tank, D. W. et al. 1983. *Proc. Natl. Acad. Sci. USA* 80: 5129–5133.

[13]Raftery, M. A. et al. 1980. *Science* 208: 1454–1457.

[14]Noda, M. et al. 1982. *Nature* 299: 793–797.

[15]Noda, M. et al. 1983. *Nature* 301: 251–255.

[16]Noda, M. et al. 1983. *Nature* 302: 528–532.

BOX 2   CLONING RECEPTORS AND CHANNELS

The application of molecular genetic techniques to the study of the nervous system has triggered extraordinarily rapid progress in the identification and characterization of proteins in neurons and their synaptic targets. What makes this approach so powerful is that the techniques are relatively straightforward, widely applicable, and exquisitely sensitive.

The first step in isolating a cDNA clone for a receptor or channel is to obtain mRNA from a tissue in which the protein is synthesized. The chance of isolating the clone of interest depends on the relative abundance of the corresponding mRNA in the starting material. Thus appropriate sources of mRNA for cloning acetylcholine receptor subunits would be electric organs of electric fish and eels or embryonic skeletal muscle. Using reverse transcriptase enzymes, complementary cDNA copies are made of total or size-selected poly(A)+ mRNA. These cDNA copies are incorporated into vectors (specially constructed bits of DNA resembling viruses) that can grow and multiply in host bacteria. The collection of viral or plasmid vectors is called a LIBRARY. To screen such a cDNA library, one exposes host bacteria to the library under conditions such that no bacterium is "infected" with more than one vector. Colonies of bacteria carrying the cDNA of interest are identified by using oligonucleotide probes that recognize the amplified cDNA or by using antibody probes that recognize the protein synthesized from it. Bacteria harboring a vector bearing the appropriate cDNA insert can then be grown in large numbers and the cDNA insert isolated and sequenced. The sequence of nucleotides is translated into the corresponding sequence of amino acids, and the deduced amino acid sequence can be examined for features such as hydrophobic transmembrane domains, α-helical regions, sites for posttranslational modifications, or stretches of amino acids similar to those in other proteins.

A variety of enzymes in addition to reverse transcriptase are crucial for these and other manipulations. These include (1) restriction endonucleases that recognize specific nucleotide sequences and cut double-stranded DNA exactly at those points, (2) exo- and endonucleases that selectively digest single or double stranded RNA or DNA, and (3) ligase enzymes that couple DNA molecules together and polymerase enzymes that replicate them.

Once the cDNA clone for a protein has been isolated, similar recombinant DNA techniques can be used to test the function of its various domains. Parts of the cDNA sequence can be deleted, then the remaining sequence inserted back into an appropriate vector and a truncated version of the protein produced. Stretches of cDNA from different proteins can be spliced together, and the resulting hybrid protein expressed and tested for activity. Using specific oligonucleotide primers to initiate DNA synthesis, individual bases in the cDNA can be changed by SITE-DIRECTED MUTAGENESIS to make proteins in which a single amino acid has been altered.

Several techniques are used to express such modified transcripts. Bacterial hosts can often be used to produce large amounts of protein; however, many eukaryotic proteins are not properly processed or glycosylated in bacteria. This drawback can be overcome by transfecting cDNA vectors into eukaryotic cell lines or by injecting mRNA synthesized from the cDNA into oocytes (see Box 4). Following this the proteins are synthesized by the host cells and can be incorporated into their membranes. Transfected cells can be fused with other cells, such as muscle cells, that are more convenient for analysis of the properties of the encoded protein.

Although the results of such experiments must be interpreted with care, the ability to make specific, discrete alterations in protein structure provides a powerful tool for understanding how proteins carry out their functions. In addition, the screening of libraries with probes derived from cDNAs encoding proteins of known function allows one to search for and isolate clones for related proteins. Small differences in homologous proteins can be identified and characterized using the POLYMERASE CHAIN REACTION (PCR) technique to amplify selected regions of mRNAs. The sensitivity and specificity of PCR is such that mRNA from a single neuron can be analyzed. Such experiments have led to the characterization of superfamilies of receptor and channel proteins and to the identification of a surprising number of isoforms of their subunits.

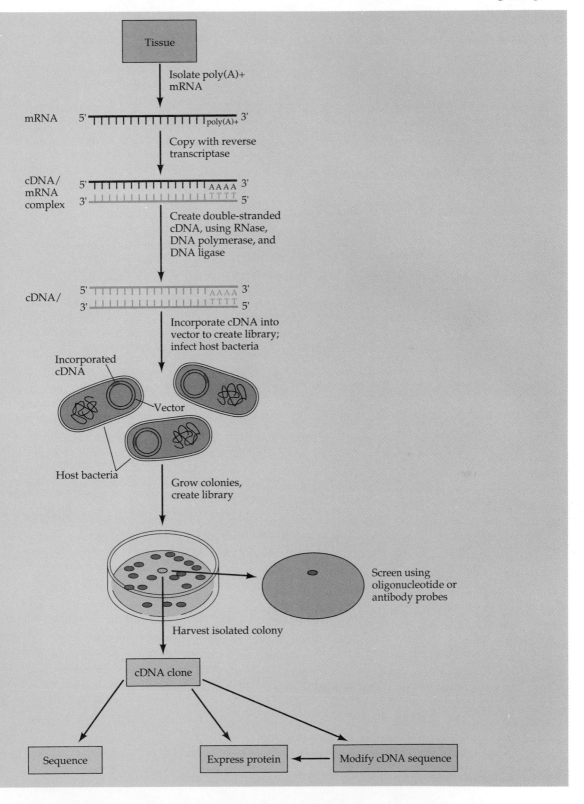

									10										20
Ser	Glu	His	Glu	Thr	Arg	Leu	Val	Ala	Asn	Leu	Leu	Glu	Asn	Tyr	Asn	Lys	Val	Ile	Arg
Pro	Val	Glu	His	His	Thr	His	Phe	Val	Asp	Ile	Thr	Val	Gly	Leu	Gln	Leu	Ile	Gln	Leu
Ile	Ser	Val	Asp	Glu	Val	Asn	Gln	Ile	Val	Glu	Thr	Asn	Val	Arg	Leu	Arg	Gln	Gln	Trp
Ile	Asp	Val	Arg	Leu	Arg	Trp	Asn	Pro	Ala	Asp	Tyr	Gly	Gly	Ile	Lys	Lys	Ile	Arg	Leu
Pro	Ser	Asp	Asp	Val	Trp	Leu	Pro	Asp	Leu	Val	Leu	Tyr	Asn	Asn	Ala	Asp	Gly	Asp	Phe
Ala	Ile	Val	His	Met	Thr	Lys	Leu	Leu	Leu	Asp	Tyr	Thr	Gly	Lys	Ile	Met	Trp	Thr	Pro
Pro	Ala	Ile	Phe	Lys	Ser	Tyr	Cys	Glu	Ile	Ile	Val	Thr	His	Phe	Pro	Phe	Asp	Gln	Gln
Asn	Cys	Thr	Met	Lys	Leu	Gly	Ile	Trp	Thr	Tyr	Asp	Gly	Thr	Lys	Val	Ser	Ile	Ser	Pro
Glu	Ser	Asp	Arg	Pro	Asp	Leu	Ser	Thr	Phe	Met	Glu	Ser	Gly	Glu	Trp	Val	Met	Lys	Asp
Tyr	Arg	Gly	Trp	Lys	His	Trp	Val	Tyr	Tyr	Thr	Cys	Cys	Pro	Asp	Thr	Pro	Tyr	Leu	Asp
Ile	Thr	Tyr	His	Phe	Ile	Met	Gln	Arg	Ile	Pro	Leu	Tyr	Phe	Val	Val	Asn	Val	Ile	Ile
Pro	Cys	Leu	Leu	Phe	Ser	Phe	Leu	Thr	Gly	Leu	Val	Phe	Tyr	Leu	Pro	Thr	Asp	Ser	Gly
Gly	Lys	Met	Thr	Leu	Ser	Ile	Ser	Val	Leu	Leu	Ser	Leu	Thr	Val	Phe	Leu	Leu	Val	Ile
Val	Glu	Leu	Ile	Pro	Ser	Thr	Ser	Ser	Ala	Val	Pro	Leu	Ile	Gly	Lys	Tyr	Met	Leu	Phe
Thr	Met	Ile	Phe	Val	Ile	Ser	Ser	Ile	Ile	Ile	Thr	Val	Val	Val	Ile	Asn	Thr	His	His

M3

									310										320
Arg	Ser	Pro	Ser	Thr	His	Thr	Met	Pro	Gln	Trp	Val	Arg	Lys	Ile	Phe	Ile	Asp	Thr	Ile
Pro	Asn	Val	Met	Phe	Phe	Ser	Thr	Met	Lys	Arg	Ala	Ser	Lys	Glu	Lys	Gln	Gln	Asn	Lys
Ile	Phe	Ala	Asp	Asp	Ile	Asp	Ile	Ser	Asp	Ile	Ser	Gly	Lys	Gln	Val	Thr	Gly	Glu	Val
Ile	Phe	Gln	Thr	Pro	Leu	Ile	Lys	Asn	Pro	Asp	Val	Lys	Ser	Ala	Ile	Glu	Gly	Val	Lys

"A"

									390										400
Tyr	Ile	Ala	Glu	His	Met	Lys	Ser	Asp	Glu	Glu	Ser	Ser	Asn	Ala	Ala	Glu	Glu		
Tyr	Val	Ala	Met	Val	Ile	Asp	His	Ile	Leu	Leu	Cys	Val	Phe	Met	Leu	Ile	Cys	Ile	Ile

M4

									430							
Gly	Thr	Val	Ser	Val	Phe	Ala	Gly	Arg	Leu	Ile	Glu	Leu	Ser	Gln	Glu	Gly

◀ **8** AMINO ACID SEQUENCE OF THE *α*-SUBUNIT of the acetylcholine receptor. Sequences in blue type indicate hydrophobic regions (M1, M2, M3, M4) capable of spanning the lipid membrane. In the M1 region, 16 out of 22 amino acids are hydrophobic (see classifications in Box 3). The other putative membrane-spanning regions are of similar composition. In the region labeled "A" (gray type), polar or charged (acidic or basic) residues are interspaced between the hydrophobic residues in such a way as to line up along one side of an *α* helix. Such a region is said to be *amphipathic*. (After Numa et al., 1983.)

sponding amino acid sequences. The amino acid sequence for the *α* subunit from *Torpedo* is shown in Figure 8 (human and bovine subunits are slightly different). The sequences for all four subunits are highly similar to one another (homologous), with various insertions and deletions in one peptide with respect to another, and discussion of the structural configuration of any one is generally applicable to the others as well. As with any very large protein, various segments of the molecule can be expected to fold into ordered secondary structures such as *α* helices. These secondary structures are then folded in some way to produce a tertiary structure in each subunit. Finally, five subunits (two *α*, one *β*, one *γ*, and one *δ*) join together to form the final quaternary structure, that is, the complete channel.

The size and orientation of the intact channel with respect to the lipid membrane has been determined by electron microscope imaging, and by X-ray and electron diffraction.[17-20] The quaternary structure is shown in Figure 9. The molecule, composed of a circular array of the five subunits around a central pore, is about 8.5 nm across at its widest point, which is in the extracellular domain, and about 11 nm long. The extracellular portion extends about 5 nm above the surface of the membrane. The central pore is about 0.7 nm in diameter, as predicted from earlier measurements of its selectivity to large cations.[21]

Although the primary structure of the subunits does not provide unique information about how the protein is arranged in the membrane, various models can be made, based on the characteristics of the amino acids in the sequence. Such models of the secondary and tertiary structures depend on a number of considerations, for example, identifying in the primary sequence extended runs of nonpolar (and hence hydrophobic) amino acid residues capable of forming *α* helices that span the membrane. In the original model proposed by Numa and his colleagues,[15] four such regions were identified (Figure 8) and the model shown in Figure 10A was postulated. The amino-terminal end, because it is known to contain the ACh-binding site, is placed extracellularly in the *α* subunit and, by analogy, in the other subunits as well. This

[17]Wise, D. S., Schoenborn, B. P. and Karlin, A. 1981. *J. Biol. Chem.* 256: 4124–4126.
[18]Kistler, J. et al. 1982. *Biophys. J.* 37: 371–383.
[19]Unwin, N., Toyoshima, C. and Kubalek, E. 1988. *J. Cell Biol.* 107: 1123–1138.
[20]Toyoshima, C. and Unwin, N. 1988. *Nature* 336: 247–250.
[21]Maeno, T., Edwards, C. and Anraku, M. 1977. *J. Neurobiol.* 8: 173–184.

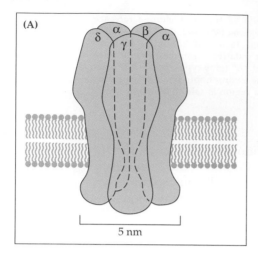

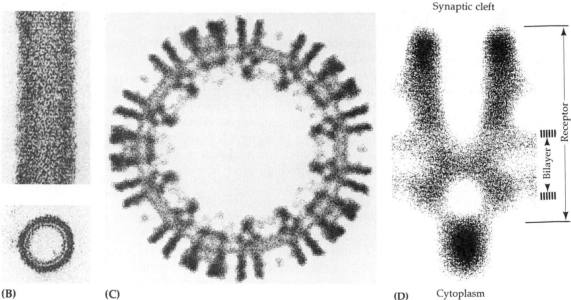

9

COMPLETE ACh RECEPTOR, (A), consists of five subunits—two *α*, one *β*, one *γ*, and one *δ*—spaced radially in increments of about 72 degrees around a central core. The α subunits contain the receptor sites for ACh. Dashed lines indicate the approximate dimensions of the internal passage in the open channel. (B) Longitudinal and transverse electron microscopic images of cylindrical vesicles from postsynaptic membrane of *Torpedo*, showing closely packed ACh receptors. (C) Reconstructed image of transverse section through tube, at higher magnification. (D) Further enlarged image of a single ACh receptor, showing position and size relative to the membrane bilayer. (A based on Stroud and Finer-Moore, 1985, and Toyoshima and Unwin, 1988; B, C, and D courtesy of N. Unwin.)

extracellular segment constitutes about half of the entire molecule, consistent with the mass distribution of the intact receptor (Figure 9). Given an even number of membrane crossings, the carboxy terminal is also extracellular.

If the membrane-spanning helices in each subunit are all hydrophobic, how do the five subunits get together to form a water-filled central pore? One might expect that such a pore, by necessity, would be lined with hydrophilic, rather than hydrophobic, regions of the molecules. One solution to this problem lies in the observation that the proposed transmembrane helices (M1–M4) are not uniformly hydrophobic: Some

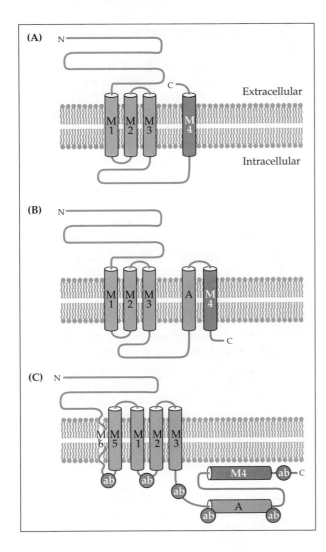

10

ALTERNATIVE MODELS OF ACh RECEPTOR SUBUNIT. (A) In the most commonly accepted model, regions M1 through M4 each form membrane-spanning helices, and both the carboxy terminus and the amino terminus of the peptide are in the extracellular space. **(B)** Additional membrane-spanning helix formed by amphipathic region A leaves the carboxy terminus on the intracellular face of the membrane. **(C)** Antibody reaction sites (ab) on the intracellular side suggest a more complex model with an intracellular carboxy terminus and two new membrane-spanning regions, one of which (M6) is too short to form a helix. (After McCrea, Popot and Engleman, 1987.)

contain amino acids with polar side chains that tend to be clustered on one side of the helix. Thus a transmembrane region could be capable of forming a portion of the pore wall by presenting a hydrophilic face to the water-filled channel. A segment of each subunit between M3 and M4, originally assigned to the cytoplasm, has such an amphipathic sequence (i.e., has regularly alternating hydrophobic and hydrophilic amino acids) and therefore could be considered a possible candidate for pore formation.[22] A model of the AChR α subunit based on this idea is shown in Figure 10B. The amphipathic region forms a fifth membrane-spanning helix (labeled A) to provide the putative pore lining. The

[22]Finer-Moore, J. and Stroud, R. M. 1984. *Proc. Natl. Acad. Sci. USA* 81: 155–159.

additional membrane crossing relocates the carboxy terminal to the cytoplasmic side of the membrane.

Models such as those in Figure 10 can be tested in a number of ways. One has already been mentioned—the location of the ACh-binding site must be extracellular. This is also true of the region that binds α-bungarotoxin. Another way to test the models is to make an antibody to a short amino acid sequence found in the subunit polypeptide and then use the antibody to determine the physical location of that sequence in the tertiary structure.[23] Such an antibody might be used, for example, to localize the carboxy terminal to the extracellular or intracellular compartment. An example of a model based on such studies is shown in Figure 10C, in which the amphipathic sequence (A) and a highly hydrophobic region (M4) are intracellular, and two new membrane-spanning sequences (M5 and M6) are added. In addition, this model differs from the other two in that the bulk of the peptide is intracellular (about 200 residues) rather than extracellular (144 residues), a feature that is less consistent with the apparent mass distribution of the channel (Figure 9).

Although antibody labeling might seem a straightforward way of localizing particular segments of the primary structure, models of the tertiary structure based on such experiments are not always consistent with those based on other, seemingly equally straightforward, biochemical observations. For example, AChRs aggregate in pairs (dimers) when closely packed. There is biochemical evidence that the dimers are formed by disulfide cross-links between cysteine residues near the carboxy terminals of the δ subunits, and that the cross-links are extracellular.[24,25] This would place the carboxy terminal of the δ subunit on the extracellular side of the membrane, contrary to the models in Figures 10B and 10C.

Receptor superfamilies

After the nicotinic AChR was sequenced, similar isolation and sequencing were carried out for subunits of a number of other ligand-activated channels, including nAChR from vertebrate brain (neuronal nAChR),[26] and subunits of channels gated by the amino acids γ-amino butyric acid (GABA$_A$ receptors), glycine, and glutamate.[27] These various subunits have amino acid sequences that show a large degree of homology with each other. Furthermore, they are highly homologous with AChR subunits from *Torpedo* and vertebrate muscle. One indication of these similarities is shown in Figure 11, in which a hydropathy index for the amino acids in the sequence (Box 3) is plotted against residue number for two GABA$_A$ receptor subunits from mammalian brain and

[23]Ratnam, M. et al. 1986. *Biochemistry* 25: 2633–2643.
[24]McCrea, P. D., Popot, J.-L. and Engleman, D. M. 1987. *EMBO J.* 6: 3619–3626.
[25]DiPaola, M., Czajkowski, C. and Karlin, A. 1989. *J. Biol. Chem.* 264: 15457–15463.
[26]Leutje, C. W., Patrick, J. and Seguela, P. 1990. *FASEB J.* 4: 2753–2760.
[27]Betz, H. 1990. *Neuron* 5: 383–392.

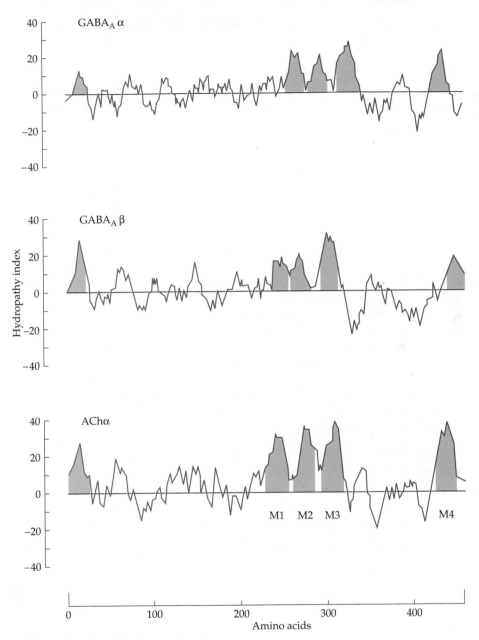

11 HYDROPATHY INDEX FOR GABA$_A$ α- and β-subunit and ACh α-subunit sequences. Indices are obtained by taking a moving average of the relative hydrophobicity of adjacent amino acids in the sequence. Regions in the positive range are hydrophobic, those in the negative range hydrophilic, on an arbitrary scale. Shaded areas indicate hydrophobic regions, four in similar positions toward the carboxy terminal end, corresponding to the M1 to M4 regions of the ACh receptor. (Amino acid numbering system is slightly different from that in Figure 8.) (After Schofield et al., 1987.)

BOX 3 CLASSIFICATION OF AMINO ACIDS

Channel subunits, like all other peptides, are composed of amino acids, and it is the amino acid side chains that determine many of the local chemical and physical properties of the subunits.

The 20 amino acids fall into three groups: basic, acidic and neutral, as shown. (Three-letter and one-letter abbreviations are given in parentheses.) Acidic and basic amino acids are hydro-

BASIC

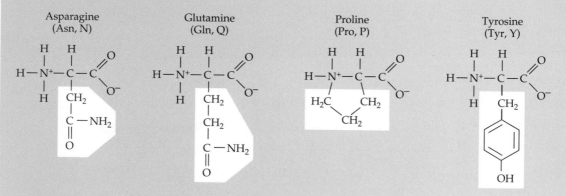

Lysine
(Lys, K)

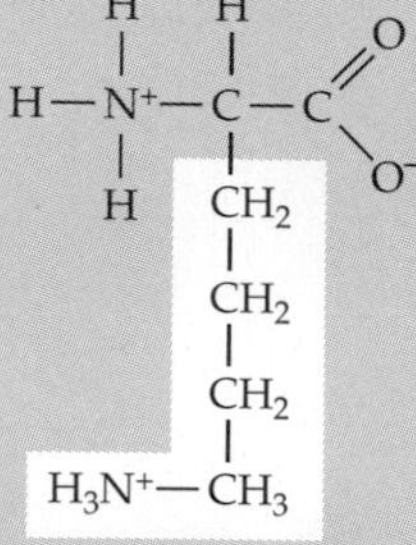

Arginine
(Arg, R)

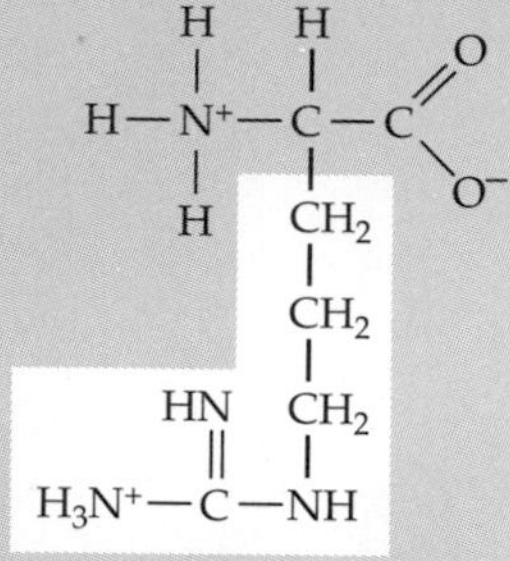

Histidine
(His, H)

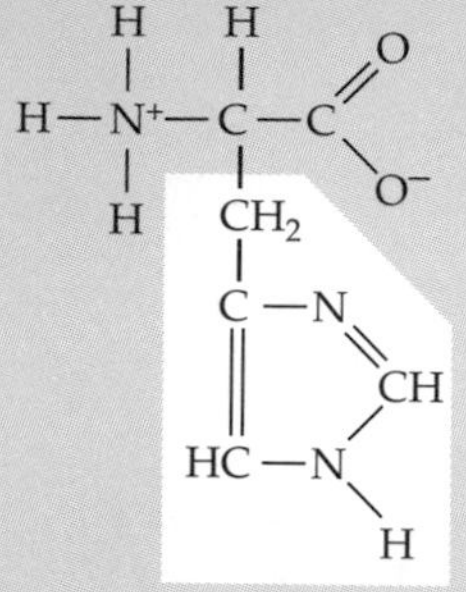

NEUTRAL

STRONGLY HYDROPHILIC

Asparagine
(Asn, N)

Glutamine
(Gln, Q)

Proline
(Pro, P)

Tyrosine
(Tyr, Y)

WEAKLY HYDROPHOBIC

Alanine
(Ala, A)

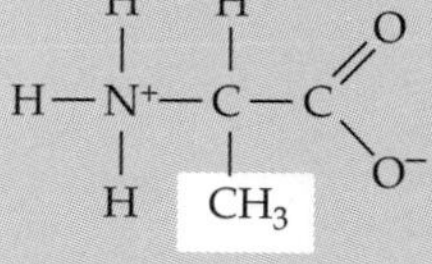

Methionine
(Met, M)

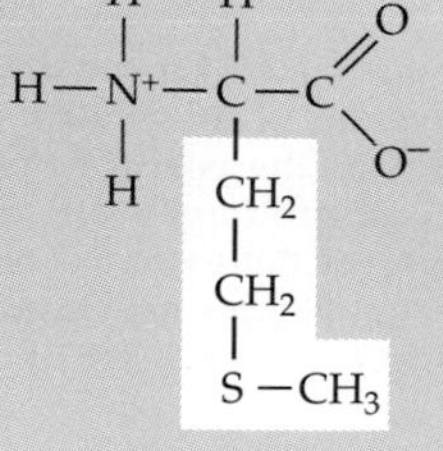

Cysteine
(Cys, C)

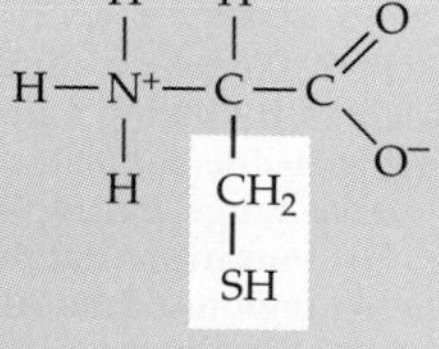

Phenylalanine
(Phe, F)

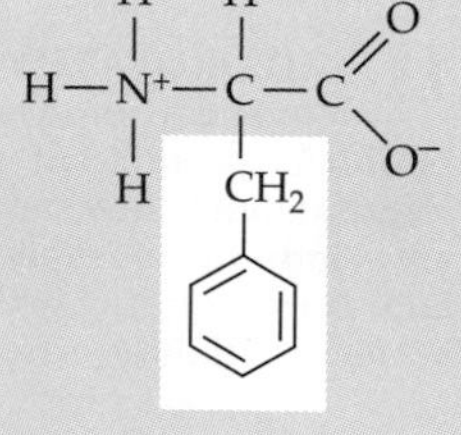

philic. The neutral amino acids are arranged according to the hydrophobicity index of Kyte and Doolittle,[28] starting with the least hydrophobic (most polar) and progressing to the most hydrophobic. Sections of a peptide are candidates for membrane-spanning regions if they contain se- quences of hydrophobic amino acids capable of forming an α helix long enough to traverse the lipid bilayer (see Figure 8).

[28]Kyte, J. and Doolittle, R. F. 1982. *J. Molec. Biol.* 157: 105–132.

ACIDIC

Aspartic acid
(Asp, D)

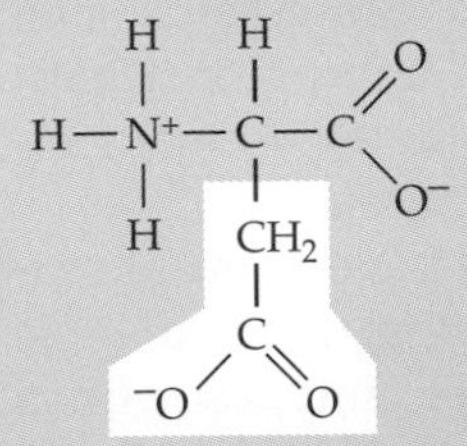

Glutamic acid
(Glu, E)

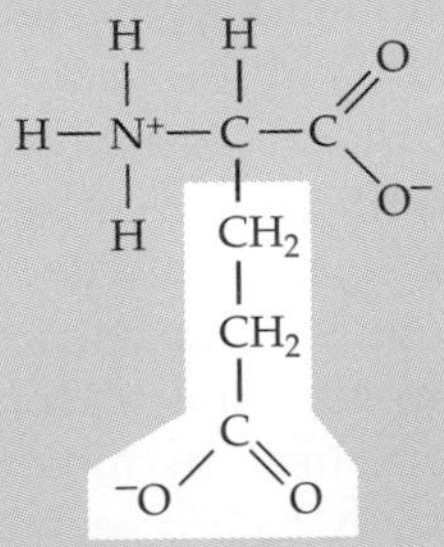

WEAKLY HYDROPHILIC

Tryptophan
(Trp, W)

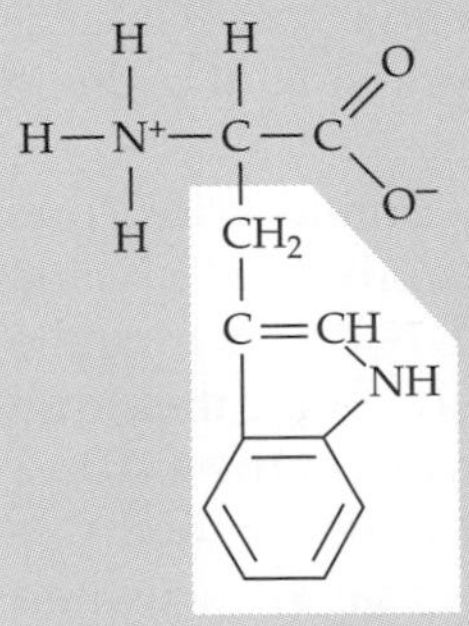

Serine
(Ser, S)

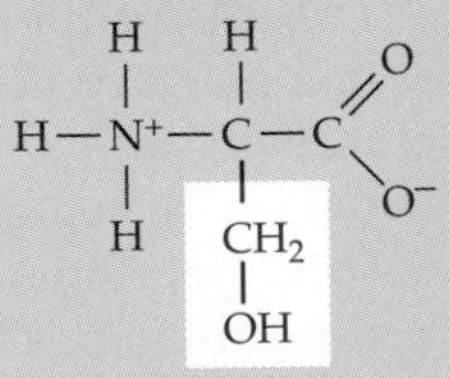

Threonine
(Thr, T)

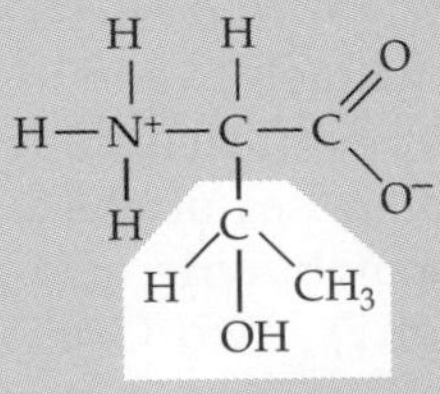

Glycine
(Gly, G)

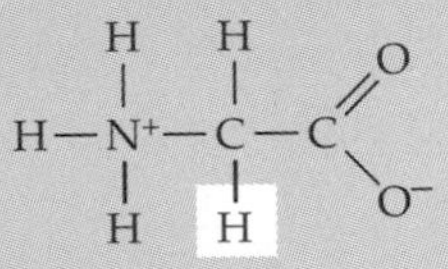

STRONGLY HYDROPHOBIC

Leucine
(Leu, L)

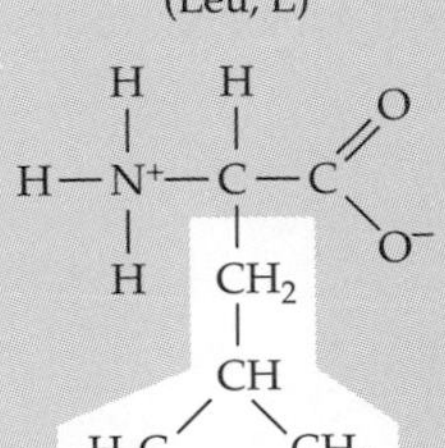

Valine
(Val, V)

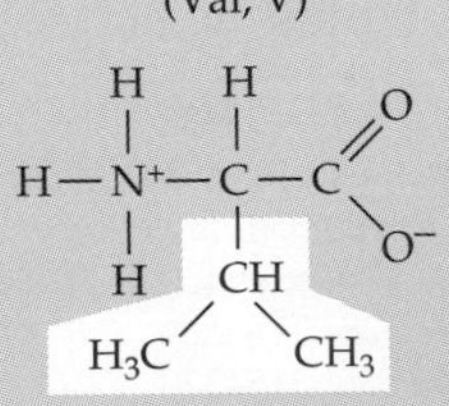

Isoleucine
(Ile, I)

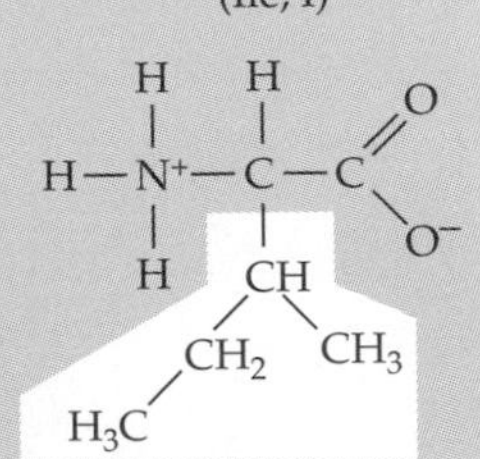

for the *Torpedo* AChR α subunit. The three peptides have similar profiles, with four hydrophobic sequences between residues 220 and 500 (shaded areas) indicating possible membrane-spanning regions.

The structural similarities between ligand-activated channels indicate that they form a SUPERFAMILY of common genetic origin. As we will see shortly, the structures of voltage-activated channels are quite different but again are similar to one another, forming a second superfamily. Membrane proteins that influence intracellular second messengers, such as the β-adrenergic receptor and the "muscarinic" ACh receptor (Chapter 8), and rhodopsin, the visual pigment protein (Chapter 16) make up yet another superfamily with highly conserved structures.[29]

Diversity of ligand-activated ion channels

Numerous subunits of ligand-activated channels have been identified in a variety of nervous system tissues, suggesting that a diverse number of gene products combine to form channel isotypes with a variety of functional properties. In most cases, the exact functional significance of this diversity (which is seen in voltage-activated channels as well) is unclear. Neuronal nAChR subunits have been classified as *a* if their amino-terminal sequence suggests the presence of an ACh-binding site (in particular if there is a pair of cysteine residues in locations analogous to those at positions 192 and 193 in Figure 8[30]). Otherwise they are classified as β or non-α. So far, at least nine such subunits have been isolated from chicken and rat brain:[26,31] $\alpha2$, 3, 4, 5, 6, and 7; and non-$\alpha1$, 2, 3 ($\beta2$, 3, 4). Injection of mRNA for any α–β combination into oocytes (Box 4) results in the formation of ACh-activated channels,[32] and in at least one instance the expression of a single subunit is sufficient for channel formation.[31]

Two classes of chloride channel are activated by amino acids, one by γ-aminobutyric acid (the GABA$_A$ receptor), the other by glycine (Chapter 7). Two GABA$_A$ receptor subunits, α and β,[34] and a single glycine receptor α subunit[35] were sequenced originally. Since then, the amino acid sequences of an increasing number of isotypes of the GABA$_A$ subunits (to date three α, two β, a γ and a δ) have been determined.[36,37]

A family of gene products forms cationic channels that are activated by glutamate and its analogues kainate, quisqualate, and AMPA (α-amino-3-hydroxyl-5-methyl-4-isoxazolepropionic acid).[38,39] These have been named variously GluR-K1, -K2, -K3, -K4, or GluA, B, C, D, or GluR1, R2, and so on. Expression of one or a combination of these "AMPA-selective" receptor subunits in oocytes or in cultured mam-

[29]Weiss, E. R. et al. 1988. *FASEB J.* 2: 2841–2848.

[30]Kao, P. N. and Karlin, A. 1986. *J. Biol. Chem.* 261: 8085–8088.

[31]Couturier, S. et al. 1990. *Neuron* 5: 847–856.

[32]Leutje, C. W. and Patrick, J. 1991. *J. Neurosci.* 11: 837–845.

[34]Schofield, P. R. et al. 1987. *Nature* 328: 221–227.

[35]Grenningloh, G. et al. 1987 *Nature* 328: 315–320.

[36]Schofield, P. R. 1989. *Trends Pharmacol. Sci.* 10: 476–478.

[37]Sigel, E. et al. 1990. *Neuron* 5: 703–711.

[38]Nakanishi, N., Schneider, N. A. and Axel, R. 1990 *Neuron* 5: 569–581.

[39]Keinanen, K. et al. 1990. *Science* 249: 556–560.

BOX 4   EXPRESSION OF RECEPTORS AND CHANNELS IN *XENOPUS* OOCYTES

Expression of messenger RNA in oocytes has been an indispensable tool for examining the properties of receptors and channels. The methods of oocyte preparation and mRNA isolation have been described in detail by Miledi and his colleagues.[33] The steps are shown schematically below.

The separated oocytes are incubated overnight in a saline solution before injection, and a further 2 to 7 days is allowed for expression of the message after injection. RNA is prepared from brain homogenate. Proteins are denatured to facilitate their separation from nucleic acids and to inactivate RNase. After injection of total brain poly(A)+ mRNA, a large variety of proteins are expressed. Partial mRNA purification, for example by size separation, serves to reduce this number, with the additional advantage that the remaining mRNAs, being more concentrated, express more of the desired protein. With mRNA derived from cDNA clones, it is possible to express only the desired receptor or channel.

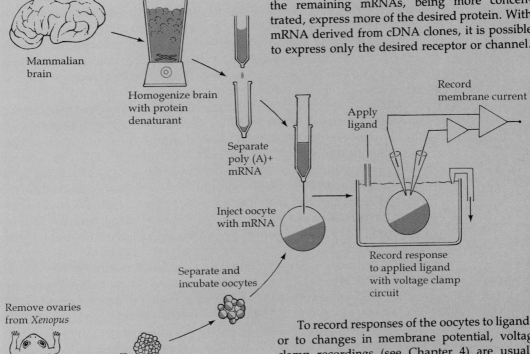

Mammalian brain

Homogenize brain with protein denaturant

Separate poly (A)+ mRNA

Inject oocyte with mRNA

Separate and incubate oocytes

Remove ovaries from *Xenopus*

Apply ligand

Record membrane current

Record response to applied ligand with voltage clamp circuit

To record responses of the oocytes to ligands, or to changes in membrane potential, voltage clamp recordings (see Chapter 4) are usually made after removing the enveloping follicular layers with gentle collagenase treatment. If, in addition, the surrounding vitelline membrane is removed (by osmotic shock), single-channel recordings can be made from the exposed oocyte membrane with patch clamp electrodes.

[33]Sumikawa, K., Parker, I. and Miledi, R. 1989. *Meth. Neurosci.* 1: 30–45.

malian cell lines leads to formation of cationic channels with very low conductances, similar to a class of glutamate-activated channels found in intact neurons. A separate group of brain polypeptides, KaiBP1 and KaiBP2 bind kainate but do not form functional ion channels.[27]

As we have already discussed, the *Torpedo* AChR is heteromultimeric; that is, it is formed as a multimer, in this case a pentamer, from different subunits. There is now strong evidence, obtained by combining mutant and naturally occurring subunits with different physiological properties (discussed later), that the neuronal nAChR is a pentamer as well, normally composed of two α and three β subunits.[40] It was proposed originally for the GABA$_A$ receptor that two α and two β subunits (Figure 12) were joined together to form a tetrameric channel in the membrane. Since then, alternative channel models have been proposed,[41] including pentamers analogous to the muscle nAChR. The combinations $\alpha_2\beta_2\gamma$, or $\alpha_3\beta\gamma$ give channels with appropriate physiological and pharmacological properties.[36] Similarly, the glycine receptor appears to be a heteropentamer, with the most likely configuration being $\alpha_3\beta_2$, deduced from the masses of the subunit types and the mass of the completely linked protein.[42]

The voltage-activated sodium channel

The methods that were successful in characterizing the behavior and molecular structure of the AChR have been applied with equal success to the voltage-activated sodium channel. Again the essential steps were biochemical isolation of the protein,[43–45] isolation of cDNA clones, and deduction of the amino acid sequence of the protein.[46] As with the AChR, an electric fish—this time the eel, *Electrophorus electricus*—provided a rich source of material, and a number of toxins were available for assaying the isolated protein, principally tetrodotoxin (TTX) and saxitoxin (STX), both of which block ion conduction in the native channels.

Subsequently, sodium channels were isolated from brain and skeletal muscle. The structural characteristics of voltage-activated channels and their molecular diversity have been discussed in several reviews.[47,48] The sodium channel purified from eel consists of a single large (260 kD) protein. The primary 260-kD (α) subunit isolated from mammalian brain is accompanied by two additional subunits of uncertain function: β_1 (36 kD) and β_2 (33 kD). Because both glial cells[49] and neurons in the brain

[40]Cooper, E., Couturier, S. and Ballivet, M. 1991. *Nature* 350: 235–238.

[41]Olson, R. W., and Tobin, A. J. 1990. *FASEB J.* 4: 1469–1480.

[42]Langasch, D., Thomas, L. and Betz, H. 1988. *Proc. Natl. Acad. Sci. USA* 85: 7394–7398.

[43]Miller, J., Agnew, W. S. and Levinson, S. R. 1983. *Biochemistry* 22: 462–470.

[44]Hartshorn, R. P. and Catterall, W. A. 1984. *J. Biol. Chem.* 259: 1667–1675.

[45]Barchi, R. L. 1983. *J. Neurochem.* 40: 1377–1385.

[46]Noda, M. et al. 1984. *Nature* 312: 121–127.

[47]Catterall, W. A. 1988. *Science* 242: 50–61.

[48]Trimmer, J. S. and Agnew, W. S. 1989. *Annu. Rev. Physiol.* 51: 401–418.

[49]Ritchie, J. M. 1987. *J. Physiol.* (Paris) 82: 248–257.

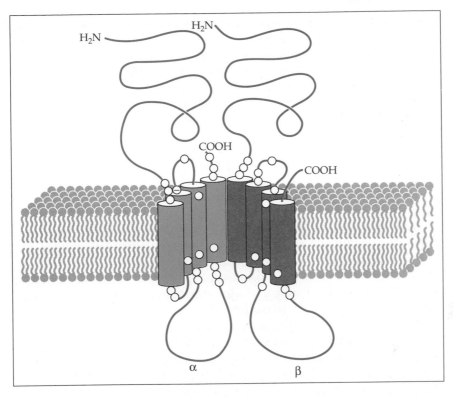

12 GABA$_A$ RECEPTOR α AND β SUBUNITS, each of which has four membrane-spanning helices (represented as cylinders). A complete channel is probably formed of two of each of these subunits, plus an additional γ or δ subunit. (After Schofield et al., 1987.)

contain an abundance of voltage-activated sodium channels, the cellular source of the channels is not clear. For the α subunit, several different mRNAs, producing a number of subtypes, have been identified. At least two additional subtypes are expressed in mammalian skeletal muscle. One, from normal muscle (RSkM1) is TTX-sensitive.[50] Another, characteristic of developing fetal muscle or denervated muscle (RSkM2), is TTX-insensitive.[51] A TTX-insensitive isotype is also found in mammalian heart.[52] After translation, the protein is heavily glycosylated; about 30 percent of the mass of the mature eel channel consists of carbohydrate chains containing large amounts of sialic acid. Removal of sialic acid from purified channel proteins in reconstituted lipid membranes results in alteration of their functional characteristics.[53]

[50]Trimmer, J. S. et al. 1989. *Neuron* 3: 33–49.
[51]Kallen, R. G. et al. 1990. *Neuron* 4: 233–342.
[52]Rogart, R. B. et al. 1989. *Proc. Natl. Acad. Sci. USA* 86: 8170–8174.
[53]Recio Pinto, E. et al. 1990. *Neuron* 5: 675–684.

The eel channel is composed of a sequence of 1832 amino acids, within which are four successive domains (I–IV) of 300 to 400 residues, with about a 50 percent sequence homology from one to the next. The domains are architecturally equivalent to the subunits of the AChR family of channel proteins, except for the genetically important difference that they are expressed together as a single protein. Each domain contains multiple hydrophobic or amphipathic sequences capable of forming transmembrane helices. The simplest of several proposed transmembrane topologies, with six such membrane-spanning segments (S1–S6), is illustrated in Figure 13A. In other models, as many as eight membrane segments in each domain are postulated.[54] As with the AChR subunits, each domain is postulated to contribute to the formation of the pore.

Other voltage-activated channels

The structure of other voltage-activated channels indicates that these, too, belong to a superfamily of like proteins. The voltage-activated calcium channel is substantially similar in amino acid sequence to the sodium channel, and, consequently, a similar secondary structure has been postulated (Figure 13B). In particular, the putative transmembrane regions, S1 to S6 (including the charged S4 helix), are highly homologous to those of the sodium channel.

A channel protein of particular interest is that associated with the voltage-activated potassium A CHANNEL (Figure 13C), expressed by *Drosophila*. In general, A channels have the property of being closed at relatively large negative membrane potentials (e.g., more negative than -60mV), opening briefly in response to depolarization, and then inactivating.[55] Although there are a number of distinct messages that give rise to a diverse family of such proteins, their amino acid sequences are all similar to those of a segment of domain IV of the eel sodium channel. Experimental evidence indicates that these single proteins assemble to form homomultimeric ion channels in the membrane.[56] In addition, it has been shown that different members of the *Drosophila* family, as well as of a corresponding family found in mammalian brain, can combine to form heteromultimeric channels.[57,58]

Ion channel structure and function

Modern methods of biochemistry, cell biology, and electrophysiology have led to a number of important observations relating channel function to structure. The examples to be discussed here involve expression of channels by *Xenopus* oocytes, induced by injection of the appropriate mRNA[59] (Box 4). In such experiments, the characteristics of single-channel currents are measured with patch clamp techniques; alternatively, other techniques are used to measure whole-cell currents (representing the behavior of the entire population of inserted channels). Oocytes normally do not express nAChRs or voltage-activated

---

[54]Guy, H. R. and Conti, F. 1990. *Trends Neurosci.* 13: 201–206.

[55]Connor, J. A. and Stevens, C. F. 1971. *J. Physiol.* 213: 21–30.

[56]Timpe, L. C. et al. 1988. *Nature* 331: 143–145.

[57]Isacoff, E. Y., Jan, N. J. and Jan, L. Y. 1990. *Nature* 345: 530–534.

[58]Ruppersburg, J. P. et al. 1990. *Nature* 345: 535–537.

[59]Miledi, R., Parker, I. and Sumikawa, K. 1983. *Proc. R. Soc. Lond. B* 218: 481–484.

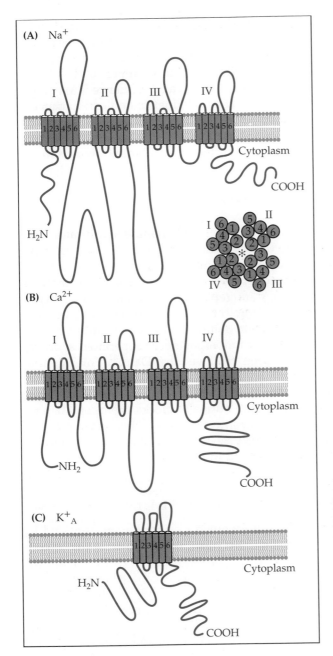

(A) Na$^+$

I II III IV

Cytoplasm

COOH

H$_2$N

I II
IV III

(B) Ca^{2+}

I II III IV

Cytoplasm

NH$_2$

COOH

(C) K^+_A

Cytoplasm

H$_2$N

COOH

13

ION CHANNEL TOPOLOGY. (A) Proposed membrane topology of the voltage-sensitive sodium channel, showing four domains (I–IV), each comprising six transmembrane helices. The inset indicates possible packing arrangement around the pore (∗). (B) Proposed structure for voltage-sensitive calcium channel is similar. (C) Potassium A channel subunit has a structure similar to a single domain of the sodium channel. Four such subunits combine to form a complete channel. (After Catterall, 1988, and Noda et al., 1984.)

sodium channels in their membranes. Yet after the appropriate message has been injected, they not only express the proteins but also assemble them in the membrane to form functionally active channels.

The first example involves the properties of ACh channels in bovine skeletal muscle. It is known that during development, the properties of the ACh channels change. In early development they are dispersed throughout the myotubes (Chapter 11). As they aggregate to form end plates, their mean open time decreases and their conductance increases. In parallel with these functional changes, the *g* subunit, present in the fetal form of the receptor, is replaced in the adult receptor by a different subunit (ε) with a similar, but distinct, amino acid sequence.[60] We might conclude, then, that the change in channel kinetics and conductance is related to the change in subunit structure of the channel. This conclusion has been confirmed by experiments in which channels were produced in oocytes by injection of different combinations of subunit mRNAs.[61] Oocytes injected with α, β, γ, and δ subunit mRNA expressed channels with fetal properties; those receiving message for α, β, ε, and δ subunits expressed channels with properties similar to those found in adult muscle.

Another technique available for functional analysis involves the construction of mutant cDNAs, with mutations directed at a particular site in the channel structure (site-directed mutagenesis), so that selected amino acids with particular properties (positively or negatively charged, or highly polar, or nonpolar) are replaced by others with different properties. The functional properties of channels expressed by oocytes, or other cells in vitro, after injection of mRNAs derived from the mutant cDNAs are then examined. A number of such experiments on the acetylcholine receptor have been reviewed by Dani.[62] For example, mutations of AChR subunits have indicated that the M2 helix forms part of the wall of the open channel. Mutations within M2 affected binding of a molecule (QX222) that normally blocks the open channel, and altered the channel conductance.[63] Specifically, M2 regions of mouse AChR α and δ subunits have the following amino acid sequences (from cytoplasm to extracellular fluid in Figure 10):

α: M T L S I S V L L S L T V F L L V I V

δ: T S V A I S V L L A Q S V F L L L I S

When the polar serines in the underlined positions were replaced by alanines (weakly hydrophobic), there was a reduction in both the conductance of the channel for outward current (from 24 to 13 pS) and the binding affinity for QX222. These effects suggest that the serine-negative dipoles, and hence all or part of the M2 helices, are in the wall of the aqueous channel and contribute directly to its functional properties.

Mutant subunits have been used to study subunit composition and stoichiometry of channels. As has already been mentioned, such experiments have provided convincing evidence in support of the notion that neuronal nAChR, like AChR from *Torpedo* and skeletal muscle, is assembled in the membrane as a pentamer with at least two α sub-

[60]Takai, T. et al. 1985. *Nature* 315: 761–764.
[61]Mishina, M. et al. 1986. *Nature* 321: 406–411.
[62]Dani, J. A. 1989. *Trends Neurosci.* 12: 127–128.
[63]Leonard, R. J. et al. 1988. *Science* 242: 1578–1581.

units.[39] Mutations at the beginning of the M2–M3 connecting loop were made in order to alter the conductance of channels formed from chicken α and non-α subunits. Coinjection into oocytes of normal and mutant α cDNA, together with native non-α, resulted in three different channel conductances. One of these was the same as the conductance of normal channels, and another the same as that of channels obtained with only mutant α cDNA. The third conductance presumably represented channels with one native and one mutant α subunit. Because there was only one such intermediate type, it was concluded that the channels contained only two α subunits. When native and mutant non-α cDNA were injected, the resulting channels showed four different conductances, indicating the presence of three non-α subunits. The deduced stoichiometry of the channel, then, was $\alpha_2(\text{non-}\alpha)_3$.

Mutant subunits have also been used to determine the stoichiometry of voltage-activated potassium A channels. It has generally been assumed that the potassium channel subunits aggregate to form a tetrameric structure analogous to that of the four-domain sodium channel protein. This assumption has been investigated by examining the sensitivity to the potassium channel blocker charybdotoxin (CTX) of heteromultimeric A channels formed in *Xenopus* oocytes by coexpression of subunits from wild-type *Drosophila* and toxin-resistant mutants.[64] When only wild-type subunits were expressed, forming homomultimeric channels, A currents in the oocyte membrane (produced by depolarization) were completely blocked by high concentrations (e.g., 500 n*M*) of CTX. Currents in oocytes with heteromultimeric channels were only partially blocked. Occupation by the toxin of a single site on the A channel is sufficient to produce block, so that at high toxin concentrations only homomultimeric channels formed from the mutant subunits should remain unaffected. The relative number of such channels can be calculated from the ratio of the injected mRNAs (assuming equal potency of expression) and the number of subunits in the channel structure. For example, if there are four subunits and if 90 percent of the RNA is mutant, then 66 percent $[(0.9)^4]$ of the channels will be composed entirely of mutant subunits. The remainder (34 percent) will have at least one wild-type subunit and be subject to block by the toxin. Block of this fraction was observed experimentally, providing direct evidence for a tetrameric structure. Block of the current by CTX should have been 27 percent had the channels been trimers, and 41 percent with pentamers.

One structure of particular interest in voltage-activated channels is the S4 region, which has attracted attention because it contains positively charged lysine or arginine side chains interspersed at regular intervals between the nonpolar residues. This feature (which is exhibited by the S4 regions of voltage-gated calcium and potassium channels as well) has led to the idea that the regularly spaced charges endow

[64]MacKinnon, R. 1991. *Nature* 350: 232–238.

the S4 segment with properties sufficient to link a transmembrane voltage change to channel opening. Thus application of a positive potential to the inside of the cell membrane (depolarization) would displace the positive charges outward, causing both an outward movement and rotation of the helix, and (by steps unknown) a consequent increase in the probability of channel opening. To test this idea, mutations have been directed at S4 regions in domains I and II of a rat brain sodium channel.[65]

In the native ("wild-type") rat brain channel, the S4 region of domain I contains the sequence

$$A \quad L \quad \mathbf{R^+} \quad T \quad F \quad \mathbf{R^+} \quad V \quad L \quad \mathbf{R^+} \quad A \quad L \quad \mathbf{K^+} \quad T \quad I$$

| Extracellular | 217 | 220 | 223 | 226 | Cytoplasmic |

with a positively charged arginine (R) or lysine (K) at every third position. The numbering is in the extracellular to cytoplasmic direction (Figure 14). In the mutants, neutral or acidic amino acids were substituted for one or more of the basic residues to determine the effect on channel activation. In oocytes injected with message for wild-type channels, a depolarization from a holding potential of -120 mV to -30 mV was required to open half the channels ($V_{1/2} = -30$ mV). Removing the positive charge at position 226 (near the cytoplasmic end of the region) by replacing the lysine with glutamine resulted in a reduction in voltage sensitivity, so that a larger depolarization, to -13 mV, was required to open half the channels in the membrane. When the positively charged lysine at position 226 was replaced by a negatively charged glutamic acid, $V_{1/2}$ was increased still further, to -3 mV. Similar effects were observed with charge removal or replacement in domain II.

The effects of charge removal and replacement near the cytoplasmic end of the helix are consistent with the idea that channel activation is associated with displacement of the S4 helices. Thus, when the charge on the helix is reduced, a greater voltage change is required to effect the displacement. However, removal of positive charges from the extracellular half of the helix resulted in a *reduction* in the depolarization required for half-maximal activation. For example, a double mutant in which the positive charges at positions 217 and 220 were both removed resulted in a $V_{1/2}$ of -51 mV.

Similar results have been obtained more recently with mutants in the S4 region of potassium A channels.[66] Overall, such experiments suggest that the S4 helix is indeed involved in channel gating, but their interpretation is difficult. The helix may translate or rotate in response to membrane depolarization, or it may provide a potential field for some other element acting as the voltage sensor. In the latter case, removal of a positive charge near the cytoplasmic face (position 226) would increase the transmembrane voltage field seen by the sensing element, and a greater depolarization would then be required for its

[65]Stühmer, W. et al. 1989. *Nature* 239: 597–603.
[66]Papazian, D. M. et al. 1991. *Nature* 349: 305–349.

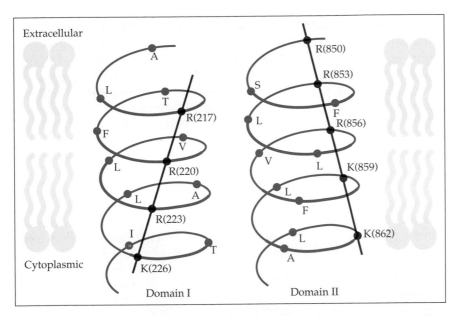

Extracellular

A
L
T
F R(217)
L
V
L R(220)
A
L
R(223)
I
T
K(226)

Cytoplasmic

Domain I

S
L
R(850)
R(853)
F R(856)
V
L K(859)
L
F
L K(862)
A

Domain II

14 S4 HELICES SPANNING THE CELL MEMBRANE in the first and second domains of the voltage-sensitive sodium channel. Every third amino acid in the
coils is a positively charged (basic) arginine (R) or lysine (K). The numbers indicate
their positions in the complete molecule. Replacing basic by neutral or acidic amino
acids affects voltage gating of the channel (see text).

activation. Removal of positive charge from near the extracellular face
(position 217) would have the reverse effect.

In summary, we have learned a great deal about the permeability
and kinetics of both ligand-activated and voltage-activated ion channels
by direct measurements of single-channel behavior. In addition, modern
methods of biochemistry and molecular biology have provided detailed
information about the molecular organization and structure of channels.
We know, for example, that channels are formed by four or more
polypeptide subunits, or domains, arranged in an orderly fashion
around a central core. Each subunit contains, in turn, four to six transmembrane segments joined by extra- and intracellular connecting loops.
Channels that are relatively selective, such as the voltage-dependent
channels, are usually (perhaps always) tetramers; larger, less selective,
ligand-activated channels are pentamers. As an extension of this principle, the very largest channels—the gap junctions—have a hexameric
structure (Chapter 5). One feature of ion channels is that their extra-
membraneous connecting loops form structures that extend considerable distances from the surfaces of the lipid bilayer (e.g., Figure 9).
Unwin pointed out that in ligand-activated cation channels, the loops
lining the walls of the projecting channel vestibules have a net excess
of negative charges; in anion channels, the excess is positive.[67] As the

[67]Unwin, N. 1989. *Neuron* 3: 665–676.

channel openings are of the order of 1–2 nm, and the effective radius for electrostatic interaction in physiological solutions is about 1 nm, excess charges in the vestibules can contribute considerably to accumulation of counterions. Such accumulation would add to both the conductance and the ion selectivity of the channels. As we will see later, such loops may be related to other specific functions, such as inactivation of voltage-activated channels (Chapter 4).

In this chapter we have presented only a brief overview of the breadth of our knowledge of ion channels. The crucial questions that we are now beginning to answer concern the relation between molecular structure and function: What are the molecular mechanisms whereby a pore opens in response to activation by a ligand, or in response to a change in voltage? What features of the channel structure determine its kinetic behavior? How do the structural elements determine its ionic selectivity? Answers to such questions are clearly close at hand.

SUGGESTED READING

General

Cold Spring Harbor Symposium on Quantitative Biology 48. 1983. pp. 1–146. [This series of papers describes original experiments on isolation, characterization, and sequencing of the nACh receptor.]

Kao, C. Y. and Levinson, S. R. (eds.). 1986. *Tetrodotoxin, Saxotoxin, and the Molecular Biology of the Sodium Channel. Ann. N.Y. Acad. Sci.* 479. [The papers in this monograph describe experiments on isolation, characterization and sequencing of the voltage-dependent sodium channel.]

Reviews

Betz, H. 1990. Ligand-gated ion channels in the brain: The amino acid receptor superfamily. *Neuron* 5: 383–392.

Dani, J. A. 1989. Site-directed mutagenesis and single channel currents define the ionic channel of the nicotinic acetylcholine receptor. *Trends Neurosci.* 12: 127–128.

Guy, H. R. and Conti, F. 1990. Pursuing the structure and function of voltage-gated channels. *Trends Neurosci.* 13: 201–206.

Leutje, C. W., Patrick, J. and Seguela, P. 1990. Nicotinic receptors in mammalian brain. *FASEB J.* 4: 2753–2760.

Olson, R. W. and Tobin, A. J. 1990. Molecular biology of $GABA_A$ receptors. *FASEB J.* 4: 1469–1480.

Unwin, N. 1989. The structure of ion channels in membranes of excitable cells. *Neuron* 3: 665–676.

Original Papers

Anderson, C. R. and Stevens, C. F. 1973. Voltage clamp analysis of acetylcholine-produced end-plate current fluctuations at frog neuromuscular junction. *J. Physiol.* 235: 655–691.

Cooper, E., Couturier, S. and Ballivet, M. 1991. Pentameric structure and subunit stoichiometry of a neuronal nicotinic acetylcholine receptor. *Nature* 350: 235–238.

Grenningloh, G., Rienitz, A., Schmitt, B., Methfessel, C., Zensen, M., Beyreuther, K., Gundelfinger, E. D. and Betz, H. 1987. A strychnine-binding subunit of the glycine receptor shows homology with nicotinic acetylcholine receptors. *Nature* 328: 215–220.

Hamill, O. P., Marty, A., Neher, E., Sakmann, B. and Sigworth, J. 1981. Improved patch-clamp techniques for high-resolution current recording from cells and cell-free membrane patches. *Pflügers Arch.* 391: 85–100.

MacKinnon, R. 1991. Determination of the subunit stoichiometry of a voltage-activated potassium channel. *Nature* 350: 232–238.

Sigel, E., Baur, R., Trube, G., Mohler, H. and Malherbe, P. 1990. The effect of subunit composition of rat brain GABA$_A$ receptors on channel function. *Neuron* 5: 703–711.

Sumikawa, K., Parker, I. and Miledi, R. 1989. Expression of neurotransmitter receptors and voltage-activated channels from brain mRNA in *Xenopus* oocytes. *Methods Neurosci.* 1: 30–45.

Unwin, N., Toyoshima, C. and Kubalek, E. 1988. Arrangement of acetylcholine receptor subunits in the resting and desensitized states, determined by cryoelectron microscopy of crystallized *Torpedo* postsynaptic membranes. *J. Cell Biol.* 107: 1123–1138.

CHAPTER THREE

IONIC BASIS
OF THE RESTING POTENTIAL

The electrical potential difference between the inside and the outside of a nerve cell membrane depends on the ionic concentration gradients across the cell membrane and the relative permeability of the membrane to the ions present. Using simple principles of physical chemistry, one can explain how resting membrane potentials arise in excitable cells. For a steady state to be maintained, the total distribution of ions on either side of a cell membrane must satisfy three major constraints: (1) The bulk solutions inside and outside the cell must each be electrically neutral; (2) the osmotic concentration of intracellular ions and molecules in solution must be equal to that in the extracellular fluid; and (3) there must be no net flux of any permeant ion across the membrane.

Each permeant ion species has quite different intracellular and extracellular concentrations and is subject to two separate gradients tending to drive it into or out of the cell: a concentration gradient and an electrical gradient. For example, potassium is more concentrated inside the cell than out, so outward movement of potassium ions along their concentration gradient would be expected. On the other hand, the inner surface of the membrane is negative with respect to the outside, tending to restrain the outward movement of positively charged ions. In normal resting cells, these concentration and electrical gradients are nearly in balance, so that the tendency of potassium ions to move out of the cell from high to low concentration is opposed almost exactly by the electrical gradient in the reverse direction. The membrane potential at which there is no net potassium flux is called the potassium equilibrium potential (E_K). The equilibrium potential for any ion, in terms of extracellular and intracellular ionic concentrations, is given by the Nernst equation.

Chloride concentration is higher outside the cell than inside, and this concentration gradient is again balanced by the membrane potential, the internal negativity tending to oppose inward movement of the negatively charged ion. The resting membrane potential is, however, determined mainly by the potassium concentration ratio because the internal chloride concentration, being low, can change to accommodate itself to changes in the resting potential.

Sodium is much more concentrated in the extracellular fluid than in the cell cytoplasm, so to oppose sodium entry the membrane would have to be positive on the inside; that is, the equilibrium potential for sodium (E_{Na}) is positive rather than negative. Thus, in a normal cell with a negative resting potential, the concentration gradient and membrane potential both favor inward movement of sodium. Although the resting membrane is only sparingly permeable to sodium, the inward sodium leak depolarizes the membrane slightly from the potassium equilibrium potential, so that there is an accompanying outward leak of potassium. To maintain a steady state in the face of these continual leaks, sodium is transported actively outward, and potassium inward, across the cell membrane.

The resting membrane potential depends on the relative permeabilities of the cell membrane to sodium and potassium. If the permeability is very much larger to potassium than to sodium, then the membrane potential will be very close to E_K. If the permeability to potassium is relatively smaller, then the membrane potential will be farther away from E_K and closer to E_{Na}. One way to express the dependence of membrane potential on cation concentrations and membrane permeabilities is the constant field equation. A more accurate description is provided by a steady-state equation that includes the contribution of the active transport processes for sodium and potassium.

These same conditions can be considered in terms of an electrical model of the membrane in which ionic equilibrium potentials are substituted for concentrations and membrane conductances for permeabilities.

The active-transport process for sodium and potassium involves a single protein, Na–K ATPase, that transports three sodium ions out of the cell and two potassium ions in for each molecule of ATP hydrolyzed. Other ions are also transported actively across the cell membrane, most of the transport processes being driven by the electrochemical gradient for sodium. In some cells chloride ions are transported outward and bicarbonate ions inward across the membrane by a process coupled to inward sodium movement. Other cells accumulate (rather than excrete) chloride, and at the same time accumulate potassium, in a similar way. Inward sodium movement is also coupled to proton excretion and to calcium extrusion. Calcium is also transported out of the cell by Ca–Mg ATPase.

Electrical signals are generated in nerve cells and muscle fibers primarily by changes in the permeability of the cell membrane to ions such as sodium and potassium, allowing them to move inward or outward across the cell membrane along established electrochemical gradients. As we have seen in the previous chapter, such changes in permeability are associated with activation of ion channels. Ions moving through the open channels change the charge on the cell membrane and hence change the membrane potential. In order to understand how such signals are generated, it is necessary to understand the nature of the established ionic gradients in the resting cell, and how such gradients come to exist.

Ions, membranes, and electrical potentials

It is useful to begin with the model cell shown in Figure 1. This cell contains potassium, sodium, chloride, and a large anion species and is bathed in a solution of sodium and potassium chloride. Other ions present in real cells, such as calcium or magnesium, are ignored for the moment, as their direct contributions to the membrane potential are negligible. The ionic concentrations inside and outside the model cell are similar to those found in frogs. In birds and mammals, ionic concentrations are somewhat higher; in marine invertebrates such as the squid, very much higher. In the model, the volume of the extracellular fluid is assumed to be infinitely large. Thus movements of ions and water into or out of the cell have no significant effect on extracellular concentrations. There are three major requirements for such a cell to exist:

1. The intracellular and extracellular solutions must each be electrically neutral. For example, a solution of chloride ions alone cannot exist; their charges must be balanced by an equal number of positive charges on cations such as sodium or potassium (otherwise electrical repulsion would literally blow the solution apart).
2. The cell must be in osmotic balance. If not, water will enter or leave the cell, causing it to swell or shrink, until osmotic balance is achieved. Osmotic balance is achieved when the total concentration of solute particles inside the cell is equal to that on the outside.
3. Finally, just as there can be no net movement of water, there must be no net movement of any particular ion into or out of the cell.

Another way of expressing the second and third conditions is to say that in the model cell, water and the permeating ions must be in EQUILIBRIUM. There is no equilibrium requirement for sodium and the internal anion, as both are impermeant. (We will see later that in real cells there is a small resting permeability to sodium and that, in general, none of the ions is in equilibrium.)

How are the equilibrium conditions satisfied for the permeant ions (potassium and chloride) and what electrical potential is developed across the cell membrane? Figure 1 shows that the two ions are distributed in reverse ratio; potassium is more concentrated on the inside of

1

IONIC DISTRIBUTIONS IN A MODEL CELL. The cell membrane is impermeable to Na^+ and to the internal anion (A^-), and permeable to K^+ and Cl^-. The concentration gradient for K^+ tends to drive it out of the cell (black arrow); the potential gradient tends to attract K^+ into the cell (blue arrow). In a cell at rest the two forces are exactly in balance. Concentration and electrical gradients for Cl^- are in the reverse directions. Ionic concentrations are expressed in mM.

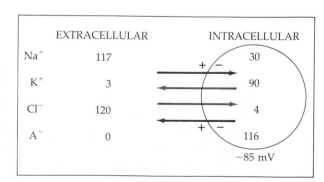

	EXTRACELLULAR	INTRACELLULAR
Na^+	117	30
K^+	3	90
Cl^-	120	4
A^-	0	116

−85 mV

the cell, chloride on the outside. Imagine first that the membrane is permeable only to potassium; the question that arises immediately is why potassium ions do not diffuse out of the cell until the concentrations on either side of the cell membrane are equal. The answer is that they cannot, because as they start to do so, a charge separation develops across the membrane; the resulting membrane potential then hinders further diffusion. As the ions leave the cell, positive charges accumulate on the outer surface of the membrane and an excess of negative charges is left on the inner surface. When the membrane potential becomes sufficiently large, further net efflux of potassium is stopped. The concentration gradient for potassium and the potential gradient across the membrane then balance one another exactly (arrows), and the potassium is said to be in electrochemical equilibrium. Individual ions may still enter and leave the cell, but no *net* movement occurs.

The conditions for potassium to be in equilibrium across the cell membrane are the same as those described in Chapter 2 for maintaining zero net flux through an individual channel in a membrane patch. There, a concentration gradient was balanced by a potential applied to the pipette. The important difference here is that the ion flux itself produces the required transmembrane potential. In other words, equilibrium in the model cell is automatic and inevitable.

Exactly how large a membrane potential is required to balance a given potassium concentration difference across the membrane? This potential is called the POTASSIUM EQUILIBRIUM POTENTIAL (E_K). One guess might be that the potential would be proportional simply to the difference between the intracellular concentration $[K]_i$ and the extracellular concentration $[K]_o$; this is not quite right. It turns out instead that the required potential depends on the difference between the *logarithms* of the concentrations:

$$E_K = k(\ln[K]_o - \ln[K]_i)$$

The constant k is given by RT/zF, where R is the thermodynamic gas constant, T the absolute temperature, z the valence of the ion (in this case $+1$), and F the Faraday (the number of coulombs of electrical charge in one mole of monovalent ion). The answer, then, is

$$E_K = (RT/zF)(\ln[K]_o - \ln[K]_i)$$

which is the same as

$$E_K = \frac{RT}{zF} \ln \frac{[K]_o}{[K]_i}$$

This is the NERNST EQUATION for potassium. The expression RT/zF has the dimensions of volts and is equal to about 25 mV at room temperature (20°C). It is sometimes more convenient to use the logarithm to the base 10 (log) of the concentration ratio, rather than the natural logarithm. Then RT/zF must be multiplied by $\ln(10)$, or 2.306, which gives a value of 58 mV. At mammalian body temperature (37°C), this value increases

The Nernst equation

to about 61 mV. For the cell shown in Figure 1, the concentration ratio for potassium is 1:30 and E_K is therefore 58 log(1/30) = −85 mV. Suppose now that, in addition to potassium channels, the membrane has chloride channels. Because for an anion $z = -1$, the equilibrium potential for chloride (E_{Cl}), in terms of the outside ($[Cl]_o$) and inside ($[Cl]_i$) concentrations, is given by

$$E_{Cl} = -58 \, \log \frac{[Cl]_o}{[Cl]_i}$$

or (from the properties of logarithmic ratios):

$$E_{Cl} = 58 \, \log \frac{[Cl]_i}{[Cl]_o}$$

In our model cell, $[Cl]_i/[Cl]_o$, like $[K]_o/[K]_i$, is 1/30 and E_{Cl} is therefore also −85 mV. As with potassium, this internal negativity balances exactly the tendency for chloride to move along its concentration gradient, in this case *into* the cell.

In summary, the tendency for potassium ions to leave the cell and for chloride ions to diffuse inward are both opposed by the membrane potential. Because the concentration ratios for the two ions are of exactly the same magnitude (1:30), their equilibrium potentials are exactly the same. As potassium and chloride are the only two ions that can move across the membrane and both are in equilibrium at −85 mV, the model cell can exist indefinitely without any net gain or loss of ions.

Electrical neutrality   The charge separation at the membrane means that there is an excess of anions at the inner surface and cations at the outer surface. This appears to violate the principle of electrical neutrality that we started out with, and indeed it does so in fact. Quantitatively, however, the charge separation produces differences in anionic and cationic concentrations so negligible that they could not possibly be measured. For example, if we consider our model cell to have a radius of 25 μm, then at a concentration of 120 mM there are 4×10^{12} cations and an equal number of anions inside the cell. At a resting membrane potential of −85 mV, we can calculate that there are approximately 4×10^7 excess negative charges on the inner surface of the membrane, which is 1/100,000 of the number of anions in free solution. Thus for every 100,000 cations inside the cell, there are 100,001 anions—a trivial difference.

The dependence of the resting potential on extracellular potassium   In neurons, and in many other cells, the membrane potential is sensitive to changes in extracellular potassium concentration but is relatively unaffected by changes in extracellular chloride concentration. To understand how this comes about it is useful to consider the consequences of such changes in the model cell. Figure 2A shows the changes in intracellular composition and membrane potential that result from increasing the extracellular potassium from 3 mM to 6 mM. This is done by replacing 3-mM NaCl with 3-mM KCl, thereby keeping the osmolarity unchanged with a total solute concentration of 240 mM. As a result of the increase in extracellular potassium concentration, the cell

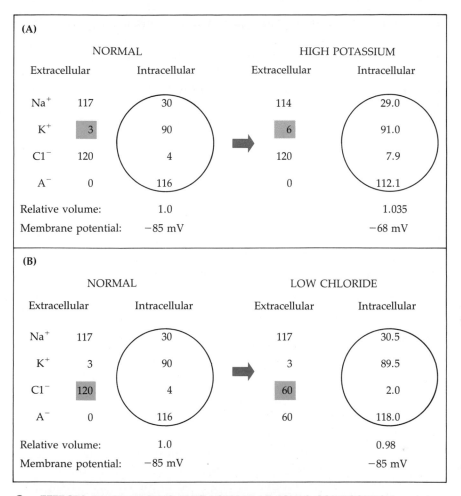

2 **EFFECTS OF CHANGING EXTRACELLULAR IONIC COMPOSITION** on intracellular ionic concentrations and on membrane potential. **(A)** Extracellular K^+ is
doubled, with a corresponding reduction in extracellular Na^+, to keep osmolarity constant. **(B)** Half the extracellular Cl^- is replaced by an impermeant anion, A^-. Ionic
concentrations are in mM, and extracellular volumes are assumed to be very large with
respect to cell volumes so that fluxes into and out of the cell do not change extracellular
concentrations.

is depolarized from −85 to −68 mV, the intracellular potassium concentration is increased slightly, and the intracellular chloride concentration
almost doubled. How does this change come about? It is achieved by
potassium and chloride entering the cell. First, when the external potassium concentration is increased, potassium is no longer in equilibrium. Consequently, potassium ions move inward. As positive charges
accumulate on the inner surface of the membrane, chloride ions, being
no longer in equilibrium, move in as well. This process of potassium
and chloride entry continues until a new equilibrium is established,

with both ions at a new concentration ratio, consistent with the new membrane potential.

Potassium and chloride entry is accompanied by entry of water to maintain osmotic balance, resulting in a slight increase in cell volume. When the new equilibrium is reached (Figure 2A), intracellular potassium has increased in concentration from 90 to 91 mM, intracellular chloride from 4 to 7.9 mM, and the cell volume has increased by 3.5 percent. At first glance it seems that more chloride than potassium has entered the cell, but think what the concentrations would have been if the cell had *not* increased in volume: The concentrations of both ions would be 3.5 percent greater than the indicated values. Thus the intracellular chloride concentration would be about 8.2 mM (instead of 7.9 mM after the entry of water), and the intracellular potassium concentration would be about 94.2 mM, both 4.2 mM higher than in the original solution. In other words, we can think first of potassium and chloride entering in equal quantities, and then of water following to achieve the final concentrations shown in the figure.

Effect of changing extracellular chloride Similar considerations apply to changes in the extracellular chloride concentration, but with a marked difference: No significant change occurs in membrane potential. The consequences of a 50 percent reduction in extracellular chloride concentration are shown in Figure 2B, in which 60 mM of chloride in the solution bathing the cell is replaced by an impermeant anion. Chloride leaves the cell because it is no longer in equilibrium and, as before, potassium and chloride move together, accompanied by water. Equal quantities of the ions leave the cell, and those remaining are concentrated slightly as the cell shrinks.

From Figure 2 we can arrive at some general conclusions: In the model cell, and in most real cells, changes in external potassium concentration result in changes in membrane potential, with the internal chloride concentration accommodating itself to the change; changes in external chloride concentration result in a similar accommodating change in internal chloride concentration without a major effect on membrane potential. This difference in the effects of a change in potassium concentration on the one hand, and a change in chloride concentration on the other, is due to the difference in their internal concentrations. When the external concentration of either is changed, the two ions move across the membrane in concert. Because the internal chloride concentration is low, relatively small ion fluxes are sufficient to adjust the chloride concentration ratio (and hence the chloride equilibrium potential) to the new conditions. In contrast, because of the high intracellular potassium concentration, small movements of KCl into or out of the cell have very little effect on the potassium concentration ratio. Thus potassium equilibrium can be restored only by an appropriate change in the membrane potential.

Membrane potentials in real cells The idea that the resting membrane potential is the result of an unequal distribution of potassium ions between the extracellular and intracellular fluids was first proposed by Julius Bernstein in 1902.[1] He

[1]Bernstein, J. 1902. *Pflügers Arch.* 92: 521–562.

could not test this hypothesis directly, however, because there was no satisfactory way of measuring membrane potential. It is now possible to measure membrane potential accurately and to see whether changes in external and internal potassium concentrations produce the potential changes predicted by the Nernst relation. The first such experiments were done on giant axons that innervate the mantle of the squid. The axons are up to 1 mm in diameter,[2] and their large size permits the insertion of recording electrodes into their cytoplasm to measure transmembrane potential directly (Figure 3A). Further, they are remarkably resilient and continue to function even when their axoplasm has been squeezed out with a rubber roller and replaced with an internal perfusate (Figure 3B)! Thus their internal as well as external ionic composition can be controlled. A. L. Hodgkin, who together with A. F. Huxley initiated many experiments on squid axon (for which they later received the Nobel prize), has said:[3]

> It is arguable that the introduction of the squid giant nerve fiber by J. Z. Young in 1936 did more for axonology than any other single advance during the last forty years. Indeed a distinguished neurophysiologist remarked recently at a congress dinner (not, I thought, with the utmost tact), 'It's the squid that really ought to be given the Nobel Prize.'

The concentrations of some of the major ions in squid blood and in the axoplasm of the squid nerves are given in Table 1 (several ions, such as magnesium and internal anions, are omitted). Experiments on isolated axons are usually done in seawater, with the ratio of intracellular to extracellular potassium concentrations being 40:1. If the membrane potential (V_m) were equal to the potassium equilibrium potential, it would be −93 mV. In fact, the measured membrane potential is considerably less negative (about −65 to −70 mV). On the other hand, the membrane potential is more negative than the chloride equilibrium potential, which is about −55 mV.

Bernstein's original hypothesis was tested not only by measuring the resting potential and comparing it with the potassium equilibrium potential, but also by examining how changes in extracellular potassium concentration affected the potential. (As with our model cell, such changes would be expected to produce no significant change in internal potassium concentration.) From the Nernst equation, changing the concentration ratio by a factor of 10 should change the membrane potential by 58 mV at room temperature. The results of such an experiment on squid axon, in which the external potassium concentration was changed, are shown in Figure 4. The external concentration is plotted on a logarithmic scale on the abscissa, and the membrane potential on the ordinate. The expected slope of 58 mV per tenfold change in extracellular potassium concentration is realized only at relatively high concentrations (solid straight line), with the slope becoming less and less

[2]Young, J. Z. 1936. *J. Microsc. Sci.* 78: 367–386.
[3]Hodgkin, A. L. 1973. *Proc. R. Soc. Lond. B* 183: 1–19.

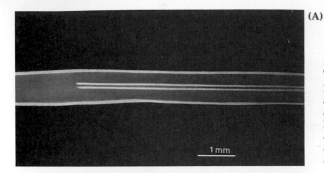

(A)

3

ISOLATED GIANT AXON of the squid (A), with axial recording electrode inside. (B) Extrusion of axoplasm from the axon, which is then cannulated and perfused internally. (C) Comparison of action potentials from a perfused and an intact axon. (A from Hodgkin and Keynes, 1956; B, C after Baker, Hodgkin, and Shaw, 1962.)

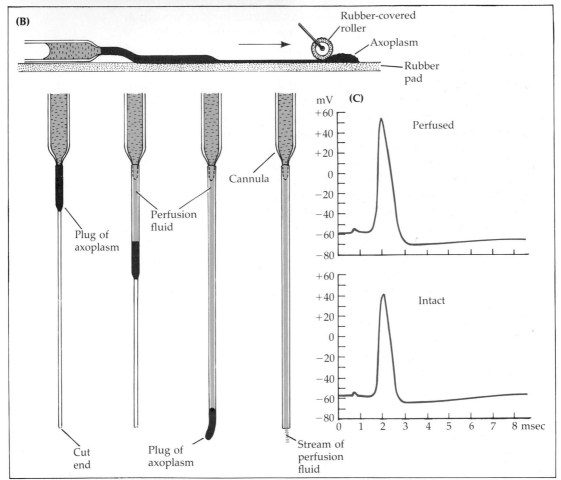

as the external potassium concentration is reduced. This result suggests that the potassium ion distribution is a major, but not the only, factor contributing to the membrane potential.

The effect of sodium permeability

From the experiments on squid axon, then, we can conclude that the hypothesis made by Bernstein in 1902 was almost correct; the membrane potential is strongly but not exclusively dependent on the potas-

TABLE 1
CONCENTRATIONS OF IONS INSIDE AND OUTSIDE FRESHLY ISOLATED AXONS
OF SQUID

Ion	Concentration (mM)		
	Axoplasm	**Blood**	**Seawater**
Potassium	400	20	10
Sodium	50	440	460
Chloride	60	560	540
Calcium	0.0001	10	10

Modified from Hodgkin (1964); ionized intracellular calcium from Baker, Hodgkin and
Ridgeway (1971).

sium concentration ratio. How do we account for the deviation from
the Nernst relation shown in Figure 4? Simply by abandoning the notion
that the membrane is impermeable to sodium. Real cell membranes
have, in fact, a permeability to sodium that ranges between 1 and 10
percent of their permeability to potassium. In the model, and in the
squid axon, the concentration gradient and the membrane potential
both tend to drive sodium into the cell. As sodium ions enter, the
accumulation of positive charge depolarizes the membrane. As a result,
potassium is no longer in equilibrium and potassium ions leave the cell.
As we will discuss later, the intracellular sodium and potassium concentrations are maintained in the face of these constant leaks by a
transport system that uses metabolic energy to pump sodium out of the
cell and accumulate potassium. Unlike our original model, the cell is
not in equilibrium; metabolic energy is used to maintain it in a STEADY
STATE. For the moment we will ignore the transport system to consider
the immediate question of how the membrane permeability to sodium
affects the resting potential.

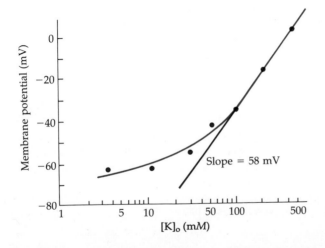

4

**MEMBRANE POTENTIAL AND EXTERNAL
POTASSIUM** concentration in squid axon, plotted on a semilogarithmic scale. The solid straight
line is drawn with a slope of 58 mV per tenfold
change in extracellular potassium concentration,
according to the Nernst equation. Because the
membrane is also permeable to sodium, the
points deviate from the straight line, especially
at low potassium concentrations (see text). (After
Hodgkin and Keynes, 1955.)

How to describe the membrane potential in a cell whose membrane is permeable to sodium, as well as to potassium and chloride, was considered originally by Goldman,[4] and later by Hodgkin and Katz.[5] The equation for the membrane potential is sometimes known, therefore, as the GHK EQUATION. The derivation of the equation is based on the idea that if the potential is constant, then the charge on the membrane must not change although ions leak across it. Thus currents carried by the inward leak of sodium, the outward leak of potassium, and any existing leak of chloride must add up to zero. Otherwise there would be a steady accumulation or loss of charge and hence a steady drift in membrane potential. Because the ionic currents depend on how readily the ions cross the membrane, the resulting equation includes the membrane permeability to each ion (p_K, p_{Na}, and p_{Cl}) as well as the ionic concentrations:

$$V_m = 58 \log \frac{p_K[K]_o + p_{Na}[Na]_o + p_{Cl}[Cl]_i}{p_K[K]_i + p_{Na}[Na]_i + p_{Cl}[Cl]_o}$$

This equation is known also as the CONSTANT FIELD EQUATION because one of the assumptions made in arriving at the expression was that the voltage gradient (or "field") across the membrane is uniform. The equation looks like the Nernst equation, but with all the ions included instead of just one. This similarity extends to the fact that, unlike the cation concentrations, the internal chloride concentration appears in the numerator and the external concentration in the denominator. The equation differs from the Nernst equation in that it includes the ionic permeabilities in addition to concentrations.

If there is no net chloride current across the membrane, then the constant field equation can be written

$$V_m = 58 \log \frac{[K]_o + b[Na]_o}{[K]_i + b[Na]_i}$$

where $b = p_{Na}/p_K$. It can be seen from the equation that if the permeability to sodium is very much smaller than that to potassium (i.e., if b is very small), then the membrane potential will be close to the potassium equilibrium potential. Conversely, if the permeability to sodium is relatively high (b is large), then the membrane potential will be near the sodium equilibrium potential. Thus the equation is consistent with what we might expect intuitively. It does not, however, provide a precise description of the resting membrane potential, because it fails to take into account the ion transport processes that maintain the intracellular sodium and potassium concentrations.

Active transport and
the steady-state
equation

Active transport of sodium and potassium across the cell membrane will be discussed shortly. For now it is sufficient to state that the major transport system for the two ions is the NA–K PUMP, which transports

[4]Goldman, D. E. 1943. *J. Gen. Physiol.* 27: 37–60.
[5]Hodgkin, A. L. and Katz, B. 1949. *J. Physiol.* 108: 37–77.

three sodium ions out of the cell for each two potassium ions carried inward; in other words the Na:K coupling ratio of the pump is 3:2. Because the pump is not electrically neutral, it contributes directly to the membrane potential and is said to be ELECTROGENIC.

The relations between ionic permeabilities, ion transport, and membrane potential were considered in detail by Mullins and Noda,[6] who used intracellular microelectrodes to study the effects of ionic changes on membrane potential in muscle. They considered their experimental results in relation to a true steady-state condition, in which the net movement of *each* ion across the membrane is equal to zero. To use potassium as an example, its inside and outside concentrations determine E_K. The difference between the membrane potential and E_K, plus the permeability of the membrane to potassium, in turn determine the rate of outward potassium leak. In a steady state, this outward leak must be exactly equal in magnitude to the rate at which potassium is being transported into the cell. Accordingly, for a given concentration gradient, permeability, and transport rate, the resulting membrane potential must be positive to E_K by just the right amount, so that potassium ions leak out of the cell at the same rate as they are pumped in. The same type of argument applies to sodium.

Such considerations lead to a relatively straightforward expression for the membrane potential:

$$V_m = 58 \log \frac{r[K]_o + b[Na]_o}{r[K]_i + b[Na]_i}$$

where b, as before, is the ratio of sodium to potassium permeability (p_{Na}/p_K) and r is the coupling ratio of the transport system (transport$_{Na}$/transport$_K$). Because the net fluxes of the pertinent ions are zero, we can refer to the relation derived by Mullins and Noda as the STEADY-STATE EQUATION. Theoretically, it provides an exact description of the resting membrane potential, provided all the other permeant ions (e.g., chloride) are in a steady state.

How does the transport ratio for sodium relative to potassium (r) affect the steady-state membrane potential? If sodium and potassium are transported at the same rate ($r = 1$), the pump is not electrogenic and has no direct effect on resting potential; the membrane potential predicted by the steady-state equation is the same as that predicted by the constant field equation. If sodium is transported more rapidly than potassium ($r > 1$) the membrane is hyperpolarized; in the reverse situation ($r < 1$) the membrane is depolarized. The size of the pump contribution depends on a number of factors, particularly the relative ionic permeabilities. For a transport ratio of 3:2, steady-state contribution to the resting membrane potential is limited to about -11 mV.[7] If the transport process is stopped, the electrogenic contribution disap-

Contribution of the transport system to membrane potential

[6]Mullins, L. J. and Noda, K. 1963. *J. Gen. Physiol.* 47: 117–132.
[7]Martin, A. R. and Levinson, S. R. 1985. *Muscle Nerve* 8: 359–362.

pears immediately, and the membrane potential then declines gradually as the cell gains sodium and loses potassium.

Under some conditions—for example, after accumulation of excess sodium inside a cell—the pump can be stimulated to transport sodium and potassium at rates that far exceed their passive leak rates. Such activity can produce a large net efflux of cations, resulting in a pronounced hyperpolarization that declines gradually as steady-state conditions are restored (see Figure 7).

Chloride distribution   How do the steady-state considerations apply to chloride? As we have already shown, chloride is able to reach equilibrium simply by an appropriate adjustment in internal concentration, without affecting the steady-state membrane potential. In many cells, however, there are transport systems for chloride as well. These are discussed in more detail later. In squid axon and in muscle, chloride is transported into the cells; in many nerve cells transport is outward. In either case, the intracellular chloride concentration is "pumped up" (or down) to a steady-state value such that the leak out of (or into) the cell matches the rate of active transport in the opposite direction.[8]

Predicted values of membrane potential   How do these considerations explain the relation between potassium concentration and membrane potential shown in Figure 4? This can be seen by using real numbers in the equations. In squid axon, the relative permeability constants for sodium and potassium are roughly in the ratio $0.04:1.0$.[5] We can use this p_{Na}/p_K ratio (b) together with the ionic concentrations given in Table 1 to calculate the resting membrane potential in seawater:

$$V_m = 58 \log \frac{(1.5)10 + (0.04)(460)}{(1.5)400 + (0.04)(50)} = -73 \text{ mV}$$

The constant field equation gives a smaller value:

$$V_m = 58 \log \frac{10 + (0.04)460}{400 + (0.04)50} = -67 \text{ mV}$$

The difference (6 mV) is the electrogenic contribution of the Na^+–K^+ transport system.

The numerical examples are useful in illustrating quantitatively why, when the extracellular potassium concentration is altered, the membrane potential does not follow the Nernst potential for potassium, as shown in Figure 4. If, in the numerator of the constant field equation, we look at the magnitude of the extracellular potassium concentration (10 mM) and the effective sodium concentration ($0.04 \times 460 = 18.4$ mM), we see that potassium contributes only about 35 percent of the total. Because of this, doubling the external potassium concentration does not double the numerator (as would happen in the Nernst equation) and, as a consequence, the effect on potential of changing the extracellular potassium concentration is less than would be expected if

[8]Martin, A. R. 1979. Appendix to G. Matthews and W. O. Wickelgren. *J. Physiol.* 293: 393–414.

potassium were the only permeant ion. When the external potassium concentration is raised to a high enough level, the potassium term becomes sufficiently dominant for the relation to approach the theoretical limit of 58 mV per tenfold change in concentration (Figure 4). This effect is further enhanced by a factor not yet discussed: The ionic permeabilities are not constant. In particular, when the membrane is depolarized, voltage-sensitive potassium channels are activated. Because of the increased permeability to potassium, the relative contribution of sodium to the membrane potential (represented in the equation by the permeability ratio *b*) is reduced.

In summary, the membranes of real nerve cells are permeable to sodium, potassium, and chloride. Sodium and potassium concentrations inside the cell are kept constant by a Na–K transport mechanism that has a transport ratio of three sodium to two potassium and is therefore electrogenic. Chloride may be in equilibrium in some nerve cells, or be transported either inward or outward in others. These features are summarized in Figure 5, which shows the relative magnitudes and directions of the passive and active cation fluxes in a neuron at rest.

So far we have discussed the resting membrane potential in terms of ionic concentrations and permeabilities. These same principles can be represented in a rather different way by an electrical model of the membrane, shown in Figure 6. Concentration ratios of the major ions are represented by their EQUILIBRIUM POTENTIALS (E_K, E_{Na}, and E_{Cl}) and ionic permeabilities are represented by CONDUCTANCES. The conductance of the membrane to a given ion (g_{Na}, g_K, or g_{Cl}) is simply the sum of the conductances of all the open channels permeable to that ion.

An electrical model of the resting membrane

In the electrical model, the inward leak of sodium through the resting cell membrane is expressed as sodium current (i_{Na}), which is proportional to the net ionic flux through the sodium channels. This current depends on two factors: (1) the sodium conductance (g_{Na}) and (2) the driving potential, which is the difference between the membrane potential and the sodium equilibrium potential ($V_m - E_{Na}$). Thus:

$$i_{Na} = g_{Na}(V_m - E_{Na})$$

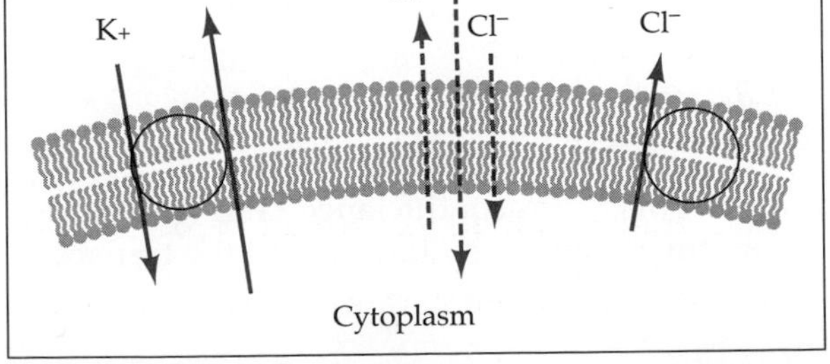

5

IONIC LEAKS AND PUMPS IN A CELL IN A STEADY STATE. Net passive ion movements across the membrane are indicated by dashed arrows, transport systems by solid arrows and circles. Lengths of arrows indicate the magnitudes of net ion movements. Total flux is zero for each ion. For example, net inward leak of sodium ions is equal to rate of outward transport. Na:K transport is coupled with a ratio of 3:2.

6

ELECTRICAL MODEL OF THE STEADY STATE
CELL MEMBRANE. E_K, E_{Na}, and E_{Cl} are the
Nernst potentials for the individual ions. The
individual ionic conductances are represented by
resistors (having a resistance $1/g$ for each ion).
The individual ionic currents i_K and i_{Na} are equal
and opposite to the currents $i_{T(K)}$ and $i_{T(Na)}$ supplied by the active transport pump, $T_{Na,K}$, so that
the net flux across the membrane of each ion is
zero. The resulting membrane potential is V_m.
For simplicity, it is assumed that $E_{Cl} = V_m$ so that
$i_{Cl} = 0$.

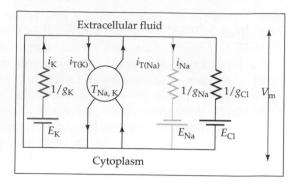

Similarly, for potassium:

$$i_K = g_K (V_m - E_K)$$

and for chloride:

$$i_{Cl} = g_{Cl}(V_m - E_{Cl})$$

These currents are equivalent to the ionic fluxes across the cell membrane shown by the arrows in Figure 5. They are represented in the
electrical model as currents produced by batteries and resistors (i.e.,
conductances) spanning the membrane. It is important to note that
under normal resting conditions, i_{Na} is negative, which is the conventional sign for *inward* current across the membrane. Normally, i_K is
outward and hence positive. As noted previously, chloride flux, and
hence i_{Cl}, can be in either direction (or zero), depending upon the
direction of transport.

To complete the electrical model in Figure 6, an active transport
mechanism ($T_{Na,K}$ acting as a battery charger) is added. This corresponds to the sodium–potassium exchange pump in Figure 5. For simplicity, chloride transport is ignored. The Na–K pump produces ionic
currents $i_{T(Na)}$ and $i_{T(K)}$. In the steady state, these must be equal and
opposite to the corresponding leak currents. In other words, for sodium
and potassium:

$$i_{T(Na)} = -g_{Na}(V_m - E_{Na})$$

$$i_{T(K)} = -g_K(V_m - E_K)$$

These two simple relations can be used to arrive at the membrane
potential of the cell represented by the electrical model. Dividing one
by the other, we write:

$$\frac{i_{T(Na)}}{i_{T(K)}} = \frac{-g_{Na}(V_m - E_{Na})}{-g_K(V_m - E_K)}$$

If we designate the sodium-to-potassium conductance ratio (g_{Na}/g_K) by
b' (numerically different from the permeability ratio b) and the transport

ratio by $-r$ (negative because the ions are transported in opposite directions), then the relation becomes:

$$-r = b' \frac{(V_m - E_{Na})}{(V_m - E_K)}$$

Rearranging, we get:

$$V_m = \frac{rE_K + b'E_{Na}}{r + b'}$$

This is the electrical equivalent of the steady-state equation derived by Mullins and Noda, with equilibrium potentials rather than concentrations, and conductances rather than permeabilities. The electrical equivalent of the constant field equation can be obtained simply by assuming $r = 1$:

$$V_m = \frac{E_K + b'E_{Na}}{1 + b'}$$

The resting conductances of membranes to sodium, potassium, and chloride have been determined in many nerve cells. It is a curious fact, however, that no precise identification of all the channels underlying these resting conductances has been made in any specific cell. A large number of potassium channel types have been identified (Chapter 4); many of them are activated by changes in membrane potential or by chemical ligands of one sort or another. Candidates for those active in the resting membrane vary from one cell to the next. Among the contributors to resting potassium conductance are channels activated by intracellular cations: sodium-activated and calcium-activated potassium channels. In addition, many nerve cells have M CHANNELS that are open at rest and closed by intracellular messengers. It is unlikely that a large fraction of voltage-activated potassium channels (DELAYED RECTIFIER and A CHANNELS) are open at rest; nevertheless only 0.1 to 1 percent of the total number would be required to account for a substantial fraction of the resting conductance.[9]

The specific source of the resting sodium conductance of nerve cells is also uncertain. Part of the conductance is due to movement of sodium through potassium channels, most of which have a sodium-to-potassium permeability ratio between 0.01 and 0.03.[10] In addition, both inward sodium and outward potassium leak may be through cation channels that show little selectivity for potassium over sodium.[11,12] Finally,

> What ion channels are associated with the resting membrane potential?

[9]Edwards, C. 1982. *Neuroscience* 7: 1335–1366.

[10]Hille, B. 1992. *Ionic Channels of Excitable Membranes*, 2nd ed. Sinauer, Sunderland, MA. p. 352.

[11]Yellen, G. 1982. *Nature* 296: 357–359.

[12]Chua, M. and Betz, W. J. 1991. *Biophys. J.* 59: 1251–1260.

tetrodotoxin has been shown to block a fraction of the resting sodium conductance, indicating a contribution by voltage-activated sodium channels.[9]

In central nervous system neurons, chloride channels may account for as much as 10 percent of the resting membrane conductance[13] and channels presumed to underlie this conductance have been described.[14]

ACTIVE TRANSPORT OF IONS

The Na–K Pump   The viability of nerve cells is maintained by the constant transport of sodium and potassium across the cell membranes against their electro-chemical gradients. This perpetual task is carried out by the Na–K pump, the required energy being obtained from hydrolysis of adenosine triphosphate (ATP). Indeed, it has been shown that the phosphatase itself is an integral part of the ion transport system. The properties of the enzyme have been summarized succinctly in a review by Skou.[15] It consists of two molecular subunits: α, with an apparent molecular mass of about 100 kD, and β, about 38 kD. The active enzyme appears to exist in the membrane as a tetramer, $(\alpha\beta)_2$. The stoichiometry of the enzyme is as expected from the transport characteristics: An average of three sodium and two potassium ions are bound for each molecule of ATP hydrolyzed. The requirement for sodium is remarkably specific. It is the only substrate accepted for net outward transport; conversely it is the only monovalent cation *not* accepted for inward transport. Thus lithium, ammonium, rubidium, cesium, and thallium are all able to substitute for potassium in the external solution but not for sodium in the internal solution. The requirement for external potassium is not absolute. In its absence the pump will extrude sodium at about 10 percent of capacity in an "uncoupled" mode. The transport system is blocked specifically by the digitalis glycosides, particularly ouabain and strophanthidin.

Both the α and β subunits have been sequenced,[16,17] and various models have been proposed for their tertiary structure. The α subunit has six major hydrophobic regions capable of forming transmembrane helices; the β subunit has only one such region. Various schemes for the transport mechanism have been proposed. All involve alternate exposure of sodium and potassium binding sites (presumably within a channel-like structure) to the extracellular and intracellular solutions. The cyclic conformational changes are driven by phosphorylation and dephosphorylation of the protein and are accompanied by changes in binding affinity for the two ions. Thus sodium is bound during intra-cellular exposure of the sites and subsequently released to the extracel-

---

[13]Gold, M. R. and Martin, A. R. 1983. *J. Physiol.* 342: 99–117.
[14]Krouse, M. E., Schneider, G. T. and Gage, P. W. 1986. *Nature* 319: 58–60.
[15]Skou, J. C. 1988. *Methods Enzymol.* 156: 1–25.
[16]Kawakami et al. 1985. *Nature* 316: 733–736.
[17]Noguchi et al. 1986. *FEBS Letters* 196: 315–320.

lular solution; potassium is bound during extracellular exposure and released to the cytoplasm.

Transport of sodium and potassium was studied in squid axon by Hodgkin and Keynes and their colleagues[18,19] and in snail neurons by Thomas.[20,21] To examine the relations among internal sodium concentration, pump current, and membrane potential, Thomas used two intracellular pipettes to deposit ions in the cell, one filled with sodium acetate and the other with lithium acetate (Figure 7A). A third intracellular pipette was used as an electrode to record membrane potential. A fourth pipette was used as a current electrode for voltage clamp experiments (Chapter 4), and a fifth, made of sodium-sensitive glass, to monitor the intracellular sodium concentration. To inject sodium, the sodium-filled pipette was made positive with respect to the lithium pipette. Thus, current flow in the injection system was between the two pipettes, with none of the injected current flowing through the cell membrane. The result of such a sodium injection is shown in Figure 7B. After a brief injection the cell became hyperpolarized by about 20 mV and gradually recovered over several minutes. Injection of lithium (by making the lithium pipette positive) produced no hyperpolarization.

Several lines of evidence showed that the potential change after sodium injection was due to the action of a sodium pump and not to changes in membrane permeability. For example, the input resistance of the cell did not decrease, as might be expected if hyperpolarization were the result of an increased permeability to potassium or chloride. The hyperpolarization could, however, be greatly reduced or abolished by addition of the transport inhibitor ouabain to the bathing solution (Figure 7C), as would be expected if it were due to pump activity. Similarly, sodium injection had little effect on potential when potassium was absent from the external solution; reintroduction of potassium after injection, however, resulted in immediate hyperpolarization (Figure 7D).

Quantitative estimates of the pump rate and the exchange ratio were obtained by voltage clamp experiments in which membrane current was measured while the membrane potential was being held constant (clamped). At the same time, intracellular sodium concentration was monitored. Sodium injection gave rise to an outward surge of current whose amplitude and duration followed the intracellular sodium concentration (Figure 7E). The total charge carried out of the cell, measured by integrating the total membrane current, was only about one-third of the charge injected in the form of sodium ions. This evidence was consistent with the idea that for every three sodium ions pumped out of the cell, two potassium ions were carried inward.

Experimental evidence that the pump is electrogenic

[18]Hodgkin, A. L. and Keynes, R. D. 1955. *J. Physiol.* 128: 28–60
[19]Baker, P. F. et al. 1969. *J. Physiol.* 200: 459–496.
[20]Thomas, R. C. 1969. *J. Physiol.* 201: 495–514.
[21]Thomas, R. C. 1972. *J.Physiol.* 220: 55–71.

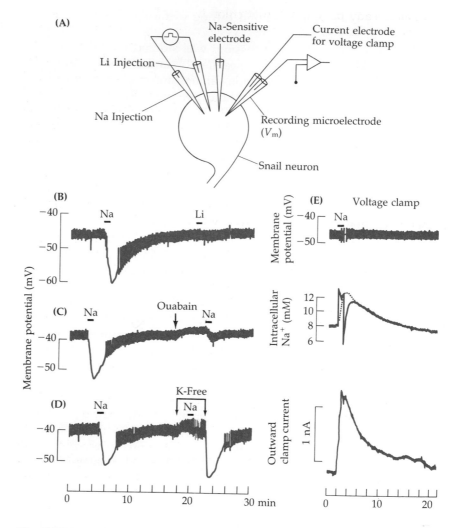

7   **EFFECT OF SODIUM INJECTION. Changes in intracellular sodium concentration,** membrane potential, and membrane current following injection of sodium into snail neurons. **(A)** Sodium is injected by passing current between two electrodes filled with sodium acetate and lithium acetate (see text). A sodium-sensitive electrode measures $[Na]_i$; two other electrodes measure membrane potential and pass current through the cell membrane to obtain the voltage clamp records in **(E)**. **(B)** Hyperpolarization of the membrane following intracellular injection of sodium. (The small rapid deflections are spontaneously occurring action potentials, reduced in size because of the poor frequency response of the pen recorder.) Injection of lithium does not produce hyperpolarization. **(C)** After application of ouabain (20 μg/ml), which blocks the sodium pump, hyperpolarization by sodium injection is greatly reduced. **(D)** Removal of potassium from the extracellular solution blocks the pump, so that sodium injection produces no hyperpolarization until potassium is restored. **(E)** Voltage clamp records. Sodium injection results in increased intracellular sodium concentration and in outward current across the cell membrane. The sharp deflections on the sodium concentration record are artifacts from the injection system. The time course of the concentration change is indicated by dashed lines. (After Thomas, 1969.)

It has already been noted that chloride may be transported into or out of nerve cells. Such transport is accompanied by movement of other ions. In neurons in which chloride is transported out of the cell, the outward movement is associated with inward transport of bicarbonate. Such a chloride–bicarbonate exchange mechanism is found in a number of cell types and has been studied by Thomas in relation to intracellular pH regulation in snail neurons.[22] Recovery from acidification of the cytoplasm (by exposure to CO_2 or intracellular injection of HCl) was prolonged when extracellular bicarbonate concentration was reduced or when intracellular chloride was depleted. Furthermore, recovery was virtually abolished when sodium was removed from the extracellular bathing solution. The recovery, then, appeared to involve inward movement of sodium and bicarbonate in exchange for chloride. Such an exchange mechanism exists in a variety of cells types; it is inhibited by 4-aceto-4'-isothiocyanostilbene-2,2'-disulfonic acid (SITS) and a related compound, DIDS. Although both anions are carried against their electrochemical gradients, the exchange mechanism is not an ATPase. Instead, the energy required appears to be obtained through passive inward movement of sodium down its electrochemical gradient.

An outward transport mechanism for chloride that is SITS-insensitive has been reported in mammalian cortical neurons.[23] The system appears to involve co-transport of chloride and potassium out of the cells. Whether or not the effluxes are coupled to inward sodium movement is not clear.

In describing the chloride–bicarbonate exchange system, we have introduced a new idea into our consideration of the properties of cell membranes, namely SECONDARY ACTIVE TRANSPORT, in which the electrochemical gradient for sodium is used to transport other ions across the membrane against their individual electrochemical gradients. Although part of the continual movement of sodium into the cell is simply through passive channels, some sodium entry is through such coupled transport mechanisms. The mechanisms require the sodium gradient for their operation and are, therefore, ultimately dependent upon maintenance of that gradient by Na–K ATPase. As a general rule, such mechanisms tend to be electrically neutral; that is, there is no net transfer of charge across the membrane. Such neutrality is achieved by the appropriate stoichiometry; for example, transport of one sodium and two bicarbonate ions into the cell for every chloride transported outward. A simpler example is the 1:1 sodium–hydrogen exchange mechanism, which also contributes to the maintenance of intracellular pH.[24] Protons are carried out of the cell against their electrochemical gradient in exchange for inward movement of sodium.

Active transport of chloride and bicarbonate

Secondary active transport

[22]Thomas, R. C. 1977. *J. Physiol.* 273: 317–338.
[23]Thompson, S. M., Deisz, R. A. and Prince, D. A. 1988. *Neurosci. Lett.* 89: 49–54.
[24]Moody, W. J. 1981. *J. Physiol.* 316: 293–308.

Inward chloride
transport

In cells such as muscle fibers and squid axon, chloride is actively accumulated. It has been shown by Russell that in squid axon inward chloride transport requires both sodium and potassium in the extracellular bathing medium.[25] It appears that for the entry of every two sodium ions, three chloride ions and one potassium ion are carried into the cell; that is, there is a Na:K:Cl ratio of 2:1:3,[26] with the sodium influx again supplying the necessary energy for transport of the other ions against their electrochemical gradients. This chloride transport system is insensitive to DIDS but is blocked by furosemide and bumetamide, substances known to block chloride transport in other tissues such as renal tubular cells.

Regulation of
intracellular calcium

As we will see later, changes in intracellular calcium concentrations play an important role in many neuronal functions such as action potential generation, release of neurotransmitters at synapses, initiation of postsynaptic conductance changes, and photoreceptor responses. In addition, calcium plays a primary role in initiation of muscle contraction. Intracellular concentrations of free calcium have been measured by injecting molecules such as aequorin[27] or FURA2[28] that emit or absorb light in the presence of ionized calcium. The absorption or fluorescence, which is dependent on calcium concentration, is then monitored with highly sensitive optical techniques. In squid axon and in various other neurons, the resting intracellular concentration of free calcium ranges from 10 to 100 nM. The extracellular calcium concentration is about 10 mM in squid blood (Table 1) and 2 to 5 mM in vertebrate interstitial fluid. Maintenance of the low intracellular concentrations requires that calcium be transported out of the cells against a very large electrochemical gradient.

Two major transport systems are responsible for extrusion of calcium from the cytoplasm across the plasma membrane.[29,30] The first is an ATPase, activated by calcium; magnesium is a necessary cofactor for ATP binding. The enzyme, then, is known as Ca–Mg ATPase. It has a high affinity for calcium ($K_{m(Ca)} < 300$ nM) and transports one calcium ion out of the cell for every ATP hydrolyzed. The distribution of the enzyme in the plasma membrane is generally sparse, however, so that the transport capacity is low. Nevertheless, it serves to maintain the low cytoplasmic calcium concentration in resting cells.

The second means of extruding calcium across the plasma membrane is sodium–calcium exchange, driven by the electrochemical gradient for sodium. The system has a lower affinity for calcium ($K_{m(Ca)} \leq 1.0$ μM), but about 50 times the transport capacity. In most cells, one calcium ion

[25]Russell, J. M. 1983. *J. Gen. Physiol.* 81: 909–925.
[26]Altamirano, A. A. and Russell, J. M. 1987. *J. Gen. Physiol.* 89: 669–686.
[27]Baker, P. F., Hodgkin, A. L. and Ridgeway, E. B. 1971. *J. Physiol.* 218: 709–755.
[28]Tsien, R. Y. 1988. *Trends Neurosci.* 11: 419–424.
[29]Blaustein, M. P. 1988. *Trends Neurosci.* 11: 438–443.
[30]Carafoli, E. 1988. *Methods Enzymol.* 157: 3–11.

is transported outward for each three sodium ions entering the cell.[31] Consequently the exchange is electrogenic, one positive ionic charge entering on each cycle. The exchange system is called into play in excitable cells when calcium influx due to electrical activity overwhelms the transport ability of the ATPase.

In general, ion exchange mechanisms can be made to run backward by altering or reversing one or more of the ionic gradients involved in the exchange. An interesting feature of sodium–calcium exchange is that such reversal can occur readily under physiological conditions, in which case calcium *enters* through the system and sodium is extruded. The direction of transport is determined simply by whether the energy provided by the entry of three sodium ions is greater than or less than the energy required to extrude one calcium ion. One factor determining this energy balance is the membrane potential of the cell. The energy dissipated by sodium entry (or required for extrusion) is simply the amount of charge moved across the membrane times the driving potential for such movement or, in other words, the charge times the difference between the sodium equilibrium potential (E_{Na}) and the membrane potential (V_m). For three sodium ions this is $3(E_{Na} - V_m)$. Similarly, for a single (divalent) calcium ion the energy is $2(E_{Ca} - V_m)$. There is no exchange when the energies are exactly equal—that is, when

Reversal of sodium–calcium exchange

$$3(E_{Na} - V_m) = 2(E_{Ca} - V_m)$$

or (by rearrangement) when

$$V_m = 3E_{Na} - 2E_{Ca}$$

Now suppose a nerve cell has internal sodium and calcium concentrations of 15 mM and 100 nM, respectively, and is bathed in a solution containing 150 mM sodium and 2 mM calcium. These are reasonable physiological values for mammalian cells. The equilibrium potential for sodium is +58 mV and for calcium +124 mV. Ion movement through the exchanger will be zero when $V_m = -74$ mV. At more negative membrane potentials, sodium will enter through the system and calcium will be extruded; at less negative potentials calcium will enter, extruding sodium. (The reader who cares to do the same calculation with the ionic distributions given in Table 1 for squid axon in seawater will find that calcium extrusion will occur only when the membrane potential is more negative than −121 mV!) The value of −74 mV is in the range of resting membrane potentials for many cells so that, in any given cell, ion movements through the exchanger may be in one direction or the other, depending on membrane potential or on whether there has been previous sodium or calcium accumulation.

In heart muscle, it appears that entry of calcium through the sodium–calcium exchange system during the action potential may contribute to

---

[31]Caputo, C., Bezanilla, F. and DiPolo, R. 1989. *Biochim. Biophys. Acta* 986: 250–256.

activation of the contractile process,[32] and that calcium extrusion through the same system after repolarization may be important for relaxation.[33] These theoretical and experimental observations indicate that the role of the sodium–calcium exchange system extends beyond merely transporting calcium out of the cell, and that the bidirectional nature of the system may have significant physiological consequences.

It is useful to keep in mind that although the cytoplasmic concentration of free calcium is very low, neurons contain substantial quantities of calcium buffered by intracellular organelles, principally the endoplasmic reticulum. In squid axon, for example, the concentration of bound calcium is about 50 μM, or about 500 times the free concentration (Table 1). A number of neuronal functions are mediated by release of calcium from these intracellular stores (Chapter 8).

In summary, intracellular ionic concentrations are governed by several different transport mechanisms. Sodium and potassium concentrations are maintained by Na–K ATPase. Similarly, Ca–Mg ATPase contributes to the maintenance of low intracellular calcium concentrations. The rest of the transport systems are driven by the electrochemical gradient for sodium rather than by direct hydrolysis of ATP. Thus they are all dependent, in the long run, upon Na–K ATPase. In the secondary transport systems, inward sodium movement is accompanied variously by chloride–bicarbonate exchange, by inward co-transport of chloride and potassium, by outward transport of protons, or by outward transport of calcium. Of these, only the sodium–calcium exchange system is electrogenic; however, the ion fluxes involved in sodium–calcium exchange are so small that their contributions to the resting membrane potential are trivial.

SUGGESTED READING

General

Junge, D. 1992. *Nerve and Muscle Excitation*, 3rd ed. Sinauer, Sunderland, MA. Chapters 1–3.
Läuger, P. 1991. *Electrogenic Ion Pumps*. Sinauer, Sunderland, MA.

Reviews

Blaustein, M. P. 1988. Calcium transport and buffering in neurons. *Trends Neurosci.* 11: 438–443.
Skou, J. C. 1988. Overview: The Na,K pump. *Methods Enzymol.* 156: 1–25.
Tsien, R. Y. 1988. Fluorescent measurement and photochemical manipulation of cytosolic free calcium. *Trends Neurosci.* 11: 419–424.

Original papers

Hodgkin, A. L. and Horowitz, P. 1959. The influence of potassium and chloride ions on the membrane potential of single muscle fibres. *J. Physiol.* 148: 127–160.

---

[32]LeBlanc, N. and Hume, J. R. 1990. *Science* 248: 372–376.
[33]Bridge, J. H. B., Smolley, J. R. and Spitzer, N. W. 1990. *Science* 248: 376–378.

Hodgkin, A. L. and Katz, B. 1949. The effect of sodium ions on the electrical activity of the giant axon of the squid. *J. Physiol.* 108: 37–77. (The constant field equation is derived in Appendix A of this paper.)

Mullins, L. J. and Noda, K. 1963. The influence of sodium-free solutions on membrane potential of frog muscle fibers. *J. Gen. Physiol.* 47: 117–132.

Russell, J. M. 1983. Cation-coupled chloride influx in squid axon. Role of potassium and stoichiometry of the transport process. *J. Gen. Physiol.* 81: 909–925.

Thomas, R. C. 1969. Membrane currents and intracellular sodium changes in a snail neurone during extrusion of injected sodium. *J. Physiol.* 201: 495–514.

Thomas, R. C. 1977. The role of bicarbonate, chloride and sodium ions in the regulation of intracellular pH in snail neurones. *J. Physiol.* 273: 317–338.

IONIC BASIS
OF THE ACTION POTENTIAL

The ionic mechanisms responsible for generating the nerve impulse in squid axons have been described quantitatively, largely through the use of the voltage clamp method to measure membrane currents produced by depolarizing pulses. Such experiments have shown that depolarization increases sodium permeability and, more slowly, potassium permeability. The activation of sodium permeability is transient, being followed by inactivation. The increase in potassium permeability persists for as long as the depolarizing pulse is maintained. It is the voltage dependence of sodium and potassium permeabilities that is responsible for the action potential. The magnitudes of the permeability changes to the two ions and their sequential timing account quantitatively for the rising and falling phases of the action potential, as well as for other phenomena such as threshold and refractory period.

Patch clamp experiments on excitable cells have been used to examine the behavior of individual sodium and potassium channels associated with the action potential. The behavior of the channels is consistent with previous voltage clamp experiments on whole cells: Depolarization increases the probability that both sodium and potassium channels will open. This increase in open channel probability follows the same time course as that of the corresponding voltage clamp currents. For example, sodium channels open most frequently near the beginning of the depolarizing pulse, and openings then become less frequent as inactivation develops.

Calcium channels are also activated by depolarization and in some tissues are responsible for the rising phase of the action potential. In addition there are a number of other potassium channels that are voltage-dependent and that play a role in determining the excitability of nerve cells.

The concepts introduced by voltage clamp experiments, together with the molecular mechanisms revealed by patch clamp recordings, explain the properties of a wide variety of action potentials seen in very different cells such as squid axons, myelinated vertebrate axons, cardiac muscle, skeletal muscle, and smooth muscle.

Sodium and the action potential

In Chapter 3 we showed that the resting potential is determined mainly by the potassium concentration ratio (as postulated by Bernstein in 1902) but is influenced as well by the concentration ratio of sodium and, to a much lesser extent, by that of chloride. At the same time that Bernstein proposed his hypothesis about the nature of the resting potential, Overton made the important discovery that sodium ions are necessary for nerve and muscle cells to produce action potentials, and suggested (somewhat hesitantly) that the action potential might come about by sodium entering the cell.[1] Further clarification of this idea again came with experiments on the squid axon. In 1939 Hodgkin and Huxley[2] and Curtis and Cole[3] showed that the action potential was more than a just a simple breakdown of the membrane to zero potential; instead, there was an "overshoot" during which the membrane potential became transiently positive on the inside. This suggested that sodium was indeed involved, because sodium entry across the membrane would continue beyond zero membrane potential until the sodium equilibrium potential (E_{Na}) was reached. Ten years later, Hodgkin and Katz showed that changes in external sodium concentration affected the amplitude of the action potential, as shown in Figure 1, and that the changes could be predicted with reasonable accuracy by the constant field equation.[4] They concluded that the action potential was the result of a large, transient increase in the sodium permeability of the membrane. We now know that this permeability increase is due to opening of a large number of voltage-activated sodium channels.

What is the effect of such an increase in sodium permeability? Recall from Chapter 3 that in squid axon at rest the permeability ratio p_K:p_{Na} was 1.0:0.04, giving a resting membrane potential of -67 mV (using the constant field equation). Hodgkin and Katz postulated that during the action potential the sodium permeability increased by a factor of about 500, giving a permeability ratio of 1.0:20. Using the ionic concentrations in squid axoplasm and seawater from Table 1 in Chapter 3, these permeability ratios predict the membrane potential at the peak of the action potential to be

$$V_m = 58 \log \frac{10 + (20)460}{400 + (20)50} = +47 \text{ mV}$$

If the external sodium concentration is reduced to one-half (230 mM) and then to one-third (153 mM), the peak potentials predicted by the equation are 30 and 20 millivolts, respectively. These predicted effects of changing sodium concentration are similar to those observed experimentally (Figure 1).

What about the falling phase of the action potential? One might expect that the membrane potential would return to the resting level if the sodium channels simply closed. Indeed, this is one factor involved. If nothing else occurred, however, the return in most cells would be much slower than that observed experimentally. This is because the resting permeability of the membrane is usually quite small, and, consequently, loss of the accumulated positive charge through resting potassium and chloride channels would take several, or even tens of,

The role of potassium ions

[1]Overton, E. 1902. *Pflügers Arch.* 92: 346–386.
[2]Hodgkin, A. L. and Huxley, A. F. 1939. *Nature* 144: 710–711.
[3]Curtis, H. J. and Cole, K. S. 1940. *J. Cell. Comp. Physiol.* 15: 147–157.
[4]Hodgkin, A. L. and Katz, B. 1949. *J. Physiol.* 108: 37–77.

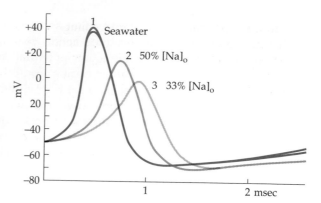

1

ROLE OF SODIUM in conduction of an action potential in squid axon. Records 1 were taken in normal seawater before and after exposure to low sodium. In record 2, external sodium was reduced to one-half and in record 3 to one-third of normal. (After Hodgkin and Katz, 1949.)

milliseconds. The return to normal is very rapid because of an additional large increase in membrane permeability—this time due to opening of voltage-activated potassium channels. The membrane potential, having raced toward E_{Na}, now returns with almost equal rapidity toward E_K. The increase in potassium permeability lasts for several milliseconds, so that in many cells the membrane is actually *hyperpolarized* beyond its normal resting potential for a time (Figure 1). The magnitude of the hyperpolarization can be predicted by the constant field equation. For example, if p_K increased tenfold, so that $p_K:p_{Na}$ became 10:0.04, the membrane potential would increase in magnitude to -89 mV from the -67 mV calculated previously for the resting potential.

To summarize, the action potential is the result of a sudden, large increase in sodium permeability of the membrane. The resulting inrush of sodium and accumulation of positive charge on the inside surface of the membrane drives the potential toward E_{Na}. Repolarization is accomplished by a subsequent increase in potassium permeability and loss of the accumulated positive charge, carried now by the efflux of potassium ions.

How many ions enter and leave during an action potential?

If the interior of the nerve gains sodium during the rising phase of the action potential and loses potassium during the falling phase, one might expect the sodium and potassium concentrations in the cytoplasm to change. In fact, as we have already noted for the charge separation associated with the resting membrane potential, the ionic movements required to charge and discharge the membrane during the action potential are negligible compared with the internal concentration of the anions. This can be shown in two ways: by calculation and by direct measurement.

The calculation requires a knowledge of the membrane capacitance, which provides a measure of how much charge movement is needed to produce the observed change in membrane potential (Chapter 5). If we assume a realistic value of 1 $\mu F/cm^2$, then we can show that to change the resting potential of a 1-cm length of squid axon, 1 mm in diameter, from a resting potential of -67 to $+40$ mV at the peak of the action potential requires the influx of about 3×10^{-13} moles of sodium. The same length of axon contains (at 50 mmol/l) 4×10^{-7} moles of sodium, so that the influx would change the sodium concentration by about one part in a million. The potassium efflux required to return the membrane potential to its resting value is the same, of course, and represents only one part in almost 10 million of the potassium content of the fiber. These calculations are supported by experimental measurements of radioactive sodium entering and radioactive potassium leaving the fiber during action potential activity[5]. The measured value was 10^{-12} moles per centimeter length of axon for each impulse. This value is somewhat higher than that calculated above, largely because the calculation takes no account of the fact that the sodium and potassium fluxes overlap in time. Thus the actual amount of sodium influx is greater than that required to charge the membrane to the peak of the action potential because potassium efflux (carrying charge in the opposite direction) begins before the peak is reached. These calculations and measurements illustrate an important point that is sometimes misunderstood: *The action potential is produced because sodium and potassium fluxes change the charge on the cell membrane, not because the fluxes change ionic concentrations in the cytoplasm.*

The ideas we have discussed so far were developed in detail by Hodgkin and Huxley (suggested readings at the end of this chapter), who carried out and analyzed in elegant detail electrophysiological experiments on the giant axon of the squid. They showed experimentally that changes in sodium and potassium conductances (and hence in permeabilities) occurred, and that these changes were timed appropriately and were of the correct magnitude to account exactly for the magnitude and time course of the action potential.

The basic feature underlying the ion movements associated with the action potential is that both the sodium and potassium conductances are voltage-dependent: The probability that the channels will open increases with depolarization. Thus depolarization increases the membrane conductance to sodium and, with a delay, to potassium as well. The effect on sodium conductance is *regenerative*. A small depolarization increases the number of open sodium channels; the resulting sodium entry along its electrochemical gradient produces still more depolarization, opening more sodium channels, leading to still more rapid sodium entry, and so on (Figure 2A). This kind of explosive process is often referred to as "positive feedback". The voltage dependence of potassium conductance, in contrast, results in "negative feedback" (Figure 2B). Depolarization increases the number of open potassium channels, resulting in efflux of potassium along its electrochemical gradient. The efflux leads to repolarization and return of the potassium conductance to its resting level.

What kinds of experiments were done to arrive at these conclusions? At first thought, it appears simple to obtain the appropriate measurements of conductance of the membrane to sodium (g_{Na}) or potassium (g_K). All that is needed is to measure the amount of current (i) flowing

[5]Hodgkin, A. L. and Keynes, R. D. 1955. *J. Physiol.* 128: 253–281.

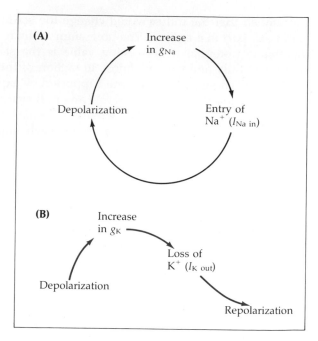

2

EFFECTS OF INCREASING g_{Na} and g_K on membrane potential. (A) Sodium entry reinforces depolarization. (B) Potassium efflux leads to repolarization.

inward or outward across the membrane at various levels of potential (V_m), since for each ion

$$g_{Na} = \frac{i_{Na}}{V_m - E_{Na}}$$

and

$$g_K = \frac{i_K}{V_m - E_K}$$

However, there are two major problems with this approach. One is that current flowing across the membrane will change the membrane potential; this, in turn, will alter the membrane conductances. The solution was to devise a method for holding steady, or clamping, the membrane potential while measuring the magnitude and time course of the membrane current. With this VOLTAGE CLAMP technique the membrane potential can be set rapidly to a preselected value and the magnitude and time course of the resulting membrane current observed without additional complicating factors. As the voltage is fixed for the period of observation, the observed changes in current will represent accurately the underlying changes in membrane conductance. The second problem is to separate the ionic currents so that their individual characteristics can be assessed. This was accomplished in a number of ways, including replacement of sodium with impermeant ions and, later, by the use of selective toxins and poisons.

BOX 1   THE VOLTAGE CLAMP

The figure shown here illustrates an experimental arrangement for voltage clamp experiments on squid axons. The axon is bathed in seawater, and two fine silver wires are inserted longitudinally into one end. One of the wires provides a measure of the potential inside the fiber with respect to that of the seawater (which is grounded) or, in other words, a measure of the membrane potential V_m. It is also connected to the *inverting* (−) input of the voltage clamp amplifier. The *noninverting* (+) input of the voltage clamp amplifier is connected to a variable voltage source, which can be set at any desired value; the value to which it is set is known as the *command potential*. As with any other amplifier, the voltage clamp amplifier will deliver current from its output whenever there is a voltage difference between the inverting and noninverting inputs. The output current flows across the cell membrane between the second fine silver wire and the seawater (arrows); it is measured by the voltage drop across a small series resistor.

Suppose that the resting potential of the fiber is −70 mV and the command potential, V_c, is set to −70 mV as well. Because the voltages at the two inputs of the voltage clamp amplifier are equal, there will be no output current. Consider now what will happen if the command potential is stepped to −65 mV. Because the noninverting input of the amplifier has been made positive to the inverting input, the amplifier will deliver positive current into the axon and across the cell membrane, driving V_m to −65 mV and thereby removing the voltage difference between the two inputs. Thus the membrane potential is kept equal to the command potential. If the circuitry is properly designed, all this happens within a few microseconds.

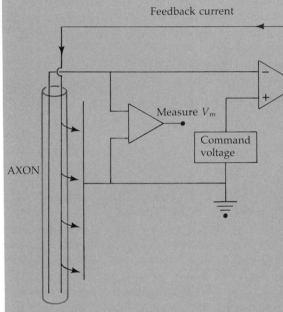

In principle, the system operates like a thermostatically controlled water bath. A thermometer measures the actual bath temperature and the thermostat control provides the command setting. If the two are equal, then no heat is supplied to the bath; if the thermostat control is turned up, heat is supplied until the bath temperature equals the command setting.

Now suppose that the command potential is stepped from −70 mV to −15 mV. We would expect that the amplifier would deliver positive current to the axon to drive V_m to −15 mV. This is indeed what happens, but only transiently (Figure 3). Then something more interesting occurs. The depolarization to −15 mV produces an increase in sodium conductance and there is a consequent flow of sodium ions *inward* across the membrane. In the absence of the clamp this would tend to depolarize the membrane still further (toward the sodium equilibrium potential); with the clamp in place, however, the amplifier provides just the correct amount of negative current to hold the membrane potential constant. In other words, the current provided by the amplifier is exactly equal to the current flowing across the membrane. Here, then, is the great power of the voltage clamp: In addition to holding the membrane potential constant, it provides an exact measure of the membrane current required to do so.

Voltage clamp
experiments

The voltage clamp was devised by Cole and his colleagues[6,7] and developed further by Hodgkin, Huxley, and Katz.[8] The experimental arrangement is described in Box 1. All we need to know to understand the experiments themselves is that the method permits us to set the membrane potential of the cell almost instantaneously at any level and hold it there (i.e., "clamp" it), while at the same time recording the current flowing across the membrane. Figure 3A shows an example of the currents that occur when the membrane potential is stepped suddenly from its resting value (in this example −65 mV) to a depolarized level (−9 mV). The current produced by the voltage step consists of three phases: (1) a brief outward surge lasting only a few microseconds while the potential moves to its new value, (2) a transient inward current, and (3) a delayed outward current.

Capacitative and
leak currents

The first component is the capacitative current, which occurs because a step from one potential to another requires that the membrane capacitance be recharged from the old potential to the new. If the clamp amplifier is capable of delivering a large amount of current, then the membrane can be recharged rapidly and this current will last only a very short time. Once the new potential is reached, there is no more capacitative current. The initial phase of the current is shown in more detail on an expanded time scale in Figure 3B. It can be seen that in practice the surge of capacitative current lasts about 20 *micro*seconds and is followed by a small, steady outward current.

This steady current is what one would expect if the resting membrane conductances were unaltered by the step depolarization. Because the cell is depolarized (i.e., more positive than at rest), there will be a net outward current. This outward ionic current is known as LEAK CURRENT and is carried largely by potassium and chloride ions. It varies linearly with voltage displacement from rest and lasts throughout the duration of the voltage step. However, during the later phases of the response, it is obscured by much larger ionic currents.

Currents carried by
sodium and potassium

Turning now to the second and third phases, Hodgkin and Huxley showed that they were due first to the entry of sodium and then to the exit of potassium across the cell membrane, and were able to deduce

[6]Marmont, G. 1940. *J. Cell. Comp. Physiol.* 34: 351–382.

[7]Cole, K. S. 1968. *Membranes, Ions and Impulses.* University of California Press, Berkeley.

[8]Hodgkin, A. L., Huxley, A. F. and Katz, B. 1952. *J. Physiol.* 116: 424–448.

3 MEMBRANE CURRENTS PRODUCED BY DEPOLARIZATION. (A) Currents ▶ measured by voltage clamp during a 56-mV depolarization of a squid axon membrane. The currents (lower trace) consist of a brief positive capacitative current, a transient phase of inward current, and a delayed, maintained outward current. These are shown separately in (B), (C), and (D). The capacitative current (B) lasts for only a few microseconds (note change in time scale). The small outward leakage current is due in part to movement of chloride. The transient inward current (C) is due to sodium entry and the prolonged outward current (D) to potassium movement out of the fiber.

(A)

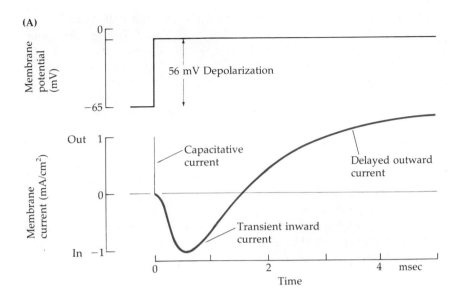

(B)

(C)

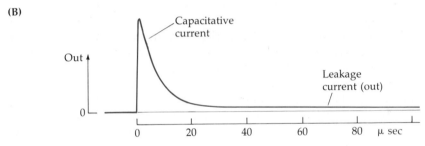

(D)

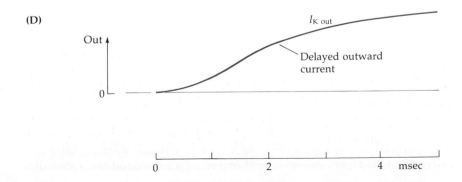

A. L. Hodgkin, 1949   A. F. Huxley, 1974

the relative size and time course of the separate currents. In particular, replacing extracellular sodium by choline abolished the inward current, revealing the underlying potassium current. The time course of the potassium current is shown in Figure 3D. Subtraction of the potassium current from the total ionic current (Figure 3A) gives the magnitude and time course of the sodium current (Figure 3C). We will see that later workers were able to separate the currents pharmacologically (Figure 6) with similar results.

Dependence of ionic currents on membrane potential

One way to obtain information about the nature of the early (inward) and late (outward) ionic currents is to determine how the magnitudes of the currents depend on the size of the depolarizing voltage step. Currents produced by various levels of depolarization from a holding potential of −65 mV are shown in Figure 4. First of all, a step *hyperpolarization* to −85 mV (lower record) produced only a small inward current, as would be expected from the resting properties of the membrane. As already shown in Figure 3, moderate depolarizing steps each produced an early inward current followed by a sustained outward current. With greater depolarizations, the early current becomes smaller, is absent at about +52 mV, and then reverses to become outward as the depolarizing step is increased still further. The current–voltage relations for the early and late currents are shown in Figure 5, in which the peak amplitude of the early current and the steady-state amplitude of the late current are plotted against the potential to which the membrane is stepped. With hyperpolarizing steps there is no separation of early and late currents; the membrane simply responds as a passive resistor, with a hyperpolarization producing the expected inward current. The late current also behaves as one would expect of a resistor in the sense that

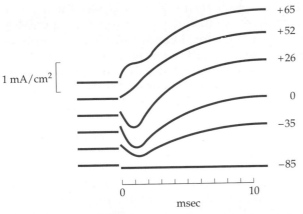

4

CURRENTS PRODUCED BY VOLTAGE STEPS from a holding potential of −65 mV to a hyperpolarized level (−85 mV) and to successively increasing depolarized levels as indicated. The late potassium current increases as the depolarizing steps increase. The early sodium current first increases, then decreases with increasing depolarization, is absent at +52 mV, and is reversed in sign at +65 mV. (After Hodgkin, Huxley and Katz, 1952.)

depolarization produces outward current, but as the depolarization is increased, the magnitude of the current becomes much greater than expected from the resting membrane properties. This is due to the voltage-activated potassium conductance, which allows more current through the membrane. The early inward current behaves in a much more complex way. As already noted, it first increases and then decreases with increasing depolarization, becoming zero at about +52 mV and then reversing in sign. The reversal potential is very near the equilibrium potential for sodium, providing one important piece of evidence that the early current is carried by sodium ions.

One point of interest in the current–voltage relation for the early

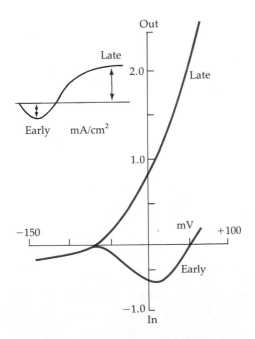

5

AMPLITUDE OF EARLY AND LATE CURRENTS plotted against the potential to which the membrane is stepped. Late outward current increases rapidly with depolarization. Early inward current first increases in magnitude, then decreases, reversing to outward current at about +55 mV (the sodium equilibrium potential). (After Hodgkin, Huxley and Katz, 1952.)

current is that between about −50 and +10 mV the slope of the relation is negative. In this region the current, $i_{Na} = g_{Na}(V_m - E_{Na})$, increases even though the driving potential for sodium entry ($V_m - E_{Na}$) is decreasing. This is because sodium conductance (g_{Na}) is increasing with depolarization, and the increase in conductance overrides the decrease in driving force. In this voltage range, the current–voltage relation is said to have a region of NEGATIVE SLOPE CONDUCTANCE.

Selective poisons for sodium and potassium channels

Since the original experiments of Hodgkin and Huxley, convenient pharmacological methods have been found for blocking sodium and potassium currents selectively. Tetrodotoxin (TTX) and its pharmacological companion saxitoxin (STX) in particular have turned out to be useful for a wide range of experiments. TTX is a virulent poison, concentrated in the ovaries and other organs of certain fish, whose potent effects have given rise to the Chinese proverb "To throw away life eat blowfish" (puffer fish). Kao has reviewed the fascinating history of TTX, beginning with the discovery of its effects by the Chinese emperor Shun Nung (2838–2698 B.C.), who personally tasted 365 drugs while compiling a pharmacopoeia and lived (for an amazingly long time) to tell the tale.[9] STX is synthesized by marine dinoflagellates and concentrated by filter-feeding shellfish, such as the Alaskan butter clam *Saxidomus*. Its virulence competes with that of TTX: Ingestion of a single clam (cooked or raw) can be fatal.

One great advantage of TTX for neurophysiological studies is that its action is highly specific. Working with squid axons, Moore, Narahashi, and colleagues have shown that it blocks the voltage-activated sodium conductance selectively.[10] When a TTX-poisoned axon is subjected to a depolarizing voltage step, no inward sodium current is seen, only the delayed outward potassium current. The poison does not change the amplitude or time course of the potassium current. Application of TTX to the inside of the membrane by adding it to an internal perfusing solution has no effect. The actions of STX are indistinguishable from those of TTX. Both toxins appear to bind to the same site in the outer mouth of the channel through which sodium ions move, thereby physically blocking ionic current through the channel.[11] Many other excitable cells are affected by TTX and STX in a similar manner, including vertebrate myelinated and unmyelinated axons and skeletal muscle fibers. The effect of TTX on sodium current in a myelinated nerve fiber is shown in Figure 6A and 6B. There are other substances that block the sodium current, for example local anesthetics such as procaine.

Just as TTX and STX block sodium channels selectively, a number of substances have been found that have similar effects on voltage-activated potassium channels. For example, in squid axons and in frog

[9]Kao, C. T. 1966. *Pharmacol. Rev.* 18: 977–1049
[10]Narahashi, T., Moore, J. W. and Scott, W. R. 1964. *J. Gen. Physiol.* 47: 965–974.
[11]Hille, B. 1970. *Prog. Biophys. Mol. Biol.* 21: 1–32.

myelinated axons, Armstrong, Hille, and others have shown that voltage-activated potassium currents are blocked by tetraethylammonium (TEA),[12] as shown in Figure 6C and 6D. In squid axon, TEA must be added to the axoplasm or to an internal perfusate and exerts its action at the inner mouth of the potassium channel; in other preparations, such as the frog node of Ranvier, it is effective at an external site as well. Other compounds, such as 4-aminopyridine (4-AP) and 3,4-diaminopyridine (DAP), block potassium currents from either the inside or the outside of the membrane.

It is apparent from the experiments of Hodgkin and Huxley and from those shown in Figure 6 that the time courses of the sodium and potassium currents are quite different. The potassium current is much delayed, compared with the onset of the sodium current, and remains high throughout the duration of the depolarizing step. The sodium current, on the other hand, rises much more rapidly but then decreases to zero, even though the membrane is still depolarized. This decline of the sodium current is called INACTIVATION.

Inactivation of the sodium current

[12]Armstrong, C. M. and Hille, B. 1972. *J. Gen. Physiol.* 59: 388–400.

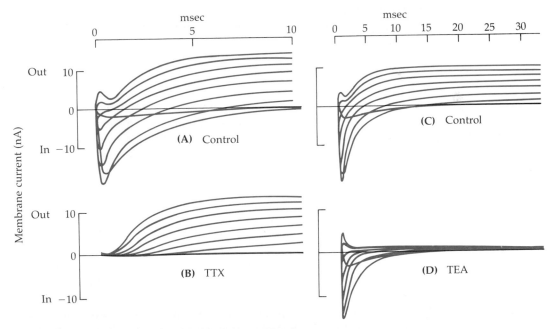

6 **PHARMACOLOGICAL SEPARATION OF MEMBRANE CURRENTS into sodium and potassium components.** The membrane potential of a frog myelinated nerve fiber is displaced to various levels between −60 and +75 mV. (A) and (C) are controls in normal bathing solution. In (B), the addition of 300 n*M* tetrodotoxin (TTX) causes the sodium currents to disappear while the potassium currents remain. In (D), the addition of tetraethylammonium (TEA) blocks the potassium currents, leaving the sodium currents intact. (After Hille, 1970.)

Hodgkin and Huxley studied the nature of the inactivation process in detail. In particular, they investigated the effect of hyperpolarizing and depolarizing prepulses on the peak amplitude of the sodium current produced by a subsequent depolarizing step. Records from such an experiment are shown in Figure 7. In Figure 7A the membrane is stepped from a holding potential of -65 mV to -21 mV, producing a peak sodium current of about 1 mA/cm^2. When the step is preceded by a hyperpolarizing prepulse of -13 mV, the peak sodium current is increased (Figure 7B). Depolarizing prepulses, on the other hand, cause a decrease in the sodium current (Figure 7C,D). These effects of hyperpolarizing and depolarizing prepulses are time-dependent; brief pulses of only a few milliseconds duration have little effect. In the experiment shown here, the prepulses were of sufficient duration for the effects to reach their maximum. The results were expressed quantitatively by plotting the peak sodium current after a conditioning prepulse as a fraction of the control current ($i_{Na(prepulse)}/i_{Na(no\ prepulse)}$) against the amplitude of the prepulse, as shown in Figure 7E. With a depolarizing prepulse of about 30 mV, the subsequent sodium current was reduced to zero; that is, inactivation was complete. Hyperpolarizing prepulses of 30 mV or larger increased the sodium current by a maximum of about 70 percent. Hodgkin and Huxley represented this range of sodium currents from zero to their maximum value with a single parameter h, varying between zero (complete inactivation) and 1 (maximum activation), as indicated on the right-hand ordinate of Figure 7E. In these experiments there was about 40 percent inactivation at the resting potential. Subsequent experiments have shown that all neurons show some degree of inactivation at rest.

The experiments with prepulses suggested that inactivation was a distinct phenomenon, separable from the activation process. A subsequent experimental observation supporting this idea was that pronase, a mixture of proteolytic enzymes, when perfused through the inside of a squid axon, led to virtual abolition of inactivation.[13] The enzyme was ineffective when applied in the same concentration to the outer surface. It appeared, then, that pronase had removed from the cytoplasmic side of the sodium channels some group of amino acids associated specifically with the inactivation process. In later experiments,[14] site-directed mutagenesis was used to delete cytoplasmic segments of rat brain sodium channel expressed in *Xenopus* oocytes (Box 4 in Chapter 2). In mutants with deletions in the intracellular loop connecting domains III and IV, inactivation of sodium currents was greatly reduced (but not eliminated). In addition, treatment with an antibody directed at the same region of the channel prolonged single-channel currents in membrane patches from rat brain neurons.[15]

[13]Armstrong, C. M., Bezanilla, F. and Rojas, E. 1973. *J. Gen. Physiol.* 62: 375–391.
[14]Stühmer, W. et al. 1989. *Nature* 339: 597–603.
[15]Vassilev, P., Scheuer, T. and Catterall, W. A. 1989. *Proc. Natl. Acad. Sci. USA* 86: 8147–8151.

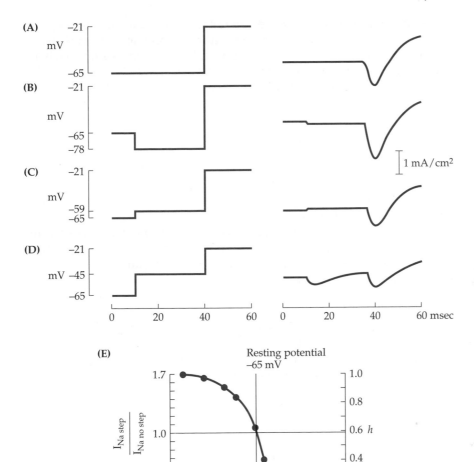

EFFECT OF MEMBRANE POTENTIAL ON SODIUM CURRENTS. (A) A depolarizing step from −65 to −21 mV produces an inward sodium current, followed by an outward potassium current. **(B)** When the depolarizing step is preceded by a 30-msec hyperpolarizing prepulse, the sodium current is increased. Prior depolarizing pulses **(C** and **D)** reduce the size of the inward current. The graph in **(E)** shows the fractional increase or reduction of the sodium current as a function of the amplitude of the preceding conditioning pulse. With a prepulse of −40 mV (hyperpolarizing), the maximum current is about 1.7 times larger than the control value. A prepulse of +40 mV reduces the subsequent response to zero. Full range of the sodium current is scaled from zero to unity (100%) by the *h* ordinate (see text).

Evidence that a particular intracellular loop of amino acids is associated with inactivation led to a new interest in the "ball-and-chain" model of inactivation, originated more than a decade earlier by Armstrong and Bezanilla to account for the effect of pronase.[16] In this model, a clump of amino acids (the ball) is tethered by a string of residues (the chain) to the main channel structure. After activation, the ball has access to a site in the mouth of the open channel, thereby producing channel block. A similar ball-and-chain model of inactivation of potassium A channels has been tested by examining the behavior of channels formed in oocytes from mutant subunits (recall that the A channel is a tetramer rather than a single polypeptide). Mutations and deletions were made in the 80 or so amino acids between the amino terminus and the first (S1) membrane helix.[17] Channels formed by mutants with deletion of residues 6 to 46 showed virtually no inactivation, suggesting that some or all of these residues were involved in the normal inactivation process. When a synthetic peptide matching the first 20 amino acids in the amino-terminal chain was simply added to the solution bathing the cytoplasmic face of the membrane, inactivation was restored with a linear dose dependence over the range of 0 to 100 μM.[18] This amazing observation provides unusually strong support for the idea that in potassium A channel subunits, the first 20 or so amino acid residues constitute a blocking particle responsible for inactivation of the channel.

The relation between sodium channel inactivation and the experimental results obtained with potassium A channels is, at this point, uncertain. There are major structural differences between the amino terminus of the A channel and the amino acid chain between domains III and IV of the sodium channel, not the least of which is the fact that the III-IV loop contains only about 50 amino acid residues and is anchored at both ends, in contrast to the 80 or more amino acids in the potassium channel, ending in a free-swinging ball. This structural dissimilarity, plus the fact that deletions and antibody treatments in the III-IV loop do not remove inactivation completely, suggests that the story of sodium channel inactivation may be far from complete.

Sodium channel activation and inactivation are also affected by a group of lipid-soluble toxins, including veratridine, an alkaloid from plants of the lily family, and batrachotoxin (BTX) from the skin of South American frogs. They virtually eliminate inactivation so that the channels remain open indefinitely.[19] In addition, the voltage dependence of activation is shifted so that the channels are open at the normal resting potential.

Sodium and potassium conductances as functions of membrane potential

The voltage clamp techniques enabled Hodgkin and Huxley to deduce the magnitude and time course of sodium and potassium currents as a function of the membrane potential, V_m, and to determine the

[16]Armstrong, C. M. and Bezanilla, F. 1977. *J. Gen. Physiol.* 70: 567–590.
[17]Hoshi, T., Zagotta, W. N. and Aldrich, R. W. 1990. *Science* 250: 533–550.
[18]Zagotta, W. N., Hoshi, T. and Aldrich, R. W. 1990. *Science* 250: 568–571.
[19]Catterall, W. A. 1980. *Annu. Rev. Pharmacol. Toxicol.* 20: 15–43.

equilibrium potentials E_{Na} and E_K. It was then a straightforward matter to deduce the magnitude and time courses of the sodium and potassium conductance changes, using the relations noted earlier:

$$g_{Na} = \frac{i_{Na}}{V_m - E_{Na}}$$

and

$$g_K = \frac{i_K}{V_m - E_K}$$

The results for five different voltage steps are shown in Figure 8. Both g_{Na} and g_K increase progressively with increasing membrane depolarization. The behavior of the sodium conductance is quite unlike that of the sodium current (Figure 5), which first increases and then decreases in magnitude as the voltage steps increase in amplitude. The progressive decrease in the current is, of course, because the larger depolarizations come progressively closer to the sodium equilibrium potential. As a result the inward current decreases, even though the sodium conductance is increasing. The relations between peak conductance and membrane potential are shown for sodium and potassium in Figure 9. The curves are remarkably similar.

In summary, the results obtained by Hodgkin and Huxley indicated that depolarization of the nerve membrane led to three distinct processes: (1) activation of a sodium conductance mechanism, (2) subsequent inactivation of the mechanism, and (3) activation of a potassium conductance mechanism.

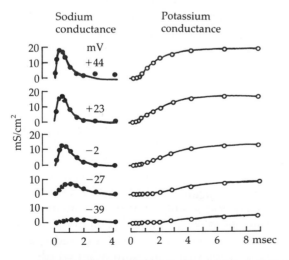

8   MEMBRANE CONDUCTANCE CHANGES produced by voltage steps from −65 mV to the indicated potentials. Peak sodium conductance and steady-state potassium conductance both increase with increasing depolarization. (After Hodgkin and Huxley, 1952a.)

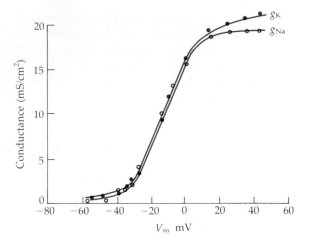

9

Na AND K CONDUCTANCES as functions of membrane potential. Peak sodium conductance and steady-state potassium conductance are plotted against the potential to which the membrane is stepped. Both increase steeply with depolarization between −20 and +10 mV. (After Hodgkin and Huxley, 1952a.)

Sodium and potassium conductances as functions of time

After obtaining the experimental results, Hodgkin and Huxley proceeded to develop a mathematical description of the precise time courses of the sodium and potassium conductance changes produced by the depolarizing voltage steps. To deal first with the potassium conductance, one might imagine that the effect of a sudden change in membrane potential would be to provide a driving force for the translocation of one or more charged regions in the voltage-activated potassium channel, which would then lead to channel opening. If a single process were involved, then the change in the overall potassium conductance might be expected to be governed by ordinary first-order kinetics; that is, its rise after the onset of the voltage step would be exponential. Instead the onset of the conductance change is S-shaped, with a marked delay. Because of this delay, and because the potassium current occurs during depolarization but not hyperpolarization (i.e., it is rectified), it is often referred to as DELAYED RECTIFIER current. Hodgkin and Huxley were able to account for the S-shaped onset of the conductance by assuming that opening of each potassium channel required the activation of four first-order processes, for example the translocation of four particles in the membrane. In other words, the S-shaped time course of activation could be fitted by an exponential raised to the fourth power. The expression for the increase in potassium conductance for a given voltage step, then, is

$$g_K = g_{K(max)}n^4$$

where $g_{K(max)}$ is the maximum conductance reached for the particular voltage step and n is a rising exponential function varying between zero and unity, given by $n = 1 - \exp(-t/\tau_n)$. As shown in Figure 9, $g_{K(max)}$ varies with voltage. As might be expected, the increase in conductance becomes more rapid with larger depolarizing steps. In other words, the exponential time constant τ_n is voltage-dependent, ranging between about 4 msec for small depolarizations and 1 msec for depolarization to zero, at a temperature of 10°C.

The time course of the rise in sodium conductance, also S-shaped, was fitted by an exponential raised to the third power. In contrast, the fall in sodium conductance due to inactivation was consistent with a simple exponential decay process. The overall time course of the sodium conductance change, then, was given by the product of the activation and inactivation processes:

$$g_{Na} = g_{Na(max)}m^3h$$

where $g_{Na(max)}$ is the maximum level to which g_{Na} would rise if there were no inactivation, and $m = 1 - \exp(-t/\tau_m)$. The inactivation process is a falling, rather than a rising, exponential and is given by $h = \exp(-t/\tau_h)$. As with the potassium conductance, $g_{Na(max)}$ is voltage-dependent, as are the activation and inactivation time constants. The activation time constant τ_m is much smaller than that for potassium, having a value at 10°C the order of 0.6 msec near the resting potential and decreasing to about 0.2 msec at zero potential. The inactivation time constant, τ_h, on the other hand, is similar in magnitude to τ_n.

Once the theoretical expressions were obtained for sodium and potassium conductances as a function of voltage and time, Hodgkin and Huxley were able to predict the entire time course of the action potential. Starting with a depolarizing step to just above threshold, they calculated what the subsequent potential changes would be at successive intervals of 0.01 msec. Thus, during the first 0.01 msec after the membrane had been depolarized to, say, −45 mV, they calculated how g_{Na} and g_K would change, what increments of i_{Na} and i_K would result, and then the amount of additional depolarization produced by the net current. Knowing the change in V_m at the end of the first 0.01 msec, they then repeated the calculations for the next time increment, and so on, all through the rising and falling phases of the action potential (a laborious exercise to undertake in the days before electronic computers were available). The result was that the procedure duplicated with remarkable accuracy the naturally occurring action potential in the squid axon. An example of such a calculated action potential and the time courses of the underlying sodium and potassium conductance changes are shown in Figure 10A. Theoretical and observed action potentials produced by brief depolarizing pulses at three different stimulus strengths are compared in Figure 10B. In order to appreciate fully the magnitude of this accomplishment, it is necessary to keep in mind that the calculations used to duplicate the action potential were based on current measurements made under completely artificial conditions with the membrane potential clamped first at one value, then at another, and with varying sodium concentrations in the external fluid.

In addition to describing the action potential accurately, Hodgkin and Huxley were able to explain in terms of ionic conductance changes the propagation of the action potential and many other properties of excitable axons such as refractory period and threshold. Further, their findings have been found to be applicable to a wide variety of other excitable tissues.

Reconstruction of the action potential

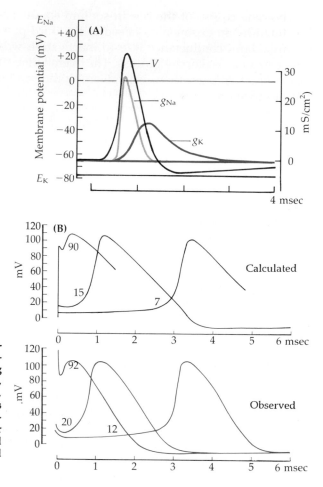

10

RECONSTRUCTION OF THE ACTION PO-TENTIAL. (A) Calculations of a propagated action potential (curve V) and the underlying changes in sodium and potassium conductances, using results from voltage clamp experiments. (B) Calculations of membrane action potentials in a segment of squid axon following brief depolarizations at three different intensities (upper traces) and observed action potentials produced by three similar stimuli. (After Hodgkin and Huxley, 1952d.)

Threshold and refractory period

How do the findings of Hodgkin and Huxley explain the threshold membrane potential at which the impulse takes off, especially when it might seem that a discontinuity like threshold would require a discontinuity in g_{Na} or g_K? The phenomenon can be understood if we imagine passing current through the membrane to depolarize it just to threshold, and then turning the current off. Because the membrane potential is farther from the potassium equilibrium potential, there will be an increase in outward potassium current (plus a small outward leak current). We will also activate some sodium channels, increasing inward sodium current. At threshold the increased currents are exactly equal and opposite, and the membrane is in a condition similar to that at rest. But there is an important difference: The sodium conductance is now unstable. If an extra sodium ion enters the cell, the depolarization is increased, g_{Na} increases, more sodium enters, and the regenerative process explodes. If, on the other hand, an extra potassium ion leaves the cell the depolarization is decreased, g_{Na} decreases, sodium current decreases and the excess potassium current causes repolarization. As the membrane potential approaches its resting level, the potassium current decreases until it again equals the resting inward sodium current. Depolarization above threshold results in an increase in g_{Na} sufficient for inward sodium movement to swamp outward potassium movement immediately. Subthreshold depolarization fails to increase g_{Na} sufficiently to override the resting potassium conductance.

And how is the refractory period explained? Two changes develop after an action potential that make it impossible for the nerve fiber to produce a second action potential immediately: (1) inactivation, which prevents any increase in g_{Na}, is maximal during the falling phase of the action potential and requires several more milliseconds to decay to zero, and (2) g_K is now very large, thus requiring a very large increase in g_{Na} to initiate any regenerative depolarization. These two factors result in an ABSOLUTE REFRACTORY PERIOD lasting throughout the falling phase of the action potential during which no amount of externally applied depolarization can initiate a second regenerative response. Following the action potential there is a RELATIVE REFRACTORY PERIOD during which the residual inactivation of the sodium conductance and the relatively high potassium conductance combine to produce an increase in threshold for action potential initiation.

From our present perspective, it was an extraordinary achievement for Hodgkin and Huxley to have provided rigorous quantitative explanations for such complex biophysical properties of membranes. The subsequent observations on single channels have provided a new depth and an understanding of the underlying molecular mechanisms, but by no stretch of the imagination would single-channel studies on their own, without the previous voltage clamp experiments and insights, have been able to account for how a nerve cell generates and conducts impulses. The older work has become enriched, rather than supplanted, by the new.

Hodgkin and Huxley suggested that sodium channel activation was associated with the translocation of charged structures, or particles, within the membrane. Such charge movements would be expected to appear as capacitative currents in response to a depolarizing voltage step. After a number of technical difficulties were resolved, these gating currents were finally seen.[20,21] An example is shown in Figure 11A. A step depolarization of an internally perfused squid axon produced an outward capacitative current, followed by an inward sodium current. The sodium current was much smaller than usual because the extracellular sodium concentration was reduced to 20 percent of normal. When TTX was added to the solution, sodium entry was blocked and only the gating current remained (outward current in Figure 11B; note the change in scale).

Gating currents

[20]Armstrong, C. M. and Bezanilla, F. 1974. *J. Gen. Physiol.* 63: 533–552.
[21]Keynes, R. D. and Rojas, E. 1974. *J. Physiol.* 239: 393–434.

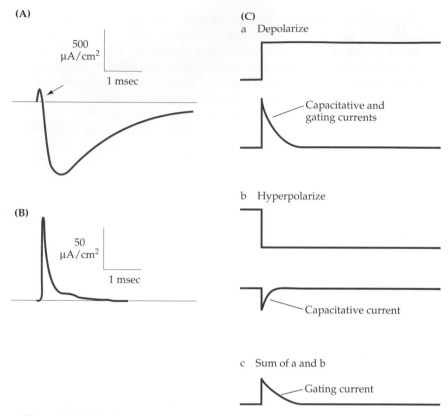

11   SODIUM CHANNEL GATING CURRENT. (A) Current recorded from squid axon in response to depolarizing pulse. The inward sodium current is reduced by reducing extracellular sodium to 20 percent of normal. The small outward current (arrow) preceding the inward current is a sodium channel gating current. (B) Response to depolarization from same preparation after adding TTX to the bathing solution, recorded at higher amplification. Only gating current remains. (C) Method of separating gating current from capacitative current. Depolarizing pulse (a) produces capacitative current in the membrane, plus gating current. Hyperpolarizing pulse of the same amplitude (b) produces capacitative current only. When the responses to a hyperpolarizing and a depolarizing pulse are summed (c), capacitative currents cancel out and only gating current remains. (A and B after Armstrong and Bezanilla, 1977.)

How is the gating current separated from the usual capacitative current expected with a step change in membrane potential (e.g., Figure 3)? Briefly, currents associated merely with charging and discharging the membrane capacitance should be symmetrical. That is, they should be of the same magnitude for depolarizing steps as for hyperpolarizing steps. On the other hand, currents associated with sodium channel activation should appear upon depolarization of, say, 50 mV, from a holding potential of −70 mV, but not upon hyperpolarization. In other words, if the channels are already closed, there should be no gating current upon further hyperpolarization. Similarly, gating currents associated with channel closing might be expected at the termination of a brief depolarizing pulse but not after a hyperpolarizing pulse. One experimental way of recording gating currents then, is to sum the currents produced by two identical voltage steps of opposite polarity, as shown in Figure 11C: The asymmetry due to gating currents is seen in parts (a) and (b) of the figure. The current at the beginning of the depolarizing pulse is larger than that produced by the hyperpolarizing pulse because of the additional charge movement associated with gating of the sodium channel. When the two currents are added (part c), the net result is the gating current (or ASYMMETRY CURRENT) alone. The evidence that asymmetry currents produced in the manner just described are, in fact, associated with sodium channel activation has been summarized by Armstrong.[22]

Patch clamp techniques have now provided detailed information about the way in which single sodium channels respond to depolarization. One such experiment is illustrated in Figure 12. The records are from a cell-attached patch (Figure 12A) on cultured rat muscle fibers.[23] To remove inactivation of sodium channels, a steady command potential was applied to the electrode, hyperpolarizing the patch membrane to -100 mV or so. On successive trials, a 40-mV depolarizing pulse was applied to the electrode for about 23 msec, as shown in Figure 12B, trace a. Single-channel currents appeared during the pulse, occurring most frequently near the onset of depolarization (trace b). The mean channel current was 1.6 pA. Assuming the sodium equilibrium potential to be $+30$ mV, the driving potential for sodium entry was about 90 mV; thus the single-channel conductance was about 18 pS. This is comparable to sodium channel conductances measured in a variety of other cells. When 300 of the individual traces were summed (trace c), the result was an inward current that followed the same time course as that expected from the whole-cell sodium current.

One major point of interest in Figure 12 is that the mean channel open time (0.7 msec) is short relative to the overall time course of the summed current. Specifically, the time constant of decay of the summed current (about 4 msec) is not a reflection of correspondingly long channel open times. Instead, it indicates a slow decay in the probability that an individual channel will open. What Hodgkin and Huxley called "activation" (m^3) and "inactivation" (h), then, do not represent actual channel opening and closing, but instead represent first an increase, and then a decrease in the *probability* that a channel will open for a brief period. Their product (m^3h) describes the time course of the probability change. It increases rapidly near the beginning of the pulse, reaches a peak, and then decreases with time. In any given trial, however, an individual channel may open immediately after the onset of the pulse, at any subsequent time during the pulse, or not at all.

Response of single sodium channels to depolarization

[22]Armstrong, C. M. 1981. *Physiol. Rev.* 61: 644–683.
[23]Sigworth, F. J. and Neher, E. 1980. *Nature* 287: 447–449.

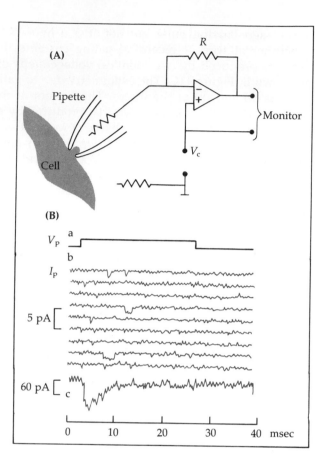

12

SODIUM CHANNEL CURRENTS recorded from cell-attached patch on cultured rat muscle cell. (A) Recording arrangement. (B) Repeated depolarizing voltage pulses applied to the patch, with waveform shown in (a), produce single-channel currents (downward deflections) in the nine successive records shown in (b). The sum of 300 such records (c) shows that most channels open in the initial 1 to 2 msec after the onset of the pulse, after which the probability of channel opening declines with the time constant of inactivation. (After Sigworth and Neher, 1980.)

A second point of interest is illustrated by referring again to Figure 11: Charge movement is virtually complete before sodium current reaches its peak; that is, before most of the channels begin to open. This means that the conformational change in the channel protein associated with the charge movement is distinct from the conformational change associated with channel opening. Thus it is not accurate to think of the charges as being connected with some kind of "handle" that opens a gate in the channel directly. Instead the charge movement simply increases the probability that the channel structure will enter the open or the inactivated state.

Kinetic models of channel activation and inactivation

From their observations that the time courses of activation of the sodium and potassium currents were best fitted by exponential functions raised to the third and fourth powers (m^3 and n^4), Hodgkin and Huxley suggested that activation could be explained by the independent displacement of three or four charged particles in the membrane. For example, we might imagine that a voltage step has to produce displacements of the S4 helices in all four domains of the potassium channel before the channel can open. Similarly, we might suppose that at least three such displacements are necessary for sodium channel activation. Further, we might suppose that in the sodium channel one or more of the displacements also leads ultimately to inactivation. A parallel model of this kind has been proposed by Keynes.[24]

The idea that four separate events (such as S4 helix displacements) are necessary for channel opening gives rise to the possibility of 16 different channel states: no displacement (one state), one displacement in any one of four domains (four possible states), two displacements in any two domains (six possible states), three displacements in any three (four possible states), and displacements in all four (one state). If the steps are independent and kinetically identical, then this reduces to five states: no displacement, displacement in any one domain, in any two, in any three, and in all four. On this basis, the transition from open to closed can be represented as follows:

$$C_5 \longleftrightarrow C_4 \longleftrightarrow C_3 \longleftrightarrow C_2 \longleftrightarrow C_1 \longleftrightarrow O$$

where C_5 represents the state of the channel at rest, C_4 and so on represent a series of closed states into which the channel can be driven by depolarization, and O is the open state. For sodium channels we must add the inactivation process. Measurements of both macroscopic and single-channel currents suggest that the channel can be inactivated whether or not it has opened previously.[25,26] Thus inactivation (I) can occur both from the open state and from one or more of the closed states:

$$C_5 \longleftrightarrow C_4 \longleftrightarrow C_3 \longleftrightarrow C_2 \longleftrightarrow C_1 \longleftrightarrow O$$
$$\searrow I \swarrow$$

Many variations of this kind of model have been proposed with more or fewer steps and with more than one inactivated state.[27,28] They differ from the original model of Hodgkin and Huxley in the sense that activation and inactivation are envisioned as sharing a number of sequential events, rather than proceeding in parallel as independent processes. In addition, although progression though any number of the steps may depend on membrane potential, the final steps leading to activation and inactivation need not themselves be voltage-dependent.[29,30]

How many states really exist? This is not known for certain, but it appears that sodium channel activation involves at least three discrete

[24]Keynes, R. D. 1990. *Proc. R. Soc. Lond. B* 240: 425–432.

[25]Bean, B. P. 1981. *Biophys. J.* 35: 595–614.

[26]Aldrich, R. W. and Stevens, C. F. 1983. *Cold Spring Harbor Symp. Quant. Biol.* 48: 147–153.

[27]Hille, B. 1992. *Ionic Channels of Excitable Membranes*, 2nd ed. Sinauer, Sunderland, MA. Chapter 18.

[28]Armstrong, C. M. and Gilley, W. F. 1979. *J. Gen. Physiol.* 74: 691–711.

[29]Aldridge, R. W. and Stevens, C. F. 1987. *J. Neurosci.* 7: 418–431.

[30]Cota, G. and Armstrong, C. M. 1989. *J. Gen. Physiol.* 94: 213–232.

charge displacements. Conti and Stühmer reached this conclusion by measuring gating currents in large cell-attached membrane patches (macropatches) on *Xenopus* oocytes injected with exogenous mRNA coding for rat brain sodium channels.[31] The size of the elementary gating charge movement was deduced by measuring the mean and variance of a large number of individual gating currents. The procedure is analogous to using noise measurements for determining the size of single-channel currents (Chapter 2): The variance-to-mean ratio gives the size of the elementary charge movements. The elementary gating charge was calculated to be 2.3 electronic charges (2.3e). The total charge transfer per channel can be estimated from the steepness of the activation curve: The more charges there are on the affected structure, the smaller the voltage increment required to change its conformation. These considerations suggested that channel activation was associated with a charge transfer between 6e and 8e—that is, three of the elementary gating charges. This finding is remarkably consistent with the activation model proposed originally by Hodgkin and Huxley. Conti and Stühmer note that it is attractive to identify the charge transfers with structural transitions in three of the four sodium channel domains and to ascribe inactivation to interaction of these three with the fourth. Such findings, of course, do not rule out additional state transitions that are electrically silent.

In summary, the kinetic models suggest that depolarization initiates a series of stepwise conformational changes that lead eventually to channel opening, with one or more alternative steps leading to inactivation. Although we can imagine in a very general way how such structural changes might occur in the protein, it is difficult to specify them precisely. An initial step has been made in relating inactivation of potassium A channels with a particular group of amino acids in the channel subunit. Further correlations will no doubt be forthcoming when the molecular anatomy is understood in greater detail.

Properties of channels associated with the action potential

The conductance of the voltage-activated sodium channel has been determined directly by patch clamp measurements (Figure 12). The density of sodium channels has been determined in a number of tissues by measuring the density of TTX binding sites. Using tritiated TTX, Levinson and Meves estimated that in squid axon an average of 553 molecules were bound to each μm^2 of membrane.[32] Values in other tissues have been found to range from a low of 2 per μm^2 in neonatal rat optic nerve[33] to 2000 per μm^2 at the node of Ranvier in rabbit sciatic nerve.[34] In skeletal muscle the sodium channel density has been found to be maximal at the end plate region, and to decrease more or less exponentially with distance toward the tendons, falling to 10 percent

[31]Conti, F. and Stühmer, W. 1989. *Eur. Biophys. J.* 17: 53–59.
[32]Levinson, S. R. and Meves, H. 1975. *Philos. Trans. R. Soc. Lond. B* 270: 349–352.
[33]Waxman, S. G. et al. 1989. *Proc. Natl. Acad. Sci. USA* 86: 1406–1410.
[34]Ritchie, J. M. 1986. *Ann. NY Acad. Sci.* 479: 385–401

or less of the maximum value.[35] In other experiments, sodium channels in the muscle fiber membrane were found to be distributed in clusters rather than uniformly.[36] In both sets of experiments, sodium channel densities were measured by depolarizing small areas of the membrane with a focal extracellular pipette. Inward sodium currents varied from one patch to the next, indicating the underlying variation in channel density.

Single-channel conductance of the potassium channels underlying the late current (delayed rectifier channels) has been measured in squid axon both by noise analysis[37] and with patch clamp experiments.[38] The latter experiments were particularly ingenious, as the squid axon membrane was patch clamped from the inside! Both methods gave a single-channel conductance of about 10 pS. In similar patch clamp experiments on cut-open squid axon, potassium channels were observed with conductances of 10, 20, and 40 pS, with the 20 pS channel predominating.[39] In frog muscle, potassium channels, like sodium channels, are distributed nonuniformly,[36] and the distributions of the two types of channel are not correlated. At nodes of Ranvier in rabbit myelinated nerve, depolarization produces no late outward current, and the delayed rectifier channels are therefore presumably absent.[40] During the action potential, repolarization is achieved by a large leak current after rapid inactivation of the sodium channels.

An ion of considerable importance, not yet discussed here, is calcium, which plays a key role in a variety of processes (Chapters 7 and 8). For example, a transient increase in intracellular calcium concentration is responsible for secretion of chemical transmitters by neurons and for contraction of muscle fibers. Calcium ions also affect excitation: A reduction in extracellular calcium concentration lowers the threshold for initiation of impulses; conversely, increasing extracellular calcium concentration raises the threshold. Frankenhaeuser and Hodgkin examined these effects in the squid axon and found not only a change in threshold but also a change in the amount of sodium and potassium current produced by depolarizing pulses.[41] When, for example, the extracellular calcium concentration was lowered, the ionic currents produced by small depolarizing pulses were increased. Thus the overall effect of a decrease in extracellular calcium concentration was to shift the current–voltage relations so that depolarizing pulses (1) reached threshold sooner and (2) produced more current. These effects were similar to those seen when a fixed voltage increment was added to each

Calcium ions and excitability

[35]Beam, K. G., Caldwell, J. H. and Campbell, D. T. 1985. *Nature* 313: 588–590.
[36]Almers, W., Stanfield, P. and Stühmer, W. 1983. *J. Physiol.* 336: 261–284.
[37]Conti, F., DeFelice, L. J. and Wanke, E. 1975. *J. Physiol.* 248: 45–82.
[38]Conti, F. and Neher, E. 1980. *Nature* 285: 140–143.
[39]Lanno, I., Webb, C. K. and Bezanilla, F. 1988. *J. Gen. Physiol.* 92: 179–196.
[40]Chiu, S. Y. et al. 1979. *J. Physiol.* 292: 149–166.
[41]Frankenhaeuser, B. and Hodgkin, A. L. 1957. *J. Physiol.* 137: 218–244.

of the pulses. A fivefold reduction in the extracellular calcium concentration was equivalent to a 10- to 15-mV voltage increment.

It was suggested that the effect might be associated with screening by calcium ions of negative charges on the outer surface of the membrane. Removal of calcium would leave such negative charges unneutralized and thus reduce the potential field across the membrane. This idea introduces a new concept with regard to membrane potential: The potential between the intracellular and extracellular solutions is determined by extracellular and intracellular ion concentrations and ion permeabilities, as discussed in Chapter 3. The potential field across the membrane itself is subject to an additional factor, namely the presence of charged molecules attached to its inner and outer surfaces. For example, externally facing portions of most membrane proteins are believed to be glycosylated with carbohydrate chains that include negatively charged sialic acid. The eel sodium channel itself has over 100 sialic acid residues.[42] These negative charges on the outer surface reduce the potential field across the membrane. In normal extracellular solutions, monovalent and divalent cations—particularly calcium—neutralize, or "screen" the surface charges so that the potential field across the membrane is approximately equal to the measured membrane potential. Removal of extracellular calcium unscreens the negative charges and reduces the potential field. Because of this mechanism, calcium removal is equivalent to depolarization. One role of calcium, then, is to stabilize the membrane, that is, to maintain a margin of safety between the potential field across the membrane and the threshold for activation of voltage-activated conductance channels.

Calcium channels

In addition to its action on excitability, calcium itself enters squid axons during the impulse. This was first demonstrated by measuring accumulation of radioactive calcium and subsequently by the use of aequorin (Chapter 3).[43] Voltage clamp experiments established that calcium enters the axon in two phases. The early phase is blocked by TTX and had a time course similar to that of the sodium current. This phase represents leak through the sodium channels; the permeability of the sodium channel to calcium is about 1 percent of the permeability to sodium. The later phase of calcium entry is not blocked by TTX or by TEA and indicates the presence of a separate voltage-activated calcium conductance.

Voltage-activated calcium channels have been shown to have a number of subtypes.[44] T-TYPE CHANNELS produce *transient* membrane currents in response to depolarization and are activated and inactivated over a voltage range similar to that of voltage-activated sodium channels. The channels are permeable to barium and have conductances of the order of 10 pS with 100 mM barium on either side of the membrane. Under physiological conditions (e.g., about 2 mM extracellular and 100

[42]Miller, J. A., Agnew, W. S. and Levinson, S. R. 1983. *Biochemistry* 22: 462–470.
[43]Baker, P. B., Hodgkin, A. L. and Ridgeway, E. B. 1971. *J. Physiol.* 218: 709–755.
[44]Tsien, R. W. et al. 1988. *Trends Neurosci.* 11: 431–437.

nM intracellular calcium), their conductances can be expected to be very much lower.

L-TYPE CHANNELS are noninactivating and therefore produce *long-lasting* currents in response to depolarization. They require larger depolarizations for activation than T-type channels, have conductances of about 25 pS in symmetrical 100 mM barium solutions, and are blocked by the dihydropyridines (DHP) nitrendipine and nifedipine. This channel, the *DHP receptor*, has been isolated from skeletal muscle and purified,[45] and the isolated protein has been shown to form L-type channels when reconstituted into lipid bilayers.[46] The channel has been cloned and sequenced, so that its full primary structure is known.[47]

N-TYPE CHANNELS have been shown to underlie currents that are *neither* long-lasting nor rapidly inactivating. They require relatively large depolarizations for activation, have conductances of about 15 pS in symmetrical 100 mM barium solutions and inactivate slowly.

These three classifications are not all-inclusive—many calcium channels do not fit neatly into one or another. Instead, the classifications are indicative of a range of calcium channel types with varying activation and inactivation thresholds and inactivation rates.

In some muscle fibers and some neurons, calcium currents through T-type channels become sufficiently large to contribute significantly to, or even be solely responsible for, the rising phase of the action potential. Because g_{Ca} increases with depolarization, the process is a regenerative one, entirely analogous to that discussed for sodium. The participation of calcium in the action potential process was first studied in invertebrate muscle fibers by Fatt and Ginsborg[48] and subsequently by Hagiwara and colleagues.[49] Calcium action potentials occur in cardiac muscle, in a wide variety of invertebrate neurons, and in neurons in the vertebrate autonomic and central nervous systems. Such action potentials occur in non-neural cells as well, including a number of endocrine cells and some invertebrate egg cells. The voltage-activated calcium currents can be blocked by adding millimolar concentrations of cobalt, manganese, or cadmium to the extracellular bathing solution. As already noted, barium can substitute for calcium as the permeant ion; magnesium, on the other hand, cannot. A particularly striking example of the coexistence of sodium and calcium action potentials in the same cell is found in the mammalian cerebellar Purkinje cell, which generates sodium action potentials in its soma and calcium action potentials in the branches of its dendritic tree.[50,51]

Calcium action potentials

[45]Campbell, K. P., Leung, A. T. and Sharp, A. H. 1988. *Trends Neurosci.* 11: 425–430.
[46]Flockerzi, V. et al. 1986. *Nature* 323: 66–68.
[47]Tanabe, T. et al. 1987. *Nature* 328: 313–318.
[48]Fatt, P. and Ginsborg, B. L. 1958. *J. Physiol.* 142: 516–543.
[49]Hagiwara, S. and Byerly, L. 1981. *Annu. Rev. Neurosci.* 4: 69–125.
[50]Llinás, R. and Sugimori, M. 1980. *J. Physiol.* 305: 197–213.
[51]Ross, W. N., Lasser-Ross, N. and Werman, R. 1990. *Proc. R. Soc. Lond.* B 240: 173–185.

Potassium channels   It is now known that in addition to the increase in g_K responsible for the delayed rectifier current, there are other voltage- and messenger-activated conductance pathways for potassium (Table 1), some that respond transiently, and others that contribute continually to the resting conductance of the membrane.[52–54] Among these are potassium channels in skeletal muscle and squid axon that are turned *off* by depolarization and are therefore called ANOMALOUS RECTIFIER or INWARD REC-

[52]Latorre, R. and Miller, C. 1983. *J. Memb. Biol.* 71: 11–30.
[53]Castle, N. A., Haylett, D. G. and Jenkinson, D. H. 1989. *Trends Neurosci.* 12: 59–65.
[54]Rudy, B. 1988. *Neuroscience* 25: 729–749.

TABLE 1
POTASSIUM CHANNEL TYPES

Channel	Properties and Function	Blocked by
Voltage-activated		
Delayed rectifier (outward rectifier)	Activated by depolarization; responsible for action potential repolarization.	TEA, aminopyridines, quinine, Cs^+, Ba^+
Anomalous rectifier (inward rectifier)	Activated by hyperpolarization; responsible for a small fraction of the resting (outward) potassium current.	TEA, Cs^+, Ba^+
A channel	Activated by depolarization, then inactivated. Largely inactivated at the resting potential, so that activation requires prior hyperpolarization.	TEA, aminopyridines
Messenger-modulated		
M channel	Activated by depolarization; little or no voltage inactivation. Contributes to resting potassium conductance. Inactivated by ACh, acting on *muscarinic* receptors, and by peptides, in both cases through intracellular messenger systems. Speeds repolarization of action potentials and synaptic potentials; messenger inactivation produces depolarization and increased excitability.	TEA, Ba^+
S channel	Like M channel, but only weakly voltage-dependent and inactivated by serotonin (5-HT). Speeds action potential repolarization in *Aplysia* cells.	TEA, Ba^+
Ion-activated		

Calcium-activated	Activated by intracellular calcium concentrations between 10 nM and 1 mM. Three subtypes:
	     Small conductance (SK—10 pS)
	     Intermediate conductance (IK—30 pS)
	     Large conductance (Maxi-K—200 pS)
	 Contribute to resting potassium conductance, and action potential repolarization and afterhyperpolarization.
		
		 Quinine, strychnine
		 Quinine, Ba^+, Cs^+
		 TEA, quinine, Ba^+

Sodium-activated	Activated by intracellular sodium concentrations between 10 and 50 mM. Contribute to resting potassium conductance.	Aminopyridines

TIFIER channels. A CHANNELS, seen in many neurons, are activated rapidly by depolarization and then inactivate, but in most cells activation occurs only after a preceding hyperpolarization; that is, they are inactivated at rest. M CHANNELS are also seen in a variety of neurons. These are similar to delayed rectifier channels in that they open in response to depolarization; they are inactivated by acetylcholine through MUSCARINIC ACh receptors (hence their name). S CHANNELS, open at rest, are inactivated by serotonin. M channels and S channels will be discussed in more detail later in relation to neuromodulation (Chapter 8).

Still other potassium channels are activated by intracellular cations. The most ubiquitous are calcium-activated potassium channels, which have at least three subtypes with very large (200 pS), intermediate (30 pS), and small (10 pS) conductances. Their presence can be demonstrated experimentally by raising intracellular calcium concentrations above normal, for example, by injection from an intracellular micropipette.[55] Following such an injection the membrane resistance of the cell decreases rapidly and the resting membrane potential approaches the equilibrium potential for potassium. The resistance and potential then return to their control levels as the excess calcium is removed from the cytoplasm by internal buffering mechanisms and outward transport. Calcium-activated potassium channels occur in a wide variety of cells (including nonexcitable cells such as erythrocytes). In many cells, inward calcium current associated with the action potential stimulates an increase in potassium conductance that contributes to repolarization and produces a subsequent hyperpolarization. Yet another type of potassium conductance is activated by increases in the intracellular sodium concentration.[56,57]

[55]Meech, R. W. 1974. *J. Physiol.* 237: 259–277.
[56]Partridge, L. D. and Thomas, R. C. 1976. *J. Physiol.* 254: 551–563.
[57]Martin, A. R. and Dryer, S. E. 1989. *Q. J. Exp. Physiol.* 74: 1033–1041.

SUGGESTED READING

Books and Reviews

Armstrong, C. M. 1981. Sodium channels and gating currents. *Physiol. Rev.* 61: 644–683.

Campbell, K. P., Leung, A. T. and Sharp, A. H. 1988. The biochemistry and molecular biology of the dihydropyridine-sensitive calcium channel. *Trends Neurosci.* 11: 425–430.

Hille, B. 1992. *Ionic Channels of Excitable Membranes*, 2nd Ed. Sinauer, Sunderland, MA.

Rudy, B. 1988. Diversity and ubiquity of K channels. *Neuroscience* 25: 725–749.

Tsien, R. W., Lipscombe, D., Madison, D. V. Bley, K. R. and Fox, A. P. 1988. Multiple types of neuronal calcium channels and their selective modulation. *Trends Neurosci.* 11: 431–437.

Original papers

Connor, J. A. and Stevens, C. F. 1971. Voltage clamp studies of a transient outward membrane current in gastropod neural somata. *J. Physiol.* 213: 21–30.

Frankenhaeuser, B. and Hodgkin, A. L. 1957. The action of calcium on the electrical properties of squid axons. *J. Physiol.* 137: 218–244.

Hodgkin, A. L. and Huxley, A. F. 1952. Currents carried by sodium and potassium ion through the membrane of the giant axon of *Loligo*. *J. Physiol.* 116: 449–472.

Hodgkin, A. L. and Huxley, A. F. 1952. The components of the membrane conductance in the giant axon of *Loligo*. *J. Physiol.* 116: 473–496.

Hodgkin, A. L. and Huxley, A. F. 1952. The dual effect of membrane potential on sodium conductance in the giant axon of *Loligo*. *J. Physiol.* 116: 497–506.

Hodgkin, A. L. and Huxley, A. F. 1952. A quantitative description of membrane current and its application to conduction and excitation in nerve. *J. Physiol.* 117: 500–544.

Hodgkin, A. L., Huxley, A. F. and Katz, B. 1952. Measurement of current-voltage relations in the membrane of the giant axon of *Loligo*. *J. Physiol.* 116: 424–448.

Meech, R. W. 1974. The sensitivity of *Helix aspersa* neurones to injected calcium ions. *J. Physiol.* 237: 259–277.

NEURONS AS CONDUCTORS OF ELECTRICITY

Impulses propagate along axons by the longitudinal spread of current. As each region of the membrane generates an all-or-nothing action potential, it depolarizes and excites the adjacent not-yet-active region and gives rise to a new regenerative impulse. For an understanding of impulse propagation, as well as of synaptic transmission and integration, we must know how electrical currents spread passively along a nerve.

As current spreads along a nerve axon or dendrite, it becomes attenuated with distance. This attenuation depends on a number of factors, principally the diameter and membrane properties of the fiber. Longitudinal current spreads farther along a fiber with large diameter and high membrane resistance. The electrical capacitance of the membrane influences the time course of the electrical signals and usually their spatial spread as well. To estimate how far a subthreshold potential change will spread, one needs to know the geometry and membrane characteristics of the neuron and, in addition, the waveform of the potential change.

The axons of many vertebrate nerve cells are covered by a high-resistance, low-capacitance myelin sheath. This sheath acts as an effective insulator and forces currents associated with the nerve impulse to flow through the membrane at intervals where the sheath is interrupted (nodes of Ranvier). The impulse jumps from one such node to the next, and thereby its conduction velocity is increased. Myelinated nerves occur in pathways in the nervous system where speed of conduction is important.

Electrical activity can also pass between neurons through specialized regions of close membrane apposition called gap junctions. Pathways for current flow in these regions are provided by intercellular channels called connexons. The structure of connexons and the amino acid composition of their constituent protein (connexin) has been characterized in several non-neural tissues, such as liver epithelium and heart muscle.

The permeability properties of nerve cell membranes and the way in which these properties produce regenerative electrical responses have been discussed in the preceding chapters. In this chapter we describe in more detail how currents spread along nerve fibers to produce local graded potentials. The *passive* electrical properties of nerves that un-

Passive electrical properties of nerve and muscle membranes

derlie such current spread are essential for signaling in the nervous system. At sensory end organs they are the link between the stimulus and the production of impulses; along axons they allow the impulse to spread and propagate; at synapses they enable the postsynaptic neuron to add and subtract synaptic potentials that arise from numerous converging inputs, some close to the cell body, others on remote dendritic sites. The discussion that follows deals primarily with the spread of current along nerve fibers of uniform diameter; that is, along cylindrical conductors. Further, throughout the discussion we will assume that in the absence of regenerative action potentials, the nerve fiber membranes are indeed passive—changes in potential below action potential threshold do not activate any voltage-sensitive channels that would change the membrane resistance.

We begin by considering how the resistive properties of the membrane and of the axoplasm together determine the size of the voltage response to an injected current and, in addition, how far this response will spread. The main requirement is that we keep in mind Ohm's law: A given amount of current i passed through a resistance r produces a voltage $V = ir$ (Appendix A). Later we consider the additional effects of capacitance. The concepts apply qualitatively to more complex systems as well, such as tapered dendrites or the convergence of currents from a number of such dendrites into a cell body; the more complicated structures require more complicated analyses. In the central nervous system, complex geometries of axonal branching patterns, and of dendritic trees with nonuniform electrical properties, play a major role in nervous system function.[1,2] As a result, electrical models of the nervous system that represent neurons as simple, well-behaved, input–output elements are almost certain to be inadequate.

A cylindrical nerve fiber has the same formal components as an undersea cable, namely a central or core conductor and an insulating sheath surrounded by a conducting medium. However, the two systems are quantitatively very dissimilar. In a cable the core conductor is usually copper, which has a very high conductance, and the surrounding insulating sheath is neoprene, plastic, or some other material of very high resistance. In addition, the sheath is usually relatively thick, so it has a very low capacitance. Voltage applied to one end of such a cable will spread an immense distance because the resistance to longitudinal current flow along the copper conductor is relatively low and virtually no current is lost through the insulating sheath. In a nerve fiber, on the other hand, the core conductor is a salt solution similar in concentration to that bathing the nerve and (compared with copper) is a poor conductor. Furthermore, the plasma membrane of the fiber is a relatively poor insulator and, being thin, has a relatively high capacitance. A voltage signal applied to one end of a nerve fiber, then, will fail to spread very far for two reasons: (1) The core material has a low con-

[1]Rall, W. 1967. *J. Neurophysiol.* 30: 1138–1168.
[2]Lev-Tov, A. et al. 1983. *J. Neurophysiol.* 50: 399–412.

ductance so that resistance to current flow down the fiber is high, and (2) current that starts off flowing down the axoplasm is lost progressively along the fiber by outward leakage through the poorly insulating plasma membrane. The analysis of current flow in cables was developed by Lord Kelvin for application to transatlantic telephone transmission and refined by Oliver Heaviside in the late nineteenth century. Heaviside was the first to consider the effect of resistive leak through the insulation, equivalent to the membrane resistance in nerve, and made many other contributions to cable theory, including the concept of what he called impedance (it was, incidently, Heaviside, not Maxwell, who formulated Maxwell's equations in the modern form seen on T-shirts worn by biophysicists). Cable theory was first applied coherently to nerve fibers by Hodgkin and Rushton, who used extracellular electrodes to measure the spread of applied current along lobster axons.[3] Later, intracellular electrodes were used in a number of nerve and muscle fibers for similar studies.

One way to gain an intuitive feeling about how current spreads in a cable or a nerve fiber is to think of the spread of heat along a metal rod surrounded by insulation and immersed in a conducting material (such as water). If one end of the rod is heated, heat is conducted along the rod and, at the same time, is lost to the surrounding medium. At progressively greater distances from the heated end, the temperature becomes less and less; similarly, the rate at which heat is lost decreases progressively with distance as the rod itself becomes less hot. Assuming that the surrounding medium is a good heat conductor, the distance the heat spreads depends primarily on (1) the conductivity of the rod and (2) the effectiveness of the insulation in preventing heat loss.

Current flow in a cable can be described in similar terms by the flow of ions: If we inject current into a nerve fiber, for example from a micropipette (Figure 1A), positive charges flowing into the axoplasm from the tip of the microelectrode repel other cations and attract anions. By far the most abundant small ion in the axoplasm is potassium, which therefore carries most of the current away from the electrode. Positive charge accumulates at the membrane and spreads laterally along the axon; as it spreads some is lost by ion movements through the membrane. (In reality, no one ion migrates very far through the axoplasm; the displacement of ions resembles more closely collisions along a series of billiard balls.) The distance the potential spreads along the axon depends on the membrane resistance relative to the internal longitudinal resistance. A low-resistance membrane has large ionic conductances, allowing current to leak out before it can spread very far. A higher-resistance membrane allows a greater portion of the current to flow laterally before escaping to the external solution.

In order to understand the quantitative factors that determine the spread of current and potential along a nerve fiber, it is useful to consider the experiment illustrated in Figure 1A in more detail. A mi-

---

[3]Hodgkin, A. L. and Rushton, W. A. H. 1946. *Proc. R. Soc. Lond. B* 133: 444–479.

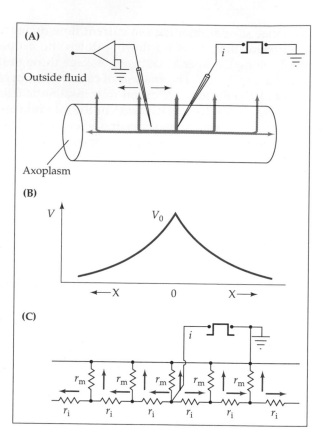

(A)

Outside fluid

Axoplasm

(B)

V

V_0

$\longleftarrow X$ 0 $X \longrightarrow$

(C)

i

r_m r_m r_m r_m r_m

r_i r_i r_i r_i r_i r_i

1

PATHWAYS FOR CURRENT FLOW IN AN
AXON. (A) Current flow across the surface membrane produced by injection of current from a
microelectrode. Thickness of arrows indicates
current density at various distances from the
point of injection. Membrane potential is recorded with a second electrode at various distances from the current electrode. (B) Potential V
measured along the axon as a function of distance
x from the point of current injection. Decay of
voltage is exponential. (C) Equivalent electrical
circuit, assuming zero resistance in the external
fluid and ignoring membrane capacitance; r_i is
longitudinal resistance of axoplasm per unit
length; r_m is membrane resistance of a unit
length.

croelectrode is placed in the middle of a large fiber (for example, a
lobster axon), and a small amount of positive current is passed into the
axoplasm. As shown by the arrows, some of the current will flow
outward across the membrane immediately adjacent to the electrode;
the remainder will spread laterally along the axon before leaving the
cell. The relative amounts of current flowing across the membrane are
indicated roughly by the thickness of the arrows. The potential change
produced across the membrane at a given distance from the electrode
will be proportional to the current flow across the membrane at that
point, in accordance with Ohm's law. There are two things we would
like to find out in an experiment of this kind. (1) For a given amount
of current injected into the pipette, how much voltage change will be
produced at the electrode? (2) How far will this voltage change spread
along the fiber? This can be done by measuring the potential change
with a second microelectrode, which we can insert at various positions
along the fiber as indicated. The results of such measurements are
shown in Figure 1B. The current produces a change in potential that is
greatest at the point of injection and falls off with distance on either
side. The decrease in potential with distance from the current electrode

is exponential, so that the potential V_x at any distance x on either side is given by

$$V_x = V_0 e^{-x/\lambda}$$

The shape of the curve is determined by two parameters, V_0 and λ. The peak potential change V_0 is proportional to the size of the injected current. The constant of proportionality is known as the INPUT RESISTANCE of the fiber, r_{input}. It is the average resistance presented by the fiber to flow of current along the axoplasm and back out through the membrane. Thus, if the amount of current injected is i, then

$$V_0 = i r_{input}$$

The parameter λ is known as the LENGTH CONSTANT of the fiber and is the distance over which the potential falls to $1/e$ (37 percent) of its maximum value. The two parameters r_{input} and λ define quantitatively for a given amount of current how much depolarization is produced and how far that depolarization spreads along the fiber.

What factors determine r_{input} and λ? As suggested by our previous qualitative discussion, they depend on the membrane resistance of the fiber, r_m, and on the internal longitudinal resistance of the axoplasm, r_i. We can see how r_m and r_i are defined by considering the equivalent electrical circuit in Figure 1C. Only resistive components are shown; the membrane capacitance is, for now, ignored. The circuit is obtained by imagining that the axon is cut along its length into a series of short cylinders. The membrane resistance r_m represents the resistance outward across such a cylinder; the longitudinal resistance r_i represents the internal resistance along the axoplasm from the midpoint of one cylinder to the midpoint of the next. Because nerves in a recording chamber are normally bathed in a large volume of fluid, the *extracellular* longitudinal resistance along the cylinders is represented as being zero. This approximation is not always adequate in the central nervous system, where nerve axons, dendrites, and glial cells (Chapter 6) are closely packed and pathways for extracellular current flow thereby restricted. For our experiment, however, the assumption is valid and serves the purpose of keeping the algebra as simple as possible. Any length can be selected for the cylinders themselves; however, the resistances r_m and r_i are always specified for a 1-cm length of axon. The dimensions of r_m are ohm-cm (Ωcm) and those of r_i are ohms per cm (Ω/cm). The dimensions of r_i seem reasonable—more axonal length means more internal resistance; those of r_m seem strange until one realizes that membrane resistance *decreases* as the fiber length increases (more channels are available for current to leak through the membrane). Thus the resistance in ohms of a given length of axon membrane is the resistance of 1 cm length (r_m, in ohm-cm) divided by the length (in cm).

The length constant of the fiber depends on both r_m and r_i:

$$\lambda = (r_m/r_i)^{1/2}$$

The expression has the dimensions of centimeters as required and, in addition, fulfills the intuitive expectation that the distance over which the potential change spreads should increase with increasing membrane resistance (which prevents loss of current across the membrane) and should decrease with increasing internal resistance (which resists current flow along the core of the fiber). Similarly, the input resistance depends on both parameters:

$$r_{input} = 0.5(r_m r_i)^{1/2}$$

Again the expression has the required dimensions (ohms) and tells us that the input resistance increases with both membrane resistance and internal resistance. The factor 0.5 arises because the axon extends both directions from the point of current injection; each half has an input resistance $(r_m r_i)^{1/2}$. In summary, the input resistance and length constant of a nerve fiber are each related to both the membrane resistance and the axoplasmic resistance of the fiber. These relations are described by the two simple equations given above.

Specific resistive characteristics of the membrane and axoplasm

Experiments such as that illustrated in Figure 1 are often done to determine the resistive properties of the membrane and axoplasm. The resistance of the membrane is of particular interest because it tells us something about its complement of channels. After we have measured r_{input} and λ experimentally, it is a simple matter to reverse the equations already given and use them to calculate

$$r_m = r_{input}\lambda$$

and

$$r_i = r_{input}/\lambda$$

These specify the resistive characteristics of a cylindrical segment of the axon 1 cm in length. They do not, however, provide precise information about the specific resistance of the membrane itself or of the axoplasmic material, because they depend on the size of the fiber. All else being equal, we would expect a 1-cm length of small fiber to have a higher membrane resistance than the same length of larger fiber, simply because the smaller fiber has less membrane surface. On the other hand, a 1-cm length of large fiber with a membrane containing very few ionic channels might have the same membrane resistance as a much smaller fiber with very many channels. If we know the fiber radius a, we can use r_m and r_i to determine the specific resistances of the membrane and axoplasmic material.

The SPECIFIC MEMBRANE RESISTANCE R_m is defined as the resistance of one square centimeter of membrane. The parameter R_m is important because it is independent of geometry and therefore enables us to compare the membrane of one cell with that of another of quite different size or shape. To determine the relation between R_m and r_m, we need to recall that membrane resistance goes down as area goes up. As a result, the membrane resistance of a 1-cm length of axon (r_m) is obtained by dividing R_m by the membrane area:

$$r_m = R_m/2\pi a$$

Turning the equation around,

$$R_m = 2\pi a r_m$$

In most neurons, R_m is determined primarily by the resting permeabilities to potassium and chloride (Chapter 3); these vary considerably from one cell to the next. The average value for R_m reported by Hodgkin and Rushton for lobster axon was about 2000 Ωcm^2; in other preparations, measurements range from less than 1000 Ωcm^2 for membranes with a large number of channels through which ions can leak to more than 50,000 Ωcm^2 for membranes with relatively few such channels.

The specific resistance of the axoplasm (R_i) is the internal longitudinal resistance of a 1-cm length of axon 1 cm^2 in cross-sectional area. It is also independent of geometry, and gives us a measure of how freely ions migrate through the intracellular space. To calculate R_i from r_i for a cylindrical axon, we recall that the resistance along the core of a cylinder increases as the cross-sectional area decreases. Therefore, the resistance of a 1-cm length of axon (r_i) is obtained by dividing R_i by the cross-sectional area of the axon:

$$r_i = R_i/\pi a^2$$

Again we can turn the equation around to get:

$$R_i = \pi a^2 r_i$$

The parameter R_i has the dimensions Ωcm. In squid nerve, it has a value of about 30 Ωcm at 20°C, or about 10^7 times that of copper. In mammals, where the ionic strength of the cytoplasm is lower, the specific internal resistance is higher (about 125 Ωcm at 37°C); in frogs, with still lower ionic strength, the specific internal resistance is even higher (about 250 Ωcm at 20°C).

Given a specific resistance R_i for the axoplasm and a specific membrane resistance R_m, how are the cable parameters r_{input} and λ influenced by fiber diameter? The answer can be obtained quantitatively from the relations presented in the preceding paragraphs. Beginning with input resistance, we know $r_{input} = 0.5(r_m r_i)^{1/2}$. In addition, $r_i = R_i/\pi a^2$ and $r_m = R_m/2\pi a$. Combining these relations we get

Effect of diameter on cable characteristics

$$r_{input} = 0.5 \left(\frac{R_m R_i}{2\pi^2 a^3} \right)^{1/2}$$

Thus as the fiber radius a increases, the input resistance decreases; the resistance varies inversely with the ½ power of the radius. The length constant, $\lambda = (r_m/r_i)^{1/2}$, is given by

$$\lambda = \left(\frac{a R_m}{2 R_i} \right)^{1/2}$$

Other properties being equal, λ increases with the square root of the fiber radius. A squid axon 1 mm in diameter with a specific internal resistance of 30 cm and a specific membrane resistance of 2000 Ωcm^2 has a length constant of almost 13 mm. The length constant of a frog muscle fiber with the same specific membrane resistance and a diameter of 100 μm would be 1.5 mm, and that of a 1-μm diameter mammalian nerve fiber only 0.2 mm.

In summary, the cable parameters r_{input} and λ can be measured experimentally. These describe the size and the steady-state distribution of potential produced by injection of current into the cable, as shown in Figure 1. In addition, their values can be used to calculate the transverse membrane resistance, r_m, of a unit length of fiber, and its internal longitudinal resistance, r_i. Using the fiber diameter, these in turn can be used to determine the specific resistances of the membrane and axoplasm. Conversely, if we know the specific resistances of the membrane and axoplasm, and the fiber diameter, we can predict the length constant and input resistance of a fiber. For given values of specific membrane resistance and specific axoplasmic resistance, the fiber's length constant increases with the square root of its diameter, and its input resistance is inversely related to the ⅔ power of the diameter.

Membrane capacitance   In addition to allowing the passage of ionic currents, the cell membrane accumulates ionic charges on its inner and outer surfaces. It is this charge separation that determines the membrane potential of the cell. Electrically, the charge separation means that the membrane has the properties of a capacitor. In general, a capacitor consists of two conducting sheets or plates separated by a layer of insulating material; in manufactured capacitors the conducting sheets are usually metallic foil and the insulator is mica or a plastic such as Mylar. The closer together the plates are, the greater their ability to separate and store charge. In the case of a nerve cell, the plates are the conducting fluids on either side of the membrane and the insulating material is the lipoprotein of the membrane itself. Because the membrane is only about 7 nm thick, it is capable of storing a relatively large amount of charge. The capacitance C of a capacitor is defined by how much charge Q it will accumulate for each volt of potential V applied to it; that is, $C = Q/V$. The capacitance C has the units coulombs per volt or farads (F). Typically nerve cell membranes have a capacitance on the order of 1 μF/cm^2. Turning the equation around, the charge stored in a capacitor is given by $Q = CV$. Thus, if a cell has a resting potential of -80 mV, the amount of charge separated by the membrane will be $(1 \times 10^{-6}) \times (80 \times 10^{-3}) = 0.08$ coulombs per cm^2, which is 5×10^{11} univalent ions per cm^2.

The *current* flowing into or out of a capacitor can be deduced from the relation between charge and voltage, and by remembering that current (i, in amperes) is the rate of change of charge with time; that is, *amperes = coulombs per second*. Thus, since $Q = CV$, we can write

$$i = \frac{dQ}{dt} = C\left(\frac{dV}{dt}\right)$$

In other words, the current is directly proportional to the rate of change of the voltage. Conversely, if a constant current is applied to the capacitor, the voltage will rise at a constant rate, $dV/dt = i/C$.

for all nerve and muscle membranes, τ provides a convenient measure for the specific membrane resistance of a cell.

How does the time constant affect current flow in a cable? As with the simple *RC* circuit (Figure 2C), the rise and fall of the potential change produced by a rectangular current pulse are slowed by the presence of the capacitance. The effects are more complicated, however, because current no longer flows into a single capacitor; instead each segment of the circuit with its capacitive and resistive elements interacts with the others. Because of this interaction, the rising and falling phases of the potential changes are not exponential. In addition, the growth and decline of the potentials become increasingly prolonged as records are made farther and farther from the point of current injection (Figure 3). Because the waveform of the potential is not exponential, the membrane time constant cannot be obtained by measuring the time to rise to 63 percent of its final value. At the point of current injection, for example, the potential rises much faster, and is 84 percent complete in one time constant. At a distance of one length constant away from the point of injection, it does in fact reach 63 percent of maximum in one time constant; at greater distances the rise is slower still (Box 1).

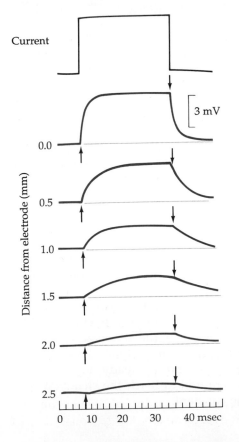

3

SPREAD OF POTENTIAL along a lobster axon, recorded with a surface electrode. Rectangular current pulse is applied at 0 mm, producing large electrotonic potential. With increasing distance from the site of current injection, the rise time of the potential change is slowed and the height of the plateau attenuated. (After Hodgkin and Rushton, 1946.)

BOX 1   ELECTROTONIC POTENTIALS AND MEMBRANE TIME CONSTANT

The electrotonic potentials shown in Figure 3, recorded at various distances along an axon from the point of current injection, do not rise and fall exponentially. Instead their waveforms are described by complicated functions of both time and distance. The potentials rise more rapidly than exponentials near the current electrode and more slowly farther away. Consequently the membrane time constant, $\tau = R_m C_m$, cannot be obtained by measuring the time for an electrotonic potential to rise to 63 percent of its final value, as with an exponential voltage change. At the point of current injection, the potential, in fact, rises to 84 percent of its maximum amplitude in one time constant. At a distance of two length constants from the current electrode, on the other hand, it reaches only 37 percent of its final value in the same time. How, then, does

one measure τ in a cable? To do this, it is necessary to know the separation between the current-passing and voltage-recording electrodes as a fraction of the length constant, λ. One then consults a table of values to find out how far the electrotonic potential will rise in one time constant at that particular electrode separation. An abbreviated version of such a table (modified from Hodgkin and Rushton, 1946) is presented here. The numbers give the amplitude V_τ reached by an electrotonic potential at one time constant after the onset of the current pulse, for various electrode separations d. The amplitudes are expressed as a fraction of the final steady-state amplitude V_∞, and the distance between the electrodes as fractions or multiples of one length constant λ.

Separation (d/λ)	0	0.2	0.4	0.6	0.8	1.0	1.5	2.0
Amplitude (V_τ/V_∞)	0.84	0.81	0.77	0.73	0.68	0.63	0.50	0.37

A further consequence of the membrane capacitance is that brief signals do not spread as far as signals of long duration. For sufficiently long pulses, when the potential can reach a steady state, its spread along the axon is unaffected by capacitance; that is, $V_x = V_0 e^{-x/\lambda}$, as before. However, for brief events, such as synaptic potentials, the current flow giving rise to the signal may be ended before the membrane capacitances become fully charged. This has the effect of reducing the spread of the potential along the fiber. In other words, for brief signals the effective length constant is less than for longer duration ones. In addition, such signals are distorted as they spread along the fiber, the peaks becoming more rounded and occurring progressively later with increasing distance.

Again, the effect of membrane capacitance can be explained in terms of ionic movements. When positive current is injected into an axon, intracellular ions (mostly potassium) on the inside spread longitudinally along the fiber. This current recharges the membrane capacitance and flows out through the membrane resistance. (Negatively charged chloride ions are, at the same time, moving in the opposite direction.) Eventually the membrane potential reaches a new steady state with the distributed capacitances fully charged and a steady ionic current through the membrane. The time required to reach the steady state is determined by the membrane time constant.

Early experiments by Hodgkin supplied the first direct evidence for the idea that action potential propagation depended solely on the flow of current through passive membrane elements ahead of the active region.[4] His demonstration that such current flow through local circuits was responsible for propagation was in marked contrast to the prevailing concepts attributing propagation to active metabolic processes.

The velocity of action potential propagation depends on the cable properties of the nerve fiber (r_{input} and λ). The effects of these passive properties on conduction are important because conduction velocity plays a significant role in the scheme of organization of the nervous system. It varies by a factor of more than 100 in nerve fibers that transmit messages of different information content. In general, nerves that conduct most rapidly (more than 100 m/sec) are involved in mediating rapid reflexes, such as those used for regulating posture. Slower conduction velocities are associated with less urgent tasks such as regulating the distribution of blood flow to various parts of the body, controlling the secretion of glands, or regulating the tone of visceral organs.

To illustrate the factors involved in impulse propagation, we can imagine the action potential frozen at an instant in time and plot its spatial distribution along the axon, as shown in Figure 4. The distance occupied depends on its duration and conduction velocity. For example, if the action potential duration is 2 msec and it is conducted at 10 m/sec (10 mm/msec), then the potential will be spread over a 20-mm length of axon. At the peak of the action potential there is a transient reversal of the membrane potential, the inside being charged positively with respect to the outside. Along the leading edge of the action potential, where the membrane potential has reached threshold, there is a rapid influx of sodium ions along their electrochemical gradient, depolarizing the cell. The current spreads through the axoplasm ahead of the active region, depolarizing the adjacent membrane and flowing out across it. Behind the peak, on the falling phase of the action potential, the membrane is still depolarized; the conductance to potassium is high, and outward potassium movement quickly restores the membrane potential toward its resting level.

If an action potential is set up by electrical stimulation in the middle of an axon, it propagates in both directions from the point of excitation. Normally such an event does not arise in nerve cells; impulses arise at one end of an axon and travel to the other. On the other hand, impulses produced at a neuromuscular junction in the middle of a muscle fiber travel away from the junction toward both muscle tendons. In any case, once initiated, an action potential cannot double back on itself, reversing its direction of propagation. This is because of the refractory period. In the refractory region, indicated in Figure 4, the sodium conductance is still inactivated and the potassium conductance is high, so that a backward-conducting regenerative response cannot occur. As the action

[4]Hodgkin, A. L. 1937. *J. Physiol.* 90: 183–210, 211–232.

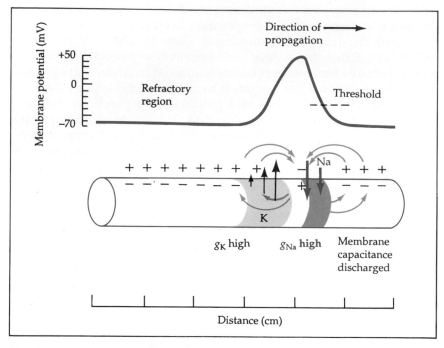

4   **CURRENT FLOW DURING A NERVE IMPULSE** at an instant in time. Current flowing into the active region spreads ahead of the impulse to depolarize the membrane toward threshold. Outward potassium current from the depolarized segment to behind the active region causes rapid repolarization.

potential leaves the region, the membrane potential returns to its resting value and excitability recovers.

The conduction velocity of the action potential depends largely on the rate at which the membrane capacitance ahead of the active region is discharged to threshold by the spread of positive charge. This, in turn, depends on the amount of current generated in the active region and on the cable properties of the fiber, in particular on the membrane capacitance c_m and the internal longitudinal resistance r_i through which the current must flow to discharge the capacitance. Because of the dependence on the longitudinal resistance of the axon core, conduction velocity is higher in large than in small fibers. In theory, other factors being equal, the velocity of propagation should vary directly with the square root of the fiber diameter.[5]

Myelinated nerves and saltatory conduction

In the vertebrate nervous system, the larger nerve fibers are myelinated. In the periphery myelin is formed by Schwann cells, and in the central nervous system by oligodendrocytes (Chapter 6), that wrap themselves tightly around axons. With each wrap, the cytoplasm of the Schwann or glial cell between the membrane pair is squeezed out so that the result is a spiral of tightly packed membranes. The number of wrappings (lamellae) ranges from a low of between 10 and 20 to a

[5]Hodgkin, A. L. 1954. *J. Physiol.* 125: 221–224.

maximum of about 160.[6] A wrapping of 160 lamellae means that there are 320 membranes in series between the plasma membrane of the axon and the extracellular fluid. Thus the effective membrane resistance is increased by a factor of 320 and the membrane capacitance is reduced to the same extent. In terms of dimensions, the myelin occupies 20 to 40 percent of the diameter of the fiber—that is, the diameter of the axon is 60 to 80 percent of the overall fiber diameter.

The myelin sheath is interrupted periodically by NODES OF RANVIER, exposing patches of axonal membrane. The internodal distance is usually about 100 times the external diameter of the fiber and ranges from 200 μm to 2 mm. The effect of the myelin sheath is to restrict membrane current flow largely to the node, because ions cannot flow easily into or out of the high-resistance internodal region, and the internodal capacitative currents are very small as well. It follows that ions enter and leave the axon only at the nodes. As a result excitation jumps from node to node, thereby greatly increasing the conduction velocity. Such impulse propagation is called SALTATORY CONDUCTION (from Latin *saltare*, "to jump, leap, or dance"). Saltatory conduction does not mean that the action potential occurs in only one node at a time. While excitation is jumping from one node to the next on the leading edge of the action potential, many nodes behind are still active. Myelinated axons not only conduct more rapidly than unmyelinated ones but also are capable of firing at higher frequencies for more prolonged periods of time.

An additional consequence of myelination is that during impulse propagation fewer sodium and potassium ions enter and leave the axon, because regenerative activity is restricted to the nodes. Consequently, less metabolic energy is required by the nerve cell to restore the intracellular concentrations to their resting levels.

Experiments that demonstrated saltatory conduction were first made in 1941 by Tasaki[7] and later by Huxley and Stämpfli,[8] who recorded current flow at nodes and internodes. Such an experiment on a single myelinated axon is illustrated in Figure 5. The nerve is placed in three saline pools, the central pool being narrow and separated from the others by air gaps of very high resistance. Electrically, the pools are connected by the external recording circuitry as shown, so that during impulse propagation currents that would otherwise be interrupted by the air gap flow instead into or out of the central pool through the resistor R. The voltage drop across the resistor provides a measure of the magnitude and direction of the currents. In the first experiment (Figure 5A), the central pool contains a node of Ranvier. Upon stimulation of the nerve, current first flows outward through the node and back toward the region of oncoming excitation (upward deflection) as the node is depolarized. This is followed by inward current at the node

[6]Arbuthnott, E. R., Boyd, I. A. and Kalu, K. U. 1980. *J. Physiol.* 308: 125–157.

[7]Tasaki, I. 1959. *In* J. Field (ed.), *Handbook of Physiology*, Section 1, Vol. I. American Physiological Society, Bethesda, MD. pp. 75–121.

[8]Huxley, A. F. and Stämpfli, R. 1949. *J. Physiol.* 108: 315–339.

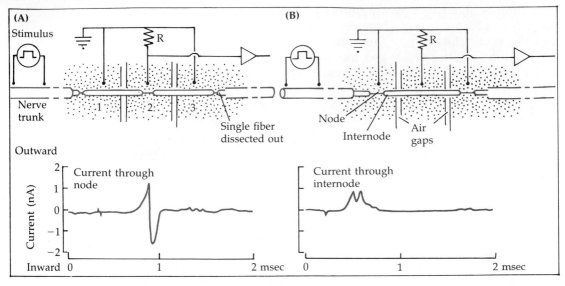

5   **CURRENT FLOW THROUGH A MYELINATED AXON.** A single myelinated axon passes through two air gaps that create three compartments not linked by extracellular fluid. During the propagated action potential, currents into and out of the center compartment (2) flow through the resistor R; the voltage drop across the resistor is a measure of the current. (A) A node of Ranvier is in compartment 2. Initially, as the action potential approaches and the node is being depolarized, current flows through the resistor from compartment 2 to compartment 1 (upward deflection); when threshold is reached at the node a large inward flux follows and the current is reversed. (B) An internode is in the center compartment and there is only outward current flow from the compartment, with no inward current, as the action potential first approaches and then leaves the internodal segment. (After Tasaki, 1959.)

(downward deflection) when threshold is reached and an action potential is generated. When the central pool contains an internode (Figure 5B), there is no inward current, only two small peaks of capacitative and resistive current flowing from the central pool toward the regions of excitation as the impulse first approaches in compartment 1 and then travels onward in compartment 3. Experiments such as this confirmed that there was no inward current, and hence no regenerative activity, in the internodal region. Sophisticated recording techniques for recording saltatory conduction in undissected mammalian axons in situ have been developed by Bostock and Sears.[9] With such techniques it is possible to measure both inward currents at the nodes and longitudinal internodal currents and thereby to estimate accurately the positions of the nodes and distances between them.

Conduction velocity in myelinated fibers

Conduction velocities of myelinated fibers vary from a few to more than 100 meters per second. In the vertebrate nervous system, peripheral nerves have been classified into groups according to conduction velocity and function (Box 3). Theoretical calculations suggest, and experimental recordings confirm, that in myelinated fibers the conduction

[9]Bostock, H. and Sears, T. A. 1978. *J. Physiol.* 280: 273–301.

velocity is proportional to the diameter of the fiber. The relation between fiber diameter and conduction velocity was first considered theoretically by Rushton[10] and has been examined in peripheral nerve in some detail by Boyd and his colleagues.[6] Experimental measurments show that large myelinated fibers have a conduction velocity in meters per second that is approximately 6 times their outside diameter in μm; for smaller fibers, below 11 μm in diameter, the constant of proportionality is about 4.5.

One theoretical point of interest is the best thickness for the myelin sheath for optimal conduction velocity, given a particular outer diameter. Obviously the increase in membrane resistance in the myelinated region will be greater with a thick sheath than with a thin one. On the other hand, as the myelin thickness increases, the cross-sectional area of the axoplasm must decrease, thereby increasing internal longitudinal resistance. The first effect would be expected to increase conduction velocity, the second to decrease it. It turns out that the optimal compromise between these opposing effects is achieved when the axon diameter is about 0.7 times the overall fiber diameter. As already noted, the observed ratio in mammalian peripheral nerve ranges between 0.6 and 0.8.

The calculated optimum internodal length for conduction is approximately that found in reality, namely about 100 times the external fiber diameter. Greater internodal distances would allow the excitation to jump farther, tending to increase conduction velocity. On the other hand, because of increased longitudinal resistance between the more distant nodes, the nodal depolarization produced by activity in a preceding node would be smaller and rise more slowly. As a result, excitation would be slowed, tending to decrease conduction velocity. Because of these opposing factors, modest variations in internodal length have little effect on conduction velocity. Of course, with a very large internodal distance, depolarization from activity in a preceding node would no longer reach threshold and conduction would be blocked.

In myelinated fibers, sodium channels are highly concentrated in the nodes of Ranvier, with potassium channels more concentrated under the paranodal sheath.[11] The properties of the axon membrane in the paranodal region normally covered by myelin were first examined by Ritchie and his colleagues.[12] To do this, the myelin was loosened by enzyme treatment or osmotic shock. Voltage clamp studies were then made of currents in the region of the node and compared with those obtained before the treatment. Such experiments showed that in rabbit nerve, nodes of Ranvier normally display only inward sodium current upon excitation. Repolarization is caused not by an increase in potassium conductance (as in other cells considered so far) but instead by a relatively large leak current and rapid sodium inactivation. After the

Distribution of channels in myelinated fibers

---

[10]Rushton, W. A. H. 1951. *J. Physiol.* 115: 101–122.

[11]Black, J. A., Kocsis, J. D. and Waxman, S. G. 1990. *Trends Neurosci.* 13: 48–54.

[12]Chiu, S. Y. and Ritchie, J. M. 1981. *J. Physiol.* 313: 415–437.

BOX 2 STIMULATING AND RECORDING WITH EXTERNAL ELECTRODES

For much physiological work in the central and peripheral nervous systems, extracellular electrodes are used to stimulate or to record from axons of various diameters. When two extracellular electrodes are used to stimulate a nerve trunk, much of the current flow is diverted through the extracellular fluid; the remainder enters individual axons under the positive electrode, flows along the axoplasm, and exits under the negative electrode. Current entering an axon produces a voltage drop across the membrane, causing local hyperpolarization (the outside is made more positive than the inside). Current flow toward the negative electrode produces an additional voltage gradient along the core of the axon. Current leaving the axon under the negative electrode causes a local depolarization. These three voltage gradients must equal the total voltage drop in the extracellular solution between the electrodes. The voltage that must be applied to bring the membrane under the negative electrode to threshold depends on fiber diameter: Large fibers require less stimulating voltage than small fibers. This is because of geometrical factors. As fiber size increases, the core resistance (which varies with the *square* of the diameter) decreases more rapidly than the membrane resistance (which varies with diameter). As a result, less of the applied voltage is dissi-

pated along the core of the fiber and a greater voltage drop is produced across the membrane. This principle is illustrated numerically in the following diagrams, showing schematic representations of current flow between a pair of stimulating electrodes 2 cm apart through an unmyelinated fiber (current flow in the extracellular fluid is omitted for clarity). Plausible values for transverse and longitudinal resistances to current flow for a 20-μm fiber are given in part (A) of the figure. A potential of 120 mV applied to the fiber between the electrodes produces a depolarization of 12 mV. For a fiber with twice the diameter (40 μm; part (B) of the figure) the transverse resistances are halved, the longitudinal resistance reduced by a factor of four, and the depolarization increased to 20 mV.

Extracellular electrodes can also be used to record action potential activity in nerve trunks. This is possible because during action potential propagation, longitudinal currents in the surrounding extracellular fluid create potential gradients along the nerve fiber. Because of their larger membrane area and lower core resistance, larger fibers generate more current, and hence larger extracellular gradients, than smaller ones. As a result, they produce larger signals at the recording electrodes.

The relation between size and threshold for

BOX 3 CLASSIFICATION OF NERVE FIBERS IN VERTEBRATES

If one stimulates a nerve bundle electrically at one end and records from it some distance away, one sees a series of peaks in the record. These are the result of dispersion of nerve impulses that travel at different velocities and therefore arrive at the recording electrodes at different times after the stimulus. Vertebrate nerve fibers were classified into groups on the basis of these

differences in conduction velocity, combined with differences in function. Unfortunately, two such classifications were developed. In the first, group A refers to myelinated fibers in peripheral nerve; these conduct at velocities ranging from 5 to 120 m/sec. Group B consists of myelinated fibers in the autonomic nervous system, which have conduction velocities in the lower part of

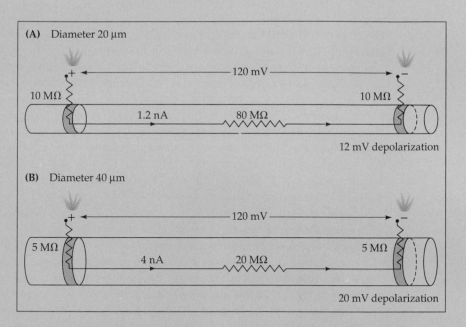

(A)   Diameter 20 μm

120 mV

10 MΩ   10 MΩ

1.2 nA   80 MΩ

12 mV depolarization

(B)   Diameter 40 μm

120 mV

5 MΩ   5 MΩ

4 nA   20 MΩ

20 mV depolarization

stimulation by external electrodes is fortunate for physiological and clinical purposes. For example, threshold and conduction velocity can be tested in motor nerves, which are relatively large, without exciting pain fibers, which are very much smaller. Just as the largest fibers are the easiest to stimulate, they are often the most difficult to block, for example by cooling, or by local anes-thetics. Again, this means that pain fibers can be blocked with anesthetic without interfering with conduction in larger sensory and motor fibers. But this relation between size and block of con-duction does not always hold: Block by localized pressure affects large axons first, then smaller axons as the pressure is increased.

the A-fiber range. Group C refers to unmyelin-ated fibers, which conduct very slowly (less that 2 m/sec). Group A fibers were further subdivided according to conduction velocity into α (80–120 m/sec), β (30–80 m/sec), and δ (5–30 m/sec). The term "γ fibers" is reserved for motor nerves sup-plying muscle spindles (Chapter 14), which have conduction velocities that span the β and lower part of the α range.

The second nomenclature applies to group A sensory fibers arising in muscle: Group I, corre-sponding to Aα, Group II (Aβ), Group III (Aδ). Group I afferent fibers were further classified into two separate groups depending on whether they conveyed information from muscle spindles (Ia) or from sensory receptors in tendons (Ib).

axon membrane in the paranodal region was exposed, excitation produced a delayed outward potassium current as well, with no increase in inward current. It seems, then, that the axon under the myelin in the region adjacent to the nodes contains voltage-activated potassium channels but no voltage-activated sodium channels. In other words, its properties are the reverse of those of the nodal membrane. One of the consequences of extensive demyelination is conduction block. However, mammalian axons that have been demyelinated chronically with diphtheria toxin can develop continuous conduction through a demyelinated region,[9] implying that after demyelination voltage-activated sodium channels appear in the exposed axon membrane. This conclusion has been confirmed by labeling demyelinated nerves with antibodies to sodium channels.[13] Voltage-activated potassium channels are present in the demyelinated region as well.[14]

Influence of neuronal shape on conduction

The simple uniform cable is an idealized structure resembling an unmyelinated axon, but is not an adequate representation of the cell body, elaborate dendritic arborization, and numerous axonal branches of a complete neuron. Branch points and abrupt changes in diameter may constitute regions unfavorable for signal transmission, leading to failure of action potential propagation. For example, current produced by an action potential in a single slender dendrite is small compared with that required for depolarization of a large cell body. As a result the action potential may fail as it reaches the cell body because the input resistance of the cell body is relatively low and its capacitance large. Similarly, where a small diameter axon splits into two larger axons, the safety factor for propagation is reduced. This arrangement is common in invertebrates (Chapter 13). Under normal circumstances an impulse propagates into both branches, but after repeated firing, propagation may become blocked at the branch point. In crustacean axons, such block is associated with an increase in extracellular potassium concentration.[15] In leech sensory cells, the block occurs because of persistent hyperpolarization, induced by increased electrogenic activity of the sodium pump (Chapter 2) and long-lasting increases in potassium permeability that increase the amount of current required for depolarization to threshold.[16,17]

In myelinated peripheral nerve, the safety factor for conduction is about 5; that is, the depolarization produced at a node by excitation of a preceding node is approximately five times larger than necessary to reach threshold. Again, this safety factor can be reduced considerably in several morphological circumstances. For example, if a myelinated nerve divides into two branches, then the current supplied by the single

[13]England, J. et al. 1990. *Proc. Natl. Acad. Sci. USA* 87: 6777–6780.

[14]Bostock, H., Sears, T. A. and Sherratt, R. M. 1981. *J. Physiol.* 313: 301–315

[15]Grossman, Y., Parnas, I. and Spira, M. E. 1979. *J. Physiol.* 295: 307–322.

[16]Yau, K. W. 1976. *J. Physiol.* 263: 513–538.

[17]Gu, X. N., Macagno, E. R. and Muller, K. J. 1989. *J. Neurobiol.* 20: 422–434.

node at the branch point is divided between two nodes beyond the branch and the safety factor for conduction along one or both may be reduced. Similarly, when the myelin sheath terminates, for example near the end of a motor nerve, the current from the last node is then distributed over a large area of unmyelinated nerve terminal membrane and, as a consequence, provides less overall depolarization than would occur at a node. It is perhaps for this reason that the last few internodes before an unmyelinated terminal are shorter than normal, so that more nodes can contribute to the depolarization of the terminal.[18]

In most circumstances, electrical signals cannot pass directly from one cell to the next. Certain cells, however, are *electrically coupled*. The properties and functional role of electrical synapses are discussed in Chapter 7. Here we describe special intercellular structures that are required for electrical continuity between cells.

Pathways for current flow between cells

The necessity for such specialized structures is illustrated in Figure 6. In the first model (Figure 6A), the ends of two cylindrical neuronal processes, A and B, are separated by a gap (as occurs, for example, at chemical synapses). Current passed into process A exits along the length of the process and out of the end. How is the current flow distributed quantitatively? This can be determined by considering the resistances involved. The cylindrical conductor is a cable extending in one direction only. Its input resistance is

$$r_{input} = \left(\frac{R_m R_i}{2\pi^2 a^3}\right)^{1/2}$$

The resistance of the circular end is

$$r_e = \frac{R_m}{\pi a^2}$$

Suppose that for both cells R_m is 2000 Ωcm^2, R_i is 100 Ωm, and a is 10 μm (all plausible values). Then r_{input} will be 3.2 MΩ, and r_e will be 637 MΩ. Because the end resistance is 200 times larger than the input resistance, only 0.5 percent of the current injected into the cylinder will flow outward through the end. Further, all of the current leaving the end will flow laterally out of the cleft, which has a low resistance (represented by r_e), rather than entering the end of the second process.

If we now butt the two ends of the processes tightly together (Figure 6B), the resistance to outward current flow along process A will be unchanged. The resistance to current flow out the end will be 1274 MΩ (two membranes in series) plus 3.2 MΩ (the input resistance of process B). As a result, current flowing through the two ends and into process B is 0.025 percent of the total. An applied current of 10 nA would depolarize cell A by 31.7 mV, cell B by only about 79 μV. Clearly, significant coupling between the two processes is excluded unless two

[18]Quick, D. C., Kennedy, W. R. and Donaldson, L. 1979. *Neuroscience* 4: 1089–1096.

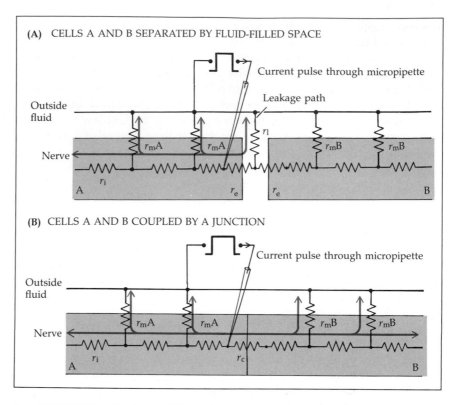

6   PATHWAYS FOR CURRENT FLOW BETWEEN CELLS. (A) Electrical model of two cell processes separated by a fluid-filled gap. Current flow out of the end of process A through the end resistance r_e escapes from gap (resistance r_i) without entering process B. (B) Cells brought into close apposition. In this model, current can flow from process A into process B. For significant coupling to occur, however, the coupling resistance r_c must be low compared with the other membrane resistances r_m (see text).

conditions are met: Current must be prevented from escaping from the region of apposition, and the intercellular resistance must be made very much smaller than that normally found in membranes.

Structural basis for electrical coupling: The gap junction

At sites of electrical coupling the intercellular current flow is through GAP JUNCTIONS.[19] The gap junction is a region of close apposition of two cells characterized by aggregates of particles distributed in hexagonal arrays in each of the adjoining membranes (Figure 7). Each particle is made up of six protein subunits arranged in a circle about 10 nm in diameter, with a 2-nm diameter central core.[20,21] Identical particles in the apposing cells are exactly paired to span the 2- or 3-nm gap in the region of contact. The unit formed in this way is called a CONNEXON.[22] The core provides the pathway for the flow of small ions and molecules

[19]Loewenstein, W. 1981. *Physiol. Rev.* 61: 829–913.
[20]Unwin, P. N. T. and Zampighi, G. 1980. *Nature* 283: 545–549.
[21]Tibbits, T. T. et al. 1990. *Biophys. J.* 57: 1025–1036.
[22]Caspar, D. L. D. 1977. *J. Cell Biol.* 74: 605–628.

between cells. The conductance of an individual channel connecting adjacent cells is on the order of 100 pS.[23]

Several varieties of the subunit protein (CONNEXIN) have been sequenced, including connexin32 (32 kD), normally present in rat liver;[24] connexin43, found in heart muscle;[25] and a 38-kD protein from *Xenopus*.[26] These three proteins are highly homologous, and hydropathy plots (see Chapter 2) suggest that they contain four membrane-spanning helices. Antibody binding studies are consistent with this model and indicate that the amino terminal (and hence the carboxy terminal as well) resides in the cytoplasm.[27]

Injection of mRNA coding for connexin into pairs of *Xenopus* oocytes results in the formation of connexons between the two cells, after a delay of about 4 hours.[28] Functional intercellular coupling occurs between connexins of different types—that is, when message for connexin32 is injected into one cell and for connexin43 in the other.[29]

Cell-to-cell channels are widely distributed phylogenetically from sponges to humans in tissues of mesenchymal and epithelial origin. Generally a given cell within a tissue is coupled to its neighbors so that whole organs or subdivisions of organs are coupled from within. Coupling may serve a variety of functions, such as tissue homeostasis among large number of cells, and transmission of regulatory signals from one cell to the next.[19] In the nervous system, cell-to-cell coupling is responsible for transmitting excitation at *electrical synapses* (Chapter 7). Smooth muscle fibers in gut, bronchi, and blood vessels are electrically coupled, as are cardiac muscle fibers. The junctions allow waves of contraction to spread through the muscle in concert. In addition to allowing the passage of current between cells, the intercellular channels allow passage of small hydrophilic molecules up to about 2 nm in diameter. Coupling is widespread in embryos,[30] even between cells with quite different future functional roles, and such cells uncouple as development progresses. Conversely, premature uncoupling leads to disruption of normal development.[31]

Uncoupling can occur under a variety of circumstances. Embryonic sensory cells in the spinal cord of tadpole larvae, which are initially coupled to one another,[32] uncouple about the time they develop the ability to produce sodium action potentials. Other cells can be uncoupled experimentally by depolarization of one or both of the coupled

Significance of coupling between cells

[23]Neyton, J. and Trautmann, A. 1985. *Nature* 317: 331–335.

[24]Paul, D. L. 1986. *J. Cell Biol.* 103: 123–134.

[25]Beyer, E. C., Paul, D. L. and Goodenough, D. A. 1987. *J. Cell Biol.* 105: 2621–2629.

[26]Ebihara, L. et al. 1989. *Science* 243: 1194–1195.

[27]Yancey, S. B. et al. 1989. *J. Cell Biol.* 108: 2241–2254.

[28]Werner, R. et al. 1985. *J. Memb. Biol.* 87: 253–268.

[29]Swensen, K. I. et al. 1989. *Cell* 57: 145–155.

[30]Sheridan, J. D. 1978. *In* J. Feldman et al. (eds.), *Intercellular Junctions and Synapses.* Chapman & Hall, London. pp. 39–59.

[31]Guthrie, S. C. and Gilula, N. B. 1990. *Trends Neurosci.* 12: 12–16.

[32]Spitzer, N. C. 1982. *J. Physiol.* 330: 145–162.

144

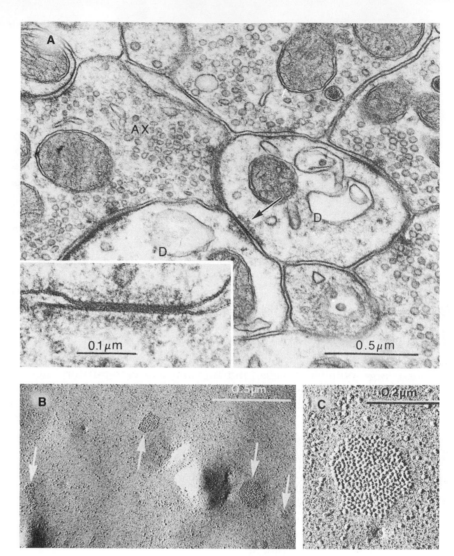

7

GAP JUNCTIONS BETWEEN NEURONS. (A)
Two dendrites (labeled D) in the inferior olivary
nucleus of the cat are joined by a gap junction
(arrow), shown at higher magnification in the
inset. The usual space between the cells is almost
obliterated in the contact area, which is traversed
by cross bridges. (B) Freeze-fracture through the
presynaptic membrane of a nerve terminal that
forms gap junctions with a neuron in the ciliary
ganglion of the chicken. A broad area of the
cytoplasmic fracture face is exposed, showing
clusters of gap junction particles (arrows). (C)
Higher magnification of one such cluster. Each
particle in the cluster represents a single connexon. (D) Sketch of gap junction region showing individual connexons bridging the gap between the lipid membranes of two apposed cells.
(A from Sotelo, Llinás and Baker, 1974; B and C
from Cantino and Mugnaini, 1975. D after Makowski et al. 1977.)

(D)

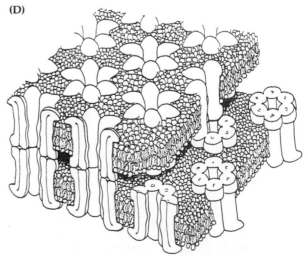

cells, or specifically by changes in potential across the junction itself. Uncoupling can also occur in response to changes in chemical composition of the cytoplasm, such as increased intracellular calcium concentrations or decreased pH.[33]

SUGGESTED READING

Reviews

Bennett, M. V. L., Barrio, L. C., Bargiello, T. A., Spray, D. C., Herzberg, E. and Saez, J. C. 1991. Gap junctions: New tools, new answers, new questions. *Neuron* 6: 305–320.

Black, J. A., Kocsis, J. D. and Waxman, S. G. 1990. Ion channel organization of the myelinated fiber. *Trends Neurosci.* 13: 48–54.

Guthrie, S. C. and Gilula, N. B. 1990. Gap junction communication and development. *Trends Neurosci.* 12: 12–16.

Original papers

Arbuthnott, E. R., Boyd, I. A. and Kalu, K. U. 1980. Ultrastructural dimensions of myelinated peripheral nerve fibres in the cat and their relation to conduction velocity. *J. Physiol.* 308: 125–157.

Hodgkin, A. L. and Rushton, W. A. H. 1946. The electrical constants of a crustacean nerve fibre. *Proc. R. Soc. Lond. B* 133: 444–479.

Huxley, A. F. and Stämpfli, R. 1949. Evidence for saltatory conduction in peripheral myelinated nerve fibers. *J. Physiol.* 108: 315–339.

Rushton, W. A. H. 1951. A theory of the effects of fibre size in medullated nerve. *J. Physiol.* 115: 101–122.

[33]Obaid, A. L., Socolar, S. J. and Rose, B. 1983. *J. Memb. Biol.* 73: 68–89.

PROPERTIES AND FUNCTIONS OF NEUROGLIAL CELLS

Most nerve cells in the central and the peripheral nervous systems are surrounded by satellite cells. These satellites are divided into two main categories: (1) neuroglial cells in the brain, which are further subdivided into oligodendrocytes and astrocytes, and (2) Schwann cells in the periphery. Taken together, the neuroglial cells make up almost one-half the volume of the brain, and they greatly outnumber neurons. Microglial cells derived from blood constitute a separate, distinctive population of nonneuronal phagocytotic cells in the nervous system.

The membrane properties of neuroglial cells differ in several essential respects from those of neurons. Glial cells behave passively in response to electric current, and their membranes, unlike those of neurons, do not generate conducted impulses. The glial membrane potential is greater than that of neurons and depends primarily on the distribution of potassium, the principal intracellular cation. Further, glial cells are linked by low-resistance connections that permit the direct passage of ions and small molecules such as fluorescent dyes between the cells. Neurons and glial cells, on the other hand, are separated from each other by narrow, fluid-filled, extracellular spaces that are about 20 nm wide and prevent currents generated by nerve impulses from spreading into neighboring cells. Neurons influence glial cells by releasing potassium in the intercellular spaces during the conduction of impulses, thereby depolarizing the glial membrane. Potassium-mediated glial depolarization creates potential changes that can be recorded from the surface of tissue; as a result, glial cells contribute to the electroencephalogram and the electroretinogram.

Neuroglial cells and Schwann cells perform a variety of functions. They form the myelin sheaths around the larger axons and speed up conduction of the nerve impulse. They are also involved in guiding axons to their targets during growth and regeneration. The fluid composition surrounding neurons is influenced by neuroglial cells and by the endothelial cells lining capillaries that constitute the blood–brain barrier. Intriguing questions that await further elucidation include how glial cells alter their performance in response to potassium-mediated depolarization, whether they supply important molecules to neurons, and what role they play in immune responses of the nervous system.

Nerve cells in the brain are intimately surrounded by satellite cells called NEUROGLIAL CELLS. From counts of cell nuclei, it has been estimated that they outnumber neurons by at least 10 to 1 and make up about one-half of the bulk of the nervous system. Studies of glial cells are in a peculiar state. From the time of their discovery to the present, the importance of these cells has been stressed. New findings and attractive speculations about their functions accumulate year by year. Evidence has accumulated for subtle, long-term influences of neuroglial cells on development and repair and on homeostasis of the fluid environment of the brain. And yet, except for their role in speeding conduction, clear-cut, essential functions of neuroglial cells for signaling by neurons remain elusive. It is remarkable that one should have to discuss the physiological activities of the nervous system in adult animals in terms of neurons only, as if glial cells did not exist. For example, there is virtually no mention of glial cells to be found in recent authoritative books that deal with drug actions on the brain or with the mammalian visual system.[1,2]

Glial cells were first described in 1846 by Rudolf Virchow, who later gave them their name. He clearly recognized that they differed fundamentally from neurons and from interstitial tissue elsewhere in the body. Several excerpts from a paper by Virchow give the flavor of his approach and thinking.[3] He pointed out many aspects of glial tissue that later became important for formulating various hypotheses.

> Hitherto, considering the nervous system, I have only spoken of the really nervous parts of it. But . . . it is important to have a knowledge of that substance also which lies *between the proper nervous parts*, holds them together and gives the whole its form. [Our italics]

Speaking of the ependyma (see later), he continued,

> This peculiarity of the membrane, namely that it becomes continuous with interstitial matter, the real cement, which binds the nervous elements together, and that in all its properties it constitutes a tissue different from the other forms of connective tissue, has induced me to give it a new name, that of *neuro glia*. [Nerve glue; our italics]

Later on he stated,

> Now it is certainly of considerable importance to know that in all nervous parts, in addition to the real nervous elements, a second tissue exists, which is allied to the large group of formations, which pervade the whole body, and with which we have become acquainted under the name of connective tissues. In considering the pathological or physiological conditions of the

[1]Hubel, D. H. 1988. *Eye, Brain and Vision*. Scientific American Library, New York.
[2]Snyder, S. 1986. *Drugs and the Brain*. Scientific American Library, New York.
[3]Virchow, R. 1859. *Cellularpathologie* (F. Chance, trans.). Hirschwald, Berlin. [Excerpts are from pp. 310, 315, and 317.]

brain or spinal marrow, the first point is always to determine how far the tissue which is affected, attacked or irritated is nervous in its nature or merely an interstitial substance. . . . Experience shows us that this very interstitial tissue of the brain and spinal marrow is one of the most frequent sites of morbid change, as for example, of fatty degeneration. . . . Within the neuroglia run the vessels, which are therefore nearly everywhere *separated from the nervous substance* by a slender intervening layer, and are not in immediate contact with it. [Our italics]

In the subsequent hundred years, neuroglial cells were intensively studied, chiefly by neuroanatomists and also by pathologists who knew them to be the most common source of tumors in the brain. This is perhaps not surprising, because normally—unlike neurons—certain glial cells can still divide in the mature animal. Among early hypotheses for glial cell functions in relation to neurons were structural support, secretion, and the prevention of "cross talk" by current spread during conduction of nerve impulses.[4] A nutritive role of neuroglia was proposed by Golgi in about 1883.[5] He wrote:

I find it convenient to mention that I have used the term connective tissue with regard to neuroglia. I would say that "neuroglia" is a better term, serving to indicate a tissue which, although connective because it connects different elements and for its own part *serves to distribute* nutrient substances, is nevertheless different from ordinary connective tissue by virtue of its morphological and chemical characteristics and its different embryological origin. [Our italics]

Coupled with Golgi's histological staining methods, these ideas seemed so reasonable and had such force that they were hardly questioned through the years.

This chapter considers glial cells from a cellular and molecular standpoint. The morphology, the molecular characterizations, and the various functions attributed to glial cells are reviewed, but the main emphasis is on their physiological properties, about which a good deal is known. Background knowledge about their ion channels, membrane potentials, and electrical signals is a prerequisite for dealing with wider issues concerning the influences of glial cells on such problems as homeostasis and axonal conduction.

Appearance and classification of glial cells

One of the most distinct structural features of neuroglial cells compared with neurons is the absence of axons, but many other differences have been demonstrated by light and electron microscopy. A representative picture of mammalian neuroglial cells is shown in Figure 1. The cytoplasmic contents suggest that glial cells are metabolically active structures containing the usual organelles, including mitochondria, endoplasmic reticulum, ribosomes, lysosomes, and often deposits of glycogen and fat. In the vertebrate central nervous system they are usually subdivided into two main groups (astrocytes and oligodendrocytes) and several subgroups.

[4]Ramón y Cajal, S. 1955. *Histologie du Système Nerveux*, Vol. II. C.S.I.C., Madrid.
[5]Golgi, C. 1903. *Opera Omnia*, Vols. I,II. U. Hoepli, Milan.

ASTROCYTES can be classified into two principal subgroups: (1) fibrous astrocytes, which contain filaments and are more prevalent among bundles of myelinated nerve fibers, the white matter of the brain; and (2) protoplasmic astrocytes, which contain less fibrous material and are more abundant in the gray matter around nerve cell bodies, dendrites, and synapses. Both types of astrocytes make contacts with capillaries and neurons. Other subtypes of astrocytes are described below.

OLIGODENDROCYTES are predominant in the white matter, where they form myelin around the larger axons (Chapter 5). This is a wrapping of glial cell processes with practically all the cytoplasm squeezed out in between so that the membranes are tightly apposed as they spiral around the axon (see Figure 13). The large numbers of smaller diameter axons (1 μm or less) that are unmyelinated are also surrounded by glial cells, singly or in bundles.

RADIAL GLIAL CELLS play an essential role in the development of the mammalian central nervous system. They stretch from the deep region of profuse cell division through the thickness of the structure (spinal cord, cerebellar or cerebral cortex) to the surface, forming elongated filaments along which developing neurons migrate to their final destinations.

EPENDYMAL CELLS that line the inner surfaces of the brain, in the ventricles, are also usually classified as glial cells. No physiological role has been assigned to them.

MICROGLIAL CELLS are distinct from the neuroglial cells in their structure, properties, and lineage. They resemble macrophages in the blood and probably arise from them.[6] One role, discussed below, is to act as scavengers for removing debris.

In invertebrates, the classification of glial cells into distinct groups is not well established. However, there is no question about the functional analogy of the various glial structures.[7,8]

In vertebrate peripheral nerves SCHWANN CELLS are analogous to oligodendrocytes in that they form myelin around the larger, fast-conducting axons (up to 20 μm in diameter). Smaller axons (usually below 1 μm in diameter), as in the brain, have a Schwann cell envelope without myelin. The glial cells of the brain and of peripheral nerves have different embryological origins: Glial cells in the central nervous system are derived from precursor cells that line the neural tube constituting the inner surface of the brain; Schwann cells arise from the neural crest.

The various types of satellite cells can now be further distinguished by sensitive immunological techniques. For example, a clear distinction has been made between fibrous and protoplasmic astrocytes. Fibrous astrocytes contain a protein against which specific antibodies have

Characterization of glial cells by immunological techniques

[6]Perry, V. H. and Gordon, S. 1988. *Trends Neurosci.* 11: 273–277.
[7]Lane, N. J. 1981. *J. Exp. Biol.* 95: 7–33.
[8]Meyer, M. R., Reddy, G. R. and Edwards, J. S. 1987. *J. Neurosci.* 7: 512–521.

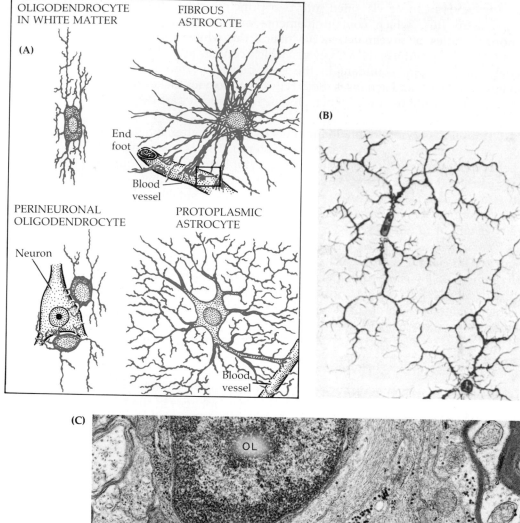

(A) OLIGODENDROCYTE IN WHITE MATTER

FIBROUS ASTROCYTE

End foot

Blood vessel

PERINEURONAL OLIGODENDROCYTE

Neuron

PROTOPLASMIC ASTROCYTE

Blood vessel

(B)

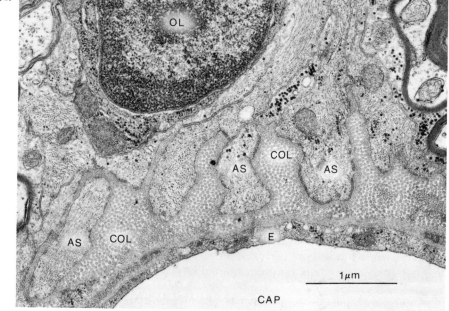

(C)

◄1   NEUROGLIAL CELLS IN THE MAMMALIAN BRAIN. (A) Neuroglial cells
stained with silver impregnation. Oligodendrocytes and astrocytes represent the
principal neuroglial cell groups in the vertebrate brain. They are closely associated with
neurons and form end feet on blood vessels. (B) Microglial cells represent small, wandering, macrophage-like cells. (C) Electron micrograph of glial cells in the optic nerve
of the rat. In the lower portion is the lumen of a capillary (CAP) lined with endothelial
cells (E). The capillary is surrounded by end feet formed by processes of fibrous astrocytes (AS). Between the end feet and the endothelial cells is a space filled with collagen
fibers (COL). In the upper portion is part of a nucleus of an oligodendrocyte (OL) and
to the right are axons surrounded by myelin wrapping. (A, B after Penfield, 1932, and
del Rio-Hortega, 1920; C from Peters, Palay and Webster, 1976.)

been prepared. When this antibody is labeled with a fluorescent marker,
the fibrous astrocytes, to which it binds selectively, can be clearly distinguished in micrographs. The protein, known as *glial fibrillary acidic
protein*, or GFAP, is present in fibrous astrocytes in all vertebrates that
have been examined and also in glial cells of the enteric neural plexus.[9]
Examples of stained fibrous astrocytes are shown in Figure 2. The
function of GFAP in the astrocytes is not yet known. Other antibodies
have been found that bind specifically to either astrocytes, oligodendrocytes, or Schwann cells.[10] Such labels can make it possible to study
not only different molecular components of various glial cell types in
adult animals but also precursor cells in the embryonic brain that will
eventually give rise to them. A recombinant retrovirus has also been
used to label glial cells and neurons as they form the cortex in mouse
embryos.[11] Precursor cells are infected at an early stage with a retrovirus
encoding a marker gene that is handed on to their descendants. These
cells can then be identified by the gene product. In this way it becomes
possible to study glial cell lineages and to pinpoint the stages at which
they diverge from neurons during development.

Another promising approach has been to combine specific antibody
labeling with tissue culture techniques. In cultures of embryonic rat
optic nerves, Raff and his colleagues have observed glial cells as they
develop from precursors.[12,13] Oligodendrocytes that make myelin start
to appear in the optic nerve at the time of birth. They can be distinguished by their morphology and by monoclonal antibodies that bind
specifically to them. Two subtypes of astrocytes, known as type I and
type II, have been described. Both react with the specific antibody that
labels fibrous astrocytes. Their morphology is, however, different; they
develop from different precursor cells at different stages, and they react
differently with other antibodies. In culture, the group II astrocytes

[9]Bignami, A. and Dahl, D. 1974. *J. Comp. Neurol.* 153: 27–38.

[10]Schachner, M. 1982. *J. Neurochem.* 39: 1–8.

[11]Luskin, M. B., Pearlman, A. L. and Sanes, J. R. 1988. *Neuron* 1: 635–647.

[12]Raff, M. C. 1989. *Science* 243: 1450–1455.

[13]Miller, R. H., ffrench-Constant, C. and Raff, M. C. 1989. *Annu. Rev. Neurosci.* 12:
517–534.

(A)

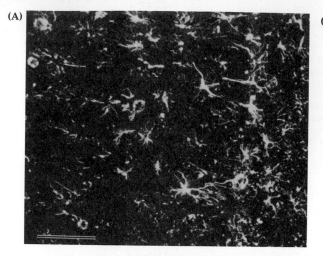

(B)

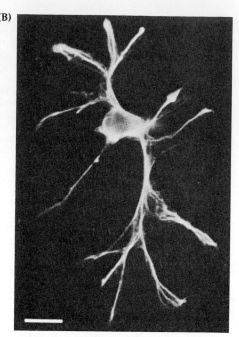

2 FIBROUS ASTROCYTES specifically labeled by an antibody against glial fibrillary acidic protein (GFAP). (A) Astrocytes labeled by antibody against GFAP in the brain of a rat. The antibody is coupled to a fluorescent dye. (B) A freshly dissociated cell from the optic nerve of a salamander. The antibody staining and the shape unequivocally identify the cell as a fibrous astrocyte. Scale in (A) is 0.1 mm; scale in (B) is 20 μm. (A after Bignami and Dahl, 1974; B from Newman, 1986.)

arise from the same precursor cell as the oligodendrocyte, but later (two weeks after birth). The common precursor cell for type II astrocytes and oligodendrocytes appears to migrate into the developing optic nerve from the brain. The classification by antibodies of astrocytes in tissue culture suggests different functional roles for type I and type II cells. It is not yet known, however, how valid this classification is for mature astrocytes in different regions of the adult CNS. Cells resembling the type I astrocytes are found in mature optic nerve, where they envelop capillaries and can influence their properties (see below). By contrast, processes of glial cells resembling type II astrocytes tend to be associated with oligodendrocytes at the nodes of Ranvier.

The embryological origin of glial cells can also be followed with precision in invertebrates such as the leech. In this animal, the large glial cells that surround the neuronal cell bodies have been shown to originate from a precursor cell that also gives rise to neurons (Chapter 13). Other glial cells that envelop axons and synapses arise from different precursor cells.[14]

Structural relations between neurons and glia

A glance at almost any electron micrograph of brain tissue brings home the difficulty of making physiological and chemical studies of neuroglial cells. Figure 3 shows an example from the cerebellum of a

[14]Weisblat, D. A., Kim, S. Y. and Stent, G. S. 1984. *Dev. Biol.* 104: 65–85.

rat. The section is filled with neurons and glial cells, which can be distinguished from each other only after considerable experience. To make the task simpler, the glial contribution is highlighted. The extracellular space is restricted to narrow clefts, about 20 nm wide, that separate all cell boundaries. Astrocytic processes generally surround neurons except where synaptic contacts are made. Many of the axons are typically grouped together, and instead of each having an individual

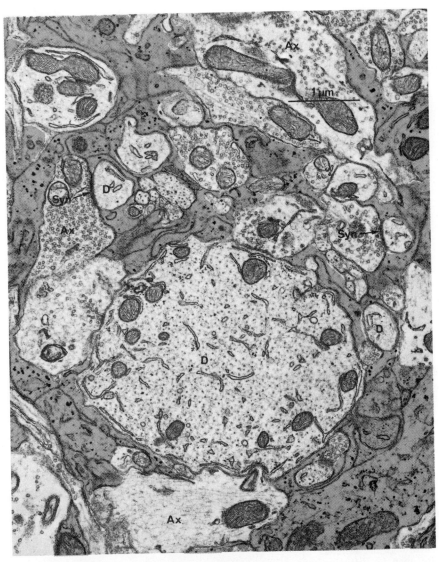

3 NEURONS AND GLIAL PROCESSES in the cerebellum of the rat. The glial contribution is lightly colored. The neurons and glial cells are always separated by clefts about 20 nm wide. The neural elements are dendrites (D) and axons (Ax). Two synapses (Syn) are marked by arrows. (After Peters, Palay and Webster, 1976.)

covering, entire bundles of axons are surrounded by a glial envelope. This arrangement is common in the central nervous system.

From comparisons of neurons and glial cells, one sees that in some regions of the brain the cross-sectional area is about equally divided between neurons and astrocytes, whereas in others, as in Figure 3, the glial contribution is smaller. Glial processes tend to be thin, at times less than 1 μm thick. Only around the glial nuclei are there larger volumes of glial cytoplasm.

Electron microscopy has clarified the neuron–glial relation by demonstrating the intimate apposition of their membranes. No special connections, such as gap junctions, are seen between the two cell types in adult CNS.[15] The clefts (several tens of nanometers wide) seen in Figure 3 always intervene between the surface membranes of neurons and glial cells. Similarly, physiological tests fail to reveal direct low-resistance pathways between the neurons and glial cells. Such pathways do, however, link glial cells to one another. The special connections are made by gap junctions (Chapter 5). One interesting junction with defined structure is the close apposition of myelin to axons at the edge of the node; specialized contacts that limit longitudinal spread of current are seen between the two structures in freeze-fracture preparations and electron micrographs.[16] The relation of glial cells, neurons, and extracellular space is diagrammed in Figure 4.

Basic to the difficulty of studying glial function is the lack of simple methods of separating neurons from glia. In spite of promising tissue culture techniques in immature, developing nervous systems, it is still difficult to separate adult brain tissue into pure glial and neuronal fractions since the tissues are so interwoven in adult brain. Exceptions are provided by preparations such as optic nerve and retina, from which individual, viable astrocytes and Müller cells have been isolated (see below).

[15]Mugnaini, E. 1982. *In* T. A. Sears (ed.), *Neuronal–Glial Cell Interrelationships*. Springer-Verlag, New York, pp. 39–56.
[16]Black, J. A. and Waxman, S. G. 1988. *Glia* 1: 169–183.

4

NEURONS, GLIA, EXTRACELLULAR SPACE, AND BLOOD. (A) The relationship between neurons and glia and between glial cells. While neurons are always separated from glia by a continuous cleft, the interiors of glial cells are linked by gap junctions. (B) Relations of capillary, glia, and neurons as seen in the light and electron microscopes. The most direct pathway from the capillary to the neuron is through the aqueous intercellular clefts that are open for diffusion. Cell dimensions are not in proportion. (After Kuffler and Nicholls, 1966.)

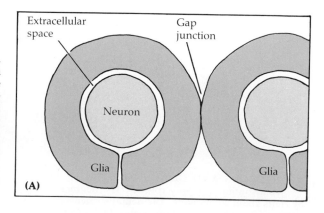

PHYSIOLOGICAL PROPERTIES OF NEUROGLIAL CELL MEMBRANES

The study of the physiological properties of neuroglia has been aided greatly by preparations in which the glial cells are large and accessible in their normal relation to nerve cells. One is the central nervous system of the leech; another is the optic nerve of the mud puppy, *Necturus*. A third preparation consists of Müller glial cells isolated from the retinas of frogs and salamanders. Such glial cells can be impaled with micro-electrodes and their membrane properties studied. These experiments illustrate once again the basic unity of the principles by which the nervous system works in higher and lower animals. Thus, for technical reasons, the membrane properties of glial cells were first determined in especially large cells of the leech brain. Once this had provided a clue to what to look for, it became simpler to investigate amphibian and

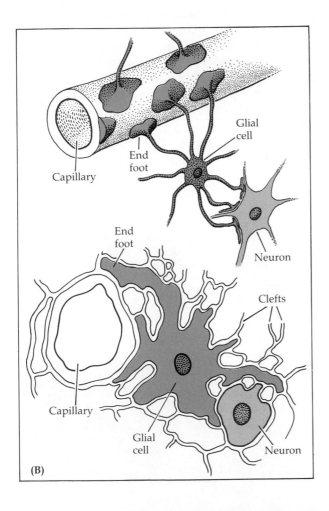

(B)

then mammalian glial cells, which share many key properties with those in the more modest nervous system of the leech. Far from being a roundabout approach, this turned out to be a shortcut.[17]

Simple preparations
used for intracellular
recording from glia

The leech nervous system, more fully described in Chapter 13, consists of a chain of ganglia joined by connectives; each ganglion is less than 1 mm in diameter, contains no blood vessels, and is quite transparent. The relative sizes of neurons and glia in this nervous system are the reverse of what is seen in vertebrates, for in the leech the glial cells are larger than the nerve cell bodies and are few in number.[18] The glial cells in the ganglion are transparent and therefore appear under the dissecting microscope as spaces between nerve cells. They can be impaled consistently by a microelectrode inserted next to a neuron.

The optic nerve of *Necturus* is about 0.15 mm in diameter and is covered by a layer of connective tissue containing blood vessels that run parallel to the surface; no blood vessels occur within the tissue.[19] The glial cells are large, with dark prominent nuclei, and intimately surround the nerve fibers. Figure 5A presents a cross section of the *Necturus* optic nerve stained with toluidine blue. The outlines of the bundles of unmyelinated axons (fiber diameters 0.1–1.0 μm) are too small to be seen in the light micrograph. The neuron–glia relation is illustrated in Figure 4; spaces about 20 nm wide exist between the cells, as in the mammalian brain (Figure 3) and the leech nervous systems.

Glial cells in both the central nervous system of the leech and the optic nerve of *Necturus* occupy 35 to 55 percent of the cross-sectional

[17]Kuffler, S. W. and Nicholls, J. G. 1966. *Ergeb. Physiol.* 57: 1–90.
[18]Kuffler, S. W. and Potter, D. D. 1964. *J. Neurophysiol.* 27: 290–320.
[19]Kuffler, S. W., Nicholls, J. G. and Orkand, R. K. 1966. *J. Neurophysiol.* 29: 768–787.

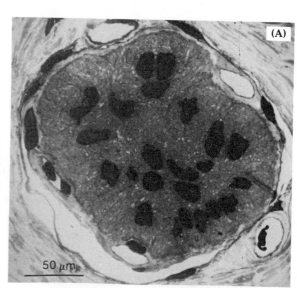

(A)

5

OPTIC NERVE OF THE MUD PUPPY (*Necturus*). (A) Cross section. Nuclei of glial cells are stained black. The outlines of the glial cytoplasm and bundles of nonmedulated axons cannot be resolved. The nerve is surrounded by connective tissue containing capillaries. **(B,C)** Two identical electron micrographs of part of the optic nerve. In one the glial processes are lightly shaded and the clefts that separate the cells are traced in black. Two cleft openings reaching the surface are marked by arrows. Axons run in closely packed bundles. (From Kuffler, Nicholls and Orkand, 1966.)

50 μm

area. For emphasis the glial cells with their processes between the nervous elements are shaded in Figure 5C, and the tortuous intercellular clefts are traced in black. Note that in two places (arrows) the clefts open to the outside. The fine structure of the leech nervous system appears strikingly similar in many respects, with narrow spaces intervening between neuronal and glial membranes.

Recordings of membrane potentials in the leech show directly that glial cells have resting potentials greater than those of the neurons they surround.[18] In vertebrates, including the frog, mud puppy, cat, and rat, the highest membrane potentials recorded for neurons are about −70 to −75 mV, whereas the values for glial cells consistently approach −90 mV. Their membrane resistance (R_m, Chapter 5) is about 1000 Ωcm^2 or more, which is comparable to the value measured in neurons.

Glial membrane potentials

To study the origin of the large resting potentials, the membrane potentials of glial cells have been measured in solutions containing different potassium concentrations (Figure 6). In isolated optic nerves of *Necturus*, the glial membrane behaves like a perfect potassium electrode; that is, its behavior accurately follows the Nernst equation[19] (Chapter 3):

Dependence of membrane potential on potassium

$$E = 59 \log \frac{[K]_o}{[K]_i}$$

Changes in sodium and chloride concentrations do not produce significant changes in potential. One can conclude that ions other than potassium make a negligible contribution to the resting membrane potential. Figure 6 shows a series of membrane potential measurements plotted against $[K]_o$ on a logarithmic scale. The solid line is the theoretical slope of 59 mV per tenfold change in concentration predicted by the Nernst equation (at 24°C), which is in excellent agreement with the

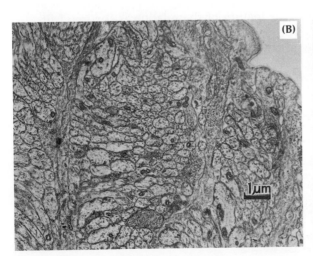

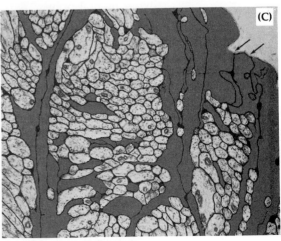

6

GLIAL MEMBRANE POTENTIAL depends on potassium concentration. (A) System of perfusing the optic nerve while recording from a glial cell. (B) Reducing the potassium concentration from the normal 3.0 mM to 0.3 mM increases the normal membrane potential of −89 mV to −125 mV; increasing the potassium concentration to 30 mM decreases it by 59 mV. (C) Various values of potassium plotted against membrane potential show that the relation predicted by the Nernst equation (solid line) accurately fits the experimental results over a wide concentration range. The membrane potential is zero when the internal and external potassium concentrations are equal to 100 mM. (After Kuffler, Nicholls and Orkand, 1966.)

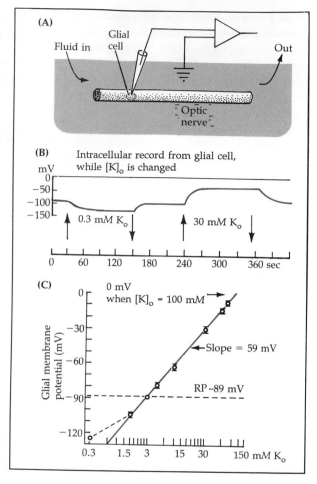

(A)

Fluid in   Glial cell   Out

Optic nerve

(B)   Intracellular record from glial cell, while $[K]_o$ is changed

mV

0.3 mM K_o 30 mM K_o

0   60   120   180   240   300   360 sec

(C)   0 mV when $[K]_o$ = 100 mM

Glial membrane potential (mV)

Slope = 59 mV

RP −89 mV

0.3   1.5   3   15   30   150 mM K_o

experimental points. A particular feature of the relation is the good fit at low concentrations of $[K]_o$, down to 1.5 mM. In this respect glial cells differ significantly from most neurons, which deviate from the Nernst prediction in the physiological range of 2 to 4 mM $[K]_o$ (Chapter 3).

The experiments illustrated in Figure 6 provide a good estimate of the internal concentration of potassium ($[K]_i$). The Nernst equation indicates that when $[K]_o$ is the same as $[K]_i$, the membrane potential is zero. This occurred when the outside potassium concentration was increased to 100 mM.

The distribution of potassium channels has been explored over the surface of Müller cells and astrocytes isolated from the retina and optic nerves of many species, such as salamanders and rabbits.[20,21] The potassium sensitivity is distributed in a characteristic pattern, being high-

[20]Newman, E. A. 1987. *J. Neurosci.* 7: 2423–2432.
[21]Brew, H. et al. 1986. *Nature* 324: 466–468.

est over the end feet and lower over the soma of the Müller cell. Figure 7 shows an isolated rabbit Müller cell and its responses to a high concentration of potassium locally applied to different regions of the surface by a pipette. A large depolarization implies a high density of potassium channels in that membrane area of the isolated cell. The possible significance of this highly nonuniform distribution of potassium currents is discussed below.

Glial cells and Schwann cells in culture were first shown by Ritchie and his colleagues to display a variety of ion channels and pumps in their membranes.

Ion channels, pumps, and receptors in glial membranes

1. As already shown, potassium channels predominate. At least two potassium currents have been distinguished, one of which is voltage-activated.[22]
2. The membranes of cultured Schwann cells and astrocytes also display sodium channels; they are voltage-activated and resemble those found in neurons.[23] The overall ratio of potassium to sodium permeability of Müller cells has been estimated to be approximately 100:1.[20]

[22]Howe, J. R. and Ritchie, J. M. 1988. *Proc. R. Soc. Lond.* B 235: 19–27.
[23]Bevan, S. et al. 1985. *Proc. R. Soc. Lond.* B 225: 299–313.

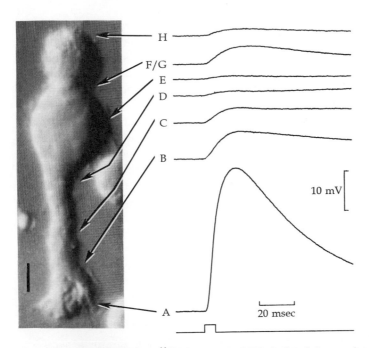

10 mV

20 msec

7 **RESPONSES OF A MÜLLER GLIAL CELL** isolated from salamander retina to potassium. Recordings were made with an intracellular microelectrode while potassium was applied to different sites, as indicated. A is the end foot and H is the distal part of the cell. The sensitivity to potassium is much greater at the end foot, suggesting a higher concentration of potassium channels in that region. Scale bar, 10 μm. (After Newman, 1987; micrograph courtesy of E. Newman.)

3. Patch clamp recordings have revealed the presence of chloride channels in Schwann cells and astrocytes.[24,25]

4. Ion pumps for transport of sodium, bicarbonate, and glutamate have been demonstrated in glial cells.[26-28]

At present it is not possible to provide a comprehensive scheme for the properties and distribution of receptors in glial cells. Oligondendrocytes, astrocytes, and Schwann cells display depolarizing and hyperpolarizing responses to the application of transmitters such as glutamate, GABA, ACh, and numerous peptides.[29] The possible functional roles of such receptors in relation to neuronal signaling are discussed below.

Absence of regenerative responses or impulses

A salient feature of glial cells is the absence of an axonlike process. This indicated to many of the earlier workers that glial cells were unlikely to conduct impulses in the manner of axons. However, the possibility remained that they might give active electrical responses, possibly slow ones. In the leech, frog, mud puppy, and mammals, it has been shown that identified glial cells do not produce impulses. Although a few examples of regenerative responses have been found in cultured glial cells,[30] action potentials cannot usually be initiated. Such properties make glial cells fundamentally different from the excitable neurons.

Electrical coupling between glial cells

Adjacent glial cells, including those of mammals, are linked to each other by gap junctions through which dyes as well as current can pass.[31] In this respect they resemble epithelial and gland cells and heart muscle fibers.[32] The functional significance of the electrical coupling of glial cells is not known. Clearly, ions can exchange directly between cells without passing through the extracellular space, and such interconnections may be useful for equalizing concentration gradients that may arise. The possibility is open that there exists some metabolic interaction between coupled cells, linked with demand that is induced by activity. It is shown later that the low-resistance coupling between the cells enables glial cells to generate currents that extracellular electrodes can record from the surface of nervous tissues.

As mentioned earlier, no gap junctions between nerve and glia have been detected. This is of physiological interest because it is natural to wonder about the manner in which neurons and glial cells may interact. Direct tests have been made in the leech nervous system, where the potentials of neurons can be changed in a controlled manner by passing

[24]Gray, P. T. and Ritchie, J. M. 1986. *Proc. R. Soc. Lond. B* 228: 267–288.

[25]Ritchie, J. M. 1987. *J. Physiol.* (Paris) 82: 248–257.

[26]Astion, M. L., Obaid, A. L. and Orkand, R. K. 1989. *J. Gen. Physiol.* 93: 731–744.

[27]Deitmer, J. W. and Schlue, W. R. 1989. *J. Physiol.* 411: 179–194.

[28]Szatkowski, M., Barbour, B. and Attwell, D. 1990. *Nature* 348: 443–446.

[29]Kettenmann, H. and Schachner, M. 1985. *J. Neurosci.* 5: 3295–3301.

[30]Newman, E. A. 1985. *Nature* 317: 809–811.

[31]Gutnick, M. J., Connors, B. W. and Ransom, B. R. 1981. *Brain Res.* 213: 486–492.

[32]Loewenstein, W. 1981. *Physiol. Rev.* 61: 829–913.

currents through them while recording from adjacent glial cells.[18] The reverse procedure has also been done—recording from nerve cells while changing the membrane potential of glial cells. Analogous tests in the optic nerve of *Necturus* show similar results: Current flow around glial cells created by synchronized nerve impulses has no significant effect on the neighboring glial membrane, and therefore an electrical interaction between nerve and glia seems quite unlikely.[19]

A SIGNALING SYSTEM FROM NEURONS TO GLIAL CELLS

Most speculation about the role of glial cells entails some kind of interaction with neurons. Such a mutual influence might be expected between two cell types that are so closely interwoven. An effect of nerve activity on glial cells can be most simply illustrated by experiments made in the brain of *Necturus*; similar results have been obtained in the leech and in mammals.

The basic observation is illustrated in Figure 8. Recordings are made

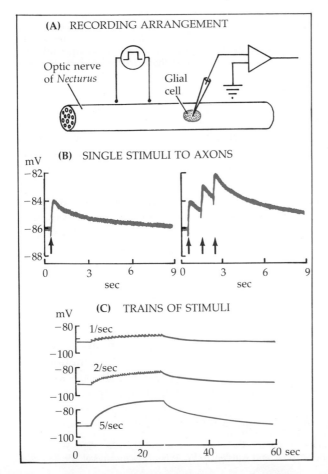

8

EFFECT OF NEURAL ACTIVITY on glial cells in the optic nerve of the mud puppy. Synchronous impulses in nerve fibers cause glial cells to become depolarized. Each volley of impulses leads to a depolarization that takes seconds to decline. The amplitude of the potentials depends on the number of axons activated and the frequency of stimulation, as shown in (B) and (C). (After Orkand, Nicholls and Kuffler, 1966.)

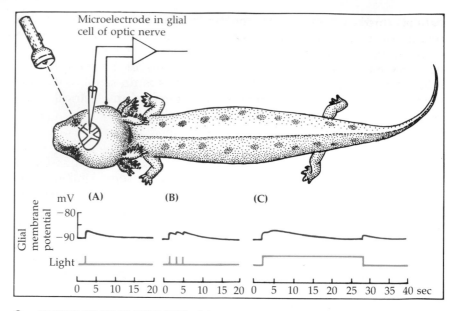

9 EFFECT OF ILLUMINATION of the eye on the membrane potential of glial cells in the optic nerve of an anesthetized mud puppy, with intact circulation. **(A)** Single flash of light 0.1 sec long. **(B)** Three flashes. **(C)** Light stimulus maintained for 27 sec. During such prolonged illumination, the initial glial depolarization declines as the nerve discharge adapts. At the end of illumination, a burst of "off" discharges initiates a renewed glial depolarization. Lower beams monitor light. (After Orkand, Nicholls and Kuffler, 1966.)

from a glial cell in the optic nerve of the mud puppy, and a volley is set up in the nerve fibers so that impulses travel past the impaled glial cell. Each volley of impulses is followed by a depolarization of the glial cell, rising to a peak in about 150 msec and declining slowly over several seconds. The size of the potential is graded, depending on the number of nerve fibers activated. With repeated stimulation, the potentials in the glial cells sum, depending on the frequency of the stimulation (Figure 8B and 8C). If stimulation is maintained, surprisingly large glial membrane depolarization of up to 48 mV can be seen. At the end of a train of stimuli, a residue of such large potentials may persist for 30 seconds or longer.[33]

Experiments were also done with natural stimulation of optic nerve fibers in anesthetized mud puppies with intact circulation. A single brief flash of light caused a glial depolarization of about 4 mV (Figure 9). Repeated flashes added distinct, but smaller, potentials. The glial potential declined progressively during maintained illumination but reappeared when the light was turned off. These results are in good agreement with the conclusion that discharges in the optic nerve are responsible for the glial potentials, since the discharge rate declines during maintained illumination but a renewed burst follows when the

[33]Orkand, R. K., Nicholls, J. G. and Kuffler, S. W. 1966. *J. Neurophysiol.* 29: 788–806.

light is turned off (Chapter 16). The depolarization of glial cells through neuronal activity can also be observed by recording optically from the optic nerves of frogs and mammals.[34,35]

In the cortex, glial cells become depolarized when neurons in their vicinity are activated by the stimulation of neural tracts, peripheral nerves, or the surface of the cortex.[36] The magnitude of the potentials recorded from glia once more depends on the strength of stimulation as additional axons conducting into the region are activated. These results suggest that glial cells within the visual cortex should become depolarized only when certain specific patterns of light are shone into the eye to activate groups of neighboring neurons. Kelly and Van Essen found such an effect in the visual cortex of the cat.[37] Glial cells identified by physiological and morphological criteria became significantly depolarized only when a bar of light with an appropriate orientation was shone into the eye (Figure 10). Illumination of both eyes was effective if corresponding areas were illuminated, but diffuse lights or inappropriate orientations produced no appreciable change in potential. These results are in good agreement with the assumption that the glial cells' response reflects the activity in the neurons they surround.

We have already noted that flow of current during nerve impulses is not the cause of glial depolarization. In any case the time courses of the events are greatly mismatched. The peak current flow caused by impulses, such as a synchronous volley in the optic nerve, has already declined when the glial depolarization starts to rise. Further, passing current directly into nerve cells in the leech does not produce detectable responses in neighboring glial cells.

Potassium release as mediator of the effect of nerve signals on glial cells

A more likely hypothesis is suggested by experiments on squid axons by Frankenhaeuser and Hodgkin, who showed that following nerve impulses, potassium accumulated in the spaces between axons and the surrounding satellite cells.[38] To determine the concentration changes produced by potassium leakage from axons, use was made of the observation that the glial membrane potential provides a quantitative assay for potassium in the environment of the cells (Figure 6). If potassium leaves axons and accumulates in the intercellular clefts, it changes the $[K]_o/[K]_i$ ratio and alters the membrane potential of glial cells in a predictable way. Accordingly, axons in the optic nerve of *Necturus* were stimulated by brief trains of stimuli. Comparison of the glial depolarization produced by a standard train in various external potassium concentrations allowed an estimate of the change in concentration to be made.

These results showed that in the physiological range of $[K]_o$, the brief train of nerve impulses released constant amounts of potassium into the intercellular clefts resulting in a concentration increase of about

[34]Konnerth, A., Orkand, P. M. and Orkand, R. K. 1988. *Glia* 1: 225–232.
[35]Lev-Ram, V. and Grinvald, A. 1986. *Proc. Natl. Acad. Sci. USA* 83: 6651–6655.
[36]Ransom, B. R. and Goldring, S. 1973. *J. Neurophysiol.* 36: 869–878.
[37]Kelly, J. P. and Van Essen, D. C. 1974. *J. Physiol.* 238: 515–547.
[38]Frankenhaeuser, B. and Hodgkin, A. L. 1956. *J. Physiol.* 131: 341–376.

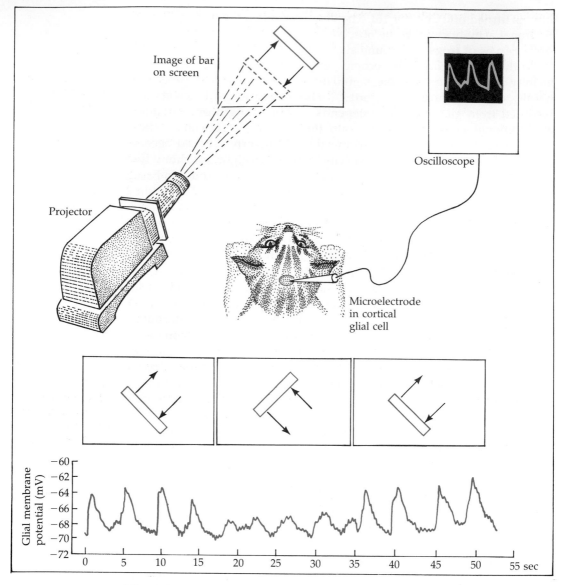

10   **DEPOLARIZATION OF A GLIAL CELL in the cortex of a cat as a result of visual stimulation. The greatest depolarization was produced by moving slits of light having a certain orientation and confined to a part of the visual field. Neurons in the region near the glial cell had the same orientation preference. (After Kelly and Van Essen, 1974.)**

2 mM. As a result of the potassium that accumulates transiently, the membrane of the glial cell becomes depolarized, and its potential returns to normal as the potassium disappears by uptake and diffusion.

The introduction of potassium-sensitive glass electrodes has enabled several groups of investigators to measure directly the accumulation of potassium in the extracellular spaces of the brain during neuronal activity.[39,40] With repetitive stimulation of neurons, the potassium concentration increases, the value being comparable to that estimated from the glial cell depolarization.

A cell generates currents if various regions of its surface are at different potentials. Nerve cells, of course, use this principle as the mechanism for conduction. Thus, current flows from inactive regions of an axon into the part that is occupied by a nerve impulse. Positive charges are thereby drawn away from the uninvolved area in front of a nerve impulse so that the region ahead becomes depolarized and eventually "active." Although most glial cells do not extend over long distances, they are linked to each other by low-resistance connections.[17,31] The conducting properties of such contiguous, coupled cells are therefore much the same as for a single elongated cell. As a result, if several glial cells become depolarized by increased potassium concentrations in their environment, they draw current from the unaffected cells, thereby creating current flow. Similarly, an elongated Müller cell that extends through the thickness of the retina will generate electrical currents if the potassium concentration increases locally at one region of its surface[20] (see Figure 11). In the region of raised $[K]_o$, potassium ions enter, while potassium ions leak out through regions of the glial cell (and other glial cells coupled to it) that are surrounded by normal $[K]_o$. Other ions to which glial membranes are relatively impermeable would contribute little to such currents.

Electroencephalography is one of the avenues available for obtaining objective information about activity in the human brain. Electroencephalograms (EEGs) are made routinely in fully conscious subjects or in animals under a variety of experimental conditions. Therefore, any method that separates the contribution of various elements, such as

Current flow and potassium movement through glial cells

Glial contribution to the electroretinogram and electroencephalogram

[39]Jendelová, P. and Syková, E. 1991. *Glia* 4: 56–63.
[40]Dietzel, I., Heinemann, U. and Lux, H. D. 1989. *Glia* 2: 25–44.

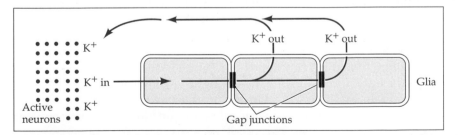

11 POTASSIUM CURRENTS IN GLIAL CELLS. The glial cells in the diagram are linked by gap junctions. Potassium supplied by active axons in one region depolarizes the glial cell and enters, causing current flow and outward movement of potassium elsewhere in the glial tissue. The concept of spatial buffering of potassium has been postulated as a mechanism for influencing neuronal function by glial cells.

neurons and glia, is of potential interest. The same applies to recordings from the eye—electroretinograms (ERGs)—in which one electrode is placed on the cornea and another indifferent electrode elsewhere on the body. In each case the potential changes represent the summed electrical activity of the underlying mass of neurons and glial cells. That glial cells can produce currents large enough to be measured by external electrodes was shown directly in the optic nerve of *Necturus.*[41]

The contribution of glial cells to the ERG has been assessed by experiments in which slow glial membrane potentials are recorded from Müller cells from the eye of the mud puppy following stimulation of neurons.[42] At the same time, potentials are monitored with extracellular electrodes on the surface of the eye and with other electrodes inserted at different depths in the retina. Illumination causes firing by optic nerve axons, buildup of potassium, and a depolarization of the Müller cells; the response slowly declines if the light is kept on. When illumination is turned off, Müller cells give "off" responses. The Müller cell potentials have a time course similar to that of one of the components (the b waves) in the ERG obtained with extracellular recordings. They are likely to be responsible for most of that component of the standard ERG in higher animals.

A quantitive answer about the contribution of glial cells to the EEG is more difficult to obtain for a number of reasons. (1) Both neurons and glial cells generate current flow, but depending on their distribution in the volume of the brain the potentials recorded on the surface of the tissue may be either positive or negative. (2) The contributions of currents by neurons and glia sum algebraically. If they occur simultaneously, their potentials may be additive; if they are in the opposite direction, they may cancel each other. (3) Slow potentials are by no means confined to glia but are common in neurons as well; thus, neurons may be hyperpolarized through conductance changes or the activity of an electrogenic pump[43] (Chapter 3) at the same time as glia are depolarized by the buildup of potassium.

Glial cells and the fluid environment of neurons

The interposition of glial cells between capillaries and neurons in the brain, the narrowness of the extracellular spaces, and the intricate anatomical relationship of the two types of cells suggest dynamic interactions. Certain clear-cut effects of glia on the fluid surrounding neurons have been established. One indirect long-term effect is mediated via the blood–brain barrier (Box 1).

Brightman and his colleagues have shown that astrocytes in culture can induce the formation of bands of tight junctions sealing endothelial cells of brain capillaries.[51,52] Grown on their own the endothelial cells are occasionally attached to each other by junctions. The presence of

[41]Cohen, M. W. 1970. *J. Physiol.* 210: 565–580.
[42]Ripps, H. and Witkovsky P. 1985. *Progr. Retinal Res.* 4: 181–219.
[43]Baylor, D. A. and Nicholls, J. G. 1969. *J. Physiol.* 203: 555–569.
[51]Tao-Cheng, J. H., Nagy, Z. and Brightman, M. W. 1987. *J. Neurosci.* 7: 3293–3299.
[52]Tao-Cheng, J. H., Nagy, Z. and Brightman, M. W. 1990. *J. Neurocytol.* 19: 143–153.

BOX 1 THE BLOOD–BRAIN BARRIER

A homeostatic system controls the fluid environment in the brain and keeps its chemical composition relatively constant compared with that of the blood plasma. This constancy seems particularly important in a system where the activity of many cells is integrated and where small variations may upset the balance of delicately poised excitatory and inhibitory influences. In contrast, at myoneural junctions, where information is generally transmitted with a good safety margin, fluctuations in the environment are of relatively small consequence. Within the brain are three different types of fluid: (1) the blood supplied to the brain through a dense network of capillaries; (2) the cerebrospinal fluid (CSF) that surrounds the bulk of the nervous system and is also contained in the internal cavities (ventricles); and (3) the fluid in the intercellular clefts (Figure I).

The intercellular spaces are generally no more than 20 nm wide; they provide the immediate environment of nerve and glial cells in the brain and constitute the main channels through which materials are distributed to reach neurons and glial cells. The BLOOD–BRAIN BARRIER keeps the environment of neural elements different from blood plasma. It depends on specialized properties of the endothelial cells of capillaries in the brain, which are far less permeable than those supplying organs in the periphery.[44,45] Charged particles, proteins, ions, and hydrophilic molecules cannot pass through; uncharged lipophilic molecules (that pass through mem-

[44]Reese, T. S. and Karnovsky, M. J. 1967. *J. Cell Biol.* 34: 207–217.

[45]Brightman, M. W. and Reese, T. S. 1969. *J. Cell Biol.* 40: 668–677.

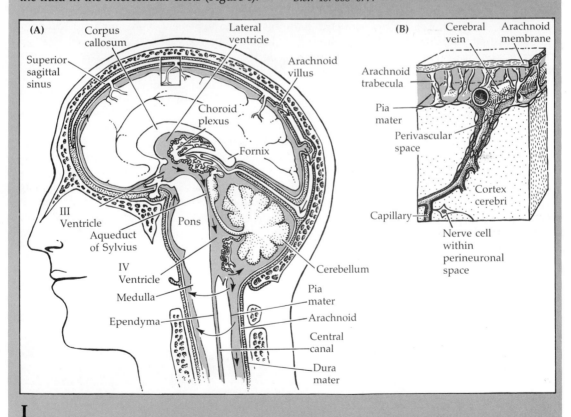

I

DISTRIBUTION OF CEREBROSPINAL FLUID and its relation to larger blood vessels and to structures surrounding the brain (A). All spaces containing CSF communicate with each other. CSF is drained into the venous system through the arachnoid villi (B).

branes) and gases can. A second essential component is the choroid plexus: Epithelial cells surrounding these capillaries secrete CSF and act as a barrier for ions and various small molecules.[46] The CSF itself is almost devoid of protein, containing only about $\frac{1}{200}$ of the amount present in blood plasma. Chemical substances such as metabolites move relatively freely from the alimentary canal into the bloodstream but not into the CSF. As a result, the levels in the blood plasma of amino acids or fatty acids fluctuate over a wide range while their concentrations in the CSF remain relatively stable. The same is true for hormones, antibodies, electrolytes, transmitters, and a variety of drugs including penicillin. Injected directly into the bloodstream they act rapidly on peripheral tissues such as the muscle, heart, or glands—but they have little or no effect on the central nervous system. When administered by way of the CSF, however, the same substances exert a prompt and strong action. The conclusion is that the substances injected into the bloodstream do not reach the CSF and the brain rapidly enough or in high enough concentrations.

For some time, around the early 1960s, it was commonly thought that the narrow intercellular spaces between cells in the brain were not functional pathways and could not play a part in the distribution of materials. It was even postulated that neuroglial cells actually constitute the extracellular space through which materials must pass to reach neurons, much as originally suggested by Golgi. Physiological experiments have shown that ions and small particles pass through the clefts and not through glia.[47] Large molecules can be seen by electron microscopy to diffuse through intercellular spaces between neurons and glia.[45] Thus ferritin, which has a diameter of 10 nm and a molecular weight of 900,000, and the enzyme horseradish peroxidase can move through these clefts. In Figure II, electron-dense molecules deposited by the peroxidase reaction are lined up in the clefts and fill the extracellular spaces after the injection of horseradish peroxidase into CSF. This result shows that large molecules can pass between the ependymal cells that line the ventricles and through intercellular clefts. In contrast, the junctions between endothelial cells lining the blood capillaries in the brain are impermeable and provide a barrier. Tracers stop spreading when they reach the capillaries. Figure IIB shows the opposite result: When enzyme was injected into the circulation, the capillary was filled with enzyme but none is seen in intercellular spaces. Physiological experiments reveal that the same sites act as barriers for the exchange of ions such as potassium and charged molecules such as neurotransmitters. During development of the central nervous system, the range of substances that can enter the CSF is qualitatively and quantitatively different.[48] For example, the protein concentration is high (up to 10 percent), containing molecules not detected in adult CSF.

The impermeability of brain capillaries in adult brain depends on the junctions linking the endothelial cells that line them. Whereas capillaries supplying muscles and organs in the periphery are loosely linked with occasional tight junctions that allow proteins and small molecules to leak out, endothelial cells in brain capillaries are tightly linked by a continuous band of tight junctions.[45] Knowledge of blood–brain barrier properties is important for understanding pharmacological actions of drugs and their effects on the body. Blood–brain barrier systems are also found in lower vertebrates and invertebrates.[49,50] The sites and cellular mechanisms are not always at the level of the capillaries and may depend instead on the glial cell properties.

[46]Cserr, H. F. 1988. *Ann. NY Acad. Sci.* 529: 9–20.
[47]Kuffler, S. W. 1967. *Proc. R. Soc. Lond. B* 168: 1–21.
[48]Møllgård, K. et al. 1988. *Dev. Biol.* 128: 207–221.
[49]Abbott, N. J., Lane, N. J. and Bundgaard, M. 1986. *Ann. NY Acad. Sci.* 481: 20–42.
[50]Bundgaard, M. and Cserr, H. F. 1981. *Brain Res.* 226: 61–73.

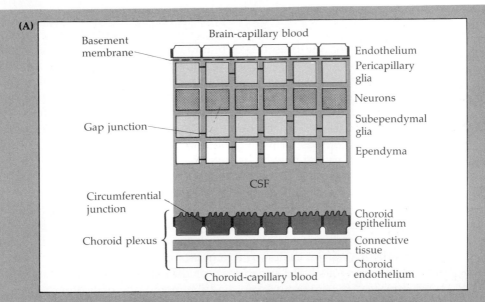

(A)

Brain-capillary blood

Basement membrane

Endothelium

Pericapillary glia

Gap junction

Neurons

Subependymal glia

Ependyma

CSF

Circumferential junction

Choroid epithelium

Choroid plexus {

Connective tissue

Choroid endothelium

Choroid-capillary blood

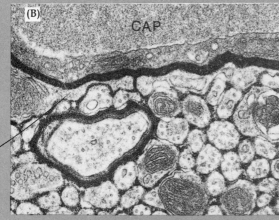

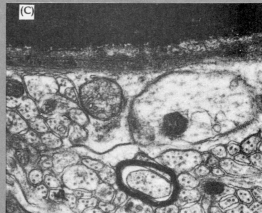

II

PATHWAYS FOR DIFFUSION IN THE BRAIN. (A) Schematic presentation of cells involved in the exchange of materials between blood, CSF, and intercellular spaces. Molecules are free to diffuse through the endothelial cell layer lining capillaries in the choroid plexus. They are, however, restrained by circumferential junctions between the choroid epithelial cells, which secrete CSF. There are no barriers between the bulk fluid of CSF and the various cell layers such as ependyma, glia, and neurons. The endothelial cells lining brain capillaries are joined by a circumferential seal, as are the choroid epithelial cells. This prevents free diffusion of molecules out of the blood. (B) Demonstration in the mouse that the enzyme microperoxidase diffuses freely from cerebrospinal fluid into the intercellular spaces of the brain, which are filled with the dark reaction product. No enzyme is seen in the capillary (CAP). (C) When injected into the circulation, the enzyme fills the capillary but is prevented by the capillary endothelium from escaping into the intercellular spaces. (B and C from Brightman, Reese and Feder, 1970.)

astrocytes triggers the formation of tight junctions resembling those seen in vivo that account for the impermeability of brain capillaries. Conversely, the presence of brain endothelial capillary cells in culture causes membrane assemblies to appear in astrocytes. These interactions are specific for astrocytes and brain capillary endothelial cells. No comparable results occur with fibroblasts or endothelial cells from peripheral arteries. The molecules released by astrocytes to cause the formation of tight junctions between brain capillary endothelial cells have been isolated.[53] Methods for promoting uncoupling and an increase in capillary permeability are now being investigated as a promising tool for circumventing the blood–brain barrier (see below) and allowing substances to enter the brain that would otherwise not be able to do so.

An obvious effect of glial cells on fluid environment is to allow the concentration of potassium in the vicinity of neurons of the brain to rise following impulses. Simply by virtue of its presence on the other side of a narrow cleft next to a neuronal membrane, the glial cell allows potassium to build up instead of diffusing into a larger extracellular space.[43] By the same token, had the membrane of another neuron been situated in place of the glial cell, it would have been subjected to a higher potassium concentration. Glial cells, therefore, separate and group neuronal processes, allowing local potassium buildup and preventing cross talk.

An attractive concept is that glial cells might regulate the potassium concentration in intercellular clefts, a process known as SPATIAL BUFFERING.[17] According to this hypothesis, glial cells act as conduits for rapid uptake of potassium from the clefts to preserve the constancy of the environment. Clearly, nonuniform depolarization of glial cells causes current to flow and results in the movement of potassium ions. Since glial cells are coupled to each other, potassium would follow the lines of current flow entering in one region and leaving at another. Evidence of such effects has been presented for Müller cells in the retina[20] (see Figure 7). These elongated glial cells have specialized areas of membrane with relatively large potassium sensitivity. That potassium will move through glial cells as a consequence of potassium buildup is inevitable. It is, however, not simple to estimate quantitatively how much potassium actually moves or how such movements alter the extracellular potassium concentration. For such calculations, numerous assumptions about geometry, conductance, diffusion, and active transport must be made.[54] The functional use of such effects are also unclear. Somehow the neurons that have been active must regain the potassium they have lost. In addition it is interesting that the membrane potential of nerve cells, in contrast to that of glial cells, is relatively insensitive to changes in potassium concentration in the environment. For example, in neurons of the leech, a fivefold increase in $[K]_o$ (from 4 to 20 mM) results in a depolarization of only 5 mV, whereas the glial membrane

[53]Rubin, L. L. et al. 1992. *Ann. NY Acad. Sci.* 633: 420–425.
[54]Odette, L. L. and Newman, E. A. 1988. *Glia* 1: 198–210.

potential changes by 25 mV.[17] The neuron's membrane potential is at least in part protected from fluctuations, presumably because it has a significant resting permeability to other ions besides potassium—in contrast to glial cells, in which potassium practically determines the full membrane potential. Nevertheless, appreciable changes can occur in conduction velocity, in excitability, and in transmitter release at synapses as a result of potassium accumulation following trains of impulses, and glial cells could act to mitigate such results.[55] Similar considerations apply to the functional significance of transmitter action, uptake, and secretion in glial cells. That glial cells can respond to, take up, and release transmitters such as GABA and glutamate is known.[56,57] How they thereby influence neuronal signaling is not.

FUNCTIONS OF NEUROGLIAL CELLS

Over the years, glial cells have been "burdened" by almost every nervous system task for which no other obvious explanation has been found. For example they have been implicated in sleep mechanisms, in learning, and in the immune responses of the brain. Even now, when much is known about the cellular and molecular properties of glial cells, numerous open questions remain about their functions. In the following sections we summarize and comment on three main types of interactions with neurons: (1) the formation of myelin, (2) the establishment of neuronal connections, and (3) the regulation of the microenvironment surrounding neurons. The properties of the blood–brain barrier, which determine the composition of the fluid in contact with neurons in the central nervous system, are closely related to the analysis of glial function.

One well-established role for oligodendrocytes and Schwann cells is the production of a myelin sheath around axons—a high-resistance covering akin to insulating material around wires. The myelin is interrupted at the nodes of Ranvier (see Figure 13), which occur at regular intervals of about 1 mm in most nerve fibers. Since the ionic currents associated with the conducted nerve impulse cannot flow across the myelin, the ions move in and out at the nodes between the insulation (Chapter 5). This leads to an increased conduction velocity and seems an ingenious solution for acquiring speed. The alternative is to make nerve fibers larger, but this is less effective. For example, in the squid the largest axons are between 0.5 and 1.0 mm in diameter and conduct at rates no faster than about 20 meters per second; a myelinated axon 20 μm in diameter conducts at 120 meters per second.

The association of satellite cells with axons to form myelin raises a number of interesting problems. For example, what are the genetic or environmental factors that enable glial cells to select the appropriate

Myelin and the role of neuroglial cells in axonal conduction

[55]Erulkar, S. D. and Weight, F. F. 1977. *J. Physiol.* 266: 209–218.
[56]Schon, F. and Kelly, J. S. 1974. *Brain Res.* 66: 275–288.
[57]Lieberman, E. M., Abbott, N. J. and Hassan, S. 1989. *Glia* 2: 94–102.

axons, to surround them at the right time, and to maintain myelin sheaths around them? What are the characteristics of some of the neurological disorders of myelin caused by genetic abnormalities?

These problems have been studied in mice by Aguayo and his colleagues.[58,59] The principle of their experiments was to remove a segment of a nerve in the leg (the sciatic or the sural nerve) and replace it with a graft. The graft consists of a segment of peripheral nerve from either the same or a different mouse. Within the grafted segment, axons, which have all been separated from their cell bodies, degenerate (Chapter 12), leaving a chain of Schwann cells. Axons from the proximal stump regenerate and grow through the grafted Schwann cells to reach the distal stump, which contains the animal's own Schwann cells. This is shown schematically in Figure 12. The experiment enables a test to be made of how effective the Schwann cells of the graft are in forming myelin, compared with those normally present in the sciatic nerve. For example, a segment of unmyelinated sympathetic nerve from the neck was grafted into a leg nerve that is normally myelinated. As axons grew into the graft, they became myelinated. Thus, a population of Schwann cells that do not make myelin in situ can do so if they come into contact with different axons of the appropriate type.[60]

In other studies, made on mutant mice, the genetic defects that are responsible for deficiency of myelin formation were analyzed. One such mutant mouse, called *Trembler*, exhibits a gross deficiency of myelin. In this inherited disorder, myelin in the peripheral nervous system is abnormally thin or totally absent. Does the defect result from Schwann cells that are unable to manufacture myelin or from axons that do not provide adequate signals to the Schwann cells? Grafts were made in normal and *Trembler* peripheral nerves, as shown in Figure 12. When a graft of *Trembler* nerve was inserted into a normal mouse sciatic nerve, regenerating axons were myelinated above and below the graft but not within it. Conversely, axons of the mutant mouse growing through a graft containing normal Schwann cells became myelinated in the graft segment but not above it or below it. In *Trembler* mice, therefore, the defect resides in the Schwann cells and not in the axons; these are capable of becoming myelinated if they come into contact with healthy satellite cells. By such techniques it has become possible to assess the sites of defects in various other disorders of myelination, including those affecting humans suffering from demyelinating diseases.

For the formation of the myelin sheath during development, complex and precise interactions occur between axons and satellite cells.[61,62]

[58]Aguayo, A. J., Bray, G. M. and Perkins, S. C. 1979. *Ann. NY Acad. Sci.* 317: 512–531.

[59]Bray, G. M., Rasminsky, M. and Aguayo, A. J. 1981. *Annu. Rev. Neurosci.* 4: 127–162.

[60]Aguayo, A. J., Charron, L. and Bray, G. M. 1976. *J. Neurocytol.* 5: 565–573.

[61]Rosenbluth, J. 1988. *Int. J. Dev. Neurosci.* 6: 3–24.

[62]Bunge, R. P., Bunge, M. B. and Bates, M. 1989. *J. Cell Biol.* 109: 273–284.

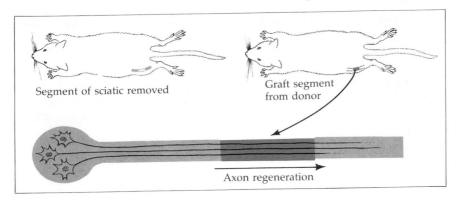

Segment of sciatic removed

Graft segment from donor

Axon regeneration

4 months

Proximal	Graft	Distal

N-N-N

N-T-N

T-N-T

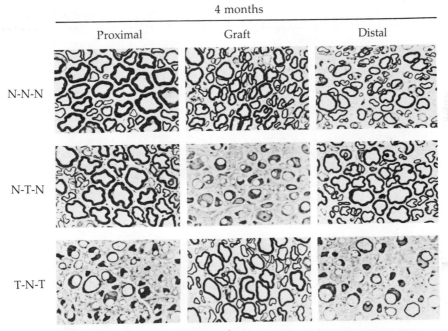

12 MYELINATION OF REGENERATING AXONS BY SCHWANN CELLS. A segment of sciatic nerve is removed from the leg of a mouse. In its place is grafted a segment of sciatic nerve from a donor mouse. Axons in the graft and in the host's sciatic nerve distal to the graft degenerate, leaving only Schwann cells and connective tissue. Axons grow back into the graft and the distal stump. N-N-N represent axons in cross section from a normal mouse in which the grafted segment was from a normal mouse: The axons are myelinated above, within, and beyond the graft. In N-T-N, the grafted segment was from a *Trembler* mouse and was inserted into a normal animal; the grafted Schwann cells did not form myelin, although distal parts of the axon became myelinated. In T-N-T, a normal segment was grafted into a *Trembler* mouse sciatic nerve; only the parts of the axons within the graft, surrounded by normal Schwann cells, became myelinated. These results show that the defect in *Trembler* resides in the Schwann cells, not in the axons. (After Aguayo, Bray and Perkins, 1979.)

The spacing of the nodes, the seals between the two cell types at the paranodal areas, and the distribution of sodium and potassium channels must be appropriately matched and positioned for rapid conduction. Figure 13 shows the principal features associated with the junctions between axons and satellite cells. At nodes within the central nervous system, a characteristic feature is the presence of astrocytic processes that contact the axon.[16] Moreover, a glycoprotein known as tenascin, or JI, is concentrated at this site.[63] Thus astrocytes as well as oligodendrocytes are associated with the myelin patterns essential for rapid

[63]ffrench–Constant, C. et al. 1986. *J. Cell Biol.* 102: 844–852.

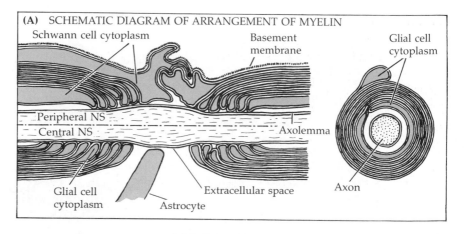

(A)   SCHEMATIC DIAGRAM OF ARRANGEMENT OF MYELIN

13

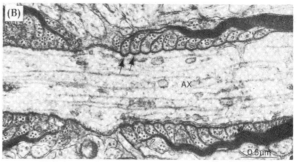

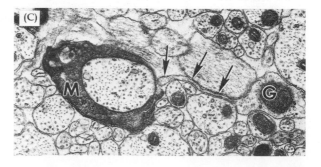

MYELIN AND NODES OF RANVIER. Oligodendroglial cells and Schwann cells form wrappings of myelin around axons. (A) At the nodes of Ranvier, the myelin covering is interrupted and the axon is exposed. The upper half of the nodal region, with a loose covering of processes, is typical of the arrangement in peripheral nerves. The lower part is representative of a node within the central nervous system. Here an astrocytic finger comes into close apposition with the nodal membrane. At that site is found the adhesion molecule tenascin or J-1 (Chapter 11). To the right is a transverse section through a myelin-covered axon portion. (B) Electron micrograph of a nodal region in a myelinated fiber in rat CNS. At the edge of the node, where the Schwann cell lamellae terminate, there occurs a specialized close contact area between the membrane of the axon (AX) and the membrane of the myelin wrapping (arrows). (C) Cross section of a myelinated axon at a node that is contacted by a process (marked with arrows) from a perinodal astrocytic glial cell (G). Myelin (M) is absent at the site of contact between the astrocyte and the node. (A after Bunge, 1968; B from Peters, Palay and Webster, 1976; C from Sims et al., 1985; micrograph courtesy of J. Black and S. Waxman.)

conduction in myelinated nerve fibers. The idea of a dynamic role for the satellites is reinforced by observations made in demyelinating or regenerating axons: Normal conduction can be restored after remyelination. Experiments by Ritchie, Waxman, Sears, Bostock, Shrager, and their colleagues have described the distribution of ion channels in nodes, paranodal areas, and internodes of normal, demyelinated, and remyelinated axons.[64–66] It has been suggested that glial cells might influence the clustering of sodium channels at sites where they contact the nodal membrane. The process would be analogous to the clustering of ACh receptors at the neuromuscular junction brought about by innervation. Astrocytic fingers in the nodal region themselves show intense labeling with saxitoxin, suggesting a high density of sodium channels in the glial membrane.[66] It has even been proposed that transfer of sodium channels might occur from satellite cells to nodes of Ranvier.[67] Direct evidence for this intriguing speculation is not available.

A role for glial cells in the growth of neurons and the formation of connections has frequently been suggested. In a comprehensive series of experiments, Rakic has studied the development of the cerebral cortex, the hippocampus, and the cerebellum in monkeys and humans.[68,69] The formation of the various cell types and their migration to their final destinations have been followed by light and electron microscopy and by labeling of the glial cells with specific antibodies. The neurons move along the glial processes during development (Figure 14). Similar experiments by Luskin and her colleagues were made with retroviral lineage markers to infect early progenitors (see above).[11] Neurons arising from a single ancestor cell were followed as they formed the layers of the cortex during development. The close association of the glia and neurons indicates that radial glia provide an initial framework around which subsequent neuronal organization takes place. Molecules responsible for the guidance of migrating neurons on radial glial fibers have been identified by Hatten and her colleagues.[70,71] In culture they have shown by video imaging that antibodies raised against the glial surface molecules can block migration as embryonic cerebellar granule cells "ride the monorail" along a strand of glia.

The role of glial cells in establishing structures during development is further illustrated by the formation of *barrel fields* in the sensory cortex of rodents. These barrels are described in Chapter 14 and consist, as their name suggests, of tubular arrays of neurons.[72] Each barrel is made

Glial cells and the formation of neuronal connections

[64]Bostock, H. and Sears, T. A. 1978. *J. Physiol.* 280: 273–301.

[65]Shrager, P. 1988. *J. Physiol.* 404: 695–712.

[66]Ritchie, J. M. et al. 1990. *Proc. Natl. Acad. Sci. USA* 87: 9290–9294.

[67]Shrager, P., Chiu, S. Y. and Ritchie, J. M. 1985. *Proc. Natl. Acad. Sci. USA* 82: 948–952.

[68]Rakic, P. 1971. *J. Comp. Neurol.* 141: 283–312.

[69]Rakic, P. 1988. *Science* 241: 170–176.

[70]Stitt, T. N. and Hatten, M. E. 1990. *Neuron* 5: 639–649.

[71]Hatten, M. E. 1990. *Trends Neurosci.* 13: 179–184.

[72]Van der Loos, H. and Woolsey, T. A. 1973. *Science* 179: 395–398.

up of those neurons responding to stimulation of one whisker on the face of the mouse or rat. During development, concentrations of glial cells appear at the cortical areas where the barrel-shaped cluster of neurons will form, before the neurons themselves have arrived and taken up their characteristic positions. Certain other prospective nuclei and groupings of neurons are also outlined by glial cells in the CNS at

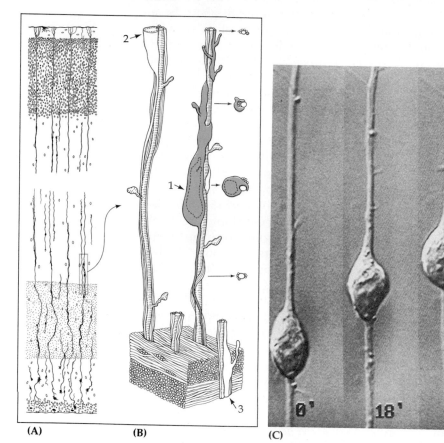

(A) (B) (C)

14 NEURONS MIGRATING ALONG RADIAL GLIA during development. (A) Camera lucida drawing of the occipital lobe of developing cortex of monkey fetus at mid-gestation. Radial glial fibers run from the ventricular zone below to the surface of the developing cortex above. (B) Three-dimensional reconstruction of migrating neurons. The migrating cell (1) has a voluminous leading process that follows the radial glia, using it as a guideline. Cell 2, which has migrated farther, retains a process still connected to the radial glia. Cell 3 at the bottom is beginning to send a process along the radial glia before migrating. (C) Migration of hippocampal neurons along cerebellar astroglial fibers in vitro. The hippocampal neuron migrates along the radial glial fiber (GF), which lies immediately underneath the neuron. As time progresses, the leading process (LP) moves farther up, with the neuronal cell body following. Times indicated at bottom represent real time taken from video photography. (A and B after Rakic, 1988; C from Hatten, 1990.)

early stages before the aggregation is apparent.[73] Comparable patterns of glial participation are observed in invertebrates during development. For example, in larvae of the moth *Manduca*, complex glomerular assemblies of neurons arise through interactions with the scaffolding provided by glial cells.[74] If the glial cells are specifically inhibited from dividing, the neurons fail to form the appropriate structures. Synapse formation can, however, occur in the absence of glial cells. In the embryonic spinal cord of the monkey and in neonatal spinal cord of the opossum, chemical synapses form before glial cells appear in the area of synapses.[75] It is also known that functioning synapses can be formed in tissue culture without satellite cells.[76]

What are the mechanisms that enable neurons to migrate along or extend processes over some glial cells but not others at different stages of development? In Chapter 11 we show that specific molecules such as laminin, fibronectin, and cellular adhesion molecules can act as neurite-promoting factors. Anchored in the extracellular matrix, they provide the substrate that triggers movement or changes in shape. In tissue culture it has been shown by Schwab and his colleagues that astrocytes and Schwann cells act as favorable substrates for neuronal adhesion and outgrowth.[77,78] Neuronal tumor cells (neuroblastoma) or embryonic optic nerve explants grow on gray matter but avoid white matter containing myelinated axons and oligodendrocytes. Treatment of the myelin with proteases or with a specific monoclonal antibody raised against myelin abolishes the inhibitory effects of oligodendrocytes on growth. Under these conditions neurons adhere to and sprout on white matter. Several lines of evidence suggest that inhibitory influences from oligodendrocytes could serve to prevent axons from sprouting laterally as they grow along a tract in the CNS toward their destination. The glial cells could thus help to delineate boundaries not to be crossed between bundles of fibers or between bundles and gray matter.

Molecular interactions of glial cells and neurons during development

Not only can glial cells be permissive or restrictive, but they can also secrete growth promoting molecules such as nerve growth factor (NGF; Chapter 11).[79] A glial neurite-promoting factor (known as GDN, or glial-derived nexin) has been isolated and shown to be a potent protease inhibitor.[80,81] Proteases secreted by growth cones can digest extracellular matrix molecules, and inhibition of the enzymes in culture by glial-derived nexin can promote sprouting. Monard has suggested that the glial-derived protease inhibitor influences the delicate balance between

[73]Cooper, N. G. F. and Steindler, D. A. 1986. *Brain Res.* 380: 341–348.
[74]Tolbert, L. P. and Oland, L. A. 1990. *Exp. Neurol.* 109: 19–28.
[75]Stewart, R. R. et al. 1991. *J. Exp. Biol.* 161: 25–41.
[76]Fuchs, P. A., Henderson, L. P. and Nicholls, J. G. 1982. *J. Physiol.* 323: 195–210.
[77]Caroni, P. and Schwab, M. E. 1988. *J. Cell Biol.* 106: 1281–1288.
[78]Schwab, M. E. 1991. *Philos. Trans. R. Soc. Lond. B* 331: 303–306.
[79]Heuman, R. (1987).*J. Exp. Biol.* 132: 133–150.
[80]Gloor, S. et al. 1986. *Cell* 47: 687–693.
[81]Monard, D. 1988. *Trends Neurosci.* 11: 541–544.

Role of satellite cells
in repair and
regeneration

neuron-derived proteases and extracellular matrix molecules that are essential for growth.[81]

Glial cells participate in scar formation, the removal of debris, and, in the periphery, provide conduits for axons to grow along and regenerate their connections after injury (Chapter 12). Microglial cells invade the CNS from blood at the site of an injury, divide, and scavenge debris from dying cells.[6,82] In invertebrates such as the leech, wandering microglial cells can play a crucial role in regeneration.[83,84] Following injury to the cockroach CNS, glial cells invade the injured region and proliferate;[85] from them are derived the new ensheathing glial cells that regulate the fluid environment surrounding neurons. A remarkable interaction of microglial cells with severed axons has been found in crustaceans.[86] After a large motor axon in the lobster has been cut, the distal stump that has been separated from the cell body and its nucleus survives for weeks or even months on its own. A puzzling question is how this axon stump continues to provide for its needs with no ribosomes or cell body. Electron micrographs show that the axoplasm contains a population of small nucleated cells resembling microglial cells. The possibility therefore arises that these invaders could provide materials essential for the axon stump's survival.

In the mammalian nervous system, astrocytes, microglia, and Schwann cells react to injury by replicating. Lesions of the ventral nerve roots containing motor axons cause pronounced changes in the glial cells surrounding motor neurons undergoing chromatolysis.[87] During regeneration of the peripheral nervous system, Schwann cells act as conduits for regenerating axons to grow along; axons follow the old trail of Schwann cells that remain in the distal stump. In addition, Schwann cells come to occupy the sites vacated by motor nerve terminals at denervated motor end plates. There they release ACh, giving rise to miniature potentials, probably the best example of transmitter release by a satellite cell under physiological conditions.[88] Schwann cells can also attract neurons to grow toward them from a distance. This has been shown in experiments on frogs in which regenerating motor axons were deprived of their target muscles.[89] At the time the motor axons were cut, the muscle they normally innervate was removed surgically, leaving a connective tissue bed. In the absence of a target, axons grew out from the cut nerve in random directions. Figure 15 shows that if a segment of nerve was placed in their vicinity, axons made an abrupt

[82]Streit, W. J., Graeber, M. B. and Kreutzberg, G. W. 1988. *Glia* 1: 301–307.

[83]McGlade-McCulloh, E. et al. 1989. *Proc. Natl. Acad. Sci. USA* 86: 1093–1097.

[84]Masuda-Nakagawa, L. M., Muller, K. J., and Nicholls, J. G. 1990. *Proc. R. Soc. Lond. B* 241: 201–206.

[85]Smith, P. J., Howes, E. A. and Treherne, J. E. 1987. *J. Exp. Biol.* 132: 59–78.

[86]Atwood, H. L. et al. 1989. *Neurosci. Lett.* 101: 121–126.

[87]Kreutzberg, G. W., Graeber, M. B. and Streit, W. J. 1989. *Metab. Brain Dis.* 4: 81–85.

[88]Dennis, M., and Miledi, R. 1974. *J. Physiol.* 237: 431–452.

[89]Kuffler, D. P. 1987. *J. Exp. Biol.* 132: 151–160.

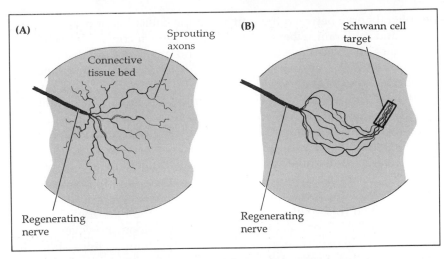

(A) Sprouting axons · Connective tissue bed · Regenerating nerve

(B) Schwann cell target · Regenerating nerve

15 GROWTH OF REGENERATING MOTOR AXONS toward a target consisting of Schwann cells and perineural tissue. **(A)** Random outgrowth of axons from regenerating muscle nerve in the absence of a target. The nerve supplies the cutaneous pectoris muscle of the frog, which had been removed in its entirety, leaving a connective tissue bed. Axons labeled with horseradish peroxidase grow in every direction. **(B)** A similar experiment in which the nerve had been cut and the muscle removed; in addition, however, a piece of nerve was placed at a distance on the connective tissue bed. Axons in the transplanted piece of nerve had degenerated, leaving Schwann cells in perineural tissue. Regenerating axons stained with horseradish peroxidase clearly head for the target. In other experiments portions of nerve enveloped by semipermeable membranes continued to attract neurite outgrowth. (After Kuffler, 1987.)

turn in a trajectory directed precisely toward it. No such growth was directed toward tendon or smooth muscle. The nerve axons in this target had degenerated, leaving only Schwann cells, connective tissue, and perineural cells. It is not uncommon for cut axons to grow for some distance to reach a distal segment that they can then grow along. Provided with a suitable substrate, axons can grow for considerable distances in the absence of glial or Schwann cells. For example, in the CNS of the leech (Chapter 13), axons grow back to reform their original connections after their processes have been severed by an injury, and such regeneration can still be achieved after glial cells surrounding them have been killed.[90]

Why is regeneration of the adult mammalian CNS so poor? Unlike leeches, fishes, or frogs, our own spinal cords or optic nerves do not regenerate after a lesion (Chapter 12). Two mechanisms could in principle account for the failure: absence of growth-promoting molecules, or active inhibition. As described in Chapter 12, Aguayo and his colleagues have shown that neurons in the adult mammalian CNS can sprout over long distances, centimeters, in an environment provided by Schwann cells.[91] The experiment entails the insertion of a graft of

[90]Elliott, E. J. and Muller, K. J. 1983. *J. Neurosci.* 3: 1994–2006.
[91]Aguayo, A. J. et al. 1990. *J. Exp. Biol.* 153: 199–224.

peripheral nerve between the retina and the colliculus or between, say, cortex and midbrain. Fibers enter the graft (which consists of Schwann cells, since all axons have degenerated) and grow into the distal tissue, where they make synapses. In the experiments of Schwab and his colleagues, the inhibitory influence of oligodendrocytes and myelin, mentioned above, has been shown to affect regeneration.[78] Antibody against myelin protein was applied to injured spinal cord in adult rats by an ingenious technique: Cells genetically engineered to produce and secrete the antibody were grafted into the lesion. As a result of the neutralization of the myelin antigen, damaged fibers regrew across the crush. By the same token, fibers in the unmyelinated spinal cord of neonatal opossum grow rapidly and profusely across a lesion.[92] The stage at which this ability for repair is lost and the role of glial cells in preventing CNS regeneration are not yet understood, however.

What direct effects do glial cells have on neuronal signaling?

Together the experiments described in preceding sections show the importance of glial cells in development, regeneration, and the formation of myelin. Detailed information is now also available about the ion channels in their membranes and about interactions at the molecular level with growing neurites. Yet it remains true that neurons survive, retain their membrane properties, and form normally functioning synapses in culture without glia. It therefore seems likely that glial cells do not play a *direct* role in neuronal signaling.

At the same time, glial cells do indicate the level of impulse traffic in their environment. The depolarization of a glial cell is graded according to the number of active neurons and the frequency at which they fire. Impulse numbers are converted into potassium concentrations in the clefts, and these, in turn, into changes in glial cell membrane potentials. Such signaling between neurons and glia is radically different from specific synaptic activity. Synaptic actions are confined to small specialized regions on the neuronal cell body and the dendrites and may be excitatory or inhibitory. In contrast, signaling by the potassium mechanism is not confined to special structures, such as synapses, but occurs along the entire length of the neuron whether an impulse liberates an excitatory or inhibitory transmitter.[17] In this respect the neuron–glia potassium signaling is nonspecific, except that glial cells are influenced preferentially by the population of neurons in their proximity. One would therefore expect the physiological role of glial cells to be generalized rather than a discriminating one. One might speculate that potassium-induced depolarization leads somehow to a stimulation of enzymes in glial cells, causing them to produce a product that the nerves need for their activity or for recovery afterward.[93] Other mechanisms of interaction that have been described involve transmitter-mediated actions of squid axons on Schwann cells; this interaction

---

[92]Treherne, J. M. et al. 1992. *Proc. Natl. Acad. Sci. USA* 89: 431–434.
[93]Orkand, P. M., Bracho, H. and Orkand, R. K. 1973. *Brain Res.* 55: 467–471.

results in release by Schwann cells of another transmitter that acts back on the Schwann cells' own membranes.[57,94,95]

We can now take account of four facts to suggest a possible role for astrocytes in the mammalian brain. First, they envelop the brain capillaries with their end feet. Indeed, it was this feature that led Golgi and so many others to suggest that they provide materials to the neurons. Second, as we show in Chapter 17, activity localized to a particular region of the brain causes a dramatic increase in the blood flow through that region, as measured by positron emission tomography (PET) scanning or optical recording. Third, astrocytes influence the formation of junctions by capillary endothelial cells in brain. Fourth, glial cells, as mentioned above, register the overall level of activity of the neurons in their vicinity. Paulson and Newman have therefore put forward the attractive idea that end feet of depolarized astrocytes might feed back onto capillaries and cause localized vasodilatation.[96] Thereby, active neurons would through glial signaling be supplied with oxygen and glucose. As for mechanisms, we have no evidence, but raised concentrations of potassium or other molecules liberated from glial end feet could influence capillary endothelial cells. Paulson and Newman's proposal is reminiscent of Golgi's original idea, which it reproduces but with signals flowing in the opposite direction. Instead of glial cells carrying nutrients through their cytoplasm from blood to neurons, neuronal activity leads to highly localized vasodilatation and increased blood flow just where it is needed.

Astrocytes and blood flow through the brain: A speculation

Until recently it was generally accepted that the tissues of the central nervous system were not patrolled by the surveillance mechanisms of the immune system. The blood–brain barrier, the absence of a lymphatic system, and the comparative ease with which grafts can be accepted all suggested the absence of immune responses to foreign antigens like those that occur in other systems and organs. Thus CNS functions are not disrupted by the massive allergic reactions to a bee sting or poison ivy. Evidence has now accumulated to show that microglia and activated T lymphocytes do enter the brain and can mediate acute inflammation of brain tissue.[6,97–99] (Sensitized, activated T cells produce responses of a tissue to specific antigens when they are presented with antigens by helper cells.) A well-known example is experimental allergic encephalitis (EAE). Antibodies are raised against a protein in myelin known as myelin basic protein. In mice and rats, activated T cells invade the brain, passing through the capillaries to cause damage in the perivascular region.[100] This is the very site at which astrocytic end feet are

Glial cells and immune responses of the CNS

[94]Villegas, J. 1981. *J. Exp. Biol.* 95: 135–151.
[95]Evans, P. D. et al. 1992. *Ann. NY Acad. Sci.* 633: 434–447.
[96]Paulson, O. B. and Newman, E. A. 1987. *Science* 237: 896–988.
[97]Wekerle, H. et al. 1987. *J. Exp. Biol.* 132: 43–57.
[98]Wekerle, H. et al. 1986. *Trends Neurosci.* 9: 271–277.
[99]Lampson, L. A. 1987. *Trends Neurosci.* 10: 211–216.
[100]Sun, D. et al. 1988. *Nature* 332: 843–845.

conspicuous. Astrocytes in culture and in situ have been shown to react with T lymphocytes, whose activity they can either stimulate or suppress. Astrocytes therefore qualify as inducible, facultative antigen-presenting cells. Wekerle and his colleagues have proposed that the immune sensitivity of the CNS is limited first by limiting the number of controlling lymphocytes to the activated T cells (which are the only ones that can cross the blood–brain barrier), and second by focusing the areas of response to limited regions around capillaries. Microglial cells and macrophages in the brain also appear in immune responses such as graft rejection. As one might expect, speculations about the role of interactions between the immune system and the central nervous system abound. They range from influences on sleep mechanisms and muscle tone to influences on mood or suspectibility to disease. Once again, glial cells have been implicated in a new field full of promise at an early stage with relatively few experimental facts.

SUGGESTED READING

General reviews

Abbott, N. J. 1992. *Glial–Neuronal Interaction. Ann. NY Acad. Sci.* Vol. 633.

Barres, B. A., Chun, L. L. and Corey, D. P. 1990. Ion channels in vertebrate glia. *Annu. Rev. Neurosci.* 13: 441–471.

Black, J. A., Kocsis, J. D. and Waxman, S. G. 1990. Ion channel organization of the myelinated fiber. *Trends Neurosci.* 13: 48–54.

Bradbury, M. 1979. *The Concept of a Blood–Brain Barrier.* John Wiley & Sons, Chichester.

Hatten, M. E. 1990. Riding the glial monorail: A common mechanism for glial-guided neuronal migration in different regions of the developing mammalian brain. *Trends Neurosci.* 13: 179–184.

Kuffler, S. W. and Nicholls, J. G. 1966. The physiology of neuroglial cells. *Ergeb. Physiol.* 57: 1–90.

Tolbert, L. P. and Oland, L. A. 1989. A role for glia in the development of organized neuropilar structures. *Trends Neurosci.* 12: 70–75.

Original papers

Aguayo, A. J., Bray, G. M., Rasminsky, M., Zwimpfer, T., Carter, D. and Vidal-Sanz, M. 1990. Synaptic connections made by axons regenerating in the central nervous system of adult mammals. *J. Exp. Biol.* 153: 199–224.

Astion, M. L., Obaid, A. L. and Orkand, R. K. 1989. Effects of barium and bicarbonate on glial cells of *Necturus* optic nerve. Studies with microelectrodes and voltage-sensitive dyes. *J. Gen. Physiol.* 93: 731–744.

Caroni, P. and Schwab, M.E. 1988. Two membrane protein fractions from rat central myelin with inhibitory properties for neurite growth and fibroblast spreading. *J. Cell Biol.* 106: 1281–1288.

Gloor, S., Odnik, K., Guenther, J., Nick, H. and Monard, D. 1986. A glia-derived neurite promoting factor with protease inhibitory activity belongs to the protease nexins. *Cell* 47: 687–693.

Kuffler, S. W. and Potter, D. D. 1964. Glia in the leech central nervous system: Physiological properties and neuron–glia relationship. *J. Neurophysiol.* 27: 290–320.

McGlade-McCulloh, E., Morrissey, A. M., Norona, F. and Muller, K. J. 1989. Individual microglia move rapidly and directly to nerve lesions in the leech central nervous system. *Proc. Natl. Acad. Sci. USA* 86: 1093–1097.

Newman, E. A. 1987. Distribution of potassium conductance in mammalian Müller (glial) cells: A comparative study. *J. Neurosci.* 7: 2423–2432.

Paulson, O. B. and Newman, E. A. 1987. Does the release of potassium from astrocyte endfeet regulate cerebral blood flow? *Science* 237: 896–898.

Rakic, P. 1988. Specification of cerebral cortical areas. *Science* 241: 170–176.

Ritchie, J. M. 1990. Voltage-gated cation and anion channels in the satellite cells of the mammalian nervous system. In *Advances in Neural Regeneration Research*, Wiley-Liss, pp. 237–252.

Smith, P. J., Howes, E. A. and Treherne, J.E. 1987. Mechanisms of glial regeneration in an insect central nervous system. *J. Exp. Biol.* 132: 59–78.

Tao-Cheng, J. H., Nagy, Z. and Brightman, M. W. 1987. Tight junctions of brain endothelium in vitro are enhanced by astroglia. *J. Neurosci.* 7: 3293–3299.

Wekerle, H., Sun, D., Oropeza-Wekerle, R. L. and Meyermann, R. 1987. Immune reactivity in the nervous system: Modulation of T-lymphocyte activation by glial cells. *J. Exp. Biol.* 132: 43–57.

PART TWO

Communication between Excitable Cells

PRINCIPLES OF SYNAPTIC TRANSMISSION

Synapses are points of contact between one nerve cell and another or between a nerve cell and an effector cell, such as a secretory cell or muscle fiber. At these points signals are passed from one cell to the next. At electrical synapses, current generated by an impulse in the presynaptic nerve terminal spreads into the next cell through low-resistance channels. More commonly, synapses are chemical. There the fluid-filled gap between the presynaptic and postsynaptic membranes prevents direct spread of current. Instead, the nerve terminal secretes a chemical, the neurotransmitter, that activates ion channels in the postsynaptic membrane. Chemical synapses may be excitatory or inhibitory.

A synapse about which detailed experimental information is available is the vertebrate neuromuscular junction, where acetyl-choline (ACh) released from the motor nerve terminals increases the permeability of the postsynaptic membrane to cations, principally sodium and potassium. This permeability increase causes a net inward movement of positive charge, and hence depolarization. The depolarization normally is large enough to produce an action potential in the muscle fiber but can be reduced below threshold by application of drugs such as curare. The permeability increase is due to activation of ACh receptors that, when open, form cation-selective channels. Similar mechanisms operate at other excitatory synapses.

At many inhibitory synapses, the effect of the inhibitory neurotransmitter is to activate postsynaptic channels permeable to anions, the major one present being chloride. The effect of this increase in permeability is to oppose depolarization. In the central nervous system, neurons receive both excitatory and inhibitory synapses and the generation of nerve impulses depends on the balance between these two influences. A second kind of inhibition is also seen—presynaptic inhibition, in which an inhibitory transmitter substance acts to prevent release of transmitter from an excitatory presynaptic terminal.

The mechanisms of transmitter release from nerve terminals have been studied in a variety of preparations, particularly the frog neuromuscular junction and the giant synapse of the squid. The stimulus for secretion of the transmitter is depolarization of the presynaptic ending, produced either by a nerve impulse or artificially, in the presence of calcium. Release occurs following calcium entry into the terminal through voltage-activated calcium channels. In low-calcium solutions, release can be reduced or completely eliminated. Invariably a delay of about 0.5 msec intervenes between the depolarization and secretion, a large part of which is due to the time taken for the calcium channels to open.

A key finding is that transmitter is secreted in multimolecular packets (quanta), each containing several thousand transmitter molecules. A quantum constitutes the fundamental physiological unit of release. Synaptic activation at the vertebrate neuromuscular junction normally involves the release of 100 to 200 quanta of ACh, released almost synchronously from the nerve terminal. Even at rest, nerve terminals release quanta spontaneously, giving rise to low-frequency miniature synaptic potentials with amplitudes in the millivolt range. At other synapses the average number of quanta released by a presynaptic action potential is generally much less— from 1 to 20 at synapses in autonomic ganglia, and about 1 at many central nervous system synapses. The average number of postsynaptic channels activated by a single quantum of transmitter also varies widely, ranging from about 10 to 1500 or more.

The number of quanta released from a nerve terminal depends on previous activity. For example, when a train of stimuli at, say, 10 per second is applied to the presynaptic nerve, successive postsynaptic potentials may increase in amplitude due to a progressive increase in the number of quanta released by each presynaptic action potential, a process known as facilitation. Following stimulation, the effect declines over a few hundred milliseconds. A longer-lasting effect can be induced by stimulating the nerve repetitively for several seconds. After such a stimulus (called a *tetanus*) the release of quanta by subsequent stimuli is enhanced for periods that can extend to more than an hour, a phenomenon known as posttetanic potentiation.

In this chapter we describe the basic mechanisms whereby chemical transmitters are released from presynaptic nerve terminals and act on postsynaptic cells to produce excitation or inhibition, and introduce the idea of synaptic plasticity. Chapters 8 and 9 will expand on these ideas, dealing in depth with factors that modulate synaptic function and neuronal membrane properties and with the metabolism, identification, and distribution of neurotransmitters.

SYNAPSES (a word coined by Sherrington) are points of contact between nerve cells, or between nerves and effector cells such as muscle fibers. It is at synapses that excitation arriving in the presynaptic nerve terminal is translated into excitatory or inhibitory electrical signals in the postsynaptic cell. It was not always clear that the two components —presynaptic terminal and postsynaptic cell—are in fact morphologically separate. In the second half of the nineteenth century there was a vigorous disagreement between proponents of the *cell theory*, who considered that neurons were independent units, and those who be-

Initial approaches

lieved that nerve cells were a *syncytium*, interconnected by protoplasmic bridges. It was not until the early part of the twentieth century that the cell theory won general acceptance and most biologists started to think of nerve cells as being independent units. It remained for electron microscopy to obtain definitive evidence that each neuron was surrounded completely by its own plasma membrane. Even so, electron microscopy and other modern techniques eventually revealed that some neurons were in fact connected by protoplasmic bridges, in the form of connexons (Chapter 5).

Henry Dale (left) and Otto Loewi, mid 1930s. (Courtesy of Lady Todd and W. Feldberg.)

Electrical and chemical transmission

　The disagreement about synaptic structure was accompanied by a parallel disagreement about function. In 1843 Du Bois-Reymond showed that flow of electric current was involved in both muscle contraction and nerve conduction, and it required only a small extension of this idea to conclude that transmission of excitation from nerve to muscle was also due to current flow.[1] Du Bois-Reymond himself favored an alternative explanation—the secretion by the nerve terminal of an excitatory substance that then caused muscle contraction. However, the idea of animal electricity had such a potent hold on people's thinking that it was more than 100 years before contrary evidence finally overcame the assumption of electrical transmission between nerve and muscle and, by extension, between nerve cells in general.

　One reason that the idea of chemical transmission seemed unattractive is the fact that transmission between nerve and muscle and between

[1]Du Bois-Reymond, E. 1848. *Untersuchungen über thierische Electricität.* Reimer, Berlin.

nerve cells in the central nervous system is extremely rapid. This difficulty appeared not to exist in the autonomic nervous system controlling glands and blood vessels, because there the actions are relatively slow and prolonged. Accordingly, results of experiments on these tissues carried little weight in the discussion. For example, Langley's observation that transmission through the mammalian ciliary ganglion was blocked selectively by nicotine[2] was generally accepted as favoring chemical transmission at ganglionic synapses, but the observation was considered irrelevant to the main argument. It is interesting that at that time an observation by Consiglio that nicotine did *not* block transmission through ciliary ganglia of birds[3] went relatively unnoticed. It was discovered 60 years later that in birds the ciliary ganglion transmits its signals electrically![4] Highlights of the experiments and ideas at the beginning of this century are contained in the writings of Dale, for several decades one of the leading figures in British physiology and pharmacology.[5] Among his many contributions were the clarification of the action of ACh at synapses in autonomic ganglia and the establishment of its role in neuromuscular transmission.

In 1921 Otto Loewi did a direct and simple experiment that established the chemical nature of transmission between the vagus nerve and the heart.[6] He perfused the heart of a frog and stimulated the vagus nerve, thereby slowing the heartbeat. When the fluid from the inhibited heart was transferred to a second unstimulated heart, it too began to beat more slowly. Apparently the vagus nerve had released, upon stimulation, an inhibitory substance into the perfusate. Loewi and his colleagues were able to demonstrate in subsequent experiments that the substance was mimicked in every way by ACh. It is an amusing sidelight that Loewi had the idea for his experiment in a dream, wrote it down in the middle of the night, but could not decipher his writing the next morning. Fortunately the dream returned and the second time Loewi took no chances; he rushed to the laboratory and performed the experiment. Later he reflected:

> On mature consideration, in the cold light of morning, I would not have done it. After all, it was an unlikely enough assumption that the vagus should secrete an inhibitory substance; it was still more unlikely that a chemical substance that was supposed to be effective at very close range between nerve terminal and muscle be secreted in such large amounts that it would spill over and, after being diluted by the perfusion fluid, still be able to inhibit another heart.

In the early 1930s the role of ACh in synaptic transmission in ganglia in the autonomic nervous system was firmly established by Feldberg

[2]Langley, J. N. and Anderson, H. K. 1892. *J. Physiol.* 13: 460–468.
[3]Consiglio. M. 1900. *Arch. Farmacol. Terap.* 8: 268–275.
[4]Martin, A. R. and Pilar, G. 1963. *J. Physiol.* 168: 443–463.
[5]Dale, H. H. 1953. *Adventures in Physiology.* Pergamon Press, London.
[6]Loewi, O. 1921. *Pflügers Arch.* 189: 239–242.

and his colleagues.[7] Then, in 1936, Dale and his colleagues demonstrated that stimulation of motor nerves to mammalian skeletal muscle caused the release of ACh.[8] In addition, when ACh was injected into arteries supplying the muscle, it caused a large synchronous contraction of the muscle fibers.

Pharmacological techniques were indispensable for such experiments. For example, collection of ACh from skeletal muscle requires that its hydrolysis be prevented by a drug (eserine) that inhibits the enzyme acetylcholinesterase, present at cholinergic synapses and in the bloodstream. Another indispensable tool was the drug curare, a South American Indian arrow poison prepared (at that time) as an extract of one or another of several tropical plants; curare was shown by Claude Bernard to block neuromuscular transmission and by Langley to have a similar action on autonomic ganglia.

ELECTRICAL SYNAPTIC TRANSMISSION

By the mid 1950s chemical transmission had become almost universally accepted as the mode of action whereby electrical signals were generated in postsynaptic cells. Then in 1959 Furshpan and Potter, using intracellular microelectrodes to record from nerve fibers in the abdominal nerve cord of the crayfish, discovered an electrical synapse.[9] They demonstrated that an action potential in a lateral giant fiber led directly to depolarization by current flow of a giant motor fiber leaving the cord. The depolarization was sufficient to initiate an action potential in the postsynaptic fiber. The electrical coupling was in one direction only: Depolarization of the postsynaptic fiber did not lead to presynaptic depolarization. In other words, the synapse *rectified*. We know now that the morphological substrate for electrical coupling at this and other electrical synapses is the gap junction and its associated connexons, which allow current to pass from one cell to the next (Chapter 5).

Electrical synaptic transmission occurs at a wide variety of synapses.[10] Electrical coupling has been demonstrated, for example, between groups of large cells in the medulla of the puffer fish.[11] Electrical synapses also occur between motoneurons in the spinal cord of the frog,[12] and the corresponding presence of gap junctions has been demonstrated.[13] Similar electrical coupling occurs in the mammalian central nervous system.[14] In the leech, electrical coupling between some sen-

[7]Feldberg, W. 1945. *Physiol. Rev.* 25: 596–642.

[8]Dale, H. H., Feldberg, W. and Vogt, M. 1936. *J. Physiol.* 86: 353–380.

[9]Furshpan, E. J. and Potter, D. D. 1959. *J. Physiol.* 145: 289–325.

[10]Bennett, M. V. L. 1974. *In* M. V. L. Bennett (ed.). *Synaptic Transmission and Neuronal Interactions.* Raven, New York, pp. 153–158.

[11]Bennett, M. V. L. 1973. *Fed. Proc.* 32: 65–75.

[12]Grinnell, A. D. 1970. *J. Physiol.* 210: 17–43.

[13]Sotelo, C. and Taxi, J. 1970. *Brain Res.* 17: 137–141.

[14]Llinás, R., Baker, R. and Sotelo, C. 1974. *J. Neurophysiol.* 37: 560–571.

sory neurons has the remarkable property that depolarization spreads readily from either cell to the other, but hyperpolarization spreads poorly;[15,16] that is, the electrical connections are *doubly rectifying*.

A surprising finding was that electrical and chemical transmission are often combined at a single synapse. Such combined electrical and chemical synapses were first found on cells of the avian ciliary ganglion,[4] where a chemical synaptic potential (produced by ACh) is preceded by an electrical coupling potential (Figure 1). Similar synapses occur widely in vertebrates, for example, on spinal interneurons of the lamprey[17] and on frog spinal motoneurons.[18] Perhaps more frequently, postsynaptic cells integrate separate chemical and electrical synaptic inputs from different converging sources. For example, in leech ganglia (Chapter 13), motor neurons receive three distinct types of synaptic input from nerve terminals signaling three different sensory modalities; one input is chemical, one electrical, and one combined chemical and electrical.[19]

One of the indications that synaptic transmission is electrical is the absence of a SYNAPTIC DELAY. At chemical synapses there is a pause of about 1 msec between the arrival of an impulse in the presynaptic terminal and the appearance of an electrical potential in the postsynaptic cell. The delay is due to the time taken for the terminal to release transmitter. At electrical synapses there is no such delay; current spreads instantaneously from one element to the next. This is illustrated in Figure 1, which shows intracellular records from a cell in the ciliary ganglion of the chick. The presence of both electrical and chemical transmission at the same synapse provides a convenient means of comparing the two modes of transmission. Stimulation of the preganglionic nerve leads to an action potential in the postganglionic cell, with very short latency (Figure 1A). When the cell is hyperpolarized slightly (Figure 1B) the action potential jumps to a later time, revealing a brief depolarization that, because the cell has been hyperpolarized, is now subthreshold. This depolarization is an electrical coupling potential, produced by current flow from the presynaptic nerve terminal into the cell. Further hyperpolarization (Figure 1C) blocks the initiation of the action potential altogether, revealing the underlying chemical synaptic potential. These cells, then, have the peculiar property that, under normal conditions, initiation of a postsynaptic action potential by chemical transmission is preempted by electrical coupling across the synapse. The coupling potential precedes the chemical synaptic potential by about 1 msec, giving us a direct measure of the synaptic delay. Additional experiments on these cells have shown that the electrical coupling is bidirectional; that is, the synapses do not rectify.

[15]Baylor, D. A. and Nicholls, J. G. 1969. *J. Physiol.* 203: 591–609.
[16]Acklin, S. E. 1988. *J. Exp. Biol.* 137: 1–11.
[17]Rovainen, C. M. 1967. *J. Neurophysiol.* 30: 1024–1042.
[18]Shapovalov, A. I. and Shiriaev, B. I. 1980. *J. Physiol.* 306: 1–15.
[19]Nicholls, J. G. and Purves, D. 1972. *J. Physiol.* 225: 637–656.

1

ELECTRICAL AND CHEMICAL SYNAPTIC TRANSMISSION in a chick ciliary ganglion cell, recorded with an intracellular microelectrode. (A) Stimulation of the preganglionic nerve produces an action potential in the cell. (B) When the cell is hyperpolarized by passing current through the recording electrode, the action potential is delayed, revealing an earlier transient depolarization. This depolarization is an electrical synaptic potential (coupling potential), caused by current flow into the cell from the presynaptic terminal, and in (A) is responsible for initiating the action potential. (C) Slightly greater hyperpolarization blocks action potential initiation, leaving a slower chemical synaptic potential (epsp). The synaptic potential follows the coupling potential with a synaptic delay of about 1 msec. (After Martin and Pilar, 1963.)

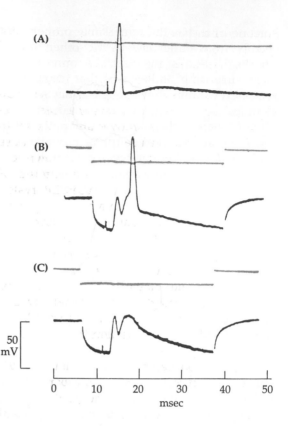

There are several possible functional advantages of electrical synaptic transmission. One is illustrated in Figure 1, namely, the absence of synaptic delay saves about 1 msec in the transmission process. This greater speed of transmission (possibly coupled with increased reliability) may be important in rapid reflexes involving escape reactions. For example, an electrical synapse on the Mauthner cell of the goldfish is involved in the strong tail flip elicited when the surface of the water is disturbed. In such a case, the saving of a millisecond may be important for surviving attack by a predator. Another possible function is the synchronization of electrical behavior of groups of cells; cells that might otherwise discharge independently may be drawn into synchrony by virtue of electrical coupling between them. Such effects depend on the degree of electrical coupling between cells in the group. This is usually expressed as a COUPLING RATIO—a ratio of 1:4 meaning that a quarter of the presynaptic voltage change appears in the postsynaptic cell. Variations in coupling between cells determine the relative degree of their electrical interactions with each other.

CHEMICAL SYNAPTIC TRANSMISSION

Synaptic structure   Chemical synapses are more complex in structure than gap junctions. Figure 2 illustrates the morphological features of the neuromuscular junction of the frog. The incoming motor nerve loses its myelin sheath and then gives off branches that run in shallow grooves in the surface of the muscle. The SYNAPTIC CLEFT between the terminal and the muscle membrane is about 30 nm thick. Within the cleft is the BASAL LAMINA, which follows the contours of the muscle surface. On the muscle side POSTJUNCTIONAL FOLDS radiate into the muscle fiber from the cleft at regular intervals. The grooves and folds are peculiar to skeletal muscle and not a general feature of chemical synapses. In muscle, this region of postsynaptic specialization is known as the MOTOR END PLATE. Schwann cell lamellae cover the nerve terminal, sending fingerlike processes around it at regularly spaced intervals. Within the cytoplasm of the terminal are mitochondria and synaptic vesicles. Many of the latter are lined up in double rows along narrow transverse bars of electron-dense material attached to the presynaptic membrane. Such a region is known as an ACTIVE ZONE. There is much evidence to indicate that the synaptic vesicles are sites of storage of ACh and that, upon excitation of the active zone, they fuse with the nerve terminal membrane to spill their contents into the synaptic cleft by EXOCYTOSIS (Chapter 9).[20-22] These details are illustrated in Figure 2B, which is an electron micrograph of a longitudinal section through a presynaptic terminal and the adjacent muscle fiber.

Synapses on nerve cells are usually made by nerve terminal swellings, called BOUTONS, which are separated from the postsynaptic membrane by a synaptic cleft. The presynaptic membrane of the bouton displays electron-dense active zones along which are clustered synaptic vesicles. Postsynaptic specializations, such as synaptic folds, are much less prominent but are often present in the form of thickening of the membrane underlying the presynaptic active zones.

The technique of freeze-fracturing has revealed many additional details of cell membranes. This procedure consists of freezing the tissue and breaking it apart before preparing replicas for scanning electron microscopy. When the tissue is broken, the lines of cleavage generally do not occur along spaces between adjacent cells; instead freeze-fracture planes occur between the bilayers of the plasma membrane. This splitting of the membrane creates two artificial surfaces, or faces, one belonging to the cytoplasmic half of the membrane leaflet, the other to the outer half. Consequently broad areas of the interior of membranes can be viewed on one face or the other.

The nature of the fracture planes is shown schematically in Figure 3. Figure 3A shows the main features of two active zones in a presynaptic terminal and the adjacent postsynaptic membrane. Two planes of separation are shown, but in practice a fracture would occur in one membrane or the other, not both at the same time. The upper portion

[20]Couteaux, R. and Pecot-Déchavassine, M. 1970. *C. R. Acad. Sci.* (Paris) 271: 2346–2349.

[21]Heuser, J. E., Reese, T. S. and Landis, D. M. D. 1974. *J. Neurocytol.* 3: 109–131.

[22]Peper, K. et al. 1974. *Cell Tiss. Res.* 149: 437–455.

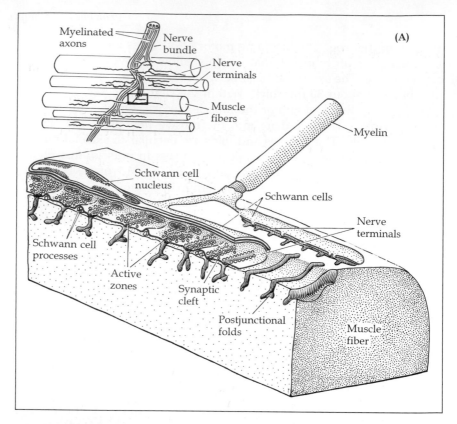

Myelinated axons

Nerve bundle

Nerve terminals

Muscle fibers

Myelin

Schwann cell nucleus

Schwann cells

Schwann cell processes

Nerve terminals

Active zones

Synaptic cleft

Postjunctional folds

Muscle fiber

(A)

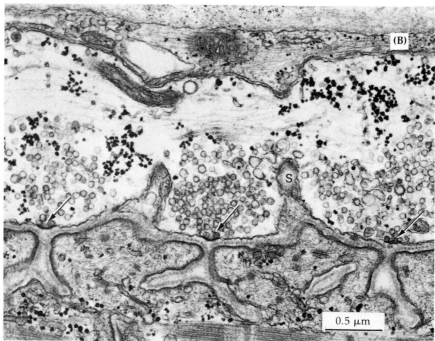

(B)

S

0.5 μm

of the figure shows the exposed surface of the cytoplasmic half of the presynaptic membrane with vesicles lined up on the cytoplasmic side. Some are represented in the process of exocytosis. There are, in addition, intramembranous particles protruding from the fracture face of the cytoplasmic leaflet; matching pits are seen on the fracture face of the outer leaflet. Similar particles and pits occur on the fracture faces of the postsynaptic membrane.

A conventional transmission electron micrograph of a horizontal section through an active zone is shown in Figure 3B. It is equivalent to looking down onto an active zone from the nerve terminal cytoplasm. An orderly row of synaptic vesicles is lined up along either side of the band of dense material. Figure 3C shows a corresponding image of the fracture face of a cytoplasmic leaflet. This is equivalent to viewing the same region from below the fracture plane (i.e., looking upward from the synaptic cleft). A row of particles, each about 10 nm in diameter, flanks the active zone on each side. Indentations, believed to indicate exocytotic openings, appear more laterally. A lower power freeze-fracture image is shown in Figure 3D. At the upper left, the first fracture face is that of the outer leaflet of the presynaptic terminal. The fracture then breaks across the synaptic cleft and exposes the face of the cytoplasmic leaflet of the postsynaptic membrane. Clusters of particles are seen around the borders of the postsynaptic folds. These are believed to correspond to ACh receptors that are concentrated in this region of the end plate.[21-23] In skeletal muscle, and other cholinergic synapses (i.e., synapses that utilize ACh as a neurotransmitter), the enzyme ACETYLCHOLINESTERASE is imbedded in the basal lamina within the synaptic cleft. The enzyme hydrolyzes ACh, thereby preventing prolonged action of the transmitter on the postjunctional receptors. Cholinesterase stains are often used as chemical markers for locating cholinergic synapses (although in theory cholinesterase is not an essential functional component of such synapses).

[23]Porter, C. W. and Barnard, E. A. 1975. *J. Membr. Biol.* 20: 31–49.

◄ **2** NEUROMUSCULAR JUNCTION OF THE FROG. (A) Low-power view (inset) of several skeletal muscle fibers and their innervation. Below is a three-dimensional sketch of part of a synaptic contact area. Synaptic vesicles are clustered in the nerve terminal in special regions opposite the openings of the postsynaptic folds. These regions, called active zones, are the sites of transmitter release into the synaptic cleft. Fingerlike processes of Schwann cells extend between the terminal and the postsynaptic membrane, separating the active zones. (B) Electron micrograph of a longitudinal section through a portion of the motor nerve terminal, showing many of the features seen in the sketch. In the nerve terminal, clusters of vesicles lie over thickenings in the presynaptic membrane—the active zones (arrows). Schwann cell processes (S) separate the clusters. In the muscle, postjunctional folds open into the synaptic cleft directly under each active zone. The line of fuzzy material in the cleft and following the contour of the postjunctional folds is the basal lamina. (B courtesy of U. J. McMahan.)

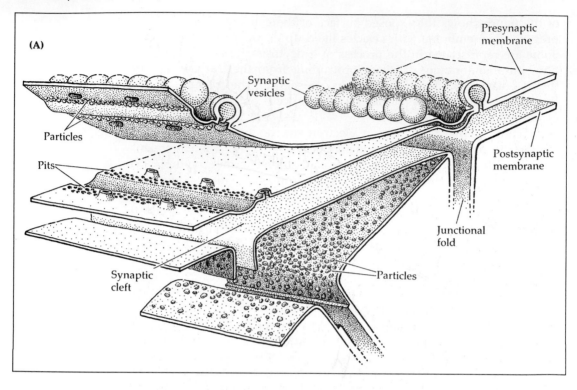

(A)

Presynaptic membrane

Synaptic vesicles

Particles

Pits

Postsynaptic membrane

Junctional fold

Particles

Synaptic cleft

Synaptic potentials at the neuromuscular junction

Much of our detailed knowledge about synaptic transmission has been obtained by experiments on the isolated vertebrate neuromuscular junction. Early studies by Göpfert and Schaefer and by Eccles and his colleagues used extracellular recording techniques to study the END PLATE POTENTIAL (epp) in muscle.[24–26] The end plate potential is the depolarization produced in the end plate region of the muscle fiber following motor nerve excitation and is produced by ACh released from the presynaptic nerve terminals. It is analogous to the EXCITATORY POST-SYNAPTIC POTENTIAL (epsp) produced by various neurotransmitters in nerve cells and in other effector cells. Normally the end plate depolarization produced in a skeletal muscle fiber by a motor nerve action potential is much greater than that needed to initiate an action potential, but when an appropriate amount of curare is added to the bathing solution the end plate potential is reduced in amplitude to below threshold, and the superimposed action potential disappears (Figure 4). The effect of curare is graded; the end plate depolarization is reduced below threshold for action potential initiation with a concentration of about 1 μM. Progressive increases in concentration reduce its amplitude still further until no detectable response remains.

[24]Göpfert, H. and Schaefer, H. 1938. *Pflügers Arch.* 239: 597–619.
[25]Eccles, J. C. and O'Connor, W. J. 1939. *J. Physiol.* 97: 44–102.
[26]Eccles, J. C., Katz, B. and Kuffler, S. W. 1942. *J. Neurophysiol.* 5: 211–230.

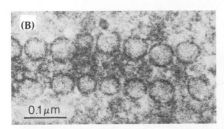

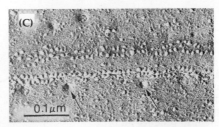

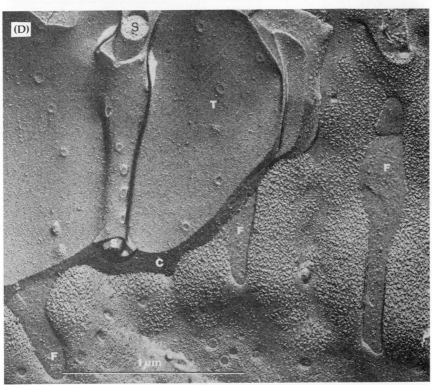

3   SYNAPTIC MEMBRANE STRUCTURE. (A) Three-dimensional view of presynaptic and postsynaptic membranes at a frog neuromuscular junction, with each membrane split along intramembranous planes as might occur in freeze-fracture. The cytoplasmic half of the presynaptic membrane at the active zone shows on its fracture face protruding particles whose counterparts are seen as pits on the fracture face of the outer membrane leaflet. Vesicles fusing with the presynaptic membrane give rise to pores and protrusions on the two fracture faces. The fractured postsynaptic membrane in the region of the folds shows a high concentration of particles on the fracture face of the cytoplasmic leaflet; these are ACh receptors. **(B)** Transmission electron micrograph of a section through the nerve terminal parallel to an active zone, showing a line of vesicles. **(C)** Fracture face of the cytoplasmic half of the presynaptic membrane in an active zone. The active zone region is delineated by particles about 10 nm in diameter and flanked by pores (arrows) caused by fusion of synaptic vesicles with the membrane. **(D)** Low-power view of fractured synaptic region. The fracture passes first through the presynaptic terminal membrane (T), showing the fracture face of the outer leaflet, then crosses the synaptic cleft (C) to enter the postsynaptic membrane. On the fracture face between the folds (F) one sees on the cytoplasmic leaflet aggregates of ACh receptors. A Schwann cell process (S) passes between the nerve terminal and the muscle. (A courtesy of U. J. McMahan; B from Couteaux and Pecot-Déchavassine, 1970; C and D from Heuser, Reese and Landis, 1974.)

4

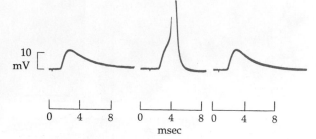

SYNAPTIC POTENTIALS recorded from a curarized mammalian neuromuscular junction. Curare concentration in the bathing solution was
adjusted so that the amplitude of the synaptic
potential was just sufficient to produce an occasional action potential in the muscle fiber (second
record). (From Boyd and Martin, 1956.)

The intracellular microelectrode[27] was used by Fatt and Katz[28] to
study in detail the time course and spatial distribution of the end plate
potential in curarized single muscle fibers. They stimulated the motor
nerve and recorded the potential at various distances from the end plate
region of the muscle (Figure 5). At the end plate the depolarization rose
rapidly to a peak and then declined slowly over the next 20 to 30
milliseconds. As they moved the recording microelectrode farther and
farther away from the end plate, the potential became progressively
smaller and its time to peak progressively longer. Fatt and Katz were
able to show that the time course of decay of the end plate potential
was consistent with the time constant of the muscle fiber membrane
and that the changes in its amplitude with distance from the end plate
were consistent with the length constant. They concluded that the
potential was generated locally at the end plate and then spread along

[27]Ling, G. and Gerard, R. W. 1949. *J. Cell. Comp. Physiol.* 34: 383–396.
[28]Fatt, P. and Katz, B 1951. *J. Physiol.* 115: 320–370.

Stephen Kuffler, John Ec-
cles, and Bernard Katz
(left to right) in Australia,
about 1941.

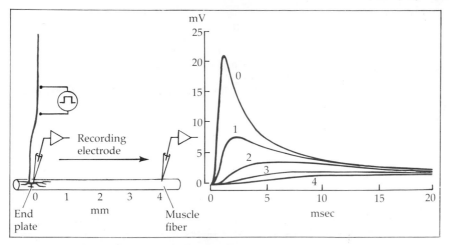

5 **DECAY OF SYNAPTIC POTENTIALS WITH DISTANCE from end plate region of a muscle fiber. Records taken by an intracellular electrode at distances of 0, 1, 2, 3, and 4 mm from the end plate show a progressive decrease in size and slowing of rise time of the synaptic potential. (After Fatt and Katz, 1951.)**

the muscle fiber and decayed in a manner determined solely by the passive electrical properties of the fiber.

Shortly after the introduction of the microelectrode for intracellular recording, glass micropipettes were also used for discrete application of ACh (and later other drugs as well) to the end plate region of muscle.[29] The technique is illustrated in Figure 6A. A microelectrode is inserted into the end plate of a single muscle fiber for recording membrane potential, while an ACh-filled micropipette is held just outside. To apply ACh to the region, a brief positive voltage pulse is applied to the top of the pipette, causing a spurt of positively charged ACh ions to leave the pipette tip. This method of ejecting charged molecules from pipettes is known as IONOPHORESIS. Using this method of application, del Castillo and Katz showed that ACh depolarized the muscle fiber only at the end plate region and only from the outside; intracellular ionophoresis was without effect.[30] The technique of ionophoresis made it possible to map with high accuracy the distribution of postsynaptic ACh receptors in muscle fibers[31] and in nerve cells.[32] When the ACh-filled pipette is in close apposition to a receptor region, the response to ionophoresis is rapid, mimicking almost exactly the effect of nerve-released ACh (Figure 6B). Movement of the pipette by only a few

Localized application of ACh

[29]Nastuk, W. L. 1953. *Fed. Proc.* 12: 102.
[30]del Castillo, J. and Katz, B. 1955. *J. Physiol.* 128: 157–181.
[31]Miledi, R. 1960. *J. Physiol* 151: 24–30.
[32]Dennis, M. J. , Harris, A. J. and Kuffler, S. W. 1971. *Proc. R. Soc. Lond. B* 177: 509–539.

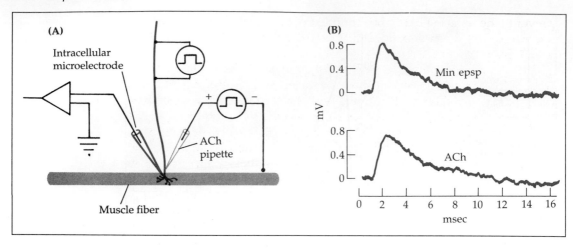

6   IONOPHORESIS OF ACh. (A) Experimental arrangement for ionophoresis of ACh. ACh-filled pipette is placed close to the neuromuscular junction, and ACh ejected from the tip by a brief positive pulse. Intracellular microelectrode records response to ionophoretic application of ACh, or response to release of ACh from the motor nerve terminal. (B) Response to a small quantity of acetylcholine released from the nerve terminal (Min epsp) is mimicked almost exactly by a pulse of ACh from a nearby micropipette. Rise time of ACh pulse is slightly slower because the location of the ACh pipette is less favorable than that of the nerve terminal. (A after Dennis, Harris and Kuffler, 1971; B from Kuffler and Yoshikami, 1975a.)

micrometers results in a reduction in amplitude and slowing of the response.

Another way to apply neurotransmitters and other substances to nerve and muscle fiber membranes is by pressure pulses. With this method, brief pulses of about 1 kg per cm^2 are applied to the top of the pipette to drive solution from the tip. The method has several advantages over ionophoresis, mainly that the applied substance need not be ionized. One disadvantage is that bulk flow of solution from the tip sometimes produces false responses due to movement of the underlying membrane.

Measurement of ionic currents produced by ACh

Except for its rapid rise, the time course of the end plate potential in muscle is determined by the cable properties of the muscle fiber. As a consequence the end plate potential provides only indirect evidence about the time course of the influx of current or, in turn, about the time course of action of ACh on the postsynaptic receptors. Synaptic currents were first measured at the frog neuromuscular junction by A. and N. Takeuchi, who used two microelectrodes to voltage clamp the end plate region of muscle fibers.[33] In addition to revealing the time course of the synaptic current, the technique enabled them to assess its ionic composition. Subsequently, similar studies were carried out by Magleby and Stevens in muscle fibers treated with hypertonic glycerol,[34] which

[33]Takeuchi, A. and Takeuchi, N. 1959. *J. Neurophysiol.* 22: 395–411.
[34]Magleby, K. L. and Stevens, C. F. 1972. *J. Physiol.* 223: 151–171.

prevents the muscle fibers from contracting (and dislodging the recording electrodes) when depolarized. Such treatment also leaves the fibers in a depolarized state.

The experimental arrangement for measuring the end plate current (epc) is shown in Figure 7A. Two microelectrodes were inserted into the end plate region of the fiber, one for recording the membrane potential, the other for current injection to clamp the potential at the desired level. In the experiment illustrated in Figure 7B, with the muscle at its resting potential (-40 mV), nerve stimulation produced an inward current of about 150 nA. At more negative holding potentials, up to -120 mV, the end plate current increased in amplitude. At depolarized levels of $+21$ and $+38$ mV, the current was outward. When the amplitude of the end plate current was plotted against holding potential (Figure 7C), the relation was nearly linear. The current changed from inward to outward near zero membrane potential. This is its REVERSAL POTENTIAL. In earlier experiments on intact muscle fibers, A. and N. Takeuchi estimated the reversal potential to be about -15 mV.[35]

The reversal potential for the end plate current gives us information about the ionic currents flowing through the channels activated by ACh in the postsynaptic membrane. For example, if the channels were permeable exclusively to sodium, then current through the channels would be zero at the sodium equilibrium potential (about $+50$ mV). The other major ions, potassium and chloride, have equilibrium potentials near the resting membrane potential (Chapter 3). None has an equilibrium potential in the range of 0 to -15 mV. What ions, then, are involved in the response? Fatt and Katz[28] suggested that ACh might open nonselective channels in the membrane, so that at the reversal potential, sodium and chloride ions would enter the cell and potassium would leave, with the total ionic currents adding up to zero. The experiments by A. and N. Takeuchi showed that the reversal potential changed when the extracellular sodium concentration was reduced or when the extracellular potassium concentration was increased. Increasing extracellular calcium concentration had a smaller effect.[36] On the other hand, the reversal potential for the end plate current was unchanged by alterations in extracellular chloride concentration. The effect of ACh, then, is to produce a general increase in *cation* permeability at the end plate.

Application of ACh opens channels in the end plate that allow sodium ions to leak inward and potassium ions outward along their electrochemical gradients. Because the calcium conductance of the channels is small, the contribution of calcium to the overall synaptic current can be ignored, as can that of other cations, such as magnesium. (It should be noted that the low calcium conductance is because of its low extracellular and intracellular concentrations; calcium *permeability* is about 20 percent of the sodium permeability.) The equivalent electrical

Significance of the reversal potential

Electrical model of the motor end plate

---

[35]Takeuchi, A. and Takeuchi, N. 1960. *J. Physiol.* 154: 52–67.
[36]Takeuchi, N. 1963. *J. Physiol.* 167: 128–140.

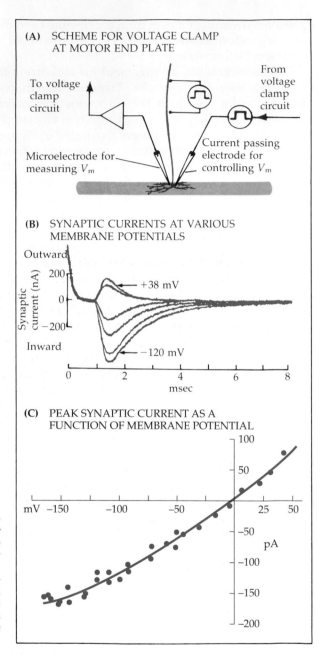

7

SYNAPTIC CURRENTS produced by nerve-released ACh. (A) Experimental arrangement for voltage clamping muscle fiber at the end plate region in order to record currents produced by ACh released from the motor nerve terminals. **(B)** Amplitude and time course of synaptic currents at membrane potentials between −120 and +38 mV. **(C)** Plot of peak end plate current (nA) against membrane potential (mV). The relation is nearly linear, with the reversal potential close to 0 mV. (After Magleby and Stevens, 1972.)

model is shown in Figure 8. The resting membrane, consisting of the usual sodium, potassium, and chloride channels, is represented by a single conductance g_m (equal to the sum of all the ionic conductances) and a single battery, V_m. It is shunted by ACh-activated channels for sodium and potassium, represented by a single conductance Δg_s and a battery whose voltage is equal to the reversal potential V_r. The synaptic conductance and reversal potential are electrically equivalent to separate conductances (Δg_{Na} and Δg_K) and driving potentials (E_{Na} and E_K) for the two ions, although the ACh receptors do not form separate pathways for sodium and potassium. It is assumed, however, that the two ions move through the channels independently, so that separate expressions can be written for the sodium and potassium currents (ΔI_{Na} and ΔI_K):

$$\Delta I_{Na} = \Delta g_{Na}(V_m - E_{Na})$$

$$\Delta I_K = \Delta g_K(V_m - E_K)$$

These equations provide a means of determining the relative conduc-

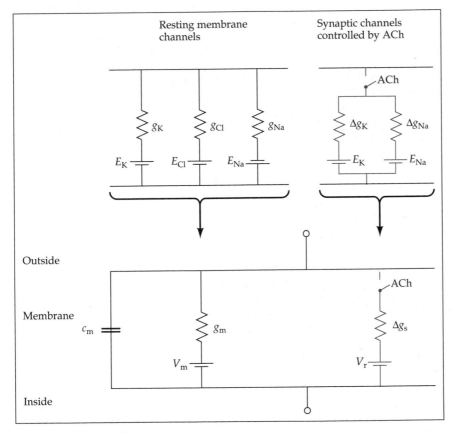

8 **ELECTRICAL MODEL OF THE SYNAPTIC MEMBRANE** activated by acetylcholine in parallel with the remaining cell membrane. The synaptic membrane has a resistance $\Delta r_s = 1/\Delta g_s$ and a reversal potential V_r. This single-current pathway is electrically equivalent to two independent pathways for sodium and potassium through the synaptic channels, as shown above. The extrasynaptic membrane is represented by a resting membrane resistance $r_m = 1/g_m$ and potential V_m. This current pathway is equivalent to the circuit above, showing separate conductance pathways for sodium, potassium, and chloride.

tance changes to sodium and potassium produced by ACh once we have determined the reversal potential V_r. This is the membrane potential at which the current is zero or, in other words, where the inward sodium current is exactly equal and opposite to the outward potassium current. So when $V_m = V_r$,

$$\Delta g_{Na}(V_r - E_{Na}) = -\Delta g_K(V_r - E_K)$$

It follows that

$$\frac{\Delta g_{Na}}{\Delta g_K} = \frac{-(V_r - E_K)}{V_r - E_{Na}}$$

The Takeuchis calculated that with $V_r = -15$ mV, $\Delta g_{Na}/\Delta g_K$ was about 1.3. The *permeabilities* of the ACh receptor to the two ions are more nearly equal.[37] If we consider the extracellular and intracellular solutions together, there are more sodium than potassium ions available to move through the channels (Chapter 3). Thus, for the same permeability change, the sodium conductance change is slightly larger (Chapter 2). Experimental measurements on single nicotinic ACh-activated channels have revealed similar sodium/potassium conductance ratios.[38]

The electrical model shown in Figure 8 illustrates an important feature of synaptic transmission, namely that the size of the synaptic potential depends not only on the synaptic conductance (i.e., how many synaptic channels are activated) but also on the resting conductance of the cell. If, for simplicity, we suppose that V_r is zero, then the membrane potential V_a during synaptic activation (when the switch is closed) is $V_a = V_m[\Delta r_s/(\Delta r_s + r_m)]$, where $\Delta r_s = 1/\Delta g_s$ (see Appendix A). The same expression can be written in terms of conductances; that is, $V_a = V_m[g_m/(\Delta g_s + g_m)]$. The amplitude of the synaptic potential itself (V_s) is the difference between the resting membrane potential and the membrane potential during synaptic activation:

$$V_s = V_m - V_a = V_m \frac{\Delta g_s}{\Delta g_s + g_m}$$

How can the amplitude of a particular synaptic potential be increased? One way is to increase the the amount of transmitter released from the nerve terminal, thereby increasing Δg_s. Another, less obvious, way is to reduce the conductance of the rest of the cell, g_m. The reduction of extrasynaptic membane conductance is an important mechanism for modulating the strength of synaptic inputs: In some cells intracellular messengers reduce the membrane conductance by closing potassium channels, thereby increasing synaptic potential amplitude (Chapter 8).

Kinetics of currents through single ACh receptor channels

One question we can ask is whether or not the behavior of the end plate current reflects the behavior of individual ACh channels. For example, is the time course of decay of the current determined by the

[37]Lassignal, N. and Martin, A. R. 1977. *J. Gen. Physiol.* 70: 23–36.
[38]Mathie, A., Cull-Candy, S. G. and Colquhoun, D. 1991. *J. Physiol.* 439: 717–750.

rate at which individual open channels close? Alternatively, is the end plate current made up of a burst of brief channel openings, with the probability of channel activation declining with time (as we saw with voltage-activated sodium channels underlying the action potential)? These kinds of questions were first examined by means of noise analysis (Chapter 2).[39] It was found that current noise produced by application of ACh to the end plate was consistent with the idea that single ACh channels opened in an all-or-nothing fashion, with a single-channel conductance of about 30 pS, and that the rate at which the channels closed determined the decay of the end plate current. This idea can be summarized by the following scheme, indicating the interaction between the transmitter molecules A (for "agonist") and the postsynaptic receptor molecules R.

$$2A + R \rightleftharpoons A_2R \underset{\alpha}{\overset{\beta}{\rightleftharpoons}} A_2R^*$$

Two ACh molecules combine with the channel (one on each α subunit), which then can undergo a change in conformation from the closed to the open state, the latter represented by A_2R^*. The rate constants for the transitions are given by α and β, as indicated. Now consider the end plate current: ACh arriving at the postsynaptic membrane interacts with a large number of channels almost simultaneously, leading to a number of conformational changes and hence open channels. Because ACh is lost rapidly from the synaptic cleft (due to hydrolysis by cholinesterase and diffusion), each channel opens only once. Once channels are open, the rate at which they close is determined by α. As with all independent or "random" events, some channels close very quickly, others stay open a long time, and the open times are distributed exponentially, with a time constant $\tau = 1/\alpha$. As the channels close, the synaptic current declines with the same time constant. The closing time constant τ is often referred to as the MEAN OPEN TIME of the channel. The two numbers are indeed the same, but this identity has nothing to do with channel properties; it is merely a property of exponential distributions: When events are distributed exponentially in time, their mean duration is the same as the time constant of the exponential.

Deductions made originally from noise analysis were confirmed with the advent of patch clamp techniques.[40] The initial experiments were done on ACh receptors that accumulate outside the end plate region in denervated muscle. This was because in normal muscle the overlying nerve terminal and Schwann cell made it extremely difficult to approach the end plate with patch electrodes. Later experiments were on perisynaptic receptors around the end plate[41] or on receptors distributed over the membranes of embryonic muscle fibers in culture. Eventually,

[39]Anderson, C. R. and Stevens, C. F. 1973. *J. Physiol.* 235: 655–691.
[40]Neher, E. and Sakmann, B. 1976. *Nature* 260: 799–801.
[41]Colquhoun, D. and Sakmann, B. 1981. *Nature* 294: 464–466.

however, patch clamp recordings were obtained from the end plate region itself, after enzymatic removal of the overlying tissues.[42] Patch clamp experiments also revealed a number of details of channel activation previously undetectable. For example, local anesthetics, such as procaine, were shown to produce brief repetitive closures of open channels, apparently by physical blockade as the molecule flickered into and out of the open pore,[43] and embryonic receptors in muscle were shown to open not only to a single main conductance state, but also to lower-conductance substates.[44]

SYNAPTIC INHIBITION

The process of inhibition by ligand-activated channels involves the same principles as those underlying synaptic excitation; only the change in postsynaptic conductance is different. Excitation is achieved by driving the membrane toward threshold, whereas inhibition is achieved by holding the membrane below threshold. The effect of inhibitory transmitters is to activate channels permeable not to cations, but instead to anions, the principal one present being chloride, which has an equilibrium potential at or near the resting potential.

Reversal of inhibitory potentials

Synaptic inhibition has been studied in detail in a number of cells, most notably the spinal motoneuron of the cat,[45] the crustacean neuromuscular junction,[46,47] and the crayfish stretch receptor.[48] Motoneurons are inhibited by sensory inputs from antagonist muscles, by way of inhibitory interneurons in the spinal cord. The effect of activation of inhibitory inputs can be studied by experiments similar to that illustrated in Figure 9A. The motoneuron is impaled with two micropipettes, one to record potential changes, the other to pass current through the cell membrane. At the normal resting potential (about −75 mV), stim-

[42]Dionne, V. E. and Leibowitz, M. D. 1982. *Biophys. J.* 39: 253–261.
[43]Neher, E. and Steinbach, J. H. 1978. *J. Physiol.* 277: 153–176.
[44]Hamill, O. and Sakmann, B. 1982. *Nature* 294: 462–464.
[45]Coombs, J. S., Eccles, J. C. and Fatt, P. 1955. *J. Physiol.* 130: 326–373.
[46]Dudel, J. and Kuffler, S. W. 1961. *J. Physiol.* 155: 543–562.
[47]Takeuchi, A. and Takeuchi, N. 1967. *J. Physiol.* 191: 575–590.
[48]Kuffler, S. W. and Eyzaguirre, C. 1955. *J. Gen Physiol.* 39: 155–184.

9 EFFECT OF INHIBITORY TRANSMITTER on cat spinal motoneuron (A), a crayfish ▶ stretch receptor (B), and a muscle fiber from a crayfish (C). In each experiment potential changes were recorded with an intracellular microelectrode. In the motoneuron and the muscle fiber, the membrane potential is set to different levels by passing current through a second intracellular electrode; in the stretch receptor the potential is altered by adjusting the amount of stretch on the dendrites. Each cell has a reversal potential at which the inhibitory stimulation causes no potential change. In (A) the reversal potential is between −74 and −82 mV; in (B) between −67 and −70 mV. In (C) membrane potential is varied continuously from −80 to −60 mV and back, with inhibitory stimulation every 2 seconds. Inhibitory potentials reverse at −72 mV (arrows). (D) Equivalent circuit of an inhibitory conductance pathway in parallel with the rest of the cell membrane. (A after Coombs, Eccles and Fatt 1955; B after Kuffler and Eyzaguirre, 1955; C after Dudel and Kuffler, 1961.)

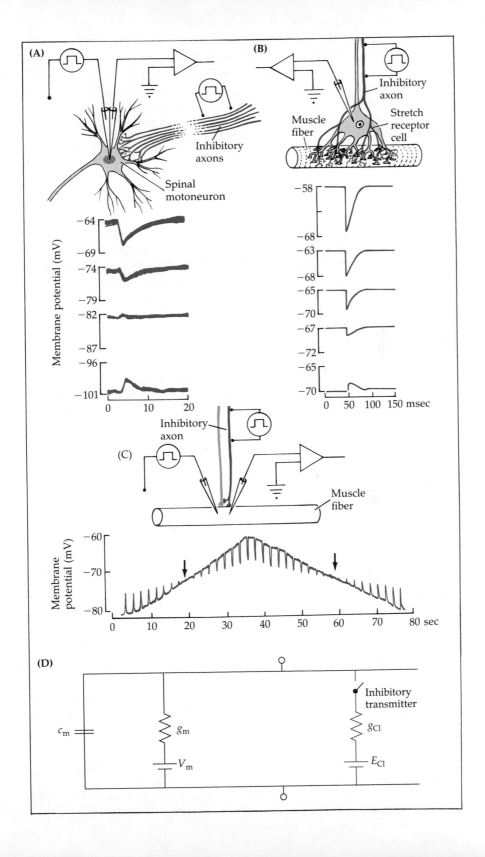

ulation of the inhibitory inputs causes a slight hyperpolarization of the cell—the INHIBITORY POSTSYNAPTIC POTENTIAL (ipsp). When the membrane is depolarized by passing positive current into the cell, the amplitude of the inhibitory postsynaptic potential is increased. When the cell is hyperpolarized to −82 mV, the inhibitory potential is very small and reversed in sign, and at −100 mV the reversed potential is increased in amplitude. The reversal potential in this experiment was about −80 mV. Similar experiments on the crayfish stretch receptor and crayfish muscle, both of which receive inhibitory innervation from ganglia in the central nerve cord, are shown in Figure 9B and 9C. No current-passing electrode is needed in the stretch receptor preparation, as its resting potential can be increased or decreased in magnitude simply by increasing or decreasing the tension on the muscle fiber in which the receptor cell dendrites are imbedded. In both preparations the inhibitory postsynaptic potential reversed at a membrane potential of about −70 mV.

Ionic basis for inhibitory potentials

The ionic selectivities of both excitatory and inhibitory channels have been reviewed by Edwards.[49] Inhibitory channels are permeable to anions, with permeabilities roughly correlated with the hydrated radius of the penetrating ion. In physiological circumstances, the only small anion present in any quantity is chloride. Thus in spinal motoneurons, for example, injection of chloride into the cell from a micropipette shifts the chloride equilibrium potential, and hence the reversal potential for the ipsp, in the positive direction. In other preparations, changes in extracellular chloride concentration also produce the expected result, but such experiments are more difficult to interpret. This is because changes in extracellular chloride concentration lead eventually to proportionate changes in intracellular concentration as well (Chapter 3), so any change in chloride equilibrium potential is only transient.

One way around this difficulty is to remove chloride entirely, as shown in Figure 10. The records are from a reticulospinal cell in the brainstem of the lamprey, in which inhibitory synaptic transmission is mediated by glycine.[50] Membrane potential was recorded with an intracellular microelectrode. A second electrode was used to pass brief current pulses into the cell; the resulting changes in potential provided a measure of the cell's input resistance. Finally, a third micropipette was used to apply glycine to the cell close to an inhibitory synapse, using brief pressure pulses. In the upper record (Figure 10A), glycine application results in a slight hyperpolarization, with a marked reduction in input resistance, as would be expected if glycine activated a large number of chloride channels. Evidence that no ions other than chloride pass through the inhibitory channels is shown in Figure 10B. Chloride was removed from the bathing solution, replaced by the impermeant ion

[49]Edwards, C. 1982. *Neuroscience* 7: 1335–1366.
[50]Gold, M. R. and Martin, A. R. 1983. *J. Physiol.* 342: 99–117.

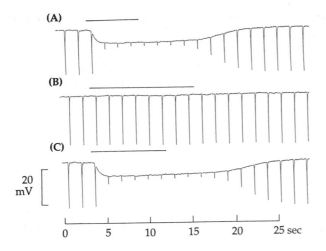

10

DEPENDENCE ON CHLORIDE of inhibitory response to glycine in lamprey brainstem neurons, recorded with an intracellular microelectrode. (A) Resting membrane potential is −63 mV. Downward voltage deflections are produced by brief 10-nA current pulses from a second intracellular pipette; their amplitudes indicate membrane resistance. On application of glycine (bar), the cell is hyperpolarized by about 7 mV and membrane resistance is reduced drastically. (B) After 20 minutes in chloride-free bathing solution, response to glycine is abolished. (C) Recovery 5 minutes after return to normal chloride solution. (From Gold and Martin, 1983b.)

isethionate. As a result, intracellular chloride was also removed by efflux through resting chloride channels. After 20 minutes, glycine application produces no detectable response. The restoration of normal extracellular chloride concentration (Figure 10C) results in restoration of the response.

Given that the inhibitory response involves an increase in chloride permeability, the reversal potential for the inhibitory current will be at the chloride equilibrium potential. It follows that the magnitude of the current is given by

$$\Delta i_{\text{inhibitory}} = \Delta i_{\text{Cl}} = \Delta g_{\text{Cl}}(V_m - E_{\text{Cl}})$$

At membrane potentials positive to E_{Cl}, the current is outward, resulting in membrane hyperpolarization; negative to E_{Cl}, inhibition causes depolarization. The equivalent electrical circuit is shown in Figure 9D.

So far we have defined excitatory and inhibitory synapses on the basis of the effect of the transmitter on the postsynaptic membrane, that is, on whether the postsynaptic permeability change is to cations or to anions. However, a number of early experiments indicated that in some instances it was difficult to account for inhibition in terms of postsynaptic permeability changes alone.[51,52] Eventually, an additional inhibitory mechanism was described in the mammalian spinal cord by Eccles and his colleagues[53] and at the crustacean neuromuscular junction by Dudel and Kuffler.[46] As shown in Figure 11, the action of the inhibitory nerve at the crustacean neuromuscular junction is exerted not only on the muscle fibers, but also on the excitatory terminals,

Presynaptic inhibition

[51]Fatt, P. and Katz, B. 1953. *J. Physiol.* 121: 374–389.
[52]Frank, K. and Fuortes, M.G.F. 1957. *Fed. Proc.* 16: 39–40.
[53]Eccles, J. C., Eccles, R. M. and Magni, F. 1961. *J. Physiol.* 159: 147–166.

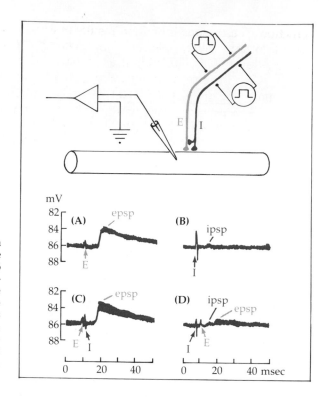

11

PRESYNAPTIC INHIBITION in a crustacean muscle fiber innervated by one excitatory and one inhibitory axon. At the resting potential (−86 mV), stimulation of the excitatory axon (E) produces a 2-mV epsp (A), while stimulation of the inhibitory axon (I) produces a depolarizing ipsp of about 0.2 mV. (C) If the inhibitory stimulus follows the excitatory one by a short interval, there is no effect on the epsp. (D) If the inhibitory stimulus precedes the excitatory one by a few milliseconds, the epsp is almost abolished. (After Dudel and Kuffler, 1961.)

Stephen W. Kuffler in 1975

reducing the output of transmitter. In the next section we shall see that transmitters are released in packets, or QUANTA, each containing several thousand molecules. The presynaptic effect of the inhibitory transmitter is to reduce the number of quanta released from the excitatory terminal. The presynaptic inhibitory effect is brief, reaching a peak in a few milliseconds and declining to zero after a total of 6 to 7 msec. For the maximum inhibitory effect to occur, the impulse must arrive in the inhibitory presynaptic terminal several milliseconds before the impulse in the excitatory terminal. The importance of accurate timing is shown in Figure 11, where parts (A) and (B) show the excitatory and inhibitory potentials following separate stimulation of the respective nerves. In Figure 11C the inhibitory impulse follows the excitatory one by 1.5 msec and arrives too late to exert its effect; in Figure 11D, however, it precedes the excitatory impulse and causes a marked reduction in the size of the excitatory postsynaptic potential. The presynaptic effect, like that on the postsynaptic membrane, is mediated by GABA and is associated with a marked increase in chloride permeability in the excitatory terminals.[47] When chloride permeability is high, the depolarizing effect of sodium influx during the rising phase of the action potential is canceled in part by an accompanying influx of chloride. As a result, the presynaptic action potential is smaller in amplitude and its effectiveness in releasing transmitter is reduced.

In the mammalian spinal cord, as at the crayfish neuromuscular junction, presynaptic inhibition results in a reduction in the number of quanta of transmitter released from excitatory nerve terminals.[54] In the nervous system in general, presynaptic and postsynaptic inhibition serve quite different functions. Postsynaptic inhibition, for example in a spinal motoneuron, reduces the excitability of the cell itself, rendering it relatively less responsive to all excitatory inputs. Presynaptic inhibition is much more specific, aimed at particular excitatory inputs and leaving the postsynaptic cell free to go about its business of integrating information from other sources. Presynaptic inhibition implies that inhibitory axons make synaptic contact with excitatory nerve terminals. Such axo-axonic synapses have been demonstrated directly by electron microscopy at the crustacean neuromuscular junction[55] and at numerous locations in the mammalian central nervous system.[56] Moreover, inhibitory nerve terminals themselves can be influenced presynaptically; the requisite ultrastructural arrangement has been reported at inhibitory synapses on crayfish stretch receptors.[57]

A novel kind of inhibition has been discovered recently between efferent fibers in the auditory nerve and cochlear hair cells of the chick.[58] Acetylcholine, released from the nerve terminals onto the postsynaptic membrane of the hair cells, opens channels that allow the entry of calcium and other cations. This effect would normally be excitatory except that the incoming calcium acts to open calcium-activated potassium channels, thereby producing inhibition. These events are illustrated in Figure 12. Patch clamp electrodes were used to obtain records of whole-cell currents from the hair cells. In Figure 12A, application of a brief pulse of ACh solution from a micropipette produced a large outward (inhibitory) potassium current, preceded by a small, brief inward current. When the same experiment was repeated with the calcium chelator BAPTA (a modified form of EGTA[59]) in the recording pipette (and hence in the cytoplasm of the cell), the outward current was blocked, revealing a substantial inward cation current. The incoming calcium, being chelated, could no longer act to open the calcium-activated potassium channels. A similar mechanism of inhibition may occur in neurons in the brain as well.[60]

The receptors mediating the response are of some interest. They have many of the characteristics common to nicotinic ACh receptors (nAChRs) in that they form ion channels when activated, and their responses to ACh are blocked by curare and bungarotoxin. However, the receptors are not activated by nicotine and have other unusual properties. Calcium flux accounts for about 2 percent of the inward

Inhibition mediated by a cation channel

---

[54]Kuno, M. 1964. *J. Physiol.* 175: 100–112.
[55]Atwood, H. L. and Morin, W. A. 1970. *J. Ultrastruct. Res.* 32: 351–369.
[56]Schmidt, R. F. 1971. *Ergeb. Physiol.* 63: 20–101.
[57]Nakajima, Y., Tisdale, A. D. and Henkart, M. P. 1973. *Proc. Natl. Acad. Sci. USA* 70: 2462–2466.
[58]Fuchs, P. A. and Murrow, B. W. 1992. *J. Neurosci* 12: 2460–2467.
[59]Tsien, R. Y. 1980. *Biochemistry* 19: 2396–2404.
[60]Wong, L. A. and Gallagher J. P. 1991. *J. Physiol.* 36: 325–346.

12

INHIBITION BY ACh-ACTIVATED CATION CHANNELS in hair cells from the chick cochlea. (A) In a whole-cell recording (inset), application of ACh near the base of a hair cell produces large outward current across the cell membrane, preceded by a small, transient inward current (arrow). In the intact cell, the outward current would be inhibitory. (B) Record from another experiment in which the calcium chelator BAPTA was added to the electrode and hence to the cell cytoplasm. Now ACh application produces only inward current due to cation influx into the cell. Outward current is seen in (A) because calcium influx through ACh-activated channels in turn opens calcium-activated potassium channels. The effect is absent in (B) because incoming free calcium is chelated. (Records courtesy of P. A. Fuchs.)

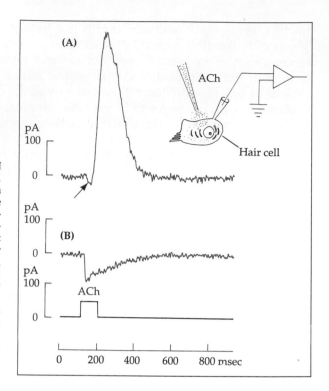

Directly acting neurotransmitters

current in most nAChR channels,[61] and this amount appears to be adequate to account for the observed inhibition.

Two major inhibitory neurotransmitters are γ-aminobutyric acid (GABA) and glycine (Chapter 10). Just as ACh acts directly to open cation channels in the postsynaptic membranes of skeletel muscle and ganglion cells, these amino acids act directly to open anion channels. An important feature of the GABA receptor is that its conductance can be influenced by other ligands, including a group of widely used drugs, the benzodiazepines (Chapter 10). Few other directly acting transmitters have been identified. One of these is glutamate, which is an important excitatory neurotransmitter in the vertebrate central nervous system.[62,63] There are a number of subtypes of glutamate-activated channels that can be grouped into two broad classifications: those that are sensitive to the glutamate agonist N-methyl-D-aspartate (NMDA receptors) and those that are not (non-NMDA receptors). Subunits of the non-NMDA receptors have been cloned and expressed in oocytes (Chapter 2). The channels have very low conductance to cations.[64] NMDA receptors have

[61]Decker, E. R. and Dani, J. A. 1990. *J. Neurosci.* 10: 3413–3420.

[62]Collingridge, G. L. and Lester, R. A. J. 1989. *Pharmacol. Rev.* 40: 143–219.

[63]Monaghan, D. T., Bridges, R. J. and Cotman, C. W. 1989. *Annu. Rev. Pharmacol. Toxicol.* 29: 365–402.

[64]Ascher, P., Nowak, L. and Kehoe, J. 1986. *In* J. M. Ritchie et al. (eds.). *Ion Channels in Neural Membranes.* Liss, New York.

an additional interesting property: Their conductance is voltage-dependent. When the receptors are activated by NMDA (or by the native transmitter, glutamate), their channels carry only a small amount of current unless the membrane is depolarized. This voltage dependence has been shown to be due to the block of resting channels by magnesium ions: Upon depolarization the magnesium is driven out of the channel and passage of other cations is permitted.[65,66] In addition, when the channel is activated by glutamate, its probability of opening is enhanced by glycine.[67] The voltage dependence of the ligand-activated channels is believed by many investigators to underlie long-term changes in synaptic efficacy (Chapter 10).

The mechanism of synaptic action illustrated in Figure 12 is different in principle from mechanisms of excitatory and inhibitory synaptic transmission described previously because the ultimate effect on the cell is mediated by channels that are not activated directly by the transmitter. An extension of this principle is found in receptors that do not form channels at all, but instead, upon activation, set in action a sequence of intracellular events that lead to modulation of ion channels or pumps that are distinct from the receptor molecules. This indirect action of neurotransmitters is discussed in detail in Chapter 8.

Indirect synaptic activation

RELEASE OF CHEMICAL TRANSMITTERS

A number of questions arise concerning the way in which the presynaptic neuron releases transmitter. Experimental answers to such questions require a highly sensitive, quantitative, and reliable measure of the amount of transmitter released, with a time resolution in the millisecond scale. In the experiments described below, this measure is provided by the membrane potential of the postsynaptic cell. Once again, the vertebrate neuromuscular junction, where ACh is known to be the transmitter, offers many advantages. However, to obtain more complete information about the release process, it is useful to be able to record from the presynaptic endings as well; for example, such recordings are needed to establish how membrane potential affects transmitter release. The presynaptic terminals at the vertebrate neuromuscular junction are usually too small to be impaled by a microelectrode (but see Morita and Barrett[68]); however, this can be done at a number of other synapses, such as the giant fiber synapse in the stellate ganglion of the squid,[69] synapses in the avian ciliary ganglion,[4] and between large axons and interneurons in the spinal cord of the lamprey.[70,71]

The stellate ganglion of the squid was used by Katz and Miledi to

Bernard Katz, 1950

[65]Mayer, M. L., Westbrook, G. L. and Guthrie, P. B. 1984. *Nature* 309: 261–263.
[66]Mayer, M. L. and Westbrook, G. L. 1987. *Prog. Neurobiol.* 28: 198–276.
[67]Johnson, J. W. and Ascher, P. 1984. *Nature* 325: 529–531.
[68]Morita, K, and Barrett, E. F. 1990. *J. Neurosci.* 10: 2614–2625.
[69]Bullock, T. H. and Hagiwara, S. 1957. *J. Gen. Physiol.* 40: 565–577.
[70]Ringham, G. 1975. *J. Physiol.* 251: 385–407.
[71]Martin, A. R. and Ringham, G. L. 1975. *J. Physiol.* 251: 409–426.

determine the precise relation between the membrane potential of the presynaptic terminal and the amount of transmitter release.[72] The preparation and the arrangement for recording from the presynaptic terminal and the postsynaptic fiber simultaneously are shown in Figure 13A. When tetrodotoxin (TTX) was applied to the preparation, the presynaptic action potential failed gradually over the next 15 minutes (Figure 13B). As it did so, the postsynaptic action potential failed as well, leaving an excitatory postsynaptic potential. Further reduction in the amplitude of the presynaptic action potential led to a continued reduction of the postsynaptic potential until the latter finally disappeared. When the amplitude of the postsynaptic potential is plotted against the amplitude of the failing presynaptic impulse, as in Figure 13C, the synaptic potential decreases rapidly as the presynaptic action potential amplitude falls below about 75 mV, and at amplitudes less than about 45 mV there are no postsynaptic responses. The fall in postsynaptic potential amplitude indicates a reduction in the amount of transmitter released from the presynaptic terminal. These experimental results suggest that there is a threshold for transmitter release at about 45 mV depolarization, after which the amount released, and hence the epsp amplitude, increases rapidly with presynaptic action potential amplitude.

Katz and Miledi used an additional procedure to explore the relation further: They placed a second electrode in the presynaptic terminal, through which they applied brief (1–2 msec) depolarizing current pulses, thus mimicking a presynaptic action potential. The relation between the amplitude of the artificial action potential and that of the synaptic potential was the same as the relation obtained with the failing action potential immediately after TTX poisoning. This result indicates that the normal sequence of permeability changes to sodium and potassium responsible for the action potential is not necessary for transmitter release, and that depolarization alone is a sufficient trigger.

Synaptic delay   One characteristic of the transmitter release process evident in Figure 13B is the delay between onset of the presynaptic action potential and that of the synaptic potential. In Katz and Miledi's experiments, which

[72]Katz, B. and Miledi, R. 1967. *J. Physiol.* 192: 407–436.

13   PRESYNAPTIC IMPULSE AND POSTSYNAPTIC RESPONSE at a squid giant ▶ synapse. **(A)** Sketch of the stellate ganglion of the squid, illustrating the two large axons that form a chemical synapse. Both axons can be impaled with microelectrodes as shown. **(B)** Simultaneous recordings from the presynaptic (dark blue records) and postsynaptic (lighter blue records) axons during the development of conduction block by TTX. Postsynaptic action potential fails after the second record, and the postsynaptic potential decreases in size as block of the presynaptic action potential progresses. **(C)** The relation between the amplitude of the presynaptic action potential and the postsynaptic potential. Closed circles are from results in **(B)**; open and half-filled circles are results obtained after complete TTX block by applying depolarizing current pulses to the presynaptic terminals. (A after Bullock and Hagiwara, 1957; B and C after Katz and Miledi, 1967c.)

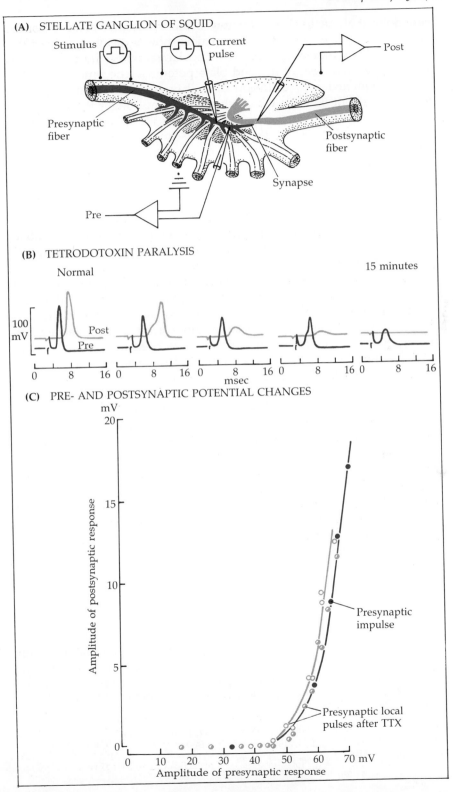

(A) STELLATE GANGLION OF SQUID

Stimulus

Current pulse

Post

Presynaptic fiber

Postsynaptic fiber

Synapse

Pre

(B) TETRODOTOXIN PARALYSIS

Normal

15 minutes

100 mV

Post

Pre

0   8   16   0   8   16   0   8   16   0   8   16   0   8   16

msec

(C) PRE- AND POSTSYNAPTIC POTENTIAL CHANGES

mV

20

15

10

5

0

Amplitude of postsynaptic response

Presynaptic impulse

Presynaptic local pulses after TTX

0   10   20   30   40   50   60   70 mV

Amplitude of presynaptic response

were done at about 10°C, the delay was 3 to 4 msec. Detailed measurements at the frog neuromuscular junction show a delay of at least 0.5 msec at room temperature between depolarization of the presynaptic terminal and the onset of the end plate potential. The time is too long to be accounted for by diffusion of ACh across the synaptic cleft (a distance of about 50 nm), which should take no longer than about 50 μsec. When ACh is applied to the junction ionophoretically from a micropipette, delays of as little as 150 μsec can be achieved, though the pipette, even with the most careful placement, can never contact the postsynaptic membrane as intimately as does the nerve terminal. Furthermore, synaptic delay is much more sensitive to temperature than would be expected if it were due to diffusion. Cooling the frog nerve–muscle preparation to 2°C increases the delay to as long as 7 msec, whereas the delay in the response to ionophoretically applied ACh is not perceptibly altered.[73] Thus the delay is largely in the transmitter release mechanism.

Evidence that calcium entry is required for release

Calcium has long been known as an essential link in the process of synaptic transmission. When its concentration in the extracellular fluid is decreased, release of ACh at the neuromuscular junction is reduced and eventually abolished.[74] The importance of calcium for release has been established at all synapses where it has been tested, irrespective of the nature of the transmitter. Its role has been generalized further to other secretory processes, such as liberation of hormones by cells of the pituitary gland, release of epinephrine from the adrenal medulla, and secretion by salivary glands.[75] As discussed below, evoked transmitter release is preceded by calcium entry into the terminal. Its effect in promoting release is antagonized by ions that block calcium entry, such as magnesium, manganese, and cobalt. Transmitter release can be reduced, then, either by removing calcium from the bathing solution or by adding a blocking ion.

Experiments by Katz and Miledi on the neuromuscular junction showed not only that calcium is essential for transmitter release to occur but also that it must be present at the time of depolarization of the presynaptic terminal. The experiment in which this was demonstrated is illustrated in Figure 14. Calcium was removed from the bathing solution so that release of ACh in response to nerve stimulation was virtually abolished. Calcium was then applied to the nerve terminal by ionophoresis, from a micropipette close to the terminal, just before the nerve was stimulated; application of calcium in this way restored transmitter release, producing an end plate potential with each trial. Calcium pulses alone were not effective in producing release, nor were calcium pulses applied after stimulation, during the synaptic delay period. Similar results were obtained in the presence of TTX, when the nerve

[73]Katz, B. and Miledi, R. 1965. *Proc. R. Soc. Lond. B.* 161: 483–495.
[74]del Castillo, J. and Stark, L. 1952. *J. Physiol.* 116: 507–515.
[75]Douglas, W. W. 1978. *Ciba Found. Symp.* 54: 61–90.

terminal was depolarized by current pulses. Again, calcium application was effective just before the depolarizing pulse. These experiments were taken as evidence that for release to occur calcium must be available during the period of depolarization.

Other experiments on nerve fibers have indicated that the calcium conductance g_{Ca} of the membrane is increased by depolarization (Chap-

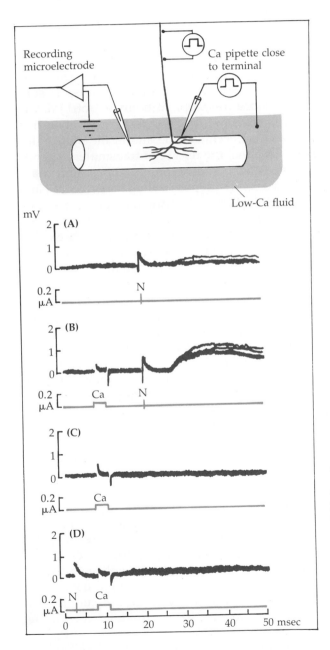

14

TIMING OF CALCIUM ACTION ON TRANS-MITTER RELEASE. (A) Repeated stimulation of the motor nerve (N) in low-calcium solution produces little or no transmitter release at the neuromuscular junction. (B) When preceded by a brief application of calcium (Ca) from a pipette close to the nerve terminal, nerve stimulation is followed by transmitter release. (C) Calcium pulse alone does not cause release. (D) Application of calcium after stimulation, but just at the beginning of the expected endplate potential, has no effect. (After Katz and Miledi, 1967b.)

ter 4) and that some calcium enters with each action potential. This idea was supported by an extension of the experiment shown in Figure 13. Katz and Miledi reasoned that if the presynaptic terminal were depolarized to E_{Ca} or beyond, then there would be no calcium entry during the pulse and hence no transmitter release. To produce large and relatively long-lasting depolarization of the squid presynaptic terminal, it was necessary to block repolarization by potassium current through the delayed rectifier channels; this was accomplished by intracellular injection of tetraethylammonium ion (TEA; see Chapter 4). The results of one such experiment are shown in Figure 15. A current pulse with a duration of about 17 msec produces a steady depolarization of the nerve terminal and a brief postsynaptic potential when the depolarization is small. At a moderate level of depolarization, the initial epsp is suppressed, and there is an additional response at the end ("off") of the pulse. With a very large depolarization, the "on" response is abolished and a large postsynaptic potential occurs at the end of the pulse. This result is consistent with the idea that the release of transmitter depends on calcium entry during depolarization. The number of calcium channels that are opened increases with depolarization, but the calcium *current* is suppressed during the pulse, when the membrane potential approaches E_{Ca}. When the pulse is terminated, the calcium channels are still open and calcium can then enter the terminal down its electrochemical gradient to trigger the transmitter release process.

Ricardo Miledi

The experiment illustrated in Figure 15, in addition to providing evidence about the role of calcium in the release process, also gives us a clue about the nature of the synaptic delay. At low and moderate depolarizations, the postsynaptic potential at the beginning of the pulse appears with a delay of about 3.5 msec. At the end of the pulse, however, the "off" response occurs with a delay of less than 1.0 msec. This difference presumably arises because the "on" response cannot occur until a sufficient number of calcium channels have opened, whereas at the termination of the pulse the "off" response can occur with little delay because the calcium channels are already open and calcium can enter immediately upon repolarization. These observations suggest that a significant fraction of the normal synaptic delay is the time taken for opening of the voltage-activated calcium channels.

Calcium entry into the presynaptic terminal was demonstrated directly by Llinás and his colleagues, using the luminescent dye aequorin (Chapter 3). In addition, they measured the magnitude and time course of the calcium current produced by presynaptic depolarization, using voltage clamp techniques.[76] An example is shown in Figure 16A. The sodium and potassium conductances associated with the action potential were blocked by TTX and TEA so that only the voltage-activated calcium channels remained. A presynaptic depolarizing pulse from -70 to -18 mV (upper record) produced an inward calcium current that increased slowly in magnitude to about -400 nA (middle record). The

[76]Llinás, R. 1982. *Sci. Am.* 247: 56–65.

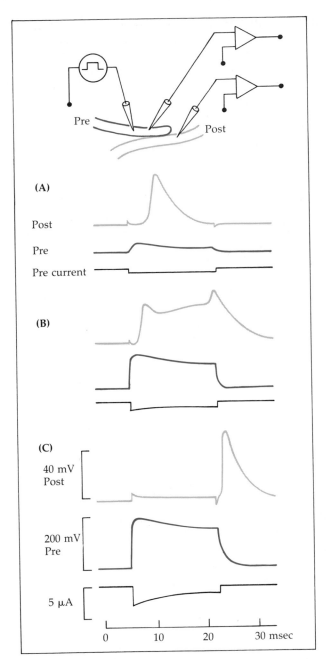

(A)

Post

Pre

Pre current

(B)

(C)

40 mV
Post

200 mV
Pre

5 µA

0   10   20   30 msec

15

EFFECT OF PROLONGED DEPOLARIZATION of the presynaptic terminal of the squid giant synapse after treatment with TTX and TEA. **(A)** Moderate presynaptic depolarizing pulse (pre) produces postsynaptic potential (post) with a delay of about 3.5 msec. **(B)** Increased depolarization reduces the amplitude of the initial postsynaptic potential slightly, and postsynaptic depolarization due to continued presynaptic transmitter release is maintained throughout the pulse. At the termination of the pulse there is an additional postsynaptic "off" response. **(C)** During a very large presynaptic pulse, transmitter release is suppressed completely, leaving only the "off" response. Suppression occurs because during the pulse the presynaptic membrane potential is near the calcium equilibrium potential and, as a consequence, there is no calcium entry until repolarization occurs. Note that in (C) the delay between repolarization and the "off" response is considerably less than that in (A) between depolarization and the "on" response. (After Katz and Miledi, 1967c.)

onset of calcium entry was followed, after a delay of more than a millisecond, by a large synaptic potential in the postsynaptic cell (lower record). When the depolarizing pulse was increased to +50 mv (Figure 16B), the calcium current was suppressed during the pulse. On repolarization, however, there was an immediate calcium current, accompanied by a postsynaptic potential with a delay of less than 0.2 msec.

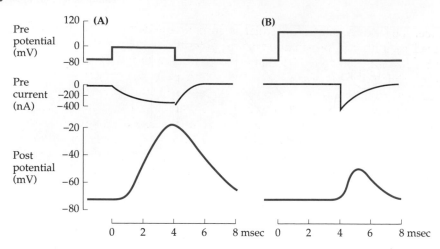

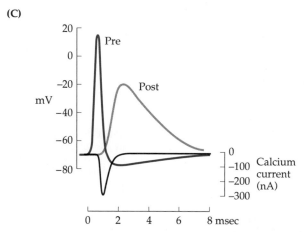

16   PRESYNAPTIC CALCIUM CURRENTS AND TRANSMITTER RELEASE at the squid giant synapse. The presynaptic terminal is voltage clamped and treated with TTX and TEA to abolish voltage-activated sodium and potassium currents. Records show potential applied to the presynaptic fiber, presynaptic calcium current, and epsp in the postsynaptic fiber. (A) Voltage pulse from −70 to −18 mV (upper trace) results in a slow inward calcium current (middle trace) and, after a delay of about 1 msec, an epsp (lower trace). (B) Larger depolarization to +50 mV suppresses calcium entry. At termination of the pulse, a surge of calcium current is followed within about 0.2 msec by an epsp. (C) A voltage waveform identical in shape to a normal action potential (pre) produces an epsp (post) indistinguishable from that seen normally. The black curve gives the magnitude and time course of the associated calcium current. (After Llinás, 1982.)

This experiment shows directly that much of the synaptic delay is associated with the time taken for activation of the calcium conductance. The effect of an artificial action potential is shown in Figure 16C. A presynaptic action potential, recorded before the addition of TTX and TEA to the preparation, is "played back" through the voltage clamp circuit to produce exactly the same voltage change in the terminal, but now without the normal changes in sodium and potassium conductance. The postsynaptic potential is indistinguishable from that produced by a normal presynaptic action potential, showing that only the voltage change, and not the changes in sodium and potassium conductances that normally accompany the action potential, is necessary for normal transmitter release. The voltage clamp technique also enabled Llinás and his colleagues to deduce the magnitude and time course of the calcium current produced by the artificial action potential (black curve). The current began near the peak of the action potential, rose rapidly to a maximum during the repolarizing phase of the action potential, and then declined over the next 1.0 msec.

Calcium currents related to synaptic transmission have also been observed at the neuromuscular junction of the mouse. Using extracellular recording with precisely placed microelectrodes, combined with discrete application of TTX and TEA to confined regions of the nerve terminal, Brigant and Mallart were able to deduce the sequence of permeability changes associated with nerve terminal depolarization.[77] Their experiments show that as the impulse arrives, there is inward sodium current restricted to the preterminal membrane (where the myelin ends), followed by inward calcium current throughout the terminal and repolarization by outward potassium current. Thus, the actual terminal appears to have few voltage-sensitive sodium channels and is specialized to promote calcium entry. On the other hand, the frog motor nerve terminal, which is much longer, has a propagating sodium action potential along most of its length.[78] More recently, calcium currents have been recorded intracellularly in presynaptic terminals in ciliary ganglia of the lizard[79] and chick.[80]

Experiments on the squid giant synapse have provided additional information about the role of calcium in release and, in particular, about the proximity of calcium channels to the sites of transmitter release.[81] In these experiments, injection of BAPTA into the presynaptic terminal resulted in a severe attenuation of transmitter release, without affecting the presynaptic action potential. On the other hand, EGTA, an equally potent calcium buffer, had little effect on release. This disparity is due to the fact that the binding of calcium to BAPTA has much faster kinetics than that of calcium to EGTA. The observation that release occurs before calcium chelation by EGTA can reach equilibrium indicates that calcium reaches its functional binding sites within 200 μsec after its influx into the terminal. This time course requires, in turn, that the calcium entry must occur within 100 nm of the calcium binding sites associated with the release process.

The close association that is required between calcium channels and release sites suggests that the channels might be represented morpho-

Localization of calcium entry sites

[77]Brigant, J. L. and Mallart, A. 1982. *J. Physiol.* 333: 619–636.
[78]Katz, B. and Miledi, R. 1968. *J. Physiol.* 199: 729–741.
[79]Martin, A. R. et al. 1989. *Neurosci. Lett.* 105: 14–18.
[80]Yawo, H. 1990. *J. Physiol.* 428: 191–213.
[81]Adler, E. M. et al. 1991. *J. Neurosci.* 11: 1496–1507.

logically by the rows of 10-nm particles seen between lines of synaptic vesicles in freeze-fracture images of presynaptic terminals (Figure 3). Results obtained with toxin-binding studies at the neuromuscular junction of the frog are consistent with this idea.[82,83] Ω-Conotoxin, which blocks neuromuscular transmission irreversibly by binding to presynaptic calcium channels, was conjugated with a fluorescent stain. Upon microscopic examination, the fluorescence was found to be concentrated in narrow bands at 1-μm intervals, the same spacing as that of the active zones in the terminal. Combined staining of postjunctional ACh receptors with fluorescent α-bungarotoxin showed that the presynaptic bands were in spatial register with the postjunctional folds.

Does depolarization play a direct role in release?

The evidence we have presented so far indicates that depolarization of the presynaptic terminal by an incoming action potential serves simply to open the voltage-activated calcium channels, and that the depolarization itself has no other role in evoking transmitter release. However, Parnas and his colleagues have suggested that depolarization itself may have a direct effect on evoked release at the crayfish neuromuscular junction.[84] The strategy behind the experiments was to establish a constant elevated calcium concentration in the presynaptic terminal and then test whether depolarization affected release. The calcium concentration was raised either by chemical interference with cytoplasmic calcium buffering or by a photochemical method involving the instantaneous release into the cytoplasm of previously injected "caged" calcium. At the same time calcium influx was inhibited by removing calcium from the extracellular solution and adding magnesium and manganese. Under those conditions, nerve stimulation caused an increase in transmitter release from the terminal. It was concluded that under the condition of increased intracellular calcium, depolarization by the incoming action potential was itself sufficient to evoke release without any accompanying entry of additional calcium. Other experiments of this nature have yielded conflicting results.[85]

Quantal release

So far the general scheme can be summarized as presynaptic depolarization → calcium entry → transmitter release. Once this general framework has been established, it remains to be shown how the transmitter is secreted from the terminals. In experiments on the frog neuromuscular junction, Fatt and Katz showed that ACh is released from the terminals in MULTIMOLECULAR PACKETS, which they called QUANTA.[86] A packet, or quantum, simply describes the smallest unit in which the transmitter is normally secreted and is now known to represent the contents of one synaptic vesicle. Quantal release means that only the contents of 0, 1, 2, 3, and so on, vesicles are released, but not the contents of 1½ or 2⅝. In general, the number of quanta released from

[82]Robitaille, R., Adler, E. M. and Charlton, M. P. 1990. *Neuron* 5: 773–779.

[83]Cohen, M. W., Jones, O. T. and Angelides, K. J. 1991. *J. Neurosci.* 11: 1032–1038.

[84]Hochner, B., Parnas, H. and Parnas, I. 1989. *Nature* 342: 433–435.

[85]Mulkey, R. M. and Zucker, R. S. 1991. *Nature* 350: 153–155.

[86]Fatt, P. and Katz, B. 1952. *J. Physiol.* 117: 109–128.

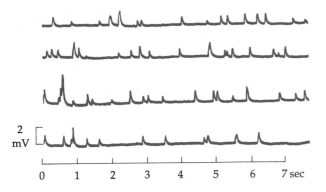

17

MINIATURE SYNAPTIC POTENTIALS occurring spontaneously at the frog neuromuscular junction. The potentials, which are due to spontaneous release of single quanta of ACh, are about 1 mV in amplitude and are confined to the end plate region of the muscle fiber. (From Fatt and Katz, 1952.)

the nerve terminal (the QUANTUM CONTENT of a synaptic response) may vary considerably, but the number of molecules in each quantum (QUANTUM SIZE) is relatively fixed (with a variance of about 10 percent).

The first evidence for packaging of ACh in multimolecular quanta was the observation by Fatt and Katz that at the motor end plate, but not elsewhere in the muscle fiber, spontaneous depolarizations of about 1 mV occurred irregularly (Figure 17). They had the same time course as the potentials evoked by nerve stimulation. The spontaneous miniature potentials were decreased in amplitude and eventually abolished by increasing concentrations of curare in the bathing solution, and they were increased in amplitude and time course by acetylcholinesterase inhibitors such as prostigmine. These two pharmacological tests suggested that the potentials were produced by the spontaneous release of discrete amounts of ACh from the nerve terminal. The possibility that they might be due to the release of single molecules of ACh was ruled out by several lines of reasoning. (1) Curare and prostigmine had graded effects on miniature potential amplitude; if the events were due to the action of single ACh molecules, their amplitude should have been fixed and the drugs would have been expected to affect their frequency of occurrence. Prostigmine, by inhibiting hydrolysis of ACh and thereby increasing its concentration in the synaptic cleft, should have increased miniature potential frequency. Curare, by occupying postsynaptic receptors and thereby decreasing the number available for interaction with ACh, should have decreased the frequency. (2) Addition of ACh to the solution bathing the muscle caused a graded depolarization of the end plate region without any increase in miniature potential frequency. Again, if the spontaneous miniature potentials were caused by single molecules, then addition of ACh to the bathing solution should have increased the number of molecule collisions with the postsynaptic membrane and hence the miniature potential frequency. Moreover, with small applications of ACh, depolarizations smaller in amplitude than a miniature potential could be produced.

Fatt and Katz made an additional important observation on the behavior of the stimulus-evoked end plate potential after synaptic transmission had been reduced by lowering the extracellular calcium concentration or adding extracellular magnesium. They found, unexpectedly, that at very low calcium levels the responses to stimulation fluctuated in a stepwise manner, as shown in Figure 18. Some stimuli produced no response at all, some a response of about 1 mV in amplitude, identical in size and shape to a spontaneous miniature potential; some responses were twice the unit amplitude, and some three times as large. This remarkable observation led Fatt and Katz to propose that the single quantal events observed to occur spontaneously also represented the building blocks for the synaptic potentials evoked by stimulation, and that the effect of a reduced extracellular calcium concentration was to reduce the quantum content of the potentials in a graded fashion, without reducing the quantal size. When the extracellular calcium concentration was sufficiently low, only a few quanta were released and stepwise fluctuations in the release process could then be observed.

Subsequent evidence confirmed in a variety of different ways that the spontaneous miniature potentials are indeed due to ACh liberated by the nerve terminal. For example, depolarization of the nerve terminal by passing a steady current through it causes an increase in frequency of the spontaneous activity,[87] whereas muscle depolarization has no effect on frequency. Botulinum toxin, which blocks release of ACh in response to nerve stimuli, also abolishes the spontaneous activity.[88] Shortly after denervation of the muscle and degeneration of its motor nerve, the miniature potentials disappear.[89] After an interim period, spontaneous potentials reappear in denervated frog muscle, but convincing evidence has been obtained that these arise because of ACh release from Schwann cells that have engulfed segments of the degenerating nerve terminals by phagocytosis.

In summary, release of ACh from motor nerve terminals at the neuromuscular junction occurs in multimolecular packets, or quanta. Quantal release occurs spontaneously at a low rate, and a large number of quanta are released in response to nerve stimulation, depending on the extracellular calcium concentration. Normally the end plate potential is made up of about 200 quantal units; in low calcium concentrations, when the number is very small, the quantum content fluctuates from trial to trial. This fluctuation in release results in stepwise fluctuations in the amplitude of the potential, as shown in Figure 18.

Statistical fluctuations of the end plate potential

Two questions arose with respect to quantal release. (1) Do the fluctuations really occur in quantal steps? (2) Why do the fluctuations occur? These questions were addressed by del Castillo and Katz, who studied the nature of the fluctuations in some detail.[90] Their experiments consisted of reducing transmitter release by reducing the extracellular calcium concentration or by adding magnesium to the bathing

[87]del Castillo, J. and Katz, B. 1954. *J. Physiol.* 124: 586–604.
[88]Brooks, V. B. 1956. *J. Physiol.* 134: 264–277.
[89]Birks, R., Katz, B. and Miledi, R. 1960. *J. Physiol.* 150: 145–168.
[90]del Castillo, J. and Katz, B. 1954. *J. Physiol.* 124: 560–573.

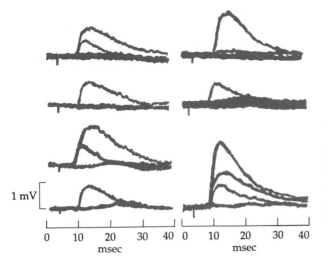

1 mV

| | | | | | | | | | |
| 0 | 10 | 20 | 30 | 40 | 0 | 10 | 20 | 30 | 40 |

msec   msec

18

FLUCTUATIONS IN SYNAPTIC RESPONSE at a neuromuscular junction. Presynaptic release of ACh was reduced by reducing the extracellular calcium concentration in the bathing solution. Each set of traces shows two to four superimposed responses to nerve stimulation. Stepwise fluctuations in amplitude are due to variations in the number of quanta of ACh released on successive trials. (From Fatt and Katz, 1952.)

solution and then recording a large number of end plate potentials evoked by nerve stimulation, as well as a large number of spontaneous miniature potentials. They plotted histograms of the amplitude distributions, similar to that shown in Figure 19. First, the spontaneous miniature potentials (inset) were not all the same size; they were distributed around their mean amplitude with a variance of about 10 percent of the mean. Because of this variance, evoked end plate potentials did not occur in discrete steps (main histogram), but pronounced peaks occurred in their amplitude distribution, occurring at precisely one, two, three, and four times the mean amplitude of the spontaneous potentials. Further, the variance of the first peak is the same as that of the spontaneous potentials, and the multiple responses have correspondingly increasing variances. This kind of experiment was later done on mammalian muscle,[91,92] autonomic ganglion cells,[93,94] spinal motoneurons,[95] and a variety of other synapses.[96] The result was always the same: Evoked release occurred in multiples of the quantal unit. The answer to the first question, then, was that release is indeed quantized.

To answer the second question, del Castillo and Katz proposed that the fluctuations were statistical, and formulated what has come to be known as the QUANTUM HYPOTHESIS:[90]

Suppose we have, at each nerve–muscle junction, a population of n units capable of responding to a nerve impulse. Suppose, further, that the average probability of responding is p . . . , then the mean number of units responding to one impulse is $m = np$.

[91]Boyd, I. A. and Martin, A. R. 1956. *J. Physiol.* 132: 74–91.
[92]Liley, A. W. 1956. *J. Physiol.* 132: 650–666.
[93]Blackman, J. G., Ginsborg, B. L. and Ray, C. 1963. *J. Physiol.* 167: 355–373.
[94]Martin, A. R. and Pilar, G. 1964. *J. Physiol.* 175: 1–16.
[95]Kuno, M. 1964. *J. Physiol.* 175: 81–99.
[96]McLachlan, E. M. 1978. *Int. Rev. Physiol. Neurophysiol. III* 17: 49–117.

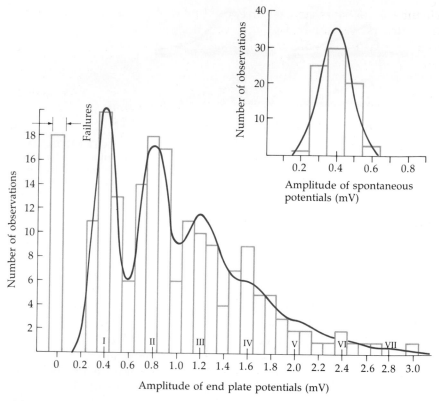

19 **AMPLITUDE DISTRIBUTION** of end plate potentials at a mammalian neuro-muscular junction in high **(12.5 m***M***)** magnesium solution. The histogram shows the number of end plate potentials observed at each amplitude. The peaks of the histogram occur at one, two, three, and four times the mean amplitude of the sponta-neous miniature end plate potentials (inset), indicating releases of one, two, three, and four quanta. The solid line represents the theoretical distribution of end plate potential amplitudes calculated according to the Poisson equation and allowing for the spread in amplitude of the quantal size (see text). Arrows indicate the predicted number of failures of the response. (From Boyd and Martin, 1956.)

What they proposed, then, was that release of individual quanta from the nerve terminal is similar to shaking marbles out of a box. If such a box has a small hole in the top and contains a large number *n* of marbles, and if each time the box is turned over and shaken each marble has a small probability *p* of falling out through the hole, then, on successive tries, sometimes one marble will fall out, other times two, other times none, and so on. (The marbles must be replaced each time so that *n* remains constant.) In a large number of trials, the mean number falling out per trial, *m*, will be given by *np*.

The hypothesis proposed by del Castillo and Katz not only explains why there are fluctuations around the mean but also provides a way of predicting how the fluctuations are distributed (i.e., how many failures, singles, doubles, and so forth to expect in a series of trials). Under such circumstances the relative occurrence of multiple events is predicted by the BINOMIAL DISTRIBUTION. (Note that the sequence in which the various responses occur is not predictable, only their number.) In N trials the expected total number of quanta released would be Nm, m being the mean number per trial. Of this total, how many failures, single releases, etc. can we expect? If we specify the number of responses containing x quanta by n_x, where x takes on the values 0, 1, 2, . . .,n, then according to the binomial distribution, n_x is given by

$$n_x = N \left(\frac{n!}{(n - x)!x!} \right) p^x q^{n-x}$$

where $q = 1 - p$. For example, if $n = 30$ and $p = 0.01$, then $m = np$ would be 0.3 and the expected number of double releases (n_2) in 1000 trials would be

$$n_2 = (1000) \left(\frac{30!}{(28!)(2!)} \right) (0.01)^2(0.99)^{28} = 32$$

In summary, the occurrence of peaks at regular intervals in the distribution tells us that release is quantized; the heights of the peaks relative to one another are explained by the quantum hypothesis.

Use of the binomial distribution requires, however, that one know the value of n (the number of marbles in the box) and p, although the only parameter that can be observed directly is m, their product. In order to deal with this difficulty, del Castillo and Katz went on:

> Under normal conditions p may be assumed to be relatively large, that is a fairly large part of the synaptic population responds to an impulse. However, as we reduce the Ca and increase the Mg concentration, the chances of responding are diminished and we observe mostly complete failures with an occasional response of one or two units. Under these conditions, when p is very small, the number of units making up the epp in a large series of observations should be distributed in the characteristic manner described by Poisson's law.

In their experiments, then, del Castillo and Katz tested the applicability of the POISSON DISTRIBUTION to the observed fluctuations in end plate potential amplitude. In the Poisson equation, there is no n or p, only m. Expected numbers of responses containing x quanta are given by

$$n_x = N \left(\frac{m^x}{x!} \right) e^{-m}$$

If, as in the previous numerical example, $m = 0.3$, then the Poisson distribution predicts that in 1000 trials $n_2 = 33$, similar to the binomial prediction.

The only requirement for the Poisson distribution to be applicable is that the quanta be released independently so that the release of one has no influence on the probability of release of the next. In general, the distribution is used to predict the frequency of occurrence of discrete events in a continuum, for example, the distribution in time of airplane crashes. One of the best-known applications is an analysis of the number of Prussian cavalry officers hurt each year by horse kicks. Some years were "failures"—none was hurt; in other years, one or two were hurt. Over a long period the distribution was described closely by the Poisson equation, using only the mean number of kicks per year (m) to describe the theoretically expected distribution. Another convenient analogy, in which the Poisson distribution is used as an approximation of the binomial distribution, is a dime slot machine that contains a very large number of dimes but, unlike ordinary slot machines, pays out at random rather than when symbols match on a set of wheels. If the owner of the machine is to make a profit, the machine must pay out with some reluctance (i.e., m must be less than 1). Most of the time the player, on pulling the handle, will suffer a failure, but sometimes he or she will receive one, two, or (in rare cases) several dimes in return. If the mean number of dimes paid out per play (m) is known, then during a long period of play it is possible to predict the number of times there is no payoff, the number of times the player receives one dime in return, and so on. Two points are to be noted about the Poisson distribution: (1) n and p have no relevance; the characteristics of the distribution depend only on m; and (2) there is no restriction on m; in the unlikely event that the machine paid out an average of 100 dimes per play, the distribution would still be applicable (provided the content of the machine was sufficiently large). In that case, however, the frequency of failures would be vanishingly small, as would the profits of the owner.

To test the applicability of the Poisson distribution, then, it is necessary only to have a measure of m, the mean number of units released per trial. For the slot machine, this could be obtained by calculating the average amount of money paid out per trial and dividing by the size of the unit (10 cents). At the neuromuscular junction the size of the unit (v_1) is provided by the average size of the spontaneous miniature potential. Thus the mean quantum content of a series of responses (m) can be obtained by dividing the mean amplitude of the evoked potentials (v) by the mean unit amplitude:

$$m = v/v_1$$

A second method is to determine m from the number of failures. When $x = 0$ in the Poisson equation, $n_0 = Ne^{-m}$. It follows, then, that

$$m = \ln(N/n_0)$$

The agreement between m estimated in these two ways is excellent. Put the other way around, if m is calculated by the first method, the value so obtained predicts accurately the observed number of failures.

A more stringent test of the applicability of the Poisson equation is to predict the entire distribution of the response amplitudes using only m and the mean amplitude of the unit potential—that is, to predict the height of the individual peaks in Figure 19 as well as their position. To do this, *m* is calculated from the ratio of the mean response amplitude to the mean spontaneous potential amplitude, and the number of expected failures is calculated (arrows at zero amplitude). The expected number of single responses is then calculated, and these are distributed about the mean unit size with the same variance as the spontaneous events (inset). Similarly, the predicted multiple responses are distributed around their means with proportionately increasing variances. The individual distributions are then summed to give the theoretical distribution shown by the continuous curve. The agreement with the experimentally observed distribution (bars) is remarkably good.

The agreement between the overall amplitude distribution and that predicted by Poisson statistics provided strong evidence that the quanta were released independently. Such experiments, however, provided incomplete information about the validity of the quantum hypothesis; there was no information about the existence of the separate parameters *n* and *p*. Although adherence of the release process to Poisson statistics was demonstrated at a large variety of vertebrate and invertebrate preparations, it was several years before the applicability of the binomial distribution was confirmed, first at the crayfish neuromuscular junction[97] and then in a number of other preparations. In summary, there is now ample evidence that transmitter is released in packets, or quanta, and that the fluctuations in release from trial to trial can be accounted for by binomial statistics, as predicted by del Castillo and Katz. When the release probability *p* is small, as in a low-calcium medium, the Poisson distribution provides an equally good description of the fluctuations.

The quantal nature of the release process bears directly on the mechanisms that operate during integration in the nervous system because the quanta provide the units that make up synaptic signals. One striking feature of the vertebrate nervous system is the reduction in mean quantum content of the response as one moves from the neuromuscular junction where the safety factor for transmission is high (*m* in the range of 100–300), to synapses on cells in the central nervous system, such as the spinal motoneuron. The motoneuron is concerned mainly with integrating a myriad of incoming information, and no one synapse has a dominating influence on the cell. A primary afferent fiber from a muscle spindle, for example, operates with a mean quantum content of only about one quantum per incoming presynaptic action potential.[98] The low mean quantum content does not mean, however, that transmission fails most of the time, as would be expected for a Poisson distribution: The quantal release is binomial with a large release probability (often up to 0.9 or so). Inhibitory inputs to the motoneuron[99] and to reticulospinal cells in the brainstem of the lamprey[100] have cor-

General significance of quantal release

[97]Johnson, E. W. and Wernig, A. 1971. *J. Physiol.* 218: 757–767.
[98]Kuno, M. 1964. *J. Physiol.* 175: 81–99.
[99]Kuno, M. and Weakly, J. N. 1972. *J. Physiol.* 224: 287–303.
[100]Gold, M. R. and Martin, A. R. 1983. *J. Physiol.* 342: 85–98.

respondingly low quantum contents and high release probabilities. Autonomic ganglion cells occupy an intermediate position, with quantum contents of 1 to 3 in mammalian sympathetic ganglia[101] and about 20 in the ciliary ganglion of the chicken.[102]

Number of molecules
in a quantum

There are a number of methods for estimating the number of molecules in one quantum of acetylcholine. The most accurate of these was devised by Kuffler and Yoshikami, who used very fine pipettes for ionophoresis of ACh onto the postsynaptic membrane of snake muscle.[103] By careful placement of the pipette, they were able to produce a response to a brief pulse of ACh that mimicked almost exactly the spontaneous miniature end plate potential. An example of such a response is shown in Figure 6. The amplitude of the response to ACh application was the same as that of the miniature synaptic potential, and the time course was only slightly longer.

Having mimicked the synaptic response by ionophoresis, Kuffler and Yoshikami then set out to measure the number of molecules released from the pipette. One method they tried was to fill the pipette with radioactive ACh, place the pipette tip in a small volume of fluid, and apply a large number of pulses. In theory, the amount of ACh released per pulse could then be determined by measuring the total amount of radioactivity released into the fluid. This method was not satisfactory for a number of technical reasons, and a second method was adopted. The ACh was released by repetitive pulses into a small (about 0.5 μl) droplet of saline under oil. The droplet was then applied to the end plate of a snake muscle fiber (Figure 20) and the resulting depolarization measured. The response was compared with responses to droplets of exactly the same size containing known concentrations of ACh. In this way the concentration of ACh in the test droplet was determined and the number of ACh molecules released per pulse was calculated. A pulse of ACh identical to that required to mimic the miniature synaptic potential contained fewer than 10,000 molecules. This represents an upper limit to the number of molecules in a quantum, as the ionophoretic pipette is in a somewhat less favorable position with respect to the subsynaptic membrane than is the nerve terminal itself and therefore must release more ACh to achieve the same effect.

Number of channels
activated by a
quantum

Given that a quantum of ACh consists of 10,000 or fewer molecules, one might expect that only a few thousand of these would actually combine with postsynaptic receptors at the neuromuscular junction, those remaining being lost because of diffusion out of the cleft or hydrolysis by cholinesterase. This expectation is correct. The conductance change associated with a miniature potential can be deduced directly by voltage clamping the end plate, measuring the amplitude of the miniature end plate current.[104] At the peak of the end plate current

[101]Blackman, J. G. and Purves, R. D. 1969. *J. Physiol.* 203: 173–198.
[102]Martin, A. R. and Pilar, G. 1964. *J. Physiol.* 175: 1–16.
[103]Kuffler, S. W. and Yoshikami, D. 1975. *J. Physiol.* 251: 465–482.
[104]Gage, P. W. and Armstrong, C. M. 1968. *Nature* 218: 363–365.

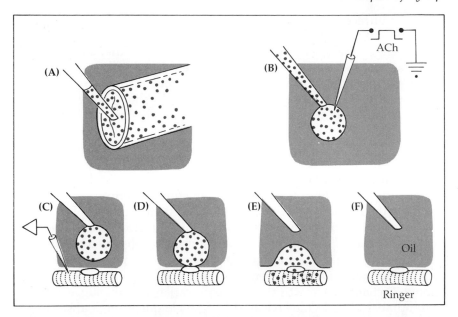

20   ASSAY OF ACh ejected from a micropipette by ionophoresis. (A) A droplet of fluid (stippled) is removed by a transfer pipette from the dispensing capillary under oil (shaded). (B) ACh is then injected into the droplet by a series of ionophoretic pulses. (C–F) After its volume has been measured, the ACh-loaded droplet is touched against the oil–Ringer interface at the end plate of a snake muscle and discharges its contents into the aqueous phase. The depolarization of the end plate is measured (not shown) and compared with that produced by droplets with known ACh concentration. Once the concentration in the test droplet is known, the amount of ACh released per pulse from the electrode can be calculated. (After Kuffler and Yoshikami, 1975b.)

the conductance change is on the order of 40 to 50 nS. As the ACh receptor has an open conductance of about 30 pS, this corresponds to about 1500 channels. This is similar to the number calculated by Katz and Miledi, who estimated the contribution of a single channel to the end plate potential from noise measurements.[105] A similar value was also obtained at glycine-mediated inhibitory synapses in lamprey brain-stem cells,[100] but considerably lower values are observed at other synapses. For example, at inhibitory synapses in hippocampal cells, a quantal response corresponds to activation of about 20 channels.[106]

Why is there such a difference between one cell and another in the number of channels activated by a quantum of transmitter? A little thought leads to the conclusion that such differences are necessary for the nervous system to function properly: The number of postsynaptic receptors activated by a quantum of transmitter released from a single presynaptic bouton must be tailored to the size of the cell. In very large cells with low input resistances, such as lamprey Müller cells, a large

[105]Katz, B. and Miledi, R. 1972. *J. Physiol.* 244: 665–699.
[106]Sakmann, B. et al. 1989. *Q. J. Exp. Physiol.* 74: 1107–1118.

number of receptors must be activated for the effect of a quantum to be significant. Activation of the same number of receptors on a very small cell, on the other hand, would overwhelm all other conductances, depolarizing the cell to a potential near zero if the synapse were excitatory, or locking its membrane potential firmly at the chloride equilibrium potential if the effect were inhibitory. In short, large cells must have an adequate number of a receptors under a bouton, small cells must not have too many.

Are the sizes of the quanta released from presynaptic boutons matched to postsynaptic cell size as well? Observations on fluctuations of miniature synaptic potentials and currents suggest they are not. Miniature end plate potentials, and miniature inhibitory synaptic currents in the lamprey cells, fluctuate around their mean amplitude with a variance of about 10 percent (Figure 19), presumably due to variations in quantal size. The fact that fluctuations are observed means that activation by one of the smaller quanta leaves some postsynaptic receptors unoccupied. Quantal events in the hippocampal cells, where only about 20 channels are activated, have no variance at all, suggesting that the number of molecules released in a single quantum is always more than sufficient to activate all the available receptors, that is, the reduction in receptor number has not been accompanied by a concomitant reduction in size of the quantal package.

Changes in mean quantal size at the neuromuscular junction

Throughout this discussion it has been implied that at any given synapse the mean quantal size is constant. This is not always true. For example, during regeneration of motor nerve terminals, the amplitudes of spontaneous miniature potentials, rather than being distributed normally (Figure 19), are skewed into the baseline noise; that is, there are large numbers of very small spontaneous potentials. The small quantal units, however, do not appear to be released in response to stimulation.[107] Small miniature potentials are also seen in normal frog and mouse muscle. It has been suggested that these might represent subunits of the usual miniature potential,[108] on the grounds that amplitude distribution of the large miniature potentials appears to be fragmented into multiples of the small miniature amplitude. The significance of the small miniature potentials is not clear. Conversely, spontaneous synaptic potentials larger than the usual miniature potentials are seen occasionally. In some preparations these appear to be due to the spontaneous release of two or more quanta simultaneously; in others their size shows no clear relation to normal quantal amplitude.[109]

Nonquantal release

In addition to leaving the motor nerve terminal in the form of individual quanta, ACh also leaks continuously from the cytoplasm into the extracellular fluid. In other words, there is a steady nonquantal "ooze" of ACh from the presynaptic terminal. This was first detected

[107]Dennis, M. J. and Miledi, R. 1974. *J. Physiol.* 239: 571–594.
[108]Kriebel, M. E., Vautrin, J. and Holsapple, J. 1990. *Brain Res. Rev.* 15: 167–78.
[109]Martin, A. R. 1977. *In* E. Kandel (ed.). *Handbook of the Nervous System*, Vol. 1. American Physiological Society, Baltimore, pp. 329–355.

electrophysiologically by Katz and Miledi at frog muscle end plates treated with anticholinesterase.[110] Under these conditions, ACh that would normally have been hydrolyzed by the cholinesterase at the synapse instead built up a steady concentration in the synaptic cleft, producing a slight depolarization. This depolarization was detected by applying large pulses of curare to the end plate region by ionophoresis from a focally placed pipette. Such application resulted in a hyper-polarization of about 50 μV. The concentration of ACh in the cleft required to account for a potential of this magnitude was calculated to be about 10 pM. The rate of leak of ACh from the terminals required to maintain such a concentration was estimated to be larger by two orders of magnitude than the spontaneous quantal release of ACh. Nonquantal release is significantly greater from mammalian motor nerve terminals.[111] The leak itself is of no significance with respect to the immediate processes involved in synaptic transmission, as its post-synaptic effect is seen only when cholinesterase is inhibited. It does, however, explain the biochemical finding that the amount of ACh collected from cholinesterase-treated resting neuromuscular preparations is about 100 times greater than that which can be accounted for by spontaneous miniature potentials.[112]

Many changes in synaptic efficacy are due to changes in quantum content. For example, at most synapses, including the neuromuscular junction of the frog and crayfish, a second impulse to a presynaptic fiber delivered shortly after the first gives rise to a larger postsynaptic potential. Similarly, a short train of impulses can result in a continuing growth of the response. This is illustrated in Figure 21A, which shows end plate potentials recorded from a frog's neuromuscular junction, produced by a short train of impulses to the motor nerve. The amplitudes of the potentials (measured from the starting point of each rising phase) increase progressively during the train. Furthermore, the effect outlasts the stimulus train for several hundred milliseconds.[113] The response to a test stimulus about 230 msec after the end of the conditioning train is still larger that the first response in the sequence. This effect, called FACILITATION, has been shown to be due to an increase in the mean number of quanta of transmitter (m) released by the presynaptic terminal.[114–116] Further statistical analyses at the crayfish neuromuscular junction have led to the suggestion that this increase in m is due, in turn, to an increase in the probability of release, p.[117] Transmitter release can also be subject to *synaptic depression* if the number of quanta released by a train of stimuli is large. In the experiment shown in Figure

Facilitation and depression of transmitter release

[110]Katz, B. and Miledi, R. 1977. *Proc. R. Soc. Lond. B* 196: 59–72.

[111]Vyskočil, F., Nikosky, E. and Edwards, C. 1983. *Neuroscience* 9: 429–435.

[112]Vizi, S. E. and Vyskočil, F. 1979. *J. Physiol.* 286: 1–14.

[113]Mallart, A. and Martin, A. R. 1967. *J. Physiol.* 193: 679–694.

[114]del Castillo, J. and Katz, B. 1954. *J. Physiol.* 124: 574–585.

[115]Dudel, J. and Kuffler, S. W. 1961. *J. Physiol.* 155: 543–562.

[116]Kuno, M. 1964. *J. Physiol.* 175: 100–112.

[117]Wernig, A. 1972. *J. Physiol.* 226: 751–759.

21

FACILITATION AND DEPRESSION at the frog neuromuscular junction. (A) Muscle bathed in low-calcium, high-magnesium solution (mean quantum content of initial response about 10). Amplitudes of end plate potentials increase progressively during a train of four impulses, due to the release of increased numbers of quanta by successive nerve stimuli. When a test stimulus is given 230 msec after the end of the train, the response amplitude is still facilitated over that of the initial response in the sequence (the arrows indicate the same amplitudes). (B) A similar experiment with a curarized preparation (mean quantum content >100). There is no facilitation after the second response in the train, and the amplitude of a response evoked 230 msec after the end of the train is depressed below that of the initial response rather than facilitated. Records are averaged potentials from a number of end plates. (After Mallart and Martin, 1968.)

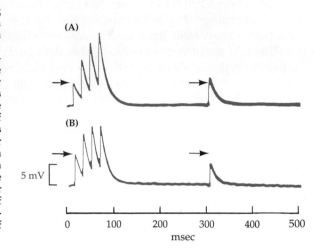

21A, the end plate potentials were reduced in amplitude by lowering the calcium concentration in the bathing solution; that is, the initial quantum content of the potentials was low (10 or less). A similar experiment on a curarized muscle is shown in Figure 21B. Here the responses have been reduced in amplitude by blocking the postjunctional ACh receptors rather than by interfering with quantal release. Thus their quantum content is relatively large (>100). During the train, facilitation of the responses is less than in Figure 21A, and a test pulse 230 msec after the train produces a response that is *smaller* than the first one in the series. This depression of the end plate potential, like facilitation, is presynaptic in origin. Its mechanism is not completely clear, but the fact that large quantal release is needed to produce depression suggests that one factor may be depletion of vesicles from the nerve terminal during the conditioning train.[118] Recovery from depression occurs over several seconds.

In summary, transmitter release from presynaptic terminals is subject to two relatively short-term modifications. The first, facilitation of release, appears to be due to an increase in efficacy of release of quanta from the presynaptic terminal. The second, depression, seems to be associated with depletion of the number of quanta available for release. These two effects interact as shown in Figure 21B. During the train of impulses, the initial effect of facilitation outweighs that of depression and the responses increase in amplitude. Later, when the test pulse is given, facilitation has partially worn off and is overridden by depression, which has a more prolonged time course.

Role of calcium in facilitation   Experimental evidence obtained by Katz and Miledi[119] suggested that facilitation of transmitter release by the second of two action po-

[118]Mallart, A. and Martin, A. R. 1968. *J. Physiol.* 196: 593–604.
[119]Katz, B. and Miledi, R. 1968. *J. Physiol.* 195: 481–492.

tentials arriving in the nerve terminal was related to a residue of calcium left over from the first. Several other possible mechanisms have been ruled out. For example, it has been shown at synapses in the chick's ciliary ganglion that facilitation is not due to an increase in amplitude or duration of the second action potential.[120] Similarly, at synapses between cultured leech neurons, facilitation of the second of two response occurs with no increase in presynaptic calcium *entry*.[121] Thus the idea that facilitation is related to residual calcium seems the most viable. With repetitive stimulation, continuing calcium accumulation would lead to a progressive increase in transmitter release, as in Figure 21A. Various theoretical approaches used to examine the relation between intracellular calcium kinetics and the time course of facilitation have been reviewed by Parnas, Parnas, and Segel.[122]

Facilitation is the shortest lived of several components of increase in quantal release following repetitive stimulation. These have been classified according to their time course.[123] Later components can be seen following longer periods of stimulation, principally POSTTETANIC POTENTIATION, which may last minutes to hours, depending on the preparation and experimental conditions. Still more persistent is LONG-TERM POTENTIATION at central nervous synapses, which will be discussed in Chapter 9.

Posttetanic potentiation (PTP) is similar to facilitation in that it describes an increase in transmitter release from the presynaptic nerve terminal. It is different in that it is produced by prior repetitive activity and it is much longer lasting. (As *tetanic* refers to muscle contraction, the phenomenon should really be called post*activation* potentiation, but the proper term has never come into general use.) The phenomenon is illustrated in Figure 22, this time by an experiment on a cell in a curarized ciliary ganglion from a chicken. The cell was hyperpolarized (long downward deflection) before each stimulus to prevent the postsynaptic potential from triggering an action potential. The first upward deflection in Figure 22A is an electrical coupling potential, the second, slower, depolarization is a chemical postsynaptic potential produced by release of ACh. It is the chemical postsynaptic potential that is of interest. Initially the postsynaptic potential was only about 4 mV in amplitude (because of curarization). The presynaptic nerve was then stimulated at 100/sec for 15 seconds (1500 stimuli), which caused a transient depression of the postsynaptic potential (not shown). Fifteen seconds later, however, a single test stimulus produced a postsynpatic potential well over 20 mV in amplitude—so large, in fact, that it exceeded threshold and produced an action potential! The amplitudes of the potentials produced by subsequent test shocks then declined, but the response was still twice the pretetanic amplitude 10 minutes after the end of the tetanic stimulation.

Posttetanic potentiation

[120]Martin, A. R. and Pilar, G. 1964. *J. Physiol.* 175: 16–30.
[121]Stewart, R. R., Adams, W. B. and Nicholls, J. G. 1989. *J. Exp. Biol.* 144: 1–12.
[122]Parnas, H., Parnas, I. and Segel, L. A. 1990. *Int. Rev. Neurobiol.* 32: 1–50.
[123]Magleby, K. L. and Zengel, J. E. 1982. *J. Gen. Physiol.* 80: 613–638.

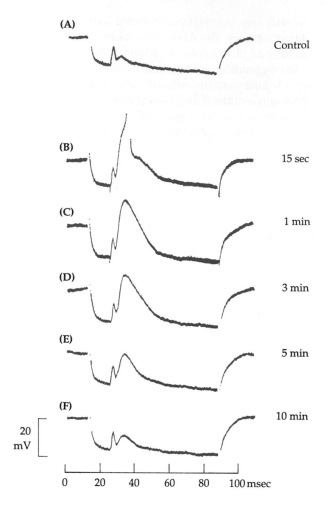

(A) Control

(B) 15 sec

(C) 1 min

(D) 3 min

(E) 5 min

(F) 10 min

20 mV

0   20   40   60   80   100 msec

22

POSTTETANIC POTENTIATION of the postsynaptic potential in a chick ciliary ganglion cell, produced by preganglionic nerve stimulation. Potentials recorded with an intracellular microelectrode. To prevent action potential initiation, end plate potential amplitude was reduced with curare and a hyperpolarizing current pulse applied through the recording electrode before each stimulus. (A) Control record shows electrical coupling potential (brief depolarization) followed by a small postsynaptic potential. (B) Response to stimulation 15 sec after end of a train of 1500 stimuli to the presynaptic nerve. The postsynaptic potential amplitude is more than six times greater than control, giving rise to an action potential. Electrical coupling is unchanged. (C–F) Test stimuli at 1, 3, 5 and 10 minutes after the tetanus show slow decline of potentiation, with the postsynaptic potential amplitude in the last record still more than twice the control value. (From Martin and Pilar, 1964b.)

Like facilitation, PTP is presynaptic in origin; that is, it is due to increased release of quanta from the presynaptic terminal. The exact mechanism underlying the phenomenon is obscure, but experiments at the neuromuscular junction of the frog have shown that it depends on calcium entry into the nerve terminal during the conditioning train of stimuli. For example, if calcium is removed from the bathing solution while the stimulation is applied, then no potentiation occurs.[124] On the other hand, sodium entry is not necessary, as PTP can be produced by trains of artificial depolarizing pulses applied to the nerve terminal in the presence of TTX.[125] In that circumstance, the magnitude of the potentiation is increased with increasing extracellular calcium concentration, and in very high calcium concentrations (83 mM), potentiation

[124]Rosenthal, J. L. 1969. *J. Physiol.* 203: 121–133.
[125]Weinrich, D. 1970. *J. Physiol.* 212: 431–446.

after 500 shocks lasts for more than two hours. While sodium is not necessary for potentiation, sodium entry nevertheless contributes to its duration, at least at the rat neuromuscular junction.[126] There, potentiation is prolonged by treatments that block extrusion by Na-K ATPase of sodium accumulated during the tetanic stimulation, such as adding ouabain or removing potassium from the bathing solution.

SUGGESTED READING

Review Articles

Edwards, C. 1982. The selectivity of ion channels in nerve and muscle. *Neuroscience* 7: 1335–1366.

Llinás, R. 1982. Calcium in synaptic transmission. *Sci. Am.* 247: 56–65.

Martin, A. R. 1977. Junctional transmission. II. Presynaptic mechanisms. *In* E. Kandel (ed.). *Handbook of the Nervous System*, Vol. 1. American Physiological Society, Baltimore, pp. 329–355.

Martin, A. R. 1990. Glycine- and GABA-activated chloride conductances in lamprey neurons. *In* O. P. Otterson and J. Storm-Mathisen (eds.). *Glycine Neurotransmission*. Wiley, New York, pp. 171–191.

McLachlan, E. M. 1978. The statistics of transmitter release at chemical synapses. *Int. Rev. Physiol. Neurophysiol. III* 17: 49–117.

Takeuchi, A. 1977. Junctional transmission. I. Postsynaptic mechanisms. *In* E. Kandel (ed.). *Handbook of the Nervous System*, Vol. 1. American Physiological Society, Baltimore, pp. 295–327.

Original Papers

CHEMICAL AND ELECTRICAL SYNAPTIC TRANSMISSION

Fatt, P. and Katz, B. 1951. An analysis of the end-plate potential recorded with an intracellular electrode. *J. Physiol.* 115: 320–370.

Furshpan, E. J. and Potter, D. D. 1959. Transmission at the giant motor synapses of the crayfish. *J. Physiol.* 145: 289–325.

Martin, A. R. and Pilar, G. 1963. Dual mode of synaptic transmission in the avian ciliary ganglion. *J. Physiol.* 168: 443–463.

IONOPHORESIS AND RECEPTOR LOCALIZATION

del Castillo, J. and Katz, B. 1955. On the localization of end plate receptors. *J. Physiol.* 128:157-181.

Dennis, M. J., Harris, A. J. and Kuffler, S. W. 1971. Synaptic transmission and its duplication by focally applied acetylcholine in parasympathetic neurones in the heart of the frog. *Proc. R. Soc. Lond.* B 177: 509–539.

Kuffler, S. W. and Yoshikami, D. 1975. The distibution of acetylcholine sensitivity at the post-synaptic membrane of vertebrate skeletal twitch muscles: Ionophoretic mapping in the micron range. *J. Physiol.* 244: 703–730.

PERMEABILITY CHANGES AT THE END PLATE

Magleby, K. L. and Stevens, C. F. 1972. The effect of voltage on the time course of end-plate currents. *J. Physiol.* 223: 151–171.

---

[126]Nussinovitch, I. and Rahamimoff, R. 1988. *J. Physiol.* 396: 435–455.

Takeuchi, A. and Takeuchi, N. 1960. On the permeability of the end-plate membrane during the action of transmitter. *J. Physiol.* 154: 52–67.

INHIBITORY PERMEABILITY CHANGES

Coombs, J. S., Eccles, J. C. and Fatt, P. 1955. The specific ion conductances and the ionic movements across the motoneural membrane that produce the inhibitory postsynaptic potential. *J. Physiol.*130: 326–373.

Takeuchi, A. and Takeuchi, N. 1967. Anion permeability of the inhibitory post-synaptic membrane of the crayfish neuromuscular junction. *J. Physiol.* 191: 575–590.

PRESYNAPTIC INHIBITION

Dudel, J. and Kuffler, S. W. 1961. Presynaptic inhibition at the crayfish neuro-muscular junction. *J. Physiol.* 155: 543–562.

Kuno, M. 1964. Mechanism of facilitation and depression of the excitatory postsynaptic potential in spinal motoneurons. *J. Physiol.* 175: 100–112.

Takeuchi, A. and Takeuchi, N. 1966. On the permeability of the presynaptic terminal of the crayfish neuromuscular junction during synaptic inhibition and the action of γ-aminobutyric acid. *J. Physiol.* 183: 433–449.

CALCIUM AND TRANSMITTER RELEASE

Adler, E. M., Augustine, G. J., Duffy, S. N. and Charlton, M. P. 1991. Alien intracellular calcium chelators attenuate neurotransmitter release at the squid giant synapse. *J. Neurosci.* 11: 1496–1507.

Katz, B. and Miledi, R. 1967. A study of synaptic transmission in the absence of nerve impulses. *J. Physiol.* 192: 407–436.

Katz, B. and Miledi, R. 1967. The timing of calcium action during neuromuscular transmission. *J. Physiol.* 189: 535–544.

QUANTAL RELEASE

Boyd, I. A. and Martin, A. R. 1956. The end-plate potential in mammalian muscle. *J. Physiol.* 132: 74–91.

del Castillo, J. and Katz, B. 1954. Quantal components of the end-plate potential. *J. Physiol.* 124: 560–573.

Fatt, P. and Katz, B. 1952. Spontaneous subthreshold activity at motor nerve endings. *J. Physiol.* 117: 108–128.

Johnson, E. W. and Wernig, A. 1971. The binomial nature of transmitter release at the crayfish neuromuscular junction. *J. Physiol.* 218: 757–767.

Kuffler, S. W. and Yoshikami, D. 1975. The number of transmitter molecules in a quantum: An estimate from iontophoretic application of acetylcholine at the neuromuscular junction. *J. Physiol.* 251: 465–482.

FACILITATION AND PTP

Magleby, K. L. and Zengel, J. E. 1982. A quantitative description of stimulation-induced changes in transmitter release at the frog neuromuscular junction. *J. Gen. Physiol.* 80: 613–638.

Mallart, A. and Martin, A. R. 1967. Analysis of facilitation of transmitter release at the neuromuscular junction of the frog. *J. Physiol.* 193: 679–697.

Mallart, A. and Martin, A. R. 1968. The relation between quantum content and facilitation at the neuromuscular junction of the frog. *J. Physiol.* 196: 593–604.

INDIRECT MECHANISMS OF SYNAPTIC TRANSMISSION

The majority of neurotransmitters do not bind directly to ligand-activated ion channels, but rather to receptors that influence ion channels and pumps indirectly through membrane-associated or cytoplasmic second messengers. At some synapses, transmission occurs solely by such indirect mechanisms. Other synapses are modulated by indirect action.

Even at a synapse as reliable and simple as the skeletal neuro-muscular junction, transmission can be modulated. Norepinephrine released from axons of the sympathetic nervous system causes a long-lasting increase in the amplitude and duration of the end plate potential, both by increasing the amount of transmitter released from motor axon terminals and by changing the properties of the muscle fiber membrane. In the sympathetic ganglion of the bullfrog two different transmitters—ACh and a peptide resembling luteinizing hormone-releasing hormone (LHRH)—modulate synapses. Both bind to receptors that cause voltage-activated M-type potassium channels to close. As a result, excitatory inputs that would normally evoke a single impulse produce instead trains of action potentials. At many synapses, indirectly acting transmitters affect channels that are not open in the resting cell. For example, norepinephrine inhibits a calcium-activated potassium channel on neurons in the rat hippocampus. This causes little or no change in the resting potential but enhances the response of the cells to excitatory inputs. Such changes in the efficacy of transmission at synapses are called neuromodulation.

The time course of the response to activation of an indirectly coupled receptor is much slower than the rapid kinetics of synaptic potentials produced by directly coupled, ligand-activated channels; the response can last for seconds, minutes, or hours rather than milliseconds. In addition, transmitters released from one axon terminal may diffuse to modulate activity in many target cells. In each cell specificity of action is determined by the nature and distribution of the transmitter receptors.

The responses to activation of indirectly coupled receptors are mediated by G proteins, so named because they bind guanosine diphosphate (GDP) and guanosine triphosphate (GTP). G proteins modify the activity of other receptor proteins, ion channels, or pumps. G proteins are trimers of three subunits, α, β, and γ. The α

subunits interact with receptor and effector proteins, whereas the $\beta\gamma$ complex acts primarily to anchor the α subunit to the membrane. GDP bound to the α subunit keeps the $\alpha\beta\gamma$ complex intact. When a G protein is activated by interaction with a transmitter receptor, GDP is replaced by GTP, and the α subunit dissociates from the $\beta\gamma$ complex. The free α subunit binds to and modulates the activity of its target— an enzyme, a channel, or a pump. G proteins are grouped into classes according to the targets recognized by the α subunit: G_t activates cyclic GMP phosphodiesterase, G_s stimulates adenylyl cyclase, G_i inhibits adenylyl cyclase, G_p activates phospholipase C, G_k activates potassium channels, and G_o acts on still other proteins. A particular G protein can activate more than one type of target, and the activity of a target protein is often modified through more than one G protein pathway.

Phosphorylation of serine and threonine residues by enzymes called protein kinases is a common mechanism by which the activity of receptors and ion channels is modified. Adenylyl cyclase, an enzyme whose activity is modulated by G_i and G_s, catalyzes the synthesis of cyclic AMP, which, in turn, increases the activity of cyclic AMP-dependent protein kinase. The activation of phospholipase C by G_p results in both the formation of diacylglycerol and an increase in intracellular calcium concentration, which together activate protein kinase C. In addition to its stimulatory effect on protein kinase C, increased intracellular calcium concentration also activates calcium–calmodulin-dependent protein kinase.

In a wide variety of cells, potassium and calcium channels are prime targets for indirect transmitter action. Changes in channel activation in axon terminals can modify calcium entry, and thereby transmitter release, both directly, through effects on calcium channels, and indirectly, by altering the duration of the action potential. Changes in calcium and potassium channels in postsynaptic cells can alter spontaneous activity and the response to synaptic inputs.

The mechanisms by which neurotransmitters produce effects in their target cells can be divided into two general categories: direct and indirect. Direct mechanisms, discussed in Chapter 7, are also referred to as "fast" or "channel-coupled." In such cases the postsynaptic receptor is itself a ligand-activated channel, which contains both the transmitter binding site and the ion channel opened by the transmitter as part of the same protein. Indirectly acting transmitters, on the other hand, bind to receptors that are not themselves ion channels, but which modify the activity of other receptor proteins, ion channels, or ion pumps so that the response of the cell is altered. Indirect mechanisms are also referred to as "slow" or "second messenger-linked." Many indirectly acting receptors produce their effects through interaction with GTP-binding, or G PROTEINS.[1]

---

[1]Dunlap, K., Holz, G. G. and Rane, S. G. 1987. *Trends Neurosci.* 10: 244–247.

At some synapses, such as those on smooth muscle of the gut, transmission may occur solely through indirectly coupled receptors. Alternatively, indirectly acting transmitters may influence the efficacy of transmission at synapses where other transmitters are released, a process referred to as NEUROMODULATION.

MODULATION OF SYNAPTIC TRANSMISSION

Before discussing the mechanisms by which indirectly coupled receptors produce their effects, it is useful to illustrate how indirect transmitter action can modify synaptic efficacy. It is somewhat ironic that one of the first synapses at which neuromodulation was described was the vertebrate skeletal neuromuscular junction. After all, the function of the neuromuscular junction, as described by Katz,[2]

> is to transfer impulses from the relatively very small motor nerve endings to the large muscle fiber and cause it to contract. At most of the myoneural junctions in vertebrate muscle, each nerve impulse is followed by a similar impulse in the muscle fibers. . . . Thus, the vertebrate myoneural junction serves a much simpler purpose than the central synapses of neurons . . . , where integration of converging signals takes place and where the effect of a single nerve impulse is generally well below the threshold of excitation of the effector cell. To put it somewhat crudely, the vertebrate nerve–muscle junction serves the purpose of a simple relay.

In spite of this apparent simplicity, direct cholinergic transmission at the neuromuscular junction is modulated in complex ways by several substances. For example, as early as 1923 it was shown that the adrenergic agents norepinephrine (noradrenaline), released from varicosities along sympathetic axons, and epinephrine (adrenaline), released into the circulation from the adrenal medulla, facilitated neuromuscular transmission.[3] For nearly half a century there was considerable confusion regarding the mechanism of this effect. Some experiments suggested a presynaptic action, others a change in the muscle fiber. Both results were found to be correct, but mediated by different receptors. Epinephrine and norepinephrine act on a variety of receptors, all of which are indirectly coupled to their effector proteins. Based on the relative potency of different agonists and antagonists, adrenergic receptors can be divided into five main classes: α_1, α_2, β_1, β_2, and β_3. The β_1 receptors predominate on cardiac tissue; β_2 and α_1 receptors are found primarily in smooth muscle, skeletal muscle, and liver cells; β_3 receptors are localized to adipose tissue; and α_2 receptors are commonly found on presynaptic nerve terminals and on postsynaptic cells in a variety of tissues.[4] At the skeletal neuromuscular junction, activation of α-adren-

Neuromodulation at the neuromuscular junction

[2]Katz, B. 1966. *Nerve, Muscle, and Synapse.* McGraw-Hill, New York.
[3]Orbeli, L. A. 1923. *Bull. Inst. Sci. Leshaft* 6: 194–197.
[4]Lefkowitz, R. J., Hoffman, B. B. and Taylor, P. 1990. *In* A. G. Gilman et al. (eds.). *Goodman and Gilman's Pharmacological Basis of Therapeutics*, 8th Ed. Pergamon Press, New York, pp. 84–121.

ergic receptors on the presynaptic nerve terminal was found to increase the number of quanta of transmitter released by an action potential. Stimulation of β-adrenergic receptors on the muscle fibers activates the Na–K pump (Chapter 3), causing hyperpolarization and a decrease in the resting membrane conductance (Figure 1).[5,6] Thus both pre- and postsynaptic adrenergic receptors contribute to the increase in the amplitude and duration of the synaptic potential.

Varicosities along sympathetic axons in skeletal muscle are not specifically juxtaposed to neuromuscular junctions. Sympathetic axons course randomly through muscle and release norepinephrine from varicosities scattered along their lengths (Figure 1A). Accordingly, norepinephrine must diffuse some distance from its site of release to reach receptors on the axon terminal and the muscle fiber. In addition, norepinephrine released from one varicosity may affect transmission at many junctions. Thus the specificity of norepinephrine's actions is achieved not by precise anatomical connections but by the nature and distribution of the receptors to which it binds. This is a common feature of neuromodulatory pathways.

Another general feature of neuromodulatory actions is that they have a prolonged time course. At the neuromuscular junction the effects of norepinephrine develop slowly over 15 to 20 minutes and decline slowly after norepinephrine is removed. This slow time course arises in two ways. First, a short delay of seconds is introduced by the time required for diffusion from the site of release to the site of action. Second, and more importantly, the indirect mechanisms by which α- and β-adrenergic receptors produce their effects involve changes in cell metabolism that have very slow kinetics compared with those of ligand-activated ion channels. Once set in motion these metabolic processes can continue long after the transmitter has diffused away from the receptors.

Neuromodulation in sympathetic ganglia

An example of how neuromodulatory effects can alter the pattern of signaling between neurons is provided by experiments of Libet, Nishi, Koketsu, Weight, and others, who described slow synaptic potentials with unusual properties in sympathetic ganglion cells of frogs and mammals. Stimulation of presynaptic inputs to the ganglion cells elicits three depolarizing potentials: a fast excitatory postsynaptic potential (epsp), a slow epsp, and a late slow epsp[7] (Figure 2A). With single presynaptic impulses, the slow epsp and late slow epsp are not evident; they are seen only after trains of action potentials at natural frequencies. Even with a train of impulses the slow and late slow epsps are not sufficient by themselves to bring the cell to threshold. However, they do alter the response of the cell to subsequent stimuli, as we shall see.

[5]Kuba, K. 1970. *J. Physiol.* 211: 551–570.
[6]Clausen, T. and Flatman, J. A. 1977. *J. Physiol.* 270: 383–414.
[7]Jan, Y. N., Jan, L. Y. and Kuffler, S. W. 1979. *Proc. Natl. Acad. Sci. USA* 76: 1501–1505.

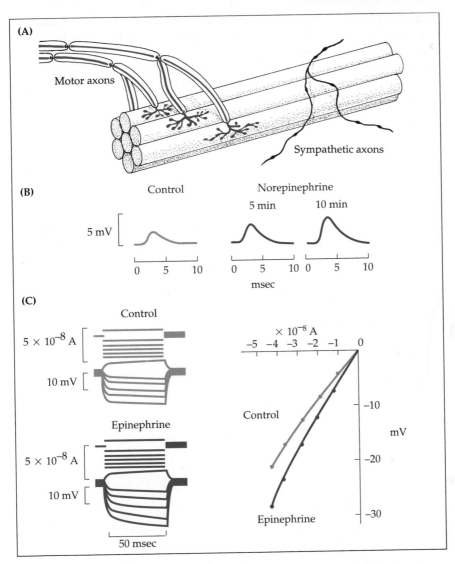

1 MODULATION OF SYNAPTIC TRANSMISSION at the skeletal neuromuscular
junction. (A) Sketch of the innervation of skeletal muscle. Motor axons form
discrete, highly specialized synapses where postsynaptic receptors and presynaptic release sites are precisely juxtaposed. Sympathetic axons course throughout the muscle.
Norepinephrine, released from varicosities scattered along their length, diffuses to reach
receptors on the muscle and axon terminal. (B) Addition of 10^{-6} g/ml norepinephrine
causes an increase in the amplitude of the synaptic potential evoked by nerve stimulation; this effect was blocked by specific antagonists of α-adrenergic receptors. The
increase is slow to develop and is due to an increase in the number of quanta released.
(Intracellular recordings from a muscle bathed in a solution containing 10^{-6} g/ml
d-tubocurare so that the synaptic potentials would be subthreshold.) (C) Addition of
5×10^{-6} g/ml epinephrine causes a decrease in resting membrane conductance. Small
current pulses were passed through an intracellular microelectrode to measure the conductance of the muscle fiber membrane. Ten minutes after epinephrine was added, the
conductance of the membrane was decreased compared with control. This effect was
blocked by β-adrenergic receptor antagonists. (After Kuba, 1970.)

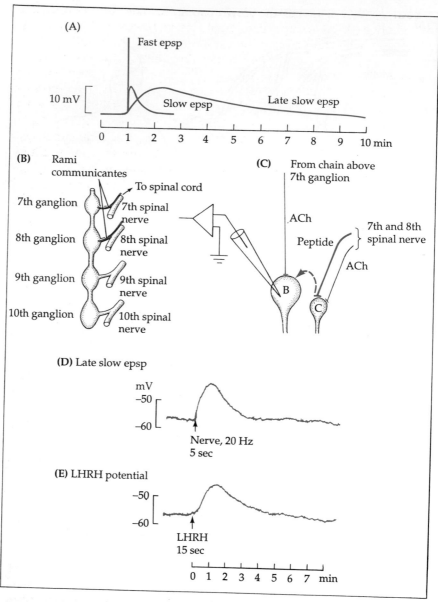

2 **SYNAPTIC POTENTIALS IN SYMPATHETIC GANGLION CELLS** in a bullfrog. (A) The fast, slow, and late slow epsp's evoked in B ganglion cells. A single stimulus to the preganglionic inputs evokes a large fast epsp; trains of stimuli (10/sec for 5 sec) are required to elicit the slow and late slow synaptic potentials. (B) Diagram of sympathetic ganglia 7, 8, 9, and 10. Nerve cells receive inputs from the spinal cord via rami communicantes, and also from the chain above. (C) The larger B cells in caudal ganglia (9 and 10) receive cholinergic input from spinal nerves rostral to ganglion 7 and receive peptide input indirectly from spinal nerves 7 and 8 via synapses onto neighboring C cells. Accordingly, the different inputs can be stimulated selectively. (D) The late slow excitatory potential evoked by stimulating spinal nerves 7 and 8 (20/sec for 5 sec). The depolarization lasts for several minutes. (E) Application of LHRH to the cell in (D) by pressure from a micropipette. The peptide mimics the action of the naturally released transmitter. (After Kuffler, 1980.)

Both the fast epsp and the slow epsp are evoked by the release of acetylcholine (ACh) from the presynaptic nerve terminals. The fast epsp results from activation of nicotinic ACh receptors. The resulting depolarization usually elicits a single action potential in the ganglion cell. The slow epsp is due to the binding of acetylcholine to muscarinic ACh receptors. Kuffler and his colleagues found that the late slow epsp in the sympathetic ganglion of the frog could be mimicked by application of the peptide luteinizing hormone-releasing hormone (LHRH) [7–10] (Figures 2D and 2E). (LHRH is a hormone released into the local circulation by neurons in the hypothalamus. It diffuses to the anterior pituitary gland and causes it to secrete luteinizing hormone, an essential hormone involved in the ovarian cycle and in the secretion of testosterone.) The dose required (1 μM or more) seemed at first too high for LHRH to be a serious transmitter candidate. With this as a clue, analogues of LHRH were tested and some were indeed found to be 100 times more potent than the hormone itself. A variety of other tests demonstrated that the late slow epsp was due to the release of an LHRH-like peptide. However, no nerve terminals containing an LHRH-like peptide were found juxtaposed to the B cells from which these recordings were made. The peptide reaches B cells by diffusion from synapses made on smaller, neighboring C cells (Figure 2C).[10] Thus transmitter liberated by a particular nerve terminal may act on more than one postsynaptic cell.

The cells of autonomic ganglia are particularly favorable for electrophysiological investigation of the mechanism of such slow neuromodulatory effects. The B cells are relatively large and nearly spherical, with no dendrites, and synaptic contacts are made directly onto the cell body. Voltage clamp studies of B cells by Adams, D. Brown, and their colleagues identified a current carried by a particular type of potassium channel that was modified by activation of muscarinic receptors, the so-called M (for muscarinic) current.[11] These M potassium channels are voltage-activated and have a threshold for activation that is near the normal resting potential. Accordingly, some of the channels are open at rest and provide a major contribution to the resting potassium conductance. Activation of muscarinic ACh receptors closes the M potassium channels (Figure 3A).

What are the consequences of the closure of potassium channels? The resting influx of sodium ions is no longer balanced by potassium efflux, so the cell depolarizes, producing the slow epsp. With trains of stimuli at natural frequencies, the depolarization produced by the slow epsp, about 10 mV, is not sufficient to bring the cell to threshold. Moreover, the change in membrane potential on its own has little effect on the response of the cell to the much larger fast epsp (Figure 3B).[12]

[8]Kuffler, S. W. 1980. *J. Exp. Biol.* 89: 257–286.

[9]Jan, Y. N., Jan, L. Y. and Kuffler, S. W. 1980. *Proc. Natl. Acad. Sci. USA* 77: 5008–5012.

[10]Jan, Y. N. et al. 1983. *Cold Spring Harbor Symp. Quant. Biol.* 48: 363–374.

[11]Adams, P. R., Brown, D. A. and Constanti, A. 1982. *J. Physiol.* 330: 537–572.

[12]Adams, P. R., Brown, D. A. and Constanti, A. 1982. *J. Physiol.* 332: 223–262.

When M-current channels are closed, however, the response of the B cell to the fast epsp is dramatically enhanced. The decrease in membrane conductance increases the amplitude of excitatory synaptic potentials (see Chapter 7), as described above for norepinephrine at the neuromuscular junction. This effect is amplified during depolarizing synaptic potentials in the following way. M channels are normally activated by depolarization and repolarize the cell, cutting short excitatory synaptic potentials. When M-current channels are kept closed by activation of muscarinic receptors, the increase in potassium conductance is prevented and depolarizations are not curtailed. The result is that the fast epsp elicited by stimulation of the preganglionic nerve, which nor-

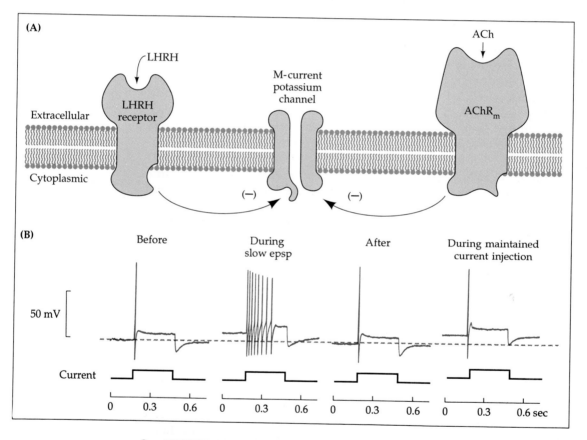

3 **INHIBITION OF POTASSIUM CURRENTS IN SYMPATHETIC GANGLION CELLS** modulates response to presynaptic stimulation. (A) Binding of ACh to muscarinic receptors (AChR$_m$) and binding of LHRH to its receptor both inhibit M-current potassium channels. (B) The effect of the decrease in the M current during the slow epsp is to increase the excitability of the B cell. Depolarizing current pulses before and after a slow epsp produce a single action potential. During the slow epsp, the same current pulse elicits a burst of action potentials. Depolarizing the B cell to the same extent as occurs during the slow epsp by injecting a maintained current has no such effect on the responsiveness of the cell. (After Jones and Adams, 1987.)

mally gives rise to one or two impulses, now becomes prolonged and evokes a train of action potentials in the B cell (Figure 3B).[13] Thus the slow epsp modifies the pattern of transmission across the ganglionic synapses. It is tempting to speculate about the role of such mechanisms in the regulation of functions that are controlled by the autonomic nervous system, such as blood pressure, gastrointestinal motility, and glandular secretion, as well as in disorders such as hypertension, glaucoma, or gastrointestinal ulcers that might result from autonomic dysfunction.

The mechanism of the peptide-evoked late slow epsp is the same as that of the slow ACh-mediated epsp; M-current potassium channels are closed (Figure 3A).[14,15] Moreover, a saturating response to prolonged application of LHRH completely occludes the response to activation of muscarinic receptors, and vice versa. Thus, although the receptors for ACh and the LHRH-like peptide are clearly distinct, they act through a common final pathway. The only difference is the time course of their effects, which appears to be determined by the time course of transmitter release, diffusion, and inactivation.

As the preceding discussion suggests, indirectly acting transmitters can modulate signaling without producing any change in the resting potential of the target cell. An example comes from studies of the effects of norepinephrine on neurons in the hippocampus by Nicoll and his colleagues. This region of the vertebrate central nervous system has been the focus of extensive investigation because of its relatively simple architecture and its importance for short-term memory (see Chapter 10).[16] Hippocampal pyramidal cells, like many other vertebrate neurons, have a slow calcium-activated potassium current.[13] The influx of calcium during an action potential causes these potassium channels to open, which in turn causes a prolonged hyperpolarization, the so-called slow after-hyperpolarization (AHP). Norepinephrine blocks this current[17] (Figure 4). Since few of these channels are open at rest, application of norepinephrine to pyramidal cells produces little or no change in resting potential. However, when the slow calcium-activated potassium current is inhibited by addition of norepinephrine, excitatory inputs that might normally produce only a few action potentials before being cut short by the after-hyperpolarization now evoke a sustained train of impulses (Figure 4B). Like the M-current channels in frog sympathetic ganglion cells, the slow calcium-activated potassium channels in hippocampal pyramidal cells are modulated by several neurotransmitters, each acting through its own receptor.[16]

Neuromodulation of hippocampal pyramidal neurons

[13]Jones, S. W. and Adams, P. R. 1987. *In* L. K. Kaczmarek and I. B. Levitan (eds.). *Neuromodulation: The Biochemical Control of Neuronal Excitability.* Oxford University Press, New York, pp. 159–186.

[14]Adams, P. R. and Brown, D. A. 1980. *Br. J. Pharmacol.* 68: 353–355.

[15]Kuffler, S. W. and Sejnowski, T. J. 1983. *J. Physiol.* 341: 257–278.

[16]Nicoll, R. A. 1988. *Science* 241: 545–551.

[17]Madison, D. V. and Nicoll, R. A. 1986. *J. Physiol.* 372: 221–244.

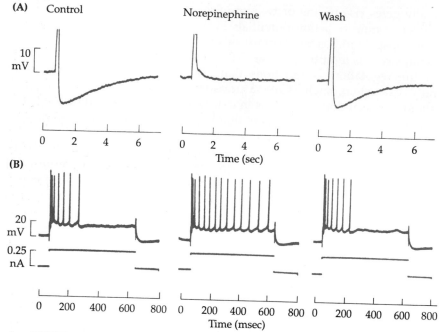

(A) Control   Norepinephrine   Wash

10 mV

0 2 4 6   0 2 4 6   0 2 4 6

Time (sec)

(B)

20 mV

0.25 nA

0 200 400 600 800 0 200 400 600 800 0 200 400 600 800

Time (msec)

4 **NOREPINEPHRINE MODULATES SIGNALING IN HIPPOCAMPAL NEURONS** by blocking calcium-activated potassium conductance. **(A)** Calcium action potentials recorded from a pyramidal cell in a slice preparation from rat hippocampus bathed in medium containing TTX and TEA. The prolonged after-hyperpolarization following each action potential is due to a slow calcium-activated potassium current. Norepinephrine blocks this current. (Action potentials in these records are truncated.) **(B)** Norepinephrine increases excitability of hippocampal pyramidal neurons. Under control conditions a depolarizing current pulse (lower trace) elicits a brief train of action potentials. In the presence of norepinephrine the same stimulus evokes a sustained response. (After Madison and Nicoll, 1986.)

INDIRECTLY COUPLED RECEPTORS AND G PROTEINS

G proteins and
G protein-coupled
receptors

All the effects described in the preceding sections are mediated by indirectly coupled receptors that exert their effects by binding to G proteins. G protein-coupled receptors constitute a superfamily of proteins that includes the visual pigment rhodopsin (see Chapter 16), muscarinic ACh receptors, α- and β-adrenergic receptors, and receptors for 5-hydroxytryptamine (5-HT, also called serotonin), dopamine, and a variety of neuropeptides.[18–21] G protein-coupled receptors are membrane proteins having seven transmembrane domains, with an extracellular amino terminus and an intracellular carboxy terminus (Figure 5A). Molecular genetic experiments have indicated that amino acid

[18]O'Dowd, B. F., Lefkowitz, R. J. and Caron, M. G. 1989. *Annu. Rev. Neurosci.* 12: 67–83.

[19]Bonner, T. I. 1989. *Trends Neurosci.* 12: 148–151.

[20]Nakanishi, S. 1991. *Annu. Rev. Neurosci.* 14: 123–136.

[21]Julius, D. 1991. *Annu. Rev. Neurosci.* 14: 335–360.

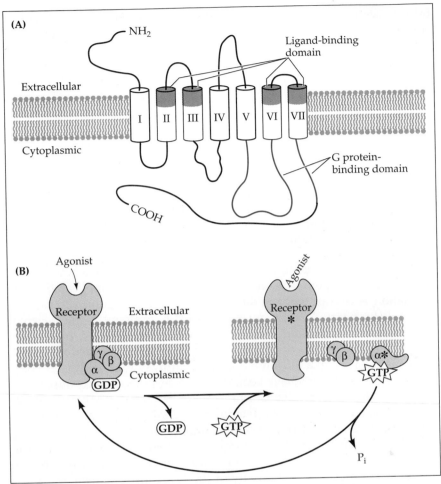

5 INDIRECTLY COUPLED TRANSMITTER RECEPTORS ACT THROUGH G PRO-TEINS. (A) The topographical structure of an indirectly coupled transmitter receptor. There are seven transmembrane domains, with an extracellular amino terminus and an intracellular carboxy terminus. The extracellular portions of transmembrane domains II, III, VI, and VII mediate ligand binding. The cytoplasmic loop between transmembrane segments V and VI together with the amino terminal region of the intracellular tail appear to mediate binding to the appropriate G protein. **(B)** G proteins are trimers of α, β, and γ subunits. Binding of an agonist to its receptor allows exchange of GTP for GDP bound to the α subunit. Thus activated (*), the α subunit dissociates from the $\beta\gamma$ complex and interacts with one of a variety of target proteins. Hydrolysis of GTP to GDP by the endogenous GTPase activity of the α subunit leads to reassociation of the $\alpha\beta\gamma$ complex and terminates the response. (A after O'Dowd et al., 1989.)

residues located in the extracellular halves of transmembrane domains II, III, VI, and VII mediate ligand binding, while the third cytoplasmic loop (between transmembrane segments V and VI) together with the amino terminal region of the carboxy tail apparently mediate binding to the appropriate G protein.[18]

G proteins, so named because they bind guanine nucleotides, are trimers of three distinct subunits, α, β, and γ (Figure 5B).[22] The α subunits bind and hydrolyze guanosine triphosphate (GTP) and interact with receptor and effector proteins. The $\beta\gamma$ complex helps anchor the α subunit to the membrane and is required for receptor–α subunit interaction. In the resting state, guanosine diphosphate (GDP) is bound to the α subunit and the three subunits are associated as a trimer. Interaction with an activated receptor allows GTP to replace GDP, resulting in dissociation of the α subunit from the $\beta\gamma$ complex. The free α subunit then binds to and modulates the activity of its target protein. There is also evidence that the $\beta\gamma$ complex may produce effects of its own through activation of phospholipase A_2.[23] The free α subunit has intrinsic GTPase activity, which results in the hydrolysis of the bound GTP to GDP. This allows the reassociation of the subunits into the G protein complex and terminates its activity. A variety of probes have been developed for identifying responses mediated by G proteins (Box 1).

Multiple forms are known for each subunit, providing a bewildering array of possible permutations. G proteins have been implicated in the regulation of at least a dozen different potassium, sodium, and calcium channels by nearly 70 different receptors.[24] G proteins couple receptor activation to modulation of channel activity either directly, by binding to the channel itself, or indirectly, by regulating the activity of an enzyme involved in a second messenger pathway that modulates channel activity. G proteins have been grouped into classes according to the targets of their α subunits: G_t (transducin, activates cGMP phosphodiesterase; see Chapter 16); G_s (stimulates adenylyl cyclase); G_i (inhibits adenylyl cyclase); G_p (couples to *p*hospholipase C); G_k (activates potassium channels); and G_o (*o*ther, targets unknown).[25] However, a particular G protein can couple to more than one effector, and different G proteins can modulate the activity of the same ion channel.

Direct modulation of
channel function
by a G protein

Loewi first showed some 70 years ago that ACh combines with muscarinic receptors on heart cells to slow the heartbeat (Chapter 7). In contrast to the closure of potassium channels by muscarinic receptors on sympathetic ganglion cells, activation of muscarinic receptors in the heart opens potassium channels, causing a hyperpolarization.[26] Results of experiments by Breitwieser, Szabo, Pfaffinger, Hille, and their colleagues indicate that activation of muscarinic receptors is coupled to opening of potassium channels by G proteins:[27,28] Intracellular GTP is required,[29] activation of potassium channels by muscarinic agonists is

[22]Casey, P. J. et al. 1988. *Cold Spring Harbor Symp. Quant. Biol.* 53: 203–208.
[23]Kim, D. et al. 1989. *Nature* 337: 557–560.
[24]Birnbaumer, L. et al. 1987. *Kidney Int.* 32 (Suppl. 23): S14–S37.
[25]Brown, D. A. 1990. *Annu. Rev. Physiol.* 52: 215–242.
[26]Sakmann, B., Noma, A. and Trautwein, W. 1983. *Nature* 303: 250–253.
[27]Brown, A. M. and Birnbaumer, L. 1990. *Annu. Rev. Physiol.* 52: 197–213.
[28]Szabo, G. and Otero, A. S. 1990. *Annu. Rev. Physiol.* 52: 293–305.
[29]Pfaffinger, P. J. et al. 1985. *Nature* 317: 536–538.

BOX 1 CHARACTERISTICS OF G PROTEIN RECEPTOR SYSTEMS

Several tests can be used to distinguish responses mediated by G proteins. For example, activation of the α subunit requires that bound GDP be replaced by GTP. Accordingly, G protein-mediated events have an absolute requirement for cytoplasmic GTP and will be blocked by intracellular perfusion with solutions that lack GTP. Two analogues of GTP, GTPγS and

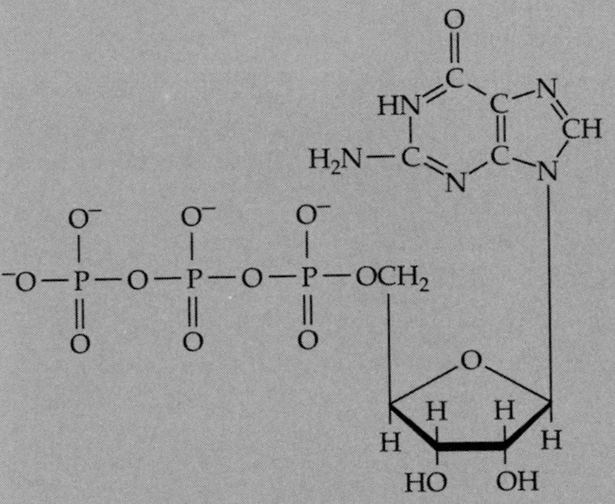

GTP
Guanosine 5'–triphosphate

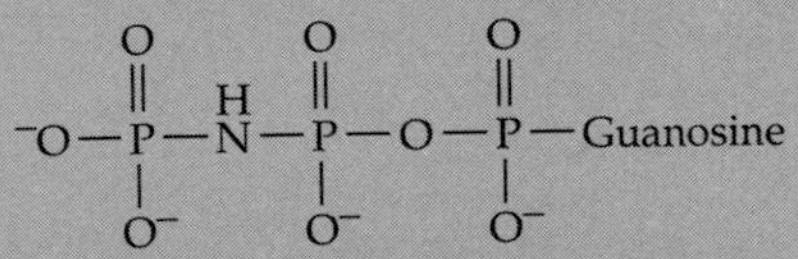

Gpp(NH)p, are useful because they cannot be hydrolyzed by the endogenous GTPase activity of the α subunit. Like GTP, they can replace GDP from the α subunit and activate it. However, because they cannot be hydrolyzed, these analogues activate the α subunit permanently. Thus intracellular perfusion with either of these analogues enhances and greatly prolongs agonist-induced activation of G protein-mediated responses, and may even initiate responses in the absence of agonist. Likewise, AlF$_4^-$ persistently activates G proteins. An analogue of GDP, GDPβS, on the other hand, binds strongly to the GDP site on the α subunit and resists replacement by GTP. Thus GDPβS inhibits G protein-mediated responses by maintaining the $\alpha\beta\gamma$ complex in the inactive state.

Two bacterial toxins are also useful for characterizing G protein-mediated processes. Each is an enzyme that catalyzes the covalent attachment of ADP–ribose to an arginine residue on the α subunit. *Cholera toxin* acts on α_s, irreversibly activating it; *pertussis toxin* acts on α_i, α_p, and α_o, irreversibly blocking their activation and so inhibiting responses mediated by the corresponding G proteins.

Gpp(NH)p
Guanosine 5'–[β,γ–imido]
triphosphate

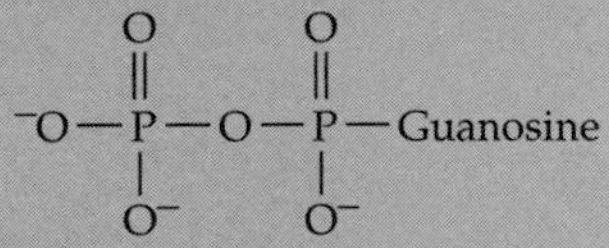

GDP
Guanosine 5'–diphosphate

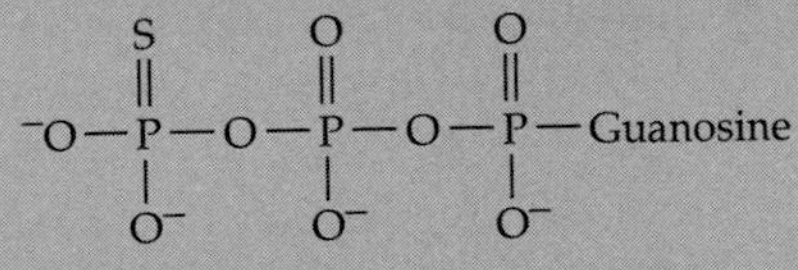

GTPγS
Guanosine 5'–O–[γ–thio]
triphosphate

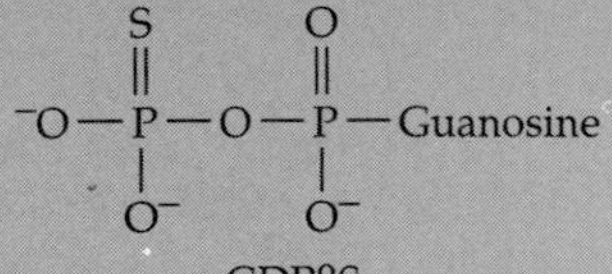

GDPβS
Guanosine 5'–O–[β–thio]
diphosphate

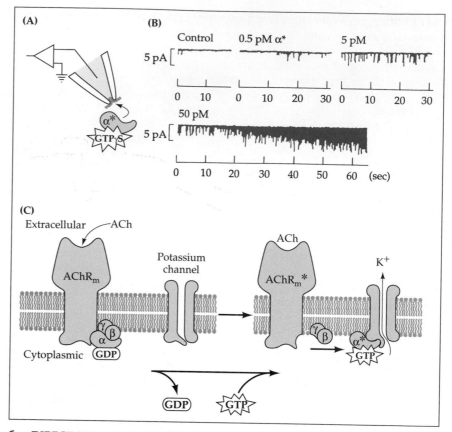

6   DIRECT MODULATION OF CHANNEL FUNCTION BY G PROTEINS. Application of G_k α subunits stably activated with GTPγS to the intracellular surface of an isolated patch of membrane from guinea pig heart atrial cells (A) results in a concentration-dependent increase in potassium channel currents (B). (C) Schematic representation of events in an intact cell. Binding of ACh to muscarinic receptors activates a G protein; the α subunit binds directly to and opens a potassium channel. (A and B after Codina et al., 1987.)

greatly prolonged by intracellular application of Gpp(NH)p (a nonhydrolyzable analogue of GTP),[30] and muscarinic activation of potassium channels is blocked by pertussis toxin,[29] which inactivates several G proteins (Box 1). Direct demonstration of the mechanism of potassium channel activation in cardiac muscle cells has come from experiments by A. M. Brown and his colleagues using inside-out excised membrane patches. When α subunits stably activated by GTPγS (another nonhydrolyzable analogue of GTP) and magnesium ion are applied to the intracellular side of the patch, potassium channels are opened, confirming that the α subunit interacts with the potassium channel itself or a very closely associated protein[31] (Figure 6). Results of similar experi-

[30]Breitwieser, G. E. and Szabo, G. 1985. *Nature* 317: 538–540.
[31]Codina, J. et al. 1987. *Science* 236: 442–445.

ments in other cells have identified a variety of potassium, sodium, and calcium channels whose activity is regulated by direct interaction with α subunits.[27]

A surprising finding is that application of the $\beta\gamma$ complex can cause opening of potassium channels in heart muscle cells. The mechanism of this effect is clearly indirect and of secondary importance for the muscarinic response.[28] It appears that the $\beta\gamma$ complex does not bind to the potassium channel itself, but activates an intracellular enzyme that, in turn, modulates channel activity. In other experiments the $\beta\gamma$ complex has been found to inhibit channel activity, apparently by reducing the concentration of endogenous α subunits.[27]

Using muscle cells dissociated from the atrium of the heart, Soejima and Noma found that potassium channel activity in cell-attached patches was increased when muscarinic agonists were added to the patch pipette solution, but not when agonists were added to the bath[32] (Figure 7). Thus, activated α subunits could not influence channels on the other side of the pipette–membrane seal. This *membrane-delimited* or *direct* mechanism of G protein action apparently reflects the limited range over which α subunits can act. In contrast to such direct effects of α subunits are those mediated by cytoplasmic intracellular second messengers (see below).

A second example of direct interaction between G proteins and ion channels comes from studies by Tsien and his colleagues on transmitter release by neurons isolated from sympathetic ganglia of adult frogs.[33]

[32]Soejima, M. and Noma, A. 1984. *Pflügers Arch.* 400: 424–431.
[33]Lipscombe, D., Kongsamut, S. and Tsien, R. W. 1989. *Nature* 340: 639–642.

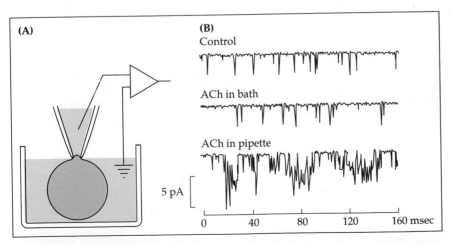

7 **DIRECT OR MEMBRANE-DELIMITED EFFECTS OF G PROTEINS** operate over short distances. **(A)** Effects of ACh were assayed by cell-attached patch clamp recording. Acetylcholine could be perfused into either the patch pipette or the bath. **(B)** Recordings of single-channel currents before and during addition of ACh. Compared with control, channel activity increased only when ACh was added to the patch pipette. (After Soejima and Noma, 1984.)

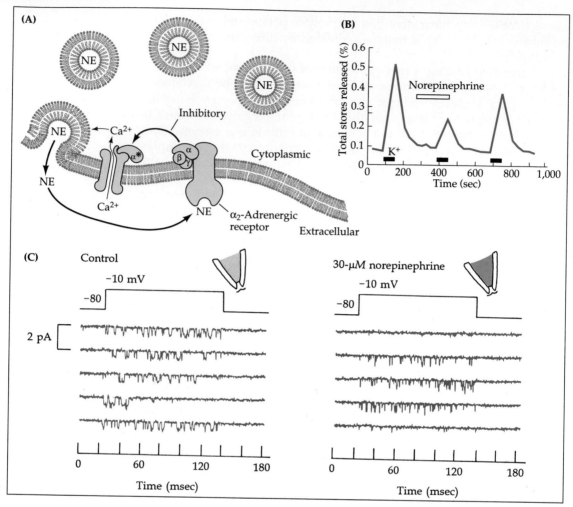

8 **NOREPINEPHRINE REDUCES TRANSMITTER RELEASE** by inhibiting calcium channel activity. **(A)** Norepinephrine released from sympathetic neurons combines with α_2-adrenergic receptors in the terminal membrane to decrease calcium influx and so reduce further transmitter release. **(B)** Norepinephrine reduces release of transmitter from sympathetic ganglia. Ganglia were loaded with radioactive norepinephrine and then enclosed in a perfusion chamber. Transmitter release was evoked by depolarization with a solution containing 50 mM potassium (dark bars). Addition of 30-μM unlabeled norepinephrine to the perfusion fluid (open bar) reduced the amount of radiolabeled transmitter released in response to potassium-induced depolarization. **(C)** Effect of norepinephrine on gating of calcium channels. Single-channel currents were recorded in cell-attached patches; channels were activated with a depolarizing pulse. When norepinephrine was included in the patch electrode, the unitary currents did not change in size, but channel openings were less frequent and of shorter duration. (B and C after Lipscombe et al., 1989.)

These neurons release norepinephrine, which binds to α_2-adrenergic receptors on their own presynaptic terminals. The α_2 receptors act to reduce transmitter release (Figure 8). Such autoinhibitory effects on transmitter release have been demonstrated in a variety of neurons.

In frog sympathetic neurons, it has been shown that norepinephrine inhibits release by reducing the probability of opening of N-type calcium channels. [Calcium channels have been grouped into three classes, called L, T, and N, which differ in their kinetic and pharmacological properties (Chapter 4). Release of transmitter from sympathetic neurons is controlled by calcium entering through N-type channels.[34]] Experiments with nonhydrolyzable derivatives of GTP (such as GTPγS) indicate that the response to norepinephrine is mediated by G proteins. Inhibition of calcium channel activity in cell-attached patches is observed only when norepinephrine is present in the recording pipette, suggesting a direct interaction between G proteins and calcium channels. Such direct interaction would provide a relatively rapid and localized mechanism for autoregulation of transmitter release.[34]

In contrast to the rapid local effects mediated by direct interaction of G proteins with their targets, many responses involve G-protein activation of cytoplasmic second messenger systems, which produce slower, more widespread effects. One of the most thoroughly studied examples of modulation via an intracellular second messenger is activation of β-adrenergic receptors in cardiac muscle cells by norepinephrine.[35,36] The changes produced by norepinephrine include an increase in the rate and force of contraction of the heart (Figure 9). The increase in contractile force is due to an increase in the height and duration of the plateau phase of the cardiac action potential. Voltage clamp studies by Reuter, Trautwein, Tsien, and others indicate that the change in the size of the action potential is due to an increase in the underlying calcium current. Single-channel recording from cardiac muscle cells, using the cell-attached mode of the patch clamp technique, confirms that β-adrenergic stimulation produces an increase in calcium channel activity. Moreover, norepinephrine added to the bathing fluid outside the patch electrode causes an increase in activity of calcium channels within the patch, a diagnostic test for responses mediated by diffusible cytoplasmic second messengers.[37,38]

Activation of β-adrenergic receptors is coupled to the increase in calcium conductance through the intracellular second messenger cyclic AMP (cAMP). It is another example of an indirectly coupled receptor acting through a G protein. As illustrated in Figure 10, binding of norepinephrine to β-adrenergic receptors activates a G protein, releasing its α subunit, which in turn binds to and activates the enzyme adenylyl cyclase. This enzyme converts ATP to cAMP, a readily diffusible intracellular second messenger that activates another enzyme, cAMP-dependent protein kinase. The catalytic subunits of this protein kinase mediate

G-protein activation of cAMP-dependent protein kinase

[34]Hirning, L. D. et al. 1988. *Science* 239: 57–61.

[35]Tsien, R. W. 1987. *In* L. K. Kaczmarek and I. B. Levitan (eds.). *Neuromodulation: The Biochemical Control of Neuronal Excitability.* Oxford University Press, New York, pp. 206–242.

[36]Trautwein, W. and Hescheler, J. 1990. *Annu. Rev. Physiol.* 52: 257–274.

[37]Reuter, H. et al. 1983. *Cold Spring Harbor Symp. Quant. Biol.* 48: 193–200.

[38]Tsien, R. W. et al. 1983. *Cold Spring Harbor Symp. Quant. Biol.* 48: 201–212.

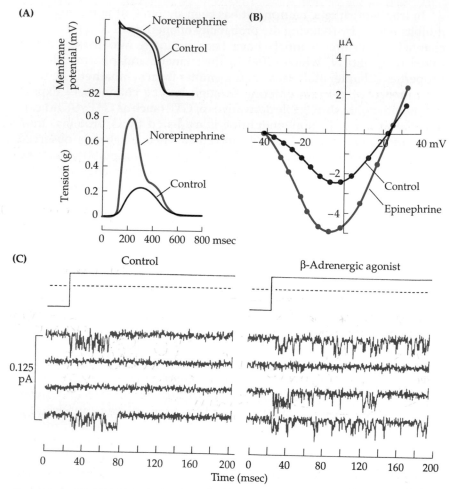

9 NOREPINEPHRINE INCREASES THE SIZE AND DURATION of the cardiac action potential by increasing calcium channel activity. **(A)** Effects of 10^{-6} *M* norepinephrine on the action potential and tension produced by cardiac muscle cells. **(B)** Epinephrine increases voltage-activated calcium current. Current–voltage relationship of calcium current measured under voltage clamp conditions in the absence and presence of 0.5 *μM* epinephrine. **(C)** Effect of *β*-adrenergic stimulation on calcium channel activity. Consecutive records from a voltage-clamped, cell-attached patch. Compared with control, 14-*μM* isoproterenol, a *β* agonist, causes an increase in the calcium channel activity evoked by the depolarizing pulse. (A after Reuter, 1974; B after Reuter et al., 1983; C after Tsien, 1987.)

the transfer of phosphate from ATP to the hydroxyl groups of serine and threonine residues in a variety of enzymes and channels, thereby modifying their activity.

Several lines of evidence are consistent with this scheme for *β*-adrenergic activation of calcium conductance in heart muscle.[36] The experimental approach takes advantage of well-established features of cAMP acting as an intracellular second messenger, as outlined in Box 2. For example, calcium channel activity is increased by forskolin, by

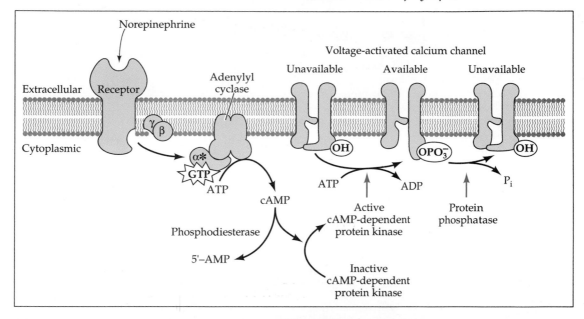

10 **EFFECTS OF NOREPINEPHRINE ON CALCIUM CHANNEL ACTIVITY are
mediated by cAMP.** Binding of norepinephrine to β-adrenergic receptors activates, through a G protein, the enzyme adenylyl cyclase. Adenylyl cyclase catalyzes the
conversion of ATP to cAMP. As the concentration of cAMP increases, it activates cAMP-
dependent protein kinase, an enzyme that phosphorylates proteins on serine and threonine residues (—OH). The response to norepinephrine is terminated by the hydrolysis
of cAMP to 5'-AMP and the removal of protein phosphate residues by protein phosphatases. In cardiac muscle cells, norepinephrine causes phosphorylation of voltage-
activated calcium channels, converting them to a form (*available*) that can be opened
by depolarization.

membrane-permeable derivatives of cAMP, by inhibitors of phosphodiesterase, and by direct intracellular injection of cAMP itself. Similarly,
intracellular injection of the catalytic subunit of cAMP-dependent protein kinase leads to an increase in calcium current, while injection of
excess regulatory subunit or inhibitors of protein kinase blocks adrenergic stimulation of calcium currents. ATPγS augments adrenergic activation of calcium channels by forming stably phosphorylated proteins,
while intracellular injection of protein phosphatases prevents or reverses adrenergic stimulation of calcium currents by rapidly removing
protein phosphate residues.

Such experiments clearly demonstrate that the effects of β-adrenergic
stimulation on calcium currents are mediated by an increase in cAMP
and activation of protein kinase, but do not identify the protein or
proteins being phosphorylated. Experiments by Catterall, Trautwein,
and their colleagues on calcium channels purified from skeletal muscle
demonstrate that the calcium channel itself is the target.[42,43] β-Adre-

[42]Curtis, B. M. and Catterall, W. A. 1986. *Biochemistry* 25: 3077–3083.
[43]Flockerzi, V. et al. 1986. *Nature* 323: 66–68.

BOX 2   CYCLIC AMP AS A SECOND MESSENGER

Experiments by Sutherland, Krebs, Walsh, Rodbell, Gilman, and their colleagues, initially aimed at understanding how the hormones epinephrine and glucagon elicit breakdown of glycogen in the liver, led to the discovery of cyclic AMP (cAMP) and the concept of intracellular second messengers.[39–41] They showed that binding of the hormone to its receptor activates a G protein that, in turn, stimulates the enzyme adenylyl cyclase. Adenylyl cyclase catalyzes the synthesis of cAMP from ATP. The increase in cAMP concentration activates cAMP-dependent protein kinase, an enzyme that phosphorylates its target proteins on serine and threonine residues. Cyclic AMP is subsequently degraded by phosphodiesterase to AMP, and the phosphate residues on

[39]Sutherland, E. W. 1972. *Science* 177: 401–408.

[40]Schramm, M. and Selinger, Z. 1984. *Science* 225: 1350–1356.

[41]Gilman, A. G. 1987. *Annu. Rev. Biochem.* 56: 615–649.

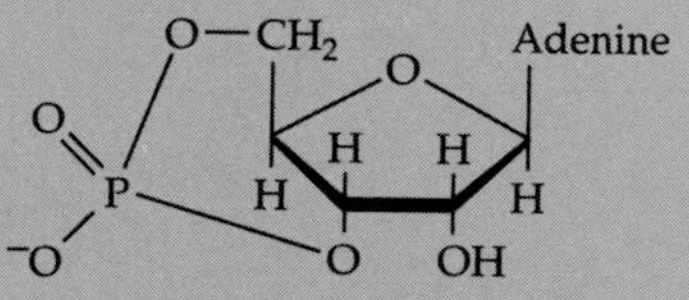

Cyclic AMP (cAMP)
Adenosine 3', 5'–monophosphate

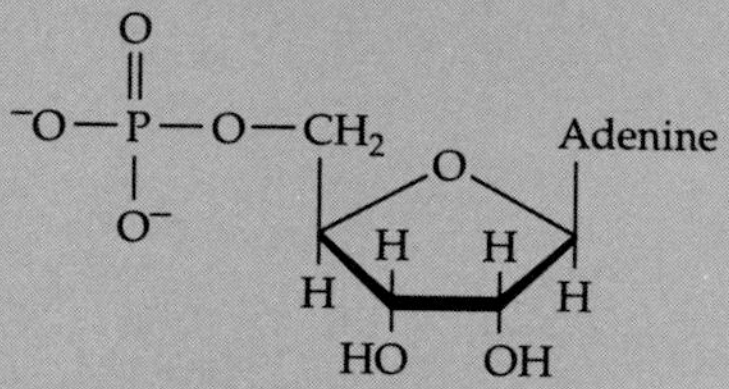

AMP
Adenosine 5'–monophosphate

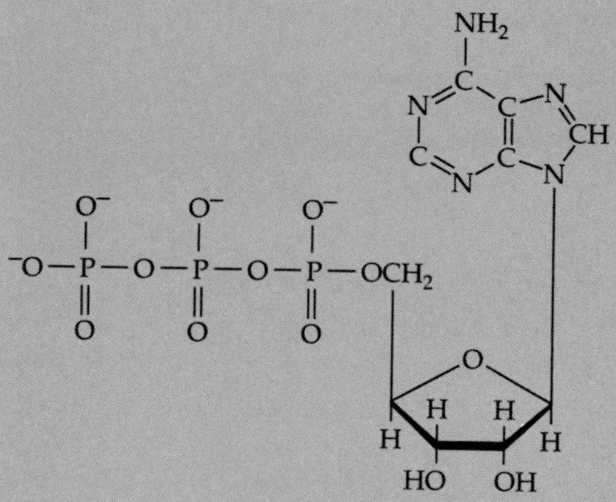

ATP
Adenosine 5'–triphosphate

the target proteins are removed by protein phosphatases (see Figure 10).

A variety of tests can be used to determine if the response to a transmitter or hormone is mediated by cAMP. Some of these depend on activating adenylyl cyclase or elevating cAMP concentrations directly. For example, intracellular injection of cAMP and addition of membrane-permeable derivatives of cAMP such as 8-bromo-cAMP or dibutyryl-cAMP mimic cAMP-mediated responses. Similarly, direct activation of adenylyl cyclase by forskolin mimics the response. Inhibitors of phosphodiesterase, such as the methylxanthines theophylline and caffeine, either mimic or enhance the response, depending on the endogenous level of cyclase activity. Other procedures test the involvement of cAMP-dependent protein kinase (also known as protein kinase A). This enzyme is composed of two reg-

nergic stimulation of heart muscle enhances the activity of L-type calcium channels. L-type calcium channels, purified from skeletal muscle on the basis of their high affinity for specific dihydropyridine inhibitors, can be incorporated into lipid vesicles[42] or planar bilayers[43] (Figure 11). If the reconstituted channels are exposed to active cAMP-dependent protein kinase and ATP, the probability that the channel will open is increased as the protein is phosphorylated. Thus β-adrenergic modu-

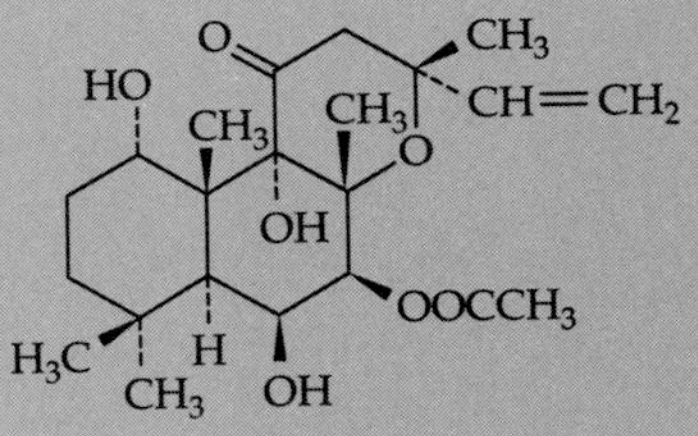

Forskolin

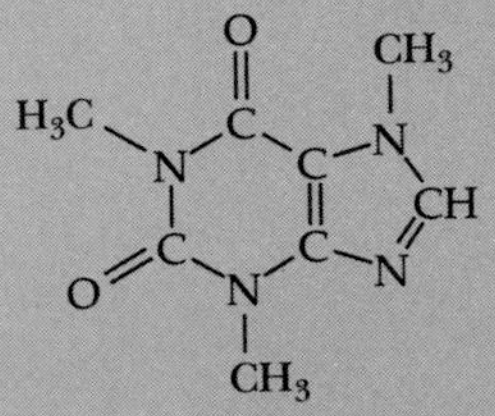

Caffeine
1,3,7–Trimethylxanthine

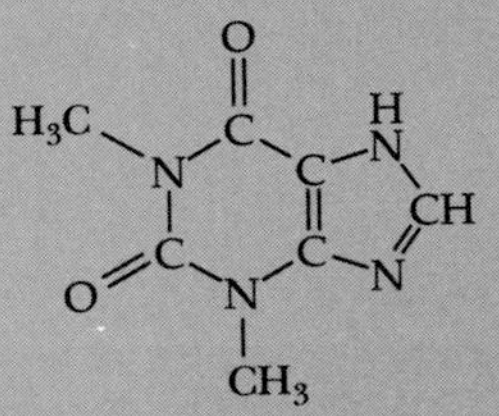

Theophylline
1,3–Dimethylxanthine

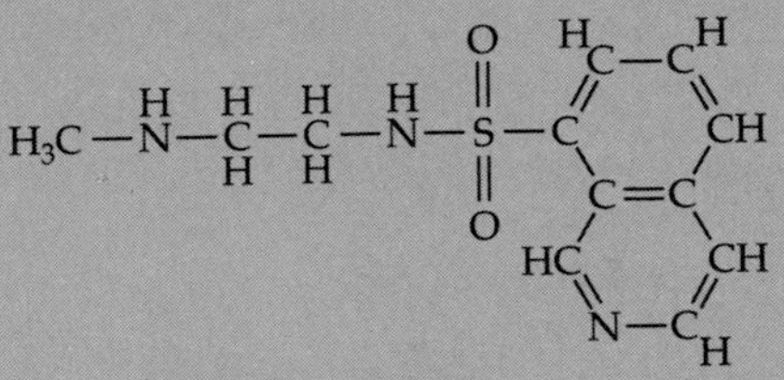

H-8
N–[2–(methylamino)ethyl]–5–
isoquinolinesulfonamide

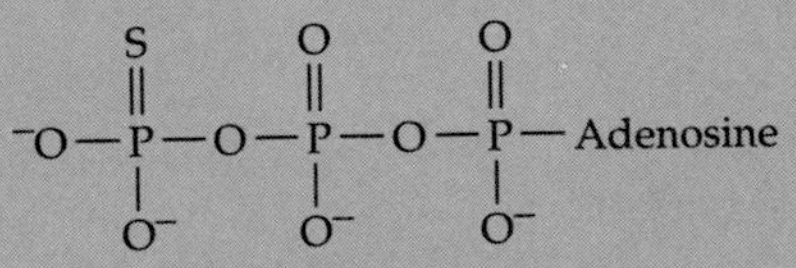

ATPγS
Adenosine 5'–O–[γ–thio]
triphosphate

excess regulatory subunits will be inhibitory. Additional inhibitors of this enzyme have been developed, including H-8 (which also inhibits several other protein serine kinases), specific peptide inhibitors, and derivatives of ATP that cannot be used by the kinase as a source for phosphate residues. These inhibitors block responses mediated by cAMP. On the other hand, treatments that inhibit protein phosphatases augment and prolong responses mediated by cAMP. These include injection of specific phosphatase inhibitors and of ATPγS, an analogue of ATP that can be used as a cosubstrate by cAMP-dependent protein kinase to form phosphoproteins with thiophosphate linkages, which are resistant to hydrolysis by protein phosphatases.

ulatory and two catalytic subunits. In the absence of cAMP, the four subunits exist as a complex, the regulatory subunits blocking the activity of the catalytic subunits. When cAMP binds to the regulatory subunits, the complex dissociates, freeing active catalytic subunits. Thus intracellular injection of purified catalytic subunit will mimic responses mediated by increased cAMP concentration, while injection of

lation of L-type calcium channels is mediated by cAMP-dependent phosphorylation of the channel protein itself.

The level of cAMP and so the activity of calcium channels in cardiac muscle cells are also influenced by another transmitter, acetylcholine. Stimulation of muscarinic receptors by ACh activates a G_i protein that inhibits the activity of adenylyl cyclase. This reduces the concentration of cAMP and so decreases calcium channel activity. In addition to

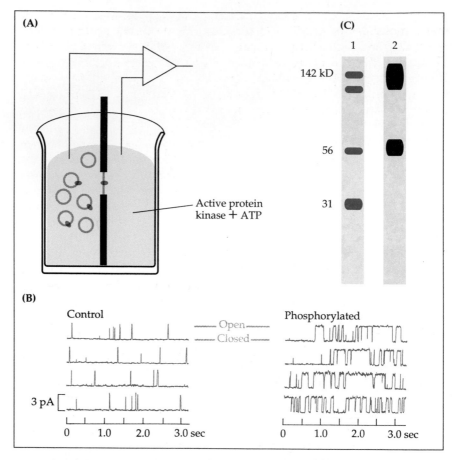

11 PHOSPHORYLATION OF ISOLATED CALCIUM CHANNELS increases their probability of opening. (A) Purified calcium channels, incorporated into small lipid vesicles (liposomes), are added to the solution on one side of a phospholipid bilayer. When liposomes fuse into the bilayer membrane, single-channel events can be recorded. (B) Single-channel records before and after addition of ATP and the catalytic subunit of cAMP-dependent protein kinase to the solution bathing one side of the bilayer. Phosphorylation increases the probability of channel opening. (C) Phosphorylation of purified calcium channels. Lane 1: An SDS polyacrylamide gel of purified calcium channels stained for protein. The calcium channel is composed of four major subunits of 142, 122, 56, and 31 kD. Lane 2: Autoradiograph of a similar gel in which purified calcium channels were exposed to the catalytic subunit of cAMP-dependent protein kinase and [32P]-ATP before electrophoresis. The 142- and 56-kD subunits are phosphorylated. (After Flockerzi et al., 1986.)

norepinephrine and ACh, various other hormones modulate the activity of the same L-type calcium channels in cardiac muscle. The effects of many of these hormones are mediated by G-protein activation or inhibition of adenylyl cyclase; others act through different protein kinases to phosphorylate calcium channels and modulate their activity.[36]

The two-step enzymatic cascade involving adenylyl cyclase and cAMP-dependent protein kinase provides tremendous amplification compared with direct opening or closing of channels by activated G proteins. Each activated adenylyl cyclase can catalyze the synthesis of many molecules of cAMP and thereby activate many protein kinase molecules, and each activated kinase can phosphorylate many proteins. Thus the activity of many molecules of a target protein at widespread sites may be modulated by the occupation of a few receptors. Moreover, in any given cell cAMP-dependent protein kinase may phosphorylate a variety of proteins and so modulate a broad spectrum of cellular processes.

Modulation of the duration of the presynaptic action potential appears to be a common mechanism by which synaptic efficacy is altered. For example, Fischbach and his colleagues made intracellular recordings from neurons dissociated from chick dorsal root ganglia and grown in cell culture. They found that a variety of neurotransmitters, including GABA, norepinephrine, 5-hydroxytryptamine, and the peptides enkephalin and somatostatin (Chapter 10), cause a decrease in the duration of the action potential and a corresponding reduction in the amount of transmitter released[44,45] (Figure 12). Voltage clamp measurements indicate that the decrease is due to a reduction in calcium current.[46] The mechanism by which norepinephrine and GABA produce their effects is beginning to be understood.[47] Binding of norepinephrine and GABA to their receptors activates, through a particular G protein, the enzyme phospholipase C. The involvement of a G protein is indicated by the findings that intracellular GDPβS, a GDP analogue that prevents activation of G proteins, blocks the decrease in action potential duration, as does treating the cells with pertussis toxin, which irreversibly inactivates G_t, G_i, G_p, and G_o. Phospholipase C hydrolyzes the membrane lipid phosphatidylinositol 4,5-bisphosphate (PIP$_2$), producing two products: inositol 1,4,5-triphosphate (IP$_3$) and diacyglycerol (DAG) (Figure 12C; Box 3). IP$_3$ releases calcium ions from the endoplasmic reticulum.[48,49] The resulting increase in cytoplasmic calcium concentration can activate a variety of intracellular enzymes. Diacylglycerol, on the other hand, activates protein kinase C, an enzyme that, like cAMP-dependent protein kinase, can phosphorylate particular serine and threonine residues on a variety of proteins. The effects of norepinephrine and GABA on action potential duration in chick sensory neuron appear to be caused by such an activation of protein kinase C.[47] Thus, direct activation of protein kinase C with diacylglycerol analogues causes a similar decrease in calcium current in these neurons; inhibitors of protein kinase C block this effect, as well as the decrease in calcium current produced by norepinephrine and GABA.

G-protein activation of phospholipase C

[44]Dunlap, K. and Fischbach, G. D. 1978. *Nature* 276: 837–839.

[45]Mudge, A. W., Leeman, S. E. and Fischbach, G. D. 1979. *Proc. Natl. Acad. Sci. USA* 76: 526–530.

[46]Dunlap, K. and Fischbach, G. D. 1981. *J. Physiol.* 317: 519–535.

[47]Dolphin, A. C. 1990. *Annu. Rev. Physiol.* 52: 243–255.

[48]Berridge, M. J. 1988. *Proc. R. Soc. Lond. B* 234: 359–378.

[49]Blumenfeld, H. et al. 1990. *Neuron* 5: 487–499.

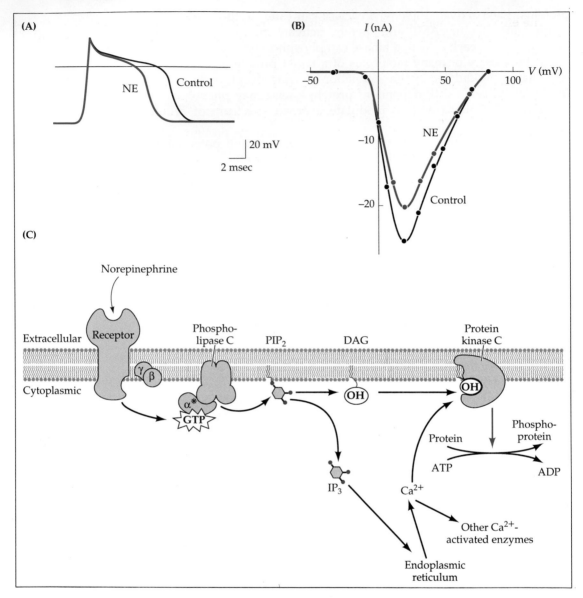

(A)

NE   Control

20 mV

2 msec

(B)

I (nA)

V (mV)

−50   50   100

NE

−10

Control

−20

(C)

Norepinephrine

Extracellular   Receptor   Phospho-lipase C   PIP$_2$ DAG   Protein kinase C

Cytoplasmic

γ β

α^*

GTP

OH

OH

Protein

ATP   ADP

Phospho-protein

IP$_3$ Ca^{2+}

Other Ca^{2+}-activated enzymes

Endoplasmic reticulum

12   **NOREPINEPHRINE DECREASES DURATION OF CALCIUM ACTION POTEN-TIALS** in chick dorsal root ganglion cells in culture. Action potentials in these neurons are unusually prolonged; a relatively large calcium current contributes to the distinct plateau during repolarization. **(A)** Norepinephrine (10^{-5} M) causes a decrease in the duration of the action potential recorded with an intracellular microelectrode from a cell bathed in saline containing 5.4 mM calcium. **(B)** Norepinephrine (10^{-4} M) decreases the amplitude of the voltage-activated calcium current. Current–voltage rela-tionship for calcium current recorded from cells bathed in solutions containing TTX, TEA, and 10-mM calcium. **(C)** The binding of norepinephrine to its receptor activates, through a G protein, the enzyme phospholipase C. This enzyme hydrolyzes the phos-pholipid PIP$_2$, releasing two intracellular second messengers: diacylglycerol (DAG) and inositol 1,4,5-triphosphate (IP$_3$). IP$_3$ releases calcium from the endoplasmic reticulum into the cytoplasm. DAG and calcium together activate protein kinase C. Activation of protein kinase C leads to a reduction in calcium content. (A and B after Dunlap and Fischbach, 1981.)

The effect of the neuropeptide FMRFamide (Phe-Met-Arg-Phe-NH$_2$) on potassium channels in sensory neurons in the central nervous system of *Aplysia* (studied by Kandel, Schwartz, Siegelbaum, and others; see Chapter 13) is mediated by yet another target of G protein action, phospholipase A$_2$. This enzyme acts on membrane phospholipids (for example, PIP$_2$) as well as on diacylglycerol to release the fatty acid arachidonic acid (Figure 13). Arachidonic acid is metabolized in two different ways: the lipoxygenase pathway, which forms products called

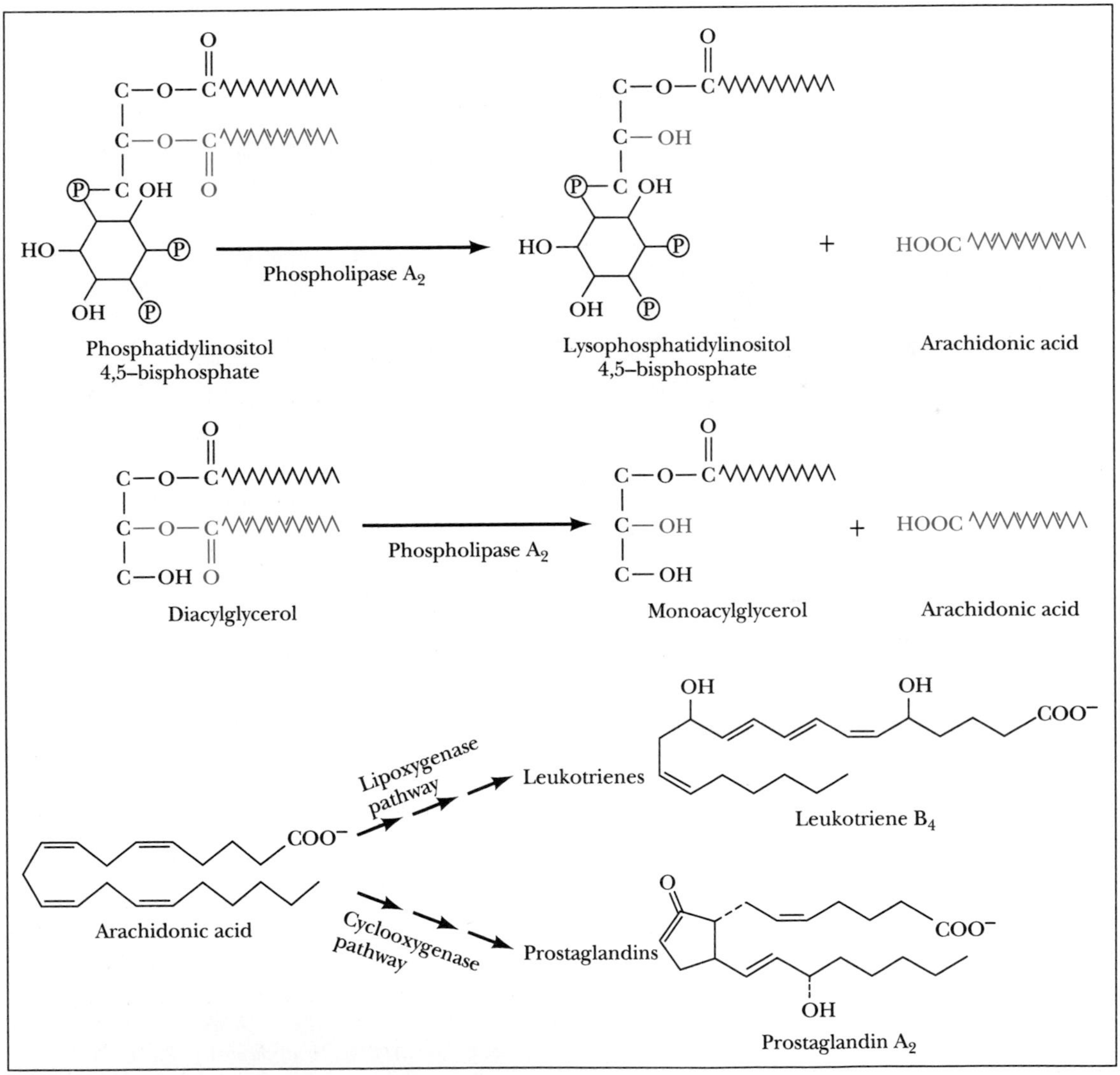

G-protein activation of phospholipase A$_2$

13 **FORMATION AND METABOLISM OF ARACHIDONIC ACID. In mammals, the metabolites of arachidonic acid—prostaglandins and leukotrienes—are found in cells throughout the body and produce a remarkably broad spectrum of effects. Anti-inflammatory drugs such as aspirin act by inhibiting the synthesis of prostaglandins.**

BOX 3   DIACYLGLYCEROL AND IP₃ AS SECOND MESSENGERS

Activation of receptors coupled through G proteins to the enzyme phospholipase C leads to the hydrolysis of the membrane lipid phosphatidylinositol 4,5-bisphosphate (PIP₂), producing two intracellular second messengers, diacylglycerol (DAG) and inositol 1,4,5-triphosphate (IP₃) (see Figure 12C).

IP₃ is water-soluble, diffuses through the cytoplasm, and binds to receptors on the endoplasmic reticulum that allow calcium ions, se-

R   Usually stearate
$(CH_3(CH_2)_{16}COO^-)$

R'   Usually arachidonate
$(CH_3(CH_2)_4(CH=CHCH_2)_4(CH_2)_2COO^-)$

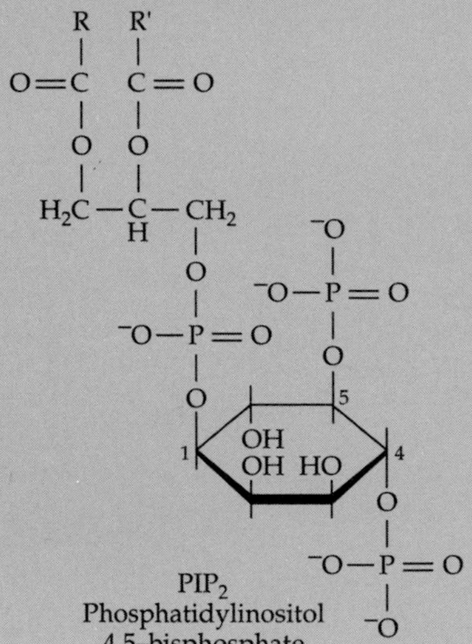

PIP₂
Phosphatidylinositol
4,5–bisphosphate

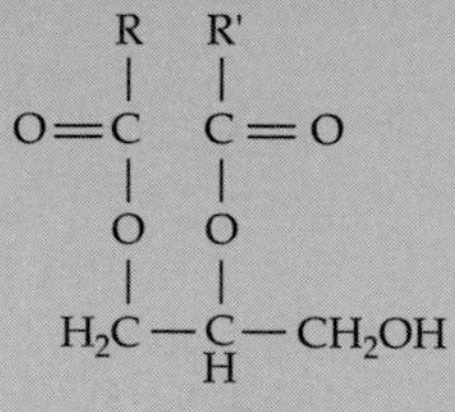

DAG
Diacylglycerol

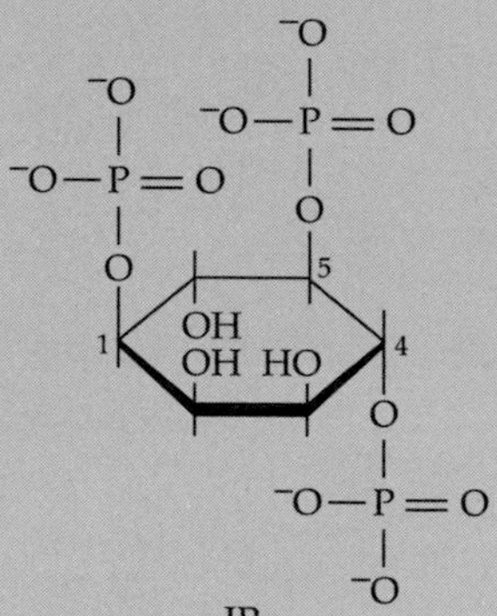

IP₃
Inositol 1,4,5–triphosphate

leukotrienes, and the cyclooxygenase pathway, which leads to the formation of prostaglandins.[50]

In *Aplysia* sensory neurons, binding of FMRFamide to its receptor activates a G protein that, in turn, activates phospholipase A₂.[51] As arachidonic acid accumulates, it is metabolized along the lipoxygenase pathway to 12-HPETE (Figure 14), which binds to S-type potassium channels (see Chapter 13) and increases their probability of opening.[52–54] This reduces the duration of the action potential, resulting in less calcium entry and less transmitter release.[49,52]

[50]Wolfe, L. S. 1989. *In* G. Siegel et al. (eds.). *Basic Neurochemistry.* Raven Press, New York, pp. 399–414.

[51]Volterra, A. and Siegelbaum, S. A. 1988. *Proc. Natl. Acad. Sci. USA* 85: 7810–7814.

[52]Piomelli, D. et al. 1987. *Nature* 328: 38–43.

[53]Belardetti, F., Kandel, E. R. and Siegelbaum, S. A. 1987. *Nature* 325: 153–156.

[54]Buttner, N., Siegelbaum, S. A. and Volterra, A. 1989. *Nature* 342: 553–555.

questered therein, to be released. Calcium then acts as a "third" messenger to regulate the activity of calcium-dependent proteins in the cell.

DAG is hydrophobic and remains associated with the membrane, where it activates protein kinase C, a protein serine kinase. In unstimulated cells, protein kinase C is found in the cytoplasm in an inactive form; in the presence of calcium and DAG it becomes associated with the membrane and is activated. The active kinase is capable of phosphorylating a variety of proteins on serine and threonine residues.

No simple tests exist to reliably identify responses mediated by IP_3. Inhibition by agents that interfere with phosphatidylinositol turnover, such as lithium, can suggest a role for IP_3, and treatments that elevate intracellular calcium concentrations can mimic effects mediated by IP_3.

On the other hand, the role of protein kinase C can be assessed using specific activators and inhibitors of this enzyme. Analogues of DAG that directly stimulate protein kinase C include synthetic diacylglycerols, such as OAG, and the tumor-promoting phorbol esters. Phospholipid analogues such as sphingosine, and protein serine kinase inhibitors such as H-7, inhibit protein kinase C and thus block responses mediated by this enzyme.

TPA
Phorbol 12–myristate 13–acetate

OAG
1–Oleoyl–2–acetylglycerol

An important generalization that emerges from such studies is that potassium and calcium channels are prime targets for modulation by transmitters acting through indirectly coupled receptors. Changes in channel activation can influence the resting potential, spontaneous activity, the response to other excitatory or inhibitory inputs, and the amount of calcium that enters during an action potential. Such effects clearly play a major role in signaling in the nervous system. The variety of indirect mechanisms that influence calcium and potassium channels make it difficult to predict how channel activity might be modulated in any given cell. For example, norepinephrine and GABA decrease calcium current in chick sensory cells via activation of protein kinase C,[47] while GABA, but not norepinephrine, acts via the cAMP pathway to lengthen the duration of calcium action potentials in lamprey sensory neurons by inhibiting calcium-activated potassium channels.[55]

Indirectly coupled receptors modulate potassium and calcium channels

[55]Leonard, J. P. and Wickelgren, W. O. 1986. *J. Physiol.* 375: 481–497.

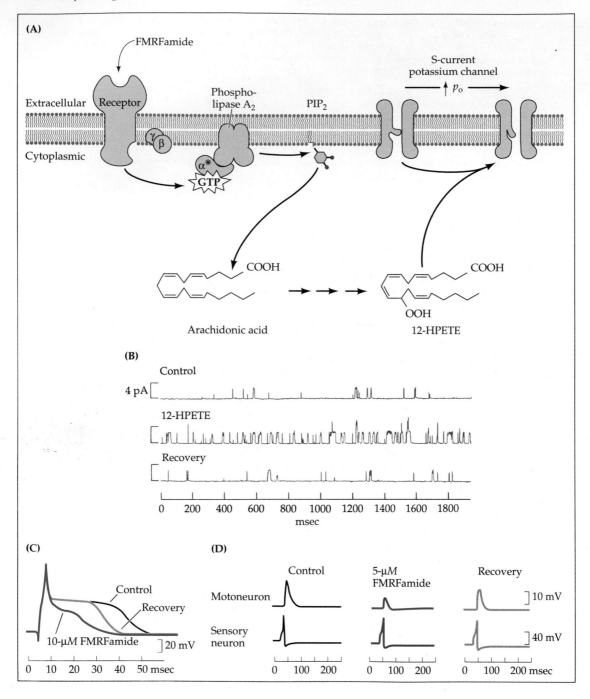

◄ **14**   THE NEUROPEPTIDE FMRFamide decreases the duration of the action potential by activation of phospholipase A_2. (A) Binding of FMRFamide to its receptor on sensory neurons in the CNS of *Aplysia* activates, through a G protein, the enzyme phospholipase A_2. This enzyme hydrolyzes membrane phospholipids, such as PIP_2, to release arachidonic acid. Arachidonic acid is converted to 12-HPETE, which binds to S-current potassium channels and increases their probability of opening (p_o). (B) 12-HPETE increases S-current potassium channel activity. Single S-channel current records from an inside-out patch, taken before, during, and after application of 20-mM 12-HPETE. (C) Intracellular recording from a sensory neuron in an intact abdominal ganglion. FMRFamide (10 μM) causes a reversible decrease in the duration of the action potential. (D) Intracellular recordings from sensory and motoneurons in an intact ganglion. The amplitude of the excitatory synaptic potential evoked by stimulation of the sensory neuron is reduced by application of 5-μM FMRFamide. The effects of FMRFamide in (C) and (D) are mimicked by application of either arachidonic acid or 12-HPETE. (B after Buttner et al., 1989; C and D after Piomelli et al., 1987.)

The complexity extends to the level of single cells; the activity of rat hippocampal pyramidal neurons is modulated by at least six different transmitters, which act on four kinds of potassium channels and two kinds of calcium channels.[16,25,47]

The intracellular concentration of calcium is determined by a complex system of regulatory mechanisms. Calcium, in turn, modulates a broad range of metabolic activities (Figure 15). One mechanism by which the concentration of calcium within cells is raised is influx through voltage-activated calcium channels (Chapter 2) and through

Calcium as an intracellular second messenger

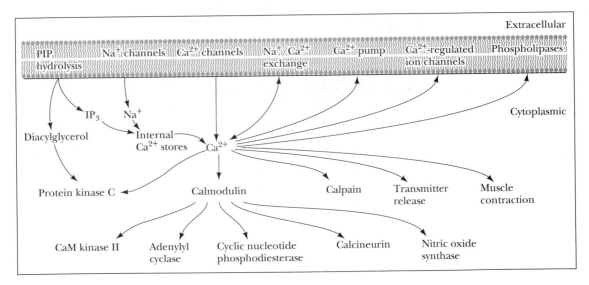

15   CALCIUM AS AN INTRACELLULAR SECOND MESSENGER. The concentration of calcium in the cytoplasm is regulated by influx through membrane channels, by the activity of calcium pumps and exchangers in the plasma membrane, by sequestration in internal stores such as the endoplasmic reticulum, and by the release from internal storage sites by sodium influx, calcium influx, and IP_3. Calcium, in turn, regulates membrane proteins (such as ion channels and phospholipases) and cytosolic proteins (including protein kinase C, calmodulin, and calpain). (After Kennedy, 1989.)

cation channels at excitatory synapses (Chapter 7). It can also be raised or lowered through the activity of calcium pumps and exchangers (Chapter 3). In addition, calcium can be released from intracellular storage sites by increases in the intracellular concentrations of calcium, sodium, and IP_3.[56,57] A variety of membrane channels are regulated directly by changes in intracellular calcium concentration, including specific potassium, cation-selective, and chloride channels.[58] Two families of membrane phospholipases are also directly regulated by cytoplasmic calcium: phospholipase C and phospholipase A_2. These enzymes, which are involved in the production of the intracellular second messengers IP_3, diacylglycerol, and arachidonic acid, are also regulated by G proteins, as described above. Within the cytoplasm, calcium activates three major targets: protein kinase C, calmodulin, and a calcium-dependent protease called calpain. Each of these proteins, in turn, influences the activity of several secondary targets. The effect of protein kinase C on the duration of the action potential in sensory neurons has already been mentioned. Calpains are a group of proteases that have been implicated in regulation of the cytoskeleton as well as a number of membrane proteins.[59] Calmodulin is a ubiquitous protein that has four calcium-binding sites.[60] When these sites are occupied, calmodulin activates a wide variety of enzymes, including calcium-calmodulin–dependent protein kinases, adenylyl cyclase, cyclic nucleotide phosphodiesterase, a protein phosphatase called calcineurin, and nitric oxide synthase. Nitric oxide, a water- and lipid-soluble gas, acts as a novel type of transmitter by diffusing into adjacent cells and activating guanylyl cyclase.[61,62]

Optical measurements of calcium signals indicate that increases in calcium concentration are often confined to particular regions of the cell.[57,63–65] The effects of changes in intracellular calcium concentration will depend on the subcellular distribution of calcium, of the calcium-activated proteins, and of their target enzymes and ion channels.

Time course of indirect transmitter action

At the skeletal neuromuscular junction, one or two milliseconds are required for acetylcholine to be released, diffuse across the synaptic cleft, bind to and open ligand-activated channels, and dissociate from its receptors. These events are much too fast to be mediated by enzymes such as adenylyl cyclase or phospholipase C, which take many milliseconds to catalyze the synthesis of a single molecule of cyclic AMP or

[56]Lipscombe, D. et al. 1988. *Neuron* 1: 355–365.

[57]Tsien, R. W. and Tsien, R. Y. 1990. *Annu. Rev. Cell Biol.* 6: 715–760.

[58]Marty, A. 1989. *Trends Neurosci.* 12: 420–424.

[59]Melloni, E. and Pontremoli, S. 1989. *Trends Neurosci.* 12: 438–444.

[60]Rasmussen, C. D. and Means, A. R. 1989. *Trends Neurosci.* 12: 433–438.

[61]Ignarro, L. J. 1990. *Pharmacol. Toxicol.* 67: 1–7.

[62]Bredt, D. S. et al. 1991. *Nature* 351: 714–718.

[63]Ross, W. N., Arechiga, H. and Nicholls, J. G. 1988. *Proc. Natl. Acad. Sci. USA* 85: 4075–4078.

[64]Lipscombe, D. et al. 1989. *Proc. Natl. Acad. Sci. USA* 85: 2398–2402.

[65]Ross, W. N., Lasser-Ross, N. and Werman, R. 1990. *Proc. R. Soc. Lond. B* 240: 173–185.

the hydrolysis of a membrane lipid. Moreover, the time course of direct activation of ligand-gated ion channels is limited by the lifetime of the transmitter. At the skeletal neuromuscular junction, transmitter lifetime is very short; as ACh dissociates from its receptor, it is hydrolyzed by acetylcholinesterase before it can activate additional receptors. Indirect synaptic mechanisms are slower, and the underlying enzymatic processes can outlive receptor activation. Thus activation of a membrane channel through binding of a G protein α subunit to the channel itself tends to have a time course of seconds, reflecting the lifetime of the activated α subunit.[22] Changes mediated by enzymatic production of diffusible cytoplasmic second messengers such as cAMP or IP$_3$ are slower still, lasting seconds to minutes, reflecting the slow time course of changes in second messenger concentration.

Yet experience tells us that changes in signaling in the nervous system can last a lifetime. How can such long-lasting changes in synaptic efficacy be produced? One answer comes from the properties of several of the protein kinases discussed above. These enzymes are themselves targets for phosphorylation. For example, when activated by calcium, calcium-calmodulin kinase II phosphorylates itself. If several of the subunits are phosphorylated, the properties of the enzyme change; it becomes constitutively active and no longer requires the presence of calcium-calmodulin for activity. This provides one mechanism by which a transient increase in calcium concentration can be translated into long-lasting activation of the kinase, which in turn can cause sustained changes in the activity of its other target proteins. For changes to persist for days or longer, protein synthesis is required (Chapter 13). Many of the second messenger systems described in this chapter have been shown to produce changes in protein synthesis.[66] Some of the most rapid effects that have been measured occur in transcription of the c-*fos* protein, one member of a family of nuclear proteins thought to be responsible for regulation of gene expression. Such alterations in gene expression can then produce metabolic or structural changes that permanently alter the response of the cell.[67]

SUGGESTED READING

General

Hall, Z. W. 1992. *An Introduction to Molecular Neurobiology.* Sinauer, Sunderland, MA.

Kaczmarek, L. K. and Levitan, I. B. (eds.). 1987. *Neuromodulation: The Biochemical Control of Neuronal Excitability.* Oxford University Press, New York.

Levitan, I. B. and Kaczmarek, L. K. 1991. *The Neuron: Cell and Molecular Biology.* Oxford University Press, New York.

Trends in Neurosciences. 1989. Special Issue: Calcium–Effector Mechanisms. *Trends Neurosci.* 12: 417–479.

[66]Morgan, J. I. and Curran, T. 1991. *Annu. Rev. Neurosci.* 14: 421–451.
[67]Rose, S. P. R. 1991. *Trends Neurosci.* 14: 390–397.

Reviews

Berridge, M. J. 1988. Inositol lipids and calcium signalling. *Proc. R. Soc. Lond. B* 234: 359–378.

Brown, D. A. 1990. G-proteins and potassium currents in neurons. *Annu. Rev. Physiol.* 52: 215–242.

Brown, A. M. and Birnbaumer, L. 1990. Ionic channels and their regulation by G protein subunits. *Annu. Rev. Physiol.* 52: 197–213.

Dolphin, A. C. 1990. G protein modulation of calcium currents in neurons. *Annu. Rev. Physiol.* 52: 243-255.

Nicoll, R. A. 1988. The coupling of neurotransmitter receptors to ion channels in the brain. *Science* 241: 545–551.

Szabo, G. and Otero, A. S. 1990. G protein mediated regulation of K^+ channels in heart. *Annu. Rev. Physiol.* 52: 293–305.

Tsien, R. W. and Tsien, R. Y. 1990. Calcium channels, stores, and oscillations. *Annu. Rev. Cell Biol.* 6: 715–760.

Original Papers

Blumenfeld, H., Spira, M. E., Kandel, E. R. and Siegelbaum, S. A. 1990. Facilitatory and inhibitory transmitters modulate calcium influx during action potentials in *Aplysia* sensory neurons. *Neuron* 5: 487–499.

Codina, J., Yatani, A., Grenet, D., Brown, A. M. and Birnbaumer, L. 1987. The α subunit of the GTP binding protein G_k opens atrial potassium channels. *Science* 236: 442–445.

Dunlap, K. and Fischbach, G. D. 1981. Neurotransmitters decrease the calcium conductance activated by depolarization of embryonic chick sensory neurones. *J. Physiol.* 317: 519-535.

Flockerzi, V., Oeken, H.-J., Hofmann, F., Pelzer, D., Cavalié, A. and Trautwein, W. 1986. Purified dihydropyridine-binding site from skeletal muscle t-tubules is a functional calcium channel. *Nature* 323: 66–68.

Jan, Y. N., Bowers, C. W., Branton, D., Evans, L. and Jan, L. Y. 1983. Peptides in neuronal function: Studies using frog autonomic ganglia. *Cold Spring Harbor Symp. Quant. Biol.* 48: 363–374.

Kuffler, S. W. 1980. Slow synaptic responses in autonomic ganglia and the pursuit of a peptidergic transmitter. *J. Exp. Biol.* 89: 257–286.

Sakmann, B., Noma, A. and Trautwein, W. 1983. Acetylcholine activation of single muscarinic K^+ channels in isolated pacemaker cells of the mammalian heart. *Nature* 303: 250–253.

CHAPTER NINE

CELLULAR AND MOLECULAR BIOCHEMISTRY OF SYNAPTIC TRANSMISSION

The identification of the transmitter at a particular synapse relies on evidence that the substance is present in the presynaptic axon terminal, is released from the terminal by nerve stimulation, and, when applied exogenously to the postsynaptic cell membrane, mimics the action of the natural transmitter. Low-molecular-weight compounds that have been identified as transmitters in the mammalian nervous system are acetylcholine, norepinephrine, γ-aminobutyric acid, dopamine, epinephrine, 5-hydroxytryptamine, histamine, adenosine triphosphate, glutamate, and glycine. In addition, 40 or more peptides are likely to act as neurotransmitters or neuromodulators. Many neurons release more than one transmitter, typically a low-molecular-weight transmitter and a neuropeptide.

A variety of mechanisms operate within nerve cells to ensure that the supply of transmitter is adequate to meet the demands for release. These mechanisms include rapid changes in transmitter synthesis (achieved by modulating the activity of rate-limiting enzymes), and long-term changes brought about by altering the number of enzyme molecules themselves. Low-molecular-weight neurotransmitters are synthesized in the axon terminal and stored there within synaptic vesicles. Neuropeptides are synthesized in the cell body, packaged into vesicles, and shipped down the axon for release. Two mechanisms move proteins along axons: Slow axonal transport carries soluble proteins and components of the axonal cytoskeleton from the cell body toward the terminal at 1 to 2 mm per day, and fast axonal transport moves vesicles and other organelles at rates up to 400 mm per day either toward the terminal (anterograde transport) or back to the cell body (retrograde transport). Fast transport is powered by two molecular "motors" that propel organelles along microtubule "tracks": kinesin in the anterograde direction and dynein in the retrograde direction.

Transmitter is released from axon terminals by exocytosis, during which a vesicle fuses with and flattens into the plasma membrane, spilling its contents into the synaptic cleft. Components of the vesicle membrane are subsequently retrieved from the surface by endocytosis and reformed into synaptic vesicles. Although several enzyme-catalyzed processes that modulate the probability of release have been identified, the detailed mechanism of calcium-induced vesicle fusion remains obscure.

The direct and indirect mechanisms by which neurotransmitters produce their effects have been described in Chapters 7 and 8. Molecular genetic experiments have revealed that directly coupled receptors constitute a superfamily of ligand-activated ion channels that share many structural features (Chapter 2). Indirectly coupled receptors are a second superfamily that produce effects through G proteins (Chapter 8). A common feature of both families is receptor desensitization, a reduction in the response during prolonged or repeated application of ligand. Directly coupled receptors tend to be concentrated in the membrane of the postsynaptic cell immediately beneath the nerve terminal; indirectly coupled receptors often have a more widespread distribution.

The final step in chemical synaptic transmission is termination of transmitter action. Typically, low-molecular-weight transmitters are either degraded after release or are taken up into the axon terminal, repackaged into vesicles, and released again. The actions of neuropeptides are terminated by diffusion. Drugs that interfere with transmitter degradation or uptake can have profound effects on signaling, indicating that the rapid termination of transmitter action is often crucial to synaptic function.

Sixty years after the idea of chemical synaptic transmission was proposed by Elliot in 1904,[1] only three compounds—acetylcholine (ACh), norepinephrine, and γ-aminobutyric acid (GABA)—had been unequivocally identified as neurotransmitters. Little was known about the biochemical properties of their receptors, which were simply classified on the basis of their pharmacology: nicotinic or muscarinic for ACh; α or β for norepinephrine. The last three decades have witnessed an explosion in biochemical research on the nervous system, fueled by rapid conceptual and technological advances in biochemistry and genetics. As a result, over 50 compounds have been shown to be neurotransmitters or neuromodulators. For many, the enzymes mediating their synthesis and degradation have been purified and cloned, the protein structure and molecular mechanisms of action of their receptors have been established, and the various intracellular pathways by which they mediate or modulate synaptic transmission have been characterized. Chapters 7 and 8 focused on the mechanisms by which transmitters influence postsynaptic cells. This chapter summarizes the biochemistry of transmitter synthesis, storage, release, and inactivation.

In Chapter 1 we emphasized that the nervous system uses stereotyped electrical signals that are endowed with meaning by virtue of the specific patterns of connections between neurons. To a considerable extent the same is true of the biochemical signals that convey information across chemical synapses. The same neurotransmitters are used over and over at synapses throughout the nervous system; the message

---

[1]Elliot, T. R. 1904. *J. Physiol.* 31: xx–xxi (Proc.).

conveyed by their release is determined by the precise way in which neurons are interconnected. For example, the release of ACh from motor nerve terminals in a gastrocnemius muscle causes extension of the foot, whereas ACh released at certain synapses in the central nervous system might enable you to remember a scene from your childhood. How such specific connections are established and maintained is discussed in Chapters 11 and 18. In the case of a few indirectly acting neuropeptides, however, meaning is encoded in the transmitters themselves. Thus diffuse application of a particular neuropeptide by injection into the cerebrospinal fluid may elicit a characteristic behavior, determined by the distribution and properties of its receptors on central neurons. In either event, to appreciate how neurons communicate requires an understanding of the chemistry of synaptic transmission. In addition, as knowledge of transmitter biochemistry increases it is possible not only to determine how drugs affect neuronal function, but also to interpret dysfunctions of the nervous system in terms of specific biochemical deficits and to define better therapeutic measures (see Chapter 10).

A first step for studying the chemistry of synaptic transmission at a particular synapse is to identify the transmitter substance(s) released. A straightforward proof of the identity of the transmitter would be to show that a particular substance, when applied to the postsynaptic cell in the same amount as is liberated from the presynaptic nerve terminal by an action potential, produces the same postsynaptic effect. It is generally considered sufficient to demonstrate that the presumptive transmitter is synthesized and stored in the axon terminal, is released by nerve stimulation, and mimics the effects of the endogenous transmitter by both physiological and pharmacological tests. Pioneering experiments by Langley, Loewi, Feldberg, and Dale helped to establish these criteria and identified ACh as the transmitter at the vertebrate skeletal neuromuscular junction and in autonomic ganglia (see Chapter 7). Definitive experiments by Cannon, von Euler, and Peart identified norepinephrine as a transmitter released by sympathetic neurons,[2] while those of Florey, Kuffler, Kravitz, and their colleagues established GABA as an inhibitory transmitter at the crustacean neuromuscular junction.[3] Such studies were usually made in readily accessible peripheral synapses where a homogeneous population of neurons or axons could be stimulated and isolated for biochemical analysis.

Meeting such requirements for the identification of transmitters at most synapses in the vertebrate central nervous system turns out to be an extraordinarily difficult task, primarily for two reasons. First, the amount of transmitter released during synaptic transmission is very small (perhaps 10^4 molecules per synapse per action potential at central synapses). Second, the anatomy of the central nervous system makes

Identification of neurotransmitters

[2]von Euler, U. S. 1956. *Noradrenaline*. Charles C. Thomas, Springfield, IL.

[3]Hall, Z. W., Hildebrand, J. G. and Kravitz, E. A. 1974. *Chemistry of Synaptic Transmission*. Chiron Press, Newton, MA.

it unfavorable for analysis. In contrast to peripheral structures such as the neuromuscular junction, the central nervous system is composed of a nonhomogeneous population of inextricably intermingled nerve and glial cell bodies and their processes. It is difficult to stimulate selectively a particular group of axons known to be excitatory or inhibitory, to record intracellularly from the postsynaptic cells, to mimic the action of nerves by applying chemicals through a micropipette, or to collect transmitter from release sites. Fortunately, some discrete areas of the CNS contain groups of cells that use the same transmitter; their axons travel in tight bundles and end in well-defined regions. Hence it becomes possible to stimulate selectively identified populations of axons and to collect the transmitter they release. Such experiments can be made in the intact CNS using, for example, microdialysis[4] or a push–pull cannula[5] to perfuse synaptic sites. Alternatively, the architecture of some areas of the CNS is such that one can cut slices 100 μm or so thick in which particular fiber tracts can be stimulated while recording from postsynaptic cells. Such slices can be maintained in vitro and transmitters applied through a micropipette or collected from the perfusate.[6]

Clues to the identities of the transmitters released at synapses can come from measuring the distribution of various proteins mediating different aspects of synaptic transmission, such as enzymes involved in transmitter synthesis and storage, transmitter receptors, and proteins mediating the termination of transmitter action. A consistent finding is that neurons contain uniquely high levels of the enzymes catalyzing the synthesis of the transmitter they release; degradative enzymes tend to have a more widespread distribution.[3] An exception is acetylcholinesterase, the enzyme that degrades ACh. It is highly concentrated at synapses where ACh is released. This concentration reflects the atypical feature of cholinergic transmission that the action of ACh is terminated by degradation (see below).

Many substances that were first characterized as neurotransmitters in invertebrates or at peripheral synapses in vertebrate nervous systems have been shown to act in the mammalian CNS. It is remarkable that the same transmitters are found so widely distributed in the animal kingdom, from leeches and insects to lampreys and mammals.

Transmitters can be divided into two groups. One group is the low-molecular-weight transmitters such as acetylcholine, norepinephrine, epinephrine, dopamine, 5-hydroxytryptamine (5-HT, or serotonin), histamine, adenosine triphosphate (ATP), and the amino acids γ-aminobutyric acid (GABA), glutamate, and glycine (Figure 1). A second group of transmitters is the neuropeptides (Table 1), more than 40 of which have been identified in the mammalian CNS. It seems highly likely that more transmitter substances remain to be discovered.

[4]Zetterstrom, T. and Fillenz, M. 1990. *Eur. J. Pharmacol.* 180: 137–143.
[5]Perschak, H. and Cuenod, M. 1990. *Neuroscience* 35: 283–287.
[6]Vollenweider, F. X., Cuenod, M. and Do, K. Q. 1990. *J. Neurochem.* 54: 1533–1540.

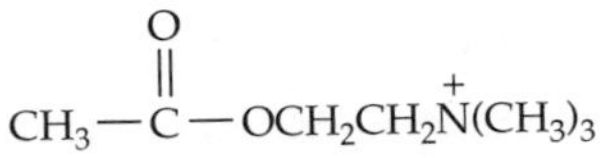

$$CH_3-\overset{\overset{\textstyle O}{\|}}{C}-OCH_2CH_2\overset{+}{N}(CH_3)_3$$

Acetylcholine (ACh)

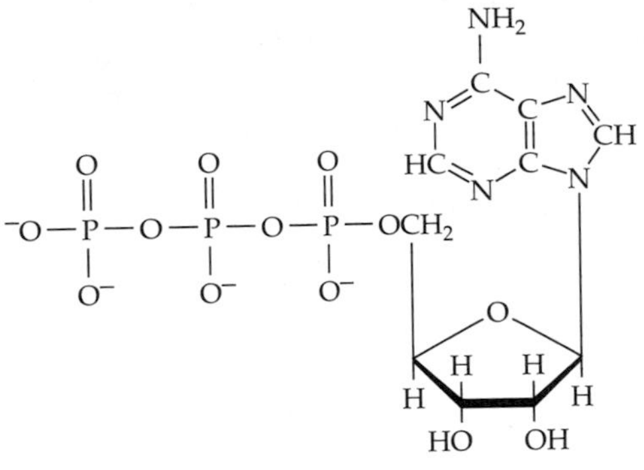

Adenosine 5'-triphosphate (ATP)

AMINO ACIDS

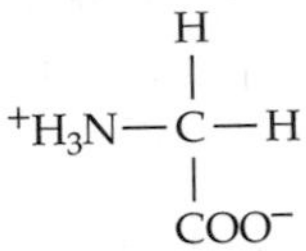

$$^+H_3N-\overset{\overset{\textstyle H}{|}}{\underset{\underset{\textstyle COO^-}{|}}{C}}-H$$

Glycine

$$^+H_3N-\overset{\overset{\textstyle H}{|}}{\underset{\underset{\textstyle COO^-}{|}}{C}}-CH_2-CH_2-COO^-$$

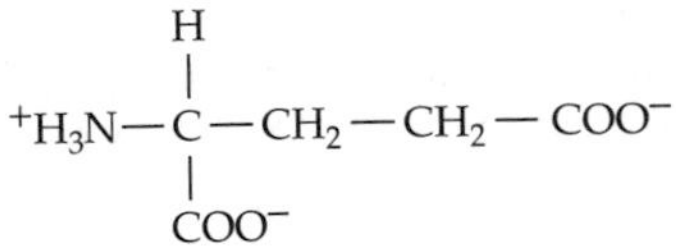

Glutamic acid

$$^+H_3N-CH_2-CH_2-CH_2-COO^-$$

γ-Aminobutyric acid (GABA)

BIOGENIC AMINES

Catecholamines

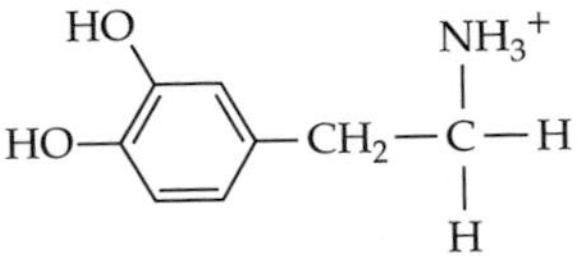

Dopamine

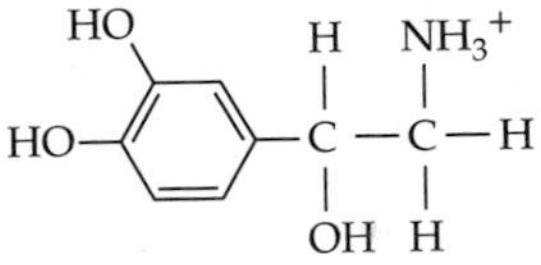

Norepinephrine (NE)
(Noradrenaline)

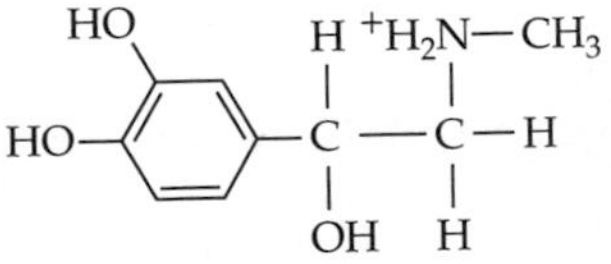

Epinephrine
(Adrenaline)

Indoleamine

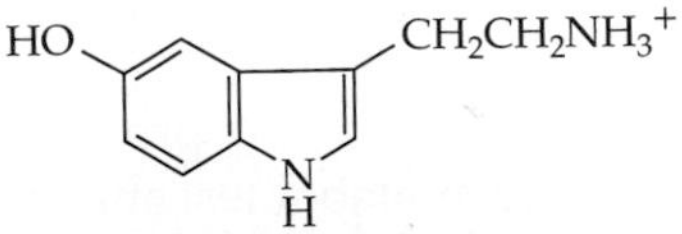

5–Hydroxytryptamine
(Serotonin, 5–HT)

Imidazole

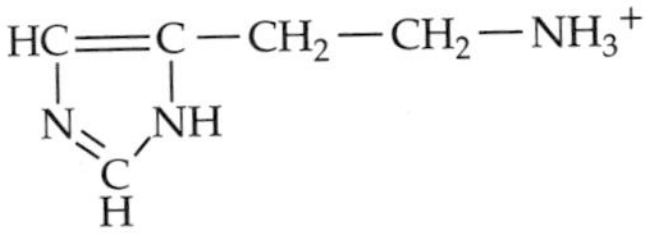

Histamine

1   STRUCTURES OF LOW-MOLECULAR-WEIGHT NEUROTRANSMITTERS.

TABLE 1
MAMMALIAN NEUROPEPTIDES

Family	Precursor	Neuropeptide
Opioid	Proopiomelanocortin (POMC)	Corticotropin (ACTH) β-Lipotropin α-MSH α-Endorphin β-Endorphin γ-Endorphin
	Proenkephalin	Met-Enkephalin Leu-Enkephalin
	Prodynorphin	α-Neoendorphin β-Neoendorphin Dynorphin A Dynorphin B (Rimorphin) Leumorphin
Neurohypophyseal	Provasopressin	Vasopressin Neurophysin II
	Prooxytocin	Oxytocin Neurophysin I
Tachykinins	α-Protachykinin A	Substance P
	β-Protachykinin A	Substance P Neurokinin A Neuropeptide K
	γ-Protachykinin-A	Substance P Neurokinin A Neuropeptide γ
	Protachykinin-B	Neurokinin B

NEUROTRANSMITTER SYNTHESIS AND STORAGE

Compared with the wealth of detailed information about the mechanisms of transmitter action, our understanding of the regulation of transmitter synthesis and storage is surprisingly rudimentary. Where are transmitter molecules synthesized and how are transmitter stores maintained and replenished? Are transmitters shipped to the nerve terminal ready-made or are they assembled there from parts provided by the cell body? The answers to such questions are different for different transmitters. Low-molecular-weight transmitters are produced within the axon terminal from common cellular metabolites and are incorporated into small synaptic vesicles (50 nm in diameter) for storage and release. Neuropeptide transmitters, on the other hand, are synthesized in the cell body, packaged in large, dense-core vesicles (100–200 nm in diameter), and shipped down the axon. An added complication

TABLE 1
MAMMALIAN NEUROPEPTIDES

Family	Precursor	Neuropeptide
Bombesin/GRP	Probombesin	Bombesin
	ProGRP	Gastrin-releasing peptide (GRP)
Secretins	—	Secretin
	—	Motilin
	Proglucagon	Glucagon
	ProVIP	Vasoactive intestinal peptide (VIP)
	ProGRF	Growth hormone-releasing factor (GRF)
Insulins	Proinsulin	Insulin
	—	Insulin-like growth factors
Somatostatins	Prosomatostatin	Somatostatin
Gastrins	Progastrin	Gastrin
	Procholecystokinin	Cholecystokinin (CCK)
Neuropeptide Y	ProNPY	Neuropeptide Y (NPY)
	ProPP	Pancreatic polypeptide (PP)
	ProPYY	Peptide YY (PYY)
Other	ProCRF	Corticotropin-releasing factor (CRF)
	Procalcitonin	Calcitonin
	ProCGRP	Calcitonin gene-related peptide (CGRP)
	Proangiotensin	Angiotensin
	Probradykinin	Bradykinin
	ProTRH	Thyrotropin-releasing hormone (TRH)
	—	Neurotensin
	—	Galanin
	—	Luteinizing hormone-releasing hormone (LHRH)

is that many neurons release more than one transmitter, typically a low-molecular-weight transmitter and one or more neuropeptides.[7] Figure 2 shows the localization of 5-HT and the neuropeptide Substance P within individual neurons in the CNS of the rat.[8] At many synapses such cotransmitters have been shown to act synergistically. For example, ACh and vasoactive intestinal peptide (VIP) are released from neurons that innervate salivary glands in cats.[9] ACh acts directly on both gland cells and blood vessels, causing secretion and vasodilation. VIP, like ACh, acts directly on blood vessels to cause vasodilation, while stimulating secretion indirectly by potentiating the response of the

[7]Kupfermann, I. 1991. *Physiol. Rev.* 71: 683–732.
[8]Schultzberg, M., Hökfelt, T. and Lundberg, J. M. 1982. *Brit. Med. Bull.* 38: 309–313.
[9]Lundberg, J. M. et al. 1980. *Proc. Natl. Acad. Sci. USA* 77: 1651–1655.

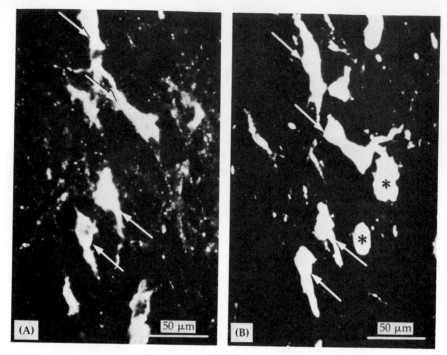

2   **5-HYDROXYTRYPTAMINE AND SUBSTANCE P in individual nerve cells of rat medulla oblongata. (A) Immunofluorescence micrograph of section labeled with antiserum to substance P. (B) Adjacent section labeled with antiserum to 5-HT. Comparison of (A) with (B) reveals many cells that contain both transmitters (marked by arrows) and other cells that do not (asterisks). (From Schultzberg, Hökfelt and Lundberg, 1982.)**

gland cells to ACh.[7] Experiments on cat salivary glands[10] and frog sympathetic ganglia[11] indicate that low-molecular-weight transmitters such as acetylcholine are released reliably by single impulses, while trains of impulses are required to release significant quantities of neuropeptide cotransmitters (see Chapter 8).

Synthesis of ACh

Transmitters are found in high concentration in neurons, especially in axon terminals. How does this accumulation come about and how are these stores maintained at rest and during periods of activity? One of the first thorough investigations of such questions was made by Birks and McIntosh in their studies of the release of ACh from the terminals of preganglionic axons in the superior cervical ganglion of the cat[12] (Figure 3A and 3B). They cannulated the carotid artery and the jugular vein, perfused the ganglion with solutions containing anticholinesterase, and analyzed the perfusate for ACh. A small amount of ACh was continually released from the ganglion at rest, amounting to 0.1 percent of the total contents per minute (Figure 3). Since the level of ACh in

[10]Lundberg, J. M. and Hökfelt, T. 1983. *Trends Neurosci.* 6: 325–333.
[11]Kuffler, S. W. 1980. *J. Exp. Biol.* 89: 257–286.
[12]Birks, R. I. and MacIntosh, F. C. 1961. *J. Biochem. Physiol.* 39: 787–827.

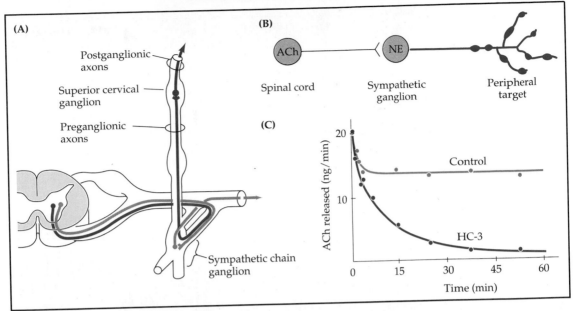

3   **MEASURING THE RELEASE OF ACh** from the terminals of preganglionic axons in the cat superior cervical ganglion. **(A)** Preganglionic axons reach the superior cervical ganglion through more caudal ganglia in the sympathetic chain. **(B)** Preganglionic neurons, whose cell bodies lie in the spinal cord, release ACh as a transmitter at synapses in sympathetic ganglia. Ganglion cells release norepinephrine from varicosities along their processes in the periphery. **(C)** Release of ACh from a cat sympathetic ganglion perfused with oxygenated plasma containing $3 \times 10^{-5} M$ eserine to inhibit acetylcholinesterase. Dashed line, resting ganglion; solid lines, preganglionic stimulation at 20/sec in control medium or medium containing $2 \times 10^{-5} M$ hemicholinium (HC-3). (After Birks and MacIntosh, 1961.)

the ganglion remained constant, this meant that at least this much ACh was continually being synthesized at rest. (Subsequently it has been shown that the ongoing rate of synthesis of ACh, determined by monitoring the incorporation of radioactively labeled choline into ACh, is 50-fold higher; an amount equal to the entire store of ACh is degraded and resynthesized within the axon terminals every 20 minutes.[13]) Birks and McIntosh then stimulated the preganglionic nerve with long trains of impulses and found that the quantity of ACh released from the ganglion increased 100-fold, so that 14 ng—an amount corresponding to 10 percent of the original content—was released each minute (Figure 3C). Remarkably, this rate of release was maintained for over an hour with no change in the level of ACh in the ganglion. The only exogenous ingredient the nerve terminals needed to maintain their stores of ACh was choline, which they cannot synthesize and so must take up from the surrounding fluid via an active transport process (Figure 4). The requirement for extracellular choline was demonstrated both by perfus-

[13]Potter, L. T. 1970. *J. Physiol.* 206: 145–166.

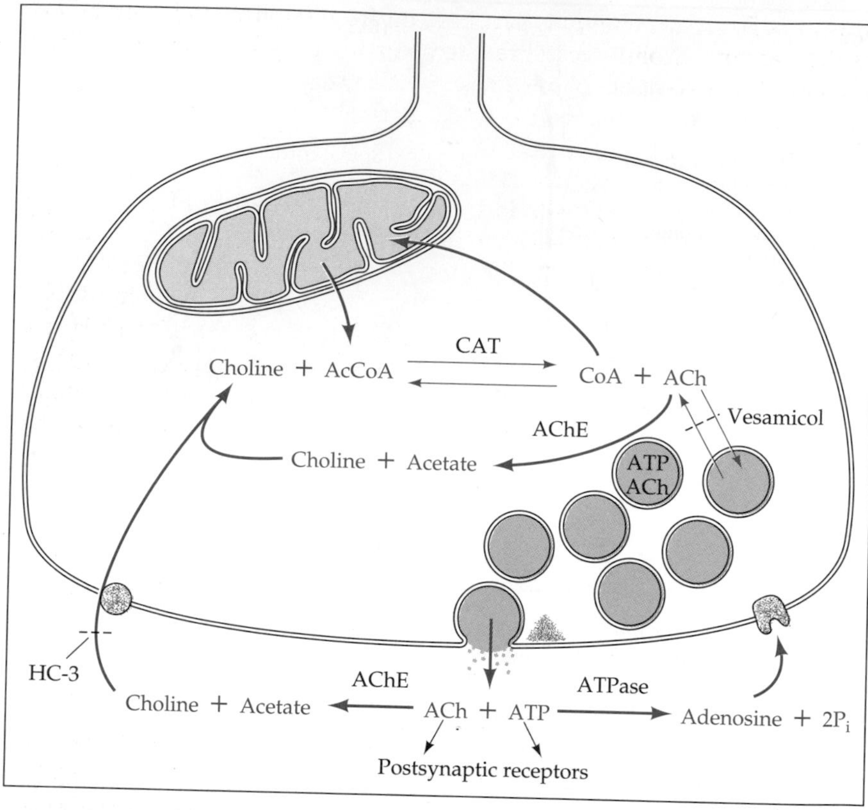

4   PATHWAYS OF ACh SYNTHESIS, STORAGE, RELEASE, AND DEGRADATION.
Acetylcholine is synthesized from choline and acetyl coenzyme A (AcCoA) by
choline acetyltransferase (CAT) and is degraded by acetylcholinesterase (AChE). AcCoA
is synthesized primarily in mitochondria; choline is supplied by a high-affinity active
transport system that can be inhibited by hemicholinium (HC-3). ACh is packaged into
vesicles together with ATP for release by exocytosis. Transport of ACh into vesicles is
blocked by vesamicol. Vesicular ACh is protected from degradation. After release, ACh
is degraded by extracellular AChE to choline and acetate. About half of the choline
transported into cholinergic axon terminals comes from the hydrolysis of ACh that has
been released. At some synapses ATP combines with postsynaptic receptors. ATP is
hydrolyzed by extracellular ATPases to adenosine and phosphate; adenosine can combine with presynaptic adenosine receptors to modulate transmitter release.

ing the preparation with solutions lacking choline and by blocking
choline uptake into the axon terminals with hemicholinium (HC-3), a
drug that inhibits the transport mechanism for choline. In both cases
the level of ACh in the ganglion and the amount released by stimulation
fell rapidly (Figure 3C). Thus, during an hour of stimulation, an axon
terminal can release an amount of transmitter equal to many times its
original content without having its stores depleted.

How is ACh synthesis controlled to meet the demands of release?
Our understanding of the mechanisms regulating ACh synthesis and
storage in cholinergic nerve terminals is surprisingly limited. The enzymatic reactions are summarized in Figure 4 and shown in detail in

Appendix B. Acetylcholine is synthesized from choline and acetyl-CoA by the enzyme choline acetyltransferase and is hydrolyzed to choline and acetate by acetylcholinesterase. Both enzymes are found in the cytosol. Because the reaction catalyzed by choline acetyltransferase is reversible, one factor controlling the level of ACh is the law of mass action. For example, a fall in ACh concentration after release would favor net synthesis until equilibrium was reestablished. However, the regulatory mechanisms at work within cholinergic axon terminals are more complex than this. For example, under resting conditions the accumulation of ACh is limited by ongoing hydrolysis by intracellular acetylcholinesterase; inhibition of acetylcholinesterase within nerve terminals causes the amount of ACh they contain to increase several fold.[12,13] Thus the level to which ACh accumulates represents a steady state between ongoing synthesis and degradation. This is a common feature of the metabolism of low-molecular-weight transmitters. Although it seems to be a waste of energy, such constant turnover may be a consequence of the mechanisms that ensure an adequate supply of transmitter is always available. For example, the supply of ACh available for release can be increased both by enhancing synthesis and by curtailing intracellular degradation. Indeed, the 100-fold increase in the *net* synthesis of ACh observed during prolonged stimulation would require only a twofold increase in the *actual rate* of ACh synthesis if intracellular degradation were blocked at the same time.

In the central nervous system, additional factors have been shown to influence the amount of ACh that accumulates within cholinergic nerve terminals; these factors include the supply of choline, the supply of the cosubstrate acetyl-CoA (made in mitochondria), and the activity of choline acetyltransferase.[14,15]

Much of the ACh in nerve terminals is sequestered in synaptic vesicles, while ACh synthesis and degradation occur in the cytosol. To influence synthesis, the release of ACh must reduce the cytoplasmic concentration of ACh, presumably by the movement of cytoplasmic ACh into newly formed vesicles. Similar interplay between cytoplasmic synthesis and vesicular release is a common feature of the metabolism of low-molecular-weight transmitters. The precise dynamics of the interaction between vesicular stores and cytoplasmic pools of transmitter is poorly understood.

Another mechanism by which the rate of synthesis of substances in cells is controlled is FEEDBACK INHIBITION, in which the rate-limiting step in a biosynthetic pathway is inhibited by the final product. A good example comes from studies by Von Euler, Axelrod, Udenfriend, and their colleagues on the synthesis, storage, and release of norepinephrine in sympathetic neurons and in secretory cells of the adrenal medulla.[16] Adrenal medullary cells resemble sympathetic neurons in many ways:

Synthesis of dopamine and norepinephrine

[14]Jope, R. 1979. *Brain Res.* Rev. 1: 313–344.
[15]Tuček, S. 1978. *Acetylcholine Synthesis in Neurons.* Chapman and Hall, London.
[16]Axelrod, J. 1971. *Science* 173: 598–606.

They share the same embryonic origin, they are innervated by cholinergic axons that originate in the central nervous system, and they release a catecholamine in response to stimulation. (The term CATECHOLAMINE is used to designate collectively the substances DOPA, dopamine, norepinephrine, and epinephrine, all of which contain a catechol nucleus—a benzene ring with two adjacent hydroxyl groups—and an amino group; see Appendix B.) Sympathetic neurons release norepinephrine; adrenal medullary cells release epinephrine as well as norepinephrine.

Norepinephrine is synthesized from the common cellular metabolite tyrosine in a series of three steps: Tyrosine is converted to DOPA by the enzyme tyrosine hydroxylase, DOPA to dopamine by aromatic L-amino acid decarboxylase, and dopamine to norepinephrine by dopamine β-hydroxylase (Figure 5; see also Appendix B). The conversion of tyrosine to DOPA and DOPA to dopamine occurs in the cytoplasm. Dopamine is then transported into synaptic vesicles, where it is converted to norepinephrine by dopamine β-hydroxylase, which is associated with the vesicle membrane. Much of the norepinephrine is stored within vesicles; some escapes into the cytoplasm, where it is susceptible to degradation by monoamine oxidase.

Some neurons release dopamine as a transmitter. Accordingly, they contain tyrosine hydroxylase and aromatic L-amino acid decarboxylase, but lack dopamine β-hydroxylase. Other neurons as well as adrenal medullary cells release epinephrine, which is derived from norepinephrine by the action of phenylethanolamine N-methyltransferase.

Evidence for the role of feedback inhibition in the regulation of the synthesis and accumulation of catecholamines came in large part from studies of enzymes isolated from the adrenal medulla. Typically it is the first enzyme in a multiple-step pathway that is rate-limiting and subject to feedback inhibition. Indeed, the activity of tyrosine hydroxylase in extracts of the adrenal medulla is two orders of magnitude lower than that of aromatic L-amino acid decarboxylase and dopamine β-hydroxylase, and tyrosine hydroxylase is inhibited by norepinephrine (and by dopamine and epinephrine as well). As norepinephrine or epinephrine accumulate, they progressively inhibit their own synthesis until a steady state is reached, at which the rate of synthesis is equal to the rate of degradation and release.

Evidence that feedback inhibition regulates the synthesis of norepinephrine in neurons came from experiments by Weiner and his colleagues on terminals of sympathetic axons innervating the smooth muscles of a duct, the vas deferens.[17] They measured the rate of norepinephrine synthesis in the terminals by bathing the preparation in radioactively labeled precursors and monitoring the accumulation of radioactively labeled norepinephrine. They found that the rate of norepinephrine synthesis was more than threefold greater if the first enzymatic step was bypassed by providing DOPA rather than tyrosine as

[17]Weiner, N. and Rabadjija, M. 1968. *J. Pharmacol. Exp. Ther.* 160: 61–71.

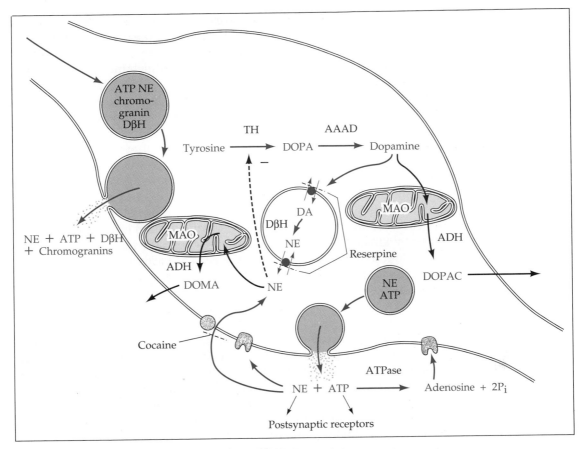

5 PATHWAYS OF NOREPINEPHRINE SYNTHESIS, STORAGE, RELEASE, AND UPTAKE. Tyrosine is converted to DOPA by tyrosine hydroxylase (TH). DOPA is converted to dopamine (DA) by aromatic L-amino acid decarboxylase (AAAD). Dopamine is transported into vesicles, where it is converted to norepinephrine (NE) by dopamine β-hydroxylase (DβH). Norepinephrine inhibits TH, thus regulating synthesis by feedback inhibition. Transport of dopamine and norepinephrine into vesicles is blocked by reserpine. Vesicles also contain ATP (large dense-core vesicles contain soluble DβH and chromogranins as well). All soluble components of vesicles are released together. NE, ATP, adenosine, and peptides derived from chromogranins can bind to pre- or postsynaptic receptors. After release, norepinephrine is transported back into the varicosity by an uptake mechanism that is blocked by cocaine. Norepinephrine in the cytoplasm can be repackaged into vesicles for release. Within the varicosity, norepinephrine is degraded to 3,4-dihydroxymandelic acid (DOMA) and dopamine to 3,4 dihydroxyphenylacetic acid (DOPAC) by monoamine oxidase (MAO) and aldehyde dehydrogenase (ADH).

the precursor. This confirmed that conversion of tyrosine to DOPA was rate-limiting. To test the idea that the rate-limiting step was controlled by feedback inhibition, they varied the concentration of norepinephrine in the cytoplasm in two ways. First, taking advantage of the fact that sympathetic axon terminals have a specific transport mechanism for norepinephrine, they added norepinephrine to the bathing fluid, which caused an increase in norepinephrine concentration in the terminals.

This decreased the rate at which norepinephrine was synthesized from tyrosine. Conversely, nerve stimulation, which lowers the concentration of norepinephrine in the cytoplasm, increased the rate of conversion of tyrosine to norepinephrine almost twofold. No such increase was seen, however, if norepinephrine was added to the bath during nerve stimulation. Apparently uptake from the medium was sufficient to maintain the level of norepinephrine in the axon terminals and so limit its biosynthesis.

A second mechanism ensures that the rate of synthesis of norepinephrine meets the demands of release. As a result of stimulation of norepinephrine-containing terminals, tyrosine hydroxylase acquires a higher affinity for its dihydropteridine cofactor (see Appendix B) and becomes less sensitive to inhibition by norepinephrine.[18] These changes appear to be the result of transient phosphorylation of tyrosine hydroxylase by kinases activated by a variety of signals, including the influx of calcium ions that occurs when the cells are active.[19]

Synthesis of 5-HT   Compared with our knowledge about acetylcholine and norepinephrine metabolism, less is known about the regulation of the presynaptic stores of other neurotransmitters. Stimulation of neurons releasing 5-HT causes an increase in the rate of conversion of tryptophan to 5-hydroxytryptophan. This reaction is catalyzed by the enzyme tryptophan hydroxylase and is the rate-limiting step in the synthesis of 5-HT (Figure 6; see also Appendix B). It has been suggested that the increase in rate is due to changes in the properties of tryptophan hydroxylase caused by calcium-dependent phosphorylation,[20] similar to the effects of stimulation on tyrosine hydroxylase.

Synthesis of amino acid transmitters   Feedback inhibition appears to regulate the accumulation of GABA in crustacean inhibitory neurons, where the activity of glutamate decarboxylase has been shown to be inhibited by GABA[21] (Figure 6). However, glutamate decarboxylase isolated from mammalian brain is not inhibited by GABA; other mechanisms must regulate the accumulation of GABA in mammalian neurons. Other amino acids that are released as transmitters, such as glutamate and glycine, are found in all cells, and there appear to be no special metabolic pathways for their synthesis in neurons. However, amino acid transmitters accumulate to higher levels in the terminals of neurons that release them than in other cells. This does not necessarily mean that cytoplasmic concentrations are higher; the excess may be represented entirely by the amount packaged in vesicles.[22]

Long-term regulation of transmitter synthesis   The regulatory mechanisms described so far operate rapidly to change the rate of synthesis within nerve terminals to meet the demands of release. They involve regulation of the activity of enzymes already

[18]Joh, T. H., Park, D. H., and Reis, D. J. 1978. *Proc. Natl. Acad. Sci. USA* 75: 4744–4748.

[19]Zigmond, R. E., Schwarzchild, M. A., and Rittenhouse, A. R. 1989. *Annu. Rev. Neurosci.* 12: 415–461.

[20]Hamon, M. et al. 1981. *J. Physiol.* (Paris) 77: 269–279.

[21]Hall, Z. W., Bownds, M. D. and Kravitz, E. A. 1970. *J. Cell Biol.* 46: 290–299.

[22]Maycox, P. R., Hell, J. W. and Jahn, R. 1990. *Trends Neurosci.* 13: 83–87.

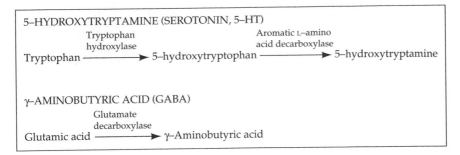

6   SYNTHESIS OF 5-HT AND GABA.

in place, rather than changes in the levels of the enzymes themselves. In addition to such short-term effects, there are also long-term regulatory mechanisms. A good example comes from the response of the sympathetic nervous system to prolonged exposure of an animal to stress. When the body is stressed, sympathetic neurons are activated. With prolonged activation, the levels of tyrosine hydroxylase and dopamine β-hydroxylase in the cell bodies and terminals of sympathetic neurons increase as much as three- to fourfold.[23,24] This increase is due to the synthesis of new enzyme molecules and is specific; the levels of other enzymes of norepinephrine synthesis and degradation, such as aromatic L-amino acid decarboxylase and monoamine oxidase, are not affected. The increase is triggered by the increased synaptic activation of sympathetic neurons by input from the central nervous system. This TRANSSYNAPTIC REGULATION provides a mechanism whereby the synthetic capability of the neurons can be matched to the demands of release.[25] Experiments on human sympathetic ganglia have demonstrated that electrical stimulation of preganglionic fibers induces a marked increase in the levels of the mRNAs for tyrosine hydroxylase and dopamine β-hydroxylase in the postsynaptic cells within 20 minutes, suggesting that the regulation of genes involved in norepinephrine synthesis is very rapid and sensitive.[26]

The levels of tyrosine hydroxylase and dopamine β-hydroxylase in sympathetic neurons also depend on signals derived from the cells they innervate. For example, nerve growth factor (NGF; see Chapter 11), produced by postsynaptic cells, is taken up into the terminals of sympathetic axons and transported back to the cell body, where it increases the synthesis of tyrosine hydroxylase and dopamine β-hydroxylase.[27]

Regulation of the stores of peptide transmitters is complicated by the separation between the sites of synthesis and release. Synthesis of

Synthesis of neuropeptides

[23]Thoenen, H., Mueller, R. A. and Axelrod, J. 1969. *Nature* 221: 1264.
[24]Thoenen, H., Otten, U., and Schwab, M. 1979. *In* F. O. Schmitt and F. G. Worden (eds.). *The Neurosciences: Fourth Study Program*. MIT Press, Cambridge, MA., pp. 911–928.
[25]Comb, M., Hyman, S. E. and Goodman, H. M. 1987. *Trends Neurosci.* 10: 473–478.
[26]Schalling, M. et al. 1989. *Proc. Natl. Acad. Sci. USA* 86: 4302–4305.
[27]Thoenen, H. and Barde, Y.-A. 1980. *Physiol. Rev.* 60: 1284–1335.

peptides occurs on ribosomes, which are found in neuronal cell bodies but not in axons or nerve terminals. Thus peptides must be synthesized in the cell body and shipped down the axon to the terminal, which limits the amount available for release. At the same time, the binding of peptides to their receptors occurs at a much lower concentration (in the range of 10^{-8} to 10^{-10} M) than the binding of low-molecular-weight transmitters, such as ACh, to their receptors (10^{-5} to 10^{-6} M) and the mechanisms by which their actions are terminated are generally slower. Moreover, neuropeptide receptors act indirectly, through intracellular pathways that can provide tremendous amplification. As a consequence, many fewer molecules of a peptide are needed to influence a postsynaptic target, and the demands of release can be met by the supply of molecules shipped from the cell body. However, this arrangement means that signals influencing the rate of synthesis of peptides must regulate events in the cell body, a relatively slow process in contrast to the rapid local control of the synthesis and storage of low-molecular-weight transmitters within the axon terminal.

Peptides are synthesized as part of larger propeptide proteins, which often contain the sequences for more than one biologically active peptide[28,29] (Table 1; Figure 7). Specific proteases cleave the propeptide into the appropriate peptide molecules. This processing can take place in the cell body, within peptide storage vesicles as they are transported down the axon to the terminal, and perhaps within the terminal itself.[30] The properties and distribution of the processing enzymes, which will determine both the amount and nature of the peptides available for release, are areas of active research.

Storage of transmitters
in synaptic vesicles
How are transmitter molecules packaged in vesicles? Peptide transmitters are synthesized in the rough endoplasmic reticulum and are

[28]Loh, Y. P. and Parish, D. C. 1987. *In* A. J. Turner (ed.). *Neuropeptides and Their Peptidases.* Ellis Horwood, New York, pp. 65–84.

[29]Sossin, W. S., Fisher, J. M., and Scheller, R. H. 1989. *Neuron* 2: 1407–1417.

[30]Gainer, H., Sarne, Y., and Brownstein, M. J. 1977. *J. Cell Biol.* 73: 366–381.

7   SYNTHESIS OF NEUROPEPTIDES. (A) Representation of the structure of bovine ▶ proopiomelanocortin. The locations of known peptide components are shown by closed boxes. Paired basic amino acid residues—common targets for processing enzymes—are indicated. (B) Steps in neuropeptide maturation. Processing usually begins with cleavage on the carboxy terminal side of the recognition site by an endoprotease. The basic residues are trimmed by carboxypeptidase E. If the peptide ends in glycine, the enzyme peptidylglycine α-amidating monooxygenase (PAM) converts the carboxy terminus to an amide. (C) Pathway for neuropeptide synthesis. Neuropeptide precursors are directed into the lumen of the endoplasmic reticulum by a signal sequence that is cleaved cotranslationally. In the endoplasmic reticulum, disulfide bonds are formed and N-linked glycosylation occurs. The propeptide is then transported through the Golgi apparatus, where further modifications, such as sulfation and phosphorylation, take place. Two packaging schemes are illustrated. On the left a propeptide is packaged into vesicles budding from the Golgi; as the vesicle matures, the propeptide is cleaved, resulting in two fragments of the precursor packaged in the same vesicle. On the right a propeptide is cleaved within the Golgi, followed by sorting of peptides into separate vesicles. (After Sossin et al., 1989.)

incorporated into large (100–200 nm diameter), dense-core vesicles as they are processed in the Golgi apparatus. They remain in vesicles during transport from the cell body to the nerve terminal.[29,30] For low-molecular-weight transmitters such as ACh and norepinephrine, on the other hand, most transmitter synthesis occurs in the axon terminal and the transmitters must be packaged into vesicles there. As seen in electron micrographs these synaptic vesicles tend to be small (50 nm in diameter) and can appear clear (e.g., ACh, amino acid transmitters) or have dense cores (e.g., biogenic amines). The concentration of trans-

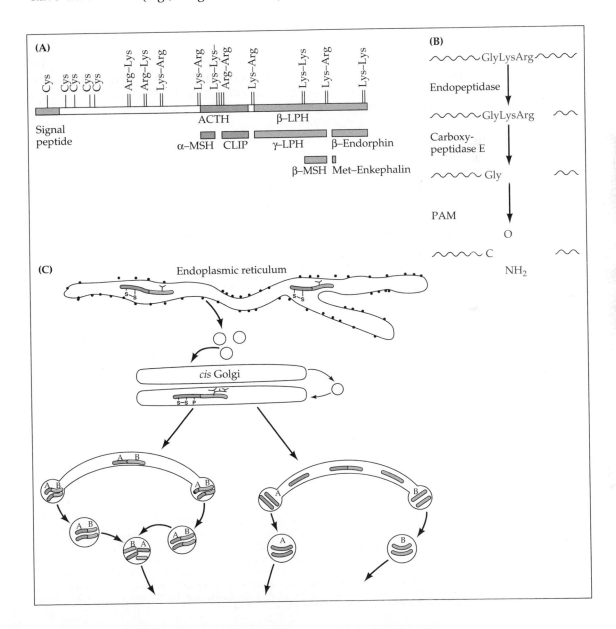

mitters in vesicles is greater than that in the surrounding cytoplasm. The accumulation of transmitter into vesicles is driven by a transmembrane proton gradient[31-33] (Figure 8). A proton-dependent ATPase in the vesicle membrane pumps protons into the vesicle, making the inside both positively charged and acidic compared with the cytoplasm; transport proteins then use the energy in this gradient to move transmitter molecules into the vesicle against their concentration gradient.[22,34] It appears that the number of different vesicle transporters is low: All biogenic amines appear to use the same transporter, GABA and glycine seem to share a common transporter, and only one transporter each for ACh and glutamate have been reported.[33] The vesicle transmitter transport mechanisms do not have the same specificity as the postsynaptic receptors. Consequently, under certain conditions, molecules that do not activate the postsynaptic receptors can accumulate in vesicles and be released from axon terminals as FALSE TRANSMITTERS.[35,36]

The first vesicles to be purified and analyzed biochemically were from the adrenal medulla. These large (200–400 nm diameter) vesicles are called chromaffin granules because of their tendency to stain with chromium salts. In addition to catecholamines, chromaffin granules contain high concentrations of ATP, a soluble form of the synthetic enzyme dopamine β-hydroxylase, and soluble proteins called chromogranins. The formation of multimolecular complexes between positively charged catecholamines, negatively charged ATP, and the chromogranins appears to aid in packing and storing the catecholamines at concentrations that might otherwise be hyperosmotic.[31] In addition, release of ATP produces effects of its own and so ATP acts as a cotransmitter.[7] Likewise, one of the proteins, chromogranin A, has been shown to serve as a precursor for a number of peptides that modulate secretion.[7]

Synaptic vesicles have been purified from cholinergic and noradrenergic axons and nerve terminals. Noradrenergic nerve terminals contain large dense-cored vesicles (70–200 nm diameter) that, like chromaffin granules, contain chromogranins and the soluble form of dopamine β-hydroxylase. The more numerous, small synaptic vesicles in catecholamine-containing nerve terminals as well as those in cholinergic nerve terminals appear to contain little soluble protein. Both cholinergic and biogenic amine-containing synaptic vesicles contain high concentrations of ATP,[37,38] which could neutralize the charge and reduce the osmotic activity of the transmitter stores. In addition, at some synapses ATP acts as a neurotransmitter, while its metabolite,

[31]Johnson, R. G., Jr. 1988. *Physiol. Rev.* 68: 232–307.

[32]Marshall, I. G. and Parsons, S. M. 1987. *Trends Neurosci.* 10: 174–177.

[33]Südhof, T. C. and Jahn, R. 1991. *Neuron* 6: 665–677.

[34]Stern-Bach, Y. et al. 1990. *J. Biol. Chem.* 265: 3961–3966.

[35]Kopin, I. J. 1968. *Annu. Rev. Pharmacol.* 8: 377–394.

[36]Luqmani, Y. A., Sudlow, G. and Whittaker, V. P. 1980. *Neuroscience* 5: 153–160.

[37]Dowdall, M. J., Boyne, A. F. and Whittaker, V. P. 1974. *Biochem. J.* 140: 1–12.

[38]De Potter, W. P., Smith, A. D. and De Schaepdryver, A. F. 1970. *Tissue Cell* 2: 529–546.

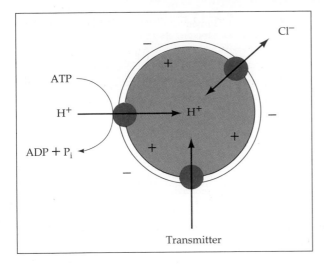

8

TRANSPORT OF TRANSMITTERS INTO SYNAPTIC VESICLES is driven by a proton electrochemical gradient. An ATP-powered pump transports protons into synaptic vesicles, making the vesicle interior acidic and positive relative to the cytoplasm. Neurotransmitters are carried into the vesicles by specific transporters, energetically coupled to the proton electrochemical gradient. Chloride movement through an anion channel maintains electrical neutrality in the bulk solution within the vesicle.

adenosine, can combine with receptors on axon terminals to modulate release.[39] The uptake of ATP into isolated amine- and ACh-containing vesicles has been characterized,[31,40] but the transporter mediating ATP uptake has not been identified. Mixed populations of synaptic vesicles have been isolated from the central nervous system and shown to contain a wide variety of transmitters, including glutamate, GABA, and glycine.[41,42]

As described above, neuropeptide transmitters and all proteins involved in transmitter synthesis, storage, release, and degradation are synthesized in the neuronal cell body and must be shipped down to the axon terminal. The first evidence for movement of material along axons came from experiments by Weiss and his colleagues, who ligated peripheral nerves and described the ballooning out of axons just proximal to the site of constriction and the subsequent movement of the accumulated material along the axons after the constriction was removed.[43] These effects suggested that normally there is a continuous bulk movement of axoplasm along the axon at the rate of 1 to 2 mm per day; this was given the name AXOPLASMIC FLOW. This idea was buttressed by later experiments using radioactively labeled amino acids to follow the movement of proteins from neuronal cell bodies along peripheral and central axons[44] (Figure 9). However, the experiments with radioactive precursors also demonstrated that some proteins were moving at much faster rates, up to 400 mm per day, which could not possibly be accounted for by bulk movements of axoplasm.[45] Thus the

Axonal transport

[39]Burnstock, G. 1990. *Ann. NY Acad. Sci.* 603: 1–17.
[40]Stadler, H. and Kiene, M.-L. 1987. *EMBO J.* 6: 2217–2221.
[41]Burger, P. M. et al. 1989. *Neuron* 3: 715–720.
[42]Burger, P. M. et al. 1991. *Neuron* 7: 287–293.
[43]Weiss, P. and Hiscoe, H. B. 1948. *J. Exp. Zool.* 107: 315–395.
[44]Droz, B. and Leblond, C. P. 1963. *J. Comp. Neurol.* 121: 325–346.
[45]Grafstein, B. and Forman, D. S. 1980. *Physiol. Rev.* 60: 1167–1283.

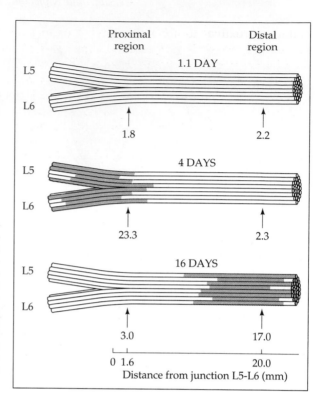

9

SLOW AXONAL TRANSPORT. Diagrams representing the sciatic nerve arising from the junction of the L5 and L6 roots (at left). Shaded areas indicate the location of radioactively labeled proteins at the indicated times after injection of [³H]-leucine into the spinal cord. The numbers below each diagram indicate the density of autoradiographic grains on cross sections of the nerve in grains/10 μm². Few radiolabeled proteins are present in the axons at 1 day; they appear in the proximal region at 4 days and have reached the distal region at 16 days, corresponding to a transport rate of approximately 1.5 mm per day. (After Droz and Leblond, 1963.)

term AXONAL TRANSPORT was adopted to encompass all the kinds of movements that occur within axons.

There is now abundant evidence for the continuous movement of substances along axons, both toward the axon terminal (ANTEROGRADE TRANSPORT) and from the terminal to the cell body (RETROGRADE TRANSPORT).[45] Different components move at different rates: Structural proteins, such as tubulin and neurofilament proteins, move at the slowest rates (1–2 mm/day); fast transport (up to 400 mm/day) is composed mainly of particulate material such as vesicles of various kinds (including synaptic vesicles) and mitochondria. A variety of approaches have been used to demonstrate movement of transmitters and enzymes along axons. For example, Geffen demonstrated the movement of norepinephrine (presumably within vesicles) along sympathetic axons after injection of radiolabeled norepinephrine into sympathetic ganglia.[46] In studies by Schwartz and his colleagues, radiolabeled precursors for ACh and 5-HT were injected intracellularly into single large neurons in *Aplysia* and the movement of labeled ACh or 5-HT followed down the axon to its terminal.[47,48] Measurement of the time course of accumulation of

[46]Livett, B. G., Geffen, L. B. and Austin, L. 1968. *J. Neurochem.* 15: 931–939.
[47]Koike, H., Kandel, E. R. and Schwartz, J. H. 1974. *J. Neurophysiol.* 37: 815–827.
[48]Shkolnik, L. J. and Schwartz, J. H. 1980. *J. Neurophysiol.* 43: 945–967.

material proximal to a constriction[49,50] or in axon terminals[51] demonstrated characteristic differences in the rates of movement within the broad spectrum of components being transported. Retrograde transport has been shown to be crucial for the movement of trophic molecules such as NGF from axon terminals back to their cell bodies (see Chapter 11). Protein tracers have been developed, such as horseradish peroxidase, and fluorescent dyes, such as Fast Blue, that are carried in both the anterograde and retrograde directions by axonal transport. Using these tracers it is possible to map synaptic connections, even over long distances, by visualizing individual axons, their terminal arborizations, and their cell bodies.[52,53]

Although early experiments demonstrated that axonal transport required metabolic energy and relied on intact microtubules, little progress was made in understanding its mechanism for 30 years. Then two technological advances triggered very rapid progress: the development of microscopic techniques that allowed direct visualization of single vesicles within living cells,[54,55] and the finding that vesicle movements persisted in cell-free systems such as extruded squid axoplasm.[56]

Microtubules and transport

Studies by Reese, Sheetz, Schnapp, Vale, and their colleagues have demonstrated that transport occurs by the attachment of organelles, such as mitochondria and vesicles, to microtubules via mechanochemical enzymes, or motors, that hydrolyze ATP and use the energy to carry organelles along the microtubule "track" (Figure 10).[57,58] Microtubules have an inherent polarity; in axons the "plus" end points toward the distal axon terminal. Anterograde transport is powered by kinesin, which moves organelles toward the plus end; retrograde transport by cytoplasmic dynein, which moves organelles toward the minus end (Figure 11).[59] There appear to be specific receptors on the surface of organelles that mediate the attachment of the kinesin and cytoplasmic dynein motors and thus regulate the direction of organelle movement.[60] Remarkably, a single kinesin motor has been shown to pull an organelle along at speeds equivalent to fast axonal transport; each molecule of ATP hydrolyzed produces a "step" of approximately 20 nm.[61] Differences in the rate of transport of different components appear to arise from differences in the proportion of time they remain "on track" and

[49]Niemierko, S. and Lubinska, L. 1967. *J. Neurochem.* 14: 761–769.
[50]Dahlstrom, A. 1971. *Philos. Trans. R. Soc. Lond.* B 261: 325–358.
[51]McEwen, B. S. and Grafstein, B. 1968. *J. Cell Biol.* 38: 494–508.
[52]La Vail, J. H. and La Vail, M. M. 1974. *J. Comp. Neurol.* 157: 303–358.
[53]Ugolini, G. and Kuypers, H. G. J. M. 1986. *Brain Res.* 365: 211–227.
[54]Inoué, S. 1981. *J. Cell Biol.* 89: 346–356.
[55]Allen, R. D., Allen, N. S. and Travis, J. L. 1981. *Cell Motil.* 1: 291–302.
[56]Brady, S. T., Lasek, R. J. and Allen, R. D. 1982. *Science* 218: 1129–1131.
[57]Vale, R. D. 1987. *Annu. Rev. Cell Biol.* 3: 347–378.
[58]Vallee, R. B. and Bloom, G. S. 1991. *Annu. Rev. Neurosci.* 14: 59–92.
[59]Schnapp, B. J. and Reese, T. S. 1989. *Proc. Natl. Acad. Sci. USA* 86: 1548–1552.
[60]Sheetz, M. P., Steuer, E. R. and Schroer, T. A. 1989. *Trends Neurosci.* 12: 474–478.
[61]Howard, J., Hudspeth, A. J. and Vale, R. D. 1989. *Nature* 342: 154–158.

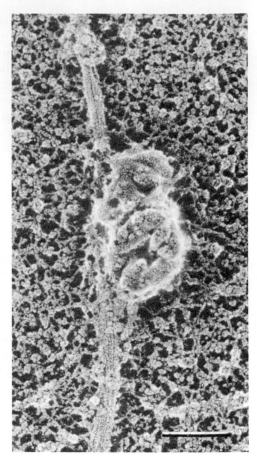

10

FAST AXONAL TRANSPORT occurs by the movement of organelles along microtubules. Electron micrograph of a vesicle moving along a microtubule. Organelles were observed by light microscopy moving along filaments in extruded squid axoplasm; then the preparation was fixed and examined by electron microscopy. The same organelles were located and identified as vesicles attached to microtubules. A layer of granular and finely filamentous material coats the glass substrate. Bar, 0.1 μm. (From Schnapp et al., 1985.)

to differences in the resistance they encounter trying to penetrate the dense network of cytoskeletal and cross-bridging elements within the axon.[58]

The mechanism of slow axonal transport, although not well understood, is clearly quite distinct, involving the turnover of components of the cytoskeleton and movement of soluble cytoplasmic proteins.[45,58]

TRANSMITTER RELEASE AND VESICLE RECYCLING

Morphological, electrophysiological, and chemical evidence indicates that transmitter is released in multimolecular packets or quanta and that release occurs by the process of exocytosis, in which the synaptic vesicle fuses with the presynaptic membrane, thereby releasing transmitters into the synaptic cleft (Chapter 7).

Vesicle contents are released by exocytosis

A prediction of the hypothesis that neurotransmitter release occurs by vesicle exocytosis is that stimulation will release the total soluble contents of synaptic vesicles: transmitter, ATP, and proteins. This pre-

(A) DYNEIN

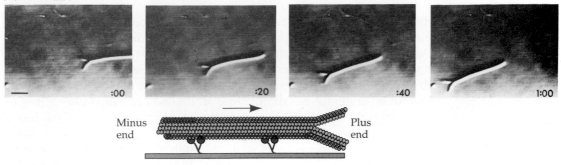

Minus end · Plus end

(B) KINESIN

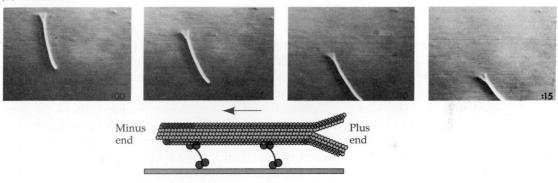

Minus end · Plus end

(C)

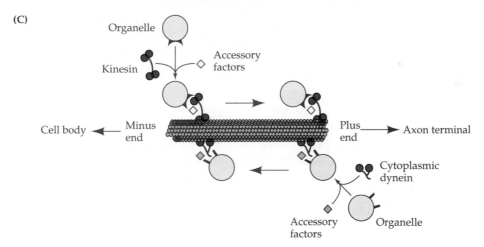

Organelle

Kinesin

Accessory factors

Cell body ← Minus end

Plus end → Axon terminal

Cytoplasmic dynein

Accessory factors

Organelle

11 FAST AXONAL TRANSPORT IS POWERED BY ENZYMATIC MOTORS. (A, B) Sequential images of the movement of microtubule fragments on purified fast-transport "motors." Time is indicated in minutes; the scale bar is 2 μm. Purified cytoplasmic dynein (A) or kinesin (B) was adsorbed to a coverslip, and fragments of microtubules were added. When the fragments contacted the surface they were propelled toward their frayed (distal, or +) end on dynein and toward their compact (proximal, or −) end on kinesin, as illustrated. (C) In the axon, the microtubules are stationary and have a polarity: The plus ends point toward the axon terminals, the minus ends toward the cell body. Cytoplasmic dynein and kinesin, together with as yet unidentified accessory factors, attach to organelles and propel them toward the cell body and axon terminal, respectively. (A and B from Paschal and Vallee, 1987; micrographs courtesy of R. Vallee; C after Vallee, Shpetner and Paschal, 1989.)

diction was first tested for adrenal medullary cells, from which chromaffin granules could be purified and their contents analyzed.[62] As described above, chromaffin granules contain not only epinephrine and norepinephrine, but also ATP, the synthetic enzyme dopamine β-hydroxylase, and chromogranins. Not only are all these components released in response to stimulation of the adrenal medulla, but they appear in the perfusate in exactly the same proportions as are found within granules. A good correspondence exists between vesicle contents and release from neurons as well, although it is difficult to isolate pure populations of synaptic vesicles from nerve terminals so as to determine their contents. For example, small synaptic vesicles in sympathetic neurons contain not only norepinephrine but also ATP; the larger, dense-cored vesicles contain, in addition, dopamine β-hydroxylase and chromogranin A. Stimulation of sympathetic axons results in the release of all these vesicle constituents.[63] Similarly, vesicles isolated from cholinergic neurons contain ATP as well as ACh and both are released by stimulation of cholinergic nerves.[64]

The idea that one quantum of transmitter corresponds to the contents of one synaptic vesicle has been examined quantitatively for cholinergic neurons. Vesicles purified from the terminals of the cholinergic electromotor neurons in the electric organ of the marine ray *Narcine brasiliensis* (a relative of *Torpedo californica*) were found to contain about 47,000 molecules of ACh.[65] The corresponding number for a synaptic vesicle at the frog neuromuscular junction, calculated assuming the same intravesicular concentration of ACh and correcting for the smaller size of frog vesicles, is 7000. This is in excellent agreement with electrophysiological estimates of the number of ACh molecules in a quantum[66] (Chapter 7).

Morphological correlates of vesicular release

An important experimental innovation developed by Heuser and Reese and their colleagues[67] enabled frog muscle to be quick-frozen within milliseconds after a single shock to the motor nerve and then to be prepared for freeze-fracture. With such an experiment it was possible to obtain scanning electron micrographs of vesicles caught in the act of fusing with the presynaptic membrane and to determine with some accuracy the time course of such fusion. To do this, the muscle is mounted on the undersurface of a falling plunger, with the motor nerve attached to stimulating electrodes. As the plunger falls, a stimulator is triggered, shocking the nerve at a selected interval before the muscle smashes into a copper block cooled to 4 K with liquid helium. An essential part of the experiment is that the duration of the presynaptic action potential is increased by addition of 4-aminopyridine (4-AP; Chapter 4) to the bathing solution. This treatment greatly increases the

[62]Kirshner, N. 1969. *Adv. Biochem. Psychopharmacol.* 1: 71–89.
[63]Smith, A. D. et al. 1970. *Tissue Cell* 2: 547–568.
[64]Silinsky, E. M. and Hubbard, J. I. 1973. *Nature* 243: 404–405.
[65]Wagner, J. A., Carlson, S. S. and Kelly, R. B. 1978. *Biochemistry* 17: 1199–1206.
[66]Kuffler, S. W. and Yoshikami, D. 1975. *J. Physiol.* 251: 465–482.
[67]Heuser, J. E. et al. 1979. *J. Cell Biol.* 81: 275–300.

magnitude and duration of quantal release evoked by a single shock and hence the number of vesicle openings seen in the electron micrographs (Figure 12A and 12B). Two important observations were made. First, the maximum number of vesicle openings occurred when stimulation preceded freezing by 3 to 5 msec. This corresponded to the peak of the postsynaptic current recorded from curarized, 4-AP-treated muscles in separate experiments. In other words, the maximum number of vesicle openings coincided in time with the peak postsynaptic conductance change determined physiologically. Second, the number of vesicle openings increased with 4-AP concentration, and the increase was related linearly to the estimated increase in quantum content of the end

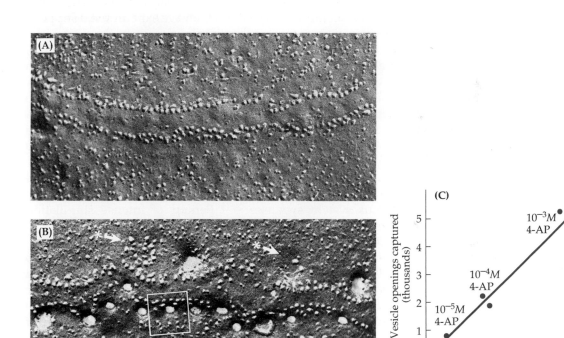

12 **VESICLES IN THE ACT OF FUSING** with the axon terminal membrane. (A) Freeze-fracture electron micrograph captures a view of the cytoplasmic half of the presynaptic membrane in a frog nerve terminal (as if observed from the synaptic cleft). The region of the active zone appears as a slight ridge delineated by membrane particles (about 10 nm in diameter). (B) Similar view of a terminal that was frozen just at the time the nerve begins to discharge large numbers of quanta (5 msec after stimulation). "Holes" (box) are sites of vesicle fusion; shallow depressions (indicated by asterisks) mark where vesicles have collapsed flat after opening. (C) Comparison of the number of vesicle openings, counted in freeze-fracture images, and the number of quanta released, determined from electrophysiological recordings. The diagonal line is the 1:1 relationship expected if each vesicle that opened released 1 quantum of transmitter. Transmitter release was varied by adding different concentrations of 4-AP (arrow indicates control, without 4-AP). (From Heuser et al., 1979; micrographs courtesy of J. E. Heuser.)

plate potentials by 4-AP, again obtained from separate physiological experiments (Figure 12C). Thus, vesicle openings were correlated both in number and in time course with quantal release. In later experiments, Heuser and Reese characterized the time course of vesicle openings in greater detail, showing that openings first increase during a 3- to 6-msec period after stimulation and then decrease over the next 40 msec.[68]

In summary, there is now much evidence that synaptic vesicles are the morphological correlate of the quantum of transmitter, each vesicle containing a few thousand molecules of transmitter. Vesicles can release their contents by exocytosis both spontaneously at a low rate (producing miniature synaptic potentials) and in response to presynaptic depolarization. This view is not universally held,[69,70] but other mechanisms proposed for quantal release, such as calcium-activated quantal gates in the presynaptic membrane, have less extensive experimental support. There is evidence that at some specialized synapses in the retina, depolarization can release transmitter by a mechanism that is nonquantal, not mediated by vesicle exocytosis, and not dependent on calcium influx.[71]

Recycling of vesicle membrane

The hypothesis that transmitter release occurs by exocytosis also predicts that during stimulation vesicles will disappear as their membrane is added to that of the axon terminal. If the processes for retrieving vesicle membrane and reforming vesicles are slow compared with the rate of exocytosis, then stimulation will lead to depletion of synaptic vesicles and expansion of the axon terminal. Such ultrastructural changes have been demonstrated at neuromuscular, ganglionic, and central nervous system synapses.[72-74] Periods of intense stimulation result in a reversible depletion of synaptic vesicles and an increase in the surface area of the axon terminal (Figure 13).

What becomes of the vesicle membrane after it has been added to the terminal plasma membrane? In elegant experiments, Heuser and Reese found that membrane is retrieved and recycled into new synaptic vesicles.[75] They studied recycling of vesicles in frog motor nerve terminals by stimulating nerve–muscle preparations in the presence of horseradish peroxidase (HRP, an enzyme that catalyzes the formation of an electron-dense reaction product). When electron micrographs of terminals fixed after short periods of electrical stimulation were examined, HRP was found primarily in coated vesicles around the outer margins of the synaptic region, suggesting that these vesicles had been formed from the terminal membrane by endocytosis and, in the process,

[68]Heuser, J. E. and Reese, T. S. 1981. *J. Cell Biol.* 88: 564–580.
[69]Tauc, L. 1982. *Physiol. Rev.* 62: 857–893.
[70]Dunant, Y. and Isreal, M. 1985. *Sci. Am.* 252: 58–66.
[71]Schwartz, E. A. 1987. *Science* 238: 350–355.
[72]Ceccarelli, B. and Hurlbut, W. P. 1980. *Physiol. Rev.* 60: 396–441.
[73]Dickinson-Nelson, A. and Reese, T. S. 1983. *J. Neurosci.* 3: 42–52.
[74]Wickelgren, W. O. et al. 1985. *J. Neurosci.* 5: 1188–1201.
[75]Heuser, J. E. and Reese, T. S. 1973. *J. Cell Biol.* 57: 315–344.

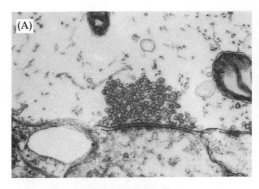

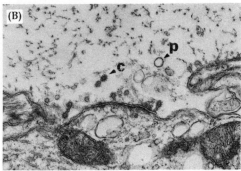

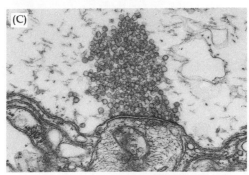

13

STIMULATION CAUSES A REVERSIBLE DE-
PLETION OF SYNAPTIC VESICLES in lamprey
giant axons. (A) Control synapse fixed after 15
minutes in saline. Synaptic vesicles are clustered
at the presynaptic membrane. (B) Synapse fixed
after stimulation of the spinal cord for 15 minutes
at 20 Hz. Note the depletion of synaptic vesicles,
the presence of coated vesicles (c) and pleo-
morphic vesicles (p), and the expanded presynaptic membrane. (C) Synapse fixed 60 minutes
after cessation of stimulation. Note similarities
to control synapse. Bar, 1 μm. (Micrographs courtesy of W. O. Wickelgren.)

had captured HRP from the extracellular space (Figure 14). HRP also
appeared, after a delay, in synaptic vesicles. Synaptic vesicles loaded
in this way with HRP could then be depleted of the enzyme by stimulation in HRP-free medium, an experimental result supporting the idea
that the recaptured membrane and enclosed HRP were recycled into
the vesicle population from which release occurred. Thus, following
stimulation, membrane is retrieved from the presynaptic terminal and
recycled into synaptic vesicles.

After particularly intense stimulation in the presence of HRP, large,
uncoated pits and cisternae containing HRP are seen.[75] The uncoated
pits and cisternae appear to be composed of components recaptured
nonselectively from the terminal membrane.[76] Synaptic vesicles are
thought to be reformed from the cisternae through the budding off of
coated vesicles and subsequent removal of the coat. These two parallel
pathways for recycling of vesicle membrane are summarized schematically in Figure 15.

The composition of synaptic vesicle membrane differs from that of
the terminal plasma membrane; nevertheless, the appropriate vesicle
membrane components are recycled[77] (Figure 16). Recapture of specific

[76]Miller, T. M. and Heuser, J. E. 1984. *J. Cell Biol.* 98: 685–698.
[77]Valtorta, F. et al. 1988. *J. Cell Biol.* 107: 2717–2727.

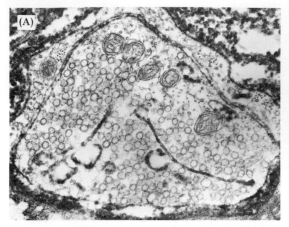

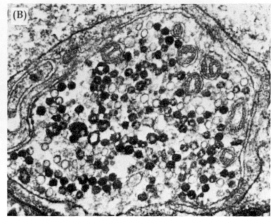

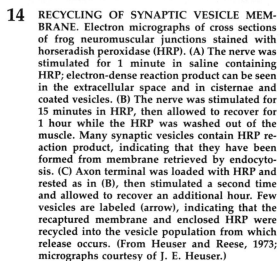

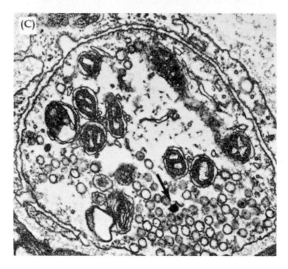

14 **RECYCLING OF SYNAPTIC VESICLE MEM-BRANE. Electron micrographs of cross sections of frog neuromuscular junctions stained with horseradish peroxidase (HRP). (A) The nerve was stimulated for 1 minute in saline containing HRP; electron-dense reaction product can be seen in the extracellular space and in cisternae and coated vesicles. (B) The nerve was stimulated for 15 minutes in HRP, then allowed to recover for 1 hour while the HRP was washed out of the muscle. Many synaptic vesicles contain HRP re-action product, indicating that they have been formed from membrane retrieved by endocyto-sis. (C) Axon terminal was loaded with HRP and rested as in (B), then stimulated a second time and allowed to recover an additional hour. Few vesicles are labeled (arrow), indicating that the recaptured membrane and enclosed HRP were recycled into the vesicle population from which release occurs. (From Heuser and Reese, 1973; micrographs courtesy of J. E. Heuser.)**

membrane proteins and lipids is mediated by formation of coated pits and endocytosis of coated vesicles.[76]

Analogous experiments have since been done using the uptake of highly fluorescent dyes to mark recycled vesicles.[78,79] This technique offers the advantage that vesicle recycling can be observed in living preparations by monitoring the stimulation-dependent accumulation and release of dye (Figure 17).

Sorting of vesicles within the nerve terminal

It appears that not all vesicles are created equal. Studies using ra-dioactive precursors to follow transmitter synthesis, incorporation into vesicles, and release suggest that there are subpopulations of vesicles

[78]Lichtman, J. W., Wilkinson, R. S. and Rich, M. M. 1985. *Nature* 314: 357–359.
[79]Betz, W. J., Mao, F. and Bewick, G. S. 1992. *J. Neurosci.* 12: 363–375.

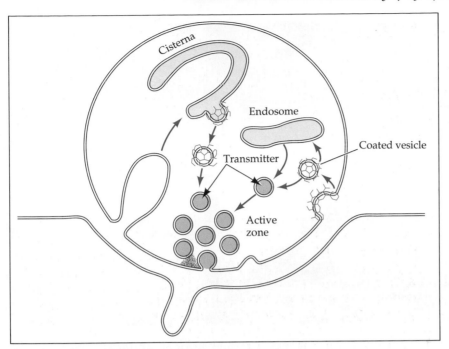

15 PROPOSED PATHWAYS FOR MEMBRANE RETRIEVAL during vesicle recycling. After exocytosis, clathrin-coated vesicles selectively recapture synaptic vesicle membrane components. New synaptic vesicles are formed from coated vesicles, either directly or through endosomes. After intense stimulation, uncoated pits and cisternae mediate the nonselective retrieval of surface membrane. Synaptic vesicles are reformed from cisternae via coated vesicles. The new synaptic vesicles formed from recycled membrane are then filled with transmitter and can be released by stimulation.

within axon terminals. In experiments made on the electric organ of the marine ray *Torpedo*, from which cholinergic synaptic vesicles can be purified, it was found that during stimulation newly synthesized ACh was not spread uniformly among the synaptic vesicle population but was localized to those vesicles recently formed by recycling.[80,81] Other studies have shown that newly synthesized transmitter molecules are preferentially released.[13,82] It is as if some vesicles fuse with the terminal membrane, release their content of transmitter, are reformed by endocytosis, refill with newly synthesized transmitter from the cytoplasm, and repopulate the active zone before other vesicles move in to take their place. Such a scheme fits with the observation that vesamicol, which specifically blocks transport of ACh into vesicles, selectively blocks release of newly synthesized ACh.[32]

Such a scheme requires that the movements of a subpopulation of vesicles be impeded in some way; vesicles free to move by Brownian

[80]Zimmermann, H. and Denston, C. R. 1977. *Neuroscience* 2: 695–714.
[81]Zimmermann, H. and Denston, C. R. 1977. *Neuroscience* 2: 715–730.
[82]Kopin, I. J. et al. 1968. *J. Pharmacol. Exp. Ther.* 161: 271–278.

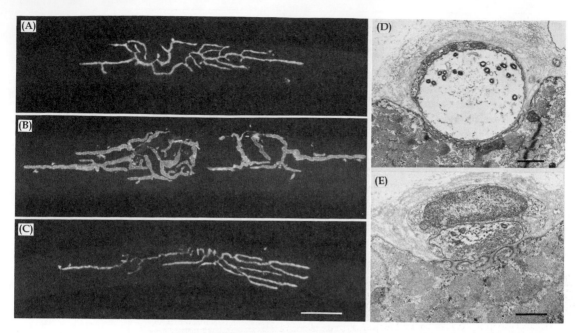

16 RECYCLING OF SPECIFIC SYNAPTIC VESICLE MEMBRANE PROTEINS. **(A–C)** Fluorescence micrographs of frog neuromuscular junction labeled with antibodies to synaptophysin, a vesicle membrane protein, and fluorescein-conjugated second antibodies. Bar, 50 μm. **(D–E)** Electron micrographs of cross sections of neuromuscular junctions. Bar, 1 μm. **(A)** Normal junction. The axon terminal membrane must be permeabilized with detergent in order for antibodies to reach synaptophysin. **(B, D)** Muscle was treated with α-latrotoxin, which causes vesicle release, in calcium-free medium, which blocks endocytosis. Under these conditions α-latrotoxin depletes nerve terminals of their quantal store of acetylcholine. Axon terminals stain without permeabilization, indicating that synaptophysin has become exposed on the terminal's surface, as expected if synaptic vesicles fuse with the axonal membrane during exocytosis. The axon terminals appear distended because retrieval of vesicle membrane is blocked by removing calcium from the bathing solution. **(C, E)** Muscle was treated with α-latrotoxin in normal saline and permeabilized with detergent before staining. Terminals have a normal appearance and can only be stained after permeabilization. Under these conditions the vesicle population is maintained by active recycling while more than two times the initial store of quanta are released. Thus, despite the active turnover of synaptic vesicles, no detectable synaptophysin remains on the terminal surface, demonstrating the specificity and efficiency of the process of synaptic vesicle membrane retrieval. (From Valtorta et al., 1988; micrographs courtesy of F. Valtorta.)

motion would quickly traverse the micrometers of distance separating them from the active zone. Greengard and his colleagues have identified a family of proteins, called synapsins, that are found in axons and axon terminals and appear to cross-link vesicles to the cytoskeleton, preventing their movement.[83,84] Interestingly, the binding of synapsins to vesicles is regulated by phosphorylation, so that addition of phosphate groups to synapsins dissociates them from vesicles. One role that re-

[83]De Camilli, P., and Greengard, P. 1986. *Biochem. Pharmacol.* 35: 4349–4357.
[84]Südhof, T. C. et al. 1989. *Science* 245: 1474–1480.

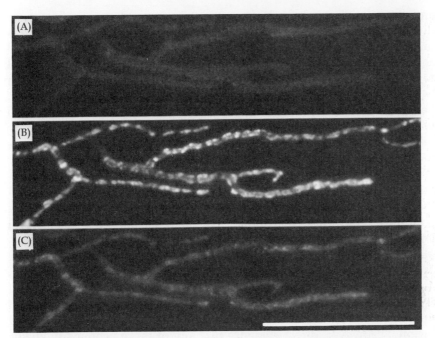

17   ACTIVITY-DEPENDENT UPTAKE AND RELEASE of fluorescent dye by axon terminals at the frog neuromuscular junction. Fluorescence micrographs of axon terminals in a cutaneous pectoris muscle. (A) Muscle was bathed for 5 minutes in fluorescent dye (2 μM FM1-43) and washed for 30 minutes. Only small amounts of the dye remain associated with the terminal membrane. (B) The same muscle was then bathed in dye for 5 minutes while the nerve was stimulated (10 Hz) and washed for 30 minutes. The fluorsecent patches are clusters of synaptic vesicles that were filled with dye during recycling. (C) The same muscle was then stimulated at 10 Hz for 5 minutes and washed for 30 minutes. Stimulation has released most of the dye. The shapes of the fluorescent patches corresponded to the distribution of synaptic vesicles seen in electron micrographs of stained terminals. Bar, 50 μm. (Micrographs courtesy of W. J. Betz.)

stricted vesicle movement might play in regulating transmitter release is illustrated by experiments by Llinás and his colleagues in which dephosphorylated synapsin I, injected directly into a presynaptic axon terminal at the squid giant synapse, caused a decrease in the amount of transmitter released by nerve stimulation.[85] When this was followed by injection of a specific protein kinase that is able to phosphorylate synapsin I, the amount of transmitter released by stimulation increased. It is tempting to speculate that the activity of endogenous kinases and phosphatases might be modulated so as to provide long-term regulation of the amount of transmitter released from axon terminals.

It is generally believed that the influx of calcium ions during an action potential promotes the fusion of vesicles with the plasma membrane at active zones, leading to transmitter release. How fusion occurs

Role of calcium in vesicle release

[85]Llinás, R. et al. 1985. *Proc. Natl. Acad. Sci. USA* 82: 3035–3039.

and the role played by calcium are not known. One complication is distinguishing between treatments that perturb regulatory pathways modulating release indirectly (for example, modifying calcium channel activity) and those that affect the release mechanism itself. It is useful to bear in mind that release occurs very rapidly after entry of calcium (within 0.1–0.2 msec), too rapidly for many enzyme-catalyzed reactions, particularly multistep cascades resulting in protein phosphorylation, to play a direct role. It seems likely that the ubiquitous calcium-binding protein calmodulin, which binds calcium with an affinity in the micromolar range and is known to mediate the effects of calcium on a variety of cellular functions, might play a role in vesicle fusion.[86] Indeed, antibodies to calmodulin and inhibitors of calmodulin action block exocytosis. In similar experiments, antibodies against a chromaffin granule-binding protein purified from adrenal medulla blocked release when injected into chromaffin cells.[87] Genetic experiments in yeast have implicated small GTP-binding proteins of the p21^{ras} superfamily in the docking and fusion of secretory vesicles; a similar protein (rab3A) is found associated specifically with synaptic vesicles.[33] Two other integral membrane proteins, synaptotagmin and synaptophysin, are localized to synaptic vesicles and may also be involved in vesicle docking and fusion.[33] However, direct evidence that any of these proteins mediates transmitter release is lacking.

A promising approach to the study of vesicle fusion is the use of dissociated nonneuronal secretory cells, such as mast cells, in which exocytosis of large, dense-cored secretory granules can be followed simultaneously by light microscopy and electrophysiological recording.[88,89] The fusion of single vesicles can been recorded as changes in electrical capacitance of the cell arising from the addition of the vesicle membrane to the cell surface (Figure 18). Similar changes in capacitance associated with transmitter release have been measured from individual nerve terminals isolated from the mammalian CNS.[90] These studies have demonstrated striking differences in the control of vesicle fusion between excitable and nonexcitable cells, particularly in the role played by calcium.[88] For example, an increase in intracellular calcium concentration is neither necessary nor sufficient for secretion to occur from mast cells; exocytosis in these nonexcitable cells appears to be under the control of an as-yet unidentified G protein. Overall, slower enzymatic processes tend to dominate stimulus–secretion coupling in nonexcitable cells, with calcium as a modulator, while in neurons the exocytotic machinery is poised so that calcium can trigger release rapidly, with slower, enzymatic processes playing a modulatory role.

[86]Reichardt, L. F. and Kelly, R. B. 1983. *Annu. Rev. Biochem.* 52: 871–926.
[87]Schweizer, F. E. et al. 1989. *Nature* 339: 709–712.
[88]Penner, R. and Neher, E. 1989. *Trends Neurosci.* 12: 159–163.
[89]Almers, W. and Tse, F. W. 1990. *Neuron* 4: 813–818.
[90]Lim, N. F., Nowycky, M. C. and Bookman, R. J. 1990. *Nature* 344: 449–451.

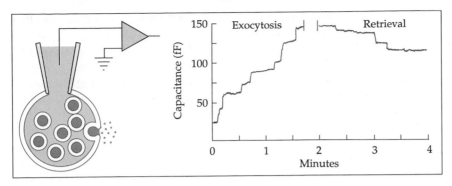

18 **RELEASE AND RETRIEVAL OF VESICLE MEMBRANE monitored by changes in membrane capacitance.** Increases in cell capacitance measured with the whole-cell patch-pipette recording technique occur in a stepwise fashion reflecting the fusion of individual vesicles with the plasma membrane. Corresponding decreases in capacitance are seen during vesicle retrieval. The recordings are from a rat mast cell, which has particularly large secretory vesicles (800 nm diameter). (After Fernandez, Neher and Gomperts, 1984.)

TRANSMITTER RECEPTORS

In Chapters 7 and 8, it was shown that transmitter receptors can be divided into two categories according to their mechanism of action: direct and indirect. Ligand-activated channels mediate direct transmitter action and belong to a superfamily of proteins that share many structural features (Chapter 2). Receptors in this family, which include nicotinic ACh receptors (nAChRs), $GABA_A$, glycine, and glutamate receptors, are oligomers of two or more different subunits. The combination of ligand-binding site and ligand-activated ion channel in the same protein enables directly coupled receptors to respond to transmitter within a fraction of a millisecond. Indirectly acting receptors, such as α- and β-adrenergic, muscarinic ACh, 5-HT, dopamine, and neuropeptide receptors, are members of a different protein superfamily. Members of this family are monomeric proteins that produce their effects through interaction with G proteins, which, in turn, modify the activity of receptors, ion channels, or pumps (Chapter 8).

Differences in the mechanism of action of directly and indirectly coupled receptors are reflected in their distribution on the surface of postsynaptic cells. Directly coupled receptors are concentrated in the postsynaptic membrane directly beneath the nerve terminal. Indirectly coupled receptors occur at lower density and are not so highly localized.

Receptor distribution

The precise localization of directly coupled receptors is exemplified by the distribution of nicotinic AChRs at the vertebrate skeletal neuromuscular junction. The existence of special properties of skeletal muscle fibers in the region of innervation has been known since the beginning of the century. For example, Langley assumed the presence of a receptive substance around motor nerve terminals, where a localized sensitivity exists for various chemical agents such as nicotine.[91] This receptive substance is now known to be the nicotinic AChR. The precise distribution of AChRs at the neuromuscular junction can be determined by measuring the chemosensitivity of the surface membrane by ionophoretic application of ACh from a micropipette onto restricted regions of the membrane (Chapter 7). This method is particularly useful with thin preparations in which the presynaptic and postsynaptic structures can be resolved with interference contrast optics and the position of the ionophoretic pipette in relation to the synapse can be determined with some precision.[92] Figure 19 shows pictures of synapses on a snake muscle fiber. The end plates in snake muscle resemble in their compactness those seen in mammals; they are about 50 μm in diameter with 50 to 70 terminal swellings that are analogous to synaptic boutons. The swellings rest in craters sunk into the surface of the muscle fiber (Figure 19B and 19C); an electron micrograph of such a synapse is shown in Figure 19B, again illustrating the characteristic features observed at all chemical synapses. The inset to the right shows an electron micrograph (at the same magnification as the rest of Figure 19B) of a typical ionophoretic micropipette. The outer diameter of the tip is about 100 nm and the opening is about 50 nm, similar in size to a synaptic vesicle. The muscle cell membrane is highly sensitive to application of ACh in the synaptic region, where the transmitter is normally released from the presynaptic boutons. The sharp delineation of sensitivity can be demonstrated dramatically in muscle fibers where the motor nerve terminal has been removed by bathing the muscle in a solution of the enzyme collagenase, which frees the terminal without damaging the muscle fiber.[93] The process of lifting off the terminals from a snake muscle is shown in Figure 19C. Each of the boutons leaves behind a circumscribed crater lined with the exposed postsynaptic membrane. This is shown in more detail in Figure 19D and 19E, in which an ACh-filled micropipette points at an empty crater. If the tip of the pipette is placed on the postsynaptic membrane, 1 picocoulomb of charge passed through the pipette releases enough ACh to cause an average 5-mV depolarization. The sensitivity of the membrane is then said to be 5000 mV per nanocoulomb (Figure 19F). In contrast, at a distance of about 2 μm, just outside the crater, the same amount of ACh applied to the extrasynaptic membrane produces a response that is 50 to 100 times smaller.[94] Along the rims of the craters the sensitivity fluctuates over a wide range.

A second way to determine the distribution of ACh receptors is to use α-bungarotoxin, a snake toxin that binds selectively and irreversibly to nicotinic AChRs, to label receptors. The distribution of bound toxin

[91]Langley, J. N. 1907. *J. Physiol.* 36: 347–384.

[92]McMahan, U. J., Spitzer, N. C., and Peper, K. 1972. *Proc. R. Soc. Lond.* B 181: 421–430.

[93]Betz, W. J. and Sakmann, B. 1973. *J. Physiol.* 230: 673–688.

[94]Kuffler, S. W. and Yoshikami, D. 1975. *J. Physiol.* 244: 703–730.

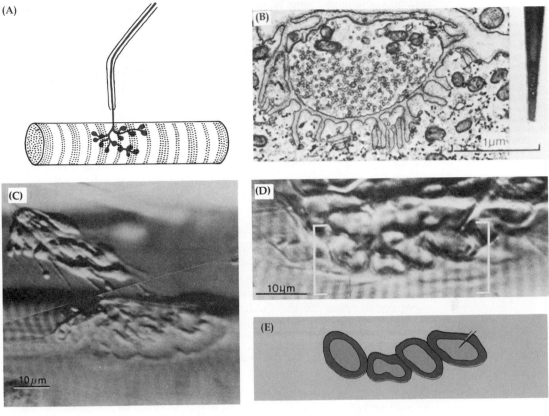

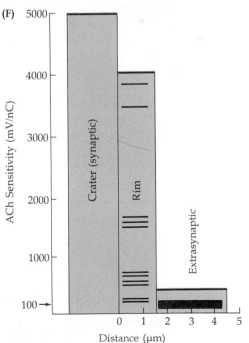

19

ACETYLCHOLINE RECEPTOR DISTRIBU-
TION at skeletal neuromuscular junctions of the
snake. (A) Sketch of an end plate on a skeletal
muscle of a snake. (B) Electron micrograph of a
cross section through a bouton. To the right is an
electron micrograph of the tip of a micropipette
used for ionophoresis of ACh; it has an outer
diameter of 100 nm and an opening of about 50
nm. (C) Removal of nerve terminal from a muscle
treated with collagenase. (D) The remaining
empty synaptic craters show exposed postsynaptic membrane. An ACh-filled pipette [entering
from upper right; compare with (E)] points to the
floor of a crater. (E) Drawing of the area bracketed
in (D). (F) Determination of sensitivity to ACh.
In the synaptic region sensitivity is uniformly
high (5000 mV/nC); in the extrasynaptic region
beyond the rims of the craters, it is uniformly
low (about 100 mV/nC). In the rim areas the sensitivity is variable. (From Kuffler and Yoshikami,
1975a.)

can be visualized using histochemical or autoradiographic techniques. For example, fluorescent markers can be attached to α-bungarotoxin and the distribution of receptors visualized by fluorescence microscopy[95] (Figure 20A); or HRP can be linked to α-bungarotoxin and the dense reaction product visualized in the electron microscope[96] (Figure 20B). These techniques confirm the results of ionophoretic application of ACh: Receptors are highly restricted to the membrane immediately beneath the axon terminal. Even more precise quantitative estimates of the concentration of ACh receptors can be obtained using radioactive α-bungarotoxin to label receptors and locating the receptor sites by autoradiography[97] (Figure 20C). By counting the number of silver grains exposed in the emulsion, the density of receptors can be determined. In muscle the density is highest along the crests and upper thirds of the junctional folds (about 10^4 per μm^2); the density in extrasynaptic regions is much lower (about 5 per μm^2).[98]

Similar techniques have demonstrated that transmitter receptors are concentrated in the postsynaptic membrane at synapses throughout the central and peripheral nervous systems. For example, $GABA_A$ and glycine receptors have been purified using specific, high-affinity ligands —benzodiazepine for the $GABA_A$ receptor, strychnine for the glycine receptor. Antibodies to the purified receptors have been used to localize the receptors at synaptic sites.[99,100] A specific snake toxin similar to α-bungarotoxin has been used to show that nicotinic AChRs are clustered beneath synaptic terminals on parasympathetic ganglion cells. However, their density is much lower than at the neuromuscular junction, approximately 600 per μm^2.[101] The lower density of receptors in neuronal postsynaptic membranes may reflect one way that nerve cells regulate their response to a quantum of transmitter. If the number of molecules of transmitter packaged in a quantum is relatively constant throughout the mammalian nervous system (synaptic vesicles are characteristically about 50 nm in diameter), then lowering the density of receptors in the postsynaptic membrane would reduce the synaptic current evoked by one quantum to a level appropriate for the size and input resistance of the cell (Chapter 7).[102]

Desensitization is a common feature of transmitter action

The response to a neurotransmitter often decreases during repeated or prolonged application, a phenomenon termed DESENSITIZATION. Desensitization is a common feature of transmitter receptor mechanisms. It was described in detail at the neuromuscular junction by Katz and Thesleff, who showed that with prolonged application of ACh the

[95]Ravdin, P. and Axelrod, D. 1977. *Anal. Biochem.* 80: 585–592.

[96]Burden, S. J., Sargent, P. B. and McMahan, U. J. 1979. *J. Cell Biol.* 82: 412–425.

[97]Fertuck, H. C. and Salpeter, M. M. 1974. *Proc. Natl. Acad. Sci. USA* 71: 1376–1378.

[98]Salpeter, M. M. 1987. *In* M. M. Salpeter (ed.). *The Vertebrate Neurosmuscular Junction.* Alan R. Liss, New York, pp. 1–54.

[99]Richards, J. G. et al. 1987. *J. Neurosci.* 7: 1866–1886.

[100]Triller, A. et al. 1985. *J. Cell Biol.* 101: 683–688.

[101]Loring, R. H. and Zigmond, R. E. 1987. *J. Neurosci.* 7: 2153–2162.

[102]Martin, A. R. 1990. *In* O. P. Ottersen and J. Storm-Mathisen (eds.). *Glycine Neurotransmission.* Wiley, New York, pp. 171–191.

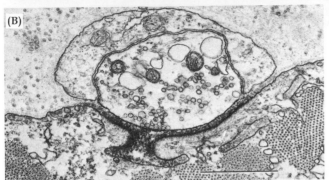

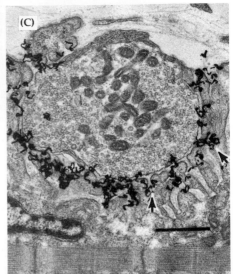

20 THE DISTRIBUTION OF ACETYLCHOLINE RECEPTORS at the neuromuscular junction. (A) Fluorescence micrograph of a frog cutaneous pectoris muscle fiber stained with rhodamine α-bungarotoxin. (B) Electron micrograph of a cross section of a frog cutaneous pectoris neuromuscular junction labeled with HRP-α-bungarotoxin. Dense reaction product fills the synaptic cleft. (C) Autoradiograph of a neuromuscular junction in a lizard intercostal muscle labeled with [125I]-α-bungarotoxin. Silver grains (arrows) show that receptors are concentrated at the tops and along the upper third of the junctional folds. Bars: A, 50 μm; B and C, 1 μm. (A courtesy of W. J. Betz; B courtesy of U. J. McMahan; C from Salpeter, 1987, courtesy of M. M. Salpeter.)

depolarizing response of the muscle fiber steadily declined (Figure 21).[103] Although the molecular mechanism by which direct ligand-activated channels become desensitized is not known, desensitization is an intrinsic molecular property of the receptor itself.[104,105] The rate at which desensitization develops, however, is modulated by receptor phosphorylation.[106] For indirectly coupled receptors, on the other hand, desensitization is a direct consequence of phosphorylation. For example, phosphorylation of the β-adrenergic receptor by at least two kinases, which are activated under different conditions, uncouples the receptor from the stimulatory G protein and so produces desensitization.[107] The phosphorylation sites on the β-adrenergic receptor are located on the intracellular loop between transmembrane domains V and VI and on the intracellular carboxy terminal tail, the same regions that have been shown, by site-directed mutagenesis studies, to mediate the interaction of the receptor with its G protein[106] (see Chapter 8).

[103]Katz, B. and Thesleff, S. 1957. *J. Physiol.* 138: 63–80.

[104]Ochoa, E. L., Chattopadhyay, A. and McNamee, M. G. 1989. *Cell Mol. Neurobiol.* 9: 141–178.

[105]Dudel, J., Franke, C. and Hatt, H. 1990. *Biophys. J.* 57: 533–545.

[106]Huganir, R. L. and Greengard, P. 1990. *Neuron* 5: 555–567.

[107]Hausdorff, W. P., Caron, M. G. and Lefkowitz, R. J. 1990. *FASEB J.* 4: 2881–2889.

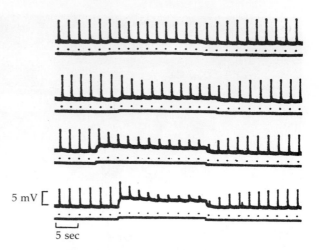

21

PROLONGED APPLICATION OF ACh CAUSES RECEPTOR DESENSITIZATION at the frog neuromuscular junction. Intracellular recordings of potential changes produced by brief iono-phoretic pulses of ACh from a micropipette (dots). Steady conditioning doses of ACh were delivered from a second pipette (upward deflections of lower traces). During the conditioning pulse the response to the test pulse decreases in amplitude as the receptors desensitize. With increasing doses the rate and extent of desensitization increase. (After Katz and Thesleff, 1957.)

5 mV

5 sec

TERMINATION OF TRANSMITTER ACTION

A variety of processes remove transmitters from the synaptic cleft and so terminate their activity. For example, the action of neuropeptides is terminated primarily by diffusion, while specific mechanisms exist to remove particular low-molecular-weight transmitters from the synaptic cleft. Physiological evidence indicates that at many synapses the rapid termination of transmitter action is crucial to synaptic function.

Termination of ACh action by acetylcholinesterase

As described in Chapter 7, the action of acetylcholine at the neuro-muscular junction is terminated by the enzyme acetylcholinesterase (AChE), which hydrolyzes ACh to choline and acetate. Much of the choline is transported back into the nerve terminal and reused for ACh synthesis (see Figure 4). Although AChE occurs at widespread locations in many tissues, it is found in particularly high concentration at cholinergic synapses[108] (Figure 22). Most of the AChE at the neuromuscular junction is bound to the synaptic basal lamina, that portion of the muscle fiber's sheath of extracellular matrix material that occupies the synaptic cleft and junctional folds.[109] There are 2600 catalytic subunits of AChE per μm^2 of synaptic basal lamina[98] (compared to the 10^4 ACh receptors per μm^2 in the postsynaptic membrane). Inhibition of AChE at the neuromuscular junction prolongs the lifetime of ACh, causing an increase in the amplitude and duration of individual synaptic potentials.[110] Under such conditions, the action of ACh is terminated solely by its diffusion from the synaptic cleft. If AChE is inhibited and axons are allowed to fire at frequencies typical of normal function, then ACh accumulates in the cleft and causes additional effects that block neuromuscular transmission. For example, the persistent depolarization of

[108]Massoulié, J. and Bon, S. 1982. *Annu. Rev. Neurosci.* 5: 57–106.
[109]McMahan, U. J., Sanes, J. R. and Marshall, L. M. 1978. *Nature* 271: 172–174.
[110]Fatt, P. and Katz, B. 1951. *J. Physiol.* 115: 320–370.

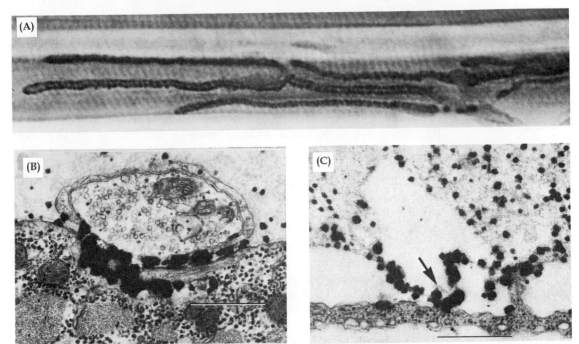

22 ACETYLCHOLINESTERASE IS CONCENTRATED in the synaptic basal lamina at neuromuscular junctions. (A) Light micrograph of a neuromuscular junction in a frog cutaneous pectoris muscle stained by a histochemical procedure for acetylcholinesterase. The dark reaction product lines the synaptic gutters and junctional folds. (B) Electron micrograph of a cross section of an axon terminal from a muscle stained for acetylcholinesterase as in (A). The electron-dense reaction product fills the synaptic cleft and the junctional folds. (C) Electron micrograph of a damaged muscle in which the nerve terminal, Schwann cell, and muscle fiber have degenerated and been phagocytized, leaving only empty basal lamina sheaths. The damaged muscle was stained for acetylcholinesterase; reaction product is associated with the synaptic basal lamina (arrow). Bar: (A), 20 μm; (B) and (C), 0.6 μm. (Micrographs courtesy of U. J. McMahan.)

the postsynaptic membrane causes inactivation of sodium channels in the muscle membrane near the end plate, preventing initiation of muscle action potentials (Chapter 4). As receptors become desensitized in the continued presence of ACh, the muscle fiber repolarizes, but transmission remains blocked because the muscle can no longer respond to ACh. The toxicity of inhibitors of AChE, such as organophosphate insecticides and nerve gases, is due in part to this form of neuromuscular blockade. Thus the rapid termination of ACh action is crucial for neuromuscular transmission under physiological conditions.

It might seem inefficient to have acetylcholinesterase situated between the axon terminal and the postsynaptic membrane, forcing molecules of ACh to traverse a minefield of degradative enzymes before having an opportunity to interact with their postsynaptic receptors. However, if the dimensions of the cleft and the rates of ACh diffusion, binding, and hydrolysis are taken into consideration, a simple scheme emerges, called the SATURATED DISK.[98,111] Following the release of one quantum, the concentration of ACh increases almost instantaneously (within microseconds) across the width of the cleft to a level high enough (0.5 mM) to saturate both ACh receptors and esterase within a disk approximately 0.5 μm in diameter centered on the release site. Binding of ACh to its receptors and to AChE is rapid compared with the rate at which AChE can hydrolyze ACh (it takes AChE 0.1 msec to hydrolyze one molecule of ACh). Therefore, the fraction of ACh molecules released that bind initially to postsynaptic receptors (rather than to cholinesterase) is determined solely by the ratio of receptors to esterase; this means that approximately 20 percent of the ACh molecules bind to AChE and 80 percent to ACh receptors. Binding causes a precipitous fall in ACh concentration. The concentration then remains low because AChE can hydrolyze ACh molecules much faster (10 per msec) than they are released from receptors as the channels close ($\tau = 1$ msec). Thus, by 0.1 msec or so after release, the concentration of ACh in the cleft has fallen to levels that make the probability of two ACh molecules being available to bind and open another receptor negligible. Such an analysis predicts that inhibition of acetylcholinesterase should have a more pronounced effect on the duration of the synaptic potential than on its amplitude, which is the case; amplitude is increased 1.5 to 2 times, while the duration is increased 3 to 5 times.[110,112] Thus the organization of the neuromuscular junction and the density and kinetic properties of ACh receptors and AChE combine to produce a synapse that is capable of very rapid responses and efficient use of ACh.

Other mechanisms for termination of transmitter action

The actions of dopamine, norepinephrine, 5-HT, and in some cases GABA are terminated not by hydrolysis but by uptake of the intact transmitter into presynaptic nerve terminals by sodium-dependent transport proteins.[113,114] Within the terminal the transmitter can be repackaged and released again. Pharmacological evidence indicates that each of these transmitters is taken up by a specific transporter.[115] Transporters for norepinephrine[116] and GABA[117] have been cloned. The cDNA clones encode highly homologous proteins that define a new protein superfamily, having 12 putative transmembrane domains, intracellular amino and carboxy termini, and three extracellular glycosylation sites. There are clear pharmacological differences between the transporters that carry neurotransmitters into nerve terminals and those that carry them into vesicles. For example, the uptake of norepinephrine into the terminal is blocked by cocaine, while its uptake into vesicles is

[111]Hartzell, H. C., Kuffler, S. W. and Yoshikami, D. 1975. *J. Physiol.* 251: 427–463.
[112]Katz, B. and Miledi, R. 1973. *J. Physiol.* 231: 549–574.
[113]Iversen, L. L. 1967. *The Uptake and Storage of Noradrenaline in Sympathetic Nerves.* Cambridge University Press, London.
[114]Bloom, F. E. and Iversen, L. L. 1971. *Nature* 229: 628–630.
[115]Ritz, M. C., Cone, E. J. and Kuhar, M. J. 1990. *Life Sci.* 46: 635–645.
[116]Pacholczyk, T., Blakely, R. D. and Amara, S. G. 1991. *Nature* 350: 350–354.
[117]Guastella, J. et al. 1990. *Science* 249: 1303–1306.

blocked by reserpine. Transmitter action can also be terminated by uptake into nonneuronal cells; for example, at the crustacean neuromuscular junction, the inhibitory transmitter GABA diffuses away from the synapse and is taken up by glial cells.[118] Glial cells in the mammalian CNS also have a high-affinity uptake mechanism for GABA and may aid in removing GABA from some CNS synapses.

Little information is available concerning the termination of action of peptide transmitters. There is no evidence for uptake of peptides into presynaptic terminals and, although enzymes that hydrolyze specific peptides are beginning to be characterized,[119] diffusion away from synaptic sites appears to be an important mechanism for termination of peptide action.

SUGGESTED READING

General

Cooper, J. R., Bloom, F. E. and Roth, R. H. 1991. *The Biochemical Basis of Neuropharmacology*, 6th Ed. Oxford University Press, New York.

Kandel, E. R., Schwartz, J. H. and Jessell, T. M. (eds.). 1991. *Principles of Neural Science*, 3rd Ed. Elsevier, New York.

Martin, J. B., Brownstein, M. J. and Krieger, D. T. (eds.). 1987. *Brain Peptides Update*, Vol. 1. Wiley, New York.

Salpeter, M. M. (ed.). 1987. *The Vertebrate Neuromuscular Junction*. Alan R. Liss, New York.

Siegel, G. J., Agranoff, B. W., Albers, R. W. and Molinoff, P. B. 1989. *Basic Neurochemistry: Molecular, Cellular, and Medical Aspects*, 4th Ed. Raven Press, New York.

Reviews

Kupfermann, I. 1991. Functional studies of cotransmission. *Physiol. Rev.* 71: 683–732.

Penner, R. and Neher, E. 1989. The patch-clamp technique in the study of secretion. *Trends Neurosci.* 12: 159–163.

Porter, M. E. and Johnson, K. A. 1989. Dynein structure and function. *Annu. Rev. Cell Biol.* 5: 119–151.

Sossin, W. S., Fisher, J. M. and Scheller, R. H. 1989. Cellular and molecular biology of neuropeptide processing and packaging. *Neuron* 2: 1407–1417.

Südhof, T. C. and Jahn, R. 1991. Proteins of synaptic vesicles involved in exocytosis and membrane recycling. *Neuron* 6: 665–677.

Trimble, W. S., Linial, M. and Scheller, R. H. 1991. Cellular and molecular biology of the presynaptic nerve terminal. *Annu. Rev. Neurosci.* 14: 93–122.

Vallee, R. B. and Bloom, G. S. 1991. Mechanisms of fast and slow axonal transport. *Annu. Rev. Neurosci.* 14: 59–92.

Zigmond, R. E., Schwarzchild, M. A. and Rittenhouse, A. R. 1989. Acute regulation of tyrosine hydroxylase by nerve activity and by neurotransmitters via phosphorylation. *Annu. Rev. Neurosci.* 12: 415–461.

[118]Orkand, P. M. and Kravitz, E. A. 1971. *J. Cell Biol.* 49: 75–89.
[119]Marcel, D. et al. 1990. *J. Neurosci.* 10: 2804–2817.

Original Papers

Birks, R. I. and MacIntosh, F. C. 1961. Acetylcholine metabolism of a sympathetic ganglion. *J. Biochem. Physiol.* 39: 787–827.

Burger, P. M., Hell, J., Mehl, E., Krasel, C., Lottspeich, F. and Jahn, R. 1991. GABA and glycine in synaptic vesicles: Storage and transport characteristics. *Neuron* 7: 287–293.

Heuser, J. E. and Reese, T. S. 1973. Evidence for recycling of synaptic vesicle membrane during transmitter release at the frog neuromuscular junction. *J. Cell Biol.* 57: 315–344.

Heuser, J. E., Reese, T. S., Dennis, M. J., Jan, Y., Jan, L. and Evans, L. 1979. Synaptic vesicle exocytosis captured by quick freezing and correlated with quantal transmitter release. *J. Cell Biol.* 81: 275–300.

Howard, J., Hudspeth, A. J. and Vale, R. D. 1989. Movement of microtubules by single kinesin molecules. *Nature* 342: 154–158.

Lim, N. F., Nowycky, M. C. and Bookman, R. J. 1990. Direct measurement of exocytosis and calcium currents in single vertebrate nerve terminals. *Nature* 344: 449–451.

Pacholczyk, T., Blakely, R. D. and Amara, S. G. 1991. Expression cloning of a cocaine and antidepressant-sensitive human noradrenaline transporter. *Nature* 350: 350–354.

Schnapp, B. J., Vale, R. D., Sheetz, M. P. and Reese, T. S. 1985. Single microtubules from squid axoplasm support bidirectional movement of organelles. *Cell* 40: 455–462.

Weiner, N. and Rabadjija, M. 1968. The effect of nerve stimulation on the synthesis and metabolism of norepinephrine from cat spleen during sympathetic nerve stimulation. *J. Pharmacol. Exp. Ther.* 160: 61–71.

IDENTIFICATION AND FUNCTION OF TRANSMITTERS IN THE CENTRAL NERVOUS SYSTEM

CHAPTER TEN

Identification of the neurotransmitters mediating synaptic interactions in the brain and spinal cord is of fundamental importance for understanding the function of the nervous system. Although it is far more difficult to establish that a particular chemical substance is a transmitter for a synapse in the central nervous system than it is for one in the periphery, considerable progress in the identification of CNS transmitters has been made. Various techniques are available for mapping the distribution of transmitter systems, based on visualizing the transmitter itself; labeling one of the proteins mediating its action, synthesis or degradation; or detecting mRNA transcripts for such proteins. Such studies not only give insight into the normal operation of the CNS, but also provide clues to biochemical deficits that underlie dysfunctions of the nervous system and suggest possible therapeutic measures.

Major transmitters in the central nervous system are the amino acids glutamate, glycine, and GABA. GABA mediates inhibitory interactions in the brain; glycine is inhibitory at many synapses in the brainstem and spinal cord. There are two classes of receptors for GABA. GABA$_A$ receptors are by far the more common. They respond to GABA with an increase in chloride conductance, and their activity is modulated by widely used anticonvulsant and antianxiety drugs such as barbiturates (phenobarbital) and benzodiazepines (diazepam and chlordiazepoxide). Glutamate is the major excitatory transmitter in the CNS. It interacts with receptors that can be divided into two broad classes: non-NMDA receptors, which are typical fast-acting directly coupled receptors that are permeable to monovalent cations; and NMDA receptors, which are both ligand-activated and voltage-dependent. Once open, NMDA channels allow calcium influx, which can trigger calcium-dependent second messenger systems in the postsynaptic cell. These properties allow NMDA receptors to mediate associative activity-dependent changes in synaptic efficacy that may underlie learning and memory; these changes include long-term potentiation (LTP) of transmission in the hippocampus.

Acetylcholine acts as a transmitter in many brain regions, largely through indirectly-coupled muscarinic receptors. Nuclei in the basal forebrain provide extensive and diffuse cholinergic innervation to the cortex and hippocampus. The effects of muscarinic agonists and antagonists suggest that this cholinergic input is important for cognitive functions. The basal forebrain cholinergic system is a conspicuous locus of degenerative changes in Alzheimer's disease, although neurons releasing other transmitters are also affected.

Peptide transmitters are widely distributed in the central, peripheral, and enteric nervous systems. Considerable interest in substance P and the opioid peptides stems from the finding that both appear to be involved in pain sensation. Substance P is released by primary afferent fibers that respond to noxious stimuli. Enkephalin, released by interneurons in the spinal cord, suppresses pain sensation by blocking release of substance P from primary afferent terminals. Other opioid peptides act at synapses within the brain to alter our perception of pain. Both substance P and opioid peptides are also found in areas of the CNS that are not involved in pain sensation.

A remarkable feature of the distribution of norepinephrine, dopamine, epinephrine, serotonin, and histamine in the CNS is that very few neurons release these amines as transmitters, but the arborizations of those cells that do are very extensive and diffuse. This morphology correlates with the physiological roles of amine-containing neurons in modulating, through indirectly coupled receptor mechanisms, synaptic activity in widespread areas of the CNS, so as to regulate global functions such as attention, arousal, the sleep–wake cycle, mood, and affect. Dysfunctions of such neuromodulatory interactions are particularly amenable to treatments based on systemic administration of drugs or neuronal transplantation.

A variety of different techniques have been devised for identifying the neurotransmitter(s) released at a particular synapse. Physiological and pharmacological tests are described in Chapter 9. In this chapter we describe other techniques for transmitter identification, some of which detect the transmitter itself, while others detect proteins associated with transmitter function or metabolism. We then summarize findings for several low-molecular-weight and neuropeptide transmitters, indicating where they are localized and what roles they may play in CNS function.

Mapping transmitter distribution   One method for identifying neurons releasing a particular transmitter is to visualize the distribution of the transmitter molecules themselves. A particularly significant advance for the determination of transmitter distribution in the CNS was the development by Falck and Hillarp of a fluorescence method that enables the recognition under the light microscope of axons or terminals containing biogenic amines such as dopamine, norepinephrine, and 5-hydroxytryptamine (5-HT).[1] When condensed with formaldehyde, each of these substances emits light of a characteristic wavelength under ultraviolet illumination (Figure 1). This fluorescence method, and subsequent modifications based on the

[1]Falck, B. et al. 1962. *J. Histochem. Cytochem.* 10: 348–354.

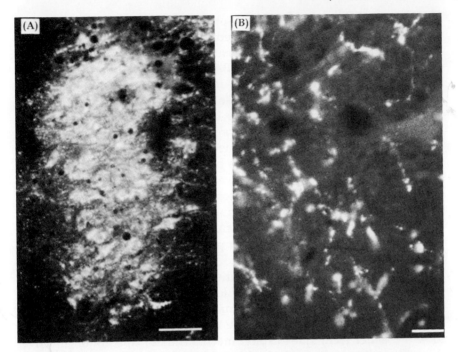

1 VISUALIZATION OF BIOGENIC AMINE-CONTAINING CELLS and their terminal arborizations by formaldehyde-induced fluorescence. (A) Norepinephrine-containing cells in the rat locus coeruleus. Bar, 100 μm. (B) The terminal arborizations of locus coeruleus cells in the cerebral cortex. Bar, 10 μm. (From Harik, 1984.)

use of glyoxylic acid,[2] paved the way for rapid progress. They opened up the entire central nervous system for easier exploration by allowing the visualization of pathways and groups of neurons whose terminals are likely to liberate amines. Another technique, immunohistochemistry, has provided an alternative way of visualizing cells containing amines and other transmitters. Antibodies have been made against small neurotransmitters, such as GABA, 5-HT, and dopamine,[3] as well as against neuropeptides. Such antibodies can be coupled, either directly or indirectly, to various markers that can be detected by light, fluorescence, or electron microscopy. The cloning of neuropeptide cDNAs and in situ hybridization techniques provide additional tools for determining the location of neurons synthesizing neuropeptides, as well as for studying the regulation of peptide expression.[4]

An alternative to labeling the transmitter itself is to make specific probes for enzymes mediating transmitter synthesis and degradation. Neurons contain uniquely high levels of enzymes catalyzing the synthesis of the transmitter they release. For example, cells using GABA

[2]de la Torre, J. C. 1980. *J. Neurosci. Methods* 3: 1–5.
[3]Wässle, H. and Chun, M. H. 1988. *J. Neurosci.* 8: 3383–3394.
[4]Mengod, G., Charli, J. L. and Palacios, J. M. 1990. *Cell Mol. Neurobiol.* 10: 113–126.

as a transmitter possess high levels of the enzyme glutamate decarboxylase. This enzyme has been purified and antibodies made. Appropriately tagged antibodies have been used to localize the enzyme in tissue sections and thereby to identify GABA-containing neurons.[5,6] Similarly, for identifying cholinergic neurons, antibodies have been made against the enzyme choline acetyltransferase.[7] Antibodies to tyrosine hydroxylase and dopamine β-hydroxylase identify cells releasing dopamine and norepinephrine.[8] The cDNAs for many of the enzymes involved in transmitter synthesis have been cloned. The localization and regulation of expression of these enzymes has been studied using in situ hybridization techniques. Enzymes involved in transmitter degradation are less reliable indicators of transmitter identity, although for ACh and GABA the distribution of their degradative enzymes has proved to be a useful marker under appropriate conditions.[9–11]

Identification of the receptors on postsynaptic cells can also provide clues about the transmitter liberated by presynaptic terminals. For instance, monoclonal antibodies against GABA receptors purified from cerebral cortex have been used to localize synapses where GABA is likely to be the transmitter.[12] Alternatively, the transmitter itself, or a specific agonist or antagonist, can be labeled and used to localize receptors.[13–15] Other techniques exploit mechanisms of transmitter inactivation. For example, cells that release biogenic amines or GABA have specific uptake systems for recapturing these transmitters. Thus, bathing tissue in radioactive norepinephrine, dopamine, 5-HT, or GABA can selectively label nerve terminals that release these transmitters.[16]

From a combination of the various approaches outlined above, it is possible to identify synapses within the central nervous system where GABA and glycine are the transmitters. For example, in an early elegant series of studies at the level of individual neurons, Otsuka, Ito, Obata, and their colleagues established that cerebellar Purkinje cells release GABA as a transmitter at inhibitory synapses on cells in the brainstem.[17–19] The visual system provides another example in which both

GABA and glycine: Major inhibitory transmitters in the CNS

[5]Matsuda, T., Wu, J.-Y. and Roberts, E. 1973. *J. Neurochem.* 21: 159–166, 167–172.
[6]Hendrickson, A. E. et al. 1983. *J. Neurosci.* 3: 1245–1262.
[7]Wainer, B. H. et al. 1984. *Neurochem. Int.* 6: 163–182.
[8]Miachon, S. et al. 1984. *Brain Res.* 305: 369–374.
[9]Shute, C. C. D., and Lewis, P. R. 1963. *Nature* 199: 1160–1164.
[10]Wallace, B. G. and Gillon, J. W. 1982. *J. Neurosci.* 2: 1108–1118.
[11]Nagai, T. et al. 1984. *In* A. Björklund, T. Hökfelt and M. J. Kuhar (eds.). *Classical Transmitters and Transmitter Receptors in the CNS*, Part 2. (*Handbook of Chemical Neuroanatomy*, Vol. 3.) Elsevier, New York, pp. 247–272.
[12]Hendry, S. H. C. et al. 1990. *J. Neurosci.* 10: 2438–2450.
[13]Kuhar, M. J., De Souza, E. B., and Unnerstall, J. R. 1986. *Annu. Rev. Neurosci.* 9: 27–59.
[14]Palacios, J. M. et al. 1990. *Ann. N.Y. Acad. Sci.* 600: 36–52.
[15]Palacios, J. M. et al. 1990. *Prog. Brain Res.* 84: 243–253.
[16]Hendry, S. H. C. and Jones, E. G. 1981. *J. Neurosci.* 1: 390–408.
[17]Obata, K. 1969. *Experientia* 25: 1283.
[18]Otsuka, M. et al. 1971. *J. Neurochem.* 18: 287–295.
[19]Obata, K., Takeda, K. and Shinozaki, H. 1970. *Exp. Brain Res.* 11: 327–342.

physiological and morphological evidence indicates that GABA is released as a transmitter (see Chapters 16 and 17). Thus in the retina, GABA is associated with certain types of horizontal and amacrine cells.[20,21] In the lateral geniculate nucleus, cells are found that contain glutamate decarboxylase, the enzyme catalyzing GABA synthesis,[6] and in the visual cortex, GABA uptake and glutamate decarboxylase are associated with several distinct types of local-circuit inhibitory neurons.[22] Local application of agents that block the action of GABA have effects on signaling; for example, bicuculline changes the receptive field organization of ganglion cells in the retina and complex cells in the cortex (Chapter 17).[23,24] There is now good evidence that GABA is the transmitter released at inhibitory synapses in widespread areas of the brain. A striking finding is the high proportion of GABA neurons. For example, in several cortical areas, one out of every five neurons contains GABA.[25] Such GABAergic neurons characteristically form local inhibitory connections. The overall importance of inhibitory interactions mediated by GABA for signaling in the CNS is demonstrated by the finding that widespread application of drugs that block GABA receptors produces convulsions.[26]

A second transmitter mediating inhibition at CNS synapses, particularly those in the brainstem and spinal cord, is the amino acid glycine. Like GABA, when applied ionophoretically glycine mimics the potentials that are produced when inhibitory pathways are stimulated. A useful distinction is in the agents that block inhibition, such as strychnine for glycine and picrotoxin or bicuculline for GABA.[27] Their actions provide diagnostic criteria for recognizing the transmitter by selectively and reliably blocking the effects of either GABA or glycine. At some synapses, such as those on neurons in the spinal cord and brainstem of the lamprey, sufficiently detailed tests have been made to indicate that glycine is the inhibitory transmitter.[27,28] A technique of growing importance used for these experiments is to measure the characteristics of conductance channels activated by transmitter candidates, either by noise analysis or by patch clamp techniques (Chapter 2). For the substance to be a viable candidate, the channels must have characteristics identical to those activated synaptically.

Two receptors for GABA have been identified on neurons in the central nervous system. GABA$_A$ receptors are the more common. They respond to GABA with a rapid increase in chloride conductance. The

GABA$_A$ and GABA$_B$ receptors

[20]Sterling, P. 1983. *Annu. Rev. Neurosci.* 6: 149–185.

[21]Lam, D. M.-K. and Ayoub, G. S. 1983. *Vision Res.* 23: 433–444.

[22]Gilbert, C. D. 1983. *Annu. Rev. Neurosci.* 6: 217–247.

[23]Caldwell, J. H. and Daw, N. W. 1978. *J. Physiol.* 276: 299–310.

[24]Sillito, A. M. 1979. *J. Physiol.* 289: 33–53.

[25]Naegele, J. R. and Barnstable, C. J. 1989. *Trends Neurosci.* 12: 28–34.

[26]Olsen, R. W. and Leeb-Lundberg, F. 1981. *In* E. Costa, G. DiChiara and G. L. Gessa (eds.). *GABA and Benzodiazepine Receptors.* Raven Press, New York, pp. 93–102.

[27]Matthews, G. and Wickelgren, W. O. 1979. *J. Physiol.* 293: 393–415.

[28]Martin, A. R. 1990. *In* O. P. Ottersen and J. Storm-Mathisen (eds.). *Glycine Neurotransmission.* Wiley, New York, pp. 171–191.

GABA$_A$ receptor belongs to the same superfamily of ligand-activated receptors as the nicotinic acetylcholine receptor (Chapter 2). Four classes of GABA$_A$ receptor subunits have been defined: α, β, γ, and δ.[29] Multiple variants exist within each class, and different combinations of subunits produce receptors with different properties. The subunit composition of native GABA$_A$ receptors is not known. Studies of the distribution of mRNAs encoding variant subunits demonstrate differences among brain regions, suggesting regional specificity in receptor subtype (Figure 2).[30] In the early 1980s, responses to GABA that were not

[29]Luddens, H. and Wisden, W. 1991. *Trends Pharmacol. Sci.* 12: 49–51.
[30]Vicini, S. 1991. *Neuropsychopharmacology* 4: 9–15.

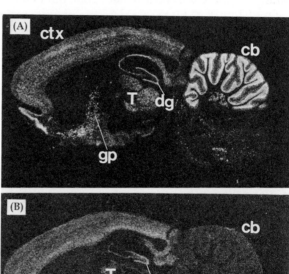
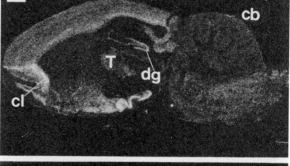

2

DISTRIBUTION OF GABA$_A$ RECEPTOR SUBUNIT mRNAs in the rat brain. Light microscopic autoradiographs of serial sections of rat brain following in situ hybridization with ^{35}S-labeled probes for the (A) α_1, (B) α_3, and (C) α_6 GABA$_A$ receptor subunit mRNAs. T, thalamus; cb, cerebellum; cl, claustrum; ctx, cortex; dg, dentate gyrus; gr, cerebellar granule cells; gp, globus pallidus. Bar, 2 mm. (From Luddens and Wisden, 1991; courtesy of W. Wisden.)

inhibited by bicuculline were observed on neurons in the CNS.[31] These effects were found to be mimicked by baclofen, which has no effect on GABA$_A$ receptors. The receptors mediating these effects are called GABA$_B$ receptors. GABA$_B$ receptors act indirectly through G proteins to block calcium channels or activate potassium channels.[32]

Every neuron in the mammalian CNS appears to have GABA receptors on its cell body, and both GABA$_A$ and GABA$_B$ receptors are found on the cell bodies of dorsal root ganglion neurons.[32] The physiological function of these receptors is unclear, since there are no GABAergic synapses on the soma of dorsal root ganglion cells. It appears that for these and many other neurons,[28] GABA receptors are inserted into membranes throughout the cell. They may, however, play a functional role only at synaptic terminals, where GABA inputs that inhibit transmitter release are located, or at synaptic sites on dendrites.

A distinctive feature of GABA$_A$ receptors is their regulation by allosteric modulation.[33] GABA$_A$ receptors exhibit binding sites for two classes of modulators, benzodiazepines and barbiturates (Figure 3). Benzodiazepines, which include diazepam (Valium) and chlordiazepoxide (Librium), are widely used as antiepileptic drugs, as antianxiety drugs, and as muscle relaxants; barbiturates, such as phenobarbital and secobarbital, are anticonvulsants. Both increase GABA-induced chloride current, benzodiazepines by increasing the frequency of channel opening and barbiturates by prolonging channel burst duration.[32] The α and β subunits have binding sites for both GABA and barbiturates; indeed, expression of either subunit alone can result in formation of a functional receptor.[34] Benzodiazepines bind to the γ_2 subunit, and this γ subunit must be coexpressed with α and β subunits to form a GABA$_A$ receptor that can be modulated by benzodiazepines.[35] The affinity and specificity

Modulation of GABA$_A$ receptor function by benzodiazepines and barbiturates

[31]Hill, D. R. and Bowery, N. G. 1981. *Nature* 290: 149–152.
[32]Nicoll, R. A., Malenka, R. C. and Kauer, J. A. 1990. *Physiol. Rev.* 70: 513–565.
[33]Costa, E. 1991. *Neuropsychopharmacology* 4: 225–235.
[34]Blair, L. A. C. et al. 1988. *Science* 242: 577–579.
[35]Pritchett, D. B. el al. 1989. *Nature* 338: 582–585.

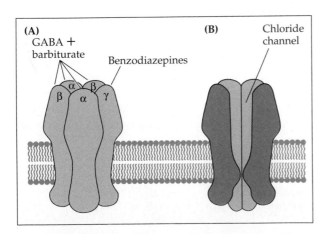

3

HYPOTHETICAL MODEL FOR THE GABA$_A$ RECEPTOR. (A) The receptor is drawn as an $\alpha_2\beta_2\gamma$ complex, by analogy with the nicotinic acetylcholine receptor; the actual number and arrangement of subunits in the native receptor are not known. GABA and barbiturates bind to different sites on the α and β subunits, each of which can form a functional homomultimeric receptor; benzodiazepines bind to the γ subunit. (B) Cross-sectional view of the receptor.

of each binding site for its ligand is determined not only by the intrinsic properties of the subunit, but also through interactions with other subunits. Assuming the native GABA$_A$ receptor is a pentamer of two α subunits, two β subunits, and one γ or δ subunit, the known diversity in subunits is sufficient to produce hundreds of different receptors.[29] Determining the distribution and properties of various receptor subtypes and identifying the endogenous ligands, if any exist, for the barbiturate and benzodiazepine binding sites are areas of active research.[33]

Glutamate: A major excitatory transmitter in the CNS

A variety of physiological tests have supported the idea that glutamate, first established as an excitatory transmitter at the locust neuromuscular junction and the squid giant axon synapse,[36] is released as the transmitter at many excitatory synapses in the central nervous system.[37] As described in Chapter 7, there are several varieties of glutamate receptors that can be grouped into two broad classes, NMDA and non-NMDA, based on their sensitivity to the glutamate agonist N-methyl-D-aspartate (NMDA).[32,38] Excitatory synaptic potentials in neurons from a variety of CNS regions have been found to consist of a fast, rapidly decaying non-NMDA receptor-mediated component and a slower NMDA receptor-mediated component. Non-NMDA receptors, which include the so-called kainate- and quisqualate-AMPA receptors, are typical directly coupled receptors that allow cations, predominantly sodium and potassium, to flow down their electrochemical gradients (Figure 4). The NMDA receptor is also a directly coupled receptor. However, this class of glutamate-activated channels carries little current unless the membrane is depolarized. This voltage-dependent block is due to magnesium ions, which clog the channel at resting levels of membrane potential but are swept out of the channel by depolarization.[39] Once opened, NMDA channels are permeable to cations, including calcium.

Excessive influx of calcium ions through NMDA receptors has been implicated in the neurotoxicity of a variety of insults to the nervous system, including anoxia, hypoglycemia, and seizure.[32] Under such conditions glutamate levels appear to remain pathologically elevated for prolonged periods, persistently activating NMDA receptors and allowing intracellular calcium to reach cytotoxic levels. Inhibitors of NMDA receptors can prevent such neuronal cell death.[40]

When neurons are at or near their resting potential, NMDA receptors make only a small contribution to the excitatory synaptic potentials produced by glutaminergic inputs. However, when neurons are depolarized—for example, during high-frequency trains of stimuli—NMDA receptors contribute significantly to excitatory synaptic poten-

[36]Kawai, N. et al. 1983. *Brain Res.* 278: 346–349.
[37]Fonnum, F. 1984. *J. Neurochem.* 42: 1–11.
[38]Betz, H. 1990. *Neuron* 5: 383–392.
[39]Mayer, M. L. and Westbrook, G. L. 1987. *Prog. Neurobiol.* 28: 197–276.
[40]Choi, D. W., Koh, J.-Y. and Peters, S. 1988. *J. Neurosci.* 8: 185–196.

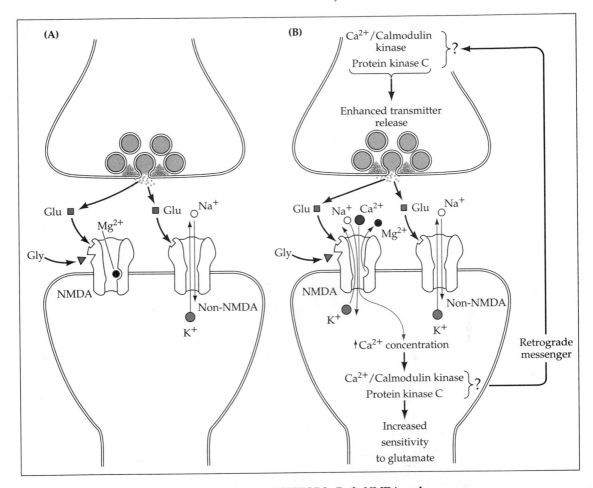

4 **NMDA AND NON-NMDA GLUTAMATE RECEPTORS. Both NMDA and non-NMDA receptors are directly coupled ligand-activated channels, and are permeable to cations. (A) When neurons are at or near their resting potential, current flows through glutamate-activated non-NMDA receptors, but glutamate-activated NMDA receptors carry little current because the channels are blocked by magnesium ions. (B) If the cell is depolarized sufficiently, for example by intense activation of non-NMDA receptors, the change in membrane potential pushes the magnesium ion from the channel of the NMDA receptor, allowing current to flow through it if it is activated by glutamate. Glycine enhances the probability of opening of NMDA receptors when they have been activated by glutamate. In neurons in the hippocampus, the influx of calcium through NMDA receptors has been shown to activate calcium-dependent second messenger systems in the postsynaptic cell, producing a long-lasting increase in the size of synaptic potentials (long-term potentiation or LTP). The increase in synaptic potential amplitude is due to an increase in the transmitter sensitivity of the postsynaptic cell and to the production of a retrograde messenger that acts on presynaptic terminals, causing a long-lasting increase in transmitter release. (After Kandel et al., 1991.)**

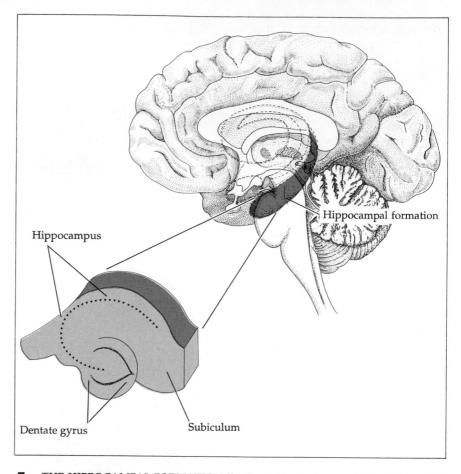

5   **THE HIPPOCAMPAL FORMATION lies buried in the temporal lobe.** Midsagittal view of the brain showing the location of the hippocampal formation, which consists of two interlocking C-shaped strips of cortex, the dentate gyrus, and the hippocampus, together with the neighboring subiculum.

tials and, perhaps more importantly, allow calcium entry. Thus, NMDA receptors provide a means by which intracellular calcium-dependent second messenger systems can be triggered in response to simultaneous presynaptic and postsynaptic depolarization, a kind of associative synaptic mechanism that appears well suited to underlie cognitive processes such as learning and memory.

Long-term changes in signaling in the hippocampus

Buried within the temporal lobe of the brain are two long, interlocking C-shaped strips of cortex known as the HIPPOCAMPUS and the DENTATE GYRUS (Figure 5). Together with the neighboring SUBICULUM, they make up the HIPPOCAMPAL FORMATION. Important insights into the function of this region came from studies of patients whose medial temporal lobes were ablated bilaterally to alleviate the symptoms of temporal lobe epilepsy.[41] Such patients displayed a particularly notice-

[41]Squire, L. R. and Zola-Morgan, S. 1991. *Science* 253: 1380–1386.

able and specific deficit in the capacity for consolidating short-term memory into long-term memory. That is, if told to remember a word or name, they could repeat it if queried immediately. However, all recollection was lost if they were distracted, even briefly. Long-term memories acquired prior to the operation remained relatively intact. The importance of the hippocampus itself for memory consolidation was established by two cases in which memory impairment was found to be associated with bilateral lesions that were confined to the hippocampus.

Such observations make it reasonable to suppose that long-lasting changes in the efficacy of signaling between neurons in the hippocampus might be involved in memory consolidation and recall. A possible cellular correlate of such long-lasting changes in synaptic efficacy was identified by Bliss and Lømo in 1973.[42] They demonstrated that high-frequency stimulation of inputs to the hippocampus produced an increase in the amplitude of excitatory synaptic potentials recorded in the postsynaptic hippocampal neurons that lasted for hours, or even for days or weeks in the intact animal (Figure 6). They called this prolonged facilitation LONG-TERM POTENTIATION (LTP).

Long-term potentiation has been shown to occur in many synaptic pathways. In each case, LTP may require a different pattern of stimulation, may decay at a different rate, and may involve different mechanisms.[43,44] For example, synapses made by axons in the Schaffer collateral/commissural tract onto pyramidal cells in a region of the hippocampus known as CA1 (Figure 7) exhibit two forms of LTP. Glutamate is released as a transmitter at these synapses, and the CA1 cells have both non-NMDA and NMDA receptors in their postsynaptic membranes. Both forms of LTP at these synapses appear to be triggered by the influx of calcium through NMDA receptors. T. H. Brown and his colleagues were the first to demonstrate LTP in this pathway.[45] They made intracellular recordings from a pyramidal cell in area CA1 and placed two extracellular stimulating electrodes in the Schaffer collateral/commissural tract so as to stimulate subpopulations of these axons that innervated two different regions on the dendritic arbor of the pyramidal cell (Figure 7). The stimulus intensities were adjusted so that electrode A evoked a large synaptic potential in the pyramidal cell, while electrode B evoked a much smaller one. A brief tetanus (100 Hz for 1 second, repeated once after 5 seconds) applied to electrode A caused long-lasting synaptic facilitation of inputs from A (not shown). This occurred presumably because temporal summation of the evoked synaptic potentials depolarized the CA1 neuron sufficiently to remove the magnesium block of NMDA receptors, allowing calcium influx and triggering LTP.

Associative LTP in hippocampal pyramidal cells

[42]Bliss, T. V. P. and Lømo, T. 1973. *J. Physiol.* 232: 331–356.

[43]Madison, D. V., Malenka, R. A. and Nicoll, R. A. 1991. *Annu. Rev. Neurosci.* 14: 379–397.

[44]Brown, T. H., Kairiss, E. W. and Keenan, C. L. 1990. *Annu. Rev. Neurosci.* 13: 475–511.

[45]Barrionuevo, G. and Brown, T. H. 1983. *Proc. Natl. Acad. Sci. USA* 80: 7347–7351.

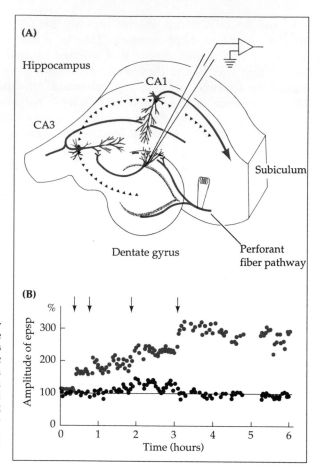

(A)

Hippocampus

CA1

CA3

Subiculum

Dentate gyrus

Perforant
fiber pathway

(B)

6

**LONG-TERM POTENTIATION OF TRANSMIS-
SION** in the hippocampus of the rat. (A) The
perforant fiber pathway from the subiculum was
stimulated and the resulting excitatory synaptic
potentials were recorded from granule cells in
the dentate gyrus. (B) Brief tetanic stimulation
(10 sec at 15 Hz) was given at the times marked
by the arrows. Each tetanus caused a long-lasting
increase in the amplitude of the synaptic potentials (filled circles); no change was seen in control
pathways not receiving tetanic stimulation (open
circles). (After Bliss and Lømo, 1973.)

This form of LTP resembles that first described in neurons in the dentate
by Bliss and Lømo. High-frequency stimulation at site A did not produce
LTP of synaptic inputs from site B because no glutamate was being
released to activate NMDA receptors at synapses from B during the
time that the CA1 cell was depolarized by the tetanus at A. Likewise,
high-frequency stimulation at site B did not produce LTP of synaptic
potentials evoked by subsequent stimulation at B because these synaptic
potentials were too small to depolarize the CA1 neuron to the extent
required to remove the magnesium block of the NMDA receptors. Simultaneous high-frequency stimulation at sites A and B together, however, did result in a prolonged increase in the size of the potentials
evoked by stimulation at B. This was called ASSOCIATIVE LTP because
depolarization of the CA1 neuron produced by synaptic *inputs from site
A* removed the magnesium from NMDA receptors postsynaptic to *inputs
from site B,* and so allowed influx of calcium through glutamate-activated
NMDA receptors at those sites. This example illustrates how the dependence of NMDA receptors on both ligand and voltage can lead to

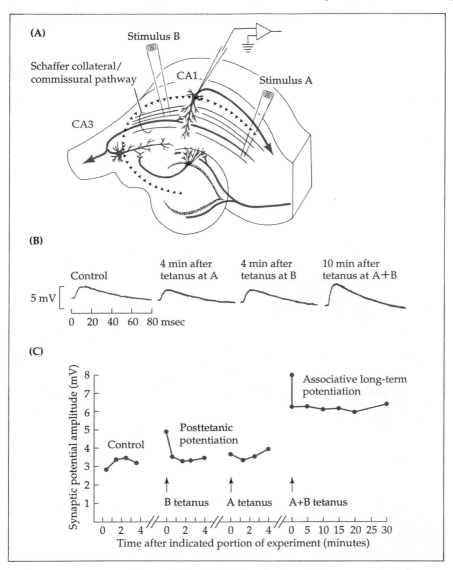

(A)

Stimulus B

Schaffer collateral/
commissural pathway

CA1

Stimulus A

CA3

(B)

Control

4 min after
tetanus at A

4 min after
tetanus at B

10 min after
tetanus at A+B

5 mV

0   20   40   60   80 msec

(C)

Synaptic potential amplitude (mV)

Associative long-term
potentiation

Posttetanic
potentiation

Control

B tetanus   A tetanus   A+B tetanus

0   2   4   0   2   4   0   2   4   0   5   10   15   20   25   30

Time after indicated portion of experiment (minutes)

7 ASSOCIATIVE LTP IN HIPPOCAMPAL NEURONS in the rat. (A) Intracellular
recordings were made from a pyramidal cell in area CA1 of the rat hippocampus.
Two distinct groups of axons in the Schaffer collateral/commissural pathway were
stimulated with extracellular electrodes; the stimulus strengths were adjusted so that
synaptic potentials evoked by stimulation at site A were five times larger than those
evoked by stimulation at site B. (B) Averaged synaptic potentials evoked by test shocks
to site B under control conditions and following brief high frequency (tetanic) stimulation (100 Hz for 1 sec, repeated once after 5 sec) at site A, site B, or both. The amplitude
of the synaptic potential was unchanged when tested 4 minutes after tetanic stimulation
at either site A or site B, but was increased almost twofold when tested 10 minutes after
tetanic stimulation at sites A and B together. (C) Summary of the time course of changes
in amplitude of synaptic potentials evoked by stimulation. Brief tetanic stimulation at
site B produced a short-lived potentiation of the subsequent test synaptic potentials
(posttetanic potentiation), tetanic stimulation at site A had no effect, while stimulation
of both A and B together produced long-term potentiation. (After Barrionuevo and
Brown, 1983.)

associative activity-dependent changes in synaptic efficacy. Activity in one pathway can make another more effective for prolonged periods.

The role of NMDA receptors in LTP in the hippocampus is demonstrated by the finding that antagonists of NMDA receptors block the induction of LTP but do not reduce LTP if applied after it has been induced.[46,47] Two lines of evidence indicate that LTP is induced by the influx of calcium ions through activated NMDA receptors. In one experiment nitr-5 (a photolabile calcium chelator that releases calcium in response to ultraviolet light) was injected into a CA1 pyramidal cell. When the injected neuron was illuminated with ultraviolet light, the increase in the level of intracellular calcium caused LTP-like synaptic enhancement.[48] Conversely, when intracellular calcium was maintained at a low concentration by injecting a calcium buffer into the postsynaptic cell, LTP was blocked.[49]

The mechanism by which increased intracellular calcium levels trigger LTP is not known; indeed, LTP may be produced by more than one mechanism. For example, there is evidence for the involvement of both calcium/calmodulin-dependent protein kinase and protein kinase C in LTP.[50,51] Moreover, the increase in size of the excitatory synaptic potentials during LTP may arise from both pre- and postsynaptic effects. The amplitude of miniature synaptic currents recorded in CA1 pyramidal cells appears to be increased during LTP, suggesting that the CA1 cells are more sensitive to transmitter.[52] On the other hand, there is evidence that quantal content is increased during LTP.[53,54] For such a presynaptic change to occur, the influx of calcium into the postsynaptic CA1 cell must act, via calcium-dependent second messenger systems, to produce a retrograde (third!) messenger that influences events in the presynaptic nerve terminal, causing a long-term increase in transmitter release (Figure 4).[55]

Some forms of LTP require a time delay after stimulation for full enhancement of transmission. For such cases, it has been speculated that calcium entry could activate splicing factors that favor the production of a glutamate channel variant that gives enhanced responses to glutamate.[56]

The first compound to be identified as a transmitter in the central nervous system was acetylcholine, which is released at synapses made

[46]Collingridge, G. L., Kehl, S. J. and McLennan, H. 1983. *J. Physiol.* 334: 33–46.
[47]Muller, D., Joly, M. and Lynch, G. 1988. *Science* 242: 1694–97.
[48]Malenka, R. C. et al. 1988. *Science* 242: 81–84.
[49]Lynch, G. et al. 1983. *Nature* 304: 719–721.
[50]Malenka, R. C. et al. 1989. *Nature* 340: 554–557.
[51]Malinow, R., Schulman, H. and Tsien, R. W. 1989. *Science* 245: 862–866.
[52]Manabe, T., Renner, P. and Nicoll, R. A. 1992. *Nature* 355: 50–55.
[53]Malinow, R. and Tsien, R. W. 1990. *Nature* 346: 177–180.
[54]Bekkers, J. M. and Stevens, C. F. 1990. *Nature* 346: 724–729.
[55]Williams, J. H. et al. 1989. *Nature* 341: 739–742.
[56]Sommer, B. et al. 1990. *Science* 249: 1580–1585.

by collaterals of spinal motoneurons onto interneurons known as Renshaw cells.[57] (These interneurons feed back onto the motoneurons and inhibit them.) This fast excitatory nicotinic synapse is, however, atypical of cholinergic synapses in the CNS. Subsequent studies of the effects of ACh in many brain regions have demonstrated a wide variety of responses, mediated by indirectly coupled muscarinic receptors.[32] These responses include an increase in a nonspecific cation conductance, increases or decreases in various potassium conductances, and a decrease in calcium conductance.

Nuclei containing the cell bodies of cholinergic neurons are scattered throughout the brain, and cholinergic axons innervate most regions of the CNS. Prominent sources of cholinergic inputs to the cortex and the hippocampus are nuclei in the basal forebrain, especially the SEPTAL NUCLEI and the NUCLEUS BASALIS OF MEYNERT (Figure 8). Cholinergic neurons in these nuclei have widespread and diffuse projections, innervating the cortex, hippocampus, amygdala, thalamus, and brainstem. Lesions of the nucleus basalis reduce the levels of choline acetyltransferase in the cortex by more than 50 percent.[58]

Considerable evidence from both animal and human studies suggests that cholinergic systems are important for learning, memory, and cognition.[58] Compounds that reversibly block muscarinic receptors,

Cholinergic neurons, cognition, and Alzheimer's disease

[57]Eccles, J. C., Eccles, R. M. and Fatt, P. 1956. *J. Physiol.* 131: 154–169.
[58]Fibiger, H. C. 1991. *Trends Neurosci.* 14: : 220–223.

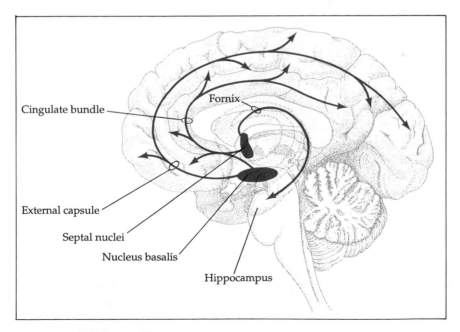

8 CHOLINERGIC INNERVATION of the cortex and hippocampus by neurons in the septal nuclei and nucleus basalis.

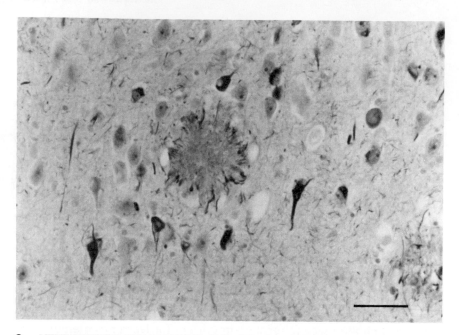

9   NEUROFIBRILLARY TANGLES AND SENILE PLAQUES characteristic of Alzheimer's disease. Scattered among cytologically normal neurons in this section of the amygdala from an Alzheimer's patient are abnormal pyramidal cells filled with darkly staining neurofibrillary tangles, formed by the accumulation of bundles of paired helical filaments. In the center, a senile plaque consists of a large, compacted deposit of extracellular amyloid surrounded by a halo of dilated, structurally abnormal neurites. Modified Bielschowsky silver stain. Bar, 50 μm. (Courtesy of D. J. Selkoe.)

such as atropine and scopolamine, can disrupt the acquisition and performance of learned behaviors, as can stereotaxic lesions of the nucleus basalis. Agents that inhibit acetylcholinesterase, such as physostigmine, can enhance performance in learning and memory tasks and can reverse some of the effects of basal forebrain lesions. However, such lesions inevitably involve cells releasing transmitters other than ACh, and treatments aimed at enhancing performance of cholinergic neurons only partially reverse the effects of such lesions. Thus, it seems certain that memory is not mediated directly by these cholinergic neurons, but rather that they, along with neurons releasing other transmitters, provide important modulatory input to cortical and hippocampal neurons.[59]

Interest in the role of basal forebrain cholinergic neurons in learning and memory was increased by the findings that declines in cognitive capacity with aging are paralleled by decreases in the levels of choline acetyltransferase in the cortex and hippocampus and by the loss of cholinergic neurons in the basal forebrain.[59] Particularly pronounced are changes in patients with Alzheimer's disease, a debilitating, pro-

[59]Dunnett, S. 1991. *Trends Neurosci.* 14: 371–376.

gressive disease characterized by loss of memory and cognitive functions. It occurs in the elderly and leads to dementia. On autopsy substantially reduced numbers of cholinergic neurons are found in the nucleus basalis. The deficits in Alzheimer's disease are not restricted to one region, however, or to neurons releasing ACh as a transmitter. Alzheimer's disease is characterized by the accumulation in neurons throughout the CNS of insoluble aggregates of modified versions of proteins normally associated with the cytoskeleton (so-called paired helical filaments) and the formation of senile plaques that contain a core region filled with insoluble fibrils composed of improperly degraded "amyloid" proteins surrounded by dystrophic neurites[60,61] (Figure 9). Although such lesions occur in cholinergic neurons in the basal forebrain and in their axon terminals in the cortex and hippocampus, they are not confined to these sites. Projection and local circuit neurons releasing many other transmitters are involved, including neurons releasing norepinephrine, dopamine, serotonin, glutamate, GABA, somatostatin, neuropeptide Y, and substance P (see below). Thus there is no direct evidence that damage to cholinergic neurons in the basal forebrain is solely responsible for the cognitive decline in Alzheimer's disease, and attempts to alleviate such cognitive deficits with drugs aimed at enhancing performance of cholinergic neurons have not been successful.

PEPTIDE TRANSMITTERS IN THE CNS

It was in the gut that the first hormone—secretin—was discovered by Bayliss and Starling in 1902.[62] Since then numerous additional intestinal hormones have been isolated and characterized. Many of the intestinal hormones, including secretin, gastrin, bradykinin, somatostatin, and cholecystokinin (CCK) were later shown to be peptides. These peptides are found in the terminals of autonomic axons that innervate the gut and in neurons of the enteric nervous system, a series of neuronal plexuses that control gut movements and secretory activities. (It is somewhat startling to realize that the enteric nervous system of the large and small intestines contains about as many neurons as the spinal cord.) It had also been known since the 1950s that certain neurons within the brain could secrete peptide hormones into the local circulation: For example, nerve cells in the hypothalamus were shown to secrete releasing factors that reached the endocrine cells of the anterior lobe of the pituitary and caused them in turn to secrete other hormones into the general circulation.[63] What was quite unexpected was the finding in the 1970s that peptide hormones identified in the enteric nervous

[60]Selkoe, D. J. 1991. *Neuron* 6: 487–498.
[61]Murrell, J. et al. 1991. *Science* 254: 97–99.
[62]Bayliss, W. M. and Starling, E. H. 1902. *J. Physiol.* 28: 325–353.
[63]Harris, G. W., Reed, M. and Fawcett, C. P. 1966. *Br. Med. Bull.* 22: 266–272.

system were also widely distributed in the brain and spinal cord.[64] Advances in immunological, cytochemical, and physiological detection techniques made it possible to demonstrate the presence of cholecystokinin, bradykinin, gastrin, vasoactive intestinal polypeptide (VIP), bombesin (first isolated from the skin of a frog, *Bombina bombina*, mentioned here for its romantic name), and other gut hormones in widespread regions of the central nervous system. In many instances, peptides can be shown to be released by stimulating appropriate regions of the intact brain or of slice preparations.[65] Conversely, peptides already known to occur in the hypothalamus were later located in the gut and pancreas, where they exert profound effects.

Substance P   An early hint of this unity of peptides occurring in the central and enteric nervous systems was provided by a transmitter known as substance P.[66,67] Substance P was first isolated in 1931 by Von Euler and Gaddum from gut and brain and was shown to cause contractions of smooth muscles. Since then the structure of the peptide has been worked out and its distribution in the nervous system established. In particular, it is present in small-diameter sensory axons concerned with nociception or pain and their endings in the dorsal layers of the spinal cord, where substance P may act as a transmitter together with glutamate (Figure 10; see also Chapter 14). The letter "P" does not refer to "pain" or "peptide" but was the name used by Von Euler and Gaddum to designate the first crude "preparation" that contained the active peptide. Like many other peptides, substance P has clear actions on smooth muscles and on neurons in the central nervous system, but its role is still not fully understood.

Opioid peptides   Interest in brain peptides increased further in the mid-1970s as a result of two sets of experiments made by Kosterlitz, Hughes, Goldstein, Snyder, and their colleagues:[68–70] First, they found in the brain and in the gut receptors to which morphine and other derivatives of opium (opiates) bound with high specificity. Second, they identified within the brain peptides that had actions similar to those of opiates. First to be characterized were the enkephalins, which are pentapeptides; one enkephalin is known as met-enkephalin and the other as leu-enkephalin, depending on whether the amino-terminal amino acid is methionine or leucine. Other key findings were that opioid peptides (peptides with opiate activity) and their receptors were concentrated in regions of the brain known to be involved in the perception of pain, that stimulation of these regions of the brain produced analgesia,[71] and that this analgesia was reversed by naloxone, a drug that blocks opiate

[64]Krieger, D. T. 1983. *Science* 222: 975–985.
[65]Iversen, L. L. et al. 1980. *Proc. R. Soc. Lond. B* 210: 91–111.
[66]Nicoll, R. A., Schenker, C. and Leeman, S. E. 1980. *Annu. Rev. Neurosci.* 3: 227–268.
[67]Maggio, J. E. 1988. *Annu. Rev. Neurosci.* 11: 13–28.
[68]Hughes, J. et al. 1975. *Nature* 258: 577–579.
[69]Teschemacher, H. et al. 1975. *Life Sci.* 16: 1771–1776.
[70]Pert, C. B. and Snyder, S. H. 1973. *Science* 179: 1011–1014.
[71]Fields, H. L. and Basbaum, A. I. 1978. *Annu. Rev. Physiol.* 40: 217–248.

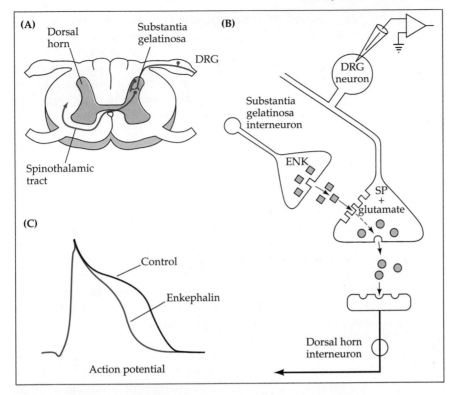

10 **PATHWAY FOR TRANSMISSION OF PAIN SENSATION in the spinal cord.**
(A,B) Dorsal root ganglion (DRG) cells responding to noxious stimuli release
substance P (SP) and glutamate at their synapses with interneurons in the dorsal horn
of the spinal cord. Enkephalin-containing interneurons in the substantia gelatinosa of
the dorsal horn block transmission by inhibiting transmitter release from terminals of
DRG cells. (C) Intracellular recordings demonstrate that enkephalin acts by causing a
decrease in the duration of the action potential in the presynaptic nerve terminal.
(C after Mudge, Leeman and Fischbach, 1979.)

receptors. Interest was spurred still further by the discovery of opioid
neurons in the spinal cord whose axons ended on the substance P-
containing terminals supposed to mediate pain sensation, and by the
finding that opiates blocked the release of substance P from the sensory
terminals[72] (Figure 10).

A clue as to how enkephalin blocks release of substance P comes
from studies of dorsal root ganglion neurons in culture.[73] When stimulated, these isolated neurons release substance P. Release is blocked
by enkephalin, which causes a decrease in the duration of the action
potential by binding to one subtype of opiate receptors called μ receptors
and activating calcium-dependent potassium channels.[74] Other opioid

[72]Jessell, T. M. and Iversen, L. L. 1977. *Nature* 268: 549–551.

[73]Mudge, A., Leeman, S. and Fischbach, G. 1979. *Proc. Natl. Acad. Sci. USA* 76: 526–530.

[74]Werz, M. A. and Macdonald, R. L. 1983. *Neurosci. Lett.* 42: 173–178.

peptides bind to a second subtype of opiate receptors, $\varkappa$ receptors, and reduce transmitter release by inhibiting voltage-dependent calcium channels.[75] Both effects appear to be mediated by indirectly coupled receptors acting through G proteins[32] (see Chapter 8). Clear evidence for the involvement of specific intracellular second messengers in the opioid peptide effects on dorsal root ganglion neurons is lacking. Evidence is available that a similar reduction in voltage-dependent calcium current caused by norepinephrine is due to activation of protein kinase C.[76]

Such studies made it clear that the characterization of the brain's own opioid peptides could be important not only for determining their possible function as transmitters but also for understanding mechanisms involved in the control of pain and addiction to drugs. As a result of extensive searches for peptides with opiate activity in the brain and gut, several additional opioid peptides, including β-endorphin, dynorphine, and neoendorphin have been discovered.[77] β-Endorphin, for example, is found in the pituitary gland, the brain, the pancreas, and the placenta. This 31-residue peptide is generated from a large molecule that also acts as a precursor for other hormones such as corticotropin (ACTH)[78,79] (see Chapter 9). The enkephalins were found to occur in the gut and adrenal glands as well as in the brain. Injection of these opioid peptides into the brain, either intraventricularly or into the central gray matter, not only mimics the analgesic and euphoric effects of opiates but also produces other profound changes of behavior, such as muscular rigidity, suggesting that the same peptides act in regions of the central nervous system that have nothing to do with pain sensation. This observation points out a characteristic feature of the organization of the nervous system: The same transmitters are often used over and over in different areas of the nervous system and in neural circuits mediating diverse physiological functions.

NOREPINEPHRINE, DOPAMINE, 5-HT, AND HISTAMINE AS REGULATORS OF CNS FUNCTION

Within the mammalian central nervous system there is evidence that norepinephrine, dopamine, 5-HT, and histamine act as transmitters. These so-called biogenic amines are found in pathways essential for sensory and motor performance as well as for higher functions. However, out of the billions of neurons in the human brain, relatively few appear to contain biogenic amines—such cells number only in the

[75]Macdonald, R. L. and Werz, M. A. 1986. *J. Physiol.* 377: 237–249.

[76]Rane, S. G. and Dunlap, K. 1986. *Proc. Natl. Acad. Sci. USA* 83: 184–188.

[77]Cooper, J. R., Bloom, F. E. and Roth, R. H. 1991. *The Biochemical Basis of Pharmacology*, 6th Ed. Oxford University Press, New York.

[78]Mains, R. E., Eipper, B. A. and Ling, N. 1977. *Proc. Natl. Acad. Sci. USA* 74: 3014–3018.

[79]Roberts, J. L. and Herbert, E. 1977. *Proc. Natl. Acad. Sci. USA* 74: 4826–4830.

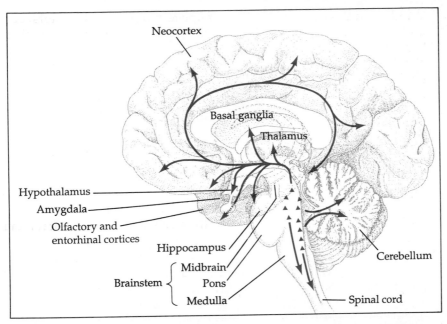

11 NEURONS CONTAINING NOREPINEPHRINE, DOPAMINE, HISTAMINE, AND 5-HT are clustered in the brainstem and have diffuse projections to widespread areas of the central nervous system.

thousands. What is more, many of the cells containing these transmitters are clustered together in a discrete region of the brain, the brainstem. Neurons in these clusters, or nuclei (which are shown schematically in Figure 11 and in Appendix B), extend axons to supply virtually all areas of the brain. In some cases these cells form synapses in which the presynaptic terminals are closely apposed to their postsynaptic targets; in other locations no obvious postsynaptic targets are seen. These anatomical characteristics suggest that an important function of amine-containing neurons may be to modulate synaptic activity simultaneously in widespread areas of the central nervous system. Consistent with such a neuromodulatory role is the finding that biogenic amines act through indirectly-coupled receptors.

Numerous comprehensive monographs and reviews describe the actions of the amines.[77,80-82] The LOCUS COERULEUS provides a good illustration of the anatomy and physiology of amine-containing cells in the central nervous system.[83] This nucleus consists of a small cluster of norepinephrine-containing cells that lies in the pons beneath the floor of the fourth ventricle (Figures 1 and 12). In the rat central nervous system, each locus coeruleus (one on either side of the brainstem)

Norepinephrine:
The locus coeruleus

[80]Fillenz, M. 1990. *Noradrenergic Neurons*. Cambridge University Press, Cambridge.
[81]Moore, R. Y. and Bloom, F. E. 1978. *Annu. Rev. Neurosci.* 1: 129–169.
[82]Moore. R. Y. and Bloom, F. E. 1979. *Annu. Rev. Neurosci.* 2: 113–168.
[83]Foote, S. L., Bloom, F. E. and Aston-Jones, G. 1983. *Physiol. Rev.* 63: 844–914.

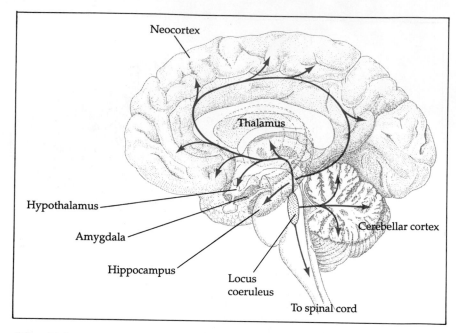

12   PROJECTIONS OF NOREPINEPHRINE-CONTAINING NEURONS in the locus coeruleus. The locus coeruleus lies in the pons just beneath the floor of the fourth ventricle. These neurons innervate widespread regions of the brain and spinal cord.

contains approximately 1500 cells; together these 3000 neurons account for approximately one-half of all norepinephrine-containing cells in the brain. Yet they have extensive projections, sending axons to the cerebellum, cerebral cortex, thalamus, hippocampus, and hypothalamus. Indeed, a single neuron in the locus coeruleus can innervate wide areas of both the cerebral and cerebellar cortices.[84] Stimulation within the locus coeruleus or application of norepinephrine can produce a variety of effects in central neurons depending on the type of receptor activated. For example, in hippocampal pyramidal cells, the most prominent effect of norepinephrine is a blockade of the slow calcium-activated potassium conductance that underlies the afterhyperpolarization following a train of action potentials[32] (Chapter 8). The response is mediated by β-adrenergic receptors that activate adenylyl cyclase, increasing the level of intracellular cAMP. How increased intracellular cAMP levels block these potassium channels is not known, although activation of a cAMP-dependent protein kinase leading to phosphorylation of the channel protein is likely. The effect of blocking the slow afterhyperpolarization is to increase dramatically the number of action potentials elicited by prolonged depolarizations.

The projections of the locus coeruleus form part of the so-called ASCENDING RETICULAR ACTIVATING SYSTEM, a functionally defined pro-

---

[84]Swanson, L. W. 1976. *Brain Res.* 110: 39–56.

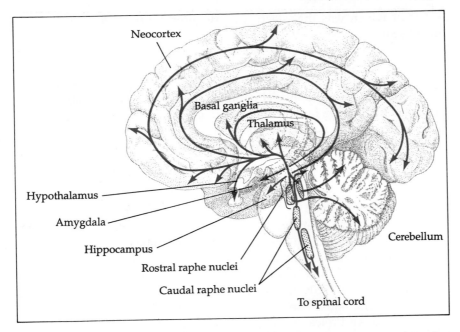

Neocortex

Basal ganglia

Thalamus

Hypothalamus

Amygdala

Hippocampus

Rostral raphe nuclei

Caudal raphe nuclei

Cerebellum

To spinal cord

13 NEURONS CONTAINING 5-HT form a chain of raphe nuclei lying along the midline of the brainstem. More caudal nuclei innervate the spinal cord, more rostral nuclei innervate nearly all regions of the brain.

jection from the brainstem reticular formation to higher brain centers that regulates, among other things, attention, arousal, and the sleep–wake cycle. Widespread projections from a few neurons, such as is the case for cells in the locus coeruleus, seem particularly well suited to carry out such global functions.

Like norepinephrine, 5-hydroxytryptamine (5-HT, also known as serotonin) is localized to a few nuclei in the brainstem.[85] These are the RAPHE NUCLEI, which lie directly along the midline of the brainstem from the midbrain to the medulla (Figure 13). (RAPHE comes from the French for "seam.") Nuclei in the medulla project to the spinal cord and modulate transmission in spinal cord pathways involved in the perception of pain (see Chapter 14). Raphe nuclei in the pons and midbrain innervate essentially the entire brain and together with projections from the locus coeruleus form part of the ascending reticular activating system. 5-HT has been implicated in the control of the sleep–wake cycle. Early experiments on cats indicated that depletion of 5-HT pharmacologically or by destruction of the raphe nuclei produced insomnia that could be reversed by administration of 5-HT or its metabolic precursor. This led to the "monoaminergic theory of sleep," one pos-

5-HT: The Raphe nuclei

[85]Steinbusch, H. W. M. 1984. *In* A. Björklund, T. Hökfelt and M. J. Kuhar (eds.). *Classical Transmitters and Transmitter Receptors in the CNS*, Part 2. (*Handbook of Chemical Neuroanatomy*, Vol. 3.) Elsevier, New York, pp. 68–125.

tulate of which was that sleep was triggered by the release of 5-HT.[86] In subsequent experiments it was discovered that the insomnia induced by depletion of 5-HT was temporary; given time, treated animals reacquired the capacity to sleep.[87] Moreover, similar surgical and pharmacological lesions in other species did not produce insomnia. Single-unit recordings in the raphe nuclei demonstrated that 5-HT-containing cells decrease their firing rate during the transition from the waking state to sleep. Thus, it is now clear that control of the sleep–wake cycle is considerably more complicated than originally proposed; several brain regions in addition to the raphe nuclei, including the locus coeruleus and surrounding structures, and several transmitters in addition to 5-HT, including norepinephrine, acetylcholine, and histamine, are known to be involved.[87–89] One conclusion that emerges from such studies is that the sleep–wake cycle, like other global changes in CNS activity, is controlled by several different groups of cells releasing a variety of neurotransmitters.

Histamine   Histamine was first identified as a natural constituent of the liver and lungs in the 1920s and was soon shown to be present in tissues throughout the body.[90] In the periphery the predominant site of histamine storage and release is the mast cell. Histamine affects a variety of peripheral tissues and is involved in diverse physiological processes, including allergic reactions, response to injury, and regulation of gastric secretions. Histamine also acts as a neurotransmitter in the brain.[91,92] The cell bodies of histaminergic neurons are concentrated in a small region of the hypothalamus, the TUBEROMAMMILLARY NUCLEUS, and extend axons that reach almost all parts of the central nervous system[93] (Figure 14). Individual histamine neurons have axon collaterals that innervate several different brain regions.[94] Like neurons releasing other biogenic amines, histaminergic neurons tend to arborize diffusely and only occasionally form classic synapses with clear pre- and postsynaptic specializations. Histaminergic processes appear to innervate not only neurons but glial cells, small blood vessels, and capillaries as well. Based on this morphology and the effects of drugs that influence histaminergic transmission, it appears that histamine neurons regulate general brain activities, such as the arousal state and energy metabolism, through indirect mechanisms mediated by receptors on neurons, astrocytes, and blood vessels.

[86]Jouvet, M. 1972. *Ergebn. Physiol.* 64: 166–307.

[87]Hilakivi, I. 1987. *Med. Biol.* 65: 97–104.

[88]Vertes, R. P. 1984. *Prog. Neurobiol.* 22: 241–288.

[89]Lin, J. S. et al. 1990. *Brain Res.* 523: 325–330.

[90]Douglas, W. W. 1980. *In* A. G. Gilman, L. S. Goodman and A. Gilman (eds.). *Goodman and Gilman's The Pharmacological Basis of Therapeutics.* Macmillan, New York. pp. 608–618.

[91]Wada, H. et al. 1991. *Trends Neurosci.* 14: 415–418.

[92]Prell, G. D. and Green, J. P. 1986. *Annu. Rev. Neurosci.* 9: 209–254.

[93]Panula, P. et al. 1990. *Neuroscience* 34: 127–132.

[94]Kohler, C. et al. 1985. *Neuroscience* 16: 85–110.

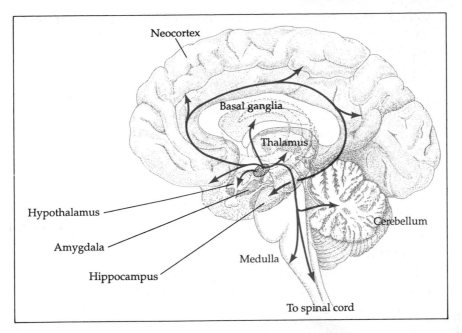

Neocortex

Basal ganglia

Thalamus

Hypothalamus

Amygdala

Hippocampus

Medulla

Cerebellum

To spinal cord

14   HISTAMINE-CONTAINING NEURONS are localized to the tuberomammillary nucleus in the hypothalamus. These neurons have diffuse projections throughout the brain and spinal cord.

There are four prominent dopamine-containing nuclei in the brainstem[81] (Figure 15). One of these lies in the ARCUATE NUCLEUS and sends processes into the median eminence of the hypothalamus, an area rich in peptide-releasing hormones. The three other clusters of dopamine cells lie in the midbrain; they project primarily to the basal ganglia, a group of nuclei in the center of the brain that are important for the control of movement (see Chapter 15). As is the case for other neurons releasing biogenic amines, a few cells with widely ramifying projections are involved. In the rat, for example, there are approximately 7000 dopamine cells in one of these midbrain clusters, the SUBSTANTIA NIGRA; each of these neurons, however, gives rise to an estimated 250,000 varicosities spread throughout its targets in the basal ganglia.[95]

It is the progressive degeneration of this group of dopaminergic neurons that is the most prominent feature of Parkinson's disease.[96] In such patients nerve cells in the PARS COMPACTA of the substantia nigra, which innervate two nuclei within the basal ganglia, degenerate. As their axon terminals disappear, the levels of dopamine in the basal ganglia fall and characteristic motor disorders appear: difficulty in ini-

Dopamine: The substantia nigra

[95]Yurek, D. M. and Sladek, J. R., Jr. 1990. *Annu. Rev. Neurosci.* 13: 415–440.
[96]Yahr, M. D., and Bergmann, K. J. 1987. *Adv. Neurol.* 45.

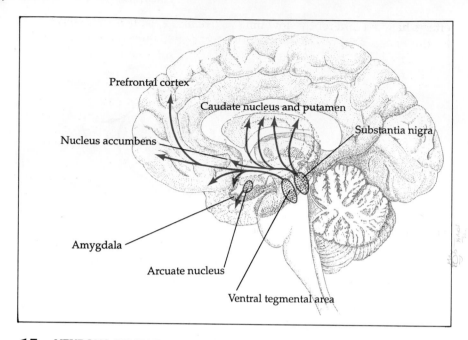

15   NEURONS CONTAINING DOPAMINE are found in nuclei in the hypothalamus and midbrain. Those in the arcuate nucleus project to the median eminence of the hypothalamus, forming the tuberoinfundibular system. Dopamine neurons in the substantia nigra project to the caudate nucleus and putamen (collectively called the striatum) of the basal ganglia, forming the nigrostriatal pathway. Dopamine neurons in the ventral tegmental area project to the nucleus accumbens, amygdala, and prefrontal cortex forming the mesolimbic and mesocortical systems.

tiating movement, muscular rigidity, and a tremor at rest. One of the early triumphs of neuropharmacology was the technique of REPLACE-MENT THERAPY for the treatment of Parkinson's disease. The idea was to attempt to alleviate the symptoms of the disease by restoring the level of dopamine in the basal ganglia. It was known that dopamine would not cross the blood–brain barrier (see Chapter 6), so the precursor for dopamine, L-DOPA, was given. Remarkably, patients receiving oral doses of L-DOPA usually showed dramatic improvement, and autopsies of patients who died from other causes while undergoing L-DOPA therapy showed near-normal concentrations of dopamine in their basal ganglia. Presumably the dopamine neurons that do remain in Parkinson's patients are able to synthesize and release sufficient transmitter if provided with extra precursor.

If one thinks in terms of neuromuscular transmission, where the precise timing and location of release is so critically important, it is difficult to imagine how excess transmitter derived from a distant surviving terminal could substitute for the absence of transmitter release from a terminal lost to degeneration. It would be like trying to restore appropriate contractions of skeletal muscles in coordinated movements by diffuse application of acetylcholine. However, if one thinks in terms of neuromodulation mediated by indirectly coupled receptors, in which effects have an inherently slow time course and specificity is determined by the nature and distribution of the receptors, then one can begin to appreciate how a generalized elevation in dopamine concentration could restore normal signaling.

The axonal projection from the substantia nigra to the basal ganglia is not the only dopaminergic pathway in the central nervous system, however. Accordingly, as you might expect, one complication of giving L-DOPA to alleviate the symptoms of Parkinson's disease is that the balance of inputs to other areas is disrupted as well. For example, several lines of evidence suggest that dopaminergic neurons of the ventral tegmental area play an important role in controlling mood and affect, and patients undergoing L-DOPA therapy often experience psychiatric disturbances. Conversely, many of the drugs given to psychiatric patients block the interaction of dopamine with its receptors and consequently produce Parkinson-like motor side effects.[97] The challenge for systemic drug therapy is to identify for a specific synapse not only the particular neurotransmitter released but also features of the receptors or of the mechanisms of transmitter synthesis, storage, release, or inactivation that set that synapse apart from others using the same transmitter, and then design specific pharmacological agents to exploit that difference.

An alternative approach is to replace degenerating cells by transplantation of appropriate embryonic neurons.[59,98] Embryonic neurons transplanted into the adult CNS have been shown to survive and make synaptic connections; however, their axons rarely extended more than a few millimeters (see Chapter 12). Therefore grafts must usually be placed in or near the target neurons. Although significant problems remain to be resolved, transplantation may provide a means by which functional interactions can be restored in one region of the CNS without the unwanted side effects of disrupting normal transmission elsewhere.

SUGGESTED READING

General

Cooper, J. R., Bloom, F. E. and Roth, R. H. 1991. *The Biochemical Basis of Pharmacology*, 6th Ed. Oxford University Press, New York.

Hall, Z. W. 1992. *An Introduction to Molecular Neurobiology.* Sinauer, Sunderland, MA.

Kandel, E. R., Schwartz, J. H. and Jessell, T. M. (eds). 1991. *Principles of Neural Science*, 3rd Ed. Elsevier, New York.

Reviews and Original Articles

Akil, H., Watson, S. J., Young, E., Lewis, M. E., Khachaturian, H. and Walker, J. M. 1984. Endogenous opioids: Biology and function. *Annu. Rev. Neurosci.* 7: 223–255.

[97]Baldessarini, R. J. and Tarsy, D. 1980. *Annu. Rev. Neurosci.* 3: 23–41.
[98]Lindvall, O. 1991. *Trends Neurosci.* 14: 376–384.

Bekkers, J. M. and Stevens, C. F. 1990. Presynaptic mechanism for long-term potentiation in the hippocampus. *Nature* 346: 724–729.

Bliss, T. V. P. and Lømo, T. 1973. Long-lasting potentiation of synaptic transmission in the dentate of the anesthetized rabbit following stimulation of the perforant path. *J. Physiol.* 232: 331–356.

Brown, T. H., Kairiss, E. W. and Keenan, C. L. 1990. Hebbian synapses: Biophysical mechanisms and algorithms. *Annu. Rev. Neurosci.* 13: 475–511.

Costa, E. 1991. The allosteric modulation of GABA$_A$ receptors. *Neuropsychopharmacology* 4: 225–235.

Cotman, C. W., Monaghan, D. T. and Ganong, A. H. 1988. Excitatory amino acid neurotransmission: NMDA receptors and Hebb-type synaptic plasticity. *Annu. Rev. Neurosci.* 11: 61–80.

Dunnett, S. 1991. Cholinergic grafts, memory and aging. *Trends Neurosci.* 14: 371–376.

Fibiger, H. C. 1991. Cholinergic mechanisms in learning, memory and dementia: A review of recent evidence. *Trends Neurosci.* 14: 220–223.

Fillenz, M. 1990. *Noradrenergic Neurons.* Cambridge University Press, Cambridge.

Lindvall, O. 1991. Prospects of transplantation in human neurodegenerative diseases. *Trends Neurosci.* 14: 376–384.

Lynch, D. R. and Snyder, S. H. 1986. Neuropeptides: Multiple molecular forms, metabolic pathways, and receptors. *Annu. Rev. Biochem.* 55: 773–799.

Madison, D. V., Malenka, R. A. and Nicoll, R. A. 1991. Mechanisms underlying long-term potentiation of synaptic transmission. *Annu. Rev. Neurosci.* 14: 379–397.

Maggio, J. E. 1988. Tachykinins. *Annu. Rev. Neurosci.* 11: 13–28.

Malinow, R. and Tsien, R. W. 1990. Presynaptic enhancement shown by whole-cell recordings of long-term potentiation in hippocampal slices. *Nature* 346: 177–180.

Manabe, T., Renner, P. and Nicoll, R. A. 1992. Postsynaptic contribution to long-term potentiation revealed by the analysis of miniature synaptic currents. *Nature* 355: 50–55.

Naegele, J. R. and Barnstable, C. J. 1989. Molecular determinants of GABAergic local-circuit neurons in the visual cortex. *Trends Neurosci.* 12: 28–34.

Nicoll, R. A., Malenka, R. C. and Kauer, J. A. 1990. Functional comparison of neurotransmitter receptor subtypes in mammalian central nervous system. *Physiol. Rev.* 70: 513–565.

Nicoll, R. A., Schenker, C. and Leeman, S. 1980. Substance P as a transmitter candidate. *Annu. Rev. Neurosci.* 3: 227–268.

Pritchett, D. B., Sontheimer, H., Shivers, B. D. S., Ymer, S., Kettenmann, H., Schofield, P. R. and Seeburg, P. H. 1989. Importance of a novel GABA$_A$ receptor subunit for benzodiazepine pharmacology. *Nature* 338: 582–585.

Selkoe, D. J. 1991. The molecular pathology of Alzheimer's disease. *Neuron* 6: 487–498.

Yurek, D. M. and Sladek, J. R., Jr. 1990. Dopamine cell replacement: Parkinson's disease. *Annu. Rev. Neurosci.* 13: 415–440.

PART THREE

Development and Regeneration in the Nervous System

NEURONAL DEVELOPMENT AND THE FORMATION OF SYNAPTIC CONNECTIONS

The way in which nerve cells acquire their unique identities and establish their orderly and precise synaptic connections during development depends on cell lineage, inductive and trophic interactions between cells, target-derived and target-independent navigational cues, specific cell–cell recognition, and ongoing activity-dependent refinement of connections. In simple invertebrates the fate of some cells is limited or even rigidly determined by autonomous, cell lineage-dependent mechanisms. The phenotype of invertebrate and vertebrate neurons and glial cells, however, is largely determined by inductive interactions.

The formation of the vertebrate nervous system begins with an inductive event: The dorsal mesoderm of the gastrula induces the overlying ectoderm to become the neural plate. The neural plate then folds to give rise to the neural tube and the neural crest. The neural crest cells form the peripheral nervous system. Neurons and glia of the central nervous system are produced by division of cells in the ventricular zone of the neural tube. Postmitotic neurons migrate away from the ventricular surface to form the gray matter of the adult nervous system.

During development of the vertebrate CNS, several of the properties of presumptive neurons appear to be determined at about the time they complete their final cell division. Other properties are regulated by subsequent interactions with their environment. Biochemical, molecular biological, and genetic studies are beginning to identify specific chemical signals that control neuronal and glial differentiation.

To establish synaptic connections with their targets, neurons extend axons tipped with growth cones that appear to explore the environment, pulling the axons to their correct destinations. Two classes of molecules have been identified as important substrates for growth cone movements. The first are the cell adhesion molecules of the immunoglobulin superfamily, such as the neural cell adhesion molecule (N-CAM). Cell adhesion molecules also mediate axonal fasciculation. The second group are the extracellular matrix adhesion molecules, such as the proteins laminin, fibronectin, and tenascin. Additional factors, as yet unidentified, have been shown to attract selectively growth cones of particular neurons and so mediate the establishment of specific connections. In other cases the trajectory of an axon to its ultimate synaptic partner is mapped out by intermediate targets, or guidepost cells. Although initial axonal projections of neurons are selective, they are often more extensive than in the adult. Thus, axonal arbors formed during development are trimmed to the adult pattern by trophic and activity-dependent mechanisms.

As axons from a particular population of neurons arrive at their destination, they innervate targets in an orderly and precise pattern. The most thoroughly studied example of target innervation is the retinotectal projection in the chick. Here the process of establishing an appropriate retinotectal map is mediated, at least in part, by repulsive interactions that discourage growth cones from entering inappropriate areas of the tectum.

Synaptic contacts are formed very rapidly. Detailed studies of synapse formation have been made at the vertebrate skeletal neuromuscular junction. Within minutes after a motor axon contacts a muscle cell, functional synaptic transmission is established. The initial contacts lack any of the specializations associated with adult junctions. The first synaptic specialization to form is the accumulation of acetylcholine receptors in the postsynaptic membrane. Over the course of several weeks, the junction matures to its adult form.

A common feature of the development of the vertebrate central nervous system is an initial overproduction of neurons followed by a period of cell death. Neuronal death appears to be regulated by competition for trophic substances released by the target tissue. The best characterized trophic substance is nerve growth factor (NGF), required by sensory, sympathetic, and certain central cholinergic neurons for their survival. NGF is one member of a family of neurotrophins, each of which sustains particular neuronal populations. The effects of neurotrophins are mediated by receptors that have an extracellular domain that binds the neurotrophin, a short transmembrane domain, and an intracellular tyrosine kinase domain.

A generalization that emerges from studies of the development of the nervous system is that few choices are absolute; neurons display a hierarchy of specificities in their interactions with their environment. In addition, although functional interactions are formed very early in development, the initial pattern of connections is not fixed; some neurons degenerate and die, and those that remain continue to make and break synaptic contacts and extend and withdraw dendritic and axonal processes. The precise pattern of connections that characterizes the adult nervous system emerges gradually as a result of these ongoing refinements and often is shaped by activity-dependent mechanisms.

The orderliness of the connections made by nerve cells with one another and to different tissues in the periphery is a prerequisite for the complex integrative processes that occur in the nervous system. The brain appears to be constructed as if each neuron had built into it an awareness of its proper place in the system. During development, the neuron moves to its appropriate site and sends one process toward its target. It ignores some cells, selects others, and makes permanent contact not just anywhere on a cell but with a specified part of it. Conversely, as described in Chapter 12, neurons behave as if they are aware when they have received their proper connections. When they lose their synaptic input, they respond in various ways. Even in the absence of denervation, simply as a result of altered patterns of activity, the effectiveness of synapses may change and new connections may be formed (Chapter 18).

Three examples illustrate the precise architecture of the nervous system. First, we describe in Chapters 1 and 14 the stretch reflex that arises as the result of impulses in the sensory nerve cell that innervates a muscle spindle. The cell body of the sensory neuron lies in a dorsal root ganglion and sends some of its processes into the periphery to appropriate regions of the intrafusal muscle fibers. It also sends processes centrally to search out and make synapses exclusively on those motoneurons that innervate the same skeletal muscle in which the sensory neuron terminates. Other branches run in the dorsal columns to end in a localized region of the dorsal column nuclei, and still other branches end on additional interneurons. A second example of specificity (Chapter 17) is provided by individual neurons in the visual cortex that selectively recognize a vertically oriented light bar shone into their receptive fields. This response is possible because the inputs are derived from selected lower-order neurons, some of which are excitatory and others inhibitory. A third example is the distribution of synapses made on Purkinje cells in the cerebellum, worked out by Ramón y Cajal, Eccles, Szentágothai, Ito, Llinás, and Palay. These large neurons provide the only known output from the cerebellar cortex and are therefore the end stations for the integrative activity of all the other cerebellar cells (Chapter 15). Each Purkinje cell may receive more than 100,000 synapses, which end on appropriate parts of the neuron. Thus, climbing fibers terminate on smooth dendrites and the basket cells on cell bodies and axons, whereas granule cells make synapses with spiny processes of Purkinje neurons.

A number of questions arise when one considers the preceding examples. What cellular mechanisms enable one neuron to select another out of a myriad of choices, to grow toward it, and to form synapses? Are both cells specified, or does the arrival of one determine the fate of the other? As for the precision of wiring, how much variability is there from animal to animal? What directs the systematic growth of nerve fibers along their well-defined paths? The answers to these questions influence thinking about the genetic blueprint for wiring up a brain containing 10^{10} to 10^{12} cells with a much smaller number of genes, 10^6 or fewer.

One approach to these problems is to study the formation of cells and connections during development. A second approach, discussed in Chapter 12, is to study the regeneration of connections by mature nerve cells after their processes have been severed by a lesion. A different but related question concerns the stability of synapses once they have been formed. How are synaptic connections influenced by use, disuse, or inappropriate use? Remarkable examples, described in Chapter 18, are the changes in the visual system produced by suturing one eye closed or by producing a squint in a kitten. Such subtle alterations of the sensory input disturb performance and disrupt pathways that had previously been effective.

The scope of all the problems relating to development, synapse formation, neural specificity, and changes in efficacy is too great for a comprehensive review. Many aspects are covered in detail elsewhere[1-3]. In this chapter we provide a brief account of neuroembryology and describe selected experimental approaches to questions of neural development. A generalization that emerges from such studies is that few if any choices are absolute; nerve cells display a hierarchy of specificities in their interactions with their environment and with one another. This imparts to the nervous system a degree of flexibility that appears to be essential for establishing, maintaining, and modifying synaptic connections.

FORMATION OF THE NERVOUS SYSTEM

Although many of the cellular mechanisms of neural development appear to be the same in invertebrates and vertebrates, the overall scheme is quite different. The formation of the nervous system in an invertebrate, the leech, is described in Chapter 13; here we summarize general features of the development of the vertebrate nervous system. During embryogenesis in vertebrates, the nervous system is induced to form from a sheet of ectoderm by the underlying dorsal mesoderm of the gastrula. The nervous system begins as a sheet of elongated neuroectodermal cells, the NEURAL PLATE (Figure 1). Over most of its length the edges of this sheet thicken and move upward, forming NEURAL FOLDS, which eventually fuse at the midline to give rise to a hollow NEURAL TUBE. Some of the cells at the lips of the neural fold are not incorporated into either the neural tube or the overlying ectoderm, but come to lie between them. These cells form the NEURAL CREST. Neural crest cells migrate away from the neural tube and give rise to a variety of peripheral tissues, including neurons and satellite cells of the sensory, sympathetic, and parasympathetic nervous systems, cells of the adrenal medulla (see Chapter 9), pigmented cells of the epidermis, and skeletal and connective tissue components of the head.

[1]Gilbert, S. F. 1991. *Developmental Biology*, 3rd Ed. Sinauer, Sunderland, MA.

[2]Patterson, P. H. and Purves, D. (eds). 1982. *Readings in Developmental Neurobiology*. Cold Spring Harbor Laboratory, Cold Spring Harbor, NY.

[3]Purves, D. and Lichtman, J. W. 1985. *Principles of Neural Development*. Sinauer, Sunderland, MA.

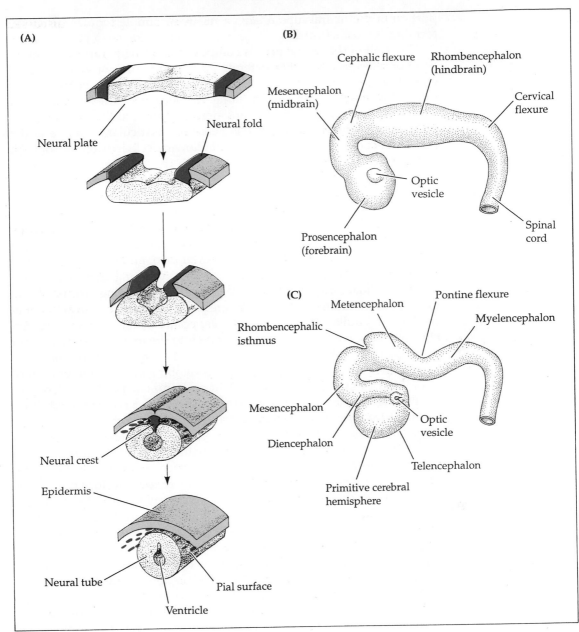

1 FORMATION OF THE NERVOUS SYSTEM. (A) Diagram of neurulation. (B, C) Sketches of the human central nervous system after 4 weeks (B) and 6 weeks (C) of development. (After Gilbert, 1991.)

NEURULATION (the formation of the neural tube) begins at the anterior (cephalic) end of the embryo; as development proceeds, this portion of the tube undergoes a series of swellings, constrictions, and flexures that form the various regions of the brain (Figure 1). The caudal portion of the neural tube retains a relatively simple tubular structure, forming the spinal cord.

The wall of the neural tube is composed of a single, rapidly dividing layer of cells, each of which extends from the luminal, or VENTRICULAR, edge to the external, or PIAL, surface. As illustrated in Figure 2, the nuclei migrate back and forth along the length of these cells. DNA synthesis occurs while nuclei are near the pial surface; during cell division (cytokinesis) the nuclei lie near the ventricular surface and the pial connections are temporarily lost. Following cell division one or both of the daughter cells may lose contact with the ventricular surface and migrate away. Once they migrate away from the ventricular zone, most neurons are postmitotic (they will never divide again). Glial cell precursors, on the other hand, can divide even after they have reached their final locations. As more and more cells are produced, the neural tube thickens and assumes a three-layered configuration: an innermost ventricular zone (where proliferation continues), an intermediate, mantle zone containing the cell bodies of the migrating neurons, and a superficial marginal zone composed of the elongating axons of the underlying neurons. This three-layered structure persists in the spinal cord and medulla. In other regions, especially in the cerebral and cerebellar cortex, some neuroblasts migrate into the marginal zone to form a cortical plate, which matures into the multilayered cortex of the adult.

Substrates for neuronal migration

In cortex and other regions of the developing brain the migration of neuroblasts and neurons is dependent on RADIAL GLIAL CELLS (Chapter 6). During development these cells maintain their contacts with both the ventricular and the pial surfaces of the neural tube. Therefore, as the walls of the neural tube thicken with the continued division of cells in the ventricular layer and the accumulation of neurons in the mantle zone and cortical plate, the radial glial cells become extremely elongated. From detailed electron microscopic studies of the development of the cerebrum and cerebellum, Rakic and his colleagues concluded that neurons move along this scaffolding of radial glial cells to reach their appropriate positions in the cortex.[4,5] Further observations by Goldowitz and Mullen[6,7] in mutant mice and experiments by Hatten and Mason[8,9] on cells maintained in culture have confirmed that neurons migrate along radial glial cells and have identified proteins involved in this process.

Movement along radially oriented glial cells is not the only mechanism by which migrating neurons are guided to their final destinations. Many neurons migrate through regions of the central nervous system in which there are no radial glial cells. Similarly, in the peripheral

[4]Rakic, P. 1971. *J. Comp. Neurol.* 141: 283–312.
[5]Rakic, P. 1972. *J. Comp. Neurol.* 145: 61–83.
[6]Goldowitz, D. and Mullen, R. J. 1982. *J. Neurosci.* 2: 1474–1485.
[7]Goldowitz, D. 1989. *Neuron* 2: 1565–1575.
[8]Hatten, M. E., Liem, R. K. H. and Mason, C. A. 1986. *J. Neurosci.* 6: 2675–2683.
[9]Edmondson, J. C. et al. 1988. *J. Cell Biol.* 106: 505–517.

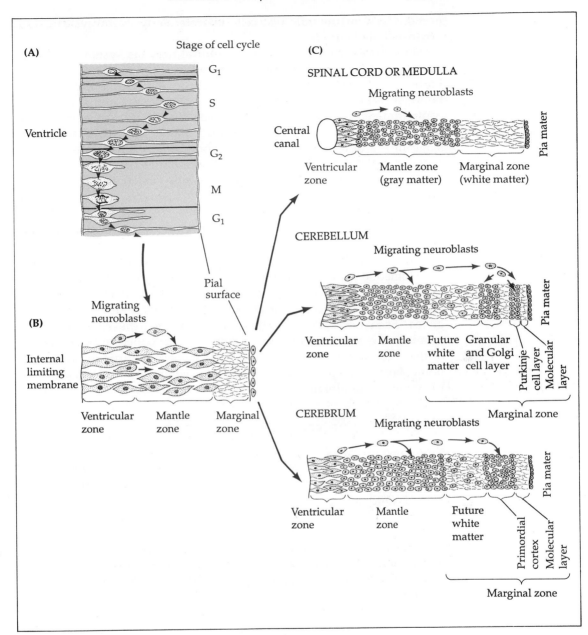

2 DIFFERENTIATION OF THE WALLS OF THE NEURAL TUBE. (A) The position of the nuclei in cells in the primitive neural tube varies as a function of the cell cycle. (B) Cells become postmitotic, migrate away from the ventricular zone, and form the mantle zone. Their processes constitute the marginal zone. (C) The three-layered organization persists in the spinal cord. In the cerebellum and cerebrum, neuroblasts migrate into the marginal zone to form a multilayered cortex. (After Gilbert, 1991.)

nervous system, neural crest cells migrate along pathways that lack organized glial structures.

Unlike the rest of the vertebrate central nervous system, the hindbrain (rhombencephalon) has a conspicuously segmented structure. From the time of their appearance early in development, segmental boundaries are not crossed by cells in expanding clones,[10] a form of compartmentalization resembling that seen in invertebrate nervous systems. Several genes have been identified that are expressed at an early stage in the development of the hindbrain and whose pattern of expression correlates with the segmental boundaries (Figure 3).[11,12] Many of these genes contain a HOMEOBOX DOMAIN similar to that found in homeotic regulator genes in *Drosophila*.[13,14] Several lines of evidence indicate that homeotic genes are "master" genes, controlling the expression of many other genes during development. For example, mutations of homeotic genes in *Drosophila* cause one body part to be replaced with another, such as extra legs to grow where antennae should be. Homeotic genes contain a conserved stretch of DNA, dubbed the HOMEOBOX. The homeobox encodes a sequence of 60 amino acids that recognize and bind to specific DNA sequences in a series of subordinate genes. Each homeotic gene thereby coordinates the expression of a number of genes that, together, determine the structure of one region of the *Drosophila* embryo. It is attractive to speculate that the vertebrate homeobox genes play a similar role, determining the identity of segments in the hindbrain.

REGULATION OF NEURAL DEVELOPMENT

Neuroanatomical techniques have revealed the startling array of highly stereotyped shapes and sizes of neurons in the central nervous system; biochemical, physiological, and immunological studies have demonstrated additional diversity in neuronal phenotype. How do neurons acquire their identities? A variety of experimental approaches indicate that while cell lineage can specify or restrict the phenotypes available to a developing neuron, especially in simple invertebrates, the fate of many neurons is determined through interactions with other cells.

Cell lineage is most readily followed in simple invertebrates like the leech,[15] the grasshopper,[16] the fruit fly,[17] or the tiny nematode *Caenorhabditis elegans*, which contains only about 300 neurons.[18] In these preparations one can follow development cell by cell and examine the ex-

[10]Fraser, S., Keynes, R. and Lumsden, A. 1990. *Nature* 344: 431–435.

[11]Keynes, R. and Lumsden, A. 1990. *Neuron* 2: 1–9.

[12]Wilkinson, D. G. and Krumlauf, R. 1990. *Trends Neurosci.* 13: 335–339.

[13]Akam, M. 1989. *Cell* 57: 347–349.

[14]DeRobertis, E. M., Oliver, G. and Wright, C. V. E. 1990. *Sci. Am.* 263(7): 46–52.

[15]Stent, G. S. and Weisblat, D. 1982. *Sci. Am.* 246(1): 136–146.

[16]Goodman, C. S. and Spitzer, N. C. 1979. *Nature* 280: 208–214.

[17]Rubin, G. M. 1989. *Cell* 57: 519–520.

[18]Horvitz, H. R. 1982. *In* J. G. Nicholls (ed.). *Repair and Regeneration of the Nervous System*. Springer-Verlag, New York, pp. 41–55.

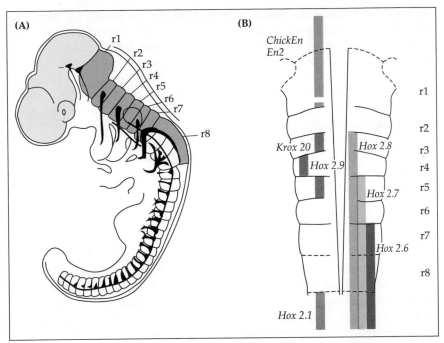

3 **SEGMENTAL EXPRESSION OF HOMEOBOX GENES in the hindbrain of the** vertebrate embryo. **(A) Diagram of a 3-day chick embryo, illustrating the segmental** arrangement of rhombomeres in the hindbrain and of somites lying alongside the spinal cord. **(B) After rhombomeres have been delineated, homeobox genes (known as *Hox*,** *Krox, ChickEn,* and *En2*) are expressed in segment-specific patterns. Data from chick and mouse are shown. **(After Keynes and Lumsden, 1990.)**

pression of characteristics such as membrane properties, transmitters, growth of axons, and branching patterns. In *C. elegans* the embryo is so small and transparent that each neuron can be identified and followed visually under the microscope. Studies by Brenner, Horvitz, Sulston, White, and their colleagues have shown that the development of the nervous system in *C. elegans* is so invariant that the sequence and pattern of cell division of each precursor cell can be predicted. Laser beams can be used to kill single cells and so determine how the fate of surrounding cells is altered. In many cases the surviving cells ignore the loss of their neighbor; their fate is determined by an autonomous, cell lineage-dependent mechanism. In some cases, however, the fate of surviving cells is modified. Thus even in animals with a rigorously stereotyped pattern of cell lineage, inductive interactions between cells can regulate their fate. In a well-known and vivid analogy, Sidney Brenner has characterized cell fate as determined by the European plan or the American plan: In the European plan who you are (as a neuron) is determined by your ancestry; in the American plan it is determined by your neighbors.

The patterned development of the compound eyes of *Drosophila* affords another system in which direct observation can identify individual cells.[19,20] *Drosophila* genetics offers an especially powerful approach to testing the effects of lineage and environmental cues on neuronal differentiation. Mutants have been isolated that lack a particular cell type, or in which the pattern of differentiation is perturbed in subtle ways, and the effects on the fate of surviving cells have been determined. Such techniques have been used in experiments by Benzer, Ready, Rubin, Tomlinson, Zipursky, and their colleagues to examine the differentiation of neurons and supporting cells in the eye.[21,22]

The adult *Drosophila* eye (Figure 4) consists of a crystalline array of repeating units, the OMMATIDIA, each of which is composed of eight photoreceptors (R1–R8) and a variety of supporting cells. The first cell to begin to differentiate in each ommatidium is one of the photoreceptors, R8. The R8 cells appear at random in the neuroepithelium, but once an R8 cell begins to differentiate, it inhibits any neighboring cell from becoming an R8 cell. R2 and R5 are recruited next, then R3 and R4, R1 and R6, and finally R7. The appearance of R7 has been particularly well studied and depends on an interaction with R8. The isolation of mutants lacking R7 and studies with genetic mosaics have led to identification of a receptor protein that has tyrosine kinase activity in the presumptive R7 cell (the product of the *sevenless* gene) that must be activated by an integral membrane protein in the R8 cell (encoded by the *boss*—bride of sevenless—gene) in order for the R7 cell to develop. These and additional studies in *Drosophila* and grasshopper support the idea that cell-specific inductive interactions are important in determining the identity of individual neurons.

An alternative approach is to mark individual cells and see what types of progeny they produce. This type of analysis, introduced by Weisblat, Stent, and their colleagues, was originally employed to trace the fates of cells in leech[23] and frog[24] embryos by injecting the enzyme horseradish peroxidase (HRP) into individual cells, allowing development to proceed, and then staining the embryos to visualize cells containing the enzyme (see Chapter 13). Fluorescently labeled tracers have been developed that can be used to follow cell lineage, even while the embryo is alive.[25] A particularly favorable vertebrate preparation for developmental studies is the zebrafish, first introduced to neurobiology by Streisinger and colleagues.[26–29] The zebrafish embryo is transparent,

[19]Ready, D. F., Handson, T. E. and Benzer, S. 1976. *Dev. Biol.* 53: 217–240.

[20]Tomlinson, A. and Ready, D. F. 1987. *Dev. Biol.* 120: 366–376.

[21]Lawrence, P. A. and Tomlinson, A. 1991. *Nature* 352: 193.

[22]Kramer, H., Cagan, R. L. and Zipursky, S. L. 1991. *Nature* 352: 207–212.

[23]Weisblat, D. 1981. *In* K. J. Muller, J. G. Nicholls and G. S. Stent (eds.). *Neurobiology of the Leech.* Cold Spring Harbor Laboratory, Cold Spring Harbor, NY, pp. 173–195.

[24]Jacobson, M. and Hirose, G. 1981. *J. Neurosci.* 1: 271–284.

[25]Bronner-Fraser, M. and Fraser, S. E. 1988. *Nature* 335: 161–164.

[26]Streisinger, G. et al. 1981. *Nature* 291: 293–296.

[27]Kimmel, C. B. and Warga, R. M. 1988. *Trends Genet.* 4: 68–74.

[28]Chitnis, A. B. and Kuwada, J. Y. 1991. *Neuron* 7: 277–285.

[29]Westerfield, M. et al. 1990. *Neuron* 4: 867–874.

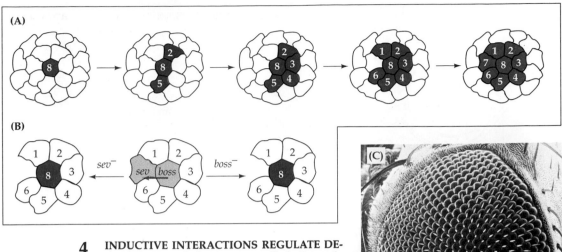

4 **INDUCTIVE INTERACTIONS REGULATE DEVELOPMENT** of photoreceptor cells in *Drosophila*. (A) Normal progression of differentiation of the eight photoreceptors in each ommatidium. (B) *Seveless* (*sev⁻*) and *bride of seveless* (*boss⁻*) mutations each prevent differentiation of R7. The product of the *seveless* gene is a tyrosine kinase that must be activated in the prospective R7 cell by the product of the *boss* gene, an integral membrane protein expressed in R8. (C) Scanning electron micrograph of a compound eye in *Drosophila*. Each facet is an ommatidium. (After Tomlinson and Ready, 1987; micrograph courtesy of D. F. Ready.)

allowing direct observation of individual cells throughout embryogenesis, development is rapid (fertilized egg to hatchling in 3 days), and the life cycle is sufficiently short (3–4 months) to allow application of genetic techniques.

Analysis of cell lineage is much more complex in the developing mammalian central nervous system, where the sheer numbers often preclude direct observation of individual identified cells. One technique that overcomes this obstacle is to map the fate of genetically marked cells in embryonic and adult chimeras.[30] Another is to inject specially engineered viruses into the CNS of developing animals[31,32] (Figure 5). These viruses are constructed so that they become permanently incorporated into the chromosomes of the host cell, replicate during cell division, and thus are passed on to the descendants of that cell. In this way the signal carried by the virus is not diluted during successive cell

Cell lineage in the mammalian CNS

[30] Rossant, J. 1985. *Philos. Trans. R. Soc. Lond. B* 312: 91–100.
[31] Turner, D. L. and Cepko, C. L. 1987. *Nature* 328: 131–136.
[32] Luskin, M. B., Pearlman, A. L. and Sanes, J. R. 1988. *Neuron* 1: 635–647.

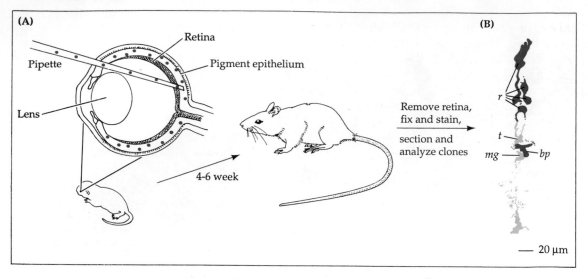

5   **CLONALLY RELATED CELLS ARE LABELED** by injecting retroviral markers into precursor cells in the rat retina. **(A)** A retrovirus encoding *β*-galactosidase is injected into the eye between the retina and the pigment epithelium early in development, infecting a few retinal precursor cells. Staining the adult retina with a histochemical reaction for *β*-galactosidase reveals clusters of labeled cells. The cells in each cluster are progeny of a single precursor cell. **(B)** Camera lucida drawing of a clone that includes five rods (r), one bipolar cell (bp) juxtaposed to the terminal of a rod cell (t), and a Müller glial cell (mg). (After Turner and Cepko, 1987.)

divisions, as are markers such as HRP. The presence of the virus can be detected at any stage by a histochemical reaction for an enzyme it encodes. Provided the number of cells infected is very low, it can be concluded that a small cluster of stained cells found later in development represents the progeny of a single infected parent cell.

When used to study the development of the mammalian cortex and retina, chick tectum, and frog retina, such techniques have demonstrated that a single neuronal precursor cell often produces progeny having a variety of neuronal phenotypes.[33] There is no evidence for specific lineages that give rise to particular types of neurons. These findings indicate the importance of interactions between postmitotic neurons and their environment in the determination of cell fate. When tracers are injected into individual neuronal precursor cells in the vertebrate hindbrain, on the other hand, the majority of clones are found to contain a single neuronal phenotype.[34] Thus, in the hindbrain, cell fate appears to be restricted at an earlier stage. Results regarding progenitors that give rise to both glial cells and neurons vary.[35] In retina, clones frequently include both a glial cell and one or more neurons, indicating that a cell's identity as a neuron or glial cell may be decided

[33]McConnell, S. K. 1991. *Annu. Rev. Neurosci.* 14: 269–300.
[34]Lumsden, A. 1991. *Philos. Trans. R. Soc. Lond. B* 331: 281–286.
[35]Sanes, J. R. 1989. *Trends Neurosci.* 12: 21–28.

only at or after the terminal division. In rodent cerebral cortex, on the other hand, clones containing both neurons and glial cells are rare, suggesting that separate populations of neuroblasts and glioblasts are present in the cortical ventricular zone. Glial cell development and the factors that regulate glial differentiation are discussed in Chapter 6.

In summary, in the nervous systems of simple organisms the lineage history of a cell limits its developmental potential. Some cells have a unique fate (for example, the sensory cells in leech ganglia; see Chapter 13), whereas others seem to have multiple potential fates depending on temporal or positional cues. In the central nervous systems of more complex organisms inductive interactions between cells are of overriding importance for determining cell fate.[33]

It requires many weeks of development to produce the vast number of neurons that make up the mammalian central nervous system. Does the time at which a presumptive neuron stops dividing and migrates away from the ventricular zone influence its fate? This question can be addressed by marking neurons that become postmitotic, or are "born," at a particular time with a labeled DNA presursor.[36] If a pulse of [³H]-thymidine is given by intrauterine or intravenous injection during development, the label will be taken up and incorporated into the DNA of cells undergoing cell division at that time. If such labeled cells continue to divide after the pulse, as would be the case for neural precursor cells in the ventricular zone and for glial cells, the label will be diluted during subsequent rounds of DNA synthesis. However, neurons that are born and migrate away from the ventricular zone during the pulse will remain heavily labeled, since such neurons are forever postmitotic. Thus, the final distribution of neurons born on a particular day can be visualized by exposing an embryo to a pulse of [³H]-thymidine on that day, allowing development to continue, and later analyzing the nervous system by autoradiography to detect labeled cells.

> The relationship between neuronal birthday and cell fate

Use of this technique has revealed that there is a systematic relationship between the time a neuron is born and its final position in the adult mammalian CNS.[37,38] This relationship is seen clearly in the cerebral cortex, where development proceeds in an inside-out fashion (Figure 6). Thus neurons of the deepest cortical layers are born first. Neurons in more superficial layers are born later and migrate through the cells of the deeper layers to assume their final positions within the cortex. A similar correspondence between time of birth and the ultimate position occupied by a neuron is found throughout the CNS, although not all regions develop in the inside-out manner of the cerebral cortex.

Experiments in developing ferrets have established with considerable precision the time at which a cortical neuron becomes committed to occupy a particular position.[39] Cells in the ventricular zone of young

[36]Angevine, J. B. and Sidman, R. L. 1961. *Nature* 192: 766–768.

[37]Rakic, P. 1974. *Science* 183: 425–427.

[38]Luskin, M. B. and Shatz, C. J. 1985. *J. Comp. Neurol.* 242: 611–631.

[39]McConnell, S. K. 1989. *Trends Neurosci.* 12: 342–349.

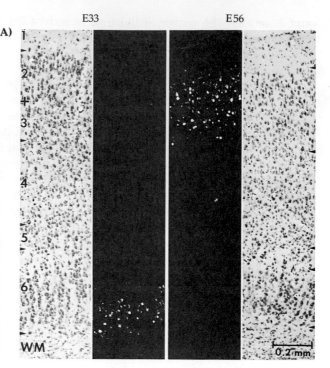

E33   E56

(A)

6

NEUROGENESIS OF THE PRIMARY VISUAL CORTEX of the cat. (A) Autoradiographs of sections of the adult visual cortex of animals injected with [³H]thymidine on embryonic day 33 (E33) or 56 (E56). Bright-field micrographs of the same sections stained with cresyl violet indicate that heavily labeled cells are located in layer 6 after the E33 injection, and in layers 2 and 3 after injection on E56. (B) Histograms showing the distribution of cells labeled on various days between E30 and E56 illustrate the inside-out pattern of neurogenesis in the visual cortex. (After Luskin and Shatz, 1985; micrograph courtesy of M. B. Luskin.)

0.2 mm

(B)

Number of cells

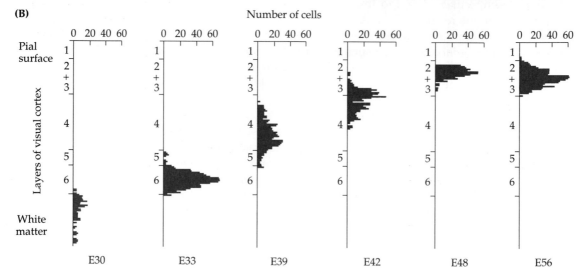

E30   E33   E39   E42   E48   E56

embryos, including embryonic cortical neuronal precursors destined to give rise to neurons of layers 5 and 6, were transplanted into the ventricular zone of older hosts, amid cells destined for layers 2 and 3. Precursor cells transplanted while in the S phase of the cell cycle (when DNA synthesis takes place) were respecified and gave rise to layer 2 and 3 neurons. However, if the transplanted cells were progenitors in the late stages of the cell cycle or were postmitotic neurons that had not yet migrated from the ventricular zone, then they maintained their original identity, migrating to layer 6 and forming connections appropriate for their birthday.

Other aspects of the identity and connections of a cortical neuron are also determined about the time it is born and can be expressed independently of position. For example, in a mutant mouse called *reeler*, neurons fail to migrate past one another in the developing cortex.[40] Thus their relative positions in the adult are inverted: Neurons born at early times end up in the most superficial layers, whereas neurons born later end up in deeper layers. In spite of their aberrant positions, the misplaced neurons acquire the morphological appearance and connections appropriate to their time of birth.

Not all properties of a cortical neuron are determined at the time it is born. Experiments in which pieces of developing cerebral cortex are transplanted to other locations in the cerebrum indicate that local influences can determine aspects of the phenotype of cortical neurons.[41–43] For example, if a section of visual cortex is transplanted into the whisker area of the somatosensory cortex in a rat, the transplanted cortex acquires the barrel structure appropriate to its new location (see Chapters 14 and 18), a phenotype not seen in the visual cortex.

Influence of local cues on cortical architecture

How rigidly fixed are the properties of neurons that descend from a particular ancestor in the peripheral nervous system of vertebrate embryos? Are all the progeny destined to form cells with specified properties, such as being autonomic rather than sensory cells, or using acetylcholine as the transmitter rather than norepinephrine? Can the transmitter released by a cell change under different influences? These questions have been studied in chick and quail nervous systems by Le Douarin, Weston, and others.[44–46] Much of the peripheral nervous system in vertebrates arises from the neural crest, the column of cells that appears between the neural tube and the overlying ectoderm early in development. In normal embryos, neural crest cells eventually give rise to a variety of cell types, including dorsal root and autonomic ganglion cells. Different regions of the crest give rise to different ganglia and neurons with different transmitters. Le Douarin transplanted cells from one region of the neural crest in the embryo of a quail to the same or a different region of a host chick embryo (Figure 7) and then followed the fates of the transplanted cells on the basis of unambiguous cytolog-

Control of neuronal phenotype in the peripheral nervous system

[40]Caviness, V. S., Jr. 1982. *Dev. Brain Res.* 4: 293–302.

[41]Stanfield, B. B. and O'Leary, D. D. M. 1985. *Nature* 298: 371–373.

[42]O'Leary, D. D. M. and Stanfield, B. B. 1989. *J. Neurosci.* 9: 2230–2246.

[43]Schlaggar, B. L. and O'Leary, D. D. M. 1991. *Science* 252: 1556–1560.

[44]Weston, J. 1970. *Adv. Morphog.* 8: 41–114.

[45]Le Douarin, N. M. 1986. *Science* 231: 1515–1522.

[46]Le Douarin, N. 1982. *The Neural Crest.* Cambridge University Press, Cambridge, England.

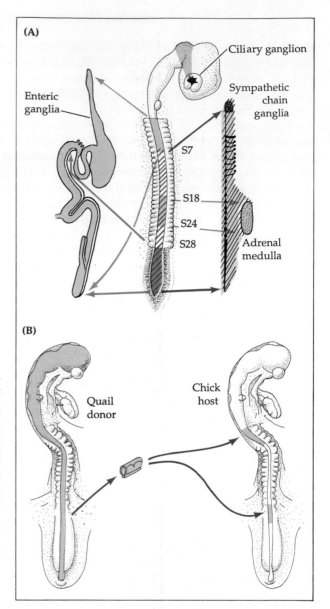

7

THE FATE OF A NEURAL CREST CELL is determined by environmental cues. (A) Neural crest cells give rise to a variety of peripheral ganglia. The ciliary ganglion is formed by cells from the mesencephalic neural crest. The ganglion of Remak and the enteric ganglia of the gut are formed by cells from the vagal (somites 1–7) and lumbosacral (caudal to S28) regions of the neural crest. The ganglia of the sympathetic chain are derived from all regions of the neural crest caudal to S5. The adrenal medulla is populated by crest cells from S18–S24. (B) If crest cells from S18–S24, which are destined to form the adrenal medulla, are transplanted from a quail donor to the vagal or lumbrosacral region of a host chick embryo, they will adopt the fate appropriate to their new location and populate the ganglion of Remak and the enteric ganglia of the gut. (After Le Douarin, 1986.)

ical differences between quail and chick cells. After transplantation, the quail cells were able to innervate structures they would never normally supply. For example, cells removed from a region that would normally become the adrenal gland could instead innervate the gut.[47] Provided that the cells are transplanted at a sufficiently early stage, they can make a new transmitter, such as ACh, instead of norepinephrine. At later stages, the cells become committed and lose the ability to differentiate in a manner determined by their environment. In some cases such a change occurs during the normal course of development. For

[47]Le Douarin, N. M. 1980. *Nature* 286: 663–669.

example, there is evidence that sympathetic neurons that innervate sweat glands initially synthesize norepinephrine, but during the second and third week of postnatal development are induced to switch to synthesizing ACh by factors associated with their target.[48]

Although the precise role played by the target tissues in determining the fates of neural crest cells is not known, results of experiments on sympathetic ganglion cells in culture have led to the identification of factors that can influence the transmitter choice made by cells of neural crest origin. When neurons are dissociated from the superior cervical ganglia of newborn rats and grown in culture in the absence of other cell types, all of them contain tyrosine hydroxylase and synthesize and store catecholamines.[49] However, if the cells are grown in the presence of certain non-neuronal cells (or medium conditioned by such cells), including target cells such as heart and skeletal muscle, the neurons stop synthesizing catecholamines and begin synthesizing choline acetyltransferase and ACh instead.[50,51] To establish unequivocally that this change occurs in individual cells, single neurons were cultured on microislands of heart cells[52] (Figure 8). The neurons rapidly extended

[48]Landis, S. C. 1990. *Trends Neurosci.* 13: 344–350.
[49]Mains, R. E. and Patterson, P. H. 1973. *J. Cell Biol.* 59: 329–345.
[50]Patterson, P. H. and Chun, L. L. Y. 1974. *Proc. Natl. Acad. Sci. USA* 71: 3607–3610.
[51]Patterson, P. H. and Chun, L. L. Y. 1977. *Dev. Biol.* 56: 263–280.
[52]Furshpan, E. J. et al. 1976. *Proc. Natl. Acad. Sci. USA* 73: 4225–4229.

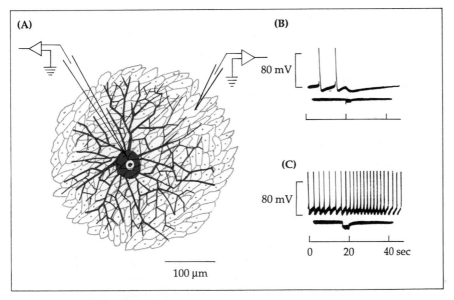

8 **SINGLE NEURONS FROM SYMPATHETIC GANGLIA** can release both acetylcholine and norepinephrine at synapses on heart cells in culture. (A) A single sympathetic neuron, grown on an island of cardiac muscle cells. (B) A brief train of impulses in the neuron (10 Hz, deflection of lower trace) produced inhibition of spontaneous myocyte activity (upper trace). (C) Addition of atropine (10^{-7} *M*) blocked the inhibitory cholinergic response, leaving the excitatory effect, which is due to the release of norepinephrine. (After Furshpan et al., 1976.)

axons and established synaptic contact with the heart cells. Initially these synapses were adrenergic. Over the course of several days, however, the neurons began to release *both* norepinephrine and ACh, and eventually transmission became cholinergic. A factor that induces cholinergic differentiation of sympathetic neurons has been purified from heart-conditioned medium, characterized, and cloned.[53] It turns out to be identical to a protein that had been characterized previously on the basis of its ability to induce differentiation of cells in the immune system. Several other molecules have been identified that can induce ACh synthesis in sympathetic neurons.[54] The roles of these various factors in determining transmitter phenotype in vivo are beginning to be investigated.

Hormonal control of development

In some regions of the central nervous system the fate of neurons is under hormonal control. This is particularly conspicuous in regions subserving sexually dimorphic behaviors. In the moth *Manduca sexta*, for example, the region of the brain concerned with olfaction has, in the male, a specialized "subsystem" of olfactory receptor cells, CNS neurons, and synaptic areas associated with the detection of female sex pheromones.[55] Songbirds provide another example.[56] In the canary, the high vocal center (nucleus HVC) plays a crucial role in the acquisition and retention of song—a uniquely male behavior. This area of the brain is more developed in males than in females. Females can be induced to sing by injections of testosterone, however, and the HVC nucleus and other structures associated with song production are enlarged in such androgenized females. In males the HVC nucleus undergoes seasonal variations in metabolic activity that depend on changes in the levels of circulating testosterone and correlate with the time of song acquisition.[57] In addition, the HVC nucleus in the adult CNS is unusual because there is within it a seasonal turnover of neurons. New neurons are added each autumn, a time when testosterone levels are increasing from a midsummer low. This is just when males modify their song for the next breeding season.

MECHANISMS OF AXON OUTGROWTH AND GUIDANCE

Nerve cells have axons that may extend a meter or more to form synapses at a particular location on the appropriate cell in a region filled with other potential target cells. How such precise connections form during development has long fascinated neurobiologists. Two general ideas regarding specificity developed over the first quarter of this century. One held that neurons and their targets were prespecified in such

[53]Yamamori, T. et al. 1989. *Science* 246: 1412–1416.
[54]Saadat, S., Sendtner, M. and Rohrer, H. 1989. *J. Cell Biol.* 108: 1807–1816.
[55]Hishinuma, A. et al. 1988. *J. Neurosci.* 8: 296–307.
[56]Nottebohm, F. 1989. *Sci. Am.* 260(2): 74–79.
[57]Kirn, J. R., Alvarez-Buylla, A. and Nottebohm, F. 1991. *J. Neurosci.* 11: 1756–1762.

a way that only appropriate synaptic connections were established.[58] The other idea was that initial connections were established more or less randomly and were then sorted out by target-induced specification of neurons, elimination of incorrect synapses, or death of inappropriately connected cells.[59] The idea that neurons could selectively innervate appropriate targets was bolstered in the 1940s and 1950s by experiments by Stone,[60] Sperry,[61,62] and their collaborators, who found that if the optic nerve was cut in a frog or a salamander, regenerating fibers grew back to the appropriate region of the brain. There they formed synapses, and eventually the animal was able to see once again. If the eye was rotated through 180° (exchanging dorsal for ventral) at the time the optic nerve was cut, then, after the nerve had regenerated, the frog behaved as though its vision were inverted. All its movements directed toward objects—as when the frog struck at a fly—were 180° out of phase; it would strike downward at a fly held over its head, getting only a mouthful of dust for its effort. The simplest interpretation of this and other behavioral experiments was that fibers had grown back from the inverted retina and selectively reinnervated their original target neurons in the tectum.

A wealth of experimental evidence now supports the idea that axonal growth is selective,[63,64] although in many instances the arborization of an axon is more extensive initially than in the adult, and the adult pattern is formed by a process of pruning.[65–68] In some cases pruning provides a mechanism for correcting mistakes;[68,69] in others it appears to reflect a strategy for establishing pathways[68,70] and for ensuring appropriate and complete innervation of a target by a particular population of neurons (see below). The importance of activity in the pruning and rearrangement of connections in the mammalian visual cortex is considered in Chapter 18.

What mechanisms enable neurons to grow to their targets? One possibility is that axons are confined to appropriate pathways by mechanical barriers in the tissues through which they are growing. Such a passive guidance mechanism is important during regeneration of peripheral axons, where regenerating axons elongate within the tubes

[58]Langley, J. N. 1895. *J. Physiol.* 22: 215–230.

[59]Weiss, P. 1936. *Biol. Rev.* 11: 494–531.

[60]Stone, L. S. 1944. *Proc. Soc. Exp. Biol. Med.* 57: 13–14.

[61]Sperry, R. W. 1963. *Proc. Natl. Acad. Sci. USA* 50: 703–710.

[62]Sperry, R. W. 1943. *J. Exp. Zool.* 92: 236–279.

[63]Stuermer, C. A. O. and Raymond, P. A. 1989. *J. Comp. Neurol.* 281: 630–640.

[64]O'Rourke, N. A. and Fraser, S. E. 1990. *Neuron* 5: 159–171.

[65]Innocenti, G. M. 1981. *Science* 212: 824–827.

[66]O'Leary, D. D. M., Stanfield, B. B. and Cowan, W. M. 1981. *Dev. Brain Res.* 1: 607–617.

[67]Ivy, G. O. and Killackey, H. P. 1982. *J. Neurosci.* 2: 735–743.

[68]Cowan, W. M. et al. 1984. *Science* 225: 1258–1265.

[69]Nakamura, H. and O'Leary, D. D. M. 1989. *J. Neurosci.* 9: 3776–3795.

[70]O'Leary, D. D. M. and Terashima, T. 1988. *Neuron* 1: 901–910.

of extracellular matrix left behind after degeneration of the original axons (Chapter 12). Although there is some evidence that mechanical barriers influence the growth of certain axons during development,[71] results of most experiments support the idea that axons actively navigate to their targets by reading chemical cues in their environment. What is the nature of those cues? Where do they come from?

Growth cones and axon elongation

The tips of growing axons expand to form growth cones. The growth cone was first recognized by Ramón y Cajal as the portion of the axon responsible for navigation and elongation toward a target (Figure 9). Time-lapse images of growth cones extending from neurons in culture demonstrate their remarkable activity.[72] Growth cones continually extend and retract broad membranous sheets called lamellipodia, and slender, spikelike protrusions termed filopodia, for distances of tens of micrometers, as if sampling the substrate in every direction. Filopodia

[71]Silver, J. and Sidman, R. S. 1980. *J. Comp. Neurol.* 189: 101–111.
[72]Smith, S. J. 1988. *Science* 242: 708–715.

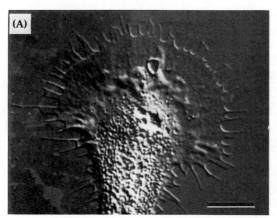

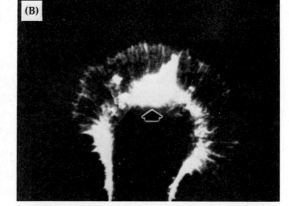

9

THE MORPHOLOGY OF GROWTH CONES.
(A) Growth cone observed by differential interference contrast microscopy. **(B)** Fluorescence micrograph showing the distribution of filamentous actin visualized with rhodamine-conjugated phalloidin. Actin filaments align with filopodia, or microspikes, in the periphery of the growth cone; randomly oriented filaments are often concentrated near the central domain (arrow). **(C)** Microtubule distribution visualized with anti-tubulin antibodies and fluorescein-conjugated secondary antibodies. Microtubules are concentrated in the axon; most terminate in the central domain of the growth cone; some (arrowhead) extend toward the growth cone margin (asterisks). Bar, 10 μm. (After Forscher and Smith, 1988; micrographs courtesy of S. J. Smith.)

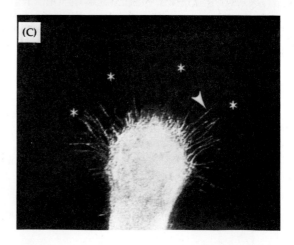

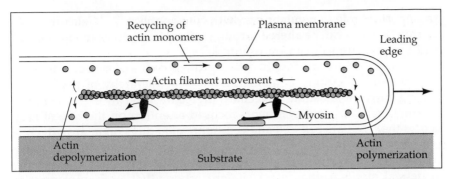

Recycling of actin monomers

Plasma membrane

Leading edge

← Actin filament movement ←

Myosin

Actin depolymerization

Substrate

Actin polymerization

10 **MODEL FOR ACTIN-BASED MOTILITY OF GROWTH CONES. Actin filaments are propelled rearward by interactions with molecules of myosin anchored to the membrane. Actin monomers are added to filaments at the leading edge, and filament depolymerization occurs in the central region of the growth cone. Extension of the leading edge may be powered by the polymerization of actin, or by the movement of myosin molecules in the region of the leading edge along actin filaments whose rearward movement has temporarily been halted by inactivity of more centrally located myosin molecules. (After Smith, 1988; Sheetz, Wayne and Pearlman, 1992.)**

appear able to adhere to the substrate and pull the growth cone in their direction.

The mechanisms by which growth cones crawl, explore, and exert force on their trailing axons have been investigated by Bray, Reuter, Kater, Smith, and their colleagues.[73,74] Actin plays a pivotal role in growth cone motility[72] (Figure 10). Both lamellipodia and filopodia are rich in filamentous actin, and agents that inhibit actin polymerization, such as the fungal toxin cytochalasin B, immobilize growth cones. The protrusion and retraction of lamellipodia and filopodia appear to be mediated by two processes: (1) ongoing polymerization and disassembly of actin filments, and (2) translocation of actin filaments away from the leading edge of the growth cone, powered by interaction with myosin. Both processes can harness the energy of ATP hydrolysis to generate force.

How do environmental cues influence the direction of axonal growth? It is thought that signals in the environment bind to surface receptors and thereby modify locally the activity of actin-binding proteins in growth cones. These proteins, in turn, control the cycle of actin polymerization and depolymerization and the interaction of actin filaments with myosin, thus modifying growth cone movement. A wide variety of actin-binding proteins have been identified in growth cones; the effects of many are known to be modulated by calcium, phosphorylation by protein kinases, or interactions with other intracellular second messenger systems. Although calcium influx is not necessary for axonal growth,[75] several lines of evidence indicate that calcium plays

[73]Bray, D. and Hollenbeck, P. J. 1988. *Annu. Rev. Cell Biol.* 4: 43–61.
[74]Lankford, K., Cypher, C. and Letourneau, P. 1990. *Curr. Opin. Cell Biol.* 2: 80–85.
[75]Usowicz, M. M. et al. 1990. *J. Physiol.* 426: 95–116.

an important role in regulating growth cone motility.[76–78] Calcium levels in growth cones can be altered through cell surface adhesion molecules[79] and by exposure to neurotransmitters.[80]

In 1975 Letourneau reported that growth cones of sensory neurons growing in cell culture demonstrated a clear hierarchy of preferences among different substrate molecules, preferring to grow on some rather than others.[81] These experiments focused attention on the role of molecules that mediate cell–cell and cell–substrate adhesion in growth cone guidance. One important group of adhesion molecules comprises transmembrane or membrane-associated glycoproteins characterized by structural motifs in their extracellular portions that are homologous to immunoglobulin constant-region domains (Figure 11). Members of this immunoglobulin superfamily include neural cell adhesion molecule (N-CAM), neuroglia CAM (NgCAM or L1), and a transiently expressed axonal surface glycoprotein called TAG-1.[82,83] These molecules mediate cell–cell adhesion, either through homophilic binding to their counterpart on other cells (e.g., N-CAM) or heterophilic binding involving a distinct receptor (e.g., TAG-1).

An additional ubiquitous cell adhesion molecule is N-cadherin (Figure 11), which mediates homophilic, calcium-dependent cell adhesion. When the effects of N-CAM or N-cadherin are assayed on cells in culture, each is found to promote not only the aggregation of nerve cells but also the extension of axons on cellular substrates and the binding together of growing axons into fascicles. Thus antibodies against N-CAM or N-cadherin reduce the outgrowth of axons from cultured nerve cells on cellular substrates and prevent axons from fasciculating into bundles, but do not block axon elongation on substrates coated with extracellular matrix molecules such as laminin. The broad distribution of N-CAM and N-cadherin in developing embryos suggests that they may play a permissive role in axon outgrowth, but are unlikely to provide specific cues for growth cone guidance. For example, antibodies to N-CAM disrupt fasciculation of developing retinotectal axons in vivo.[84,85] This does not prevent the axons from reaching their target tissue, although it degrades the precision of the projection. In addition, several members of the immunoglobulin superfamily of cell adhesion molecules are important in axon–Schwann cell interactions, including growth of regenerating axons[86] and myelination.[87]

[76]Anglister, L. et al. 1982. *Dev. Biol.* 94: 351–365.

[77]Lankford, K. L and Letourneau, P. C. 1989. *J. Cell Biol.* 109: 1229–1243.

[78]Bentley, D., Guthrie, P. B. and Kater, S. B. 1991. *J. Neurosci.* 11: 1300–1308.

[79]Schuch, U., Lohse, M. J. and Schachner, M. 1989. *Neuron* 3: 13–20.

[80]Kater, S. B. et al. 1988. *Trends Neurosci.* 11: 315–321.

[81]Letourneau, P. C. 1975. *Dev. Biol.* 44: 92–101.

[82]Dodd, J. and Jessell, T. M. 1988. *Science* 242: 692–699.

[83]Furley, A. J. et al. 1990. *Cell* 61: 157–170.

[84]Thanos, S. Bonhoeffer, F. and Rutishauser, U. 1984. *Proc. Natl. Acad. Sci. USA* 81: 1906–1910.

[85]Fraser, S. E. et al. 1984. *Proc. Natl. Acad. Sci. USA* 81: 4222–4226.

[86]Fawcett, J. W. and Keynes, R. J. 1990. *Annu. Rev. Neurosci.* 13: 43–60.

[87]Wood, P. M., Schachner, M. and Bunge, R. P. 1990. *J. Neurosci.* 10: 3635–3645.

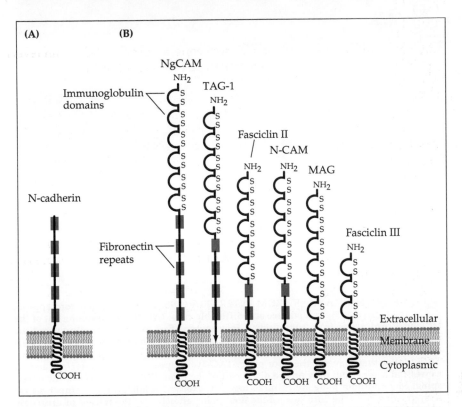

11 TWO CLASSES OF NEURAL CELL ADHESION MOLECULES. (A) N-cadherin mediates homophilic, calcium-dependent cell adhesion. (B) Members of the immunoglobulin superfamily are characterized by multiple repeats of disulfide-linked loops homologous to domains first characterized in the constant region of immunoglobulin molecules. Many of these cell adhesion molecules also contain multiple domains similar to the type III repeat of fibronectin (rectangles).

There is suggestive evidence that other members of this group may play a more instructive rule in axon guidance. For example, early in embryogenesis, TAG-1 is expressed on the surface of the axons of a particular group of spinal neurons, the commissural interneurons, as they grow ventrally through the developing spinal cord[83] (Figure 12). Upon reaching the floor plate at the base of the spinal cord, these axons make an abrupt right angle turn and grow longitudinally along the cord. At the same time, TAG-1 expression ceases, the L1 adhesion molecule is induced (but only in the distal segment of the axon), and the distal segments form fascicles as they grow along their longitudinal pathway. Other cell adhesion molecules, for example N-CAM, are expressed uniformly over the surface of the axons throughout their trajectory. Similarly, in the developing insect nervous system, changes in the temporal and spatial expression of homologous proteins, called fasciclins, in the developing insect nervous system appear to contribute to guidance of growing axons by selective fasciculation with existing fiber tracts.[88]

[88]Harrelson, A. L. and Goodman, C. S. 1988. *Science* 242: 700–708.

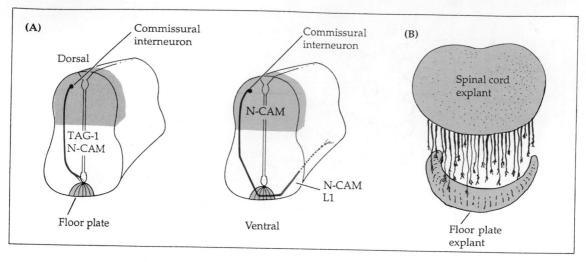

12   **GROWTH OF AXONS FROM INTERNEURONS** in the rat spinal cord appears to be directed by cell surface and diffusible cues. **(A)** During development of the rat spinal cord axons from commissural interneurons first extend ventrally toward the floor plate and then make an abrupt turn caudally. These axons express N-CAM on their surface throughout their length. Before they reach the floor plate they also express TAG-1; after they traverse the floor plate, TAG-1 disappears from their surface and L1 expression begins, but only on the distal part of the axon. **(B)** When pieces of the dorsal spinal cord and floor plate are cultured together, axons of commissural interneurons are seen growing from the cord toward the floor plate, attracted by the release of diffusible factors. (After Tessier-Lavigne et al., 1988; Furley et al., 1990.)

Extracellular matrix
adhesion molecules

A second important group of adhesion molecules are those that are associated with the extracellular matrix and mediate cell–substrate adhesion.[89,90] This group includes laminin, fibronectin, tenascin (also known as J1 or cytotactin), and thrombospondin (Figure 13). These large extracellular glycoproteins consist of two or more identical or similar polypeptide chains held together by disulfide bonds. Each subunit is characterized by repeated structural motifs, some of which are shared among several members of the group. These extracellular matrix proteins interact with cells via a family of receptors known as integrins, which vary in their subunit composition and ability to bind different extracellular matrix adhesion proteins. Laminin is a favorable substrate for neurite extension in culture. Immunohistochemical studies have shown that laminin is present along several pathways that growing axons follow in developing embryos. Laminin is generally considered to be a nonselective promoter of neurite outgrowth. However, in some regions laminin is transiently and selectively expressed precisely along pathways taken by axons to their peripheral targets, suggesting that in such cases it might play a role in growth cone guidance. Extracellular

[89]Lander, A. D. 1989. *Trends Neurosci.* 12: 189–195.
[90]Reichardt, L. F. and Tomaselli, K. J. 1991. *Annu. Rev. Neurosci.* 14: 531–570.

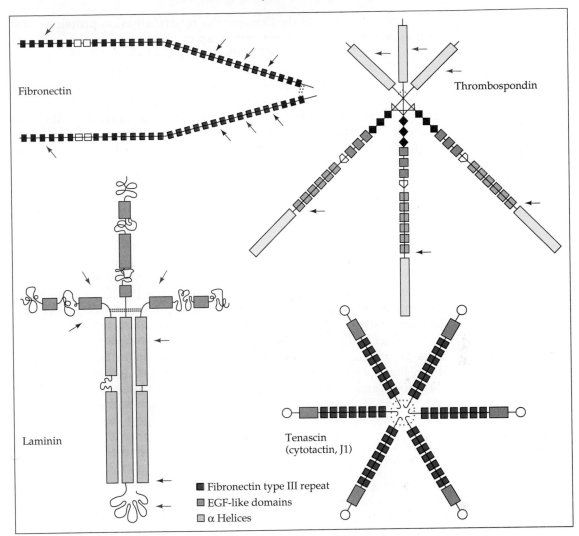

13 **EXTRACELLULAR MATRIX PROTEINS** that mediate cell–substrate adhesion. Schematic representations of extracellular matrix glycoproteins. Fibronectin and tenascin contain multiple "fibronectin type III" repeats; laminin, tenascin and thrombospondin contain multiple cysteine-rich EGF (epidermal growth factor)-like domains. Positions of interchain disulfide bridges (dotted lines), cell-binding domains (black arrows), and heparin-binding sites (blue arrows) are indicated. (After Lander, 1989.)

matrix adhesion proteins also appear to be involved in migration of neural crest cells away from the spinal cord.[91] Crest cells migrate in cell culture on substrates coated with extracellular matrix adhesion molecules, their movement is disrupted by agents that block binding of extracellular matrix components to integrin, and antibodies to integrin block migration of cranial neural crest cells in vivo.

[91]Bronner-Fraser, M. 1985. *J. Cell Biol.* 101: 610–617.

A striking feature of the extracellular matrix adhesion proteins is the multifunctionality of binding they display. For example, laminin not only binds to integrin, its cell surface receptor, but also binds to other molecules of laminin and to heparin, heparan sulfate proteoglycans, collagen type IV, entactin, and glycolipids, as well as other unidentified receptors.[89] Moreover, a general finding is that different proteins, or groups of proteins, support neurite outgrowth on different cell types. Thus, to inhibit all axonal growth on Schwann cells, antibodies to L1 (NgCAM), N-cadherin, and integrins must be applied together; none completely prevents growth by itself.[92,93] It is also clear from recent molecular biological studies that many of the proteins mediating adhesion have a variety of isoforms that may endow them with greater selectivity than previously thought.[90] For example, 12 distinct α and 6 distinct β subunits of integrin have been identified; each $\alpha\beta$ combination is likely to produce a receptor with distinctive binding properties. There are at least three isoforms of laminin, two of fibronectin, and several of tenascin. Overall, the distributions of these cell adhesion molecules in developing embryos and the effects of antibodies that block their function suggest that they are important in facilitating axon elongation and, in many pathways, may provide specific guidance cues as well.

Another indication of how extracellular matrix molecules can influence growing axons comes from experiments on identified cells isolated from the leech central nervous system and grown in culture[94] (Figure 14). Substrates that contain tenascin or laminin not only support rapid extension of neurites from leech neurons, but also influence the pattern of neurite outgrowth and the distribution of calcium channels on the

[92]Bixby, J. L., Lilien, J. and Reichardt, L. F. 1988. *J. Cell Biol.* 107: 353–361.

[93]Seilheimer, B. and Schachner, M. 1988. *J. Cell Biol.* 107: 341–351.

[94]Masuda-Nakagawa, L. M. and Nicholls, J. G. 1991. *Philos. Trans. R. Soc. Lond. B* 331: 323–335.

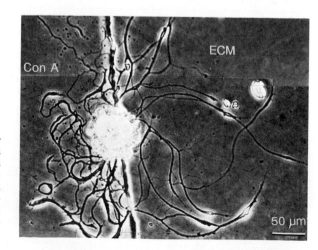

14

EXTRACELLULAR MATRIX MOLECULES DETERMINE THE PATTERN of neurite outgrowth from leech neurons in culture. A single neuron grown in cell culture on a patterned substrate. On the left side of the border the plate was coated with concanavalin A, on the right with extracellular matrix-containing extracts of leech ganglion capsules. (From Grumbacher-Reinert, 1989; courtesy of S. Grumbacher-Reinert.)

cells. Different neurons respond to particular extracellular matrix molecules in individual ways, thus providing an economical scheme by which a few molecules can exert diverse effects.

An additional class of proteins that have pronounced effects on neurite outgrowth from cells in culture are protease inhibitors,[95] such as glia-derived nexin (GDN). GDN is a serine protease inhibitor isolated from rat glioma cell cultures that promotes the outgrowth of neurites from neuroblastoma cells, rat hippocampal neurons, and chick sympathetic neurons in culture.

A mechanism often postulated for active guidance is for the growth cone to navigate along a gradient of some molecule released by the target. Such a CHEMOATTRACTANT MODEL of growth cone navigation, originally proposed by Ramón y Cajal, appears to account for directed axon outgrowth when the distance from the nerve cell body to its peripheral target is very short. For example, Lumsden and Davies have studied the growth of axons from the trigeminal ganglion in the head of the mouse into the adjacent epithelial tissue, a distance less than 1 millimeter.[96] These axons ultimately give rise to the sensory innervation of the whiskers (see Chapter 14). If the developing trigeminal ganglion is placed in cell culture near explants from several peripheral tissues, neurites grow from the ganglion toward their appropriate target but ignore other potential target tissues. Moreover, explants of target epithelium have this effect on axon outgrowth only if they are taken from embryos at the time innervation normally occurs. Thus at the appropriate developmental stage these epithelial cells appear to produce some molecule that attracts the growth cones of trigeminal ganglion cells.

A second example comes from the innervation of epaxial muscles in chick embryos[97] (Figure 15). The developing epaxial or trunk musculature lies immediately above the spinal cord. Shortly after the motor axons destined to innervate epaxial muscles exit from the spinal cord, they extend dorsally, directly toward the dermamyotome from which the epaxial muscles will develop. If the dermamyotome is surgically removed from one segment, the axons emerging from the spinal cord in that segment grow toward the closest intact dermamyotome. If several segments of dermamyotome are removed, some epaxial nerves fail to form altogether, as if the presence of a potential target cannot be sensed over distances greater than 150 μm. This effect is specific; the majority of motor axons exiting the spinal cord grow into the limb and ignore the dermamyotome altogether. An important distinction that needs to be made in such cases is whether the target supplies a diffusible cue that alters growth cone navigation or selectively stabilizes those axons that happen to contact it. In either event, pathfinding by these axons appears to rely on cues emanating from their targets.

Target-derived chemoattractants

[95]Farmer, L., Sommer, J. and Monard, D. 1990. *Dev. Neurosci.* 12: 73–80.
[96]Lumsden, A. G. S. and Davies, A. M. 1986. *Nature* 323: 538–539.
[97]Tosney, K. W. 1991. *BioEssays* 13: 17–23.

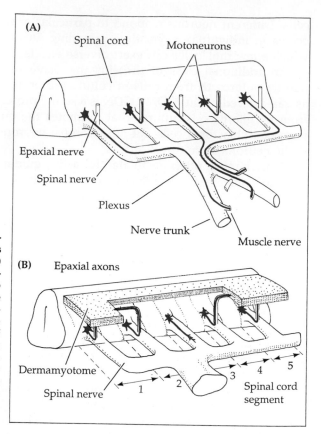

(A) Spinal cord   Motoneurons

Epaxial nerve

Spinal nerve

Plexus

Nerve trunk

Muscle nerve

(B)   Epaxial axons

Dermamyotome

Spinal nerve

1   2   3   4   5   Spinal cord segment

15

TARGET-DERIVED CUES SELECTIVELY AT-TRACT AXONS from appropriate motoneurons during development of the chick spinal cord. (A) Most of the motor axons exiting through the ventral roots enter a plexus at the base of the limb and sort out into individual muscle nerves. Those destined to innervate muscles of the trunk (epaxial muscles) turn dorsally soon after leaving the spinal cord, forming the epaxial nerves. (B) If segments of the overlying dermamyotome, which gives rise to the epaxial musculature, are removed early in development, then axons follow novel trajectories to reach muscles in adjacent segments; if no dermamyotome is present in the adjacent segment, no dorsal outgrowth of epaxial axons occurs (segment 3). (After Tosney, 1991.)

The development of axonal projections from cortical pyramidal neurons to the spinal cord and pons illustrates an additional mechanism by which target-derived cues influence axon outgrowth (Figure 16). Early in development, pyramidal cells in all cortical areas send axons to the spinal cord.[70] As development proceeds, collaterals are induced to sprout from the corticospinal fibers to innervate the nearby pontine nuclei. Some pyramidal cells, such as those in the motor cortex, maintain both axon branches in the adult; others, such as those in the visual cortex, maintain connections with the pons, but the distal portion of the original corticospinal axon degenerates. Experiments in which explants of cortex and pons are juxtaposed in tissue culture demonstrate the existence of diffusible factors released from the pons that induce and direct the growth of collaterals from pyramidal cell axons.[98] These observations suggest an overall strategy by which corticospinal and corticopontine connections are established. Thus, general "directions" are given to all cortical pyramidal cells to project to the spinal cord and to respond to sprouting factors from the pons; then appropriate axon

[98]Heffner, C. D., Lumsden, A. G. S. and O'Leary, D. D. M. 1990. *Science* 247: 217–220.

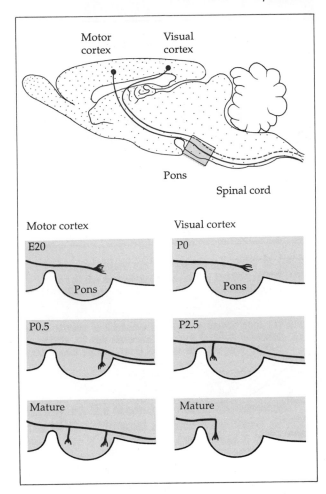

Motor
cortex

Visual
cortex

Pons

Spinal cord

Motor cortex

Visual cortex

E20

Pons

P0

Pons

P0.5

P2.5

Mature

Mature

16

**ADULT PATTERN OF ARBORIZATION is es-
tablished by selective elimination of axon
branches during the development of cortical pyramidal cells in the rat. Pyramidal neurons in
both the visual and the motor cortex initially
send axons into the spinal cord. The release of
factors from cells in the pons induces the corticospinal axons to sprout collateral branches.
Later in development the spinal segments of axons from neurons in the visual cortex degenerate.
(After O'Leary and Terashima, 1988; Heffner,
Lumsden, and O'Leary, 1990.)**

branches are selected from this generic structure and maintained by
trophic interactions (see below).

Position-dependent
navigation

For situations in which nerve cell bodies and their targets are separated by large distances, alternative active guidance mechanisms come
into play by which growth cones follow cues in their environment that
are not derived from their ultimate target. An example comes from
experiments on the developing innervation of limb musculature in chick
embryos. Detailed studies by Landmesser and her colleagues demonstrated that each muscle in the limb is innervated by a pool of moto-
neurons that extend over a well-defined set of spinal cord segments[99–101] (Figure 17). Axons from these motoneurons grow out from the spinal
cord through the adjacent segmental ventral roots, which anastomose

[99]Lance-Jones, C. and Landmesser. L. 1980. *J. Physiol.* 302: 581–602.
[100]Lance-Jones, C. and Landmesser, L. 1981. *Proc. R. Soc. Lond. B* 214: 1–18.
[101]Lance-Jones, C. and Landmesser, L. 1981. *Proc. R. Soc. Lond. B* 214: 19–52.

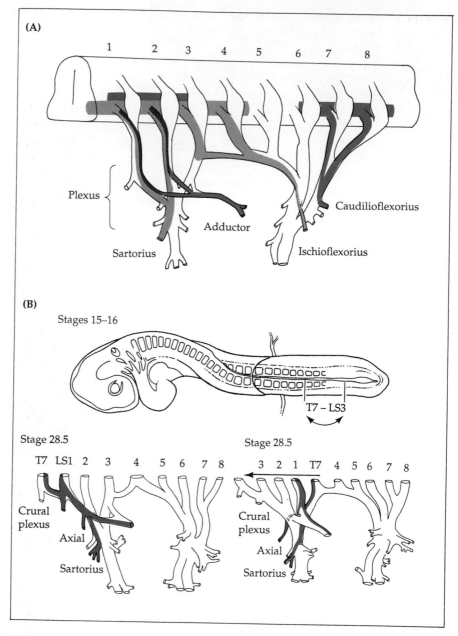

17 **LOCAL CUES GUIDE AXONS OF MOTONEURONS** toward their peripheral targets. **(A)** Motoneurons that will innervate each muscle are grouped together in the spinal cord and send their axons out via one or more adjacent spinal nerves. Axons from overlapping populations of motoneurons sort themselves out in the plexus at the base of the limb and enter the appropriate peripheral nerves. **(B)** If a region of spinal cord is reversed rostrocaudally early in development, axons from displaced motoneurons will take novel paths through the plexus to reach their appropriate peripheral nerve. Thus motoneurons specified to innervate a particular muscle follow cues in the plexus to reach the appropriate region in the periphery. (After Lance-Jones and Landmesser, 1980.)

to form a plexus at the base of the limb from which nerves to the individual muscles emerge. To test how motor axons reach their targets, sections of the spinal cord of 4-day-old embryos were cut out, rotated head-to-tail 180°, and replaced in the embryo. Axons from motor neurons followed novel pathways to reach their original target muscles; they did not innervate targets appropriate to their new position along the spinal cord. Thus, motoneurons are specified to innervate particular muscles very early in development. However, the initial extension of axons, formation of the plexus, and formation of the muscle nerves occur before myoblasts in the developing limb have fused to form muscle fibers. This finding indicates that the ability of axons to grow to the appropriate region in the limb cannot depend on the presence of their targets. This was confirmed by removing the somites, the source of the myoblasts that give rise to limb musculature, early in development.[102] Motor axons extended from the spinal cord, formed a limb plexus, and grew into appropriate muscle nerves in the absence of muscle. Thus, the factors that guide motor axons to their correct destinations in the limb plexus are not supplied by the muscles that the axons ultimately innervate.

One strategy for axonal guidance that does not rely on factors produced by a neuron's ultimate synaptic partner is to mark the appropriate pathway with intermediate targets. For example, as described above, early in development the axons of commissural interneurons in the rat spinal cord grow ventrally through the neuroepithelium, then make an abrupt turn and grow longitudinally along the spinal cord toward their synaptic targets (Figure 12). The ventral growth of these axons is directed by a chemoattractant that is released by cells of the FLOOR PLATE, a specialized set of midline neural epithelial cells.[103,104] Another example is provided by growth cones arising from sensory cells in the limbs of developing grasshoppers, which make abrupt changes in the direction of growth as they extend toward the central nervous system[105] (Figure 18). These turns occur when the growth cones contact so-called GUIDE-POST CELLS, suggesting that interactions with guidepost cells, which are often immature neurons, are responsible for redirecting the growth cones. That this is so can be demonstrated by removing the guidepost cells by laser ablation before the growth cone arrives, in which case the appropriate change in trajectory is not made. Similar kinds of experiments in both the central and the peripheral nervous systems of a variety of invertebrates and vertebrates have demonstrated the importance of guidepost cells in establishing axonal projections.[106]

In other cases, neurons have been found to make transient synaptic contacts during development that appear to be an essential intermediate

Guidepost neurons and subplate cells

[102]Phelan, K. A. and Hollyday, M. 1990. *J. Neurosci.* 10: 2699–2716.
[103]Tessier-Lavigne, M. et al. 1988. *Nature* 336: 775–778.
[104]Tessier-Lavigne, M. and Placzek, M. 1991. *Trends Neurosci.* 14: 303–310.
[105]Bentley, D. and Caudy, M. 1983. *Cold Spring Harbor Symp. Quant. Biol.* 48: 573–585.
[106]Kuwada, J. Y. 1986. *Science* 233: 740–746.

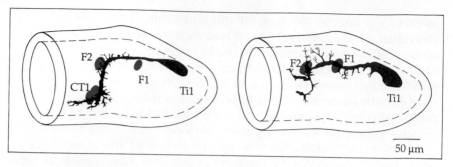

50 µm

18   GROWTH CONES OF PERIPHERAL NEURONS rely on guidepost cells to navigate through the limb of the grasshopper. In normal embryos, the axon of the Ti1 neuron encounters a series of guidepost cells on its route to the central nervous system: F1, F2, and two CT1 cells. If the CT1 cells are killed early in development, the Ti1 neuron forms several axonal branches at the site of cell F2, with growth cones extending in abnormal directions. (After Bentley and Caudy, 1983.)

stage in the innervation of their eventual target. In the mammalian visual system, for example, axons from neurons in the lateral geniculate nucleus reach the developing cortical plate before their ultimate synaptic targets, the pyramidal cells of layer 4, have been born.[107] The geniculate axons form synaptic connections with SUBPLATE NEURONS, cells that are produced very early during embryogenesis, lie beneath the developing cortical plate, and are destined to disappear shortly after birth.[33] After a few weeks, when the layer 4 pyramidal cells have reached their position in the cortex, geniculate axons abandon their connections with the subplate neurons and invade the cortex to establish the adult pattern of innervation. If the subplate neurons are eliminated early in development by local application of neurotoxins, lateral geniculate axons grow past the developing visual cortex and fail to form synaptic contacts with their targets.[108]

TARGET INNERVATION

Ever since the pioneering studies of Stone and Sperry, the retinotectal systems of frogs, goldfish, and chicks have been used to investigate how specific synaptic connections are established. The precision with which retinal ganglion cells innervate their targets in the optic tectum during development has been studied by Fujisawa, Harris, Holt, Stuermer, Bonhoeffer, O'Leary, Fraser, and others. Although results vary somewhat from species to species and between different areas of the retina, a consistent finding is that the initial pattern of the projection is less precise than that found in the adult. In frogs and fish the tectum is small and immature at the time retinotectal axons arrive. Although

[107]Luskin, M. B. and Shatz, C. J. 1985. *J. Neurosci.* 5: 1062–1075.
[108]Ghosh, A. et al. 1990. *Nature* 347: 179–181.

axons from some regions of the retina are topographically segregated throughout retinotectal development,[109] the terminal arbors of axons from cells in other regions of the retina often overlap considerably.[63,64,110] As the tectum grows, the arbors sort themselves out in a retinotopic fashion. The relationships between retinal axons and tectal neurons continue to shift throughout adulthood, as new neurons are added to the retina in adult amphibians and to both the retina and the tectum in adult goldfish.[111] In chicks and rats the tectum is much larger at the time the first retinal axons arrive.[69,112,113] Nevertheless, the initial projections of retinal axons are more diffuse than in the adult; increased map precision is brought about by course corrections and the elimination of inappropriate side branches and aberrant arbors.

What signals enable the precise pattern of innervation to be achieved? In an elegant series of experiments, Bonhoeffer and his colleagues have identified a glycoprotein that provides positional information in the chick tectum.[114] During development, axons from temporally located retinal ganglion cells innervate neurons in anterior tectum; those from nasal ganglion cells innervate posterior tectum (Figure 19). This process appears to be mediated by repulsive interactions that discourage growth cones of temporal axons from penetrating into inappropriate regions in the posterior tectum. When ganglion cells from the temporal retina are placed in culture adjacent to surfaces coated with membranes purified from either the anterior or the posterior tectum, their axons grow on the membranes from their target and avoid membranes from posterior tectum.[115,116] Retinal axons elongate rapidly on either substrate when not presented with a choice. Experiments with denatured proteins indicate that axons from temporal ganglion cells are normally repelled by a component of the membranes from the posterior tectum. The factor that repels the growth cones was found to be a 33-kD glycoprotein that, when incorporated into lipid vesicles and added to the medium bathing a growing axon, caused the growth cone to detach from the substrate and retract.[117,118] This protein is expressed in the tectum during the time retinotectal connections are being formed, and its concentration increases in a graded manner from anterior to posterior in the tectum. Glycoproteins that cause collapse of growth cones have been isolated from other areas of the developing nervous system, where they appear to play a similar role in growth cone guid-

[109]Stuermer, C. A. O. 1988. *J. Neurosci.* 8: 4513–4530.

[110]Fujisawa, H. 1987. *J. Comp. Neurol.* 260: 127–139.

[111]Easter, S. E., Jr. and Stuermer, C. A. O. 1984. *J. Neurosci.* 4: 1052–1063.

[112]Simon, D. K. and O'Leary, D. D. M. 1989. *Dev. Biol.* 137: 125–134.

[113]Thanos, S. and Bonhoeffer, F. 1987. *J. Comp. Neurol.* 261: 155–164.

[114]Walter, J., Allsopp, T. E. and Bonhoeffer, F. 1990. *Trends Neurosci.* 13: 447–452.

[115]Walter, J. et al. 1987. *Development* 101: 685–696.

[116]Walter, J. Henke-Fahle, S. and Bonhoeffer, F. 1987. *Development* 101: 909–913.

[117]Cox, E. C., Muller, B. and Bonhoeffer, F. 1990. *Neuron* 4: 31–47.

[118]Stahl, B. et al. 1990. *Neuron* 5: 735–743.

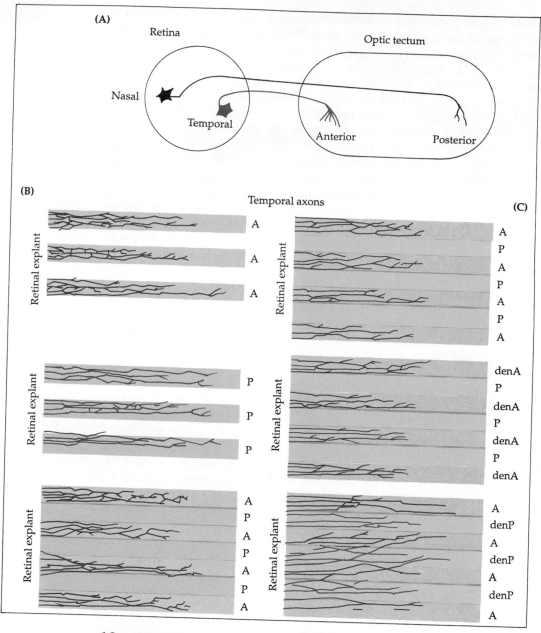

19 PATTERN OF INNERVATION OF THE OPTIC TECTUM by axons from retinal ganglion cells is mediated by repulsive interactions between growth cones and inappropriate targets. (A) Ganglion cells in the nasal retina innervate neurons in the posterior tectum, ganglion cells in the temporal retina innervate neurons in the anterior tectum. (B) In cell cultures, axons from neurons in the temporal retina grow equally well on lanes coated with membranes isolated from anterior (A) or posterior (P) tectum. However, when given a choice, they prefer to grow on anterior membranes. (C) If given a choice between denatured anterior (denA) membranes and intact posterior membranes, growth cones from temporal retina still prefer anterior; when presented with native anterior membranes and denatured posterior (denP) membranes they display no preference, indicating that normally they are repelled by components of the posterior membranes. (After Walter et al., 1987, 1990.)

ance.[119,120] The mechanisms by which a protein that causes detachment in one assay can sustain or redirect growth in other assays are not yet clear.

Certainly a single anterior–posterior gradient is not sufficient to enable retinal axons to reach their appropriate positions in the tectum. There is evidence that other molecules exhibit a topographically graded distribution in the retinotectal system,[82,121–123] although the functional roles of these gradients have yet to be defined. Activity-dependent mechanisms also play a role in establishing precise patterns of retinotectal innervation[124] (Chapter 18).

SYNAPSE FORMATION

On reaching its destination, a growth cone must establish synaptic contact with the correct cell, and often with a specific location on that cell. The formation of synaptic connections has been studied extensively at the vertebrate skeletal neuromuscular junction, particularly with regard to the distribution and properties of ACh receptors and acetylcholinesterase. Recall from Chapter 9 that ACh receptors are highly concentrated in the plasma membrane of the muscle fiber directly beneath the nerve terminal, along the crests of the junctional folds. Acetylcholinesterase, on the other hand, is associated with the synaptic portion of the sheath of extracellular matrix material, known as BASAL LAMINA, that surrounds each muscle fiber. Many other cytoplasmic, membrane, and extracellular matrix components, whose functions have yet to be determined, have been shown to be concentrated at adult neuromuscular junctions.

How does the complex synaptic apparatus arise during development? Studies by Fischbach, Cohen, Changeux, Salpeter, Steinbach, Poo, Kidokoro, and others[125] have shown that as myoblasts fuse to form myotubes they begin to synthesize ACh receptors. ACh receptors are distributed diffusely over the surface of uninnervated myotubes at a density of a few hundred receptors per μm^2. As the growth cone of a motor axon approaches a myotube, depolarizing potentials due to the release of ACh from the growth cone can be recorded in the myotube. Within minutes of contact, the rate of spontaneous release of quanta of ACh increases, as does the size of the synaptic potential evoked by stimulating the axon. Thus, within minutes a functional synaptic connection has formed. These initial contacts are very simple, bearing none

Initial synaptic contacts

[119]Davies, J. A. et al. 1990. *Neuron* 2: 11–20.

[120]Raper, J. A. and Kapfhammer, J. P. 1990. *Neuron* 2: 21–29.

[121]Trisler, D. and Collins, F. 1987. *Science* 237: 1208–1209.

[122]Constantine-Paton, M. et al. 1986. *Nature* 324: 459–462.

[123]Rabacchi, S. A., Neve, R. L. and Drager, U. C. 1990. *Development* 109: 521–531.

[124]Constantine-Paton, M., Cline, H. T. and Debski, E. 1990. *Annu. Rev. Neurosci.* 13: 129–154.

[125]Salpeter, M. M. (ed.). 1987. *The Vertebrate Neuromuscular Junction.* Alan R. Liss, New York.

of the structural specializations that characterize the adult neuromuscular junction.

Accumulation
of ACh receptors

The first synaptic specialization to form is the accumulation of ACh receptors beneath the axon terminal.[126] This begins within hours of the initial contact, and within a day or two the density of receptors beneath the terminal is several thousand per μm^2. At about the same time, acetylcholinesterase begins to accumulate at synaptic sites, and wisps of basal lamina are visible within the synaptic cleft. Further differentiation of the neuromuscular junction occurs gradually over the course of the next several weeks of development. In many mammalian species the γ subunit of the ACh receptor is replaced by an ε subunit, converting the embryonic form of the receptor to the adult form (see Chapter 2). The distribution of the receptor also changes; the concentration beneath the axon terminal continues to increase to adult levels of approximately 10^4 receptors per μm^2, while the density of receptors in nonsynaptic portions of the muscle fiber decreases to less than 10 receptors per μm^2. The metabolic stability of ACh receptors also changes. Before innervation, receptors in the membrane have a half-life of approximately 1 day; ACh receptors in innervated fibers are remarkably stable, being degraded with a half-life of 10 days. Changes also occur within the axon terminal, leading over the course of several weeks to the formation of active zones.

What triggers the accumulation of ACh receptors at synaptic sites? To begin to address this question nerve and muscle cells have been grown in cell culture where their interactions can be monitored continuously (Figure 20). Myotubes that form in culture in the absence of neurons develop clusters of ACh receptors on their surface that resemble those seen at early synaptic contacts. Do such sites represent targets for innervation by growth cones? Detailed morphological and physiological experiments demonstrate that growth cones contact muscle cells at random positions on their surface, ignoring preexisting ACh receptor clusters and rapidly inducing formation of new receptor aggregates.[127,128] Thus, axon terminals must release a signal that induces ACh receptor accumulation in the muscle cell. The signal is specific to cholinergic neurons; when noncholinergic neurons grow over muscle cells, they do not induce changes in ACh receptor distribution. The signal does not appear to be ACh itself, however, as ACh receptors accumulate beneath axon terminals in cultures grown in the presence of drugs that block the interaction of ACh with its receptor. Experiments originally aimed at identifying signals controlling regeneration of the neuromuscular junction (see Chapter 12) have identified a protein, called agrin, that is released by motor nerve terminals and triggers the accumulation of ACh receptors, cholinesterase, and other components of the postsynaptic apparatus at synaptic sites.[129] Other factors—including calci-

[126]Scheutze, S. M. and Role, L. W. 1987. *Annu. Rev. Neurosci.* 10: 403–457.
[127]Frank, E. and Fischbach, G. D. 1979. *J. Cell Biol.* 83: 143–158.
[128]Anderson, M. J. and Cohen, M. W. 1977. *J. Physiol.* 268: 757–773.
[129]McMahan, U. J. and Wallace, B. G. 1989. *Dev. Neurosci.* 11: 227–247.

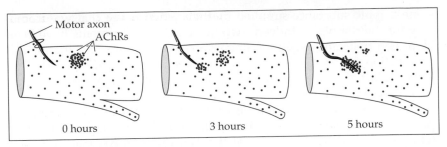

20 **MOTOR AXON TERMINALS INDUCE ACCUMULATION OF ACh RECEP-TORS** at synapses developing in culture. The density of ACh receptors was assayed by measuring the sensitivity of the myotube to focal applications of acetylcholine. AChRs accumulated beneath the terminal within 3 hours of the time the growth cone made contact with the myotube. Growth cones did not appear to be attracted to preexisting aggregates of ACh receptors. (After Frank and Fischbach, 1979.)

tonin gene-related peptide (CGRP), ACh receptor-inducing activity (ARIA), and ascorbic acid—that may regulate the synthesis of ACh receptors have been identified.

These and many other studies indicate that the formation of a synapse such as the neuromuscular junction is not a single all-or-none event. Although functional synaptic transmission is established very rapidly, the differentiation of pre- and postsynaptic specializations is a protracted process, occurring over several weeks of development, and relies on the exchange of a variety of molecular signals between the nerve terminal and muscle fiber.

COMPETITIVE INTERACTIONS DURING DEVELOPMENT

A conspicuous feature of the development of the nervous system is that many neurons are born to die. In invertebrates, neuronal cell death accompanies the sweeping changes that occur during metamorphosis and is under hormonal regulation.[130] In the developing vertebrate nervous system, however, neuronal death occurs in the absence of such gross mophological changes.[131] Experiments by Hamburger and Levi-Montalcini first documented neuronal death in vertebrate embryos and demonstrated that its extent can be influenced by manipulating the size of the target tissue.[132] They and their colleagues showed, for example, that in the developing limb of chick embryos, at about the same time as synaptic connections are first being formed on myofibers, 40 to 70 percent of the motoneurons that had sent axons into the limb died. Implantation of a supernumerary limb reduced the fraction of motoneurons that died, while removal of the limb bud exacerbated the death of motoneurons.[133] It appeared that motoneurons were competing for

Neuronal cell death

[130]Truman, J. W. 1984. *Annu. Rev. Neurosci.* 7: 171–188.
[131]Oppenheim, R. W. 1991. *Annu. Rev. Neurosci.* 14: 453–501.
[132]Hamburger, V. 1939. *Physiol. Zool.* 12: 268–284.
[133]Hollyday, M. and Hamburger, V. 1976. *J. Comp. Neurol.* 170: 311–320.

some trophic substance supplied by their target tissue that was necessary for their survival. Indeed, two muscle-derived proteins have been identified that can sustain motoneurons in cell culture, cholinergic differentiation factor (CDF) and insulin-like growth factor 1 (IGF1).[134,135] When injected into embryos, these proteins rescue motoneurons that otherwise would have died.

An overproduction of neurons and subsequent cell death during the period of synapse formation is a common finding throughout the vertebrate nervous system. Some of the neurons that die may not have made any synapses, or may have innervated an inappropriate target. In such cases cell death contributes to the specificity of innervation. Most of the cells that die, however, appear to have reached and innervated their correct targets. Thus, cell death is primarily a mechanism by which the size of the neuronal input is matched to that of its peripheral target.

Polyneuronal innervation

A more intricate control of innervation appears to come into play once the population of neurons innervating a particular target has been restricted through cell death. In this process the surviving neurons compete with one another for synaptic territory within the target. A clear example of this competition occurs in developing skeletal muscle.[136] In the adult, each motoneuron innervates a group of up to 300 muscle fibers, forming a motor unit (see Chapter 15), but each muscle fiber is innervated by only one axon. In developing muscle, however, following the period of motoneuronal cell death, the surviving motoneurons branch extensively such that each muscle fiber becomes innervated by axons from several motoneurons. As development progresses, axon branches are eliminated until the adult pattern is formed.[137,138] This elimination does not involve cell death, only a reduction in the number of muscle fibers innervated by each motoneuron until motor units of the normal adult size are formed. This removal of polyneuronal innervation appears to involve competition between axons of different motoneurons for synaptic space on the muscle cell. The clearest example of this comes from studies of a small muscle in the toe of the rat.[139] When all but one of the motor axons innervating this muscle were cut early in development, the remaining axon spread to innervate many fibers within the muscle. During the period when motor units normally are reduced to their adult size through elimination of polyneuronal inputs, however, no synapses were lost. In the absence of competition the surviving motoneuron maintained contacts with every myofiber it

[134]Caroni, P. and Grandes, P. 1990. *J. Cell Biol.* 110: 1307–1317.

[135]McManaman, J. L., Haverkamp, L. J. and Oppenheim, R. W. 1991. *Adv. Neurol.* 56: 81–88.

[136]Betz, W. J. 1987. *In* M. M. Salpeter (ed.). *The Vertebrate Neuromuscular Junction.* Alan R. Liss, New York. pp. 117–162.

[137]Redfern, P. A. 1970. *J. Physiol.* 209: 701–709.

[138]Brown, M. C., Jansen, J. K. S. and Van Essen, D. 1976. *J. Physiol.* 261: 387–422.

[139]Betz, W. J., Caldwell, J. H. and Ribchester, R. R. 1980. *J. Physiol.* 303: 265–279.

had innervated. Similar retraction of multiple inputs has been shown to occur in the autonomic ganglia of neonatal rats and guinea pigs.[140] Each ganglion cell is initially supplied by multiple inputs—about five —but by about 5 weeks after birth, only one usually remains. Experiments by Purves, Lichtman, and their colleagues have provided vivid images of this process by visualizing living nerve terminals in animals with vital dyes and observing changes in synaptic structure during the course of synapse elimination.[141,142]

Physiological experiments indicate that activity plays a role in synapse elimination, influencing both the rate and the outcome of the competition between axon terminals.[136] Increased electrical activity by electrical stimulation of the nerve via implanted wire electrodes increases the rate of synapse elimination within the muscle.[143,144] If activity is reduced, by applying tetrodotoxin in a cuff around the nerve to block action potentials or by inhibiting synaptic transmission, synapse elimination is slowed.[145,146] Muscles that receive input from axons that run in two different nerves allow the more interesting experiment of blocking impulses in one nerve and not the other.[147,148] In such cases inactive axons are clearly at a competitive disadvantage; axons in the blocked nerve innervate smaller than normal motor units, while those in the active nerve innervate more fibers than usual. Domination by the active nerve is not complete, however, suggesting that other factors are involved. The molecular basis of this competition and the mechanism by which activity contributes to synapse elimination are not known.

Similar competition for synaptic targets occurs during development of CNS pathways.[149] One example is the formation of the ocular dominance columns in visual cortex (Chapter 18), where axons from the lateral geniculate nucleus conveying information from the two eyes initially overlap extensively in layer 4 of the cortex and then sort out into left eye–right eye columns. Here the pattern of activity in the terminals from the two eyes plays a decisive role in determining the outcome of the competition.

DISCOVERY OF GROWTH FACTORS

A pivotal series of experiments that opened the door to investigation of the molecular basis of competitive interactions among neurons during development was provided by the work of Levi-Montalcini, Cohen, and

Identification of nerve growth factor

[140]Purves, D. and Lichtman, J. W. 1983. *Annu. Rev. Physiol.* 45: 553–565.
[141]Balice-Gordon, R. J. and Lichtman, J. W. 1990. *J. Neurosci.* 10: 894–908.
[142]Balice-Gordon, R. J. and Lichtman, J. W. 1990. *Soc. Neurosci. Abstr.* 16: 456.
[143]O'Brien, R. A. D., Ostberg, A. J. C. and Vrbova, G. 1978. *J. Physiol.* 282: 571–582.
[144]Thompson, W. 1983. *Nature* 302: 614–616.
[145]Thompson, W., Kuffler, D. P. and Jansen, J. K. S. 1979. *Neuroscience* 4: 271–281.
[146]Brown, M. C., Hopkins, W. G. and Keynes, R. J. 1982. *J. Physiol.* 329: 439–450.
[147]Ribchester, R. R. and Taxt, T. 1983. *J. Physiol.* 344: 89–11.
[148]Ribchester, R. R. and Taxt, T. 1984. *J. Physiol.* 347: 497–511.
[149]Shatz, C. J. 1990. *Neuron* 5: 745–756.

Rita Levi-Montalcini in 1985.

their colleagues,[150] who found a factor that supported the growth and survival of sympathetic and sensory neurons. These studies provided the framework for approaching many of the problems raised in this chapter; the course of the investigations also illustrates the manner in which research can progress in the hands of extraordinarily perceptive investigators. The search for the growth factor is a remarkable sequence of coincidences, false but profitable leads, and extraordinary and apparently fortunate choices—all leading to an important development in the area of the study of nerve growth.

Transplanting an extra leg onto the back of a tadpole, a lizard, or a newt not only rescues motoneurons that would otherwise have died, as described above, but also causes the outgrowth of nerve fibers from the central nervous system.[132] To follow up on the idea that there must be substances in transplanted limbs capable of attracting nerve fibers, it was reasonable to test the effect of rapidly growing tissues on the growth of neurons. The initial experiments were made by implanting onto chick embryos a connective tissue tumor (sarcoma) obtained from mice. On the side where the sarcoma had been implanted, there was a profuse outgrowth of sensory and sympathetic nerve fibers from the embryo into the tumor. To show that the effect was caused by a humoral factor, sarcomas were grafted onto the chorioallantoic membrane, a tissue that surrounds the embryo. There was no direct contact between the embryo and the tumor, but once again the dorsal root ganglia and sympathetic neurons on the side of the implant grew profusely.[151] Next, it was shown that sarcoma cells produce a similar dramatic effect on tissue-cultured chick ganglia, providing a simple and reliable bioassay (Figure 21).

The active factor in the sarcoma initially appeared to be a nucleoprotein. To see if nucleic acids were essential components of the growth-promoting factor, tumors were incubated with a snake venom whose action would hydrolyze the nucleic acids and thereby render the tumor fraction inactive. With venom present, however, the growth, far from being inactivated, was further increased. In fact, the control experiment of adding snake venom without the sarcoma extract revealed, surprisingly, that the venom itself was a far richer source of growth factor than was the sarcoma.[152] This, in turn, gave rise to the speculation that since venom is secreted by the salivary gland, salivary glands from other animals might also contain a similar factor.

The animal selected was the mouse.[152,153] Extracts of salivary glands of adult male mice were found to be potent in causing the growth of neurites from sensory ganglion explanted into culture (Figure 21). It was fortunate that adult male mice were chosen because the salivary

[150]Levi-Montalcini, R. 1982. *Annu. Rev. Neurosci.* 5: 341–362.

[151]Levi-Montalcini, R. and Angeletti, P. U. 1968. *Physiol. Rev.* 48: 534–569 [See this article for references to earlier work.]

[152]Cohen, S. 1959. *J. Biol. Chem.* 234:1129–1137.

[153]Cohen, S. 1960. *Proc. Natl. Acad. Sci. USA* 46:302–311.

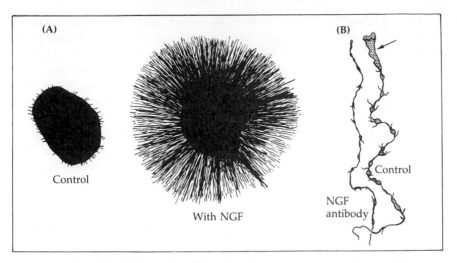

21 **EFFECT OF NERVE GROWTH FACTOR** on neurons in sensory ganglia from a 7-day-old chick embryo. The ganglia were kept in culture for 24 hours. (A) Ganglion in control medium is on the left. To the right, a ganglion maintained in medium supplemented with nerve growth factor shows prolific growth. (B) Thoracic sympathetic chain of ganglia from a control animal (mouse) is on the right. Arrow points to the stellate ganglion. To the left is a ganglion chain, much smaller in size, from a mouse injected 5 days after birth with antiserum to nerve growth factor. (A after Levi-Montalcini, 1964; B after Levi-Montalcini and Cohen, 1960.)

glands of female or immature mice contain far less growth factor, as is also true for salivary glands of other animals that have since been tried. The substance extracted from snake venom and salivary glands of mice was called NERVE GROWTH FACTOR (NGF). The functional role of NGF activity in salivary glands in the animal is still not clear, especially since removal of these glands from young animals has only minor effects on nerve growth.

An early observation by Levi-Montalcini was that neurites tended to grow toward regions containing high concentrations of nerve growth factor. A dramatic example was provided by experiments in which NGF was injected into the neonatal rat brain.[154] Axons arising from sympathetic ganglia entered the spinal cord and ascended to the site of injection in the midbrain. Studies on sensory neurons in culture showed that when NGF is slowly released from a pipette close to one side of a growth cone, the growth cone will bend toward the region of higher concentration.[155] These observations lend support to the hypothesis that NGF might act as a chemoattractant to guide sensory and sympathetic axons to their targets during normal development. However, the amount of NGF synthesized by targets in developing embryos is so low

The influence of nerve growth factor on the growth of neurites

[154]Menesini-Chen, M. G., Chen, J. S. and Levi-Montalcini, R. 1978. *Arch. Natl. Biol.* 116: 53–84.
[155]Gundersen, R. W. and Barrett, J. N. 1980. *J. Cell Biol.* 87: 546–554.

that for a long time it was difficult to detect, and when careful experiments by Davies demonstrated that the target tissue of sensory axons did in fact synthesize NGF, it was found that the synthesis did not begin until after axons had arrived.[156] Moreover, before sensory axons reached their target tissue they did not express receptors for NGF. Thus growing sensory axons do not appear to be guided to their appropriate destination by NGF secreted by the target tissue.

Nerve growth factor may play a role, however, in regulating the distribution of axon processes within the target. An example of the local control NGF can exert on individual branches of a neuron came from experiments in which sympathetic neurons were grown in three-compartment culture chambers[157] (Figure 22). Initially the medium in all three compartments contained NGF. Cells were added to the central compartment and extended axons into each of two lateral compartments. NGF was then removed from one of the lateral compartments. The axons projecting into the compartment containing NGF survived, but the axons that had extended into the compartment from which NGF was removed degenerated. Thus, within the target tissue, local sources of NGF may shape the terminal arbors of sensory and sympathetic axons by attracting and sustaining some branches while others degenerate.

Trophic effects of NGF during embryonic development

Nerve growth factor was found not only to enhance the outgrowth of neurites from sensory and sympathetic neurons, but to be essential for their survival as well. When antiserum to NGF was injected into newborn mice, their sympathetic nervous system failed to develop[158] (Figure 21). The parasympathetic nervous system was not affected, and the dorsal root ganglia were only slightly smaller than normal. The animals lived normally but responded poorly to stress conditions. In subsequent experiments it was shown that exposing fetuses to NGF antibodies by the clever tactic of immunizing the mother resulted in the failure of dorsal root ganglion cells in the fetus to survive.[159] In adults, antibodies to NGF were much less effective on either cell population. Thus each cell type displayed a critical period during development when its survival depended on a supply of NGF.

Uptake and retrograde transport of NGF

Although the effects of NGF on the direction of growth and survival of neurites might be mediated by mechanisms confined to neurites and their growth cones, the dependence on NGF for cell survival suggests an action on the cell soma. Studies made in adult animals with radiolabeled NGF have shown that it is taken up into nerve terminals and actively transported back to the soma, where it regulates (among other things) the synthesis of norepinephrine by inducing two enzymes required for its synthesis: tyrosine hydroxylase and dopamine β-hydrox-

[156]Davies, A. M. and Lumsden, A. 1990. *Annu. Rev. Neurosci.* 13: 61–73.
[157]Campenot, R. B. 1982. *Dev. Biol.* 93: 13–21.
[158]Levi-Montalcini, R. and Cohen, S. 1960. *Ann. N.Y. Acad. Sci.* 85: 324–341.
[159]Dolkart-Gorin, P. and Johnson, E. M. 1979. *Proc. Natl. Acad. Sci. USA* 76: 5382–5386.

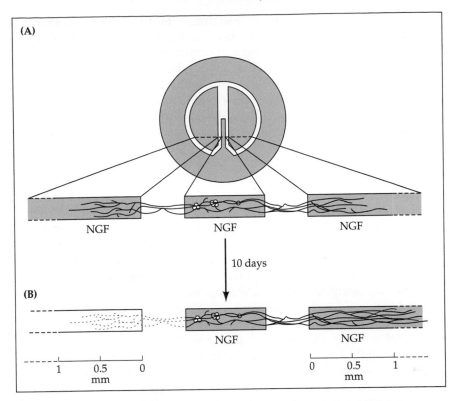

(A)

NGF   NGF   NGF

10 days

(B)

NGF   NGF

1   0.5   0   0   0.5   1
    mm   mm

22 NGF CAN REGULATE THE SURVIVAL OF AXON BRANCHES from sympathetic ganglion cells grown in cell culture. (A) Neurons dissociated from neonatal sympathetic ganglia plated in the central compartment send neurites under a Teflon divider and into the adjacent compartments, both of which contain nerve growth factor. (B) Removal of NGF from the left compartment causes the neurites entering it to degenerate, while those in the compartment containing NGF remain. (After Campenot, 1982.)

ylase.[160,161] If NGF transport in adult neurons is impaired, the levels of these enzymes fall. When embryonic sympathetic neurons are grown in three-compartment culture chambers, as described above, the central chamber in which the neurons are placed must contain NGF for the cells to survive.[162] However, after neurites have reached the side chamber, NGF can be removed from the central chamber and the cells will remain alive. This suggests that the trophic effects of NGF on developing neurons are also mediated by retrograde transport of NGF from nerve terminals to the cell soma.

Sensory neurons not only innervate targets in the periphery but make connections in the central nervous system as well. It is natural to

The molecular biology of growth factors

[160]Hendry, I. A. et al. 1974. *Brain Res.* 68: 103–121.
[161]Black, I. B. 1978. *Annu. Rev. Neurosci.* 1: 183–214.
[162]Campenot, R. B. 1977. *Proc. Natl. Acad. Sci. USA* 74: 4516–4519.

ask whether factors similar to NGF are important in establishing central connections. Indeed, a protein has been identified in the central nervous system, called BRAIN-DERIVED NEUROTROPHIC FACTOR (BDNF), that promotes the survival of dorsal root ganglion neurons in culture and rescues them in vivo if administered to embryos during the period of natural neuronal death.[163] The majority of dorsal root ganglion sensory neurons respond to both BDNF and NGF at early stages of their development; as development proceeds, BDNF- and NGF-dependent neurons become distinct. Other sensory neurons are not supported by either NGF or BDNF, suggesting the existence of additional neurotrophic factors, specific for the structures innervated by these neurons (see below).

These observations set the stage for a molecular analysis of the mechanism of action of nerve growth factor by a number of groups, including those of Levi-Montalcini, Shooter, Thoenen, and Barde.[164–166] Their studies explored such questions as the fraction of the protein that is most effective in inducing growth, the receptors on the membrane that interact with NGF, and the metabolic events that subsequently occur. NGF was first to be purified, sequenced, and cloned. In salivary glands, NGF is present as a complex made up of three types of subunits, α, β, and γ. The β subunit is responsible for promoting nerve growth and survival;[167–169] it consists of two identical peptide chains each containing 118 amino acids and 3 disulfide bridges. Purification and characterization of BDNF revealed that it has a high degree of homology to NGF, indicating that NGF and BDNF are members of a family of growth factors, given the name NEUROTROPHINS. Molecular genetic experiments have in fact identified two additional neurotrophins, NT-3 and NT-4. NT-3 can support survival and neurite outgrowth of several types of neurons. NT-4 is specifically localized to immature egg cells in the ovary; its role is unknown.

The neurotrophins interact with two types of receptors on the surface of their target neurons.[170,171] All neurotrophins bind with relatively equal and low affinity ($K_d = 10^{-9}$ M) to a membrane receptor referred to as the low-affinity–fast NGF receptor, or p75^{NGFR}, which is found on both neurons and non-neuronal cells. There are also high-affinity ($K_d = 10^{-11}$ M) receptors for the neurotrophins. Results of bioassays indicate that the effects of neurotrophins on cell survival and neurite outgrowth

[163]Barde, Y.-A. 1989. *Neuron* 2: 1525–1534.

[164]Rodriguez-Tebar, A., Dechant, G. and Barde, Y.-A. 1991. *Philos. Trans. R. Soc. Lond. B.* 331: 255–258.

[165]Welcher, A. A. et al. 1991. *Philos. Trans. R. Soc. Lond. B.* 331: 295–301.

[166]Thoenen, H. 1991. *Trends Neurosci.* 14: 165–170.

[167]Angeletti, R. H., Hermodson, M. A. and Bradshaw, R. A. 1973. *Biochemistry* 12: 100–115.

[168]Angeletti, R. H., Mercanti, D. and Bradshaw, R. A. 1973. *Biochemistry* 12: 90–100.

[169]Greene, L. A. et al. 1971. *Neurobiology* 1:37–48.

[170]Sutter, A. et al. 1979. *J. Biol. Chem.* 254: 5972–5982.

[171]Ragsdale, C. and Woodgett, J. 1991. *Nature* 350: 660–661.

are mediated by binding to the high-affinity receptors. Although the high-affinity receptor for NGF is normally found only on neurons, it was originally identified in human colon carcinoma cells as part of the product of the TRK oncogene (oncogenes are genes involved in mediating cell transformation). The oncogene encodes a receptor-like protein containing a tyrosine kinase domain fused to tropomyosin sequences; hence it was called "tropomyosin receptor kinase" or *trk*. The counterpart of the *trk* oncogene found in normal cells encodes a 140-kD protein that is referred to as p140^{Prototrk}, or simply trk. The structure of the trk protein, predicted from its deduced amino acid sequence, consists of an extracellular domain containing the neurotrophin-binding site, a short transmembrane segment, and an intracellular domain encoding a tyrosine kinase. There are at least three members of the *trk* family of proto-oncogenes, each acting as the high-affinity receptor for one or more neurotrophins. The protein trkA is the receptor for NGF; trkB appears to be the receptor for BDNF. Among the earliest events detected after binding of neurotrophins to their high-affinity receptors is an increase in tyrosine phosphorylation of proteins within the neuron. The function of the low-affinity receptor, which lacks any intracellular domain, is not known. On some cells it may interact with the high-affinity receptor during the binding of neurotrophins; on others, especially those that lack high-affinity receptors, it may provide a mechanism for restricting diffusion and establishing high local concentrations of neurotrophins.

Of particular interest is the finding of a population of NGF-sensitive cells within the central nervous system.[172] These cholinergic neurons are located in the basal forebrain and innervate several structures, including the hippocampus, a region of the CNS thought to be involved in learning and memory (see Chapter 10). If the axons of these neurons are cut in the adult rat, the cells die. However, if nerve growth factor is infused into the CNS, then these neurons survive axotomy. The number of these cells that stain with markers for cholinergic function declines with age, as does the ability of rats to learn a maze or other spatial memory tasks.[173] If NGF is infused into aged rats, the number of cells that can be stained increases and the rat's performance in spatial memory tasks improves.[174] These observations indicate that survival and growth of neurons within the CNS are likely to rely on the same or similar factors as have been identified for peripheral neurons. At the same time such findings provide a way of thinking, in molecular terms, about defects that might give rise to mental deficits and how they might be ameliorated.

NGF in the central nervous system

The question of neural specificity—how nerve cells find their appropriate targets—is just one facet of a general property of the regulation of cell growth. Other more mundane cells in the body also appear

General considerations of neural specificity

[172]Gage, F. H. et al. 1988. *J. Comp. Neurol.* 269: 147–155.
[173]Fischer, W., Gage, F. H. and Björklund, A. 1989. *Eur. J. Neurosci.* 1: 34–45.
[174]Fischer, W. et al. 1991. *J. Neurosci.* 11: 1889–1906.

to know in what direction to grow and when to stop. The restitution of skin tissue after partial removal, the appropriate closing of a wound, and regrowth of an injured organ such as the liver to its proper size are related phenomena. Just as a denervated muscle seems to attract new nerve fibers, so does a transplanted muscle attract new capillaries that grow into it by splitting off from an adjoining vascular bed. Yet the problem of establishing the vast number of specific connections required for the integrative activity of the central nervous system seems so much more complex!

Perhaps a rather commonplace analogy may be encouraging. Let us assume that we are ignorant about the workings and design of the postal system. A chapter from a book on the nervous system, without its illustrations, is posted in Basel, Switzerland and addressed to a publisher in Sunderland, Massachusetts, where it arrives a few days later. How does it get there? The writer knows only the closest letterbox and is unaware even of the location of the post office in his district. The postal worker who empties the letterbox knows where the post office is; there the clerk who handles the mail may not know where Sunderland is but does know how to direct the package to the airport —and so on, to the right country, city, street, building, and eventually to the correct person. If this were not enough, the illustrations that complement the chapter are posted by separate mail from Denver to the same destination, where they arrive almost simultaneously with the chapter from Basel. All the while other mail is moving through the same letterboxes and post offices in different directions to different final destinations.

The comforting feature of this lengthy analogy is that the problem seems altogether baffling at first sight. Yet one can solve the postal puzzle by following the mail step by step to its destination. This would reveal some of the logic and design of postal organization (albeit without disclosing the identity of the designer). At any one step, only a limited number of instructions are followed and a limited number of mechanisms operate.

Some aspects of neural specificity may not be too different. A retinal ganglion cell sends its axon toward the back of the eye, where it makes a turn to enter the optic nerve together with fibers from other regions of the retina. The optic chiasm presents the next choice point, where the decision to enter the optic tract leading to one lateral geniculate nucleus or the other may be made based on local chemical signals. Within the geniculate, retinal axons may arrange themselves and innervate their targets according to gradients of repulsive molecules. Axons of geniculate neurons likewise follow a fairly simple path to their targets in the cortex, stopping along the way to form transient connections with subplate neurons. Thus the seemingly complex task of forming specific connections between retinal ganglion cells and neurons in the visual cortex can be broken down into a series of relatively simple, independent events.

There remain formidable gaps in our understanding of the development of the nervous system: How is the expression of signal molecules and their receptors regulated? How can growth cones recognize and respond to gradients of chemoattractants? How are pathways of extracellular matrix molecules laid down? How do guidepost cells arrive at their appointed positions? Nevertheless, it seems that the problem of neural specificity can be broken down into analyzable parts, and that the tools at our disposal at present, representing a combination of genetic, tissue culture, molecular, and developmental approaches at the cellular level, are adequate for the task.

SUGGESTED READING

General

Gilbert, S. F. 1991. *Developmental Biology*, 3rd Ed. Sinauer, Sunderland, MA.

Patterson, P. H. and Purves, D. (eds). 1982. *Readings in Developmental Neurobiology*. Cold Spring Harbor Laboratory, Cold Spring Harbor, NY.

Purves, D. and Lichtman, J. W. 1985. *Principles of Neural Development*. Sinauer, Sunderland, MA.

Reviews

Bentley, D. and Caudy, M. 1983. Navigational substrates for peripheral pioneer growth cones: Limb-axis polarity cues, limb segment boundaries, and guidepost neurons. *Cold Spring Harbor Symp. Quant. Biol.* 48: 573–585.

Cowan, W. M., Fawcett, J. W., O'Leary, D. D. M. and Stanfield, B. B. 1984. Regressive events in neurogenesis. *Science* 225: 1258–1265.

DeRobertis, E. M., Oliver, G. and Wright, C. V. E. 1990. Homeobox genes and the vertebrate body plan. *Sci. Am.* 263(7): 46–52.

Dodd, J. and Jessell, T. M. 1988. Axon guidance and the patterning of neuronal projections in vertebrates. *Science* 242: 692–699.

Lander, A. D. 1989. Understanding the molecules of neural cell contacts: Emerging patterns of structure and function. *Trends Neurosci.* 12: 189-195.

Levi-Montalcini, R. 1988. *In Praise of Imperfection*. Basic Books, New York.

McConnell, S. K. 1991. The generation of neuronal diversity in the central nervous system. *Annu. Rev. Neurosci.* 14: 269–300.

Ragsdale, C. and Woodgett, J. 1991. *trking* neurotrophic receptors. *Nature* 350: 660–661.

Rodriguez-Tebar, A., Dechant, G. and Barde, Y.-A. 1991. Neurotrophins: Structural relatedness and receptor interactions. *Philos. Trans. R. Soc. Lond. B* 331: 255–258.

Sanes, J. R. 1989. Analyzing cell lineage with a recombinant retrovirus. *Trends Neurosci.* 12: 21–28.

Smith, S. J. 1988. Neuronal cytomechanics: The actin-based motility of growth cones. *Science* 242: 708–715.

Thoenen, H. 1991. The changing scene of neurotrophic factors. *Trends Neurosci.* 14: 165–170.

Walter, J., Allsopp, T. E. and Bonhoeffer, F. 1990. A common denominator of growth cone guidance and collapse? *Trends Neurosci.* 13: 447–452.

Original Papers

FORMATION OF THE NERVOUS SYSTEM

Bronner-Fraser, M. and Fraser, S. E. 1988. Cell lineage analysis reveals multipotency of some avian neural crest cells. *Nature* 335: 161–164.

Le Douarin, N. M. 1980. The ontogeny of the neural crest in avian embryo chimeras. *Nature* 286: 663–669.

Luskin, M. B., Pearlman, A. L. and Sanes, J. R. 1988. Cell lineage in the cerebral cortex of the mouse studied in vivo and in vitro with a recombinant retrovirus. *Neuron* 1: 635–647.

Patterson, P. H. and Chun, L. L. Y. 1977. The induction of acetylcholine synthesis in primary cultures of dissociated rat sympathetic neurons. I. Effects of conditioned medium. *Dev. Biol.* 56: 263–280.

Turner, D. L. and Cepko, C. L. 1987. A common progenitor for neurons and glia persists in rat retina late in development. *Nature* 328: 131–136.

MECHANISMS OF AXON GUIDANCE

Bixby, J. L., Lilien, J. and Reichardt, L. F. 1988. Identification of the major proteins that promote neuronal process outgrowth on Schwann cells in vitro. *J. Cell Biol.* 107: 353–361.

Ghosh, A., Antonini, A., McConnell, S. K. and Shatz, C. J. 1990. Requirement for subplate neurons in the formation of thalamocortical connections. *Nature* 347: 179–181.

Heffner, C. D., Lumsden, A. G. S. and O'Leary, D. D. M. 1990. Target control of collateral extension and directional axon growth in the mammalian brain. *Science* 247: 217–220.

Lance-Jones, C. and Landmesser, L. 1981. Pathway selection by embryonic chick motoneurons in an experimentally altered environment. *Proc. R. Soc. Lond. B* 214: 19–52.

Lumsden, A. G. S. and Davies, A. M. 1986. Chemotropic effect of specific target epithelium in the developing mammalian nervous system. *Nature* 323: 538–539.

O'Leary, D. D. M. and Terashima, T. 1988. Cortical axons branch to multiple subcortical targets by interstitial axon budding: Implications for target recognition and "waiting periods." *Neuron* 1: 901–910.

Phelan, K. A. and Hollyday, M. 1990. Axon guidance in muscleless chick wings: The role of muscle cells in motoneuronal pathway selection and muscle nerve formation. *J. Neurosci.* 10: 2699–2716.

Tessier-Lavigne, M., Placzek, M., Lumsden, A. G. S., Dodd, J. and Jessell, T. M. 1988. Chemotropic guidance of developing axons in the mammalian central nervous system. *Nature* 336: 775–778.

TARGET INNERVATION

Cox, E. C., Muller, B. and Bonhoeffer, F. 1990. Axonal guidance in the chick visual system: Posterior tectal membranes induce collapse of growth cones from the temporal retina. *Neuron* 4: 31–47.

Nakamura, H. and O'Leary, D.D.M. 1989. Inaccuracies in initial growth and arborization of chick retinotectal axons followed by course corrections and axon remodeling to develop topographic order. *J. Neurosci.* 9: 3776–3795.

O'Rourke, N. A. and Fraser, S. E. 1990. Dynamic changes in optic fiber terminal arbors lead to retinotopic map formation: An in vivo confocal microscopic study. *Neuron* 5: 159–171.

Stahl, B., Muller, B., von Boxberg, Y., Cox, E. C. and Bonhoeffer, F. 1990. Biochemical characterization of a putative axonal guidance molecule of the chick visual system. *Neuron* 5: 735–743.

Walter, J., Kern-Veits, B., Huf, J., Stolze, B. and Bonhoeffer, F. 1987. Recognition of position-specific properties of tectal cell membranes by retinal axons in vitro. *Development* 101: 685–696.

SYNAPSE FORMATION

Anderson, M. J. and Cohen, M. W. 1977. Nerve-induced and spontaneous redistribution of acetylcholine receptors on cultured muscle cells. *J. Physiol.* 268: 757–773.

Falls, D. L., Harris, D. A., Johnson, F. A., Morgan, M. M., Corfas, G. and Fischbach, G. D. 1990. M_r 42,000 ARIA: A protein that may regulate the accumulation of acetylcholine receptors at developing chick neuromuscular junctions. *Cold Spring Harbor Symp. Quant. Biol.* 55: 397–406.

Frank, E. and Fischbach, G. D. 1979. Early events in neuromuscular junction formation in vitro: Induction of acetylcholine receptor clusters in the post-synaptic membrane and morphology of newly formed synapses. *J. Cell Biol.* 83: 143–158.

McMahan, U. J. and Wallace, B. G. 1989. Molecules in basal lamina that direct the formation of synaptic specializations at neuromuscular junctions. *Dev. Neurosci.* 11: 227–247.

CELL DEATH AND TROPHIC SUBSTANCES

Campenot, R. B. 1982. Development of sympathetic neurons in compartmentalized cultures. II. Local control of neurite survival by nerve growth factor. *Dev. Biol.* 93: 13–21.

Fischer, W., Björklund, A., Chen, K. and Gage, F. H. 1991. NGF improves spatial memory in aged rodents as a function of age. *J. Neurosci.* 11: 1889–1906.

Hollyday, M. and Hamburger, V. 1976. Reduction of the naturally occurring motor neuron loss by enlargement of the periphery. *J. Comp. Neurol.* 170: 311–320.

POLYNEURAL INNERVATION

Betz, W. J., Caldwell, J. H. and Ribchester, R. R. 1980. The effects of partial denervation at birth on the development of muscle fibers and motor units in rat lumbrical muscle. *J. Physiol.* 303: 265–279.

Brown, M. C., Jansen, J. K. S. and Van Essen, D. 1976. Polyneuronal innervation of skeletal muscle in new-born rats and its elimination during maturation. *J. Physiol.* 261: 387–422.

DENERVATION AND REGENERATION OF SYNAPTIC CONNECTIONS

CHAPTER TWELVE

In adult animals, most neurons are postmitotic; only in a few instances do neuroepithelial stem cells persist. Thus neurons lost to injury or disease cannot be replaced. Nerve cells can, however, regenerate severed axonal and dendritic processes to reestablish synaptic connections. In this chapter we consider the consequences of denervation and the ability of neurons to regenerate.

When deprived of their synapses by denervation, vertebrate skeletal muscle fibers develop new chemoreceptors over the entire muscle surface and, consequently, an increased sensitivity to acetylcholine. Direct electrical stimulation of denervated super-sensitive muscles causes the chemosensitivity to shrink back to the original end plate area. The distribution of chemoreceptors in the muscle membrane is controlled in part by the level of muscle activity, in part by additional factors. Supersensitivity goes hand in hand with the ability of muscle fibers to accept innervation and to induce nerve terminals to sprout new branches. When muscles are denervated, even foreign nerves will form connections.

The basal lamina (an extracellular, protein-containing matrix that ensheaths the muscle, nerve terminals, and Schwann cells) plays a key role in the differentiation of the nerve terminal and the postsynaptic membrane during synapse regeneration. Agrin, a synaptic basal lamina protein synthesized by motoneurons, mediates the nerve-induced formation of postsynaptic specializations during development and regeneration.

The ability of regenerating axons to locate and innervate appropriate targets varies widely among species, from neuron to neuron, and with developmental age. Precise synaptic connections can be reestablished following lesions to the peripheral and central nervous systems of invertebrates and some lower vertebrates, such as salamanders and newts. Neurons in fetal or neonatal mammals also are able to reestablish appropriate synaptic connections with targets in the periphery after injury. Axons in the adult mammalian peripheral nervous system can regenerate, but show less specificity in contacting peripheral targets. In the adult mammalian CNS, successful regeneration is often blocked by proteins on the surfaces of astrocytes and oligodendroglia that inhibit axon growth. However, regenerating neurons that do reach their targets can reestablish synaptic connections. Techniques in which conduits of peripheral nerve sheaths or groups of embryonic neurons are grafted into regions of the adult CNS damaged by lesions demonstrate the regenerative capacities of vertebrate CNS neurons and hold promise for recovery from trauma and the amelioration of neurodegenerative diseases.

If a neuron is disconnected from its target by severing its axon, a characteristic sequence of changes usually occurs[1] (Figure 1). The distal portion of the axon degenerates together with a short length of axon proximal to the site of the lesion. The glial cells that had formed the myelin sheath of the distal segment of the nerve dedifferentiate, proliferate, and, together with invading microglia and macrophages, phagocytize the axonal and myelin remnants. This degenerative response is called WALLERIAN DEGENERATION after the nineteenth-century anatomist Augustus Waller, who first described it. In addition there are changes in the appearance of the damaged neuron, many of which reflect altered patterns of protein synthesis. The cell body and its nucleus swell, the

Effects of axotomy

[1]Grafstein, B. 1983. *In* F. J. Seil (ed.). *Nerve, Organ, and Tissue Regeneration: Research Perspectives*. Academic Press, New York, pp. 37–50.

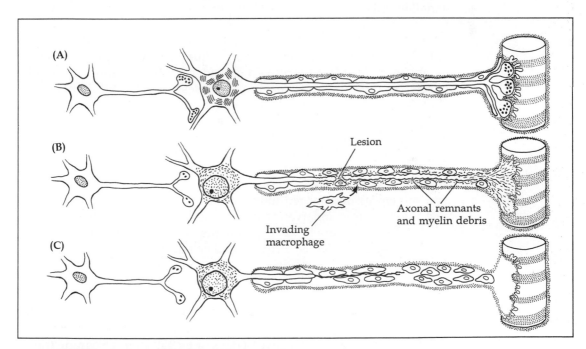

1 DEGENERATIVE CHANGES AFTER AXOTOMY. (A) A typical motoneuron in an adult vertebrate. **(B)** After axotomy the nerve terminal, the distal segment of the axon, and a short length of the proximal segment of the axon degenerate. Schwann cells dedifferentiate, proliferate, and, together with invading microglia and macrophages, phagocytize the axonal and myelin remnants. The axotomized neuron undergoes chromatolysis, presynaptic terminals retract, and degenerative changes may occur in pre- and postsynaptic cells. **(C)** The axon regenerates along the column of Schwann cells within the endoneurial tube and sheath of basal lamina that had surrounded the original axon.

nucleus moves from its typical position in the center of the cell soma to an eccentric location, and the ordered arrays of endoplasmic reticulum, called NISSL SUBSTANCE, disperse. As the Nissl substance stains prominently with commonly used basic dyes, its dispersal following axotomy causes a decrease in staining that is referred to as CHROMATOLYSIS. Within a few hours, new axonal sprouts emerge from near the tip of the proximal stump and begin regenerating. If the neuron successfully reestablishes synaptic contact with a target, the cell body usually regains its original appearance.

A variety of changes occur in adult neurons that fail to reestablish synaptic contact with their targets. In autonomic ganglia, axotomized ganglion cells become less sensitive to acetylcholine, shrink, and may eventually die. Axotomized sensory and motor neurons survive but atrophy and lose some of their differentiated properties. Retinal ganglion cells die if their axons in the optic nerve are severed. Neurons in the thalamus survive axotomy, but shrink in size.

Axotomy can also lead to marked changes in presynaptic cells that innervate the damaged neuron. The changes have been studied in detail in the autonomic ganglia of chicks and guinea pigs. Synaptic inputs to ganglion cells become less effective, not only because the ganglion cells become less sensitive to ACh, but also because many presynaptic terminals retract from the axotomized cells and those terminals that remain release fewer quanta of transmitter.[2-4] Thus, damage to a neuron alters its ability to maintain its properties and has transsynaptic retrograde effects on other neurons making synapses upon it. Rotshenker has shown an additional transsynaptic effect on spinal motoneurons in the frog and mouse.[5] When motor axons are cut on one side of the animal, axon terminals of the intact, undamaged motoneurons innervating the corresponding muscle on the other side sprout and form additional synapses after a delay of a few weeks. The evidence suggests that a signal spreads from the axotomized neurons, crosses the spinal cord, and influences undamaged motoneurons on the other side of the animal. Motoneurons innervating other muscles are not affected.

Results of a variety of experiments indicate that the effects of axotomy—chromatolysis, atrophy, and cell death—result from the loss of trophic substances produced by the target tissue and transported retrogradely along the axon to the cell body. The clearest example comes from studies of the effects of nerve growth factor (NGF) on sensory and sympathetic neurons, discussed in Chapter 11. For example, in the guinea pig the effects of axotomy are mimicked by injecting antibodies to NGF subcutaneously for several days or by blocking retrograde transport in postganglionic nerves and are largely prevented by application of NGF to the ganglion.[6]

[2]Purves, D. 1975. *J. Physiol.* 252: 429–463.
[3]Brenner, H. R. and Johnson, E. W. 1976. *J. Physiol.* 260: 143–158.
[4]Brenner, H. R. and Martin, A. R. 1976. *J. Physiol.* 260: 159–175.
[5]Rotshenker, S. 1988. *Trends Neurosci.* 11: 363–366.
[6]Nja, A. and Purves, D. 1978. *J. Physiol.* 277: 53–75.

If neurons rely on retrograde transport of trophic factors from their targets, how can they ever survive axotomy and regenerate? Experiments in which the sciatic nerve is lesioned have shown that as the peripheral portion of the axon degenerates following axotomy, the Schwann cells not only proliferate but also begin to synthesize NGF.[7] Thus the Schwann cells may temporarily supply regenerating sensory and sympathetic axons with NGF as they grow back to their peripheral targets, sustaining the neurons and perhaps providing guidance for growth cone navigation. It is interesting that these "denervated" Schwann cells also express low-affinity NGF receptors on their surface, perhaps to hold the NGF they produce along the path regenerating axons should take.[8] As regeneration progresses, the Schwann cells cease production of NGF and once again ensheath the axons.

EFFECTS OF DENERVATION ON THE POSTSYNAPTIC CELL

Neuromuscular synapses have provided a useful model for mechanisms of synaptic transmission between neurons in higher centers. Similarly, changes occurring in denervated muscles are relevant for thinking about the consequences of disruption and regeneration of neural connections in general.[9]

The denervated muscle membrane

Some phenomena in denervated skeletal muscles were originally described toward the end of the last century; these were first noted in muscle fibers that could be easily seen, such as those in the tongue. Some time after severance of their nerve supply, individual muscle fibers start to exhibit spontaneous, asynchronous contractions called FIBRILLATION. Fibrillation is caused by changes in the muscle membrane itself and is not initiated by acetylcholine,[10] although most of the spontaneous action potentials producing fibrillation originate in the region of the former end plate.[11] The onset of fibrillation may be as early as 2 to 5 days after denervation in rats, guinea pigs, or rabbits, or well over a week in monkeys and humans.

Before or at the start of fibrillation, mammalian muscle fibers become supersensitive to a variety of chemicals. This means that the concentration of a substance required to produce depolarization, or shortening of a muscle, is reduced by a factor of several hundred to a thousand. For example, a denervated mammalian skeletal muscle is about a thousand times more sensitive to ACh, applied either directly in the bathing fluid or injected into an artery supplying the muscle, than is a normally innervated muscle.[12] The increase in chemosensitivity is not restricted

[7]Heumann, R. et al. 1987. *J. Cell Biol.* 104: 1623–1631.

[8]Johnson, E. M. Jr., Taniuchi, M. and DiStefano, P. S. 1988. *Trends Neurosci.* 11: 299–304.

[9]Cannon, W. B. and Rosenblueth, A. 1949. *The Supersensitivity of Denervated Structures: Law of Denervation.* Macmillan, New York.

[10]Purves, D. and Sakmann, B. 1974. *J. Physiol.* 239: 125–153.

[11]Belmar, J. and Eyzaguirre, C. 1966. *J. Neurophysiol.* 29: 425–441.

[12]Brown, G. L. 1937. *J. Physiol.* 89: 438–461.

to the physiological transmitter—ACh—but occurs for a wide variety of chemical substances and even makes the muscle more sensitive to stretch or pressure.[13] The action potentials in denervated muscles also change, becoming more resistant to tetrodotoxin (TTX), the puffer fish poison that blocks sodium channels (Chapter 2). This change is due to the reappearance of TTX-resistant sodium channels that are the prevailing form in immature muscle.[14] Other changes occur in denervated muscle, such as a gradual atrophy or wasting of muscle fibers, but will not be discussed here.[15–17]

Appearance of new ACh receptors

Supersensitivity to acetylcholine is explained by an altered distribution of ACh receptors in denervated muscles. This has been demonstrated by applying ACh to small regions of the muscle surface by ionophoretic release from an extracellular micropipette while recording the membrane potential with an intracellular microelectrode. As explained in Chapters 7 and 9, in a normally innervated frog, snake, or mammalian muscle, only the end plate region—where the nerve fiber makes a synapse—is sensitive to ACh; the rest of the muscle membrane has a very low sensitivity. After denervation, the area of muscle membrane sensitive to ACh increases. When a nerve to a mammalian muscle is cut, the chemosensitivity increases day by day, until by about 7 days the surface of the muscle is almost uniformly sensitive to ACh[18] (Figure 2). In frog muscle, the changes are relatively small and take several weeks to develop.[19] The receptors that appear in extrasynaptic areas do not simply drift away from the original end plate. This was first shown by Katz and Miledi in experiments in which frog muscles were cut in two; nucleated fragments that were physically separated from the original end plate survived and developed increased sensitivity to ACh.[20] Thus, new ACh receptors appear in extrajunctional regions of denervated muscles.

Synthesis and degradation of receptors in denervated muscle

A valuable technique for studying the distribution and turnover of ACh receptors is to label them with radioactive α-bungarotoxin, which binds to ACh receptors strongly and with a high degree of specificity. The method used by several workers has been to bathe normal and denervated muscles in toxin and compare toxin binding at the end plate and end plate-free areas. As expected, the number and distribution of toxin-binding sites were changed after denervation.[21–23] Estimates of the

[13]Kuffler, S. W. 1943. *J. Neurophysiol.* 6: 99–110.
[14]Kallen, R. G. et al. 1990. *Neuron* 4: 233–242.
[15]Guth, L. 1968. *Physiol. Rev.* 48: 645–687.
[16]Gutmann, E. 1976. *Annu. Rev. Physiol.* 38: 177–216.
[17]Spector, S. A. 1985. *J. Neurosci.* 5: 2189–2196.
[18]Axelsson, J. and Thesleff, S. 1959. *J. Physiol.* 147: 178–193.
[19]Miledi, R. 1960. *J. Physiol.* 151: 1–23.
[20]Katz, B. and Miledi, R. 1964. *J. Physiol.* 170: 389–396.
[21]Fambrough, D. M. 1979. *Physiol. Rev.* 59: 165–227.
[22]McArdle, J. J. 1983. *Prog. Neurobiol.* 21: 135–198.
[23]Salpeter, M. M. and Loring, R. H. 1985. *Prog. Neurobiol.* 25: 297–325.

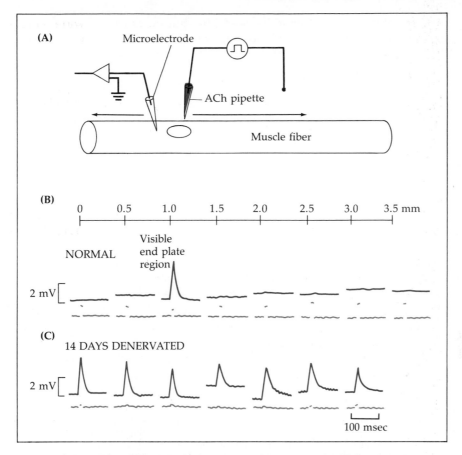

2 NEW ACh RECEPTORS APPEAR in a muscle of the cat after denervation. (A) Pulses of ACh are applied from an ACh-filled pipette onto different positions along the surface of a muscle fiber, while the membrane potential is recorded with an intracellular microelectrode. (B) In a muscle fiber with intact innervation, a response is seen only in the vicinity of the end plate. (C) After 14 days of denervation, a muscle fiber responds to ACh along its entire length. (After Axelsson and Thesleff, 1959.)

density of binding sites at postsynaptic areas of the muscle are of the order of 10^4 per μm^2 compared with fewer than 10 per μm^2 at end plate-free areas. After denervation, however, receptor sites in the extrasynaptic regions increase to about 10^3 per μm^2, with little change in density in the synaptic region.

The increase in receptors is attributable to enhanced synthesis and not to reduced degradation.[21,24] For example, substances that block protein synthesis (such as actinomycin and puromycin) prevent the increase in extrasynaptic receptor density in muscles maintained in organ culture, and measurements of the rate of appearance of new ACh receptors demonstrate a marked increase in ACh receptor synthesis in

[24]Scheutze, S. M. and Role, L. M. 1987. *Annu. Rev. Neurosci.* 10: 403–457.

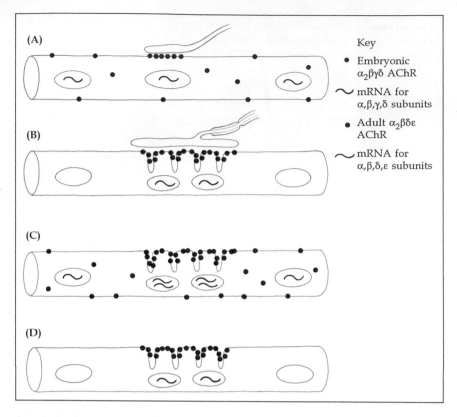

3   **SYNTHESIS AND DISTRIBUTION OF ACh RECEPTORS in muscles of the rat.**
(A) In fetal muscles, mRNAs for the α, β, γ, and δ subunits of the ACh receptor
are expressed in nuclei all along the length of the myofiber. The embryonic $\alpha_2\beta\gamma\delta$ form
of the receptor is found over the entire surface of the myofiber and accumulates at the
site of innervation. (B) In adult muscles, mRNAs for the α, β, δ, and ε subunits are
expressed only in nuclei directly beneath the end plate. The adult $\alpha_2\beta\delta\varepsilon$ form of the
receptor is highly localized to the crests of the junctional folds. (C) In denervated adult
muscles, nuclei directly beneath the end plate express α, β, γ, δ, and ε subunits; all
other nuclei reexpress the fetal pattern of α, β, γ, and δ subunits. Embryonic ACh
receptors are found all over the surface of the myofiber (producing denervation supersensitivity), including the postsynaptic membrane; the adult form of the receptor is
restricted to the end plate region. (D) If denervated muscles are stimulated directly, the
pattern of ACh receptor expression resembles that in innervated myofibers. (After Witzemann, Brenner and Sakmann, 1991.)

denervated muscles. Northern blot and in situ hybridization techniques,
which measure the levels and distribution of the mRNAs that encode
subunits of the ACh receptor, have confirmed these findings[25–27] (Figure
3). In adult muscle, only a few nuclei located immediately beneath the
end plate are synthesizing ACh receptor subunit mRNAs, whereas ACh
receptor genes are transcribed by nuclei all along the length of dener-

[25]Merlie, J. P. and Sanes, J. R. 1985. *Nature* 317: 66–68.
[26]Bursztajn, S., Berman, S. A. and Gilbert, W. 1989. *Proc. Natl. Acad. Sci. USA* 86:
2928–2932.
[27]Fontaine, B. and Changeux, J.-P. 1989. *J. Cell Biol.* 108: 1025–1037.

vated muscle fibers. Moreover, the new receptors that appear in denervated adult muscles are the embryonic type of receptor, containing a γ
rather than an ε subunit[28] (see Chapters 2 and 11).

Experiments in which receptor turnover was measured by using
labeled α-bungarotoxin show that denervation increases the rate of
receptor degradation. Receptors in neonatal rat muscles, including those
at the end plate, have a rapid turnover, with a half-life of approximately
1 day. As the muscle matures, the half-life of receptors in junctional
and extrajunctional regions increases from 1 to 10 days.[29] Following
denervation, the half-life of receptors remaining at the end plate decreases to 3 days; new receptors synthesized while the muscle is denervated (whether synaptic or extrasynaptic) resemble those in embryonic muscle, turning over with a half-life of 1 day.[30]

What mechanism causes the appearance of new receptors? Is it
inactivity of the muscle, loss of "trophic" factors, or both? Lømo and
Rosenthal investigated this problem by blocking conduction in rat
nerves with a local anesthetic or diphtheria toxin.[31] The substances were
applied by means of a cuff to a short length of the nerve some distance
from the muscle. With this technique, the muscles became completely
inactive because motor impulses failed to conduct past the cuff. Occasional test stimulation of the nerve distal to the block produced a twitch
of the muscle as usual, and miniature end plate potentials still occurred
normally, showing that synaptic transmission was intact. And yet after
7 days of nerve block, the muscle had become supersensitive (Figure
4). Other experiments have shown that new extrajunctional receptors
also appear when neuromuscular transmission is blocked by long-term
application of curare or α-bungarotoxin to a muscle.[32] All these results
show that "denervation" supersensitivity can be produced without interrupting the nerve; blockage of synaptic activation of the muscle is
sufficient.[33]

The role of muscle activity itself as an important factor in controlling
supersensitivity was further shown in other experiments in which supersensitive denervated muscles in the rat were stimulated directly
through electrodes permanently implanted around the muscle. Repetitive direct stimulation of muscles over several days caused the sensitive
area to become restricted, so that once again only the synaptic region
was sensitive to ACh (Figures 3 and 5).[31] The frequency of stimulation
and the length of the quiescent intervals were important variables in
the development or reversal of supersensitivity. This explains why denervated mammalian muscle fibers develop supersensitivity in spite of
the ongoing contractions associated with fibrillation. Sampling the activity of individual fibers showed that fibrillation is cyclical, periods of

[28]Mishina, M. et al. 1986. *Nature* 321: 406–411.

[29]Salpeter, M. M. and Marchaterre, M. 1992. *J. Neurosci.* 12: 35–38.

[30]Shyng, S.-L. and Salpeter, M. M. 1990. *J. Neurosci.* 10: 3905–3915.

[31]Lømo, T. and Rosenthal, J. 1972. *J. Physiol.* 221: 493–513.

[32]Berg, D. K. and Hall, Z. W. 1975. *J. Physiol.* 244: 659–676.

[33]Witzemann, V., Brenner, H.-R. and Sakmann, B. 1991. *J. Cell Biol.* 114: 125–141.

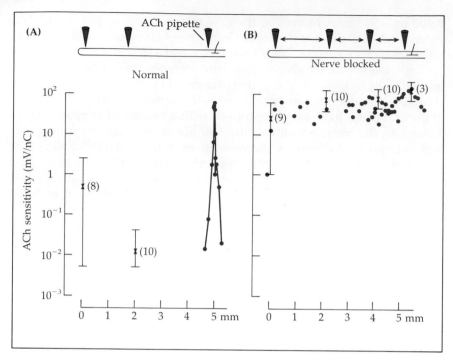

4 NEW ACh RECEPTORS in a rat muscle after block of nerve conduction. **(A) In the normal muscle the ACh sensitivity is restricted to the end plate region (near the 5-mm position). (B) After the nerve to the muscle was blocked for 7 days by a local anesthetic, the ACh sensitivity is distributed over the entire muscle fiber surface. Sensitivity is expressed numerically in millivolts depolarization per nanocoulomb of charge ejected from the pipette. The crosses and bars represent the mean and range of sensitivities of a number (in parentheses) of adjacent muscle fibers. (From Lømo and Rosenthal, 1972.)**

activity alternating with inactivity. The level of spontaneous activity is, however, below that required to reverse the effects of denervation on the distribution of ACh receptors.[21,34]

Like the development of supersensitivity, the turnover of junctional receptors is regulated by muscle activity.[35] Similar increases in the turnover rate occur in muscles paralyzed by denervation and in those paralyzed by continuous application of tetrodotoxin to the nerve. Direct electrical stimulation of denervated muscle also restores normal stability to ACh receptors at synaptic sites; receptors have a half-life of 10 days in stimulated denervated muscle. The effect of muscle activity on the degradation rate appears to be mediated by calcium influx: It is mimicked by treating inactive muscles with the calcium ionophore A23187, and activity-dependent ACh receptor stabilization is prevented by calcium channel blockers.[36] Elevation of intracellular cAMP levels also

[34]Purves, D. and Sakmann, B. 1974. *J. Physiol.* 237: 157–182.
[35]Fumagalli, G. et al. 1990. *Neuron* 4: 563–569.
[36]Rotzler, S., Schramek, H. and Brenner, H. R. 1991. *Nature* 349: 337–339.

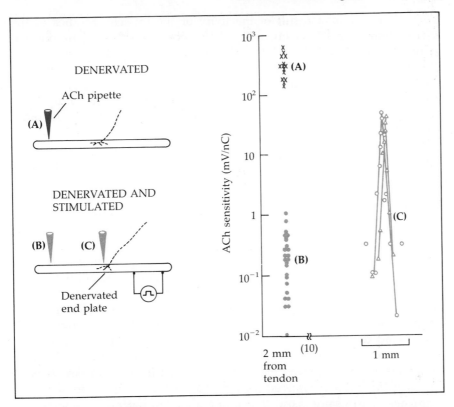

5 REVERSAL OF SUPERSENSITIVITY in a denervated muscle of the rat by direct stimulation of the muscle fibers. (A) Increased sensitivity in the nerve-free portion of a muscle fiber after 14 days of denervation. (B) Sensitivity in the nerve-free region of a muscle that had been denervated for 7 days without stimulation, and then stimulated intermittently for another 7 days. This treatment reversed the denervation supersensitivity. (C) ACh sensitivity in two stimulated fibers of the same muscle near their denervated end plate regions. The high sensitivity is confined to this region in the stimulated muscle. (After Lømo and Rosenthal, 1972.)

causes stabilization of ACh receptors in inactive muscles, suggesting that calcium influx may cause receptor stabilization through activation of adenylyl cyclase.[37]

There remains the question whether nerves provide the muscle with some products, other than transmitters, that keep the muscles normal. Experiments in which slowly developing changes occur without activity per se playing an obvious role have been made on partially denervated muscles. Many fibers in the frog sartorius muscle are innervated at more than one site along their length. If some of these synapses are denervated by cutting intramuscular branches of the nerve, leaving other axons to the same muscle fibers intact, supersensitivity develops in the denervated portions of the muscle fibers; yet these fibers have kept contracting all along.[19] Presumably such effects cannot be ex-

[37]Shyng, S.-L., Xu, R. and Salpeter, M. M. 1991. *Neuron* 6: 469–475.

plained simply as a result of inactivity of the muscle. Studies of the distribution of mRNAs encoding ACh receptor subunits in normal, surgically denervated, and toxin-paralyzed rat muscles suggest that there are at least two neural factors that regulate ACh receptor expression independently of muscle activity; one induces the expression of the adult ε subunit in nuclei at the end plate, and the other suppresses expression of the γ subunit and to a lesser extent mRNAs for the other subunits.[33]

In principle, crustacean muscles that are supplied by one inhibitory and two excitatory axons can also be used to study the effects of partial denervation. Such experiments are difficult to perform, however, because after section of a crustacean axon its distal segment can survive for weeks or months without degenerating. Accordingly, Parnas and his colleagues destroyed single inhibitory or excitatory axons by injecting proteolytic enzymes into them near their terminals and examined transmission at the remaining synapses.[38,39] This procedure has been shown to damage only the injected axon and not its neighbors. Some days after the inhibitory axon was killed, the synaptic currents recorded in response to stimulation of the excitatory axon were prolonged, due to a prolongation of the channel open times. In other experiments, it was found that when one of the excitatory axons was killed, the remaining excitatory axon released more transmitter. However, only those terminals of the remaining excitatory axon that innervated targets with reduced innervation showed an increase in transmitter release; other branches of the same axon that innervated normal muscles were not affected. Thus the signal for synaptic strengthening is generated and acts locally.

Distribution of receptors in nerve cells

When part of their synaptic input is destroyed, neurons, like muscles, undergo changes. The consequences of denervation in neurons include altered responses to injected drugs[9] and reductions in amino acid incorporation into protein, neuron number, and soma size.[40]

Alterations of the neuronal surface membrane after loss of synapses have been studied in autonomic ganglion cells in frogs and chicks. In the frog heart, parasympathetic neurons can be seen in the transparent interatrial septum. Like skeletal muscle fibers, innervated neurons are highly sensitive to the transmitter ACh at selected spots on their surfaces, immediately under the presynaptic terminals (Figure 6A; see also Chapter 9). The restriction of receptors to synaptic sites in these neurons is not as complete as in skeletal muscle, however. Experiments in which the distribution of ACh receptors is assessed by a histochemical reaction suggest that approximately 20 percent of the receptors may be extrasynaptic.[41] If the two vagus nerves to the heart are cut, synaptic trans-

[38]Parnas, I. 1987. *J. Exp. Bio.* 132: 231–247.
[39]Dudel, J. and Parnas, I. 1987. *J. Physiol.* 390: 189–199.
[40]Born, D. E. and Rubel, E. W. 1988. *J. Neurosci.* 8: 901–919.
[41]Sargent, P. B. and Pang, D. Z. 1989. *J. Neurosci.* 9: 1062–1072.

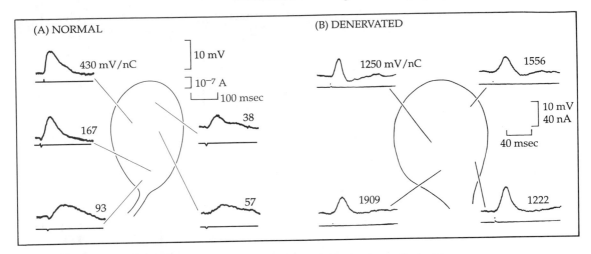

6 DEVELOPMENT OF SUPERSENSITIVITY in parasympathetic nerve cells in the heart of the frog after denervation. (A) In a normal neuron, the high ACh sensitivity is confined to synaptic regions. The large numbers indicate high sensitivity (expressed in millivolts per nanocoulomb). When ACh is applied to the extrasynaptic region, more must be released to have an effect. Such responses rise relatively slowly because ACh has to diffuse to a nearby sensitive synaptic spot. (B) After 21 days of denervation, the sensitivity of the neuronal surface is high wherever ACh is applied. (After Kuffler, Dennis and Harris, 1971; Dennis and Sargent, 1979.)

mission between vagal nerve terminals and ganglion cells fails rapidly, starting on the second day after denervation.[42,43] At the same time, the area of the neuronal surface membrane sensitive to ACh starts to increase. By 4 to 5 days, ACh causes a membrane depolarization when applied anywhere on the cell surface (Figure 6B). In other respects the cells are normal; there are no obvious changes in the resting membrane potential or excitability of the cells. Somewhat surprisingly, the number of ACh receptors on denervated neurons appears to decrease.[44] However, the remaining receptors are more widely distributed following denervation, which may account for the observed change in chemosensitivity. A normal distribution of chemosensitivity reappears if the original nerve is allowed to grow back into the heart.[43] As in muscle (see later), the sensitive area becomes restricted once more to the vicinity of the synapses.

Not all neurons respond to denervation in the same way. In chick parasympathetic ganglia and frog sympathetic ganglia, neurons have a high concentration of ACh receptors beneath presynaptic terminals.[45–47]

[42]Kuffler, S. W., Dennis, M. J. and Harris, A. J. 1971. *Proc. R. Soc. Lond. B* 177: 555–563.

[43]Dennis, M. J. and Sargent, P. B. 1979. *J. Physiol.* 289: 263–275.

[44]Sargent, P. B. and Pang, D. Z. 1988. *Neuron* 1: 877–886.

[45]McEachern, A. E., Jacob, M. H. and Berg, D. K. 1989. *J. Neurosci.* 9: 3899–3907.

[46]Loring, R. H. and Zigmond, R. E. 1987. *J. Neurosci.* 7: 2153–2162.

[47]Dunn, P. M. and Marshall, L. M. 1985. *J. Physiol.* 363: 211–225.

But in these ganglia, denervation has little or no effect on the number or distribution of ACh receptors. For example, in denervated frog sympathetic ganglia the sensitivity of neurons to ionophoretically applied ACh remains unchanged. If the sensitivity to ACh is measured by recording the gross extracellular response to bath-applied ACh, however, an 18-fold increase in sensitivity is seen. This is due entirely to the loss of acetylcholinesterase activity associated with preterminal and terminal portions of the preganglionic axons; the enzyme normally prevents ACh from reaching cells buried within the ganglion. Thus, the mechanisms that regulate ACh receptor distribution appear to be different in nerve and muscle cells, and to vary among neuronal cell types as well.

DENERVATION SUPERSENSITIVITY, SUSCEPTIBILITY TO INNERVATION, AND AXONAL SPROUTING

Synapse formation in denervated muscles

Some clues about the possible significance of supersensitivity come from studying the changes a muscle undergoes in the process of reinnervation. In the adult, an innervated muscle fiber will not accept innervation by an additional nerve.[48] Thus, if a cut motor nerve is placed on an innervated muscle, it will not "take" to form additional new end plates on the muscle fibers. In contrast, nerve fibers do grow out and reinnervate a denervated or injured muscle. Unlike the situation during development, where growth cones contact muscle fibers at random locations, reinnervation usually occurs at the site of the original end plate. However, if a cut nerve is placed far enough away from the original end plate, or if an axon is in some way prevented from reaching it, then an entirely new end plate can be formed. This is a remarkable result. It means that nerve fibers can grow out to an adult muscle fiber and form synapses in a region that had never been innervated. Thus, like the situation at developing neuromuscular junctions, regenerating axons appear capable of releasing signals that trigger postsynaptic differentiation, and inactive muscle fibers are able to respond to those signals.

What are the conditions that enable a denervated muscle to accept a nerve? Several tests have been made to see whether a correlation exists between increased chemosensitivity and synapse formation. In one of these, botulinum toxin was used to produce neuromuscular block, not by destroying nerve terminals, but by preventing them from releasing transmitter. Despite the presence of intact-looking nerve terminals, the muscle membrane developed supersensitivity in previously extrasynaptic areas and accepted additional innervation.[49,50] Similarly, after a rat muscle was made supersensitive as a result of blocking

[48]Jansen, J. K. S. et al. 1973. *Science* 181: 559–561.
[49]Thesleff, S. 1960. *J. Physiol.* 151: 598–607.
[50]Fex, S. et al. 1966. *J. Physiol.* 184: 872–882.

impulse transmission in the nerve, a foreign nerve was able to form additional synapses.[48] In such experiments, individual muscle fibers then showed synaptic potentials and contractions in response to stimulation of each of the two nerves. Conversely, when a denervated muscle was stimulated directly, its ability to accept extra innervation was lost together with its supersensitivity.

Is it a normal prerequisite that the muscle be supersensitive for innervation to occur? As described in Chapter 11, fetal and neonatal rat muscles are sensitive to ACh along their length.[51] Similarly, myofibers grown in cell culture are both sensitive to ACh and amenable to innervation throughout their length (Chapter 11).[52] After innervation of muscle fibers in vivo, the ACh-sensitive area shrinks over a period of about two weeks to a region around the end plate.[51,53] Thus, both initial innervation and reinnervation occur when the muscle fibers are supersensitive. Several experiments indicate, however, that synapse formation is not dependent upon the ACh receptor itself, or at least not on that part of it to which α-bungarotoxin or curare binds, since reinnervation still occurs in denervated rat and toad muscles in the presence of these inhibitors.[54,55]

Not only are denervated muscles amenable to innervation, but they actively induce undamaged nerves to sprout new terminal branches. For example, if a muscle is partially denervated, the remaining axon terminals will sprout and innervate the denervated fibers (Figure 7).[56] As with regulation of ACh receptor synthesis and degradation, muscle inactivity appears to play a key role in this inductive process. Sprouting and hyperinnervation occur if muscle activity is prevented by blocking action potential propagation in the nerve with a cuff impregnated with tetrodotoxin,[57] or if neuromuscular transmission is blocked with botulinum toxin or α-bungarotoxin.[58,59] Denervation-induced sprouting has also been observed for sensory axons innervating skin and spinal cord, for preganglionic axons in autonomic ganglia, and for several axonal projections in the brain.[60]

Denervation-induced axonal sprouting

The molecular signals that induce sprouting have not been identified; however, the signals are quite selective. In leech skin, for example, killing a particular sensory or motor neuron by injecting it with pronase induces axon sprouting into the denervated territory, but only of axons

[51]Diamond, J. and Miledi, R. 1962. *J. Physiol.* 162: 393–408.
[52]Frank, E. and Fischbach, G. D. 1979. *J. Cell Biol.* 83: 143–158.
[53]Bevan, S. and Steinbach, J. H. 1977. *J. Physiol.* 267: 195–213.
[54]Van Essen, D. and Jansen, J. K. 1974. *Acta Physiol. Scand.* 91: 571–573.
[55]Cohen, M. W. 1972. *Brain Res.* 41: 457–463.
[56]Brown, M. C., Holland, R. L. and Hopkins, W. G. 1981. *Annu. Rev Neurosci.* 4: 17–42.
[57]Brown, M. C. and Ironton, R. 1977. *Nature* 265: 459–461.
[58]Duchen, L. W. and Strich, S. J. 1968. *Q. J. Exp. Physiol.* 53: 84–89.
[59]Holland, R. L. and Brown, M. C. 1980. *Science* 207: 649–651.
[60]Purves, D. and Lichtman, J. W. 1985. *Principles of Neural Development.* Sinauer, Sunderland, MA.

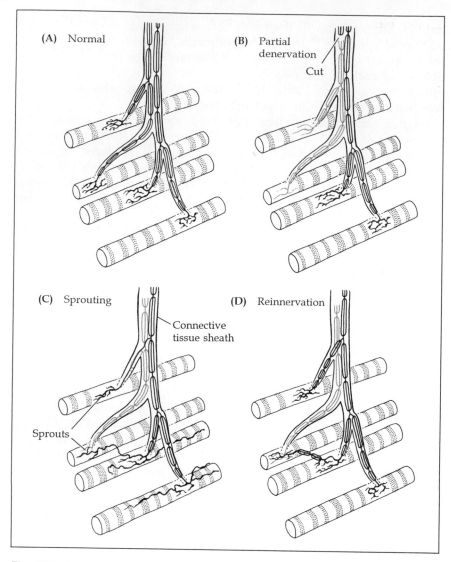

7 **NERVE TERMINALS SPROUT in response to partial denervation of a mammalian skeletal muscle. (A) Normal pattern of innervation. (B) Some fibers are denervated by cutting a few of the axons innervating the muscle. (C) Axons sprout from the terminals and from nodes along the preterminal axons of undamaged motoneurons to innervate the denervated fibers. (D) After 1 or 2 months, sprouts that have contacted vacant end plates are retained, while other sprouts disappear. (After Brown, Holland and Hopkins, 1981.)**

of cells of the same sensory or motor modality.[38,61,62] In the cat central nervous system, when the ipsilateral projections from sensory and motor cortex to the red nucleus are cut, the fibers from the contralateral cortex sprout to innervate the denervated cells. They do so—not ran-

[61]Bowling, D., Nicholls, J. G. and Parnas, I. 1978. *J. Physiol.* 282: 169–180.
[62]Blackshaw, S. E., Nicholls, J. G. and Parnas, I. 1982. *J. Physiol.* 326: 261–268.

domly, but in the correct topological pattern, via the appropriate
interneurons—and end on the correct dendrites.[63,64]

A factor that plays a key role in the regeneration of synapses between
nerve and muscle is the SYNAPTIC BASAL LAMINA, a specialized region
of the extracellular matrix. Lying between the nerve terminal and the
muscle membrane, this material constitutes a densely staining matrix
made up of proteoglycans and glycoproteins, including collagen, laminin, fibronectin, and cholinesterase. As shown in Figure 8A, basal
lamina surrounds the muscle, the nerve terminal, and the Schwann cell
and dips into the folds in the postsynaptic membrane. McMahan and
his colleagues have made a series of systematic and elegant studies on
the physiological and structural effects that molecules in this noncellular
material have on differentiation of nerve and muscle.[65–68] The key to
their analysis was to use a thin, nearly transparent frog muscle—the
cutaneous pectoris—in which regeneration occurs rapidly and the end
plates have a highly ordered arrangement. As a first step, cells in the
region of innervation were killed by cutting the nerve and muscle fibers
or by repeated application of a brass bar cooled in liquid nitrogen (Figure
8B). Within days the portion of the muscle fibers in the damaged region
together with the nerve terminals degenerated and were phagocytized,
but the basal lamina sheaths remained intact (Figure 8C and 8D). The
location of the original neuromuscular junctions could be recognized by
the distinctive morphology of the basal lamina sheaths of the muscle
and Schwann cell at the junctional sites, and because cholinesterase
remained concentrated in the basal lamina of the synaptic cleft and
folds for weeks following the operation. Two weeks after the muscles
were damaged, new myofibers had formed within the basal lamina
sheaths and were contacted by regenerated axon terminals, which
evoked muscle twitches when the nerves were stimulated. Nearly all
of the regenerated synapses were located precisely at original synaptic
sites, as marked by cholinesterase. Thus signals associated with the
synaptic basal lamina specified where regenerating synapses formed.

To investigate further the nature of the signals associated with the
synaptic basal lamina, muscles were damaged and the nerve was
crushed, but muscle fiber regeneration was prevented by X-irradiation.
Regenerating axons grew to the former synaptic sites on the basal
lamina—as marked by cholinesterase—and formed active zones for
release precisely opposite portions of the basal lamina that had projected
into the junctional folds—all this without a postsynaptic "target" (Figure 8E). In a parallel series of experiments McMahan and his colleagues
demonstrated that synaptic basal lamina in the adult also contains factors that trigger differentiation of postsynaptic specializations in regen-

Role of basal lamina at regenerating synapses

[63]Tsukahara, N. 1981. *Annu. Rev. Neurosci.* 4: 351–379.
[64]Katsumaru, H. et al. 1986. *J. Neurosci.* 6: 2864–2874.
[65]Sanes, J. R., Marshall, L. M. and McMahan, U. J. 1978. *J. Cell Biol.* 78: 176–198.
[66]Burden, S. J., Sargent, P. B. and McMahan, U. J. 1979. *J. Cell Biol.* 82: 412–425.
[67]McMahan, U. J. and Slater, C. R. 1984. *J. Cell Biol.* 98: 1452–1473.
[68]Anglister, L. and McMahan, U. J. 1985. *J. Cell Biol.* 101: 735–743.

8

BASAL LAMINA AND REGENERATION OF SYNAPSES. (A) Electron micrograph of a normal neuromuscular synapse in the frog, stained with ruthenium red to show the basal lamina that dips into the postsynaptic folds and surrounds the Schwann cell (S) and nerve terminal (N). (B) Diagram of the cutaneous pectoris muscle, showing on the right the region frozen to damage muscle fibers. (C) Damage causes all cellular elements of the neuromuscular junction to degenerate and be phagocytized, leaving only the basal lamina sheath of the muscle fiber and Schwann cell intact. New neuromuscular junctions are restored by regenerating axons and muscle fibers. (D) Nerve and muscle were damaged, and regeneration of muscle fibers was prevented by X-irradiation. Regenerating axons have contacted the original synaptic sites, which are marked by cholinesterase stain (arrows), on the basal lamina sheaths. (E) Formation of active zones at original synaptic sites by axons regenerating in the absence of muscle fibers. The tongue of basal lamina that had extended into the junctional fold (arrow) marks the site of the original synapse. (After McMahan, Edgington and Kuffler, 1980.)

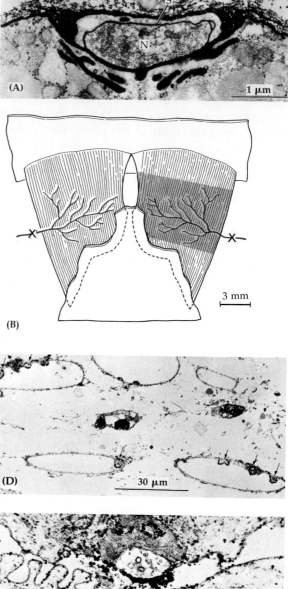

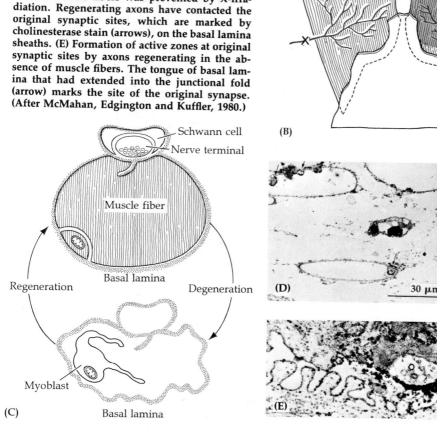

erating myofibers. Muscles were damaged as described above, but reinnervation was prevented by removing a long segment of the nerve. When new muscle fibers regenerated within the basal lamina sheaths, they formed junctional folds and aggregates of acetylcholine receptors and acetylcholinesterase precisely at the point where they came in contact with the original synaptic basal lamina (Figure 9). Thus signals

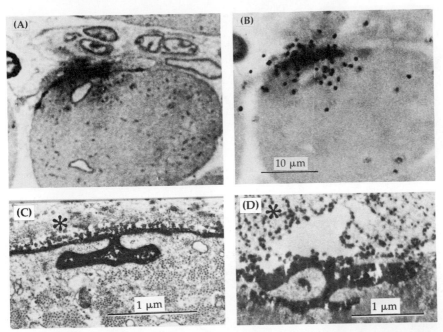

9 **ACCUMULATION OF ACh RECEPTORS AND ACETYLCHOLINESTERASE** at original synaptic sites on muscle fibers regenerating in the absence of nerve. The muscle was frozen as in Figure 8B, but the nerve was prevented from regenerating. New muscle fibers formed within the basal lamina sheaths. (A,B) Light microscope autoradiography of a regenerated muscle incubated with radioactive α-bungarotoxin to label ACh receptors [silver grains are in focus in (B)] and stained for cholinesterase to mark the original synaptic site. (C) Electron micrograph of the original synaptic site in a regenerated muscle labeled with HRP-α-bungarotoxin. The distribution of ACh receptors is indicated by the dense HRP reaction product, which lines the junctional folds (asterisk). (D) Electron micrograph of the original synaptic site in a regenerated muscle stained for cholinesterase. The original cholinesterase was permanently inactivated at the time the muscle was frozen. Thus the dense reaction product is due to cholinesterase synthesized and accumulated at the original synaptic site by the regenerating muscle fiber. (A and B from McMahan, Edgington and Kuffler, 1980; C from McMahan and Slater, 1984; D from Anglister and McMahan, 1985; micrographs courtesy of U. J. McMahan.)

stably associated with synaptic basal lamina can trigger the formation of synaptic specializations in both regenerating myofibers and regenerating nerve terminals.

In order to identify the signal in synaptic basal lamina that triggers postsynaptic differentiation, basal lamina-containing extracts were prepared from the electric organ of the marine ray *Torpedo californica*, a tissue derived embryologically from muscle that receives very dense cholinergic innervation. When added to myofibers in culture, these extracts mimicked the effects of synaptic basal lamina on regenerating muscle fibers; that is, they induced the formation of specializations at which ACh receptors accumulated, together with several other components of the postsynaptic apparatus[69] (Figure 10). The active com-

[69]McMahan, U. J. and Wallace, B. G. 1989. *Dev. Neurosci.* 11: 227–247.

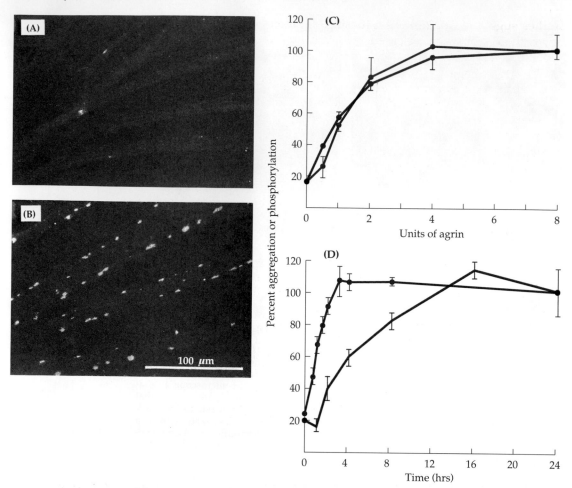

10   **AGRIN CAUSES AGGREGATION AND PHOSPHORYLATION** of ACh receptors in chick myotubes in culture. **(A,B)** Fluorescence micrographs of myotubes labeled with rhodamine-conjugated α-bungarotoxin to mark ACh receptors. **(A)** Receptors are distributed over the surface of control myotubes. **(B)** Overnight incubation with agrin causes the formation of patches at which ACh receptors accumulate. **(C,D)** Comparison of agrin-induced ACh receptor aggregation (blue traces) and tyrosine phosphorylation (gray traces) of the β subunit of the ACh receptor. **(C)** When cultured myotubes are incubated overnight with increasing amounts of agrin, the extent of aggregation and the extent of phosphorylation increase and then reach a plateau along the same dose response curve. **(D)** Agrin-induced ACh receptor phosphorylation increases rapidly after addition of agrin, reaching a steady state within 4 hours. Receptor aggregation proceeds more slowly, reaching a steady state 16 hours after addition of agrin. (After Wallace, 1986, 1992.)

ponent in the extracts, called agrin, has been purified and characterized, and cDNAs encoding it have been cloned from chick and rat brain libraries.[70-73] Results of in situ hybridization and immunohistochemical

[70]Nitkin, R. M. et al. 1987. *J. Cell Biol.* 105: 2471–2478.
[71]Tsim, K. W. K. et al. 1992. *Neuron* 8: 677–689.
[72]Ruegg, M. A. et al. 1992. *Neuron* 8: 691–699.
[73]Rupp, F. et al. 1991. *Neuron* 6: 811–823.

studies suggest that agrin is synthesized by motoneurons, transported down their axons, and released to induce differentiation of the postsynaptic apparatus at developing neuromuscular junctions.[74] Agrin apparently then becomes incorporated into the synaptic basal lamina where it helps maintain the postsynaptic apparatus in the adult and trigger its differentiation during regeneration.

The mechanism by which agrin induces the formation of postsynaptic specializations is not known. Several lines of evidence suggest that agrin binds to a receptor on the myotube surface, triggering a catalytic process within the myotube that leads to the formation of a specialization at which ACh receptors and other components of the postsynaptic apparatus accumulate.[69] Agrin has been shown to cause a rapid increase in tyrosine phosphorylation of the β subunit of the ACh receptor, suggesting that activation of a protein tyrosine kinase may be an early event in ACh receptor aggregation[75] (Figure 10). As described in Chapter 11, receptors for many growth factors, including NGF, BDNF, and NT-3, have been shown to be protein tyrosine kinases.

Agrin is not the only signal controlling the differentiation of the postsynaptic apparatus.[33] As described above, muscle activity, evoked by the release of ACh from axon terminals, has been shown to regulate the rate of synthesis of ACh receptors and the extent to which acetylcholinesterase accumulates in synaptic basal lamina.[24,76] How muscle activity increases the synthesis of acetylcholinesterase and its accumulation in synaptic basal lamina and suppresses synthesis of ACh receptors in all but subsynaptic nuclei remains an area of active investigation.[77]

SELECTIVITY IN REGENERATION

The ability of denervated skeletal muscle to be innervated by foreign nerves allows investigation of a number of questions regarding specificity of synapse formation and the way in which nerve and muscle cells influence each other that are difficult to address during development. For example, what properties must the alien nerve have in order to be accepted? Does it alter the properties of the muscle? Is the nerve itself altered as a result of innervating the wrong muscle?

Neuronal regulation of muscle fiber properties

Observations concerning these questions date back to 1904, when Langley and Anderson showed that muscles of the cat could become innervated by cholinergic preganglionic sympathetic fibers,[78] which normally make synapses on nerve cells in ganglia. The properties of synapses formed by vagus nerves on the frog sartorius muscle have been studied by Landmesser.[79,80] In these experiments there was no evidence

[74]McMahan, U. J. 1990. *Cold Spring Harbor Symp. Quant. Biol.* 50: 407–418.
[75]Wallace, B. G., Qu, Z. and Huganir, R. L. 1991. *Neuron* 6: 869–878.
[76]Dennis, M. J. 1981. *Annu. Rev. Neurosci.* 4: 43–68.
[77]Changeux, J.-P. 1991. *New Biol.* 3: 413–429.
[78]Langley, J. N. and Anderson, H. K. 1904. *J. Physiol.* 31: 365–391.
[79]Landmesser, L. 1971. *J. Physiol.* 213: 707–725.
[80]Landmesser, L. 1972. *J. Physiol.* 220: 243–256.

to suggest that the properties of the nerve were changed or that the abnormal type of innervation had altered the electrical characteristics of the muscle.

The properties of other muscles have been shown to become markedly changed with foreign innervation. Slow muscle fibers in the frog are quite distinctive: They are diffusely innervated, have a characteristic fine structure, and do not normally give regenerative impulses or twitches.[81] After denervation, slow fibers can become reinnervated by nerves that normally innervate twitch muscles at discrete end plates. Under these conditions, slow fibers become able to give conducted action potentials and twitches.[82] Eccles, Buller, Close, and their colleagues cut and interchanged the nerves to rapidly and more slowly contracting skeletal muscles in kittens and rats. Both these types of mammalian muscle fibers give conducted action potentials; they are called SLOW-TWITCH and FAST-TWITCH fibers. After the muscles were reinnervated by the inappropriate nerves, the functional and biochemical properties of the slow-twitch muscles came to resemble those of the fast-twitch fibers, and the properties of fast-twitch ones became slower.[83] Experiments by Lømo and his colleagues have shown that a major factor in this transformation is the pattern of impulses in the nerve and the resulting pattern of muscle contractions.[84,85] Motoneurons innervating slow- and fast-twitch muscle fibers tend to fire at different frequencies. If a slow-twitch muscle is denervated and then stimulated directly with the "slow" pattern of impulses, it will maintain its slow properties; however, if it is stimulated with the "fast" pattern, many of its properties will come to resemble those of fast-twitch muscle. Likewise, a denervated fast-twitch muscle stimulated with the "fast" pattern will remain fast, but will come to resemble a slow-twitch muscle if stimulated with the "slow" pattern. Thus the type of use to which a muscle is subjected can influence its biochemical and physiological properties.

Specificity of regeneration in the vertebrate peripheral nervous system

Classic experiments by Langley, now confirmed by single-cell analysis, demonstrated that regenerating mammalian preganglionic autonomic axons reinnervate the appropriate postganglionic neurons.[60] This selectivity apparently results, at least in part, from positional cues that bias synapse formation between neurons and target cells on the basis of rostrocaudal position. Similar cues appear to be used by the sympathetic and motor systems. Thus both sympathetic ganglia and muscles transplanted from different segments to a common site in the neck are reinnervated by different subsets of preganglionic axons, more caudally derived targets tending to be innervated by more caudally derived

[81]Kuffler, S. W. and Vaughan Williams, E. M. 1953. *J. Physiol.* 121: 289–317.

[82]Miledi, R., Stefani, E. and Steinbach, A. B. 1971. *J. Physiol.* 217: 737–754.

[83]Close, R. I. 1972. *Physiol. Rev.* 52: 129–197.

[84]Lømo, T., Westgaard, R. H. and Dahl, H. A. 1974. *Proc. R. Soc. Lond.* B 187: 99–103.

[85]Ausoni, S. et al. 1990. *J. Neurosci.* 10: 153–160.

axons.[86,87] In young rats, a similar degree of positional selectivity has been shown during reinnervation of myofibers in long muscles that span several segments.[88] The degree of selectivity observed in such transplantation experiments is small, however; virtually any part of the spinal cord can reinnervate muscles or ganglia from any segmental level.

Selectivity in the reinnervation of muscles by motoneurons varies with species and age. Appropriate neuromuscular connections are restored following transection of motor nerves in neonatal rats, larval frogs, and adult newts.[89] One mechanism for reestablishing connections selectively is competition between axons; in salamander muscles that have been innervated by inappropriate axons, the foreign synapses are eliminated after the normal nerve reestablishes its connection.[90] On the other hand, in adult mammals, regenerating sensory, motor, and postganglionic autonomic axons show little ability to navigate back selectively to their original targets, and foreign nerves can be as effective as the original ones in innervating fibers.[89] Further, in mammalian and fish muscle fibers that have become dually innervated, both the original and foreign synapses can function simultaneously.[48,91–93] There is evidence to show that a foreign nerve can even displace the original motor axons in intact adult rat muscles, suggesting that the maintenance of synaptic structure is a dynamic process.[94]

The probability of successful regeneration following peripheral nerve injury in adult mammals is greatly increased if the nerve is crushed rather than cut. This leaves the endoneurial tubes and Schwann cell basal lamina that surround the axons intact[89] (see Figure 1). Under such conditions, axons regenerate within their parent tubes and are guided back to their original targets. However, if the endoneurial tubes are disrupted, as when a nerve is cut, then regenerating axons enter tubes in the distal portion of the nerve at random and frequently are guided to and make synapses with inappropriate targets.

The evidence that regenerating axons can grow to predestined targets raises the question of how accurately nerve cells can reconnect in a system where individual identified neurons, rather than whole populations of cells, can be examined. A convenient preparation for a study of the precision of regenerating nerve fibers is the leech central nervous system (Chapter 13), where individual cells can be recognized without ambiguity and their connections in normal animals traced.

Regeneration in the invertebrate central nervous system

[86]Purves, D., Thompson, W. and Yip, J. W. 1981. *J. Physiol.* 313: 49–63.

[87]Wigston, D. J. and Sanes, J. R. 1985. *J. Neurosci.* 5: 1208–1221.

[88]Laskowski, M. B. and Sanes, J. R. 1988. *J. Neurosci.* 8: 3094–3099.

[89]Fawcett, J. W. and Keynes, R. J. 1990. *Annu. Rev. Neurosci.* 13: 43–60.

[90]Dennis, M. J. and Yip, J. W. 1978. *J. Physiol.* 274: 299–310.

[91]Frank, E. et al. 1975. *J. Physiol.* 247: 725–743.

[92]Scott, S. A. 1975. *Science* 189: 644–646.

[93]Kuffler, D. P., Thompson, W. and Jansen, J. K. S. 1980. *Proc. R. Soc. Lond. B* 208: 189–222.

[94]Bixby, J. L. and Van Essen, D. C. 1979. *Nature* 282: 726–728.

To observe regeneration in the leech, the procedure is to sever the axons that link two ganglia and test whether the connections become reestablished.[95] For example, individual sensory cells in one ganglion are known to cause synaptic potentials in specific motor cells in the next ganglion. Such connections between identified neurons can, in fact, be successfully reestablished, showing that individual cells can discriminate among many targets so as to interact once more with particular neurons in preference to others. There is also anatomical evidence for accurate regeneration in the leech CNS. By examining the time course of regeneration and the pathways taken by regenerating axons of touch sensory neurons, Muller, Macagno, and their colleagues found that regenerating axons grow preferentially along their normal pathway (as marked by axons of their homologues) and on occasion fuse with their severed distal segment to reestablish appropriate synaptic contacts[96,97] (Figure 11). A second example is provided by the S cell.[98] Within each ganglion there is a single S cell, which sends an axon toward each of the neighboring ganglia. This axon can be clearly recognized in cross sections of the connectives that link adjacent ganglia since it has by far the largest diameter. In the midregion of the connective, each S cell axon forms an electrical connection exclusively with the axon from the S cell situated in the adjacent ganglion. When the connective is cut or crushed, the regenerating portion of the axon reforms electrical synapses selectively, so that transmission becomes reestablished with the S cell in the neighboring ganglion. Interestingly, the initial contact is often made with the axon's own surviving distal stump and only later with the axon of the other S cell. The main point is the precision with which the target is recognized. No extraneous connections are made. An axon can also become reconnected to its old distal stump in the peripheral nervous system of crustaceans.[99] There is no evidence that repair by such mechanisms can occur in vertebrates, although in cold-blooded vertebrates, and even in mammals under certain circumstances, the distal stumps of severed axons can survive for weeks.[100]

Other invertebrates, such as the cricket, provide additional examples of regeneration with a high degree of precision.[101–104] In these creatures it has been shown anatomically and physiologically that mechanorecep-

[95]Muller, K. J. and Nicholls, J. G. 1981. *In* K. J. Muller, J. G. Nicholls and G. S. Stent (eds.). *Neurobiology of the Leech.* Cold Spring Harbor Laboratory, Cold Spring Harbor, NY.

[96]DeRiemer, S. A. et al. 1983. *Brain Res.* 272: 157–161.

[97]Macagno, E. R., Muller, K. J. and DeRiemer, S. A. 1985. *J. Neurosci.* 5: 2510–1521.

[98]Scott, S. A. and Muller, K. J. 1980. *Dev. Biol.* 80: 345–363.

[99]Hoy, R., Bittner, G. D. and Kennedy, D. 1967. *Science* 156: 251–252.

[100]Bittner, G. D. 1991. *Trends Neurosci.* 14: 188–193.

[101]Edwards, J. S. and Palka, J. 1976. *In* J. Fentress (ed.). *Simpler Networks and Behavior.* Sinauer, Sunderland, MA, pp. 167–185.

[102]Anderson, H., Edwards, J. S. and Palka, J. 1980. *Annu. Rev. Neurosci.* 3: 97–139.

[103]Murphey, R. K., Johnson, S. E. and Sakaguchi, D. S. 1983. *J. Neurosci.* 3: 312–325.

[104]Chiba, A. and Murphey, R. K. 1991. *J. Neurobiol.* 22: 130–142.

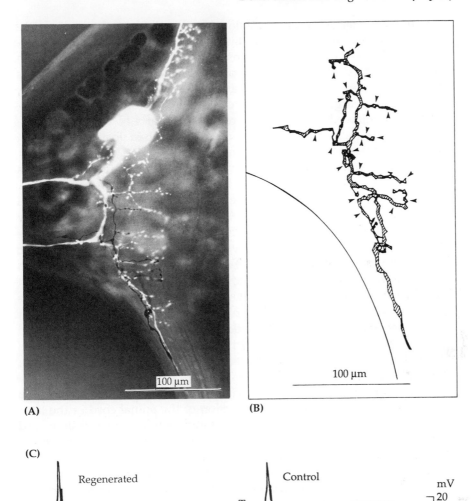

11 PARTIAL REGENERATION OF A TOUCH SENSORY CELL AXON in the central nervous system of the leech. Normally each touch sensory cell makes synaptic connections with its homologues in adjacent ganglia. Two months after a touch sensory cell axon was severed, the lesioned touch sensory cell was injected with horseradish peroxidase, and a target touch sensory cell in the adjacent ganglion was injected with a fluorescent dye. (A) The regenerated axon, filled with dark HRP reaction product, grew along the axon of the target cell, as is seen in normal ganglia. (B) Drawing of the regenerated axon, showing 22 points of apparent contact (arrowheads) with the process of the target cell. In normal ganglia, there are 40 to 60 points of contact. (C) Intracellular recordings from another pair of touch sensory cells. The regenerated axon (T_{12}) evoked a synaptic potential in the target cell (T_{11}) similar to that seen in control ganglia. (After Macagno, Muller and DeRiemer, 1985; micrograph courtesy of K. J. Muller.)

tor fibers regenerate to form functional connections on the appropriate neurons within the central nervous system. Even axons from supernumerary sensory neurons in transplanted appendages regenerate in an orderly manner to share appropriate targets in the central nervous system with their normal counterparts. These experiments also provide evidence for reorganization of the central connections when regeneration is not allowed to occur after removal of parts of the sensory apparatus. The neurons within the centers that had lost their normal inputs now receive innervation from elsewhere.

Regeneration in the CNS of lower vertebrates

The regenerative powers of neurons in the central nervous systems of cold-blooded anamniotic vertebrates, such as fish and amphibians, were first demonstrated by Matthey, who in the 1920s sectioned the optic nerve of a newt and found that vision was restored within a few weeks.[105] Beginning in the 1940s, Sperry, Stone, and their colleagues took advantage of this regenerative capacity to explore how specific connections formed within the nervous system. Their experiments on regenerating retinotectal connections provided support for the idea that neurons innervated targets selectively, rather than making connections at random that were later reorganized (see Chapter 11). Subsequent experiments have shown that selectivity is not achieved by the rigid point-to-point matching that Sperry espoused, but rather through graded preferences that allow continual reorganization of connections, even in the adult.[60]

The ability to regenerate specific functional connections has also been demonstrated following transection of the spinal cord in the lamprey.[106] Regenerating axons grow across the site of spinal lesions and make functional synaptic connections with appropriate targets, restoring coordinated behavior. In this case, however, regeneration is anatomically and physiologically incomplete; axons regenerate only short distances past the site of the lesion, and regenerated connections are less robust than normal.

Regeneration in the mammalian CNS

It has been generally believed that regrowth of cut axons in the adult mammalian central nervous system is quite restricted, largely because transection of tracts is not followed by regeneration and restitution of function. However, it has become apparent through the work of Aguayo and his colleagues that axons in the central nervous system can grow for distances of several centimeters under suitable circumstances.[107-109] Of prime importance for axonal regeneration is the immediate environment encountered by growth cones, an environment provided by Schwann cells in the periphery and astrocytes and oligodendrocytes in the central nervous system (Chapter 6). Clues to the role of glial cells are provided by several types of experiment. First, it is well known that

[105]Matthey, R. 1925. *C.R. Soc. Biol.* 93: 904–906.

[106]Cohen, A. H., Mackler, S. A. and Selzer, M. 1988. *Trends Neurosci.* 11: 227–231.

[107]David, S. and Aguayo, A. J. 1981. *Science* 214: 931–933.

[108]Aguayo, A. J. et al. 1990. *J. Exp. Biol.* 153: 199–224.

[109]Aguayo, A. J. et al. 1991. *Philos. Trans. R. Soc. Lond. B* 331: 337–343.

motor neurons, whose cell bodies lie within the spinal cord, can regenerate severed peripheral axons. However, motor axons lesioned within the central nervous system sprout little if at all. Similarly, sensory axons regrow to their targets in the periphery, and if a sensory root running toward the spinal cord is cut close to the dorsal root ganglion, axons start to grow from the cell bodies toward the cord. But they stop growing once they reach the surface of the central nervous system. Central nervous system glial cells can also prevent growth of peripheral axons.[110] Thus, although regenerating axons of the mouse sciatic nerve grow back to the periphery through a chain of remaining Schwann cells, they fail to enter chains of astrocytes and oligodendrocytes placed in their path by transplantation of a segment of optic nerve (the retina and optic nerve are part of the CNS). Together these findings suggest that (1) glial cells can actively inhibit growth, or (2) Schwann cells can provide factors that stimulate the growth of injured neurons.

Both mechanisms have been shown to play a role. As described above, following damage to a peripheral nerve the Schwann cells phagocytize the degenerating axons, multiply, and begin to synthesize NGF and NGF receptors. As a result NGF becomes bound on the surface of the Schwann cells, where it is available to help support regenerating axons. On the other hand, Schwab and his colleagues found that glial cells from the mature central nervous system have molecules on their surface that inhibit neurite outgrowth in culture.[111] They identified one such factor, raised antibodies to it, and found that in the presence of these antibodies axons could successfully regenerate across a lesion in the CNS[112] (Figure 12). Unfortunately, the extent of successful regen-

[110]Aguayo, A. J. et al. 1978. *Neurosci. Let.* 9: 97–104.
[111]Schwab, M. E. and Caroni, P. 1988. *J. Neurosci.* 8: 2381–2393.
[112]Schnell, L. and Schwab, M. E. 1990. *Nature* 343: 269–272.

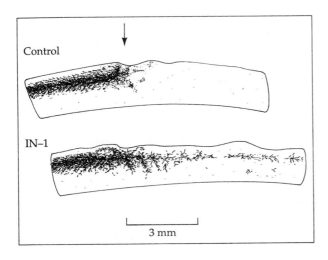

12

PROTEINS ON THE SURFACE OF CNS GLIAL CELLS INHIBIT REGENERATION of axons in the spinal cord of the rat. Camera lucida reconstructions of longitudinal sections of spinal cords showing regenerating corticospinal axons 2 to 3 weeks after the corticospinal tract was severed. Regenerating axons had grown around the lesion (arrow) and descended long distances down the spinal cord in rats treated with antibodies to a myelin protein that inhibits axon growth (IN-1); no such elongation was observed in rats treated with control antibody (control). (After Schnell and Schwab, 1990.)

eration under such conditions is still meager, suggesting that additional factors restrict regeneration in the adult vertebrate CNS.[113]

The immature CNS appears to provide a more favorable environment for regeneration. Thus, if the spinal cord of a neonatal opossum is crushed, axons grow across the lesion and conduction through the damaged region is restored within a few days, even when the spinal cord is removed from the animal and maintained in culture.[114] A striking feature of the opossum spinal cord at this age is the absence of myelin and the small number of glial cells it contains.

An additional approach to studying the ability of CNS neurons to regenerate connections has been to transplant Schwann cells, which produce a favorable environment for growth, into the central nervous system. For example, when a segment of sciatic nerve is grafted between the cut ends of the spinal cord in a mouse, fibers grow across and fill the gap.[115] (Such a graft is composed of Schwann cells and connective tissue; the axons will degenerate.) Similarly, cultures of Schwann cells implanted into the spinal cord promote growth. A dramatic effect is observed by the use of "bridges" of the type shown in Figure 13. One end of a sciatic nerve graft is implanted into the spinal cord, the other into a higher region of the nervous system (upper spinal cord, medulla, or thalamus).[107] Bridges have even been made to extend from cortex to another part of the CNS or to muscle. After several weeks or months, the graft resembles a normal nerve trunk filled with myelinated and unmyelinated axons. These neurons fire impulses and are electrically

[113]Schwab, M. E. 1991. *Philos. Trans. R. Soc. Lond. B* 331: 303–306.

[114]Treherne, J. M. et al. 1992. *Proc. Natl. Acad. Sci. USA* 89: 431–434.

[115]Richardson, P. M., McGuinness, U. M. and Aguayo, A. J. 1980. *Nature* 284: 264–265.

13

BRIDGES BETWEEN MEDULLA AND SPINAL CORD enable neurons within the central nervous system to grow for prolonged distances. The graft consists of a segment of adult rat sciatic nerve in which axons have degenerated, leaving Schwann cells. These act as a conduit along which central axons can grow. (A) Sites of insertion of the graft. Neurons are labeled by cutting the graft and applying HRP to the cut ends. (B) Positions of 1472 neuronal cell bodies labeled by retrograde transport of HRP in seven grafted rats. Most of the cells sending axons into the graft are situated close to its points of insertion. (After David and Aguayo, 1981.)

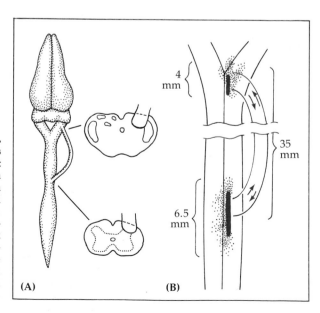

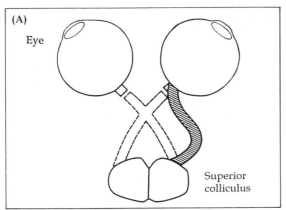

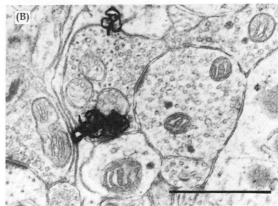

14 RECONNECTION OF THE RETINA AND SUPERIOR COLLICULUS through a peripheral nerve graft in an adult rat. (A) The optic nerves were severed, and one was replaced by a 3- to 4-cm segment of the peroneal nerve (cross-hatched). Regeneration was tested by injecting anterograde tracers into the eye or recording responses of superior colliculus neurons to light flashed onto the retina. (B) Electron microscope autoradiogram of a regenerated retinal ganglion cell axon terminal in the superior colliculus. [³H]-Labeled amino acids were injected into the eye two days before the brain was fixed and sectioned; silver grains exposed by radiolabeled proteins transported from the injected eye identify ganglion cell axon terminals. The regenerated terminal resembles those seen in control animals; it is filled with round synaptic vesicles and forms asymmetric synapses. Bar, 1 μm. (After Vidal-Sanz, Bray and Aguayo, 1991; micrograph courtesy of A. J. Aguayo.)

excited or inhibited by stimuli applied above or below the sites of implantation. By cutting the bridge and dipping the cut ends into horseradish peroxidase or other markers, the cells of origin become labeled and their distribution can be mapped. Examples such as those in Figure 13 show that the axons that have grown over distances of several centimeters arise from neurons whose cell bodies lie within the central nervous system. Usually only those neurons with cell bodies not more than a few millimeters from the bridge send axons into it. Similarly, axons leaving the bridge to enter the central nervous system grow only a short distance before terminating.

Can axons regenerating in the central nervous systems of vertebrates locate their correct targets and make functional synapses? Experiments on regenerating retinal ganglion cell axons indicate that they can.[109] If the optic nerve is cut and a bridge inserted between the eye and the superior colliculus, regenerating retinal ganglion cell axons grow through the bridge, extend into the target, arborize, and form synapses (Figure 14). The regenerated synapses are formed on the correct regions of their target cells, have a normal structure when visualized by electron microscopy, and are functional, in that the postsynaptic cells can be driven by illumination of the eye.

An important finding from these studies is that many ganglion cells in the retina die shortly after section of their axons in the optic nerve. Some cells are prevented from dying when axons regenerate through

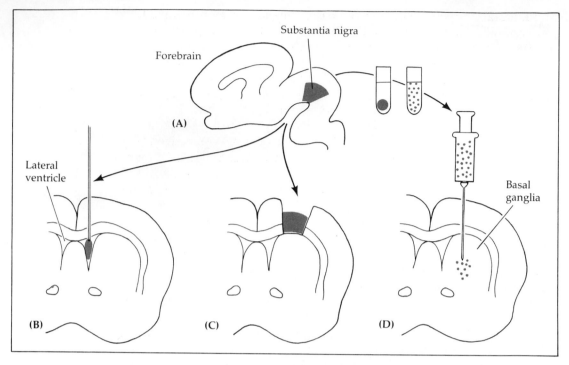

15   **PROCEDURES FOR TRANSPLANTING EMBRYONIC TISSUE into adult rat brain. Tissue rich in cells containing dopamine is dissected from the substantia nigra (A) and is injected into the lateral ventricle (B) or grafted into a cavity in the cortex overlying the basal ganglia (C). Alternatively, a suspension of dissociated substantia nigra cells can be injected directly into the basal ganglia (D). Such embryonic cells survive, sprout, and secrete transmitter. (After Dunnett, Björklund and Stenevi, 1983.)**

the Schwann cell bridge. However, even after axons have regenerated through the bridge and into the CNS, ganglion cell death continues. Apparently the retinal ganglion cells require additional trophic substances for their survival.

A different approach used by Björklund, Gage, Dunnett, Lund, Sotelo, Vrbova, Lindvall, and their colleagues to study regeneration in the adult mammalian central nervous system has been to transplant embryonic nerve cells into the adult brain.[116] Unlike neurons from adult central nervous system, which die following transplantation, explants of brain tissue or dissociated cells taken from fetal or neonatal animals survive and grow after being inserted into the adult central nervous system (Figure 15). There they differentiate, extend axons, and release transmitters. An example of this is provided by experiments in which neurons were transplanted into the basal ganglia of rats after destruction of dopamine-containing neurons in the substantia nigra.[117] In normal animals, the dopaminergic neurons in the substantia nigra (a region in

[116]Björklund, A. 1991. *Trends Neurosci.* 14: 319–322.
[117]Ridley, R. M. and Baker, H. F. 1991. *Trends Neurosci.* 14: 366–370.

the midbrain) innervate cells in the basal ganglia (a region involved in programming movements; see Chapters 10 and 15 and Appendix C). If a lesion of this dopamine pathway is made on one side of a rat, a disorder of movement results; the animal turns toward the side of the lesion spontaneously or in response to stress. This asymmetry of movement can be counteracted by grafts of dopamine-containing neurons taken from the substantia nigra of immature animals and placed in appropriate sites on the lesioned side. Fibers grow into the host tissue from the graft and the animals cease to rotate in response to the stressful stimuli.[118] Ultrastructural studies show that axons originating in neuronal grafts form synapses with host neurons. Dopamine released by the donor cells acts in a modulatory manner at synapses on host neurons; for such a diffuse "garden sprinkler" type of action, specific one-to-one cell contacts probably are not required (Chapters 8 and 10).

The prospects for functional recovery following grafts seem greatest for neurons that exert their effects through such neuromodulatory mechanisms, rather than those that rely on reconstitution of precise synaptic connections. However, integration of transplanted neurons into the appropriate synaptic circuitry has been shown to occur after grafting embryonic tissue into lesioned adult cortex, hippocampus, and striatum.[116] For instance, a fetal retina transplanted into a neonatal rat brain is capable of making specific functional connections, thereby restoring appropriate visual reflexes.[119] A particularly remarkable example is the anatomical and functional integration of transplanted cerebellar Purkinje cells in the adult PCD (Purkinje-cell-degeneration) mouse (Figure 16).[120] Sotelo and his colleagues have grafted dissociated cells or solid pieces of the cerebellar primordium into the cerebellum of adult mutant mice, whose cerebellar Purkinje cells degenerate shortly after birth. During the first week following transplantation, donor Purkinje cells migrate out of the graft across the cerebellar surface and then turn and migrate radially into the molecular layer toward the positions originally occupied by the degenerated Purkinje cells. Within two weeks many transplanted cells have formed dendritic arbors that resemble those of normal Purkinje cells; climbing fibers have formed synapses, first on the cell body and then on the proximal dendrites, and parallel fibers have innervated the distal dendrites. Characteristic synaptic potentials can be recorded following stimulation of the climbing fiber and mossy fiber inputs. However, the implanted cells rarely succeed in establishing synaptic connections with their normal targets in the deep cerebellar nuclei of the host, innervating instead nearby donor deep nuclear neurons that survive in the remnant of the graft. Nevertheless, these experiments demonstrate that transplanted cells can become incorporated into the synaptic circuitry of an adult host to a remarkable extent.

[118]Björklund, A. et al. 1980. *Brain Res.* 199: 307–333.
[119]Klassen, H. and Lund, R. D. 1990. *J. Neurosci.* 10: 578–587.
[120]Sotelo, C. and Alvarado-Mallart, R. M. 1991. *Trends Neurosci.* 14: 350–355.

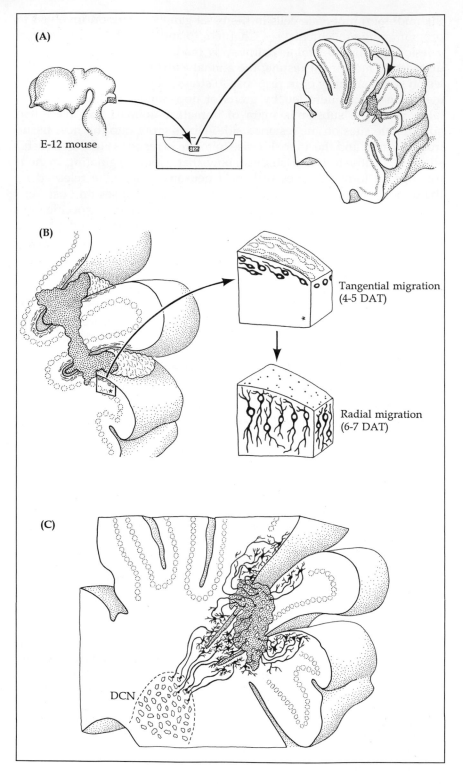

(A)

E-12 mouse

(B)

Tangential migration
(4-5 DAT)

Radial migration
(6-7 DAT)

(C)

DCN

◀ **16**   RECONSTRUCTION OF CEREBELLAR CIRCUITS by transplantation of embryonic Purkinje cells into an adult *pcd* mouse. (A) Solid pieces of cerebellar primordium from a 12-day embryo were injected into the cerebellum of a 2- to 4-month old *pcd* mouse. (B) Purkinje cells migrate out of the graft (stipled region), first tangentially along the cerebellar surface (4–5 days after transplantation, DAT) and later radially inward, penetrating the host molecular layer (6–7 DAT). (C) Axons of Purkinje cells distant from the deep cerebellar nuclei (DCN) of the host innervate deep nuclear neurons in the graft remnant (stipled region). Axons of Purkinje cells within 600 μm of the deep cerebellar nuclei innervate the host nuclei, where they make synaptic contacts on their specific targets. (After Sotelo and Alvarado-Mallart, 1991.)

It is clear that neurons in the vertebrate central nervous system retain in the adult an ability to grow and establish appropriate synaptic connections. In addition, transplantation of embryonic cells and neural tissues into adult CNS has become an accepted tool for the study of mechanisms mediating the development and regeneration of neuronal connections in experimental animals. Such work provides hope for the amelioration of functional deficits resulting from CNS lesions and neurodegenerative diseases.

SUGGESTED READING

General Reviews

Aguayo, A. J., Rasminsky, M., Bray, G. M., Carbonetto, S., McKerracher, L., Villegas-Perez, M. P., Vidal-Sanz, M. and Carter, D. A. 1991. Degenerative and regenerative responses of injured neurons in the central nervous system of adult mammals. *Philos. Trans. R. Soc. Lond. B* 331: 337–343.

Björklund, A. 1991. Neural transplantation: An experimental tool with clinical possibilities. *Trends Neurosci.* 14: 319–322.

Brown, M. C., Holland, R. L. and Hopkins, W. G. 1981. Motor nerve sprouting. *Annu. Rev. Neurosci.* 4: 17–42.

Cohen, A. H., Mackler, S. A. and Selzer, M. 1988. Behavioral recovery following spinal transection: Functional regeneration in the lamprey CNS. *Trends Neurosci.* 11: 227–231.

Fawcett, J. W. and Keynes, R. J. 1990. Peripheral nerve regeneration. *Annu. Rev. Neurosci.* 13: 43–60.

Johnson, E. M. Jr., Taniuchi, M. and DiStefano, P. S. 1988. Expression and possible function of nerve growth factor receptors on Schwann cells. *Trends Neurosci.* 11: 299–304.

Kass, J. H. 1991. Plasticity of sensory and motor maps in adult mammals. *Annu. Rev. Neurosci.* 14: 137–167.

McMahan, U. J. 1990. The agrin hypothesis. *Cold Spring Harbor Symp. Quant. Biol.* 50: 407–418.

Purves, D. and Lichtman, J. W. 1985. *Principles of Neural Development.* Sinauer, Sunderland, MA.

Rotshenker, S. 1988. Multiple modes and sites for the induction of axonal growth. *Trends Neurosci.* 11: 363–366.

Scheutze, S. M. and Role, L. M. 1987. Developmental regulation of nicotinic acetylcholine receptors. *Annu. Rev. Neurosci.* 10: 403–457.

Schwab, M. E. 1991. Nerve fibre regeneration after traumatic lesions of the CNS; Progress and problems. *Philos. Trans. R. Soc. Lond. B* 331: 303–306.

Sotelo, C. and Alvarado-Mallart, R. M. 1991. The reconstruction of cerebellar circuits. *Trends Neurosci.* 14: 350–355.

Tsukahara, N. 1981. Synaptic plasticity in the mammalian central nervous system. *Annu. Rev. Neurosci.* 4: 351–379.

Original Papers

EFFECTS OF DENERVATION ON THE POSTSYNAPTIC CELL

Axelsson, J. and Thesleff, S. 1959. A study of supersensitivity in denervated mammalian skeletal mucle. *J. Physiol.* 147: 178–193.

Fumagalli, G., Balbi, S., Cangiano, A. and Lømo, T. 1990. Regulation of turnover and number of acetylcholine receptors at neuromuscular junctions. *Neuron* 4: 563–569.

Lømo, T. and Rosenthal, J. 1972. Control of ACh sensitivity by muscle activity in the rat. *J. Physiol.* 221: 493–513.

Miledi, R. 1960. The acetylcholine sensitivity of frog muscle fibres after complete or partial denervation. *J. Physiol.* 151: 1–23.

Mishina, M., Takai, T., Imoto, K., Noda, M., Takahashi, T., Numa, S., Methfessel, C. and Sakmann, B. 1986. Molecular distinction between fetal and adult forms of muscle acetylcholine receptor. *Nature* 321: 406–411.

Sargent, P. B. and Pang, D. Z. 1988. Denervation alters the size, number, and distribution of clusters of acetylcholine receptor-like molecules on frog cardiac ganglion neurons. *Neuron* 1: 877–886.

Witzemann, V., Brenner, H.-R. and Sakmann, B. 1991. Neural factors regulate AChR subunit mRNAs at rat neuromuscular synapses. *J. Cell Biol.* 114: 125–141.

DENERVATION SUPERSENSITIVITY, SUSCEPTIBILITY TO INNERVATION, AND SPROUTING

Blackshaw, S. E., Nicholls, J. G. and Parnas, I. 1982. Expanded receptive fields of cutaneous mechanoreceptor cells after single neurone deletion in leech central nervous system. *J. Physiol.* 326: 261–268.

Bowling, D., Nicholls, J. G. and Parnas, I. 1978. Destruction of a single cell in the central nervous system of the leech as a means of analyzing its connexions and functional role. *J. Physiol.* 282: 169–180.

Jansen, J. K. S., Lømo, T., Nicholaysen, K. and Westgaard, R. H. 1973. Hyperinnervation of skeletal muscle fibers: Dependence on muscle activity. *Science* 181: 559–561.

Katsumaru, H., Murakami, F., Wu, J.-Y. and Tsukahara, N. 1986. Sprouting of GABAergic synapses in the red nucleus after lesions of the nucleus interpositus in the cat. *J. Neurosci.* 6: 2864–2874.

ROLE OF BASAL LAMINA AT REGENERATING SYNAPSES

Anglister, L. and McMahan, U. J. 1985. Basal lamina directs acetylcholinesterase accumulation at synaptic sites in regenerating muscle. *J. Cell Biol.* 101: 735–743.

McMahan, U. J. 1990. The agrin hypothesis. *Cold Spring Harbor Symp. Quant. Biol.* 50: 407–418.

McMahan, U. J. and Slater, C. R. 1984. The influence of basal lamina on the accumulation of acetylcholine receptors at synaptic sites in regenerating muscle. *J. Cell Biol.* 98: 1452–1473.

Nitkin, R. M., Smith, M. A., Magill, C., Fallon, J. R., Yao, M. Y.-M., Wallace, B. G. and McMahan, U. J. 1987. Identification of agrin, a synapse organizing protein from *Torpedo* electric organ. *J. Cell Biol.* 105: 2471–2478.

Sanes, J. R., Marshall, L. M. and McMahan, U. J. 1978. Reinnervation of muscle fiber basal lamina after removal of muscle fibers. *J. Cell Biol.* 78: 176–198.

SELECTIVITY IN REGENERATION

Aguayo, A. J., Bray, G. M., Rasminsky, M., Zwimpfer, T., Carter, D. and Vidal-Sanz, M. 1990. Synaptic connections made by axons regenerating in the central nervous system of adult mammals. *J. Exp. Biol.* 153: 199–224.

Björklund, A., Dunnett, S. B., Stenevi, U., Lewis, N. E. and Iversen, S. D. 1980. Reinnervation of the denervated striatum by substantia nigra transplants: Functional consequences as revealed by pharmacological and sensorimotor testing. *Brain Res.* 199: 307–333.

David, S. and Aguayo, A. J. 1981. Axonal elongation into peripheral nervous system "bridges" after central nervous system injury in adult rats. *Science* 214: 931–933.

Macagno, E. R., Muller, K. J. and DeRiemer, S. A. 1985. Regeneration of axons and synaptic connections by touch sensory neurons in the leech central nervous system. *J. Neurosci.* 5: 2510–2521.

Schnell, L. and Schwab, M. E. 1990. Axonal regeneration in the rat spinal cord produced by an antibody against myelin-associated neurite growth inhibitors. *Nature* 343: 269–272.

Schwab, M. E. and Caroni, P. 1988. Oligodendrocytes and CNS myelin are nonpermissive substrates for neurite growth and fibroblast spreading in vitro. *J. Neurosci.* 8: 2381–2393.

PART FOUR

Integrative Mechanisms

CHAPTER THIRTEEN

LEECH AND *APLYSIA*: TWO SIMPLE NERVOUS SYSTEMS

Invertebrates perform various complex tasks such as flying or swimming, escaping from danger, or feeding. Yet their nervous systems are made up of relatively few neurons. With large, identified cells of known function it becomes possible to study signaling, reflexes, and the coordinated behavior of the animal as a whole at the cellular and molecular level to an exceptionally full extent. The nervous systems of invertebrates also lend themselves well to following the formation of ganglia and specific connections during development and repair of the nervous system.

Two animals—the leech (a worm) and the sea hare *Aplysia* (a mollusk)—are used to illustrate this approach. Their central nervous systems consist of discrete aggregates of neurons—the ganglia— which are linked to each other and to the periphery by bundles of axons. In *Aplysia* the ganglia contain several thousand nerve cells, whereas those in the leech are smaller and contain only about 400 nerve cells. In both animals the nerve cells can be seen clearly under the dissecting microscope and can be impaled with microelectrodes. Individual sensory and motor nerve cells (as well as interneurons) have been identified in adult animals and in embryos, their synaptic connections traced, and the fields they innervate in the periphery defined. This information is an essential requirement for tracing the normal wiring diagram of the nervous system, for determining how connections are made, and for studying how the properties of the neurons change with repeated use.

In ganglia of the leech and *Aplysia*, one can correlate reflex movements of the animal with the physiological characteristics of synapses between identified sensory and motor nerve cells. For example, when impulses are initiated at their normal frequencies in sensory neurons by natural stimulation of the skin, certain chemical synapses show sequentially both facilitation and depression. The reflexes mediated by cells connected through such synapses reflect this sequence to produce sequential motor acts that can be observed readily. At higher levels of integration, coordinated movements and hormonally regulated events, such as swimming and egg laying, have been worked out in terms of the individual cells involved. Persistent changes in *Aplysia* synapses that last for hours or even days are produced by regular stimuli applied to the skin. The depression or augmentation of synaptic transmission can explain the way in which the animal becomes habituated or sensitized to stimuli. Moreover, a detailed, coherent account of molecular mechanisms is available from experiments in which second messengers, gene regulation, and protein synthesis have been analyzed. Many of the phenomena observed in the animal are faithfully reproduced by small circuits formed by identified neurons in tissue culture.

Throughout this book invertebrate neurons are used to illustrate mechanisms that are significant for understanding signaling in higher animals. From the squid axon, for example, were derived principles of nearly universal validity for the conduction of impulses; cells of other invertebrates were used to illustrate the channel mechanisms responsible for synaptic inhibition and excitation in vertebrates. The reasons for choosing invertebrate preparations are usually technical. Certain problems can be solved more easily in invertebrate nervous systems, which survive well in isolation; in addition, if their cells are large and readily accessible, they can be recognized and studied with electrical recording methods and biochemical techniques.

Curiosity about the nervous systems of invertebrates, however, goes beyond basic mechanisms of conduction and synaptic transmission. The way in which an ant navigates, a bee dances, a cricket sings, or a fly flies are all problems of first-order interest.[1] Invertebrates provide the opportunity for following the thread from development of a single, particular identified cell through the elaboration of its branching pattern and the formation of its connections; analyzing the biophysical properties of the cell and observing how they are correlated with the coordinated behavior of the whole animal; and discovering the molecular events occurring in that cell as the animal modifies its behavior in response to the outside influences playing upon it. Some behavioral responses are performed by relatively few neurons, whereas analogous responses in mammals require many thousands of neurons. Each type of preparation has its own set of advantages for approaching specific problems. The neural circuits for coordinated elementary units of behavior, such as postural reflexes, feeding, circadian rhythms, escape reactions, and swimming, have been traced in crayfish and snails.[2-4] Central nervous systems of insects have been used to study a variety of problems including development,[5,6] flight,[7] navigation,[8] walking,[9] and regeneration,[10] as well as communication by sound.[11]

[1]Wehner, R. and Menzel, R. 1990. *Annu. Rev. Neurosci.* 13: 403–414.

[2]Elliott, C. J. H. and Benjamin, P. R. 1989. *J. Neurophysiol.* 61: 727–736.

[3]Sahley, C. L., Martin, K. A. and Gelperin, A. 1990. *J. Comp. Physiol. (A)* 167: 339–345.

[4]Aréchiga, H. et al. 1985. *Amer. Zool.* 25: 265–274.

[5]Bentley, D. and Toroian-Raymond, A. 1989. *J. Exp. Zool.* 251: 217–223.

[6]Palka, J. and Schubiger, M. 1988. *Trends Neurosci.* 11: 515–517.

[7]Reichert, H. and Rowell, C. H. 1985. *J. Neurophysiol.* 53: 1201–1218.

[8]Wehner, R. 1989. *Trends Neurosci.* 12: 353–359.

[9]Ramirez, J. M. and Pearson, K. G. 1988. *J. Neurobiol.* 19: 257–282.

[10]Edwards, J. S., Reddy, G. R. and Rani, M. U. 1989. *J. Neurobiol.* 20: 101–114.

[11]Hoy, R. R. 1989. *Annu. Rev. Neurosci.* 12: 355–375.

The animals themselves and the scope of the problems are so varied that a comprehensive review is impossible. Indeed, monographs abound on *Aplysia*,[12] leech,[13,14] the marine snail *Hermissenda*,[15] crayfish[16] and isolated invertebrate neurons in culture.[17] Consequently, we have singled out for fuller discussion the nervous systems of just two invertebrates: the leech (mentioned in Chapters 6 and 12 in relation to glial cells and regeneration), and *Aplysia* (mentioned in Chapters 8 and 9 in relation to synaptic plasticity). These compact nervous systems contain relatively few cells arranged in an orderly manner; yet they provide useful preparations for the study of problems of neural organization found in the brains of higher animals. At the same time, the ganglia use synaptic mechanisms similar to those that have been identified in vertebrates. Figures 1, 2, 3, and 4 show the principal features of the animals and of their central nervous systems. In the leech, the central nervous system consists of a chain of stereotyped ganglia; in *Aplysia* the ganglia are considerably larger and are specialized in different regions of the animal, thereby controlling different functions.

Working at the level of single cells, one can identify the individual neurons that mediate reflexes, trace their synaptic connections, and establish whether they are chemical or electrical, excitatory or inhibitory. This sets the stage for studying how neural components act in concert to produce coordinated stereotyped behavior. As a next step, one can investigate how the properties of neurons and synapses change as a result of repeated natural sensory stimuli and how these changes are reflected in the performance of the animal. The advantage of using natural stimuli is that only in this way is one assured that impulses are arising in the appropriate fibers at frequencies similar to those in the intact animal. Knowledge of the circuitry in the adult nervous systems of the leech and *Aplysia* makes it possible to study development at the cellular and molecular level, trace cell lineage, follow the growth of individual axons, measure changes in membrane properties, and assess the specificity and mechanisms involved in synapse formation. The embryos are transparent and are made up of large cells that develop rapidly—in days rather than weeks—into ganglia. In addition, the nervous system of the leech can repair itself after a lesion (discussed later).

Aplysia   The gastropod ("belly-footed") mollusk *Aplysia* has a body with well-defined parts: a head, a foot, a body, and a mantle that consists of a skirtlike flange of tissue that covers the mass of viscera.[12] Among its

[12]Kandel, E. R. 1979. *Behavioral Biology of Aplysia*. W. H. Freeman, San Francisco.

[13]Muller, K. J., Nicholls, J. G. and Stent, G. S. (eds.). 1981. *Neurobiology of the Leech*. Cold Spring Harbor Laboratory, Cold Spring Harbor, NY.

[14]Nicholls, J. G. 1987. *The Search for Connections: Studies of Regeneration in the Nervous System of the Leech*. Sinauer, Sunderland, MA.

[15]Alkon, D. 1987. *Memory Traces in the Brain*. Cambridge University Press, Cambridge.

[16]Atwood, H. L. and Sandeman, D. C. (eds.). 1982 *Biology of Crustacea*, Vol. 3. Academic Press, New York.

[17]Beadle, D. J., Lees, G. and Kater, S. B. 1988. *Cell Culture Approaches to Invertebrate Neuroscience*. Academic Press, London.

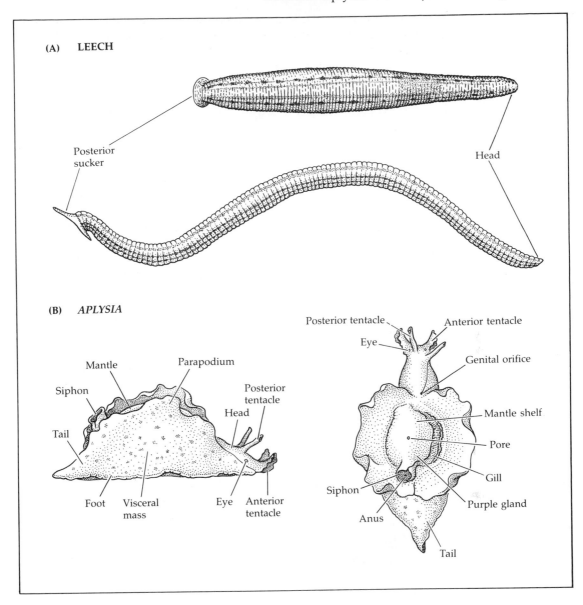

(A)   LEECH

Posterior
sucker

Head

(B)   *APLYSIA*

Posterior tentacle

Anterior tentacle

Eye

Genital orifice

Mantle shelf

Pore

Gill

Purple gland

Siphon

Anus

Tail

Mantle

Parapodium

Siphon

Posterior
tentacle

Head

Tail

Foot   Visceral
mass

Eye   Anterior
tentacle

1   ANIMALS WITH SIMPLE NERVOUS SYSTEMS. (A) The leech, *Hirudo medicin-alis*, has a segmented body with a sucker at each end. The animal can measure 5 inches in length after feeding. **(B)** *Aplysia californica*, the sea hare, viewed from the side and from above, with the parapodia retracted so as to show mantle, gill, and siphon. **(B** modified from Kandel, 1976.)

specialized organs are tentacles, eyes, gills, and a gland that squirts jets of inky purple fluid when the animal is disturbed. The *Aplysia* lives in seawater and its behavior, like that of the leech, is limited compared with that of other invertebrates such as bees, ants, or crickets. *Aplysia* walk, withdraw defensively in response to noxious stimuli, make res-

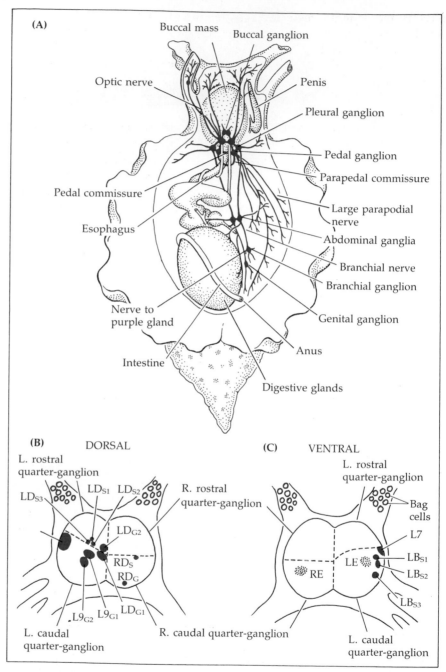

(A)

Buccal mass
Buccal ganglion
Optic nerve
Penis
Pleural ganglion
Pedal ganglion
Parapedal commissure
Pedal commissure
Large parapodial nerve
Esophagus
Abdominal ganglia
Branchial nerve
Branchial ganglion
Nerve to purple gland
Genital ganglion
Intestine
Anus
Digestive glands

(B) DORSAL

L. rostral quarter-ganglion
LD$_{S3}$
LD$_{S1}$
LD$_{S2}$
R. rostral quarter-ganglion
LD$_{G2}$
RD$_S$
RD$_G$
L9$_{G2}$
L9$_{G1}$
LD$_{G1}$
L. caudal quarter-ganglion
R. caudal quarter-ganglion

(C) VENTRAL

L. rostral quarter-ganglion
Bag cells
L7
LB$_{S1}$
RE
LE
LB$_{S2}$
LB$_{S3}$
L. caudal quarter-ganglion

2 CENTRAL NERVOUS SYSTEM OF *APLYSIA*. (A) The various ganglia and their connections are shown in relation to the internal organs. Experiments described in this chapter were made largely on the abdominal ganglia. (B, C) Identified cells and cell clusters of abdominal ganglion involved in gill and siphon withdrawal reflexes. The groups of cells labeled RE and LE are sensory; the other cells are motoneurons. The connections of these cells are shown in Figure 11. The bag cells secrete peptides that induce egg laying. (Modified from Kandel, 1976.)

piratory movements, feed, lay eggs, squirt ink, and indulge in group sex. The name *Aplysia* comes from *aplytos,* meaning "unwashed." Since the tentacles protruding from the body can make it look like a rabbit, Pliny called the animal *Lepus marinus* (sea hare). He considered it to be venomous, and Bacon in 1626 for some reason believed "that the sea hare hath an antipathy with the lungs (if it cometh near them) and erodeth them" (surely an unlikely event).

The account that follows of the behavior and neurobiology of *Aplysia* is based largely on the pioneering studies of Castellucci, Kandel, Levitan, Carew, Byrne, and their colleagues.[12,18–20] The appeal and usefulness of the large nerve cells in *Aplysia* ganglia for neurobiology were first recognized by Arvanitaki in 1941; she showed that they could be identified, and she analyzed their electrical activity and biochemical properties.[21] Later Strumwasser showed that it was possible to record continuously from an individual cell for days on end in the animal and in culture and thereby analyze the circadian rhythm exhibited by the bursts of action potentials.[22] Tauc,[23] Kehoe,[24] Hochner,[25] Siegelbaum,[26] and others have made detailed studies of transmitter release and the properties of postsynaptic chemoreceptors, as well as of the complex events under neurohumoral control, for example, egg laying, which is regulated by specific secretory neurons (called *bag cells;* see Figure 2).[27] Thanks to the large amount of cytoplasm, biochemical analyses that would be quite impossible in leech neurons have been made in single cells of *Aplysia.*

The buccal, cerebral, pleural, pedal, and abdominal ganglia of *Aplysia* each contain more than 1000 neurons. Some of the individual cells, such as R2, R14, and R15, are more than 1 millimeter in diameter, bigger than an entire leech ganglion. Many of the identified neurons have been studied in detail, particularly in the abdominal ganglion, which controls the mantle and the viscera. Some of the cells are distinctively colored—some yellow, others orange—and the ganglion as a whole has an orange hue. In *Aplysia,* but not in the leech, neuronal circuits are present in the periphery as well as in the central nervous system and can give rise to reflex responses after the ganglia have been removed. | Central nervous system of *Aplysia*

Since the days of ancient Greece and Rome, leeches have been applied by physicians to patients suffering from diverse diseases such | The leech

[18]Adams, W. B. and Levitan, I. B. 1985. *J. Physiol.* 360: 69–93.

[19]Carew, T. J. 1989. *Trends Neurosci.* 12: 389–394.

[20]Byrne, J. H. 1987. *Physiol. Rev.* 67: 329–439.

[21]Arvanitaki, A. and Cardot, H. 1941. *C. R. Soc. Biol.* 135: 1207–1211.

[22]Strumwasser, F. 1988–1989. *J. Physiol.* (Paris) 83: 246–54.

[23]Baux, G., Fossier, P. and Tauc, L. 1990. *J. Physiol.* 429: 147–168.

[24]Kehoe, J. 1985. *J. Physiol.* 369: 439–474.

[25]Hochner, B. et al. 1986. *Proc. Natl. Acad. Sci. USA* 83: 8794–8798.

[26]Siegelbaum, S. A. et al. 1986. *J. Exp. Biol.* 124: 287–306.

[27]Arch, S. and Berry, R. W. 1989. *Brain Res. Rev.* 14: 181–201.

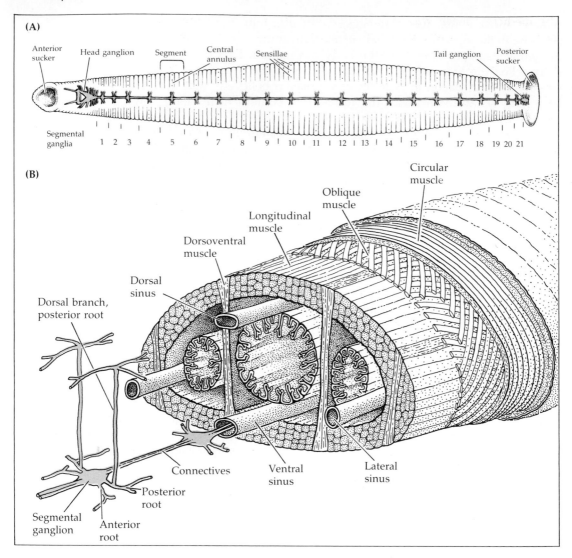

3   **CENTRAL NERVOUS SYSTEM OF THE LEECH. (A)** The CNS of the leech consists of a chain of 21 segmental ganglia, a head ganglion, and a tail ganglion. Over most of the body five circumferential annuli make up each segment; the central annulus is marked by sensory end organs responding to light and touch (the sensillae). **(B)** The nerve cord lies in the ventral part of the body within a blood sinus. Ganglia, which are linked to each other by bundles of axons (the connectives), innervate the body wall by paired roots. The muscles are arranged in three principal layers: circular, oblique, and longitudinal. In addition there are dorsoventral muscles that flatten the animal, and fibers immediately under the skin that raise it into ridges.

as epilepsy, angina, tuberculosis, meningitis, and hemorrhoids—an unpleasant treatment that almost certainly did more harm than good to the countless unfortunate victims.[28] By the nineteenth century, use of

[28]Payton, W. B. 1981. *In* K. J. Muller, J. G. Nicholls and G. S. Stent (eds.). *Neurobiology of the Leech.* Cold Spring Harbor Laboratory, Cold Spring Harbor, NY, pp. 27–34.

the medicinal leech was so prevalent that the animal became almost extinct in western Europe, forcing Napoleon to import about 6 million leeches from Hungary in one year to treat his soldiers. This mania for leeching had at least one lasting benefit for contemporary biology: The medicinal application of leeches stimulated basic research on their reproduction, development, and anatomy. Thus in the late nineteenth century, founders of experimental embryology, such as Whitman, chose the leech to follow the fates of early embryonic cells. Similarly, its nervous system was extensively studied by a roster of distinguished anatomists, including Sanchez, Ramón y Cajal, Gaskell, Del Rio Hortega, Odurih, and Retzius.[13] Interest in the leech thereafter declined, to be rekindled in 1960 when Stephen Kuffler and David Potter first applied modern neurophysiological techniques to its nervous system.[29]

Both body and nervous system are rigorously segmented and consist of a number of repeating units (segments) that are similar throughout the length of the leech (Figures 3 and 4). Since the animal has no limbs, its behavior consists of a relatively simple repertoire of movements, such as swimming, walking like an inchworm, and shortening, performed by layered groups of muscles. Each segment is innervated by a stereotyped ganglion that is much like the others within one animal and those in other animals. Even the specialized head and tail "brains" consist of fused ganglia, in which many characteristic features of segmental ganglia are still recognized. The present account focuses mainly on the work of Muller, Stent, Weisblat, Kristan, Friesen, Macagno, Parnas, Blackshaw, Fernandez, Drapeau, Calabrese, and their colleagues.

Each ganglion contains only about 400 nerve cells, which have distinctive shapes, sizes, positions, and branching patterns.[30] A ganglion innervates a well-defined territory of the body by way of paired axon bundles (ROOTS), and it communicates with neighboring and distant parts of the nervous system through another set of bundles (CONNECTIVES). Integration thus occurs in a succession of clear-cut steps:

Leech ganglia: Semiautonomous units

1. Each segmental ganglion receives information from a circumscribed body segment, the performance of which it directly regulates.
2. Neighboring ganglia influence each other by direct interconnections.
3. The coordinated operation of the whole nerve cord and the animal is governed by the brains at each end of the leech.

The distinct segmental subdivisions can be studied on their own or together.

Perhaps the main appeal of the leech is the beauty of the ganglion as it appears under the microscope, with its 400 or so neurons so recognizable and so familiar from segment to segment, from specimen to specimen, from species to species. As one looks at these limited aggregates of cells laid out in an orderly pattern, one cannot but marvel

---

[29]Kuffler, S. W. and Potter, D. D. 1964. *J. Neurophysiol.* 27: 290–320.
[30]Coggeshall, R. E. and Fawcett, D. W. 1964. *J. Neurophysiol.* 27: 229–289.

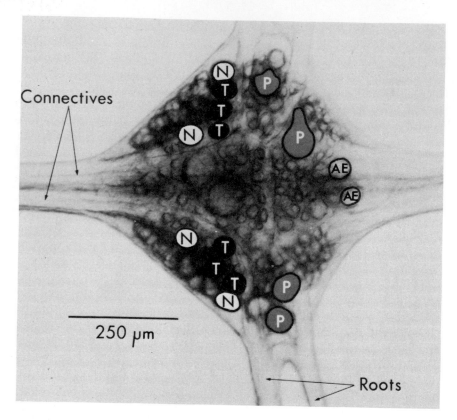

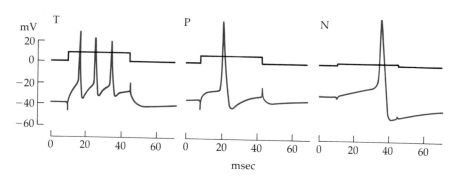

4   VENTRAL VIEW OF LEECH SEGMENTAL GANGLION. Individual cells are clearly recognized. The three sensory cells responding to touch (T) and the pairs of cell types responding to pressure (P) or noxious (N) mechanical stimulation of the skin are labeled. Each type of cell gives distinctive action potentials, as shown by the traces below. Impulses in T cells are briefer and smaller than those in P or N cells. Current injected into cells through the microelectrode is monitored on the upper traces. The cells outlined in the posterior part of the ganglion are the annulus erector (AE) motoneurons. (After Nicholls and Baylor, 1968.)

at how they, on their own, being the brain of the creature, are responsible for all its movements, hesitations, avoidance, mating, feeding, and sensations. In addition to the esthetic pleasure provided by the preparation, there is the intellectual excitement in trying to solve the circuitry and logic of a finite, well-organized nervous system, one cell at a time.

Questions concerning the way in which vertebrate neurons develop, send out their axons, form specific synapses, and establish the wiring diagram of the nervous system are considered in Chapters 6, 11, and 18. Detailed and comprehensive developmental studies have been made in embryos of a tiny nematode worm, *Caenorhabditis elegans*, the adult nervous system of which contains only 302 neurons.[31] In this nematode and in the fruit fly *Drosophila*, genetic techniques provide a powerful tool for analyzing molecular mechanisms in development. Genes have been cloned and proteins sequenced that determine the overall plan of the nervous system or the production of a single identified cell of crucial importance for function.[32] Such techniques cannot be simply applied to the larger invertebrates with their slower reproductive cycles of weeks or months.

Development of the nervous system

In leeches and *Aplysia*, as in *Drosophila*, crickets, and locusts, it is possible to follow cell division, migration, and formation of ganglia from egg to adult.[13,19,33] In leeches, which have no larval stage, four individual cells have been identified that on their own directly give rise to virtually the entire central nervous system of the animal. Moreover, cell lineage can be followed visually by injecting one of the precursor cells with a label such as horseradish peroxidase that does not interfere with the process of development yet is transferred from cell to cell during division. Figure 5, for example, shows leech eggs, a developing leech embryo, and an embryo in which one precursor cell had been inoculated with horseradish peroxidase. As a result, the nervous system on that side contains the enzyme and stains black.

Such problems as how ganglia are assembled in each segment, how the muscles in the periphery adopt their form and position, and how the individual neurons originate and establish their connections have now been tackled in leech embryos. For example, there are now many detailed experiments that show how identified precursor cells divide and migrate to form ganglia. In brief, four large ectoderm precursor cells called N, O, P, and Q TELOBLASTS develop at an early stage. Four bands of daughter cells originating from the N, O, P, and Q cells spread over the surface of the spherical egg, as shown in Figure 6. The two outside bands of N-derived cells come together and fuse at the midline. This fused line of bandlet cells will give rise to the CNS by dividing. The O, P, and Q cells also contribute certain specific nerve and glial

Generation of ganglia, target organs, and individual neurons

[31]Wood, W. B. (ed.) 1988. *The Nematode Caenorhabditis elegans.* Cold Spring Harbor Laboratory, Cold Spring Harbor, NY.
[32]Jan, Y. N. and Jan, L. Y. 1990. *Trends Neurosci.* 13: 493–498.
[33]Fernández, J. and Stent, G. 1982. *J. Embryol. Exp. Morphol.* 72: 71–96.

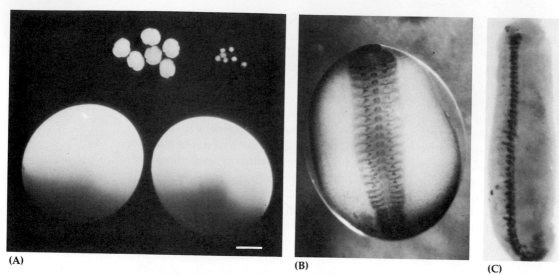

5   LEECH EGGS AND EMBRYOS. (A) Comparison of eggs from different species. The smallest eggs are from the medicinal leech (*Hirudo medicinalis*), the intermediate-size eggs are from *Helobdella triserialis* (which feeds on snails), and the large pair at the bottom are from *Haementeria ghilianii* (which lives in Guyana, attains monstrous proportions, and feeds on crocodiles). Scale is 0.5 mm. **(B)** *Haementeria* embryo, about 2 mm long at 11 days, stained with hematoxylin. Ganglia are already apparent. **(C)** *Helobdella* embryo, about 1.5 mm long. One cell (known as the N teloblast), which gives rise to much of the central nervous system, was injected with horseradish peroxidase 6 days beforehand. Note that the hemiganglia on one side, which derive from this single precursor, are stained. (Photographs courtesy of D. A. Weisblat; from Weisblat, 1981.)

cells. Stent, Fernández, Weisblat, and their colleagues have followed the division and migration step by step as cells cluster to form the ganglionic primordia.[34] Marking and ablation techniques have been essential for analyzing mechanisms. At a precise stage of development the N, O, P, or Q teloblast cell on one side can be killed by injecting a lethal enzyme or dye. Thereafter the killed cell produces no progeny but its homologue on the other side continues to do so. As a result the embryo develops with a chain of ganglia that are normal at the anterior end (the first ones to be formed before the lethal injection). Those "ganglia" developing later are made up of only the complement of cells stemming from the surviving teloblasts. A photo of a ganglion in which the contribution from the N teloblast on one side has been labeled is shown in Figure 6.

An oversimplification in the previous description of leech ganglia was that they were all much the same from head to tail within an animal and from animal to animal. As a first approximation this is correct. There are, however, interesting specializations in various segments along the length of the animal. One pronounced example is in segments 5 and 6 of the leech, which contain the genitalia, male and female, of

[34]Weisblat, D. A. and Shankland, M. 1985. *Philos. Trans. R. Soc. Lond. B* 312: 39–56.

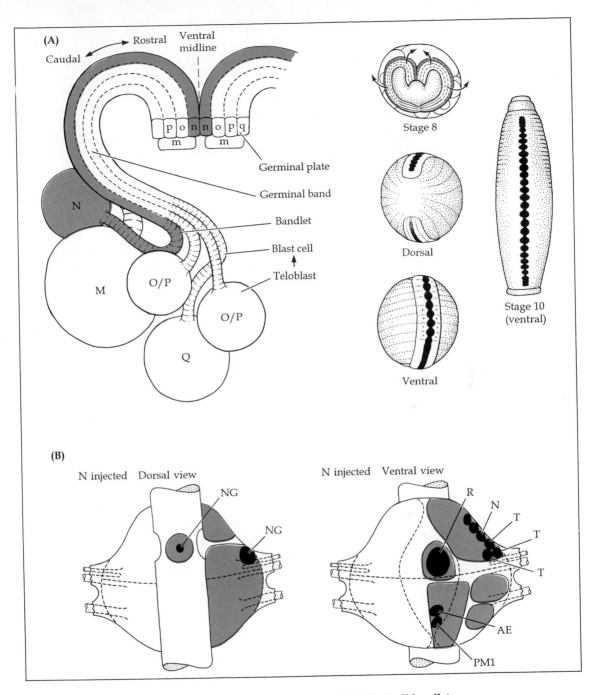

6 LEECH DEVELOPMENT. (A) Relation of the teloblasts and the blast cell bandlets that come together to form the germinal band and the germinal plate. The two bandlets derived from N come together at the ventral midline and give rise to most of the central nervous system. By stage 8 the ganglia are already identifiable. **(B)** Areas of leech ganglia (shaded) and identified cells derived from the N teloblast. The N teloblast was injected with a fluorescent peptide that was carried from cell to cell during subsequent divisions. The neuropil glial cell (NG) arises from the same teloblast as neurons seen on the dorsal and the ventral surfaces. These include Retzius (R), touch (T), nociceptive (N), and AE cells mentioned in the text. O, P, and Q teloblasts give rise to fewer cells in leech ganglia. (After Ho and Weisblat, 1987; Kramer and Weisblat, 1985.)

this hermaphrodite animal. Ganglia 5 and 6 of the nerve cord contain many more cells than their counterparts elsewhere, about 700 compared with 400.[35] The two large and distinctive Retzius cells, described below, are also different in these ganglia, having smaller cell bodies and unusual arborizations. Macagno, Kristan, and their colleagues have shown by transplantation in embryos that the target organs influence cell proliferation and cell structure.[36,37] When the sex organs were displaced to segments 4 or 7 from their normal home in segments 5 and 6, those ganglia (4 and 7) showed the typical sex specializations, with extra neurons and unusual Retzius cells. Ganglia in segments 5 and 6 from which the targets had been removed then came to resemble normal segmental ganglia.

The steps by which target organs are formed during development have been followed by Jellies and Kristan for a body wall muscle in the leech.[38] The crisscrossed oblique muscles are shown in Figure 7. These two layers of muscle fibers are sandwiched between circular and longitudinal muscles and develop later than them. By what mechanisms are these two precisely oriented orthogonal sheets laid down? An extraordinary cell shaped like a comb precedes the formation of the oblique muscles. Figure 7 shows the comb cells, one on each side, which are mirror images of each other. Each comb has about 35 fine processes running obliquely at 45° to the long axis of the animal, at right angles to the other. By labeling these cells and muscle fibers with intracellular stains and monoclonal antibodies, it was shown that the developing fibers orient themselves along the comb processes to form fascicles. Killing one comb cell prevents the formation in that segment of the fibers with that orientation. The fibers crossing from the other side, which grew along the normal undamaged cell, developed normally.

Many similarities exist between development of the nervous systems of leeches and of higher animals. Cells migrate to their final destinations; produce overabundant processes that then retract; can undergo cell death; and can compete for targets. These events were described in Chapter 11. In leeches, however, it becomes possible to describe how single identified neurons descended from an identified ancestor take up their final form. Certain lines of development are fixed and specified; others are less determined. Thus, the sensory cells for touch (T cells), the Retzius cells (which are serotonergic), other serotonergic cells, annulus erector motor cells (AE cells), and the giant neuropil glial cells are all derived exclusively from the N teloblast. The P sensory cells and dopaminergic cells can arise from P as well as O teloblast cells. If the O teloblast is killed, the P takes over.

[35]Macagno, E. R. 1980. *J. Comp. Neurol.* 190: 283–302.
[36]Baptista, C. A. and Macagno, E. R. 1988. *J. Neurobiol.* 19: 707–726.
[37]Jellies, J., Loer, C. M. and Kristan, W. B., Jr. 1987. *J. Neurosci.* 7: 2618–2629.
[38]Jellies, J. and Kristan, W. B., Jr. 1988. *J. Neurosci.* 8: 3317–3326.

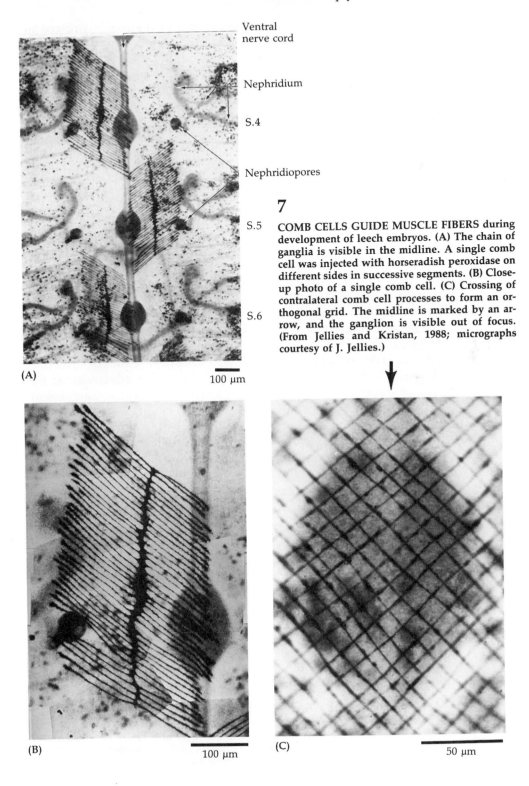

Ventral
nerve cord

Nephridium

S.4

Nephridiopores

S.5

S.6

7

COMB CELLS GUIDE MUSCLE FIBERS during development of leech embryos. **(A)** The chain of ganglia is visible in the midline. A single comb cell was injected with horseradish peroxidase on different sides in successive segments. **(B)** Close-up photo of a single comb cell. **(C)** Crossing of contralateral comb cell processes to form an orthogonal grid. The midline is marked by an arrow, and the ganglion is visible out of focus. (From Jellies and Kristan, 1988; micrographs courtesy of J. Jellies.)

(A)

100 μm

(B)

100 μm

(C)

50 μm

Cell death is exemplified by a pair of serotonergic cells situated in the posteromedial region of the embryonic ganglia. During development, one of the two cells dies.[39] Interestingly, the surviving cell is situated on alternating sides in successive ganglia (left, right, left, etc.). Competition for survival of one of the two posteromedial cells in a ganglion seems to be determined by the pattern in the neighboring ganglion. Macagno and Stewart have shown that if one of the two cells is killed by injecting it, the other one lives on, while in the neighboring ganglia its ipsilateral homologues die. Additional evidence for competition, resembling that seen in vertebrate CNS development, is provided by the outgrowth patterns of identified motor AE cells. These neurons in the adult innervate subcutaneous muscles and erect the annuli of the skin into ridges[40] (see Figure 12). Their axons leave the ganglia through the contralateral nerve roots to reach the body wall. In the adult, axons of AE motor cells in midbody ganglia do not project through connectives to adjacent ganglia. But at early embryonic stages the AE cells have a larger arborization.[41] The AE cell in an immature ganglion sends axons through connectives at either end into adjacent ganglia, where they arborize extensively. Later these axons retract. If, however, the AE cell in the neighboring ganglion is killed the axons remain in the connectives throughout development to adult life. Comparable experiments at the single-cell level would be impossible in a vertebrate embryo.

In *Aplysia*, Carew and his colleagues have taken advantage of the extensive information about behavior to correlate ganglionic development with the appearance of habituation, dishabilitation, and sensitization at different embryonic stages.[19] These behaviors, which are described below, are adaptive responses of the animal to familiar or novel stimuli. They appear in immature *Aplysia* according to different developmental timetables: Sensitization appears several days later than habituation at a stage of intense neuronal proliferation.

ANALYSIS OF REFLEXES MEDIATED BY INDIVIDUAL NEURONS

Sensory cells in leech and *Aplysia* ganglia

When one strokes, presses, or pinches the skin of a leech or an *Aplysia*, a sequence of movements follows. In the leech, one segment or more shortens abruptly, and the skin becomes raised into a series of distinct ridges. Subsequently the animal swims away or writhes. Similarly, in *Aplysia* a touch or a jet of water applied to the siphon or mantle leads to a withdrawal. In living leech and *Aplysia* ganglia, one can reliably identify the sensory and motor cells that mediate these reflexes according to their shapes, sizes, positions, and electrical characteristics.[12,40,42]

[39]Macagno, E. R. and Stewart, R. R. 1987. *J. Neurosci.* 7: 1911–1918.
[40]Stuart, A. E. 1970. *J. Physiol.* 209: 627–646.
[41]Gao, W. Q. and Macagno, E. R. 1987. *J. Neurobiol.* 18: 295–313.
[42]Blackshaw, S. 1981. *In* K. J. Muller, J. G. Nicholls and G. S. Stent (eds.). *Neurobiology of the Leech.* Cold Spring Harbor Laboratory, Cold Spring Harbor, NY, pp. 51–78.

Figures 2 and 4 show the distribution of the identified sensory cells in the abdominal ganglion of *Aplysia* and in a leech ganglion. The 14 leech neurons labeled T, P, and N in Figure 4 are all sensory and represent three sensory modalities. Each cell responds selectively to touch (T), pressure (P), or noxious (N) mechanical stimulation of the skin. Figure 8 illustrates the responses of leech and *Aplysia* sensory cells to various forms of cutaneous stimuli. Leech T cells give transient responses to light touch of the skin surface. Their sensory endings consist of small dilatations situated between epithelial cells on the surface of the skin.[43] T cells adapt rapidly to a maintained step indentation and usually cease firing within a fraction of a second. The P cells respond only to a marked pressure or deformation of the skin and show a slowly adapting discharge. The N cells require still stronger mechanical stimuli, such as radical deformation produced by pinching the skin with blunt forceps or scratching it with a pin. The signaling characteristics and action potential shapes of T, P, and N sensory cells can be accounted for in terms of the numbers and types of ion channels they possess; in particular the complement of potassium channels is distinctive for each type of cell.[44] The specificity of these leech sensory cells is remarkable —for example, each of the two N cells has different chemosensitivity to transmitters.[45] Moreover, a monoclonal antibody can be made that selectively binds to molecules in one but not the other N cell, and that does not bind at all to T, P, Retzius, motor, or any other identified cells in the segmental ganglia.[46]

Sensory responses similar to those in the leech can be recorded from groups of identified neurons in *Aplysia* ganglia following comparable mechanical stimuli to the surface.[47] The number of cells is, however, considerably larger. For example, in the abdominal ganglion, 25 distinctive cells—50 μm in diameter, orange in color, and with a dark rim (labeled LE in Figure 2)—send their axons out through the siphon nerve to innervate the skin of the siphon and the mantle shelf. They respond selectively to mechanical stimulation: some to touch, others to pressure. Unlike those in the leech, however, all these cells have similar electrical properties. A second group of 20 sensory cells, labeled RE in Figure 2, innervate the mantle shelf and the purple gland. In general, the sensitivity and responses of *Aplysia* mechanoreceptor cells are similar to those of P cells in the leech (Figure 8), except for one cell (L18), which responds only to severe mechanical deformation of the skin and resembles the N sensory neurons in the leech.

The modalities and responses of sensory neurons in leech and *Aplysia* resemble those of mechanoreceptors in the human skin, which also

Receptive fields

[43]Blackshaw, S. 1981. *J. Physiol.* 320: 219–228.
[44]Stewart, R. R., Nicholls, J. G. and Adams, W. B. 1989. *J. Exp. Biol.* 141: 1–20.
[45]Sargent, P. B., Yau, K.-W. and Nicholls, J. G. 1977. *J. Neurophysiol.* 40: 446–452.
[46]Zipser, B. and McKay, R. 1981. *Nature* 289: 549–554.
[47]Byrne, J., Castellucci, V. F. and Kandel, E. R. 1974. *J. Neurophysiol.* 37: 1041–1064.

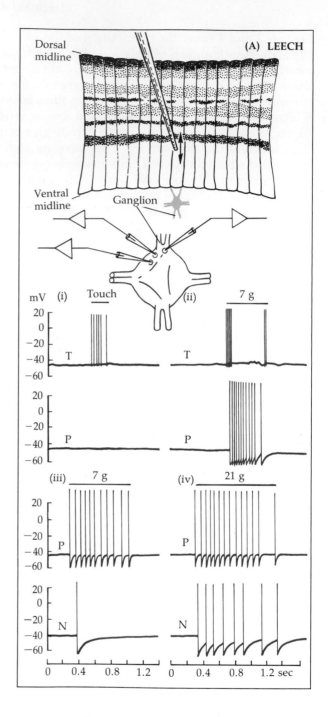

distinguish between touch, pressure, and noxious or painful stimuli. In the invertebrate, however, a single nerve cell does the job of many neurons supplying our own skin in a densely innervated region such as the fingertip. Examples of receptive fields for *Aplysia* and leech me-

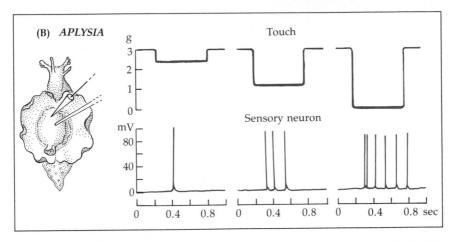

8 RESPONSES TO SKIN STIMULATION by neurons in leech and *Aplysia*. **(A)** Leech. Intracellular records of T, P, and N sensory cells. The preparation consists of a piece of skin and the ganglion that innervates it. Cells are activated by touching or pressing their receptive fields in the skin. (i) A T cell responds to light touching that is not strong enough to stimulate the P cell. (ii) Stronger, maintained pressure evokes a prolonged discharge from the P cell and a rapidly adapting "on" and "off" response from the T cell. (iii, iv) Still stronger pressure is needed to activate the N cell. **(B)** Intracellular records from an *Aplysia* sensory neuron responding to pressure applied to the siphon. In their properties, the *Aplysia* sensory neurons in general resemble P cells. **(A after Nicholls and Baylor, 1968; B modified from Kandel, 1976.)**

chanoreceptor neurons are shown in Figure 9. The fields in *Aplysia* vary considerably in size from cell to cell and show considerable overlap.[47]

In the leech each sensory cell innervates a defined territory of the periphery and responds only to stimuli applied within that area. The territory that a particular cell supplies is mapped by recording from it while applying mechanical stimuli to the skin. The boundaries can be conveniently identified by landmarks, such as segmentation or the coloring of skin, so that one can predict reliably which cell will fire when a particular area is touched, pressed, or pinched. Thus one of the touch-sensitive T cells innervates dorsal skin, another ventral skin, and a third laterally situated skin. Similarly, the two P sensory cells divide the skin into roughly equal dorsal and ventral areas, whereas the two N cells respond to noxious stimulation over roughly equivalent overlapping areas.[48] The elaborate and stereotyped branching pattern of a T sensory cell in the leech is shown in Figure 9. The T cell responds to touch outside its own segment by axons conducting through the neighboring ganglia and connectives. It will be shown later that conduction block occurs in these fine branches.

In a system that has such clear-cut boundaries in the periphery, it is possible to study the dynamics of the formation and maintenance of

[48]Blackshaw, S. E., Nicholls, J. G. and Parnas, I. 1982. *J. Physiol.* 326: 261–268.

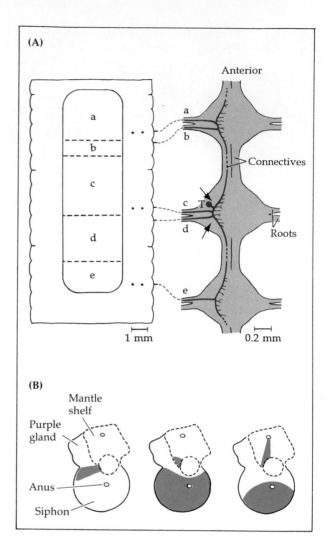

(A)

Anterior

a

b

Connectives

c   T

d

Roots

e

1 mm   0.2 mm

(B)

Mantle
shelf

Purple
gland

Anus

Siphon

receptive fields. How does the receptive field become established? What factors determine exactly where the endings of a particular cell can and cannot be formed? How is the degree of overlap by endings from the same cell and from different cells regulated? It has been shown that during development the axons of an individual sensory cell can be followed as they grow out to innervate skin for the first time. They head for the correct area to form an appropriately shaped and located field of innervation of the type seen in the adult.[49] The dimensions and position of the receptive field are not, however, immutably fixed throughout the life of the leech. Thus removal of the innervation from one area of skin causes an uninjured cell to expand its receptive field

[49]Kuwada, J. Y. and Kramer, A. P. 1983. *J. Neurosci.* 3: 2098–2111.

to occupy the vacant territory.[14,48] Interestingly, such sprouting is modality-specific: Selective killing of N cells in the leech causes uninjured N cells in the same ganglion (but not P or T sensory cells) to expand into the denervated territory. Similarly, selectively killing T cells causes the remaining uninjured T cell in the ganglion to sprout without influencing N or P cells. These results suggest the presence of a dynamic relationship between a neuron and the target that it innervates—a problem considered again in this and other chapters. Similar competition and sprouting into denervated territory has been clearly demonstrated for axons supplying crustacean muscles.[50,51]

Individual motor cells in *Aplysia* and leech ganglia are shown in Figures 2 and 4. The criterion for showing that a cell is indeed motor is that each impulse in the cell gives rise to a conducted action potential in its axon leading to the muscle and then to a synaptic potential in the muscle fiber.[40] In addition, deletion of a single cell can give rise to an obvious deficit in behavior. For example, in each ganglion there are only two annulus erector motor cells (labeled AE in Figure 4), which cause the skin of the leech to be raised into ridges like a concertina (Figure 12). One AE cell can be killed by injecting it with a mixture of proteolytic enzymes in an otherwise intact animal. When tests are made after the animal has been allowed to recover, the region of skin formerly innervated exclusively by the killed cell fails to become erect in response to appropriate sensory stimuli.[52] In the leech, more than 20 pairs of motor cells supplying the various muscles that flatten, lengthen, shorten, and bend the body, as well as the motor cell that controls the heart, have been identified in the segmental ganglia. Muscles can receive inhibitory and modulatory peptidergic inputs in addition to the excitatory inputs that are mediated by acetylcholine.[53] In *Aplysia*, a large number of motor neurons are involved in the production of reflex movements such as gill and siphon withdrawal.[12] Thirteen identified central motor neurons and about 30 peripheral motor neurons innervate muscles moving the siphon, the mantle shelf, and the gill. If the motoneurons are prevented from firing by hyperpolarizing them, reflex contractions of the muscles are abolished. Some of the motoneurons have a restricted action, causing only small movements, whereas others have a broad action, sometimes contracting completely separate effector organs, such as the gill and mantle shelf.

Here, then, are systems in which one can infer which sensory cells are activated when a mechanical stimulus is applied to a particular area of skin, and which motor cells are firing when the animal performs a movement.

In addition to sensory and motor cells, interneurons, neurosecretory cells, and modulatory cells have been identified in *Aplysia* and leech

Motor cells

Interneurons and neurosecretory cells

[50]Parnas, I. 1987. *J. Exp. Biol.* 132: 231–247.
[51]Krasne, F. B. 1987. *J. Neurobiol.* 18: 61–73.
[52]Bowling, D., Nicholls, J. G. and Parnas, I. 1978. *J. Physiol.* 282: 169–180.
[53]Kuhlman, J. R., Li, C. and Calabrese, R. L. 1985. *J. Neurosci.* 5: 2310–2317.

central nervous systems. Interneurons, whose processes are by definition confined to the CNS, confer plasticity, allow switching of pathways, and can give rise to rhythmical coordinated behavior (see below). Examples of secretory neurons that regulate an essential function in *Aplysia* are the BAG CELLS. About 300 of these large neurons are clustered in each anterior connective of the abdominal ganglion. Kupfermann first showed that an extract of these cells induced complex egg-laying behavior.[54] A battery of neuropeptides that induce the animal to lay eggs have now been purified and sequenced. The physiological properties of the bag cells, the mechanisms of peptide synthesis, and the induction of egg laying have been worked out in detail by Arch, Mayeri, Blankenship, Scheller, and their colleagues.[27,55–57] These peptide hormones, derived from a large precursor molecule, are stored in dense-core vesicles, from which they are secreted. The active peptides induce egg laying by acting directly on the target organ in the periphery. They also influence the properties of neurons within the CNS, such as L15. The bag cells are electrically coupled and fire synchronous bursts of action potentials to release their hormones. This system provides one of the best examples of an essential function that has been studied at the molecular level, from genes, intracellular messengers, and regulatory mechanisms to behavior. In invertebrates are also found cells with interesting unusual properties but no known function. An example is a huge cell (1 mm in diameter) known as R15. This cell's biophysical properties and channel characteristics have been intensely studied in relation to the generation of spontaneous rhythmical bursting patterns.[18]

Morphology of synapses

In the nervous systems of invertebrates, most (but not all)[58] of the synapses between neurons are situated not on the cell bodies but on fine processes within a central region of the ganglion (the neuropil).[12,13,30] The neuropil is highly complex and resembles that in the vertebrate brain. Despite its complexity, the neuropil is organized in an orderly manner. This fact was revealed by the experiments of Muller and McMahan, who first devised the technique of intracellular injection of horseradish peroxidase using identified nerve cells in the leech.[59] The enzyme penetrates into all the branches of the cell. The branching patterns of sensory and motor cells within the neuropil are characteristic, each cell displaying its own configuration. Examples of the typical ramifications of identified neurons in the leech and in *Aplysia* are shown in Figures 9 and 10. Horseradish peroxidase can also be detected in electron micrographs, as in Figure 10, in which characteristic chemical synapses are marked by arrows. In this manner, then, one can identify the synapses made by identified cells in the neuropil and characterize

[54]Kupfermann, I. 1967. *Nature* 216: 814–815.
[55]Brown, R. O., Pulst, S. M. and Mayeri, E. 1989. *J. Neurophysiol.* 61: 1142–1152.
[56]Nagle, G. T. et al. 1990. *J. Biol. Chem.* 265: 22329–22335.
[57]Newcomb, R. W. and Scheller, R. H. 1990. *Brain Res.* 521: 229–237.
[58]French, K. A. and Muller, K. J. 1986. *J. Neurosci.* 6: 318–324.
[59]Muller, K. J. and McMahan, U. J. 1976. *Proc. R. Soc. Lond. B* 194: 481–499.

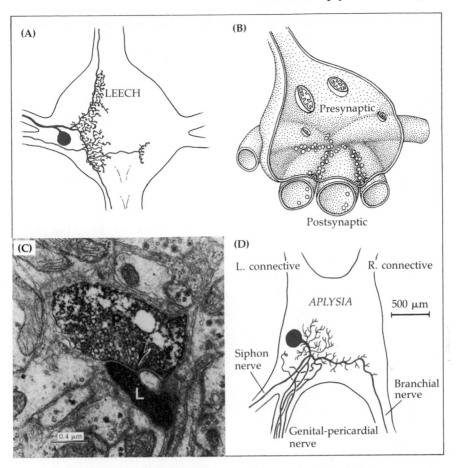

10 **SHAPES AND SYNAPSES** of leech and *Aplysia* neurons. (A) Arborization of a
pressure cell in a leech ganglion after injection with horseradish peroxidase. The
cell body sends a single process to the neuropil, where all the synaptic connections of
the ganglion are made. Axons run to neighboring ganglia through connectives and to
the body wall through roots. Small processes within the neuropil form synaptic contacts.
(B) Reconstruction made from serial sections of synaptic specializations seen in leech
neuropil. Typically, a single presynaptic process containing numerous vesicles makes
synapses on two or more postsynaptic elements. This arrangement is seen also in the
CNS of *Aplysia*, with similar thickenings and widened intercellular clefts. (C) Synapse
made by a P cell on an L motoneuron in the neuropil of a leech ganglion. Both cells
were injected with horseradish peroxidase for identification. Arrow marks presynaptic
density. (D) Cell L7 in an *Aplysia* abdominal ganglion injected with cobalt. This is a
motor cell supplying the mantle shelf. (A and B from Muller, 1981; C after Macagno,
Muller and Pitman, 1987; D after Kandel, 1976.)

their fine structure.[60,61] The synaptic structures are highly complex. A
single sensory cell supplies many postsynaptic targets, and its presynaptic endings are themselves supplied by numerous terminals arising
from other neurons that modulate its release of transmitters.

[60]Macagno, E. R., Muller, K. J. and Pitman, R. M. 1987. *J. Physiol.* 387: 649–664.
[61]Bailey, C. H. and Chen, M. 1988. *J. Neurosci.* 8: 2452–2459.

Electrical synapses in leech and *Aplysia* ganglia appear as contacts between neurons with a narrowed space 4 to 6 nm separating the two membranes; as at other gap junctions, bridges span the cleft in these regions. Fluorescent dyes such as Lucifer Yellow injected into one cell will usually, but not always, cross such electrical junctions and spread into other coupled cells. The pattern of connections revealed in this way can be more complex than one might guess from physiological evidence. For example, intracellular recordings reveal that the three touch sensory T cells on one side of a leech ganglion are all weakly coupled to those on the other.[62] This coupling, however, is not direct but is mediated by electrical synapses through two specific coupling interneurons. If these two cells are killed by injection of protease, coupling between the T cells is abolished.

Synaptic connections
of sensory and
motor cells

Synaptic potentials originating in the neuropil spread into the nearby cell body, where they appear as excitatory and inhibitory potentials. Currents injected into the cell body can influence synaptic potentials and the release of transmitter. Unfortunately, the properties of the pre- and postsynaptic terminals themselves remain inaccessible to direct physiological investigation since they are too small to be impaled by microelectrodes.

The sensory cells responding to mechanical stimuli make excitatory connections on the motoneurons used for gill and siphon withdrawal in *Aplysia* or for shortening in the leech. Several lines of evidence including electron microscopy have shown that the connections are direct (i.e., that there are no known intermediary cells).[12,13,60,61] This fact is important, because only if each constituent and its properties are known can one pinpoint the sites at which any interesting modifications in signaling take place. The sensory cells that make excitatory chemically mediated connections on motoneurons also make excitatory synapses on other cells (interneurons) that in turn synapse on the motoneurons. A diagram of the connections that mediate gill withdrawal in *Aplysia* is shown in Figure 11.

In the leech the T, P, and N sensory cells all converge on one motor cell, the L motoneuron, which innervates longitudinal muscles and produces shortening. A remarkable feature is that the mechanism of transmission onto the L motor cell is characteristically and consistently different for each of the sensory cells.[63] The N cells act mainly through

[62]Muller, K. J. and Scott, S. A. 1981. *J. Physiol.* 311: 565–583.
[63]Nicholls, J. G. and Purves, D. 1972. *J. Physiol.* 225: 637–656.

11   **WITHDRAWAL REFLEX OF GILL AND SIPHON** in *Aplysia*. **(A, B)** Water squirted onto the siphon causes it and the gill to withdraw. **(C)** Sensory cells, interneurons, motoneurons, and the connections mediating the withdrawal reflex. A single sensory neuron out of the population of 24 or more is shown. Sensory neurons make monosynaptic and also indirect connections with the pool of motoneurons (M). Excitatory (L22, L23) and inhibitory (L16) interneurons mediate indirect actions. Motor cells situated in the periphery (PS) also contribute to the reflex. (After Kandel, 1976.)

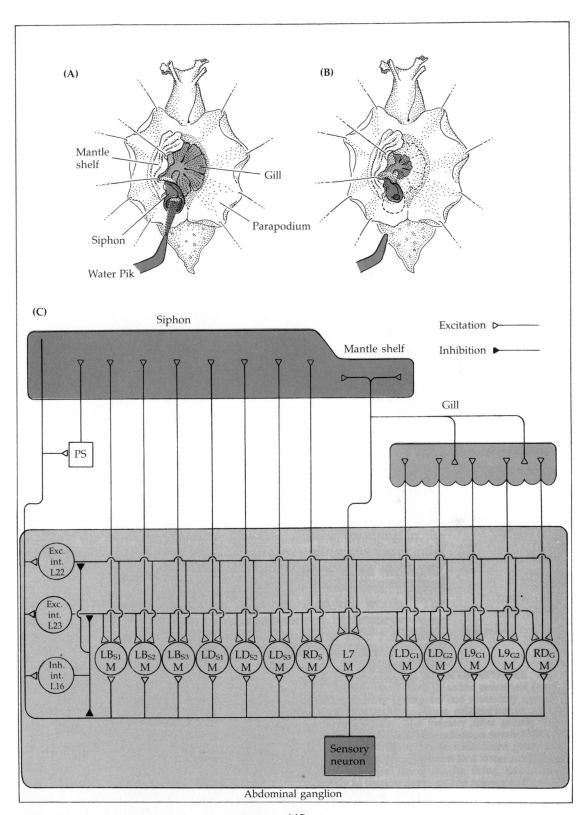

chemical synapses (with only a hint of electrical coupling), the T cells through rectifying electrical synapses, and the P cells by a combination of both mechanisms. Chemical and electrical mechanisms of transmission have been distinguished from one another by (1) observing the latency of the synaptic potentials in the motoneuron, (2) bathing the ganglion in high concentrations of magnesium ions, which block chemical synaptic transmission, (3) changing the membrane potential of the presynaptic and postsynaptic neurons, and (4) observing the synapses in electron micrographs.

Much is now known about the transmitters used by leech and *Aplysia* neurons, including ACh, GABA, dopamine, serotonin and a variety of peptides including the tetrapeptide FMRFamide.[12,13,64,65] Certain important transmitters are still not known, however, particularly those liberated by P and N cells in leech and sensory cells in *Aplysia* innervating gill and siphon.

Short-term changes in synaptic efficacy

For simplicity, the initial analysis of signaling mechanisms is usually made by observing the effects of single impulses. However, while an animal is swimming in water or moving on a surface under normal conditions, stimulation of the skin causes trains of impulses. As a second step in analysis of reflexes and behavior and to differentiate between the effects of chemical and electrical synapses, the performance of the motoneurons is tested by repetitive stimulation. From the material presented in Chapter 7 on the characteristics of transmission processes, the general expectation is that electrical synapses will remain relatively stable in their performance, whereas chemical synapses will be much more variable.

Figure 12A shows the difference between the two forms of transmission when leech nociceptive and touch cells fire. For example, when an N cell fires in response to stimulation or to pinching of the skin, the chemically evoked synaptic potentials recorded in the L cell during a train first increase (facilitation) and then decrease (depression, not shown). In sharp contrast is the synaptic effect of the touch neuron (the T cell) that makes an electrical synapse on the L cell. With repeated stimulation or stroking of the skin under the same conditions, the synaptic potentials evoked in the L cell remain unchanged.[63]

In the leech CNS, the variable effectiveness of different chemical synapses permits sequential activation or inactivation of different postsynaptic structures as a single presynaptic sensory neuron activates first one and then another postsynaptic cell. This type of differential effect adequately explains how pressing the skin of a leech leads first to shortening of the animal and then to erection of its annuli. These two reflexes follow different time courses: The shortening of the body wall occurs abruptly and is poorly maintained, whereas the annuli become erect more slowly and stay erect longer. This behavior correlates well

[64]Lent, C. M. et al. 1991. *J. Comp. Physiol.* 168: 191–200.
[65]Ocorr, K. A. and Byrne, J. H. 1985. *Neurosci. Lett.* 55: 113–118.

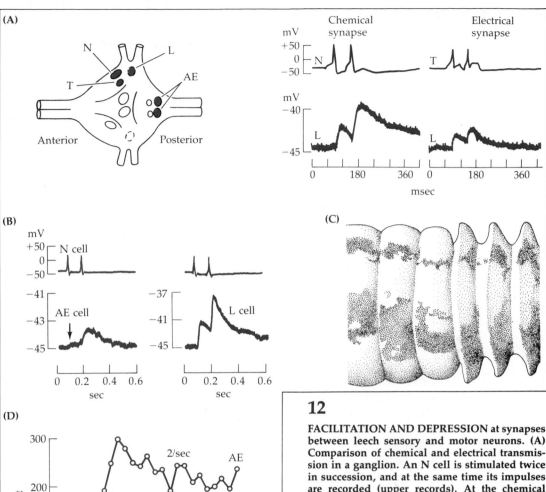

12

**FACILITATION AND DEPRESSION at synapses
between leech sensory and motor neurons. (A)**
Comparison of chemical and electrical transmission in a ganglion. An N cell is stimulated twice
in succession, and at the same time its impulses
are recorded (upper records). At the chemical
synapse between N and L cells, facilitation occurs, so a second impulse leads to a larger synaptic potential (bottom left). In contrast, two potentials evoked by T-cell impulses in an L cell
cause similar postsynaptic potentials with double
stimulation. This is typical of electrical synapses.
(B) Characteristics of transmitter release at different synapses made by a presynaptic neuron.
An N cell is stimulated and responses are recorded in L and AE cells. Facilitation is greater
at the N–AE synapse. (The small first synaptic
potential in the AE cell is marked by an arrow.)
(C) Impulses in an AE cell produce annular erection (arrows). **(D)** When the N cell is stimulated
twice per second, the synaptic potentials in the
AE cell are facilitated to more than double their
original size, whereas those in the L cell decrease
in amplitude. The abscissa indicates the number
of the synaptic potential recorded in the train.
The ordinate gives the proportional height of the
synaptic potentials compared with the average
value before the train (100 percent). (A after Nicholls and Purves, 1972; B, C, and D after Muller
and Nicholls, 1974.)

447

with the synaptic potentials shown in Figure 12 that facilitate at different rates and trigger the two events in sequence.[66] Although the N and P sensory cells make chemical synapses not only on the L cell but also on the AE motor cell, the synaptic potential recorded in the AE motor cell at first is considerably smaller (Figure 12). With trains of impulses in a single sensory cell, the synaptic potentials in the AE and L motoneurons undergo phases of facilitation and depression; the facilitation, however, is characteristically greater and longer lasting at synapses upon the AE motoneuron. A number of indirect experiments suggest that the differences in synaptic transmission can be accounted for by variations in the amount of transmitter released at presynaptic N-cell terminals, rather than by differences in the postsynaptic cells. Direct quantal analysis to determine whether such changes are pre- or postsynaptic in origin is impossible at these synapses (as indeed at most other synapses in the CNS of invertebrates or vertebrates). Difficulties arise from the complex geometry, the absence of well-defined miniature potentials, and the small unit size. A major advantage of the isolated invertebrate cell preparation in culture described below is that a fuller and more detailed analysis of transmission, facilitation, and modulation becomes possible (see Figure 20).

In addition to the direct through pathway from N cells to motor neurons shown in Figure 12, there are parallel delayed lines through interneurons. These can have pronounced effects on the strength of sustained responses to mechanical stimuli. Interestingly, a single identified interneuron may be involved in a number of different movements.[67]

Short-term facilitation and depression of the sort seen at leech synapses or frog neuromuscular junctions are less evident at *Aplysia* synapses used for withdrawal reflexes, but there, as shown below, long-term changes persisting for hours, or even days, have been observed. Moreover, a single interneuron in *Aplysia* can excite one postsynaptic target while it inhibits another.[68] For example, in Figure 13, impulses in cell L10 evoke excitatory potentials in cell R15 and inhibitory potentials in cell L3, both mediated chemically by monosynaptic pathways. At both synapses the transmitter is acetylcholine (ACh), but the effect depends on the presence of different types of channels, opened by the ACh receptors, in the two cells. The same cell—L10—also evokes either excitatory or inhibitory responses in one postsynaptic target (L7), the response depending on the frequency of firing. At certain synapses in *Aplysia*, the duration of excitatory or inhibitory potentials can be extremely long, persisting for minutes or even hours, following one or a few impulses.[69] Acetylcholine, as well as dopamine and 5-hydroxytryptamine (5-HT, or serotonin) can produce both conductance decreases

[66]Muller, K. J. and Nicholls, J. G. 1974. *J. Physiol.* 238: 357–369.
[67]Lockery, S. R. and Kristan, W. B., Jr. 1990. *J. Neurosci.* 10: 1816–1829.
[68]Wachtel, H. and Kandel, E. R. 1971. *J. Neurophysiol.* 34: 56–68.
[69]Parnas, I. and Strumwasser, F. 1974. *J. Neurophysiol.* 37: 609–620.

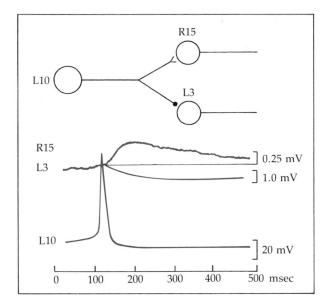

13

EXCITATION AND INHIBITION mediated by a single cell (L10) onto two postsynaptic neurons (L3 and R15) in *Aplysia*. Acetylcholine is the transmitter at both synapses, but the permeability changes produced are different. (After Wachtel and Kandel, 1971.)

and increases, and these changes persist after the transmitter has been removed, like the potentials described in autonomic ganglia (Chapter 8).

In addition to conventional frequency-dependent facilitation and depression of transmitter release, a variety of novel presynaptic mechanisms that influence the efficacy and duration of chemical synaptic transmission have been described in *Aplysia*, in the leech, and in other invertebrates. For example, at chemical synapses in the nervous systems of crustaceans, insects, leeches, and mollusks, transmitter is released tonically by presynaptic neurons that do not give impulses.[70,71] Release occurs continuously at rest and is increased by depolarization and decreased by hyperpolarization of the presynaptic terminal. Similar "nonspiking" interneurons have been observed within the mammalian central nervous system—for example, the photoreceptor, horizontal, and bipolar cells in the retina (see Chapter 16). At such synapses, the tonic release of transmitter allows finely graded, maintained action of one neuron on another.

Presynaptic depolarization and hyperpolarization have other subtle effects on chemical transmission that can markedly alter reflexes. For example, at certain synapses in *Aplysia* and the leech, maintained depolarization of the presynaptic terminals enables an impulse to release more transmitter; conversely, if the membrane of the presynaptic terminal is hyperpolarized, the impulse releases less transmitter.[72] A shift

Unusual synaptic mechanisms

[70]Burrows, M. and Siegler, M. V. 1985. *J. Neurophysiol.* 53: 1147–1157.
[71]Blackshaw, S. E. and Thompson, S. W. 1988. *J. Physiol.* 396: 121–137.
[72]Thompson, W. J. and Stent, G. S. 1976. *J. Comp. Physiol.* 111: 309–333.

in resting potential of as little as 5 millivolts (from −40 to −35 mV) can cause a threefold increase in the amount of transmitter released by an impulse. The amplitude and the duration of the presynaptic impulse are not obviously altered by such small changes in resting potential. Similar mechanisms operate at the squid giant synapse. The effects on reflexes are illustrated in the leech by a circuit in which the presynaptic terminals of certain interneurons controlling the heartbeat show naturally occurring hyperpolarization. This is produced cyclically by inhibitory inputs from other identified interneurons. As a result of the hyperpolarization of the terminal, the numbers of quanta liberated by each impulse invading it are reduced.[73] Here, then, is an instance of presynaptic inhibition that is mediated by hyperpolarization of the terminals, and in which the mechanism has been demonstrated directly by quantal analysis. This is one of the rare examples of a synapse within the CNS at which detailed quantal analysis is possible owing to simplified geometry and a relatively large unit size. In the well-defined circuit of neurons controlling the heart of the leech, such presynaptic inhibition and cyclical modulation of the membrane potential of the presynaptic terminal have been shown to be essential for establishing the rhythmicity.[74] It is not yet known whether similar mechanisms may also be involved in the presynaptic inhibition observed in the mammalian spinal cord (Chapter 15).

Failure of impulse conduction represents a further mechanism for altering the synaptic action of one cell upon its postsynaptic targets. In crustacean motor axons, in the central nervous system of the cockroach, and in the leech, repeated trains of impulses occurring at natural frequencies can produce aftereffects that lead to conduction block at branch points. In crustacean axons,[75] the changes are associated with an increase in extracellular potassium concentration. In sensory neurons of the leech, the mechanism depends on the hyperpolarization caused by the electrogenic sodium pump and by long-lasting changes in a calcium-activated potassium conductance. Thus repeated stroking or pressing the skin of the leech causes trains of impulses and a maintained hyperpolarization of T and P sensory cells; as a result, propagation of impulses becomes blocked at certain branch points where the geometry for impulse conduction is unfavorable, with a small fiber feeding into a larger one (Chapter 5).[76] Under these condiftions, the P or T cell terminals within the ganglion no longer invaded by impulses fail to release transmitter and fail to influence their postsynaptic target, while other branches of the same neuron continue to transmit.[60,77] This, therefore, represents a nonsynaptic mechanism that temporarily disconnects the neuron from one set of its postsynaptic targets whether the synapses

[73]Nicholls, J. G. and Wallace, B. G. 1978. *J. Physiol.* 281: 157–170.

[74]Arbas, E. A. and Calabrese, R. L. 1987. *J. Neurosci.* 7: 3953–3960.

[75]Grossman, Y., Parnas, I. and Spira, M. E. 1979. *J. Physiol.* 295: 307–322.

[76]Yau, K.-W. 1976. *J. Physiol.* 263: 513–538.

[77]Gu, X. 1991. *J. Physiol.* 441: 755–778.

are chemical or electrical. Alternatively, when a few, but not all, of the presynaptic fibers connected to a cell fail to conduct or release transmitter, the efficacy of transmission may be reduced. An example of conduction block in a P cell is shown in Figure 14; the synaptic potentials recorded in the postsynaptic motor cell become reduced in amplitude when impulses become blocked at the point marked by the arrow. In this way, and by making lesions with a laser at specific selected branch point sites, Gu has assessed the contribution made by a part of the sensory P cell arborization to its synaptic action on the motor cell.[77]

One aim of the studies on an invertebrate is to analyze how complex behavioral acts are built up from simple, elementary reflexes. The smooth, coordinated movements of a leech are produced through interactions of individual ganglia with their neighbors, with distant ganglia, and with the head and tail brains. The leech, with its rigorously segmented ganglia and small number of neurons, has been particularly valuable for tracing the circuits and for identifying one by one the individual cells that act in concert to produce swimming. This complex movement has been examined by Stent, Kristan, Friesen, and their colleagues.[78–80] The individual muscle groups that participate and the motor cells that control them have been identified and the central connections traced. The key problems are, first, to analyze the pattern of synaptic interconnections and the mechanisms that allow the waves of contraction to travel repeatedly from head to tail along the body; and second, to explain the source of the rhythm and factors that modulate it. The head and tail brains are not essential for the swimming movements, which can occur in a few segments or even a single segment of the animal. As in other invertebrates (such as the cockroach, the locust, and the cricket) in which central motor programs involving a small number of individual cells have been shown to control complex patterns of movement, in the leech the basic rhythm is established by synaptic interactions within the ganglia; the role of peripheral receptors is to trigger, enhance, depress, or halt the swimming.

In order to swim, a leech first flattens and elongates its body. Next the animal bends up and down as one wave spreads from head to tail, followed by another bending wave. Cinephotography shows the leech's body to bend in the form of a single sine wave as the crest and trough spread along its length. The wave is produced by alternate contraction and relaxation of ventral and dorsal muscles. Knowing the motor neurons that innervate those muscles, one can infer the pattern of activity that they generate, alternately firing and becoming quiescent. Thus, in a midbody ganglion, the motor neurons innervating ventral muscle fire, bending the tail downward, and then become inhibited as the dorsal

Higher levels of integration

[78]Stent, G. S. and Kristan, W. B., Jr. 1981. *In* K. J. Muller, J. G. Nicholls and G. S. Stent (eds.). *Neurobiology of the Leech*. Cold Spring Harbor Laboratory, Cold Spring Harbor, NY, pp. 113–146.

[79]Nusbaum, M. P. et al. 1987. *J. Comp. Physiol.* A. 161: 355–366.

[80]Brodfuehrer, P. D. and Friesen, W. O. 1986. *Science* 234: 1002–1004.

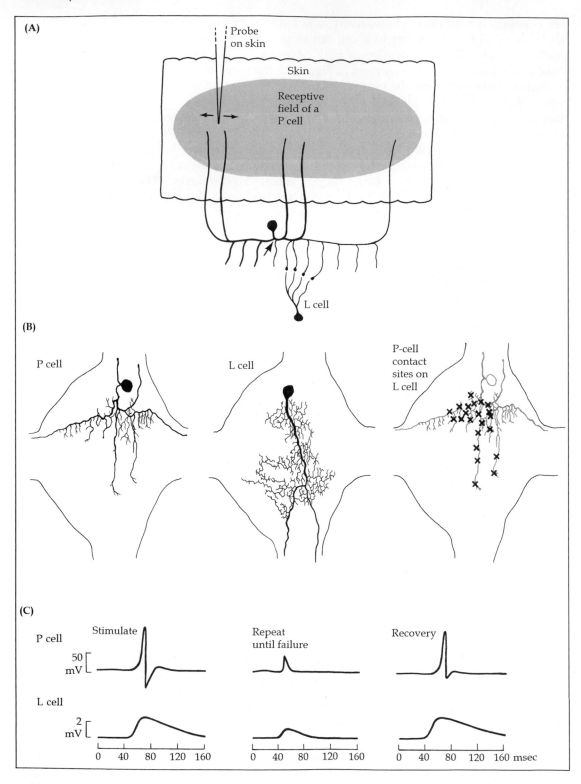

(A)

Probe on skin

Skin

Receptive field of a P cell

L cell

(B)

P cell

L cell

P-cell contact sites on L cell

(C)

P cell

Stimulate

Repeat until failure

Recovery

50 mV

L cell

2 mV

0   40   80   120   160

0   40   80   120   160

0   40   80   120   160 msec

◄ **14** FAILURE OF CONDUCTION IN A PRESSURE SENSORY NEURON following repeated stimulation. (A) The receptive field of the P neuron, which resembles that of the T cell shown in Figure 9. The skin is pressed by the probe, causing action potentials to travel along the fine branches to the soma of the P cell. The arrow indicates the point at which impulses fail as a consequence of the hyperpolarization that is produced by impulse traffic. The blockage of conduction occurs at this site because there is considerable mismatch in size between the small-diameter incoming process and the larger neurite connected to the cell body. Once block has occurred, branches distal to the failure will no longer be invaded by impulses. (B) Camera lucida drawing of P and L cells to show their arborizations and the sites at which synapses are made. Normally under resting conditions, impulses invade all the branches of the P cell. After conduction block only some of the synapses will be activated. (C) Recordings made from a P cell and an L motor neuron before and after repeated stimulation. At first the impulse in the P cell gives rise to a large synaptic potential. Once the impulse fails the postsynaptic potential becomes smaller. Subsequently recovery occurs. Such experiments, together with lesions made at selected sites of a neuron, show how its synaptic output can be graded by invasion failure. (After Macagno, Muller and Pitman, 1987; Gu, Macagno and Muller, 1989; Gu, 1991.)

motor neurons fire to bend the tail upward. In practice, it is not necessary to record intracellularly from motor neurons. Instead, it is simpler to monitor extracellularly the activity of the axons that run in specific peripheral nerve branches.[78] In a chain of ganglia removed from the body, sensory feedback is eliminated, yet the swimming rhythm remains intact, as evidenced by the firing pattern of the motor neurons. Even a single ganglion will, under certain conditions, "swim" rhythmically—that is, send appropriate commands down the appropriate axons to nonexistent muscles.

The basis for the rhythm depends on a series of inhibitory connections made by a few interneurons on each other and on the motor neurons. The scheme outlined in Figure 15 can account quantitatively for the alternation of dorsal and ventral contractions spreading as a wave from head to tail. Certain specific unpaired neurons that have been identified initiate the swimming cycles. An interesting modulatory role is played by 5-HT. Quiescent, sluggish (!), nonswimming leeches have lower blood levels of 5-HT than do active leeches. Moreover, stimulation of cells known to secrete 5-HT promotes an increase in its concentration in the blood as well as swimming in the animal. When 5-HT is depleted in embryos by means of a specific chemical (5,6-dehydroxytryptamine) that selectively destroys the 5-HT neurons in the developing ganglia, the adult leeches do not swim spontaneously but will do so if immersed in a weak solution of 5-HT.[81,82]

For the swimming of the leech, the individual nerve cells and their connections are now largely known. Other neurally mediated rhythms in invertebrates are produced by different mechanisms. For example, identified neurons controlling the heartbeat of the leech possess inherent rhythmicity—they fire spontaneously in bursts, with an independently generated depolarization and repolarization that do not require

[81]Willard, A. 1981. *J. Neurosci.* 1: 936–944.
[82]Glover, J. C. and Kramer, A. P. 1982. *Science* 216: 317–319.

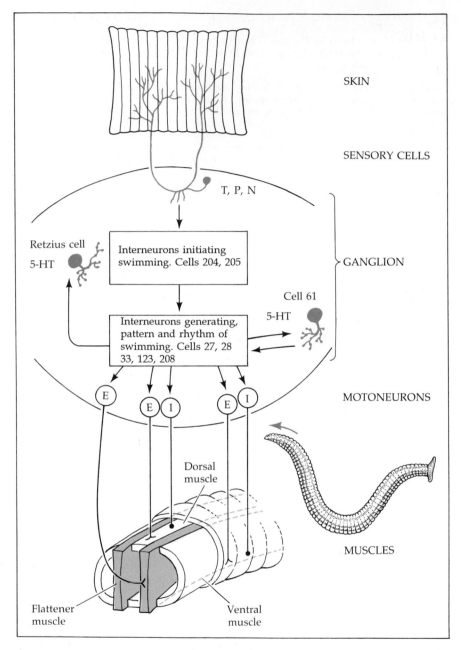

15 **SIMPLIFIED DIAGRAM OF NEURONAL CONNECTIONS** involved in swimming by the leech. Interneurons within each ganglion that generate the swimming rhythm have been identified (cells 27, 28, 33, 123, and 208); their connections, which are largely inhibitory, have been traced within the ganglion and between ganglia. These cells act on the motoneurons that control the muscles used for swimming (flatteners, dorsal and ventral longitudinal muscles). Mechanosensory cells can activate identified interneurons (cells 204, 205) that excite the pattern generating circuit. The system is also influenced by 5-HT either diffusely from the Retzius cells or at synapses made by cell 61. E, excitatory motoneuron; I, inhibitory motoneuron. (After Kristan, 1983.)

synaptic inputs to turn the cells on or off.[72,74] Synaptic inputs, however, modulate and coordinate the rhythm in several such cells.

In a series of experiments, Kandel, Castellucci, Byrne, Carew, Hochner, and their colleagues have analyzed the role played by various biophysical and molecular mechanisms on the production of complex, long-lasting behavioral responses of *Aplysia*.[20,25,83] The gill withdrawal reflex becomes progressively weaker when water jets are applied to the mantle repeatedly at regular intervals (Figure 16). For example, if con-

Habituation in *Aplysia*

[83]Carew, T. J., Castellucci, V. F. and Kandel, E. R. 1979. *Science* 205: 417–419.

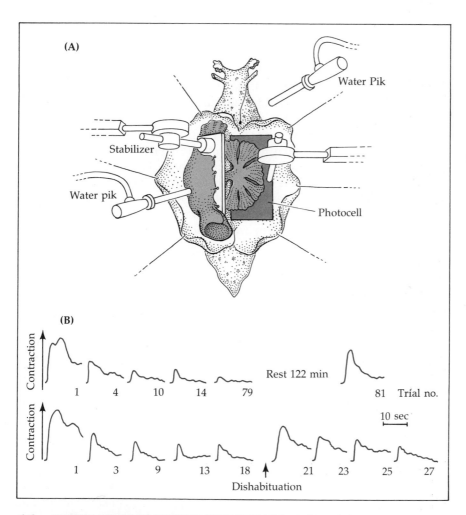

16   HABITUATION OF WITHDRAWAL REFLEX in *Aplysia*. **(A) The animal is immobilized in circulating seawater with its mantle pinned. Contractions of the gill are monitored by a photocell. The stimulus is a jet of water. (B) Traces of gill contractions in response to standard stimuli. The depression is reversed after a rest of 122 minutes, or if a strong mechanical stimulus is applied to the head (arrowhead). (After Kandel, 1976.)**

stant stimuli are used to elicit the reflex 15 times at intervals of 10 minutes, the strength of the gill withdrawal to the same stimulus becomes reduced to about 50 percent of its original value. It then recovers only slowly over a period of 30 minutes to several hours. The animal is thereby able to respond less vigorously to a stimulus that had previously been effective. The time course and the properties of this phenomenon resemble those of *habituation* in higher animals: A stimulus such as a sudden light tap on the shoulder may elicit a startle reaction, but upon repetition will cease to do so. A characteristic of habituation is that, in addition to spontaneous recovery (if the stimulus is not presented for a period), immediate recovery occurs following a strong generalized stimulus of a different type; a habituated *Aplysia* will once again withdraw its gill in response to a water jet after a strong shock to the tail.

To determine the mechanisms involved in habituation, recordings have been made from the sensory cells, the motor cells, and the interneurons in animals, and in isolated preparations that show properties similar to those of whole animal. *Aplysia* are immobilized, their ganglia exposed, and stimuli delivered to the skin while intracellular recordings and muscle tension measurements are made. Alternatively, depression of synaptic transmission can be seen when a sensory cell is repeatedly stimulated by an intracellular microelectrode: Synaptic potentials in the motoneuron become smaller. Experiments have been made that show the reduced responsiveness to be due to a long-lasting depression of transmitter release at the terminals of the sensory neurons upon the motor neurons.[84] The sensory response to the peripheral stimulus and the properties of the motor neurons show no comparable changes. Intracellular recordings made from the sensory neuron indicate that the presynaptic action potential becomes shortened in duration. This gives rise to a smaller inward calcium current and therefore to a smaller amount of transmitter release. Since the presynaptic terminals are too small to be impaled by microelectrodes, inferences about their membrane properties are drawn from records made in the soma.[20,25]

Sensitization in *Aplysia*   Sensitization represents a second behavioral modification that has been analyzed to an exceptionally full extent in *Aplysia*. After a strong, noxious, and potentially dangerous stimulus, animals react more vigorously to a mild stimulus that would normally produce a weak or negligible response. Thus in *Aplysia* a single electrical shock to the tail increases the strength of the siphon withdrawal reflex for an hour or more (Figure 17). This provides a clear example of NEUROMODULATION: A group of fibers, when stimulated, modifies the efficacy of a pathway in which they are not directly involved. How is sensitization produced by strong electrical stimulation? The suggestion is that the fibers mediating this effect release 5-HT and small peptides. These transmitters can give rise to similar sensitization when applied to the ganglion. The scheme proposed by Kandel and his colleagues[85] is shown in Figure 18

[84]Castellucci, V. F. and Kandel, E. R. 1974. *Proc. Natl. Acad. Sci. USA* 71: 5004–5008.
[85]Kandel, E. R. 1989. *J. Neuropsychiatr.* 1: 103–125.

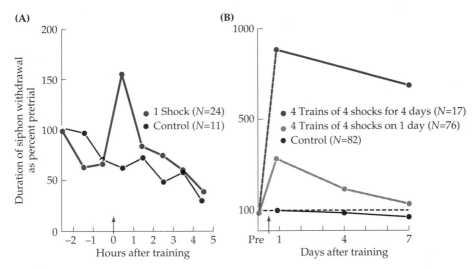

(A)

(B)

- 1 Shock (*N*=24)
- Control (*N*=11)

- 4 Trains of 4 shocks for 4 days (*N*=17)
- 4 Trains of 4 shocks on 1 day (*N*=76)
- Control (*N*=82)

17 SENSITIZATION OF WITHDRAWAL REFLEX in *Aplysia.* (A) Short-term sensitization produced by a single strong electrical shock to the tail or the neck of *Aplysia* at the time shown by the arrow. Marked potentiation of the siphon withdrawal occurs, compared with the control group. The effect lasts for more than 1 hour. (B) Long-term sensitization lasting for days is produced by trains of shocks for 4 days preceding tests. (After Kandel, 1989.)

(see also Chapter 8). Appropriate nerve stimuli or the application of 5-HT lead to an increase in the concentration of intracellular cAMP in the presynaptic cell; this activates a cAMP-dependent protein kinase. The action of the protein kinase is to phosphorylate the potassium channel protein; this decreases the potassium conductance of the membrane, which thereby allows the presynaptic action potential to become prolonged in duration. Substances that specifically block the protein kinase prevent sensitization. Conversely, injection of the catalytic subunit of the protein kinase into the appropriate sensory cell causes prolongation of the action potential and increased transmitter release as expected. The potassium channel that responds both to 5-HT and to the peptide FMRFamide is known as the *S channel*, and it gives rise to the *S current*. A detailed analysis of these S channels, their properties, their susceptibility to modulation, and their functional significance has been made by Siegelbaum and his colleagues using patch clamp.[26,86] Interestingly, 5-HT has the opposite effect on a different *Aplysia* neuron—R15. In R15, cAMP also mediates the effect of 5-HT; but in this cell it causes an increase rather than a decrease of potassium conductance as well as an increase in calcium conductance.[18]

In related experiments, Kandel, Byrne, and their colleagues have studied associative conditioning in *Aplysia.*[20] A chemosensory conditioning stimulus (shrimp extract applied to the bath) that normally causes feeding is paired with noxious stimulation (the unconditioned

Eric Kandel, 1989

[86]Belardetti, F. and Siegelbaum, S. A. 1988. *Trends Neurosci.* 11: 232–238.

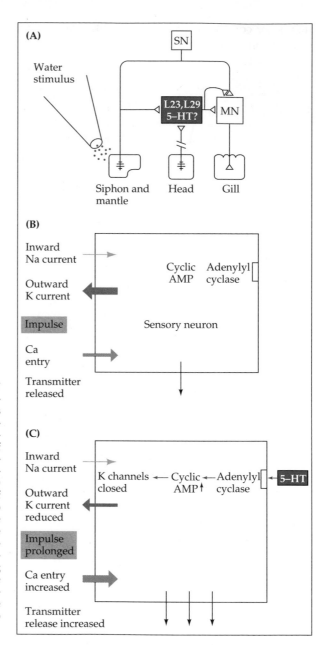

18

SCHEME TO EXPLAIN DISHABITUATION AND SENSITIZATION in *Aplysia*. **(A)** Simplified diagram of sensory (SN) and motor cells (MN) mediating withdrawal reflex. 5-HT application mimics stimuli that produce dishabituation. **(B, C)** Hypothesis for mode of action of 5-HT. **(B)** The impulse in the sensory nerve terminals allows calcium to enter; the increased calcium concentration causes transmitter release. **(C)** The effect of 5-HT is to prolong the duration of the impulse, thereby allowing more calcium to enter and more transmitter to be liberated. The steps involve 5-HT activation of adenylyl cyclase and an increase in intracellular cAMP. This increased concentration of cAMP in turn, by closing S channels and decreasing g_K, would prolong the impulse (slower repolarization). Because the terminals are small and buried within the complex neuropil, recordings and injections are made at the cell body. (Modified from Klein, Shapiro and Kandel, 1980.)

stimulus, an electric shock applied to the animal). After a sufficient number of paired trials, the animal shows defensive responses instead of feeding when the chemosensory stimulus is applied. Similarly, weak touches (conditioned stimulus) to the siphon have been coupled with strong shocks to the tail (unconditioned stimulus). Interestingly, the region of skin touched can be discriminated so that responses from neighboring regions do not become conditioned. The cellular mechanisms for conditioning, like those for sensitization, involve presynaptic changes in cAMP concentrations and channel phosphorylation, in this case brought about by the pairing of stimuli. An essential component for the detailed analysis of molecular mechanisms has been to study comparable changes that occur in even simpler preparations. These consist of pairs of identified cells or triads in culture that form synapses showing depression and facilitation (described below).

A major extension of the studies of sensitization has been to describe the origin of longer-lasting changes that persist for many days. Figure 17 shows long-term sensitization of the gill withdrawal reflex following a series of training shocks to the tail or the neck of *Aplysia*. Again the changes in synaptic efficacy can be mimicked by applying repeated doses of 5-HT to the ganglion. And again the experimental evidence indicates that the changes occur through phosphorylation and closure of S potassium channels, with enhanced action potential duration and calcium entry into sensory presynaptic cells.[85,86] Cyclic AMP injection can also induce long-term sensitization. A major difference from short-term sensitization has, however, been found: Long-term sensitization requires protein synthesis and RNA synthesis. Long-term, but not short-term, sensitization is selectively blocked by inhibition of protein synthesis and by actinomycin D (which blocks RNA synthesis) applied for about 1 hour while 5-HT is present. When given at later times, these drugs do not prevent the development of long-term sensitization following 5-HT stimulation. Long-term sensitization is prevented from developing if a specific oligonucleotide is injected into the nucleus of a sensory cell activated by 5-HT. This oligonucleotide blocks the activation of cAMP-inducible genes.[87] The proteins themselves and their turnover in the sensory cells have been studied. The scheme proposed by Kandel is shown in Figure 19. The principal feature of this hypothesis is that similar mechanisms are involved in both long- and short-term sensitization. In the long-term sensitization, however, regulatory genes are influenced to produce a cascade of events that results in transcription-dependent persistence in protein phosphorylation. A second difference from short-term sensitization is that there are anatomical correlates in synaptic structure that go hand in hand with physiological events, as described by Bailey and Chen.[61,88] It is not clear, however, whether these changes are confined to synapses made upon motor cells, since only sensory cells and not their targets were labeled for electron microscopy.

Inevitably, with such complex processes and intricate mechanisms numerous questions remain. One alluded to earlier concerns the applicability of studies made in the soma to events occurring in presynaptic terminals. Another concerns the circuit itself and the role of interneurons. In leech ganglia it has been shown that an identified interneuron can perform highly complex switching functions and that its functional

Long-term sensitization

---

[87]Dash, P. K., Hochner, B. and Kandel, E. R. 1990. *Nature* 345: 718–721.
[88]Bailey, C. H. and Chen, M. 1989. *J. Neurobiol.* 20: 356–372.

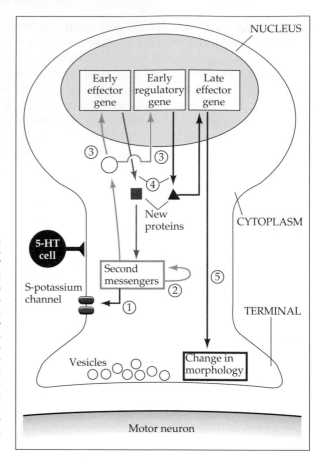

19

SCHEME PROPOSED BY KANDEL to explain differences between short- and long-term memory mechanisms in *Aplysia*. The actions of 5-HT on S channels and g_K (labeled as pathways 1 and 2) are similar to those shown to those in Figure 18. The second messengers can act back on themselves (2) and can act to modify regulators (open circle). These regulators influence early effector and early regulatory genes (pathway 3). The acquisition of long-term memory that lasts more than 1 day depends on the induction of new proteins (square and triangle, pathway 4) initiated by second messengers involved in short-term memory. Agents that block protein synthesis therefore prevent the induction of long-term memory. The regulatory genes can also trigger late effector genes, leading to changes in morphology (pathway 5). (After Kandel, 1989.)

roles may be dissimilar from trial to trial.[67] Moreover, in a pioneering study, Zecevic, Cohen, and their colleagues have studied a large population of neurons in the *Aplysia* ganglion during sensory stimulation, habituation, and sensitization.[89] By recording optically they have established that a single sensory stimulus can activate as many as 300 neurons and suggest that it may be difficult to determine the role of each cell in the behavioral responses of the animal. A further difficulty is that while all the available evidence points to a presynaptic locus in the sensory cell, it is hard to rule out additional postsynaptic mechanisms initiated by 5-HT or peptides on the motoneurons.

A natural question concerns the general significance of these results in *Aplysia* for understanding long-term changes in performance such as learning or memory in higher animals or even in other invertebrates. Genetic analysis of *Drosophila* by Benzer, Dudai, and their colleagues

[89]Zecevic, D. et al. 1989. *J. Neurosci.* 9: 3681–3689.

has revealed mutants with reduced ability to learn.[90,91] In an extensive series of experiments on the marine crab *Hermissenda*, Alkon and his colleagues have used natural stimuli to produce classic conditioning. The persistent changes in synaptic efficacy again invoke second messengers—but the site is postsynaptic, not presynaptic.[15,92] Similarly, several lines of evidence implicate changes in receptors in the postsynaptic cells as essential for long-term changes in the mammalian CNS (Chapter 10). At the very least, however, whatever variations occur, the coherent and comprehensive schemes produced for sensitization and conditioning in the invertebrates have provided a framework for analyzing these highly complex phenomena.

Neurons in the central nervous system of adult leeches retain their ability to sprout and can reform appropriate connections with their targets after an injury. Somehow the damaged cell is able to send its axons in the correct direction to the original region they innervated, there to form highly specific connections on some targets but not others.[13,14] Normal function can be restored in the animal after its nervous system has been transected. Again, an invertebrate provides the opportunity for studying, at the cellular and molecular level, mechanisms involved in an important problem: repair of the nervous system after injury. By examining one cell at a time, one can follow the course taken by a growing axon. For example, a regenerating touch cell grows back to reform its original arborization pattern in the next ganglion (see Chapter 12). T cells also reestablish appropriate and selective connections onto the "correct" target cells with virtually no errors.[93] Similar, accurate regeneration of connections also occurs when the ganglia are maintained outside the animal in organ culture.

In a detailed and technically difficult series of experiments, Muller and his colleagues analyzed the steps occurring during leech CNS regeneration. At the site of a crush an initial event is the appearance of numerous microglial cells that migrate rapidly from elsewhere in the leech CNS[94] The role that these wandering macrophage-like cells play in regeneration is not established. In addition to phagocytosing debris they are closely associated with laminin, a growth-promoting molecule (Chapter 11) at the crush site. Whether they actually make or transport this protein is not yet known.[95] After injury the severed axons sprout. Growing axons can be traced as they traverse the distance to the next ganglion and there form synapses on their targets. On occasion the severed axon has been observed to reconnect with its own old, surviving distal stump. (The part of the axon that has been severed does not

Regeneration of synaptic connections in the leech

[90]Benzer, S. 1971. *J.A.M.A.* 218: 1015–1022.

[91]Dudai, Y. 1989. *The Neurobiology of Memory.* Oxford University Press, N.Y.

[92]Alkon, D. L. 1989. *Sci. Am.* 260(7): 42–50.

[93]Macagno, E. R., Muller, K. J. and DeRiemer, S. A. 1985. *J. Neurosci.* 5: 2510–2521.

[94]McGlade-McCulloh, E. et al. 1989. *Proc. Natl. Acad. Sci. USA* 86: 1093–1097.

[95]Masuda-Nakagawa, L. M., Muller, K. J. and Nicholls, J. G. 1990. *Proc. R. Soc. Lond. B* 241: 201–206.

usually degenerate in invertebrates.) Alternatively, synapses can be formed at the crush site. Although the wandering microglial cells seem somehow to be involved, the sprouting, growth, and formation of connections can occur in the absence of the large glial cell that normally ensheathes the axons linking ganglia. Elliott and Muller[96] showed that this glial cell can be selectively killed without damaging neurons and without preventing accurate regeneration. An interesting observation was that destruction of the glial cell on its own caused accumulation of microglia and laminin and profuse sprouting of undamaged axons. This result bears on the suggestion (made in Chapter 6) that glial cells might serve to delineate tracts and prevent exuberant sprouting in adult, fully connected central nervous systems.

Identified neurons in culture: Growth, synapse formation, and modulation

A promising preparation for studying the mechanisms that play a part in growth and cell–cell recognition is provided by identified neurons that have been isolated from the ganglion and maintained in tissue culture. Such cells retain their membrane properties, sprout, and form specific chemical and electrical synapses in vitro that resemble in their properties those in the ganglion.[97,98] In a dish, the factors that influence sprouting, navigation toward the target, and synapse formation can be investigated in turn. An additional advantage is that the geometry becomes far simpler. Voltage clamp and patch clamp techniques have been used to measure the properties and distribution of ion channels over the neuronal surface and to analyze facilitation and modulation of transmitter release[99] (see above).

Both leech and *Aplysia* neurons isolated from the CNS sprout profusely under suitable conditions. Soluble growth factors play a key role in the outgrowth of neurites from *Aplysia*. By contrast, identified leech neurons extend processes in culture when bathed in Ringer's fluid provided they are plated on an appropriate substrate. For example laminin, a large protein molecule extracted from the extracellular matrix surrounding leech CNS (see Chapter 11) promotes rapid and profuse outgrowth.[95] What is interesting is that this substrate promotes different branching patterns and rates of outgrowth in different identified cells. The outgrowth is blocked by antibodies that can also be used to locate the distribution of laminin in the intact CNS (this is how it was shown that laminin accumulates at sites of injury). Laminin also influences the distribution of voltage-sensitive calcium channels over the surfaces of the growing neurons.[100] A different extracellular matrix molecule used as a substrate, say tenascin (Chapter 11), again produces a distinctive pattern and rate of outgrowth for each cell, different from that seen on laminin. These experiments indicated that a large molecule such as

[96]Elliott, E. J. and Muller, K. J. 1983. *J. Neurosci.* 3: 1994–2006.
[97]Liu, Y. and Nicholls, J. G. 1989. *Proc. R. Soc. Lond. B* 236: 253–268.
[98]Schacher, S., Montarolo, P. and Kandel, E. R. 1990. *J. Neurosci.* 10: 3286–3294.
[99]Stewart, R. R., Adams, W. B. and Nicholls, J. G. 1989. *J. Exp. Biol.* 144: 1–12.
[100]Ross, W. N., Aréchiga, H. and Nicholls, J. G. 1988. *Proc. Natl. Acad. Sci. USA* 85: 4075–4078.

laminin or tenascin, anchored in extracellular matrix, may give signals to growing neurites—signals with different meanings and outcomes depending on the identity of the cell. Moreover, laminin appears in developing or regenerating CNS at exactly the regions through which neurons will grow. This illustrates a powerful feature of the cultured cell preparation: It is possible to compare results obtained in the dish with those normally occurring in the animal.

With pairs or triads of identified cells in culture, one can examine the specificity of cell–cell recognition as well as synaptic mechanisms. Figure 20 shows synaptically coupled cells in culture from leech and from *Aplysia*. The way in which appropriate types of cells recognize each other in the absence of glia or any outside cues has been studied with leech and *Aplysia* neurons.[97,98,101] For example, serotonergic leech Retzius cells make chemical synapses on P sensory cells in culture, as in the animal. While P cells do not form chemical or electrical synapses

[101]Kleinfeld, D. et al. 1990. *J. Exp. Biol.* 154: 237–255.

20

(A)

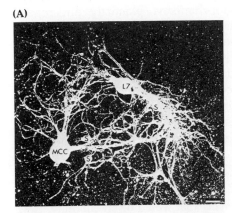

(B)

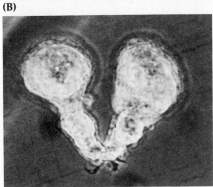

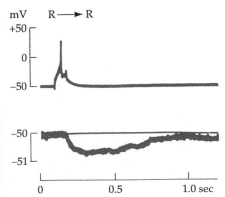

SIMPLIFIED PREPARATIONS of identified leech and *Aplysia* cells in culture. (A) Reconstitution of connections used in the gill-withdrawal reflex of *Aplysia* by isolated identified cells. A sensory neuron (S), a motor neuron (L7), and a facilitating serotonergic cell (MCC) are cultured together. When stimulated repetitively, such triads show properties similar to those observed in ganglia. Bar, 100 μm. (B) Specific connections of identified Retzius cells isolated from the central nervous system of the leech. Retzius cells, which secrete 5-HT, form chemically mediated synapses upon each other in culture in approximately 8 hours. This synapse is bidirectional and shows quantal release and facilitation, depression, and modulation similar to that seen in leech ganglia. Electrical synapses develop later. (A from Rayport and Schacher, 1986, courtesy of S. Schacher; B from Liu and Nicholls, 1989.)

on Retzius cells, they do make rectifying electrical junctions in the dish with L motor cells. Retzius cells make nonrectifying electrical synapses on such L motor cells. The Retzius cells make both chemical and electrical synapses with each other. These connections in the dish resemble those seen in the animal. Since the synapses form rapidly (in about 8 hours), reliably, and at a preappointed place on the leech neurons (Figure 20), one can study their properties as they form and mature.[97] Interestingly, the chemical synapse between Retzius cells develops before the electrical. In addition it has been shown that 5-HT receptors change in their distribution as the presynaptic cell makes contact with its postsynaptic target. Drapeau and his colleagues have shown that the apposition of a Retzius cell (presynaptic) to a P cell (postsynaptic) causes the disappearance of one type of 5-HT receptor response while leaving another unaffected.[102] At most other developing synapses it is not possible to make such tests. An open question concerns the molecules that enable the cells to recognize specific partners.

With triplets of identified *Aplysia* neurons such as those shown in Figure 20, Schacher, Hochner, Kandel, and their colleagues have been able to reproduce many of the modulatory phenomena seen in the animal and to demonstrate the role of 5-HT and peptides in sensitization.[103]

This chapter illustrates the varied uses of simple preparations for exploring development, integrative mechanisms, plasticity, and regeneration at the cellular and molecular level. Although analysis of brains composed of so few cells may seem to present a well-defined, finite problem, many essential questions remain to be explored. Some questions can be answered simply by doing the appropriate experiments. Others must await technical advances. Finally, as always, one expects a simple system to provide a stepping-stone toward an understanding of complex nervous systems. We suspect that Nietzsche anticipated our problems when he wrote (in *Thus Spake Zarathustra*):

> 'Then you must be a scientist whose field is the leech,' said Zarathustra, 'and you must pursue the leech to its last rock-bottom, you conscientious man!'
>
> 'Oh Zarathustra,' answered the man, 'that would be an enormity. How could I take up such a huge task! What I am the master and connoisseur of is the *brain* of the leech: that is *my* field! And it is a whole universe!'

SUGGESTED READINGS

General Reviews

Alkon, D. L. 1989. Memory storage and neural systems. *Sci. Am.* 260(7): 42–50.

Arch, S. and Berry, R. W. 1989. Molecular and cellular regulation of neuropeptide expression: The bag cell model system. *Brain Res. Rev.* 14: 181–201.

[102]Drapeau, P. 1990. *Neuron* 4: 875–882.

[103]Schacher, S. et al. 1990. *Cold Spring Harbor Symp. Quant. Biol.* 55: 187–202.

Belardetti, F. and Siegelbaum, S. A. 1988. Up- and down-modulation of single K$^+$ channel function by distinct second messengers. *Trends Neurosci.* 11: 232–238.

Byrne, J. H. 1987. Cellular analysis of associative learning. *Physiol. Rev.* 67: 329–439.

Dudai, Y. 1989. *The Neurobiology of Memory.* Oxford University Press, New York.

Fernandez, J., Tellez, V. and Olea, N. 1992. Hirudinea. *In* F. W. Harrison (ed.). *Microscopic Anatomy of Invertebrates,* Vol. 7, *Annelida.* Wiley-Liss, New York, pp. 323–394.

Kandel, E. R. 1979. *Behavioral Biology of Aplysia.* W. H. Freeman, San Francisco.

Kandel, E. R. 1989. Genes, nerve cells, and the remembrance of things past. *J. Neuropsychiatr.* 1: 103–125.

Kristan, W. B. 1983. The neurobiology of swimming in the leech. *Trends Neurosci.* 6: 84–88.

Muller, K. J., Nicholls, J. G. and Stent, G. S. (eds.). 1981. *Neurobiology of the Leech.* Cold Spring Harbor Laboratory, Cold Spring Harbor, NY.

Nicholls, J. G. 1987. *The Search for Connections: Studies of Regeneration in the Nervous System of the Leech.* Sinauer, Sunderland, MA.

Wehner, R. 1989. Neurobiology of polarization vision. *Trends Neurosci.* 12: 353–359.

Original Papers

Baxter, D. A. and Byrne, J. H. 1990. Differential effects of cAMP and serotonin on membrane current, action potential duration, and excitability of pleural sensory neurons of *Aplysia. J. Neurophysiol.* 64: 978–990.

Brodfuehrer, P. D. and Friesen, W. O. 1986. From stimulation to undulation: A neuronal pathway for the control of swimming activity in the leech. *Science* 234: 1002–1004.

Dash, P. K., Hochner, B. and Kandel, E. R. 1990. Injection of the cAMP-responsive element into the nucleus of *Aplysia* sensory neurons blocks long-term facilitation. *Nature* 345: 718–721.

Drapeau, P. 1990. Loss of channel modulation by transmitter and protein kinase C during innervation of an identified leech neuron. *Neuron* 4: 875–882.

Gu, X. 1991. Effect of conduction block at axon bifurcations on synaptic transmission to different postsynaptic neurones in the leech. *J. Physiol.* 441: 755–778.

Hayashi, J. H. and Hildebrand, J. G. 1990. Insect olfactory neurons in vitro: Morphological and physiological characterization of cells from the developing antennal lobes of *Manduca sexta. J. Neurosci.* 10: 848–859.

Hochner, B., Klein, M., Schacher, S. and Kandel, E. R. 1986. Additional component in the cellular mechanism of presynaptic facilitation contributes to behavioral dishabituation in *Aplysia. Proc. Natl. Acad. Sci. USA* 83: 8794–8798.

Liu, Y. and Nicholls, J.G. 1989. Steps in the development of chemical and electrical synapses by pairs of identified leech neurones in culture. *Proc. R. Soc. Lond. B* 236: 253–268.

Masuda-Nakagawa, L. M., Muller, K. J. and Nicholls, J. G. 1990. Accumulation of laminin and microglial cells at sites of injury and regeneration in the central nervous system of the leech. *Proc. R. Soc. Lond. B* 241: 201–206.

Muller, K. J. and McMahan, U. J. 1976. The shapes of sensory and motor neurones and the distribution of their synapses in ganglia of the leech: A study using intracellular injection of horseradish peroxidase. *Proc. R. Soc. Lond. B* 194: 481–499.

Schacher, S., Glanzman, D., Barzilai, A., Dash, P., Grant, S. G. N., Keller, F., Mayford, M. and Kandel. E. R. 1990. Long-term facilitation in *Aplysia*: Persistent phosporylation and structural changes. *Cold Spring Harbor Symp. Quant. Biol.* 55: 187–202.

Weisblat, D. A. and Shankland, M. 1985. Cell lineage and segmentation in the leech. *Philos. Trans. R. Soc. Lond. B* 312: 39–56.

Zecevic, D., Wu, J. Y., Cohen, L. B., London, J. A., Hopp, H. P. and Falk, C. X. 1989. Hundreds of neurons in the *Aplysia* abdominal ganglion are active during the gill-withdrawal reflex. *J. Neurosci.* 9: 3681–3689.

TRANSDUCTION AND PROCESSING OF SENSORY SIGNALS

CHAPTER FOURTEEN

The nervous system receives information through a variety of receptors that respond to light, to sound, to direct mechanical stimuli, to a variety of chemicals, and to stimuli that produce pain. In all such receptors the appropriate stimulus produces a receptor potential that may be either depolarizing or hyperpolarizing. In a short receptor, this change in membrane potential modulates the release of neurotransmitter onto the next neuron in line, thereby influencing the signals it sends to the central nervous system. In a long receptor, depolarization results in the generation of a train of action potentials; the frequency of discharge and duration of the train provide information about the intensity and duration of the initial stimulus.

Stretch receptors in invertebrate and vertebrate muscle are examples of long receptors that respond to increases in muscle length by depolarization. They are classified by the rate at which they adapt to a change in length. In slowly adapting receptors an increase in length produces a receptor potential that is well maintained for the duration of the stretch. Rapidly adapting receptors respond briskly to an increase in length, but the response declines, or *adapts*, as the stretch is maintained. Adaptation in stretch receptors is due to a combination of electrical and mechanical events. A good example of a mechanical factor in adaptation is found in the Pacinian corpuscle (a vibration detector), which adapts very rapidly when intact but only slowly when its onionlike capsule is removed from around the nerve terminal. By contrast, adaptation of photoreceptors in the retina is a chemical event.

Muscle stretch receptors receive efferent innervation from the nervous system, through which their sensitivity is controlled. The crayfish stretch receptor is provided with both excitatory and inhibitory axons. In vertebrate skeletal muscle, the rapidly and slowly adapting receptors (muscle spindles) receive a complex pattern of efferent innervation. Among other functions, this γ MOTOR SYSTEM causes receptor muscles to contract, thereby maintaining their tension, when the bulk of the muscle shortens. In this way receptor sensitivity is maintained at all muscle lengths.

Whereas muscle spindles provide information about muscle length, receptors in muscle tendons—called Golgi tendon organs—signal muscle tension. In addition, receptors within the joints respond to joint position. These three receptors together provide information about limb position, movement, and load.

467

A variety of long receptors in the skin convey information about touch and pressure. These are distributed nonuniformly over the body surface, for example occurring in high density on the fingertips and sparsely in the middle of the back. Their axons enter the spinal cord, giving off collateral branches, and ascend mostly in the dorsal columns to the thalamus, where they end on second-order neurons that project to the somatosensory area of the cerebral cortex. A topographic representation of the body surface is maintained throughout the pathways; in the cortex this representation is highly distorted, in accordance with the density of innervation. Thus the face and hand occupy disproportionately large areas of cortex compared with the back and the arm. Important sensory regions, such as the whiskers on the muzzle of a mouse, are represented in precise topographical detail.

The receptive field of a somatosensory cell is that area of skin over which its discharge rate can be altered. Rapidly adapting touch receptors in the fingertips have small, punctate fields; slowly adapting receptors in the forearm can be excited by skin deformation over a relatively large area. One characteristic of higher-order cells in the somatosensory system is that their firing rates may be either increased or decreased by activation of receptors in the skin. Thus the receptive field of a cortical cell may consist of an excitatory center with a surrounding inhibitory region. This organization sharpens the localization of stimuli; that is, the cortical cell will respond maximally when touch is applied only to the center, rather than over the whole surface of the field.

Separate sets of nerve endings respond to noxious stimuli leading to the perception of pain. These are free nerve endings without specialized accessory structures, connected to small myelinated axons and unmyelinated C fibers. The myelinated fibers, when excited, are responsible for a transient sensation of "fast" pain, the C fibers for slow, burning pain. The fibers terminate in the dorsal horns of the spinal cord, where they make synapses on second-order cells that send their axons to the contralateral side to ascend in the lateral and ventral spinothalamic tracts. These pathways are shared with fibers carrying temperature information. Activity in the pain pathways can be modulated at all levels, beginning at the segment of entry into the spinal cord, by activity in adjacent somatosensory pathways and by descending influences. Pain modulation is related to the action of naturally occurring opiatelike neurotransmitters—enkephalin and endorphin. Chronic stimulation of descending pathways has been used clinically for relief of intractable pain.

Transduction of auditory signals is by hair cells in the inner ear, which respond to bending of bundles of cilia on their apical surface as the basilar membrane vibrates. Hair cells do not produce action potentials. Bending in one direction causes depolarization, the other hyperpolarization. These changes in membrane polarity modulate neurotransmitter release onto the auditory nerve terminals, and hence impulse activity in the nerves. In higher vertebrates, frequency selectivity is determined by the mechanical properties of the basilar membrane, which responds to low frequencies near its apical (wide) end and to high frequencies near its basal tip. In lower vertebrates the distribution of frequency sensitivity is the same, but the tuning is an intrinsic property of the hair cells themselves, determined by the interaction of currents through calcium and potassium channels. The hair cells receive efferent innervation that inhibits their responses and reduces the sharpness of their tuning.

Just as deformations of hair bundles produce changes in membrane potential, changes in potential cause hair cells to change shape. Specifically, hyperpolarization causes elongation and depolarization shortening of isolated hair cells. There is evidence that such changes in length in one group of cochlear hair cells (the outer hair cells) are involved in a positive feedback mechanism to reinforce the movement of the basilar membrane and sharpen its frequency selectivity.

In the auditory cortex, some cells respond to simple tones. Their frequency response is more sharply tuned than that of primary fibers. This sharpening is due to the nature of their input from the basilar membrane: As in somatosensory receptive fields, the region producing maximum excitation is flanked by regions that are inhibitory, so that the region (and therefore the frequency) to which the cell responds best is limited. Other cells in the cortex respond to more complicated combinations of sounds related to speech. At earlier stages of auditory processing, beginning in the brainstem, comparison of binocular inputs is used for sound localization. This ability varies widely, being acute in bats and only moderately developed in humans. Differences between the two ears in both timing and intensity are utilized for localization. Some species of owl use these two features separately to locate sound in both the vertical and horizontal directions.

Taste and olfaction involve responses to the presence of a variety of molecules in the buccal or nasal mucosa. In the case of taste, such molecules act on receptor regions at the tips of accessory cells to produce depolarizations that, in turn, modify the discharge of afferent nerve fibers. Olfaction is mediated by specialized endings on the dendrites of olfactory cells themselves, without the participation of accessory receptor cells. Hence taste is mediated by short receptors, olfaction by long receptors. Taste bud cells respond to classes of molecules corresponding to the perceptions of sweet, sour, bitter, and salty. Electrical recording from single cells in taste buds indicate that sweet and bitter stimuli produce depolarization by reducing potassium permeability, through an intracellular second-messenger system utilizing cyclic AMP. Sour stimuli also reduce potassium permeability, either through a similar mechanism or by blocking of potassium channels by protons. Depolarization by sodium appears to occur simply by its entry through channels in the tip of the receptor. Unlike taste stimuli, odorants have not been classified into groups on the basis of functional responses. Their general mechanism of action is to produce depolarization, and hence action potential discharge, through a second-messenger system that activates relatively nonspecific cation channels.

In this chapter we provide an overview of how signals are transduced, how sensory information is transmitted along afferent pathways to the nervous system, and how it is analyzed. We begin with invertebrate and vertebrate muscle receptors that signal muscle length. These are used to illustrate two phenomena common to sensory receptors: receptor adaptation, and control of sensitivity by efferent fibers from the nervous system. This is followed by a discussion of the somatosensory receptors that subserve the sensations of touch, pressure, vibration, and pain, and a glimpse of how the signals they transmit are processed in the central nervous system. We then focus on mechanisms of transduction and processing in the auditory system and, finally, on the transduction of chemosensory signals by receptors for taste and olfaction. The principles underlying the processing and analysis of sensory signals are discussed in Chapters 16 and 17 in relation to the visual system, where they have been studied in considerable detail.

Sensory nerve endings as transducers

Sensory receptors are the gateways through which we perceive the outside world. Right at the outset the receptors set the stage for all the analyses of sensory events that are subsequently made by the central nervous system. They define the limits of sensitivity and determine the range of stimuli that can be detected and acted upon. With rare exceptions, each type of receptor is specialized to respond preferentially to only one type of external energy, called the ADEQUATE STIMULUS—rods and cones in the eye respond to light, nerve endings in the skin to touch, receptors on the tongue to taste. Nevertheless the stimulus, whatever its modality, is always converted (or TRANSDUCED) to an electrical signal. In general, the strength and duration of the stimulus are coded in the electrical signal itself; its modality and location are preserved anatomically. Thus a touch receptor in the foot has its own pathway into the nervous system, quite distinct from that of a vibration receptor in the leg.

For many signals there is a great deal of amplification at the receptor level, so that very small external stimuli provide a trigger to release stored charges that appear as electrical potentials. For example, odors from only a few molecules of specific substances (pheromones) are needed to act as sex attractants for moths and ants. Similarly, a few quanta of light trapped by receptors in the retina are sufficient to produce a visual sensation. This extreme sensitivity extends to the inner ear, where mechanical displacements of 10^{-10} meters are detectable.[1]

---

[1]Bialeck, W. 1987. *Annu. Rev. Biophys. Biophys. Chem.* 16: 455–478.

Equally remarkable are electroreceptors in some fish that can detect electric fields of a few nV/cm.[2,3] This is smaller than the field that would be produced if two wires connected to either pole of an ordinary flashlight battery could be dipped into the Atlantic Ocean, one at Bordeaux, the other at New York! Apart from their sensitivities, many sensory receptors have a well-defined range of stimuli to which they respond. For example, our auditory system is arranged to respond to sound within a limited bandwidth of about 20 to 20,000 Hz. Sound waves outside this range produce no response. The response by receptors in our retina to electromagnetic radiation is similarly restricted to wavelengths between about 400 and 750 nanometers. Shorter (near-ultraviolet) and longer (near-infrared) wavelengths go undetected. Restrictions of this kind are not usually because of insurmountable physical limitations. Instead, each system is tuned to the particular needs of the organism—whales and bats can hear much higher frequencies than humans, snakes can detect infrared, and moths ultraviolet, radiation.

The receptor potential generated by the transduction process reflects the intensity and duration of the original stimulus. In some receptors that do not have long processes, such as rods and cones in the retina, receptor potentials spread passively from the sensory region to the synaptic region of the cell. Such receptors are sometimes called SHORT RECEPTORS, and the passage of information from the receptor end to the synaptic end of the cell does not require the intervention of action potentials. In some cells, passive spread of the receptor potential can reach a surprisingly long distance. For example, in some crustacean[4] and leech[5] mechanoreceptors, and in photoreceptors in the barnacle eye,[6] the receptor potential spreads passively over a distance of several millimeters. In such cells, the membrane resistance, and hence the length constant for spread of passive depolarization, is unusually high. While receptor potentials are usually depolarizing, many short receptors respond to their adequate stimulus with a hyperpolarizing potential change. This occurs, for example, in photoreceptors of the vertebrate retina (Chapter 16). Cochlear hair cells have both hyperpolarizing and depolarizing responses, depending on which way the hairs are deflected. Whatever the polarity of the receptor potential, such receptors release neurotransmitter substances tonically from their synaptic regions; depolarization increases or hyperpolarization decreases the rate of release. In other receptors, called LONG RECEPTORS, such as those in skin or muscle, information from single receptors must be sent over a much longer distance (for example, from the big toe to the spinal cord). In order to accomplish this, the receptor must perform a second transformation process: Receptor potentials give rise to action potential trains whose duration and frequency code information about the duration and

Short and long receptors

[2]Kalmidjn, A. J. 1982. *Science* 218: 916–918.
[3]Heiligenberg, W. 1989. *J. Exp. Biol.* 146: 255–275.
[4]Roberts, A. and Bush, B. M. H. 1971. *J. Exp. Biol.* 54: 515–524.
[5]Blackshaw, S. E. and Thompson, S. W. 1988. *J. Physiol.* 396: 121–137.
[6]Hudspeth, A. J., Poo, M. M. and Stuart, A. E. 1977. *J. Physiol.* 272: 25–43.

Mechanoelectrical transduction in stretch receptors: The receptor potential

intensity of the original stimulus. These then carry the information to the synaptic terminals of the cell.

The way in which sensory receptors generate electrical signals has been studied extensively in stretch receptor neurons that register muscle length, both in crustaceans and in vertebrates. The first observations of this kind were made by Adrian and Zotterman,[7] who studied the relation between muscle stretch and action potential discharge in sensory nerve fibers arising in vertebrate muscle spindles. The general outline of the way in which muscle spindles respond was worked out by B. H. C. Matthews in the early 1930s.[8–10] For many years his experiments provided one of the best attempts to describe comprehensively a sensory end organ and its control. Matthews was able to detect impulses in single nerve fibers from individual spindles in frogs and cats. Recordings were made by an oscilloscope he designed for the purpose (no mean feat in 1930). Much later, Katz[11] showed that stretch

[7]Adrian, E. D. and Zotterman, Y. 1926. *J. Physiol.* 61: 151–171.
[8]Matthews, B.H.C. 1931. *J. Physiol.* 71: 64–110.
[9]Matthews, B. H. C. 1931. *J. Physiol.* 72: 153–174.
[10]Matthews, B. H. C. 1933. *J. Physiol.* 78: 1–53.
[11]Katz, B. 1950. *J. Physiol.* 111: 261–282.

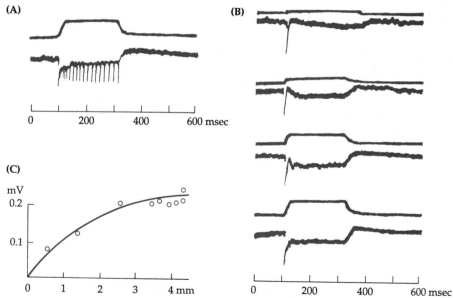

1   RECEPTOR POTENTIALS RECORDED EXTRACELLULARLY from a sensory nerve fiber supplying a muscle spindle. Recording electrode is placed as close as possible to the receptor. Downward deflection of the voltage record (lower trace) indicates receptor depolarization. (A) Stretching the muscle (upper trace) produces a receptor potential, upon which is superimposed a series of action potentials (lower trace). (B) Four stretches of increasing magnitude applied to the muscle after procaine has been added to the bathing solution. Action potentials (except for the first) are abolished by the procaine, but the receptor potentials remain. (C) Plot of receptor potential amplitude against increase in muscle length. (After Katz, 1950.)

produced a "spindle potential," or receptor potential, in the nerve terminal, that gave rise to action potentials. When the receptor potential was recorded in isolation by blocking the nerve discharge with procaine, a local anesthetic, its amplitude could be seen to increase in a graded fashion with muscle stretch (Figure 1).

The stretch receptor in crayfish muscle, first described by Alexandrowicz,[12] was used for further studies of this kind by Eyzaguirre and Kuffler.[13] This preparation is particularly useful because its cell body lies in isolation—not in a ganglion, but on its own in the periphery, where it can be seen in live preparations (Figure 2A), and is sufficiently large to permit penetration by intracellular microelectrodes. The cell

[12]Alexandrowicz, J. S. 1951. *Q. J. Microsc. Sci.* 92: 163–199.
[13]Eyzaguirre, C. and Kuffler, S. W. 1955. *J. Gen. Physiol.* 39: 87–119.

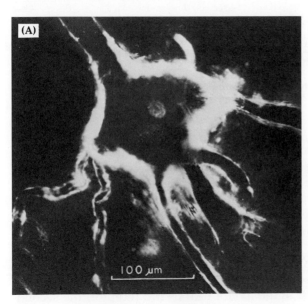

2

CRUSTACEAN STRETCH RECEPTOR. (A) Living receptor neuron viewed with dark-field illumination. Distal portions of six dendrites insert into the receptor muscle, which is not visible. (B) Relation between stretch receptor neuron and muscle, indicating method of intracellular recording. Excitatory fiber to the muscle produces contraction; inhibitory fiber innervates neuron. Two additional inhibitory fibers are not shown. (After Eyzaguirre and Kuffler, 1955.)

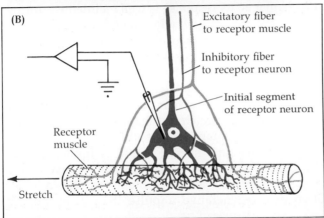

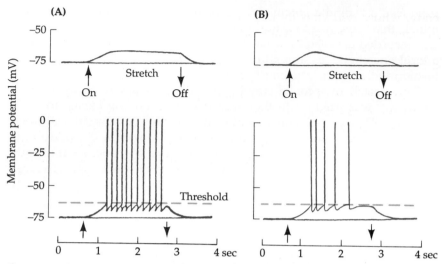

3   RESPONSES OF STRETCH RECEPTOR NEURONS to increases in muscle length,
recorded intracellularly as indicated in Figure 2. In a slowly adapting receptor (A),
a weak stretch for about 2 seconds produces a subthreshold receptor potential that
persists throughout the stretch (upper record). With a stronger stretch, a larger receptor
potential sets up a series of action potentials (lower record). In a rapidly adapting
receptor (B) the receptor potential is not maintained (upper record) and during the large
stretch the action potential frequency declines. (After Eyzaguirre and Kuffler, 1955.)

inserts its dendrites into a fine muscle strand nearby and sends an axon
centrally to a segmental ganglion (Figure 2B). In addition, the receptor
receives inhibitory innervation from the ganglion, and the muscle fibers
into which it inserts receive excitatory innervation.

There are two types of crustacean stretch receptors with distinct
structural and physiological characteristics. These have characteristic
appearances and their dendrites are embedded in different types of
muscle. One responds well at the beginning of a stretch, but its response
quickly wanes. This decrease in response to a steady stimulus is called
ADAPTATION. In contrast to the RAPIDLY ADAPTING receptor, the second
type is SLOWLY ADAPTING; that is, its response is well maintained during
prolonged stretch. The responses of a slowly adapting and a rapidly
adapting stretch receptor are shown in Figure 3A and 3B. In the slowly
adapting receptor, mild stretch of the muscle produces a depolarizing
receptor potential of 5–10 mV, lasting for the duration of the stretch. A
larger stretch produces a larger potential that depolarizes the cell to
above threshold and produces a train of action potentials that propagate
centrally along the axon. A similar stretch to the muscle produces only
transient responses in the rapidly adapting receptor.

The conducted impulses start in a special region of the axon close
to the cell body.[14] This region is called the axon hillock or (in myelinated
fibers) the initial segment. There the neuronal membrane has a lower
threshold for initiation of a regenerative impulse than does the cell

[14]Edwards, C. and Ottoson, D. 1958. *J. Physiol.* 143: 138–148.

body, whose dendrites may not conduct action potentials at all. Once initiated, impulses propagate not only to the central ganglion but also back into the cell body. It appears to be a common property of neurons that impulses are initiated in the region where the axon emerges from the soma.

As in other receptors, the intensity of the stimulus is expressed by the frequency of impulses. This relation between stimulus intensity and impulse frequency is established through the interaction of the maintained receptor current from the dendrites and the conductance changes associated with the action potential. At the end of each action potential, the increased potassium conductance that occurs on the recovery phase drives the membrane in the hyperpolarizing direction, toward E_K. This increase in potassium conductance is transient, while the receptor current is maintained by stretch and depolarizes the membrane once more to firing level. The stronger the receptor current, the sooner the firing level is reached again, and the higher the impulse frequency.

It has been shown with voltage clamp experiments that the current underlying the receptor potential in crayfish stretch receptors is associated with an increase in sodium permeability. However, its reversal potential is closer to zero than to E_{Na} because of an accompanying increase in potassium permeability.[15] Permeability to divalent cations is increased as well,[16] as is that to larger cations such as Tris and arginine. As with other cation channels, such as nicotinic ACh receptors, the increase in conductance produced by stretch is unaffected by tetrodotoxin.[17] Similarly, the receptor potential in the vertebrate muscle spindle is associated with an increased cation permeability.[18]

Ionic mechanisms underlying the receptor potential

The fact that distortion of the stretch receptor dendrites produces depolarization implies that there are channels in the dendritic membrane that are sensitive to distortion or "stretch" of the membrane. Single channels activated by membrane distortion were first observed in membrane patches from embryonic chick muscle cells[19] and in other cell membranes having nothing to do with sensory transduction.[20] Such channels were relatively silent in resting patches and activated by suction applied to the patch electrode. Similar channels have now been observed in patches made on the primary dendrites of the crayfish stretch receptor.[21] They are permeable to cations and their relative conductances to sodium, potassium, and calcium are consistent with previous observations on the whole cell.

The structural and functional properties of mammalian muscle spindles have been summarized in a concise review by Hunt.[22] An essential basis for physiological experiments on the spindles is an analysis of

Muscle spindle organization

[15]Brown, H. M., Ottoson, D. and Rydqvist, B. 1978. *J. Physiol.* 284: 155–179.
[16]Edwards, C. et al. 1981. *Neuroscience* 6: 1455–1460.
[17]Nakajima, S. and Onodera, K. 1969. *J. Physiol.* 200: 161–185.
[18]Hunt, C. C., Wilkerson, R. S. and Fukami, Y. 1978. *J. Gen. Physiol.* 71: 683–698.
[19]Guharay, R. and Sachs, F. 1984. *J. Physiol.* 352: 685–701.
[20]Sachs, F. 1988. *CRC Crit. Rev. Biomed. Eng.* 16: 141–169.
[21]Erxleben, C. 1989. *J. Gen. Physiol.* 94: 1071–1083.
[22]Hunt, C. C. 1990. *Physiol. Rev.* 70: 643–663.

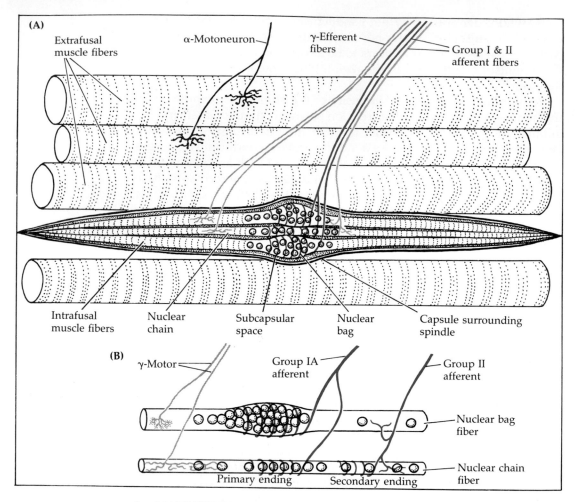

4   MAMMALIAN MUSCLE SPINDLE. (A) Scheme of mammalian muscle spindle innervation. The spindle, composed of small intrafusal fibers, is embedded in the bulk of the muscle, which is made up of large muscle fibers supplied by *α* motoneurons. Gamma motor (fusimotor) fibers supply the intrafusal muscle fibers, and group I and group II afferent fibers carry sensory signals from the muscle spindle to the spinal cord. (B) Simplified diagram of intrafusal muscle types and their innervation (see also Figure 9). (B after Matthews, 1964.)

their pattern of innervation. Figure 4 illustrates schematically the sensory apparatus of spindles in leg muscles of the cat. The spindle consists of 8 to 10 modified muscle fibers (called intrafusal fibers) running within a capsule. In the central, or equatorial, region there is in each fiber a large aggregation of nuclei. Their arrangement provides a basis for the classification of intrafusal fibers as bag or chain fibers, depending on whether the nuclei are grouped together in a swollen protuberance or are arranged linearly. This classification is refined further by subdividing the nuclear bag intrafusal muscle fibers into two groups on the basis of structural and functional differences (to be discussed later).

Two types of sensory neurons innervate each muscle spindle. The large nerve fibers, group Ia afferents, have diameters of 12–20 μm and conduct impulses at velocities up to 120 meters per second. (For a summary of the fiber classifications referred to in this chapter, see Box 2 in Chapter 5.) Their terminals are coiled around the central parts of both bag and chain fibers to form the primary endings. Smaller sensory nerves (group II afferents) are 4 to 12 μm in diameter and conduct more slowly. Their terminals are mainly in the less central region of the chain fibers, where they form the secondary endings.

When a rapid stretch is applied to a muscle and thereby to the spindles within it, receptor potentials and bursts of impulses arise in both types of sensory fibers. There is, however, a clear difference in the characteristics of the discharges in the two endings. The primary endings, connected to the larger group I axons, are sensitive mainly to the rate of change of stretch. The frequency of discharge is therefore maximal during the dynamic phase while stretch is increasing and subsides to a lower steady level while stretch is maintained. The secondary endings connected to the smaller group II fibers are relatively unaffected by the rate of stretch but are sensitive to the level of static tension.[23] This behavior is illustrated in Figure 5. The group Ia (dynamic) and group II (static) afferents are analogous to the rapidly adapting and slowly adapting receptors in the crayfish muscle.

We have noted that the responses of crayfish stretch receptors and vertebrate muscle spindles to stretch may be either slowly adapting or rapidly adapting. Adaptation to a continuing stimulus is a property common to all sensory receptors. We gradually become less aware of a constant pressure, an increased temperature on the skin, or contact with clothing or shoes. Some of this decreased awareness is undoubtedly due to processes in the central nervous system (we stop "paying attention"), but at least part is due to adaptation of the receptors themselves; as the stimulus is maintained, the frequency of firing in the primary afferent fibers declines. There are great differences in the degree and rate of such adaptation. Some mechanoreceptors adapt very rapidly, firing only a few impulses at the beginning of a maintained

Mechanisms of adaptation in sensory receptors

[23]Matthews, P. B. C. 1981. *J. Physiol.* 320: 1–30.

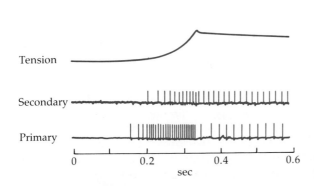

Tension

Secondary

Primary

| | | | | |
0   0.2   0.4   0.6
sec

5

DIFFERENCES IN MUSCLE SPINDLE RESPONSES. Recordings of action potentials from single primary (group Ia) and secondary (group II) sensory afferent fibers originating in a cat muscle spindle. The primary fiber greatly increases its discharge rate as tension develops during the stretch; during the maintained phase of the stretch it quickly adapts to a lower rate. The secondary fiber increases its firing rate more slowly as tension develops and maintains its discharge during the steady stretch. (After Jansen and Matthews, 1962.)

stimulus. In others there is very little adaptation. Still other receptors start at a high rate but maintain a lower rate of impulses almost indefinitely.

A number of factors can contribute to the adaptation of sensory responses. In muscle spindles, the viscoelastic properties of intrafusal fibers or slippage in the attachments of the nerves to muscle fibers may allow a gradual decrease in deformation of the sensory terminals.[24] The relative contribution of such mechanical factors to adaptation of muscle spindles and other mechanoreceptors is not clear. A variety of other processes have been shown to contribute to inactivation in crustacean stretch receptors.[16,25–27] In the slowly adapting stretch receptor, trains of impulses lead to an increase in internal sodium concentration and activation of the sodium pump. The net outward transport of positive charges by the pump reduces the amplitude of the receptor potential and hence the discharge frequency. Yet another factor contributing to inactivation is an increase in potassium conductance. For example, during a train of impulses in crayfish stretch receptors, calcium entry through voltage-activated channels in turn opens calcium-activated potassium channels. The effect of this increase in potassium conductance is to "short out" the receptor potential, again reducing its amplitude and the frequency of the sensory impulses. The rapidly adapting crayfish stretch receptor shows prompt adaptation of its firing rate even when a steady depolarizing current is applied experimentally. Because the depolarization is electrical, mechanical factors cannot be responsible, and the discharge ceases before there is any significant sodium or calcium accumulation. The response is terminated after only a few impulses by activation of voltage-sensitive potassium channels by depolarization, possibly coupled with partial sodium channel inactivation.

Adaptation in the Pacinian corpuscle

The Pacinian corpuscle is perhaps the most rapidly adapting of all receptors. Its nerve terminal is enclosed in an onionlike capsule. Pressure applied slowly to the capsule produces no response at all; more rapidly applied pressure produces only one or two action potentials. However, the receptors are exquisitely sensitive to vibration up to frequencies of 1000 per second. Although they are found generally in subcutaneous tissue, Pacinian corpuscles are particularly common around footpads and claws of mammals and in the interosseus membranes bridging the bones of the leg and forearm, possibly as sensitive detectors of ground vibration.[28] A similar structure, the Herbst corpuscle, is found in the legs, bills, and cutaneous tissue of birds (and in the tongues of woodpeckers!). Speculation about their physiological function includes detection by the duck's bill of aquatic vibrations due to

[24]Fukami, Y. and Hunt, C. C. 1977. *J. Neurophysiol.* 40: 1121–1131.
[25]Nakajima, S. and Takahashi, K. 1966. *J. Physiol.* 187: 105–127.
[26]Nakajima, S. and Onondera, K. 1969. *J. Physiol.* 200: 187–204.
[27]Sokolove, P. G. and Cooke, I. M. 1971. *J. Gen. Physiol.* 57: 125–163.
[28]Quilliam, T. A. and Armstrong, J. 1963. *Endeavour* 22: 55–60.

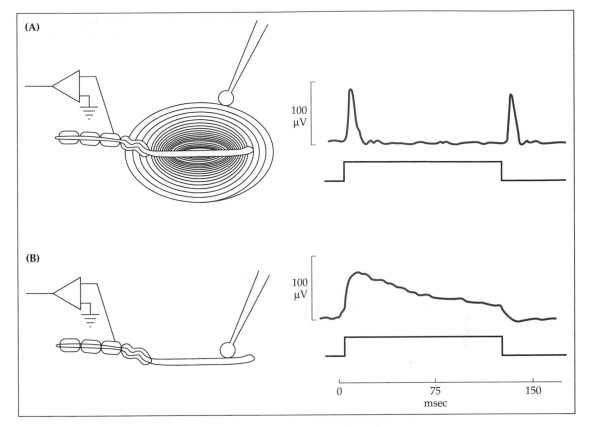

6 ADAPTATION IN A PACINIAN CORPUSCLE. (A) A pressure pulse applied to the body of the corpuscle (lower trace) produces a rapidly adapting receptor potential (upper trace), due to a transient wave of deformation that travels through the capsule to the nerve terminal. A similar response occurs on removal of the pulse. (B) After removal of the capsule layers, pressure applied to the nerve terminal produces a receptor potential that lasts for the duration of the pulse. (After Loewenstein and Mendelson, 1965.)

swimming fish, and (in soaring birds) detection of vibration of flight feathers due to improper aerodynamic trim.[29]

The mechanism of adaptation in the Pacinian corpuscle was studied in detail by Loewenstein and his associates. In particular, they showed that the adaptation was due, in part, to the dynamic properties of the capsule.[30] When a mechanical pulse was applied to an isolated, intact corpuscle, a brief receptor potential appeared at the onset and withdrawal of the pulse (Figure 6A). The transient responses were due to fluid waves propagated through the capsule at the onset and release of the deformation. After the capsule was stripped carefully from the nerve ending, the receptor potential decayed only slowly during the pulse

[29]McIntyre, A. K. 1980. *Trends Neurosci.* 3: 202–205.
[30]Loewenstein, W. R. and Mendelson, M. 1965. *J. Physiol.* 177: 377–397.

(Figure 6B). Nonetheless, even when the receptor potential was prolonged there was still only a brief burst of action potentials in the afferent axon (not shown); that is, the properties of the axon itself are matched to those of the receptor.

CENTRIFUGAL CONTROL OF MUSCLE RECEPTORS

The central nervous system not only receives information from sensory receptors but also acts back upon them to modify their responses. The brain, therefore, has the built-in ability not only to edit or censor but also to adjust the flow of information that reaches it. This feedback control is executed through pathways leading centrifugally to the peripheral sense organ. Such centrifugal control has a role in many sensory systems, for example, the auditory system (see later), and has been analyzed in detail in mammalian muscle spindles and crustacean stretch receptors.

The elements of centrifugal control in the crustacean stretch receptor are shown in Figure 7A. The receptors are innervated by excitatory and inhibitory motor axons from the central nervous system. Electrical excitation causes the ends of the receptor muscles to contract and thereby to stretch the centrally located dendrites of the sensory cells, producing a depolarizing receptor potential. It follows that the membrane potential of the receptor neuron, and hence the frequency of the sensory impulses, can be influenced not only by external stretch but also by active contraction of the receptor muscle in response to excitatory signals from the nervous system.[11,31] The excitatory control ensures that the receptors can be made to signal even when the body muscles in which they are embedded shorten, or their ongoing signals can be accelerated when they are already discharging. The effects of the excitatory stimulation on a slowly adapting receptor are shown in Figure 7B. When the muscle is stretched, the sensory discharge is greatly accelerated by a contraction of the slow receptor muscle fiber. In contrast, stimulating the inhibitory input inhibits the receptor discharge directly, as shown in Figure 7C.[32,33]

In summary, the information that leaves the crustacean stretch receptors is highly controlled and adjustable over a wide range. The signals are determined by the excitatory action of graded amounts of external stretch, by contraction of the receptor muscles, and by graded amounts of inhibition. These processes interact and exert their influence on the initial segment of the axon, where impulses start. The interaction of excitation and inhibition in the stretch receptor provides an example of multiple excitatory and inhibitory actions characteristic of neurons within the central nervous system. The depolarizing receptor potential drives the membrane potential beyond its firing level, while the inhi-

---

[31]Kuffler, S. W. 1954. *J. Neurophysiol.* 17: 558–574.
[32]Kuffler, S. W. and Eyzaguirre, C. 1955. *J. Gen. Physiol.* 39: 155–184.
[33]Jansen, J. K. S. et al. 1971. *Acta Physiol. Scand.* 81: 273–285.

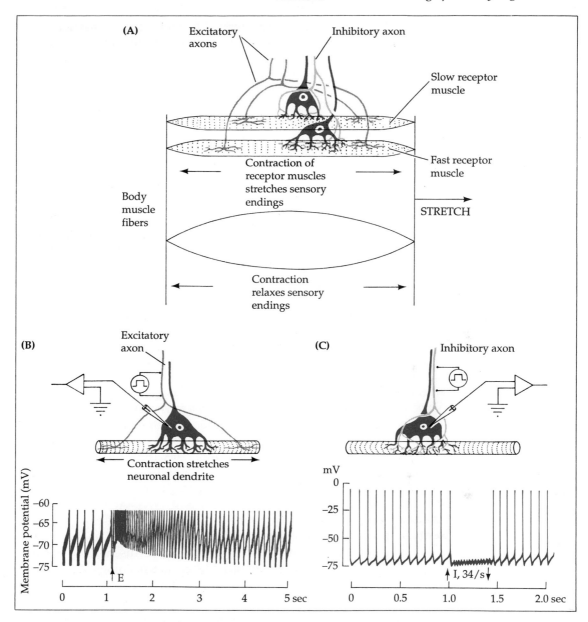

7   **EFFERENT REGULATION OF CRAYFISH STRETCH RECEPTOR NEURONS. (A)**
Slow- and fast-receptor muscle strands both receive excitatory innervation that can
make them contract, thereby stretching the dendrites of the receptor neurons. The
dendrites also receive inhibitory innervation that, when activated, counteracts the effect
of stretch. The main muscle mass is drawn separately for clarity; when it contracts,
stretch on the receptors is reduced. **(B)** A slow receptor gives a maintained sensory
discharge when stretched (only the lower parts of the action potentials are shown); the
discharge is accelerated by two closely spaced motor stimuli (E, at arrow). The duration
of the increase in frequency reflects the time course of the muscle contraction. **(C)**
Stimulation of the inhibitory axon at 34 per second (I, between arrows) suppresses
spontaneous discharge for the duration of the stimulus train. **(B and C after Kuffler and
Eyzaguirre, 1955.)**

Centrifugal control of
muscle spindles

bitory synaptic action tends to keep the membrane potential below threshold. The balance of these two competing influences determines whether the cell discharges or remains quiescent.

The motor control of mammalian muscle spindles is illustrated in Figure 8A. It is analogous to that of crustacean stretch receptors (compare with Figure 7A), except for the absence of inhibitory innervation. The efferent nerve supply to the spindles was described originally by Eccles and Sherrington,[34] and was studied in more detail by Leksell.[35] It consists of a distinct group of small motor nerves (2 to 8 μm in diameter), now known as FUSIMOTOR- or γ-FIBERS. The role of fusimotor fibers was firmly established on limb muscles of the cat in a series of technically difficult and definitive experiments by Kuffler, Hunt, and Quilliam.[36] The procedure was to record in the dorsal root the activity in an individual afferent fiber coming from a spindle in an anesthetized cat while stimulating a fusimotor fiber in the ventral root going to the same spindle. Fusimotor stimulation produced an increase in sensory activity, but no increase in muscle tension. Trains of impulses in the fusimotor neurons accelerated the sensory discharge if the muscle was stretched, or initiated a sensory discharge in a shortened muscle.

Further studies on the effects of intrafusal contraction in the cat have established the existence of two main classes of γ-efferent motor fibers to mammalian spindles.[37,38] One class ends on bag fibers and produces an increase in the dynamic discharge; the other ends mainly on chain fibers and has its major effect on the static component of the discharge. They are known accordingly as DYNAMIC EFFERENT (γ_d) and STATIC EFFERENT (γ_s) axons. Their effects on sensory discharge are shown in Figure 8B and 8C.

Although the description above illustrates the basic properties of muscle spindle regulation, the details are still more complex. These details are shown schematically in Figure 9. Bag fibers are further subdivided into two types, known as BAG$_1$ and BAG$_2$. Activation of bag$_1$ fibers affects the dynamic component of the sensory discharge, as described previously. Bag$_2$ fibers, on the other hand, are like chain fibers inasmuch as their activation increases the static discharge. The two types of bag fibers can be distinguished by their histological characteristics, by contractions they undergo, and by the type of γ-efferent nerve supply they receive. The details of fusimotor innervation and function summarized in Figure 9 have been determined by Boyd, Barker, Laporte, and their colleagues,[39–41] who have made direct observation of

[34]Eccles, J. C. and Sherrington, C. S. 1930. *Proc. R. Soc. Lond. B* 106: 326–357.
[35]Leksell, L. 1945. *Acta Physiol. Scand.* 10(Suppl.31): 1–84.
[36]Kuffler, S. W., Hunt, C. C. and Quilliam, J. P. 1951. *J. Neurophysiol.* 14: 29–51.
[37]Jansen, J. K. S. and Matthews, P. B. C. 1962. *J. Physiol.* 161: 357–378.
[38]Emonet-Dénand, F., Jami, L. and Laporte, Y. 1980. *Prog. Clin. Neurophysiol.* 8: 1–11.
[39]Barker, D. et al. 1980. *Brain Res.* 185: 227–237.
[40]Arbuthnott, E. R. et al. 1982. *J. Physiol.* 331: 285–309.
[41]Arbuthnott, E. R., Gladden, M. H. and Sutherland, F. I. 1989. *J. Anat.* 163: 183–190.

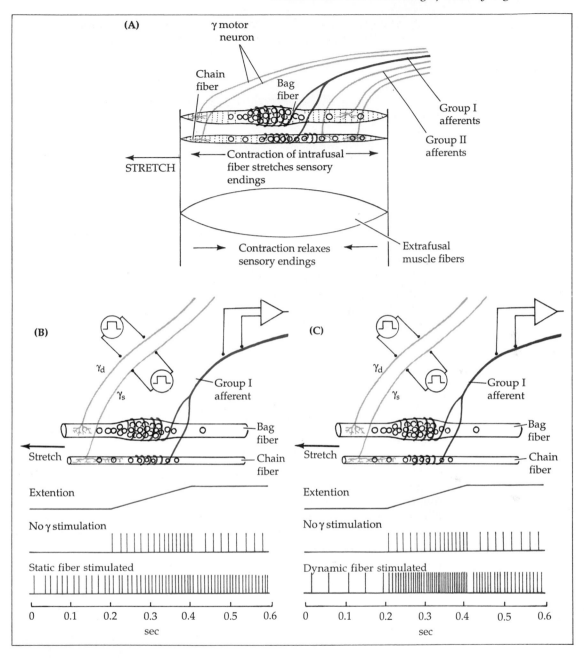

8 **EFFERENT REGULATION OF MUSCLE SPINDLES. (A) Mammalian muscle spindles are arranged in a manner similar to the crayfish stretch receptors; bag and chain fibers receive innervation by γ-efferent fibers that, when activated, cause the intrafusal fibers to contract. The mass of extrafusal muscle fibers is drawn separately; when it contracts, stretch on the intrafusal fibers is reduced. (B) Extension of the whole muscle causes discharge of group Ia afferent fibers from the spindle; stimulation of the static γ motor fiber ($γ_s$) throughout the period causes an increase in frequency both before and throughout the period of stretch. (C) Stimulation of the dynamic γ motor fiber ($γ_d$) causes an increase in sensory discharge that is most marked during the transient phase of the stretch. (B and C after Crowe and Matthews, 1964.)**

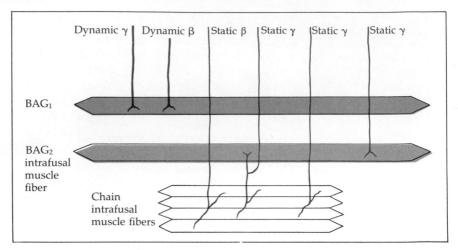

9   FUSIMOTOR INNERVATION OF BAG AND CHAIN FIBERS. The intrafusal bag₁
    fiber is supplied by dynamic γ and β axons. Bag₂ fibers are supplied by static γ
    axons, and chain fibers by static γ and β axons. Evidence for this pattern was obtained
    from living spindles by observing contractions of individual intrafusal muscle fibers
    produced by stimulation of single fusimotor fibers. (After Matthews, 1981; Arbuthnott
    et al., 1982.)

the intrafusal fibers and of the effects of γ-efferent stimulation in living spindles, dissected out of the animal and maintained in vitro. By using high-resolution Nomarski microscopy, it has become possible to measure precisely the deformation of the primary and secondary endings as stimulation of a single γ-efferent axon causes intrafusal contraction of a dynamic bag₁, a static bag₂, or a chain muscle fiber.

Figure 9 shows a further type of motor supply to the spindle. It has long been known that in frogs and toads some of the motoneurons that cause contraction of the main mass of the muscle also supply intrafusal fibers.[42] Laporte and his colleagues have since shown that such fibers also exist in the cat and the rabbit.[40] Some are dynamic, supplying bag₁ fibers, others static. Such fibers are usually called β because their diameters and conduction velocities correspond to the smaller and slower range of extrafusal motor axons. (In fact the term β is a misnomer because the fibers fall technically in the α range of conduction velocities; see Chapter 5.) Beta fibers appear to represent a system for maintaining a spindle discharge automatically when the extrafusal muscle fibers are activated, but the extent of their role in maintaining muscle spindle tone is not known.

The control
of stretch receptors
during movement

One role played by the efferent innervation of intrafusal muscle fibers in the spindle becomes apparent when sensory discharges during muscle movement are considered. Because the spindles are in parallel with the main body of the muscle (Figure 8A), muscle contraction

[42]Katz, B. 1949. *J. Exp. Biol.* 26: 201–217.

reduces the tension on the sensory element. Therefore, in the absence of efferent control, the rate of sensory discharges would be reduced or cease during muscle contraction, because of unloading of the spindle. However, activation of the γ efferents shortens the intrafusal muscle elements to take up the slack and restores the tension on the sensory terminals. The part played by efferent nerves is to adjust the sensitivity of the measuring instrument—the spindles—so that they can perform over a wide range of muscle lengths. Spindles, therefore, can be made to maintain their discharge frequencies even when the external stretch on them is reduced during contraction of muscles.

Evidence for such a mechanism has been obtained in a variety of experiments. For example Sears, von Euler, and others[43–45] have recorded discharges in afferent fibers from spindles in intercostal muscles used for respiration. In Figure 10A, records taken under normal conditions show afferent discharges from an inspiratory muscle being greatest during inspiration, when the muscle is shortest. A simple explana-

[43]Sears, T. A. 1964. *J. Physiol.* 174: 295–315.
[44]Critchlow, V. and von Euler, C. 1963. *J. Physiol.* 168: 820–847.
[45]Greer, J. J. and Stein, R. B. 1990. *J. Physiol.* 422: 245–264.

(A)   NORMAL

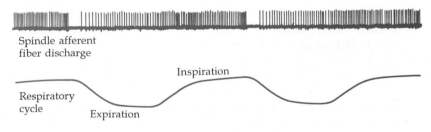

(B)   AFTER FUSIMOTOR PARALYSIS

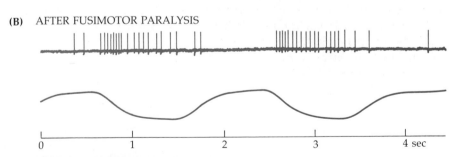

10   **REFLEX CONTROL OF SPINDLE ACTIVITY** during breathing in the cat. In each pair of records the upper traces show muscle spindle afferent activity, lower traces respiratory cycle. **(A)** Normally the sensory discharge from muscle spindles in inspiratory muscles is highest during inspiration, even though the muscles are shortening. **(B)** After fusimotor fibers are blocked selectively with procaine, the spindles behave passively, with increased discharges during expiration, when the muscles are stretched, and cessation of activity on inspiration, when the muscles are shortened. (After Critchlow and von Euler, 1963.)

tion for this unexpected result is that activation of the extrafusal muscle fibers is accompanied by activation of the intrafusal fibers through the γ efferents, and that the γ excitation more than overcomes the slack created by extrafusal contraction. This explanation is confirmed in Figure 10B, which shows the responses of the afferent fiber after the γ efferents were paralyzed by local application of procaine. After the γ fibers are blocked, afferent discharges occur only on expiration, when the intrafusal fibers are being stretched. Other examples of such parallel activation include records of muscle spindle activity accompanying finger movements in humans.[46] During voluntary isometric contractions, the majority of spindle afferents increased their discharge even though the length of the muscle was not increasing.

The skeletal muscle fibers that do the work of contraction are innervated directly by groups of large motoneurons (α MOTONEURONS) in the ventral part of the spinal cord (Chapter 15). All the neural apparatus that influences movement must converge on these neurons. In the soleus muscle of the cat, 100 α motoneurons diverge to innervate about 25,000 muscle fibers, a ratio of one neuron to 250 muscle fibers. In contrast, the 50 muscle spindles in the soleus (containing a total of about 300 intrafusal muscle fibers) are innervated by 50 γ motoneurons. Thus about one-third of the motor axons supplying the muscle are concerned with regulation of movement rather than with its execution. Sensory fibers comprise 50 group Ia and 50 group II spindle afferents, in addition to 40 Ib fibers supplying Golgi tendon organs (see later). The more finely honed the movement, the more nervous machinery is devoted to its control. Thus, the muscles of our hands are more densely supplied with spindles than are the soleus and other large limb muscles.

Golgi tendon organs   Muscle spindles, being in parallel with the main mass of the muscle, measure muscle length, not tension. Another group of receptors, the Golgi tendon organs, consist of a fine arborization of nerve terminals encapsulated within the muscle tendons.[47] These are highly sensitive to tension produced by muscle contraction. Discharges from the tendon organs contain both static and dynamic components and are carried to the central nervous system by group I afferent fibers, called GROUP IB to distinguish them from muscle spindle afferents.

Joint position   It is not clearly understood how information about joint position is derived. Mechanoreceptors in a joint such as the knee, which were once thought to signal joint angles over discrete ranges, now appear not to serve this purpose. Instead, the receptors respond best to extreme and potentially damaging ranges of flexion, extension, or torque.[48] Signals from tendon organs provide ambiguous information about position: Their static and dynamic components signal muscle tension and the rate at which it is developing. The information lacks precision because

---

[46]Vallbo, A. B. 1990. *J. Neurophysiol.* 63: 1307–1313.
[47]Barker, D. 1962. *Muscle Receptors.* Hong Kong University Press, Hong Kong.
[48]Matthews, P. B. 1988. *Can. J. Physiol. Pharmacol.* 66: 430–438.

both components of the signal depend not only on muscle contraction but also on the load against which the muscle is working. The problem of analyzing signals from muscle spindles is still more complex. The nervous system must distinguish between sensory discharges from spindles evoked on the one hand by stretch and on the other by intrafusal contraction. Presumably the analysis needed to determine joint position involves comparison of (1) the discharges causing contraction of the bulk of the muscle, (2) the discharges initiating or maintaining sensory responses, and (3) the incoming signals received from the spindles and tendon organs.

Do spindle discharges contribute to our conscious sense of position or movement? For many years it was thought that sensory information from muscle spindles was concerned solely with reflex regulation of muscles and did not reach the cerebral cortex. Position sense was attributed entirely to receptors situated in joints. It is now known that information from spindles does reach the cerebral cortex and consciousness.[49,50] Patients and volunteers have reported sensations of movements, of changes in position of a limb or digit, or of stiffness when muscles were vibrated or stretched after all sensation in joint receptors had been abolished.

SOMATIC SENSATION

Our sensations of touch, pressure, and vibration depend on the presence of appropriate receptors in the skin and deep tissues. The morphological types present in glabrous (nonhairy) skin are sketched in Figure 11. The most superficial are Meissner's corpuscles and Merkel's disks, thought to correspond respectively to rapidly adapting and slowly adapting afferent fibers. Deeper in the tissue are rapidly adapting Pacinian corpuscles and slowly adapting Ruffini's capsules. Not shown are receptors for pain and for hot or cold stimuli, all of which lack distinct morphological structures and are therefore classified as FREE NERVE ENDINGS. In hairy skin, nerve terminals surround the hair follicles, giving rapidly adapting responses to bending of the hairs. The skin of the human hand, which is densely innervated, contains 15,000 to 20,000 mechanoreceptors, about evenly divided between slowly adapting and rapidly adapting types. The density of the more superficial mechanoreceptors in the human fingertip is about 100 per cm^2, three to four times greater than in the palm.

Somatic receptors

Figure 12A shows schematically the main somatosensory pathways for carrying sensations of touch, pressure, and vibration to the brain (more details are given in Appendix C). Fibers from skin, deep tissue, muscle, and joints enter the dorsal roots, give off axon collaterals to make synapses on spinal neurons, and ascend in the dorsal columns to end on second-order cells in the cuneate and gracile nuclei. The axons

Central pathways

[49]McCloskey, D. I. 1978. *Physiol. Rev.* 58: 763–820.
[50]Matthews, P. B. C. 1982. *Annu. Rev. Neurosci.* 5: 189–218.

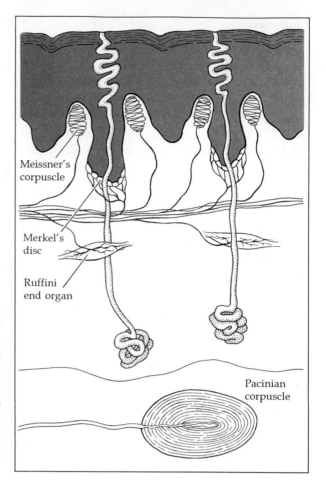

Meissner's
corpuscle

Merkel's
disc

Ruffini
end organ

Pacinian
corpuscle

11

MECHANORECEPTORS IN SKIN. Rapidly adapting Meissner's corpuscles and slowly adapting Merkel's discs respond to skin indentation within small, well-defined fields. Ruffini end organs are slowly adapting and have wider fields with poorly defined borders. Pacinian corpuscles are very rapidly adapting and sensitive to vibration over a wide area of skin. All are connected to axons with diameters in the Aβ range. Not shown are the rapidly adapting hair follicle receptors that respond to hair movement (Aβ fibers) and free nerve endings mediating temperature, itch, and pain sensations (Aδ and C fibers). (After Johansson and Vallbo, 1983.)

12   SENSORY PATHWAY SUBSERVING TOUCH SENSATION. (A) A transverse ▶
section of the spinal cord and medulla, leading to a coronal section of the brain behind the central sulcus. Sensory fibers (mostly Aβ), with cell bodies in the dorsal root ganglia, enter the spinal cord, giving off collaterals that form synapses at segmental levels, and ascend to the medulla in the dorsal columns. At progressively ascending levels in the spinal cord, incoming fibers are added laterally, so that those from the leg are medial to those from the arm and a somatotopic order is maintained. In the medulla, second-order fibers cross the midline to ascend in the medial lemniscus to the ventral posterior lateral nucleus of the thalamus. Third order fibers project to the sensory region of the cerebral cortex. Touch fibers also ascend in the spinothalamic tract (Figure 16). (B) Somatotopic representation in the human sensory cortex (the sensory "homunculus") on a coronal section of the brain. Densely innervated regions, such as the face, have a correspondingly large representation. (B after Penfield and Rasmussen, 1950.)

of the second-order cells cross the midline and ascend in a bundle called the medial lemniscus to end on cells in the nucleus ventroposterolateralis (VPL) of the thalamus. The third-order cells project to a region of the cerebral cortex in the postcentral gyrus known as the primary so-

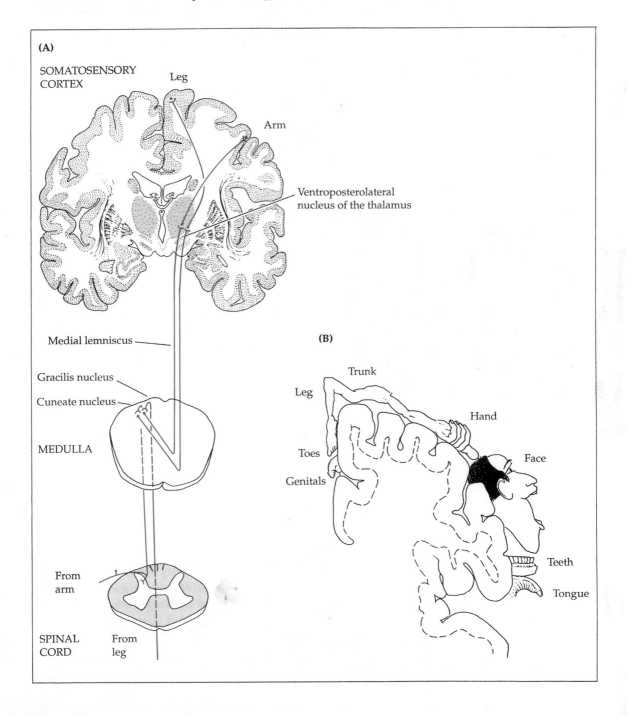

(A)

SOMATOSENSORY CORTEX

Leg

Arm

Ventroposterolateral nucleus of the thalamus

Medial lemniscus

Gracilis nucleus

Cuneate nucleus

MEDULLA

From arm

SPINAL CORD

From leg

(B)

Trunk

Leg

Hand

Toes

Face

Genitals

Teeth

Tongue

matosensory area. At each successive level there is an orderly map of the body correlated with the modalities of touch, pressure, and joint position. Since the pathways are crossed, the left side of the body is represented on the right side of the brain. As in the visual system, where a small region of retina—the fovea—is represented most extensively within the visual cortex (Chapter 17), so there exists within the sensory cortex comparable distortion of the representation of the body.[51] In monkeys and human beings, areas within the central nervous system concerned with hands, fingers, and lips are larger than those concerned with trunk or legs (Figure 12B). In different animals, various other regions of the body predominate—the whiskers for a mouse (Figure 15) or the snout for a pig, which as Adrian states "is its chief executive organ, spade as well as hand."[52] It has been shown that these representations are not immutably fixed and can be altered by lesions that modify the sensory input to the cortex or by peripheral stimulation[53] (see Chapter 18).

Receptive fields　　The receptive field of a cell in the somatosensory system is that area of skin within which a mechanical stimulus elicits a response. Part of the field may be excitatory, producing an increase in firing rate of the cell; another part may be inhibitory, decreasing its firing rate. The size and complexity of receptive fields vary as we progress through successive levels of the sensory system. Primary afferent fibers have uncomplicated fields that are excitatory and whose location and size are determined simply by the location of the receptor itself and how its sensitivity decreases with distance from the stimulus. Fibers supplying the superficial receptors (Meissner's corpuscles and Merkel's disks) usually have relatively restricted receptive fields; receptive fields of those supplying the more deeply located Ruffini end organs and Pacinian corpuscles are usually larger. An example of primary afferent receptive fields for rapidly adapting touch receptors is shown in Figure 13. The fields are a few millimeters in diameter and contain several "hot spots" of increased sensitivity, presumably indicating the location of single corpuscular endings made by the fiber.

In the primary somatosensory cortex (Figure 12B), a fine grain of the representation has been revealed by recording from individual neurons. Indeed, it was in this region of the cortex that Mountcastle, Powell, and their colleagues first demonstrated columnar organization,[54] preceding comparable work in the visual system (Chapter 17). When a microelectrode penetration perpendicular to the cortical surface was made through the thickness of the somatosensory cortex, as shown in Figure 14A, each neuron encountered shared a number of receptive field properties with the other neurons above and below in the same

[51]Penfield, W. and Rasmussen, T. 1950. *The Cerebral Cortex of Man. A Clinical Study of Localization of Function.* Macmillan, New York.

[52]Adrian, E. D. 1946. *The Physical Background of Perception.* Clarendon Press, Oxford.

[53] Jenkins, W. M. et al. 1990. *J. Neurophysiol.* 63: 82–104.

[54]Powell, T. P. S. and Mountcastle, V. B. 1959. *Bull. Johns Hopkins Hosp.* 105: 133–162.

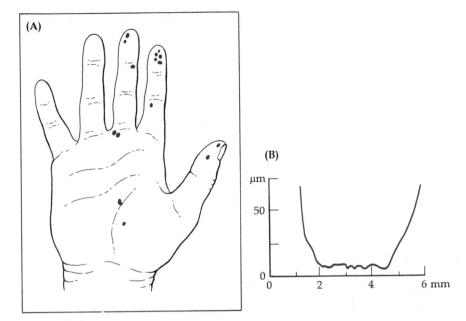

13 RECEPTIVE FIELDS OF RAPIDLY ADAPTING TOUCH RECEPTORS (probably Meissner's corpuscles) in a human hand. Fields were mapped by recording discharges from single fibers in a sensory nerve, using tungsten electrodes inserted through the skin. (A) Receptive fields consist of small, well-defined areas on the fingers and palm. Each dot represents the receptive field of one fiber. (B) Indentation (in μm) required to produce a response is plotted against location within a receptive field. The region of maximum sensitivity is about 3 mm in radius, within which indentations of only a few micrometers are sufficient to produce a response. Points of maximum sensitivity within the region presumably correspond to the position of individual receptors, all connected to the same afferent fiber. (After Johansson and Vallbo, 1983.)

track. For example, all the cells encountered in one penetration were driven by light touch applied to a small area of the skin of the hand (Figure 14B). Penetrations through an area of cortex a millimeter or so away would result in a shift of the receptive field and perhaps of the modality, all the cells now responding to pressure or to displacement of the joint of the fingertip. Still farther away—with a shift of receptive field position to the forearm or elbow—the size of the receptive field of the cortical neuron would be considerably larger, measured in centimeters instead of millimeters. Figure 14C shows a receptive field on the forearm for a cortical neuron and illustrates an important property of receptive field organization: The cell not only discharges when an area of skin is touched; that discharge is inhibited by touching any part of a large surrounding area. This INHIBITORY SURROUND makes possible greater discrimination between two nearby tactile stimuli applied to the skin.

The degree of precision with which parts of the body are represented in the somatosensory cortex is well illustrated by the projection of

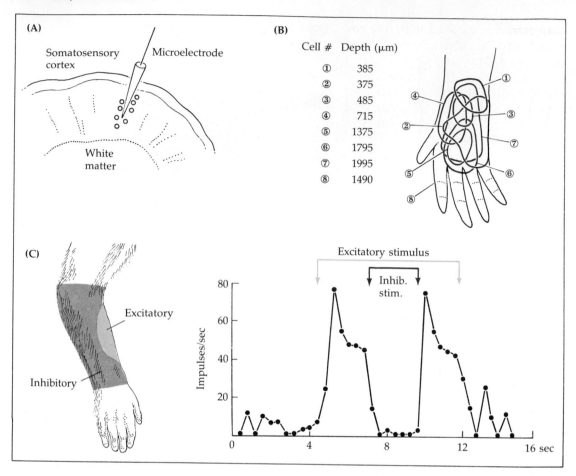

14 **RECEPTIVE FIELDS OF NEURONS** in the somatosensory cortex of the monkey.
(A) Electrode is inserted through the cortex at right angles to the surface. (B)
Each cell encountered by the electrode responds to touch over roughly the same region
of the hand. (C) Receptive field of a cell in another region of the cortex. Excitatory
region on the ventral surface of the forearm is surrounded by an inhibitory region that
suppresses the response of the cell. (B After Powell and Mountcastle, 1959; C after
Mountcastle and Powell, 1959.)

sensory innervation from whiskers of the mouse. Histological studies
by Van der Loos and his colleagues[55–57] have shown that the mouse
somatosensory cortex contains characteristic groups of nerve cells clustered in the form of cylinders that lie perpendicularly within the cortex.
These assemblies have been termed BARRELS from their shapes, which
were determined by serial reconstructions. Each barrel is 100 to 400 μm
in diameter and is composed of a ring of cells that surrounds a central
hollow containing fewer cells. The array of barrels in the mouse is

[55]Woolsey, T. A. and Van der Loos H. 1970. *Brain Res.* 17: 205–242.
[56]Nussbaumer, J. C. and Van der Loos, H. 1984. *J. Neurophysiol.* 53: 686–698.
[57]Walker, E. and Van der Loos H. 1986. *J. Neurosci.* 6: 3355–3373.

consistently organized with five rows. Van der Loos and Woolsey realized that this pattern corresponded exactly with the rows of whiskers, or vibrissae, on the mouse's face, one barrel for each whisker (Figure 15C and 15D). The functional role of the barrels is confirmed by recording electrically from individual cells.[58,59] Each cell responds to movement only of the corresponding vibrissa, some firing when the whisker is moved in one direction, others when it is moved in the opposite direction. The mouse uses the vibrissae as sensitive antennae, waving them backward and forward as it walks along to detect objects on either side of its path. The synaptic connections and overall design of circuitry within the barrels are not yet known.

In addition to the primary somatosensory cortex, multiple representations of the body occur in other, secondary, somatosensory areas (see Appendix C).[60,61] In area 5 of the parietal cortex, neurons with more

[58]Welker, C. 1976. *J. Comp. Neurol.* 166: 173–190.
[59]Simons, D. J. and Woolsey, T. A. 1979. *Brain Res.* 165: 327–332.
[60]Merzenich, M. M. et al. 1978. *J. Comp. Neurol.* 181: 41–74.
[61]Kaas, J. H. 1983. *Physiol. Rev.* 63: 206–231.

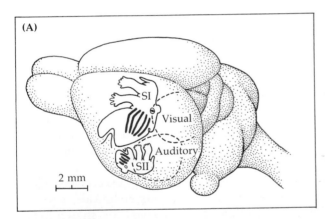

15

BARRELS IN A MOUSE SOMATOSENSORY CORTEX corresponding to vibrissae. (A) Diagram of a mouse brain to show the somatotopic representation on the sensory cortex ("musculus" corresponding to the "homunculus" in Figure 12B), particularly the face and vibrissae. SII is the second somatosensory area. **(B)** Horizontal section through the cortex in the area representing the vibrissae, showing barrels in cross section. **(C)** Close-up of whiskers. **(D)** Schematic diagram of barrel arrangement from (B), showing correspondence with vibrissae. (After Woolsey and Van der Loos, 1970.)

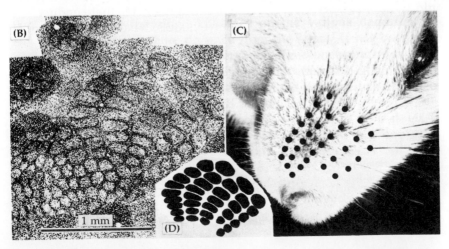

complex response properties have been found; such neurons are driven only by movements involving several joints, for example movement of the entire limb in one direction only. Other columns of neurons that have been found respond to stroking of the skin in one direction but not the other.[62,63] What at present remains elusive is the way in which the brain synthesizes information from diverse areas into a complete body image.

Temperature sensation

At a skin temperature of about 33°C we are usually unaware of any temperature sensation. Raising or lowering skin temperature above this neutral point produces a sensation of warming or cooling. Surprisingly, there are two kinds of temperature receptors in the skin—one signaling warmth, the other cold—even though in theory one would be sufficient. You can readily demonstrate this on the back of your hand: If a room-temperature probe (such as a dull pencil point) is pressed against the skin at various points, it will occasionally produce a punctate sensation of cold. Outside of such spots, only touch is felt. A warm (e.g., 45°C) metal probe can be used to find other points, spatially distinct from the cold spots, where sensations of warmth are felt, although such spots are fewer and require extensive searching. Small myelinated (Aδ) and unmyelinated fibers from thermal receptors enter the spinal cord to form synapses on second-order cells in the dorsal horn. The pathways carrying information about temperature then run with those subserving pain sensation (Figure 16) rather than in the dorsal columns.

NOCICEPTIVE SYSTEMS AND PAIN

Information about noxious or painful stimuli is conveyed to higher centers by specific receptors and by pathways largely distinct from those used for position sense, touch, or pressure. Nociceptive terminals appear as free nerve endings in the skin and the viscera; characteristically, they respond only to strong stimuli, such as pricking, excessive stretching, and extremes of temperatures, or to various chemicals, such as histamine and bradykinin.[64] The afferent axons have been shown to fall into two classes: (1) Aδ axons 1 to 4 μm in diameter that conduct at 6 to 24 meters per second and (2) unmyelinated C axons, 0.1 to 1 μm in diameter, that conduct more slowly, at 0.5 to 2 meters per second.[65,66] A single afferent fiber, whether Aδ or C, responds to noxious stimuli applied to its receptive field in a characteristic manner: The rate of adaptation is slow, and the discharge continues after removal of the stimulus. The two fiber types, however, mediate different pain sensations. In humans a brief, intense stimulus delivered to a distal limb gives rise first to a sharp and relatively brief pricking sensation (first

[62]Hyvärinen, J. and Poranen, A. 1978. *J. Physiol.* 283: 523–537.

[63]Costanzo, R. M. and Gardiner, E. P. 1980. *J. Neurophysiol.* 43: 1319–1341.

[64]Ottoson, D. 1983. *Physiology of the Nervous System.* Oxford University Press, New York, Chapter 31.

[65]Kruger, L., Perl, E. R. and Sedivic, M. J. 1981. *J. Comp. Neurol.* 198: 137–154.

[66]Torebjork, H. E. and Ochoa, J. 1980. *Acta Physiol. Scand.* 110: 445–447.

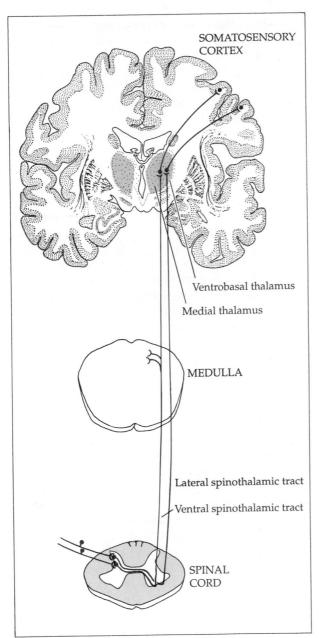

SOMATOSENSORY
CORTEX

Ventrobasal thalamus

Medial thalamus

MEDULLA

Lateral spinothalamic tract

Ventral spinothalamic tract

SPINAL
CORD

16

SENSORY PATHWAYS MEDIATING PAIN AND TEMPERATURE, shown on cross sections of the spinal cord and medulla and a coronal section of the brain. Small myelinated (Aδ) and unmyelinated (C) fibers enter the spinal cord to form synapses with second-order neurons. The second order axons cross the midline to ascend in the ventral and lateral spinothalamic tract to end in the medial and ventrobasal thalamus. Third-order cells then project to the cerebral cortex. Spinothalamic tracts also carry touch fibers (not shown). Cells in the ventral spinothalamic tract give off collateral branches in the medulla, and some terminate at this level (the spinoreticular tract; not shown).

pain), followed by a dull, prolonged burning sensation (second pain). Electrophysiological experiments have shown that first pain is related to activation of small myelinated fibers and second pain to C-fiber activation.[67,68]

[67]Collins, W. F. Jr., Nulsen, F. E. and Randt, C. T. 1960. *Arch. Neurol.* 3: 381–385.
[68]Torebjork, H. E. and Hallin, R. G. 1973. *Exp. Brain Res.* 16: 321–332.

Nociceptive pathways within the central nervous system are shown in Figure 16. Entering fibers form synapses on cells in the dorsal horn. It is not certain which neurotransmitters mediate the effects of the afferent fibers on the spinal neurons; substance P and glutamate are favored (Chapter 10), along with somatostatin and VIP (vasoactive intestinal polypeptide).[69,70] Fibers from second-order cells cross the midline and ascend through two major pathways: the lateral spinothalamic tract and the ventral spinoreticular thalamic pathway. The ascending fibers end in the ventrobasal and medial nuclei of the thalamus. Cells in these nuclei project to somatosensory cortex and other widespread areas of the nervous system. Unlike other interneurons in the somatosensory system, those in the thalamus and cortex receiving nociceptive inputs have large, ill-defined receptive fields, often covering wide areas of the body, contralateral as well as ipsilateral.

Although the afferent systems for nociception have their own through lines, it is clear that no noxious stimulus can fail to activate other receptors responding to touch, pressure, displacement, stretch, cooling, heating, and so on. Numerous experiments have shown that the two systems interact. In particular, light touch or stroking of the skin can influence the discharge of nociceptive neurons within the central nervous system. For example, Poggio and Mountcastle[71] showed that the activation (by stroking) of somatosensory afferents with nearby receptive fields had an inhibitory effect on the nociceptive discharges of cortical and thalamic neurons. The subject of modulation of pain sensation received considerable attention with the discovery of opiate receptors in the central nervous system and with the identification of naturally occurring opiatelike peptides in specific neurons concerned with nociceptive function[72] (see discussion of peptide neurotransmitters in Chapter 10).

As in other sensory systems, descending influences from higher centers can dramatically modify the flow of sensory information about pain that reaches consciousness.[73,74] A discussion of the specific pathways subserving these influences can be found in an informative monograph by Fields on the neurophysiological and clinical aspects of pain.[75] The known pathway starts in the midbrain with a group of cells in the periaqueductal gray matter. In some cases stimulation of this area through implanted electrodes has been found to produce a selective reduction of severe clinical pain.[76] The cells involved are believed to be enkephalinergic and project to serotonergic neurons in the rostral me-

[69]Basbaum, A. I. 1985. *In* H. L. Fields et al. (eds.). *Advances in Pain Research and Therapy*, Vol. 9. Raven, New York.

[70]Lynn, B. and Hunt, S. P. 1984. *Trends Neurosci.* 7: 186–188.

[71]Poggio, G. F. and Mountcastle, V. B. 1960. *Bull. Johns Hopkins Hosp.* 106: 266–316.

[72]Hughes, J. (ed.). 1983. *Br. Med. Bull.* 39: 1–106.

[73]Basbaum, A. I. and Fields, H. L. 1979. *J. Comp. Neurol.* 187: 513–522.

[74]Fields, H. L. and Besson, J.-M. (eds.). 1988. *Pain Modulation. Prog. Brain Res.* 77.

[75]Fields, H. L. 1987. *Pain.* McGraw-Hill, New York.

[76]Fields, H. L. and Basbaum, A. I. 1978. *Annu. Rev. Physiol.* 40: 217–248.

dulla. The medullary neurons then send descending fibers along the dorsolateral funiculus of the spinal cord to end in the dorsal horns. There they form synapses with interneurons and terminals of afferent fibers to modulate transmission in the pain pathway. The descending fibers are accompanied by noradrenergic fibers belonging to a second pathway originating in the dorsolateral pons that is also involved in pain modulation.

Pain itself remains an elusive and difficult concept, beyond the scope of this book. In sharp contrast to the analysis of visual, auditory, or somatosensory systems, a discussion of "pain," with its high emotional content, of necessity deals with subjective matters—feelings akin to "anguish" and "suffering" that cannot at present be expressed in the language of neurobiology (just as "seeing a sunset" or "feeling warm" cannot be considered in those terms). Nevertheless, certain correlates between neural activity and pain are apparent. Thus the common experience of a sharp, well-localized stab of pain followed by a dull, gradually swelling, poorly localized, burning ache can be attributed to the activity of the A and the C fibers. Similarly, the analgesic effect of stroking the skin can be accounted for by synaptic interactions at the level of the spinal cord, the thalamus, and the cortex. Particularly appealing is the idea that the level of pain perceived may be reduced by descending influences that involve the action of morphinelike peptides, enabling a soldier charging into battle to ignore injuries that would be excruciatingly painful if they were inflicted while he sat in a chair.

Adrian clearly enunciated a key problem for pain, namely, its function.[52]

Why should we suffer pain, what is its purpose, and why should it be so unpleasant? Our ancestors may have believed the moralists (and the doctors) who told them that pain could be a valuable experience, not to be avoided by such unnatural means as anesthetics for childbirth, but we have been made less hardy, and we need biological as well as moral grounds for the existence of such an evil. The chief biological excuse for pain is that it is a danger signal. The argument is convincing to a point. A successful type of animal, one which can look after itself, must have a sensory mechanism which will signal events likely to damage it and the signals must have priority over all others. . . . It must be a sensation to which we cannot manage to remain inattentive and one which we feel compelled to bring to an end as soon as possible. . . . The danger signal may get out of order, the warning sometimes sounds, although there is no chance of injury. . . . Nonetheless, the danger signal seems to wear thin when it comes to explaining the pain which you may have to suffer when the injury does not come from without so that there is some chance of avoiding it but from something happening inside the body, a slowly growing tumour or the movements of a renal calculus. Such pain may be a warning to call in the surgeon, but a few hundred years ago . . . calling in the surgeon would not have made much difference. . . . Medical science has already done so much to lessen pain that we need not be ashamed to confess that there is still a great deal more to do before we can fully understand how and why it arises.

E. D. Adrian

THE AUDITORY SYSTEM

Two of our sensory systems, unlike those for touch, temperature and pain, enable us to detect and localize events at great distances beyond the range of our ordinary physical contact with the environment. These are the visual and auditory systems. The fact that distant events can be perceived by these systems arises from the physical properties of the stimuli. Light waves reflected from objects around us, or emitted from distant stars, travel to eyes to be absorbed by retinal rods and cones, and sound waves propagate through the atmosphere to enter our outer ears, setting up local mechanical vibrations that ultimately excite hair cells in the cochlea. Visual transduction and perception are discussed in Chapters 16 and 17. Here we describe how the auditory system captures and transduces sound waves.

Auditory transduction   In terrestrial vertebrates, sound waves in the air enter the outer ear, strike the TYMPANIC MEMBRANE (eardrum), and, by a series of mechanical couplings in the middle ear, are converted to fluid waves in the COCHLEA (Figure 17A). The fluid waves, in turn, cause vibrations in the BASILAR MEMBRANE, on which sit sensory HAIR CELLS. It is the hair cells that transduce the mechanical vibrations into electrical signals and re-

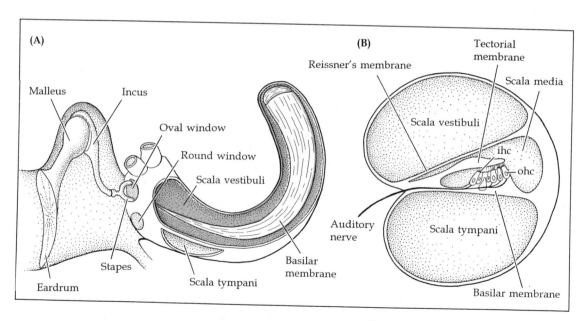

17 **STRUCTURE OF THE COCHLEA. (A)** The middle ear and cochlea, showing the eardrum and its bony connections to the oval window. The cochlea is represented as being unwound and cut open to show the major internal compartments (scala vestibuli and scala tympani) and overlying structures removed to reveal the shape of the basilar membrane, which is narrow near the base and broad at the apex. **(B)** Cross section, showing the scala media (which contains endolymph, a high-potassium solution) and the structural relations between the basilar membrane, inner (ihc) and outer (ohc) hair cells, and tectorial membrane. Hair cells form synapses on the terminals of auditory nerve fibers that have their cell bodies in the spiral ganglion.

lease transmitter onto the afferent nerve terminals that carry information to the brain. The process has been summarized admirably by Aldous Huxley:[77]

> Pongileoni's blowing and the scraping of the anonymous fiddlers had shaken the air in the great hall, had set the glass of the windows looking on to it vibrating; and this in turn had shaken the air in Lord Edward's apartment on the further side. The shaking air rattled Lord Edward's *membrana tympani*; the interlocked *malleus*, *incus*, and stirrup bones were set in motion so as to agitate the membrane of the oval window and raise an infinitesimal storm of fluid in the labyrinth. The hairy endings of the auditory nerve shuddered like weeds in a rough sea; a vast number of obscure miracles were performed in the brain, and Lord Edward ecstatically whispered 'Bach!'

The cochlea is divided into three compartments (Figure 17B). The SCALA MEDIA contains a high-potassium solution, the endolymph. It is separated from the overlying SCALA VESTIBULI by Reissner's membrane and from the underlying SCALA TYMPANI by intercellular tight junctions between the apical ends of the hair cells and their surrounding supporting cells. There are two distinct groups of hair cells, INNER HAIR CELLS and OUTER HAIR CELLS. Each hair cell extends a bundle of stereocilia into the high-potassium environment of the scala media. The cilia are arrayed in order of progressive length, like organ pipes, with each attached to the next at their apical tips, and the longest attached to the overlying TECTORIAL MEMBRANE.

The physical arrangement between the hair cells, the basilar membrane, and the tectorial membrane is such that up and down movement of the membranes causes the cilia to move back and forth with respect to the hair cells (Figure 18). This mechanical arrangement, which appears at first to be relatively inefficient, is highly effective, in part because of the extreme sensitivity of the hair cells, which can signal bundle displacements of a few tenths of a nanometer. Experiments on lateral line organs by Flock indicated that bending the ciliary bundle of hair cells toward the longest cilia caused depolarization, while bending in the reverse direction caused hyperpolarization.[78] (Lateral line organs in fish are analogous to our ears and detect water vibrations.) This was confirmed subsequently by Corey and Hudspeth by direct measurements on individual hair cells of the vestibular apparatus[79] (Box 1). Voltage changes are achieved by the opening and closing of cationic channels that appear to be located at or near the tips of the stereocilia, and conductance measurements suggest that each stereocilium may contain only one channel.[80] One mechanism proposed for channel activation lies in the observation that the tip of each cilium appears to be connected to the wall of the adjacent, longer one by a thin strand. It has been suggested that these strands act as "gating springs" that are

Electrical transduction by hair cells

[77]Huxley, A. 1928. *Point Counter Point*. HarperCollins, New York, Chapter 3.
[78]Flock, A. 1964. *Cold Spring Harbor Symp. Quant. Biol.* 30: 133–145.
[79]Hudspeth, A. J. and Corey, D. P. 1977. *Proc. Natl. Acad. Sci. USA* 74: 2407–2411.
[80]Howard, J. and Hudspeth, A. J. 1988. *Neuron* 1: 189–199.

BOX 1   THE VESTIBULAR APPARATUS

Changes of the orientation of the head in space are detected by the vestibular apparatus, an organ of extreme importance in maintaining balance as well as orientation in relation to the visual world. When we walk, or run, or swim, information is sent continuously from the vestibular apparatus to the central nervous system. This vestibular information, combined with sensory input from muscle receptors in the neck that tell us how the body is oriented in relation to the head, is essential for the maintenance of balance and position. The illustrations indicate the essential components of the vestibular apparatus and how they work.

Shown in Part I are the five components of the system, as seen from the front. The three semicircular canals are filled with fluid that, because of its inertia, flows within them when the head is rotated. The canals are oriented in three planes more or less at right angles to one another. The anterior canal, A, is in a vertical plane angled forward to about 45° off the midline; the posterior canal, P, is angled backward the same amount. (The relations of the planes to the head are shown in the inset.) The horizontal canal, H, is in a plane that is nearly horizontal. Two internal structures, the utricle and the saccule (colored regions), are responsible for detection of linear acceleration.

The mechanism responsible for detection of rotation is illustrated in Part II, which shows a cross section through the fluid-filled ampulla of the horizontal canal. When the head is rotated clockwise, the fluid, because of its inertia, moves in the reverse direction within the rotating canal (arrow). The fluid pushes on the gelatinous mass called the cupula, in the base of which are embedded the stereocilia of hair cells. As in the auditory system, movement of the stereocilia toward the long kinocilium causes depolarization of the cells. If the rotation is continued (as when sitting on a rotating stool), the fluid will catch up with the body rotation and then cease to move inside the canal. When rotation is stopped, transient fluid movement in the reverse direction will occur, bending the cilia in the reverse direction and causing hyperpolarization. It can be seen that the system detects only changes in velocity of rotation (i.e., rotary acceleration). In the same way the anterior and posterior semicircular canals signal rotations of the head in any vertical plane.

Hair cells in the utricle and saccule respond to linear accelerations by a different mechanism. The utricle is oriented in a horizontal plane, and the saccule is oriented vertically (Part III). As in the ampullae, hair cell cilia are embedded in an overlying gelatinous material. However, this material also contains crystals of calcium carbonate called OTOCONIA ("ear dust") or OTOLITHS ("ear stones") that make it heavier than the surrounding fluid. Thus when the head is moved suddenly forward, the cilia will be bent backward in both the utricle and the saccule, again because of inertia. Sudden upward acceleration, as in an elevator (or "lift"), will bend the cilia of the saccule hair cells downward. When the structures are moved, all the hair cells associated with them have their cilia bent in the same direction. However, the hair cells themselves have different orientations, so a movement that depolarizes some will have no effect on, or will hyperpolarize, others. In this way the direction of movement is defined. Hair cell orientation varies in a systematic manner, indicated by the arrows in Part III, which show the direction of movement that causes depolarization of hair cells in each region. As with rotary movement, the system responds only to change in velocity—that is, to acceleration. Although the acceleration of gravity exerts

stretched or relaxed when the bundle is tilted toward or away from the larger cilia, thereby opening a channel or allowing it to close.[81] Because the cilia are exposed to the high-potassium endolymph, channel open-

[81]Hudspeth, A. J. 1989. *Nature* 341: 397–404.

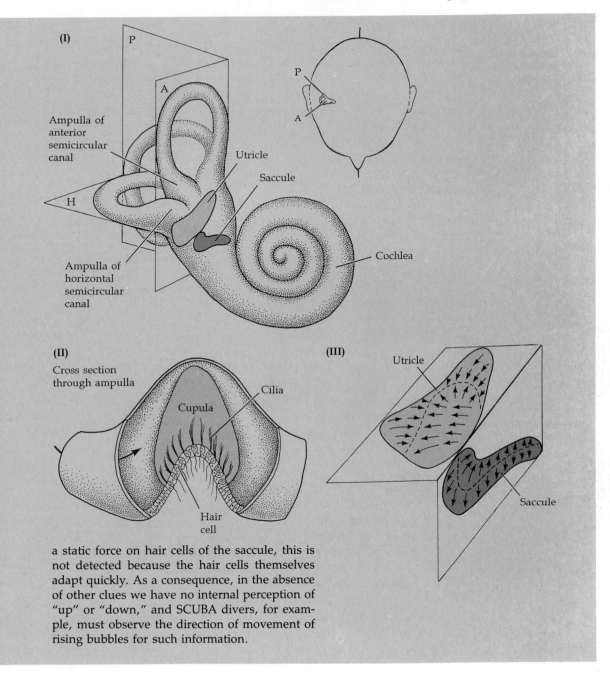

(I)

Ampulla of
anterior
semicircular
canal

Utricle

Saccule

H

Ampulla of
horizontal
semicircular
canal

Cochlea

(II)

Cross section
through ampulla

Cupula

Cilia

Hair
cell

(III)

Utricle

Saccule

a static force on hair cells of the saccule, this is
not detected because the hair cells themselves
adapt quickly. As a consequence, in the absence
of other clues we have no internal perception of
"up" or "down," and SCUBA divers, for example, must observe the direction of movement of
rising bubbles for such information.

ing results in inward potassium current and depolarization of the hair
cell.

How do the hair cells transmit signals to the auditory nerve? As in
other short receptors, modulation of their membrane potential results
in modulation of transmitter release onto the sensory nerve terminals,

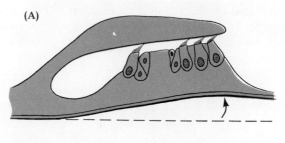

(A)

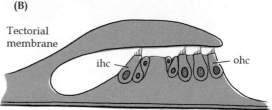

(B)

Tectorial
membrane

ihc   ohc

Basilar membrane

18

DISPLACEMENT OF HAIR CELL BUNDLES
produced by movement of the basilar membrane.
(A) As the basilar membrane and overlying tectorial membrane move upward, the resulting
shearing motion tilts the hair cell bundles off the
cells' axis, the direction of motion being toward
the longest bundles. Such movement depolarizes
the hair cells. **(B)** The membrane is at rest and
the hair cell bundles undistorted. **(C)** Downward
movement displaces the hair cell bundles in the
opposite direction and produces hyperpolarization.

(C)

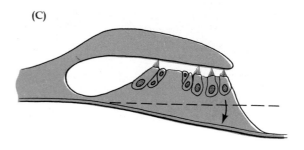

depolarization causing an increase, and hyperpolarization a decrease in
release. The transmitter, which is excitatory to the sensory terminals,
has not yet been identified, but may be glutamate or aspartate.[82]

*Frequency
discrimination in the
auditory system*

The auditory system must be able to discriminate between the song
of a bird and the growl of a tiger; an important first step is an analysis
of the frequency components of the incoming sounds. It was discovered
by von Bekesy that in higher vertebrates much of the analysis is done
mechanically by the cochlea.[83] When he observed the vibrations of the
basilar membrane in animals and in human cadavers he found that the
wider, more flexible region at the apical end of the cochlea was most
sensitive to low frequencies, while the narrower, stiffer region at the
basal end, close to the oval window, vibrated best in response to high
frequencies (Figure 19A). In other words (as predicted by Helmholtz in
the last century), sound frequency is converted into position on the
basilar membrane; activation of hair cells at a particular location indi-

[82]Klinke, R. 1986. *Hearing Res.* 22: 235–243.
[83]von Bekesy, G. 1960. *Experiments in Hearing.* McGraw-Hill, New York.

cates that the corresponding frequency is present in the incoming sound waves. In the living cochlea, the tuning of individual afferent fibers is considerably sharper than would be expected from the passive properties of the cochlear membrane (Figure 19B), possibly because some of the cochlear hair cells themselves provide a mechanical "boost" to cochlear membrane movement (see later). Primary afferent tuning, which is still relatively broad, is sharpened by further processing in the central nervous system (discussed later in this chapter).

In lower vertebrates such as turtles and chicks, the cochlea is too abbreviated in length to be tuned mechanically over the frequency range to which the animals are sensitive. Instead, individual hair cells are tuned electrically.[84] Intracellular records from a cell in the cochlea of a turtle are shown in Figure 20. In response to an acoustic click, the cell gives a brief oscillatory response at a frequency of about 350 Hz. When a rectangular current pulse is applied to the cell, a similar response is produced at both the onset and termination of the pulse. The conclusion from this type of experiment is that the acoustical tuning is a property of the hair cell itself, rather than of the cochlear membrane to which it is attached.[85] This conclusion has been confirmed by whole-cell recording with patch clamp electrodes from hair cells that have been isolated from the cochlea of turtles[86] and chicks.[87] Electrical resonance provides

Electrical tuning in cochlear hair cells

[84]Fettiplace, R. 1987. *Trends Neurosci.* 10: 421–425.
[85]Crawford, A. C. and Fettiplace, R. 1981. *J. Physiol.* 312: 377–412.
[86]Art, J. J. and Fettiplace, R. 1987. *J. Physiol.* 385: 207–242.
[87]Fuchs, P. A., Nagai, T. and Evans, M. G. 1988. *J. Neurosci.* 8: 2460–2467.

(A)

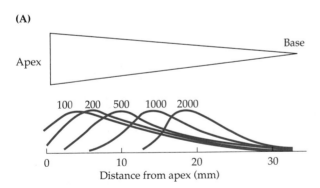

(B)

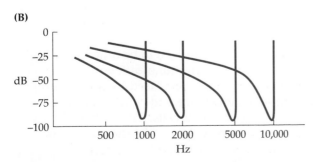

19

COCHLEAR TUNING. (A) Location of maximum displacement of the basilar membrane in the cochlea by sound waves depends on frequency. Curves represent relative displacement at the indicated frequencies (100 to 2000 Hz). At low frequencies maximum displacement is near the wider apical membrane; higher frequencies produce maximal displacement near the base. (B) Typical tuning curves of four individual eighth nerve fibers innervating different locations on the cochlea. Sound intensity in decibels (dB) needed to produce discharges in a fiber is plotted against frequency of the auditory stimulus (note logarithmic frequency scale). The best frequencies for the fibers (i.e., the frequencies requiring the least stimulus intensity) are 1000, 2000, 5000 and 10,000 Hz. (A after von Bekesy, 1960; B after Katsuki, 1961.)

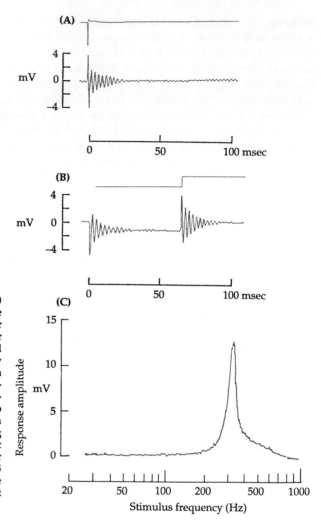

20

HAIR CELL TUNING in the turtle cochlea. (A) The effect of an acoustic click (indicated by the upper trace) on the membrane potential of a hair cell (lower trace), recorded with an intracellular microelectrode. The resting membrane potential was about −50 mV. The click produces a rapidly damped oscillation in membrane potential at a frequency of about 350 Hz, with an initial peak-to-peak amplitude of about 8 mV. (B) A hyperpolarizing current pulse (upper trace) applied to the same cell produces similar oscillations at both the onset and termination of the pulse, indicating that the frequency of oscillation is an intrinsic electrical property of the hair cell. (C) In another cell, stimulated with pure tones ranging from 25 to 1000 Hz, the peak-to-peak amplitude of the oscillatory response has a sharp maximum at about 310 Hz. (After Fettiplace, 1987.)

relatively sharp tuning, as indicated in Figure 20C by the frequency response curve of another cell with a resonant frequency near 310 Hz. Even though determined electrically rather than mechanically, the tuning is still tonotopically arranged in the cochlea. Cells isolated from near the apical end have relatively low resonant frequencies (e.g., 10 Hz), while those from the basal end resonate at much higher frequencies (e.g., 1000 Hz or greater).

What properties of the cell membrane allow it to be tuned electrically and how do these properties vary to determine different tuning frequencies? First, the oscillatory response is produced by interaction between voltage-activated calcium conductance and calcium-activated potassium conductance in the body of the hair cells. These channels were first reported as a basis of oscillatory behavior in the saccula of the bullfrog.[88] Depolarization opens calcium channels, resulting in inward current and further depolarization; as calcium accumulates near the inner surface of membrane, potassium conductance is increased, resulting in outward current and repolarization toward E_K and thereby reduced calcium influx. As long as a depolarizing stimulus is maintained, the cycle repeats with a characteristic frequency, gradually dying out (Figure 21). The characteristic frequency is determined in a remarkably simple and straightforward way, namely by the density and kinetic properties of the calcium-activated potassium channels.[80,89,90] In cells responding to lower frequencies the potassium conductance is relatively small and slow to activate, and thus gives rise to relatively slow oscillations. In higher frequency cells it is larger and more rapidly activating. The density and activation rate of the calcium-activated potassium channels increases monotonically from apex to base of the cochlea, thereby imparting a corresponding gradation of resonance frequencies to the hair cells. This dependence of tuning on calcium-activated potassium conductance has been confirmed during development: Extension of the upper frequency limit to hearing in chicks coincides with the appearance of calcium-activated potassium currents in the developing hair cells.[91]

[88]Hudspeth, A. J. and Lewis, R. S. 1988. *J. Physiol.* 400: 237–274.
[89]Fuchs, P. A. and Evans, M. G. 1990. *J. Physiol.* 429: 529–551.
[90]Fuchs, P. A., Evans, M. G. and Murrow, B. W. 1990. *J. Physiol.* 429: 553–568.
[91]Fuchs, P. A. and Sokolowski, B. H. A. 1990. *Proc. R. Soc. Lond.* B 241: 122–126.

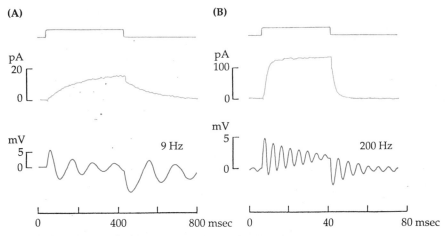

21   **TUNING FREQUENCY AND POTASSIUM CONDUCTANCE in isolated turtle hair cells** measured by whole-cell recording with a patch clamp electrode. **(A)** Middle record shows outward current, carried mainly by potassium, produced by a small depolarizing voltage pulse (duration indicated in top record). Current rises slowly to a maximum of 15 pA with a time constant of about 200 msec. A small current step of the same duration produces oscillatory voltage responses (lower record) with a resonant frequency of 9 Hz. **(B)** In another cell a similar depolarizing pulse produces a much larger, rapidly rising inward current (middle record; note changes in current and time scales), indicating a greater density of potassium channels with faster kinetics. The oscillatory response to a small current pulse reveals a concomitant increase in tuning frequency to 200 Hz (lower record). (After Fettiplace, 1987.)

Efferent regulation of
hair cell responses

Experiments have shown that auditory receptors, like muscle spindles, are subject to efferent regulation.[92] This regulation is through synapses formed by efferent fibers on the base of the hair cells. In the auditory system of the turtle, efferent fiber activity reduces both the sensitivity and the frequency selectivity of the hair cells.[93,94] The effect is mediated by acetylcholine, which acts indirectly to increase the potassium conductance of the postsynaptic membrane (Chapter 7). An example is shown in Figure 22A. A short train of eight shocks to the efferent nerve hyperpolarizes the cell and severely attenuates the response to an acoustic stimulus at 220 Hz (near the cell's characteristic frequency). At lower and higher frequencies of acoustic stimulation, activation of the efferent fibers still results in hyperpolarization, but the

[92]Galambos, R. 1956. *J. Neurophysiol.* 19: 424–427.
[93]Art, J. J. et al. 1985. *J. Physiol.* 360: 397–421.
[94]Art, J. J., Fettiplace, R. and Fuchs, P. A. 1984. *J. Physiol.* 356: 525–550.

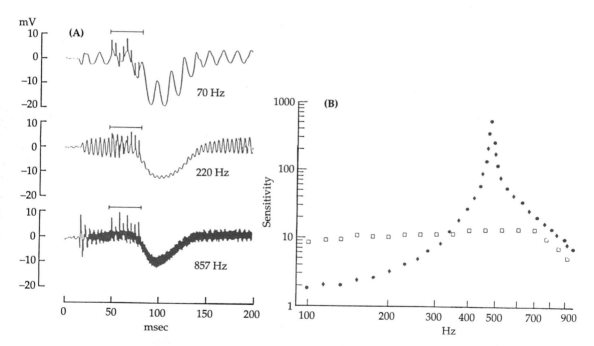

22   **EFFECT OF EFFERENT STIMULATION** on cochlear hair cell responses to acoustic stimuli. **(A)** Center record: Response of a cell to an acoustic stimulus of 220 Hz (its resonant frequency) is inhibited by a brief train of efferent stimuli (indicated by the bar), and the cell is hyperpolarized. Upper record: An acoustic stimulus at 70 Hz has been adjusted to produce a response of similar amplitude. Efferent stimulation produces a hyperpolarization but an increase rather than a decrease in the acoustic response. Lower record: Response to an 857-Hz stimulus (again adjusted to produce a response similar to that at the resonant frequency) is unchanged by efferent stimulation. **(B)** Sensitivity of another cell (in millivolts per unit of sound pressure) as a function of frequency in the absence (closed symbols) and presence (open symbols) of efferent stimulation. Efferent inhibition reduces the response at the resonant frequency and disrupts tuning specificity. (A after Art, Crawford, Fettiplace and Fuchs, 1985; B after Fettiplace, 1987.)

effects on the acoustic oscillations are different: They are increased at the lower frequency and unchanged at the higher. This preferential effect on the acoustic responses results in broadening of the frequency response of the cell (Figure 22B).

So far we have made no distinction between inner and outer hair cells in the cochlea of higher vertebrates (Figures 17B and 18). The outer hair cells outnumber the inner ones by about 3:1, yet give rise to only about 5 percent of the afferent fibers carrying information to the brain. This relative sparsity of afferent innervation suggests that the outer cells may have some other role. There is both experimental and theoretical evidence to suggest that in higher vertebrates the mechanical properties of the basilar membrane alone are not adequate to explain the sharpness of its frequency selectivity, and that the outer hair cells may play an active role as force generators in cochlear tuning.[95,96] The general scheme is that local mechanical displacement of the basilar membrane produces voltage changes in the outer hair cells, and that these voltage changes in turn produce changes in hair cell length that reinforce the original displacement. In such a system, attenuation of the electrical response of the hair cell by efferent inhibition would reduce the mechanical response and broaden the mechanical tuning. Various experiments lend support to this idea. Current injection into turtle hair cells, for example, has been shown to cause hair bundle displacement.[97] Mechanical responses have also been demonstrated in isolated outer hair cells from the guinea pig cochlea.[98] When cells were held at their base with a patch clamp electrode in the whole-cell recording configuration, hyperpolarization produced elongation and depolarization shortening of individual cells, over a total range of about 4 percent of the cell's length (4 μm for 100-μm apical cells). Cells were able to follow mechanically sinusoidal voltage changes at greater than 1 kHz. Unlike hair cells from lower vertebrates, they were not tuned electrically to any particular frequency.

In summary, the frequency analysis of incoming auditory signals in higher vertebrates is accomplished mechanically by the resonance characteristics of the basilar membrane. The mechanical tuning appears to be sharpened by electromechanical transduction in the outer hair cells. The signals sent to the central nervous system arise primarily from inner hair cells. In lower vertebrates, frequency analysis is achieved by the hair cells themselves, through their electrical tuning properties. These properties depend largely on the density and kinetic properties of calcium-activated potassium channels. In both cases, the frequency sensitivity progresses tonotopically along the cochlea, the lower-frequency responses arising near the apex and the higher-frequency responses near the base.

> Electromechanical tuning of the basilar membrane

[95]Sellick, P. M., Patuzzi, R. and Johnstone, B. M. 1982. *J. Acoust. Soc. Am.* 72: 131–141.

[96]de Boer, E. 1983. *J. Acoust. Soc. Am.* 73: 567–573.

[97]Crawford, A. C. and Fettiplace, R. 1985. *J. Physiol.* 364: 359–379.

[98]Ashmore, J. F. 1987. *J. Physiol.* 388: 323–347.

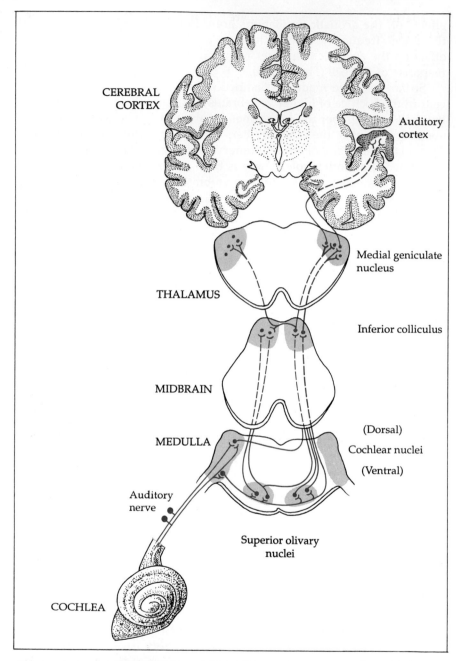

23   **CENTRAL AUDITORY PATHWAYS shown schematically on transverse sections of the medulla, midbrain, and thalamus, and a coronal section of the cerebral cortex. Auditory nerve fibers end in the dorsal and ventral cochlear nuclei. Second-order fibers ascend to the contralateral inferior colliculus; those from the ventral cochlear nucleus also supply collaterals bilaterally to the superior olivary nucleus. Further bilateral interaction occurs at the level of the inferior colliculus. Fibers then ascend to the medial geniculate nucleus of the thalamus and hence to the auditory cortex. (After Berne and Levy, 1988.)**

The main auditory pathways are illustrated in Figure 23. Auditory fibers of the eighth nerve travel centrally and send branches to both the dorsal and ventral cochlear nuclei. Second-order axons ascend in the contralateral medial lemniscus to innervate cells in the inferior colliculus. Those in the dorsal nucleus also provide collateral branches to both the ipsilateral and contralateral olivary nuclei. Third-order cells in the olivary nuclei, in turn, also send ascending fibers to the inferior colliculus. The pathway continues through the medial geniculate nucleus of the thalamus to the auditory region of the cerebral cortex in the temporal lobe.

Central auditory processing

As one ascends higher in the auditory system, pure tones become less and less important as stimuli for individual cells. Instead cells in the auditory cortex respond to combinations of tones, for example sweeping upward or downward in frequency, to binaural but not monaural sounds, and to other complex auditory stimuli. When cells are found that do respond to pure tones, their frequency selectivity often is enhanced over that of the primary afferent fibers. If we compare the tuning curve (sensitivity vs. frequency) of a primary afferent fiber with that of a cortical neuron, the latter is usually sharper (Figure 24A). This sharpening of frequency response corresponds to sharpening of spatial localization on the basilar membrane and is accomplished by lateral inhibition, just as in the somatosensory system. Thus the "receptive field" of the cortical neuron is an excitatory strip on the basilar membrane flanked on either side by inhibitory bands. This translates to a narrow frequency range that excites the cell, flanked by higher and lower frequencies that are inhibitory (Figure 24B).

Frequency sharpening

The processing of auditory signals is complex. Not only must the frequency content of incoming sounds be analyzed in some way, but so must their sequence in time (playing a tape recorder backward produces gibberish). In addition, such processing in any animal species must deal not only with frequency discrimination and timing related to sounds from the surrounding world, but also with vocalization within the species. In humans the basic elements of speech, called PHONEMES, are common to all languages and are the sounds first babbled by babies before particular ones are selected to be combined into words.[99] The basic sounds can be analyzed into combinations of frequency–time relations—for example, a continuous component at 1000 Hz accompanied by a second, frequency-modulated component starting at 5000 Hz and descending rapidly to 500 Hz. The components are called FORMANTS. By analogy with the visual system, which contains cells that respond to slits, corners, edges, and other geometrical forms (Chapter 17), we might expect to find higher-order cells in the human auditory cortex that respond to particular formants, or perhaps even phonemes. This principle holds for lower animals. For example, some cells in the auditory cortex of the mustached bat respond to specific combinations

Sound elements

[99]De Boysson-Bardies, B. et al. 1989. *J. Child Lang.* 16: 1–17.

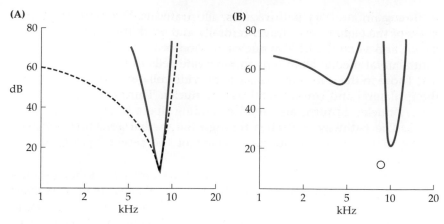

24 LATERAL INHIBITION OF A NEURON IN THE AUDITORY CORTEX. (A)
Tuning curve for a neuron in the auditory cortex of a cat (continuous line). Best
frequency is about 7 kHz. Dashed line shows, for comparison, the broader tuning curve
of a mammalian auditory nerve fiber with the same best frequency (Figure 19). (B)
Lateral inhibition of the same cortical neuron. A response is produced by a continuous
tone at the best frequency (open circle); an additional tone is then added, which inhibits
the best-frequency response. Curves indicate the sound pressure required for the additional tone to reduce the response by 20 percent. (After Arthur, Pfeiffer and Suga,
1971.)

of constant frequency and frequency-modulated tones that are equivalent to the bat's own ultrasonic vocalizations.[100] At the same time, we
might expect to search in vain for cells responsive to human sound
elements in the auditory system of the ground squirrel. Such higher-
order cells *are* found, however, in mynah birds that have been trained
to talk.[101]

Sound localization   One distinctive feature of the auditory system is that pathways from
the two ears form bilateral connections almost immediately, thus providing an opportunity for binaural interactions. Such interactions presumably provide a basis for sound localization by permitting the comparison of auditory signals arriving at the two ears. Sound localization
is a relatively important function for vertebrates, more for some than
for others.[102] Bats have a complex system of echolocation whose accuracy in localizing distant objects approaches that of the human visual
system.[103,104] Humans, on the other hand, can determine the direction
of a sound source with only moderate accuracy, approaching one degree
of arc in the horizontal plane at higher frequencies. Psychophysical
studies have shown that localization is accomplished by using differences between the two ears in time of arrival and/or intensity of the

[100]Tsuzuki, K. and Suga, N. 1988. *J. Neurophysiol.* 60: 1908–1923.
[101]Langer, G., Bronke, D., and Scheich, H. 1981. *Exp. Brain Res.* 43: 11–24.
[102]Lewis, B. (ed.). *Bioacoustics: A Comparative Approach.* Academic Press, New York.
[103]Griffin, D. R. 1958. *Listening in the Dark.* Yale University Press, New Haven.
[104]Neuweiler, G. 1990. *Physiol. Rev.* 70: 615–641.

incoming sound.[105] Thus, if clicks are presented through earphones with varying delays, the sound is localized toward the ear in which the click arrives first. If the clicks are simultaneous, but of different intensity, the localization is to the side with the loudest click. These two effects can cancel, so that if a click arrives earlier at one ear but is of a lesser intensity, the sound may be located directly ahead. This is called TIME–INTENSITY TRADING.

Barn owls are able to locate sound sources with exceptional accuracy with respect to both horizontal and vertical position.[106,107] Vertical localization is made possible by the anatomy of the external ears, which makes the right ear more sensitive to sounds from above, and the left ear to sounds from below. Sound collected by the facial ruff feathers is directed around preaural flaps into the ears. The flaps are asymmetrical so that the right ear trough points upward and emerges above the flap high in the facial plane, the left ear downward and below the flap. Intensity and time differences are analyzed separately. Intensity differences indicate vertical localization, a greater intensity in the right ear being interpreted as "up," in the left ear as "down." Phase shifts are used to indicate horizontal location. In accordance with the behavioral observation, Konishi and his colleagues have found cells in the auditory midbrain nucleus of the owl, corresponding to the inferior colliculus in mammals, that respond only to sounds from a particular position in the owl's frontal plane. When sounds were presented separately to each ear through earphones, the cells responded only when both the intensity and phase differences were appropriate for that position. Such neurons are arranged in an orderly manner, so that the auditory space in the owl's frontal plane is mapped topographically on the nucleus, just as visual space is represented topographically in the visual system. Such organization suggests that for the barn owl the perceptual experience of hearing the squeak of a mouse in a quiet barn may be not too different from our own when we see a flash of light in a dark room, and that the owl has at all times an ongoing two-dimensional "picture" of auditory space. One developmental aspect of auditory localization is that it apparently depends initially on visual clues. Thus baby barn owls fitted with prismatic glasses that offset the visual field (Figure 25) develop similar offsets in their auditory field localization.[108]

OLFACTION AND TASTE

Discussion of the sensations of taste and olfaction are often combined because the stimuli arriving from the outside world are similar: Unlike sound waves, or light, both are chemical. However, that is almost the

[105]Blauert, J. 1983. *Spatial Hearing: The Psychophysics of Human Sound Localization*. MIT Press, Cambridge.

[106]Knudsen, E. I. and Konishi, M. 1979. *J. Comp. Physiol.* 133: 13–21.

[107]Moiseff, M. 1989. *J. Comp. Physiol.* 164: 637–644.

[108]Knudsen, E. I. and Knudsen, P. K. 1989. *J. Neurosci.* 9: 3306–3313

25

BABY BARN OWL WITH GLASSES consisting of Fresnel prisms that offset the owl's visual field 34° to the right. The auditory field is offset in the same direction. (Courtesy of E. I. Knudsen.)

full extent of the similarity. The neuroanatomical arrangements of the two systems are quite different (Figure 26), as are their central projections. Odors are detected directly by long receptors—sensory cells of the olfactory nerve—embedded in the mucosa of the nasal passages. Taste, on the other hand, is transduced by short receptors, assembled into small groups within the taste buds, and the transduced signal is passed on to adjacent nerve endings.

Perhaps the most remarkable collection of facts related to taste and olfaction has been accumulated by organic chemists who, by and large, can identify the chemicals that make (for example) grapefruit taste like grapefruit and not like papaya. The chemistry is also interesting in another sense: It is highly stereospecific.[109] Thus D-carvone smells like caraway, whereas its enantiomeric form (L-carvone) smells like spearmint. Aspartane, which is L-aspartic acid-L-phenylalanine methyl ester, is very sweet. If the L-aspartate is replaced by D-aspartate, the compound is not sweet at all. The bases for these differences at the chemoreceptor level are still unknown but studies at the molecular level are beginning to yield information on this point (see below).

Taste transduction   There appear to be four basic taste receptor responses, corresponding to the taste perceptions SALT, ACID, BITTER, and SWEET. A fifth classification—UMAMI—has been proposed, induced selectively in gustatory afferent fibers of rats by substances like monosodium glutamate.[110] In addition, catfish have receptors that respond specifically to amino acids.[111]

How are taste responses generated? Experimental evidence indicates

[109]Pickenhagen, W. 1989. *In* J. G. Brand et al. (eds.). *Chemical Senses*, Vol. 1. *Receptor Events and Transduction in Taste and Olfaction*. Marcel Dekker, New York, pp. 505–509.

[110]Kawamura, Y. and Kare, M. R. 1987. *Umami: A Basic Taste*. Marcel Dekker, New York.

[111]Caprio, J. 1978. *J. Comp. Physiol.* 132: 357–371.

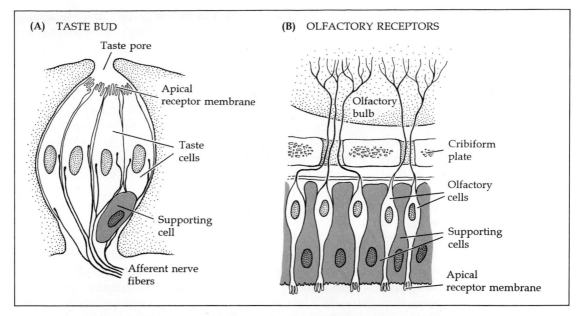

(A) TASTE BUD

Taste pore

Apical
receptor membrane

Taste
cells

Supporting
cell

Afferent nerve
fibers

(B) OLFACTORY RECEPTORS

Olfactory
bulb

Cribiform
plate

Olfactory
cells

Supporting
cells

Apical
receptor membrane

26   **RECEPTORS FOR TASTE AND OLFACTION. (A) Drawing of a taste bud. Receptor (apical) ends of taste cells sample fluid through a small taste pore. The basal end receives innervation from afferent nerve fibers. (B) Olfactory neurons with apical receptors projecting from the nasal mucosa send axons through the bony cribiform plate into the olfactory bulb. There they make synapses with second-order neurons (not shown).**

that the receptor cells respond selectively to one or another of the basic stimuli at their apical end by depolarization. The depolarization may either evoke action potentials,[112,113] or else spread passively to the synaptic end. In either case, voltage-sensitive calcium channels are activated to release neurotransmitter onto the sensory nerve terminals. Nerve impulses, either generated or modulated in frequency by the transmitter, carry the information to the central nervous system. Evidence for the nature of the receptor response and the subsequent steps has been reviewed by Kinnamon[114] and Roper.[115] The scheme itself is reasonable inasmuch as it coincides in general with transduction in other systems with specialized receptor cells (e.g., the auditory system). However, many details remain to be clarified, particularly with respect to the synaptic events. Putative synapses between the receptor cells and the afferent nerve endings are often poorly defined morphologically and are frequently of ambiguous polarity, raising the possibility of efferent innervation as well. This lack of clear morphological definition is possibly related to the relatively brief life span of the receptor cells:

[112]Roper, S. D. 1983. *Science* 220: 1311–1312.
[113]Avenet, P. and Lindemann, B. 1987. *J. Membr. Biol.* 95: 265–269.
[114]Kinnamon, S. C. 1988. *Trends Neurosci.* 11: 491–496.
[115]Roper, S. D. 1989. *Annu. Rev. Neurosci.* 12: 329–353.

Each lives for only a few days. The rapid turnover may militate against formation of stable, well-defined synaptic relations. In any case, no specific neurotransmitters for the taste synapses have yet been identified with certainty, although acetylcholinesterase has been demonstrated histochemically in the taste buds, responses to extrinsically applied norepinephrine have been recorded, and dense core vesicles in the taste bud cells have been reported to contain serotonin.

What are the steps leading to depolarization of the receptor? Whole-cell recording with patch clamp electrodes and single-channel recording techniques have been used to approach this question.[116,117] Taste cells do not rely on a single mechanism of transduction. Instead, evidence for at least two different mechanisms has been obtained: (1) a reduction in potassium conductance and (2) entry of cations (other than potassium) through channels that are mostly open at rest. A variety of potassium channels are found selectively at the apical membrane of the receptor cells.[118,119] Reduction in potassium permeability by bitter and sweet stimuli is believed to be mediated by an intracellular second-messenger system leading to phosphorylation of normally open apical potassium channels. The channels close upon phosphorylation, leading to the reduction in conductance. Evidence in favor of this scheme includes the direct observation that potassium conductance of isolated frog taste receptors is reduced by cyclic AMP,[120] and that amino acid stimulation of taste epithelium in catfish increases cyclic AMP synthesis.[121] Sour stimuli also produce a reduction in potassium conductance, either through a similar second messenger system or by direct block of the apical potassium channels by protons.[122] Finally, it is proposed that depolarization of salt receptors occurs simply by sodium or other cations diffusing into the cell through normally open channels.[123]

Olfaction — Detection of olfactory stimuli is by long receptors that send their axons into the olfactory bulb (Figure 26). One major difference between olfaction and taste is that there appear to be no groups of odorants to which various olfactory receptors respond selectively, such as "flowery," "acrid," "burnt," "putrid," and so forth. This apparent lack of a finite number of classes of stimuli seems to imply that the population of olfactory receptors (if not each individual cell) must be endowed with a wide variety of receptor proteins, each responsive to one or a few odorants. Alternatively, the full range of smell sensations may involve

[116]Avenet, P. and Lindemann, B. 1987. *J. Membr. Biol.* 97: 223–240.

[117]Kinnamon, S. C. and Roper, S. D. 1987. *J. Physiol.* 383: 601–614.

[118]Kinnamon, S. C., Dionne, V. E. and Beam, K. G. 1988. *Proc. Natl. Acad. Sci. USA* 85: 7023–7027.

[119]Roper, S. D. and McBride D. W. 1989. *J. Membr. Biol.* 109: 29–39.

[120]Avenet, P., Hofmann, F. and Lindemann, B. 1988. *Nature* 331: 351–354.

[121]Kalinsoki, D. L. et al. 1989. *In* J. G. Brand et al. (eds.). *Chemical Senses*, Vol. 1. *Receptor Events and Transduction in Taste and Olfaction*. Marcel Dekker, New York, pp. 85–101.

[122]Kinnamon, S. C. and Roper, S. D. 1988. *Chem. Senses* 13: 115–121.

[123]Avenet, P. and Lindemann, B. 1988. *J. Membr. Biol.* 105: 245–255.

activation of combinations of cells. One odorant-binding protein has been identified and cloned,[124] but its diffuse localization and willingness to bind a wide variety of odorants with little selectivity[125] suggest that it is not a specific receptor protein. Instead it may serve a transport role, carrying odorants (many of which are hydrophobic) through the mucous layer of the olfactory epithelium to the receptor surface of the olfactory cells. Its putative tertiary structure—a hydrophilic exterior surrounding a hydrophobic binding pocket—is consistent with this role.

In addition, a number of members of a very large multigene family whose expression is restricted to olfactory epithelium have been cloned.[126] The genes encode proteins with seven putative transmembrane domains, three of which are more divergent than the rest, suggesting that they might represent diversified odorant-binding sites. Further experiments are necessary to decide whether they indeed mediate responses to specific odorants.

The first electrical responses to olfactory stimuli were measured by Ottoson,[127] who recorded the transepithelial potential change (electro-olfactogram, or EOG) produced in the nasal mucosa by odor application. Since then, evidence has accumulated that odorant molecules interact with the apical membrane of the receptor cells to produce an increase in conductance that results in depolarization and the generation of action potentials that travel to the central nervous system. The depolarization is mediated by a second-messenger system utilizing cyclic AMP.[128,129] Patch clamp techniques have been used to record odorant-induced currents from isolated olfactory cells,[130] and to record the precise time course and localization of the odor response.[131] An example of such an experiment on a cell isolated from the olfactory mucosa of the salamander is shown in Figure 27. The membrane potential of the cell is held at −65 mV by the patch clamp pipette, and a solution containing a mixture of odorant molecules (approximately 0.1 mM) in 100 mM KCl is applied from a second pipette by a brief (35 msec) pressure pulse, first to the soma of the cell and then to the distal portion of the dendrite and the cilia. Pipette solution applied to the soma produces a rapid inward current, due to the local increase in potassium concentration. The time course of the potassium response provides a

Olfactory transduction

[124]Pevsner, J. et al. 1989. *In* J. G. Brand et al. (eds.). *Chemical Senses*, Vol. 1. *Receptor Events in Transduction in Taste and Olfaction*. Marcel Dekker, New York, pp. 227–242.

[125]Pelosi, P. and Tirindelli, R. 1989. *In* J. G. Brand et al. (eds.). *Chemical Senses*, Vol. 1. *Receptor Events in Transduction in Taste and Olfaction*. Marcel Dekker, New York, pp. 207–226.

[126]Buck, L. and Axel, R. 1991. *Cell* 65: 175–187.

[127]Ottoson, D. 1956. *Acta Physiol. Scand.* 35 (Suppl.122): 1–83.

[128]Lancet, D. 1986. *Annu. Rev. Neurosci.* 9: 329–355.

[129]Lancet, D. 1988. *In* F. L. Margolis and T. V. Getchell (eds.). *Molecular Biology of the Olfactory System*. Plenum, New York, pp. 25–50.

[130]Trotier, D. 1986. *Pflügers Arch.* 407: 589–595.

[131]Firestein, S., Shepherd, G. M. and Werblin, F. 1990. *J. Physiol.* 430: 135–158.

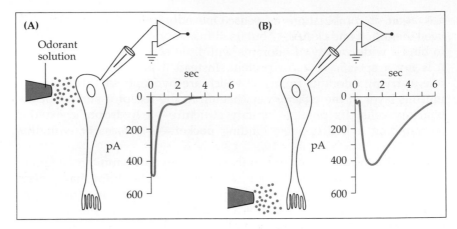

27 **RESPONSES OF ISOLATED OLFACTORY CELLS** from the salamander. Patch clamp electrode is used to record whole-cell current. Solution containing 0.1 m*M* odorant mixture in 100 m*M* KCl is applied to the cell by a brief (35 msec) pressure pulse. **(A)** When solution is applied to the cell body, there is a rapid transient inward current due to the increased KCl concentration, followed by a small, slower current as the odorant reaches the apical end of the dendrite. The time course of the fast inward current is indicative of the time course of application and dissipation of the electrode solution. **(B)** When the solution is applied to the dendrite, there is only a small rapid current due to the KCl; but a large current, due to the odorant, lasts for several seconds after the electrode solution has washed away. (After Firestein, Shepherd and Werblin, 1990.)

measure of the speed of application and subsequent dissipation of the solution by diffusion into the surrounding bath. A second smaller and slower inward current appears when the odorants reach the apical dendrite. Solution applied to the apical dendrite and cilia produces only a small potassium response, while the odorant produces a large inward current that outlasts the time of application of the solution by several seconds. The experiment clearly indicates that the region of sensitivity to the odorants is the distal dendrite and cilia, and the prolonged time course of the dendritic response is consistent with the idea that the conductance change is produced by a second messenger system, rather than by a direct action of the odorants themselves.

SUGGESTED READINGS

Reviews and Original Papers

GENERAL

Bialeck, W. 1987. Physical limits to sensation. *Annu. Rev. Biophys. Biophys. Chem.* 16: 455–478.

Heiligenberg, W. 1989. Coding and processing of electrosensory information in gymnotiform fish. *J. Exp. Biol.* 146: 255–275.

Kalmidjn, A. J. 1982. Electric and magnetic field detection in elasmobranch fishes. *Science* 218: 916–918.

McIntyre, A. K. 1980. Biological seismography. *Trends Neurosci.* 3: 202–205.

MECHANORECEPTORS

Erxleben, C. 1989. Stretch-activated current through single ion channels in the abdominal stretch receptor organ of the crayfish. *J. Gen. Physiol.* 94: 1071–1083.

Eyzaguirre, C. and Kuffler, S. W. 1955. Processes of excitation in the dendrites and in the soma of single isolated sensory nerve cells of the lobster and crayfish. *J. Gen. Physiol.* 39: 87–119.

Hunt, C. C. 1990. Mammalian muscle spindle: Peripheral mechanisms. *Physiol. Rev.* 70: 643–663.

Katz, B. 1950. Depolarization of sensory nerve terminals and the initiation of impulses in the muscle spindle. *J. Physiol.* 111: 261–282.

Kuffler, S. W., Hunt, C. C. and Quilliam, J. P. 1951. Function of medullated small-nerve fibers in mammalian ventral roots: Efferent muscle spindle innervation. *J. Neurophysiol.* 14: 29–54.

Loewenstein, W. R. and Mendelson, M. 1965. Components of adaptation in a Pacinian corpuscle. *J. Physiol.* 177: 377–397.

CENTRAL PROCESSING AND PAIN

Fields, H. L. 1987. *Pain*. McGraw-Hill, New York.

Kaas, J. H. 1983. What if anything is SI? Organization of first somatosensory area of cortex. *Physiol. Rev.* 63: 206–231.

HAIR CELL TRANSDUCTION AND TUNING

Ashmore, J. F. 1987. A fast motile response in guinea-pig outer hair cells: The cellular basis of the cochlear amplifier. *J. Physiol.* 388: 323–347.

Crawford, A. C. and Fettiplace, R. 1981. An electrical tuning mechanism in turtle cochlear hair cells. *J. Physiol.* 312: 377–412.

Fettiplace, R. 1987. Electrical tuning of hair cells in the inner ear. *Trends Neurosci.* 10: 421–425.

Fuchs, P. A. and Evans, M. G. 1990. Potassium currents in hair cells isolated from the cochlea of the chick. *J. Physiol.* 429: 529–551.

Fuchs, P. A., Evans, M. G. and Murrow, B. W. 1990. Calcium currents in hair cells isolated from the cochlea of the chick. *J. Physiol.* 429: 553–568.

Hudspeth, A. J. 1989. How the ear's works work. *Nature* 341: 397–404.

Hudspeth, A. J. and Corey, D. P. 1977. Sensitivity, polarity, and conductance change in the response of vertebrate hair cells to controlled mechanical stimuli. *Proc. Natl. Acad. Sci. USA* 74: 2407–2411.

Hudspeth, A. J. and Lewis, R. S. 1988. Kinetic analysis of voltage- and ion-dependent conductances in saccular hair cells of the bull-frog, *Rana catesbeiana*. *J. Physiol.* 400: 237–274.

ECHOLOCATION

Knudsen, E. I. and Knudsen, P. K. 1989. Vision calibrates sound location in developing barn owls. *J. Neurosci.* 9: 3306–3313.

Knudsen, E. I. and Konishi, M. 1979. Mechanisms of sound location in the barn owl (*Tyto alba*). *J. Comp. Physiol.* 133: 13–21.

Neuweiler, G. 1990. Auditory adaptations for prey capture in echolocating bats. *Physiol. Rev.* 70: 615–641.

CHEMICAL SENSES

Buck, L. and Axel, R. 1991. A novel multigene family may encode odorant receptors: A molecular basis for odor recognition. *Cell* 65: 175–187.

Firestein, S., Shepherd, G. M. and Werblin, F. 1990. Time course of the membrane current underlying sensory transduction in salamander olfactory receptor neurones. *J. Physiol.* 430: 135–158.

Kinnamon, S. C. 1988. Taste transduction: A diversity of mechanisms. *Trends Neurosci.* 11: 491–496.

Lancet, D. 1986. Vertebrate olfactory reception. *Annu. Rev. Neurosci.* 9: 329–355.

Roper, S. D. 1989. The cell biology of vertebrate taste receptors. *Annu. Rev. Neurosci.* 12: 329–353.

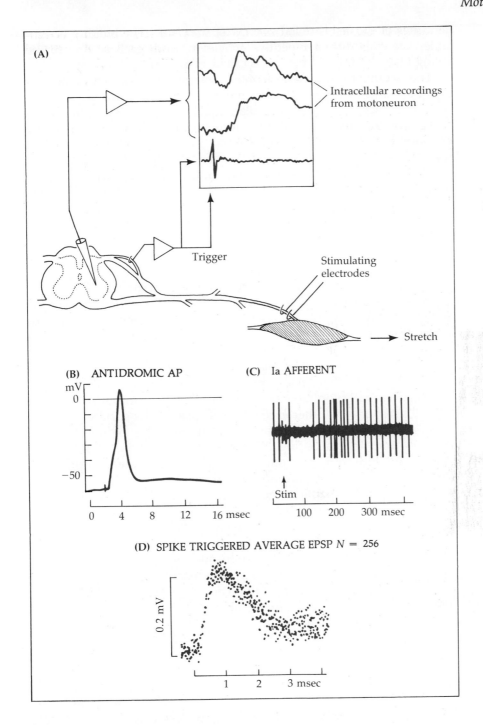

(A)

Intracellular recordings from motoneuron

Trigger

Stimulating electrodes

Stretch

(B) ANTIDROMIC AP

mV

0

−50

0   4   8   12   16 msec

(C) Ia AFFERENT

Stim

100   200   300 msec

(D) SPIKE TRIGGERED AVERAGE EPSP *N* = 256

0.2 mV

1   2   3 msec

dendrites of the cell is called SPATIAL SUMMATION. The buildup of synaptic potentials during repetitive activation, with each adding to the falling phase of the one before, is called TEMPORAL SUMMATION.

Synaptic integration

The fact that the axon hillock of the motoneuron is the site of impulse initiation, and therefore the focal point of integrative activity, has important consequences. In particular, the locations of synaptic inputs determine their relative contribution to excitation or inhibition: Those on the soma, close to the axon hillock, have relatively greater effect than those on the distal parts of the dendritic tree. Apart from their proximity to the site of impulse initiation, the effects of synaptic inputs on one another depend on their relative locations. For example, inward currents at excitatory synapses in the dendrites spread passively to the axon hillock and, if their sum is sufficient, depolarize the cell to threshold, producing an impulse and subsequent muscle contraction. Postsynaptic inhibition opposes impulse initiation in two ways: (1) The inhibitory conductance change provides a path for outward current (i.e., anion influx), thereby tending to reduce the excitation; and (2) if the inhibitory conductance increase occurs between the excitatory synapse and the cell body, spread of excitation to the axon hillock is attenuated. These principles are illustrated in Figure 3: Inhibition near the tip of a dendrite is relatively less effective than inhibition close to the cell body. Clearly, then, *the sites of the synapses on a neuron* play a key role in establishing their effectiveness. Finally, through presynaptic inhibition, particular excitatory influences can be attenuated without influencing the cell's response to other excitatory synapses on its surface.

An important principle that bears on problems of development emerges from the preceding discussion of integration. In the motor system, and indeed throughout the entire nervous system, appropriate synaptic interactions require not only that neurons make connections with specific postsynaptic targets, but also that these connections be formed in the appropriate locations on the postsynaptic neurons, on particular dendrites or regions of the cell body.

Muscle fiber properties

Muscle fibers are not homogeneous: Some are faster in their contractions than others, and their contractile mechanisms fatigue more rapidly. The slowly contracting, fatigue-resistant fibers depend on oxidative metabolism for energy production, the fast, rapidly fatiguing fibers on glycolysis. The differences in speed of contraction are related to the presence in the fibers of different isoenzymatic forms of myosin. In a detailed series of experiments on the range of muscle fiber properties, Burke and his colleagues have shown that the fibers can be grouped into four main classes, extending from "slow-twitch, fatigue-resistant" to "fast-twitch, fast-fatiguing," each with distinctive histochemical characteristics.[12] Interestingly, in any given motor unit all the muscle fibers belong to the same class. As we will discuss in more detail, the pattern of activation of any given motoneuron is matched appropriately to the properties of its muscle fibers (see Chapter 11).

[12]Burke, R. E. 1978. *Am. Zool.* 18: 127–134.

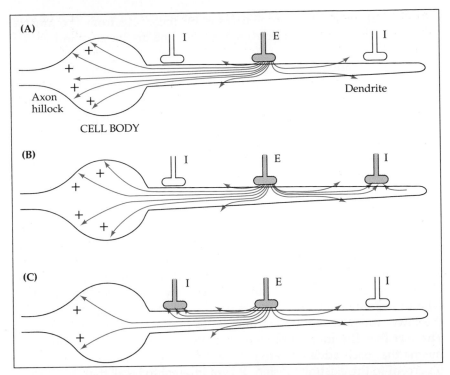

3 **EFFECT OF SYNAPTIC LOCATION** on synaptic interactions in a dendrite. **(A)** Activation of an excitatory synapse results in inward current that spreads toward the soma and the distal end of the dendrite. The inward current spreads along the proximal dendrite to depolarize the soma and axon hillock. The current toward the distal portion of the dendrite is small because of high input resistance, due to a decreasing cross-sectional area and the relatively small area of the surface membrane. **(B)** Activation of the distal inhibitory synapse provides an additional path for outward current in the distal dendrite but has little effect on current flowing toward the soma. **(C)** Activation of a proximal inhibitory synapse provides an alternate pathway for outward current, reducing the depolarization of the soma and axon hillock.

The size principle

An important property of motor units is that they are not of uniform size. Instead their size is graded. Some motoneurons supply many muscle fibers, and impulses from such a motoneuron give rise to a large increase in tension. Other motoneurons innervating relatively few muscle fibers generate less tension. Further, the smallest motor units are usually made up of slowly contracting, fatigue-resistant fibers, the largest of fast, rapidly fatiguing fibers. There is evidence that the number of muscle fibers in a motor unit is related directly to the size of the motoneuron cell body and of its axon. The relation of motor unit size to axon diameter makes it possible to study the behavior of motor units of various sizes by recording action potentials from the ventral roots. When records are made from electrodes placed on the roots, the extracellular impulses from large axons are larger than those from small axons (Chapter 5). Motor units of different sizes can therefore be distinguished by differences in spike amplitudes. A second way of distinguishing different-sized motor units is by recording from the muscle itself. In such recordings (electromyograms), the discharge of individual motor units can be recognized.

Henneman and his colleagues have used such recordings to examine how motor units are activated during graded muscle contractions.[13] They have shown that, in general, when a contraction occurs, small motor units fire first, producing small increments in tension. As the strength of the contraction increases, larger and larger units are recruited, each contributing progressively more tension. A nice control is thereby achieved, enabling fine or coarse movements to be appropriately graded. It is clearly efficacious for smaller units to fire at the lower end of the contractile range, rather than near the maximum, when the percentage tension increment they produce would be far less. Thus, in the soleus muscle of the cat, the firing of a small motoneuron may give rise to an increase in tension of about 5 grams, whereas a larger unit may contribute more than 100 grams; and the maximum contraction brought about by all the motor units firing may reach over 3.5 kilograms. Plainly a small motor unit would be relatively ineffective if brought in when the contraction was near its maximum, and a large unit firing in the lower range would be detrimental to fine gradation of movement. The fact that the motor units are recruited in order of increasing size means that each adds a relatively fixed percentage increment of about 5 percent to the existing tension. It is of interest to recall that the muscle fibers of small motor units, which are recruited first and are therefore always involved in muscle contractions, are fatigue-resistant, while fibers in the large motor units, which are recruited less frequently, fatigue more rapidly. There is some evidence that in humans the largest motoneurons are so inexcitable that they are recruited only rarely during feats of exceptional strength.

The principle that motor unit recruitment adds a relatively fixed fraction, rather than a fixed increment, to existing muscle tension is reflected in the sensory aspects of motor activity as well. For example, we judge weights by the force needed to support them, and can easily distinguish the difference between 2 and 3 grams but not between 2002 and 2003 grams. Again, it is the *relative* change that is important. Indeed, much of our perception of the world is determined in the same way, and we act accordingly. You would not mind paying $2003 for an item normally costing $2002 but would be outraged at paying $3 for a $2 item.

The stretch reflex   Extensor muscles of vertebrate limbs can be described generally as those that open, or extend, the joints and (in quadrupeds) oppose the force of gravity. Flexor muscles close, or flex, the joints, thereby opposing the action of extensors. Flexors and extensors are thus said to be ANTAGONIST muscles. The simplified diagram in Figure 4A outlines connections to extensor and flexor leg muscles involved in the stretch reflex. This reflex pathway is the simplest of many converging onto the

[13]Henneman, E., Somjen, G. and Carpenter, D. O. 1965. *J. Neurophysiol.* 28: 560–580.

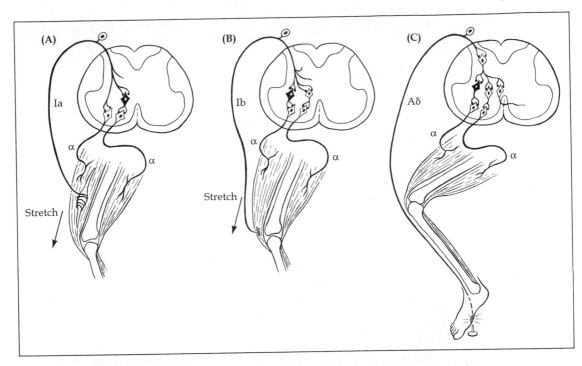

4 **ORGANIZATION OF SYNAPTIC CONNECTIONS** for reflex actions in the spinal cord. The spinal cord is shown in transverse section, with inhibitory interneurons in black. **(A)** In the *myotatic reflex*, stretch of the muscle spindle generates impulses that travel along group Ia afferent fibers to the spinal cord and produce monosynaptic excitation of α motoneurons to the same muscle. Impulses also excite interneurons that, in turn, inhibit motoneurons supplying the antagonist muscles. **(B)** Activation of *Golgi tendon organs* produces impulses in group Ib afferent fibers that, through interneuronal connections, provide inhibition to motoneurons supplying the same muscle and excitation to antagonist motoneurons. (This is sometimes called the *inverse myotatic reflex.*) **(C)** The *flexor reflex* is a limb-withdrawal reflex, produced in this example by stepping on a tack. Excitation of Aδ pain fibers results in elevation of the thigh (synaptic connections not shown) and flexing of the knee joint by polysynaptic excitation of flexor motoneurons and inhibition of extensors. Also not shown are contralateral connections that subserve extension of the opposite leg for support.

motoneuron and represents only a small part of what is known about spinal reflexes. When the reflex is activated by muscle stretch—for example, by tapping the patellar tendon to produce a knee-jerk (Chapter 1)—the primary sensory endings in the muscle spindle are deformed and initiate impulses in group IA afferent fibers. These impulses produce monosynaptic excitation of the α motoneurons going back to the muscle that has been extended, resulting in a reflex contraction and extension of the leg. The leg extension is further helped by a simultaneous inhibition of the α motoneurons that innervate the antagonistic flexor muscles. This principle of one group of muscles being excited while its antagonists are inhibited was first described by Sherrington, who called it RECIPROCAL INNERVATION.

For the sake of simplicity, a number of pathways are omitted from

Figure 4A. For example, discharges from the smaller group II afferents reinforce the reflex largely by way of interneurons, but also monosynaptically.[14] Intraspinal connections have been worked out in detail by Lundberg, Jankowska and their colleagues.[15,16] One type of interneuron, the Renshaw cell, mediates inhibition within populations of α motoneurons supplying the same muscle. Motoneuron discharge produces excitation of Renshaw cells (mediated by acetylcholine), which in turn causes inhibition of α motoneurons within the population. The precise role played by this recurrent inhibition in regulating movement is not understood. Like other interneurons involved in coordination, Renshaw cells receive descending inputs from higher centers.

The Golgi tendon organs near the tendon–muscle junctions are in series with contracting skeletal muscles (Chapter 14). They can be made to discharge impulses by passive stretch, but contraction of muscle fibers, to which they are extremely sensitive, is the principal stimulus that activates their firing. A contraction of one or two muscle fibers, leading to a tension increase of less than 100 mg, can cause a brisk discharge.[17,18] Axons from tendon organs activate interneurons that, in turn, inhibit α motoneurons supplying their muscle of origin (Figure 4B).[18,19] In addition, the information that they provide about muscle tension is relayed to higher centers.

The flexor reflex   More widespread spinal reflexes that extend beyond the muscles of origin involve groups of motoneurons—flexors and extensors on the same (ipsilateral) and opposite (contralateral) sides of the body and, in quadrupeds, flexors and extensors in the other pair of limbs. They illustrate the high degree of complexity and variability of responses that can occur at the spinal level without immediate input from higher motor centers. The most familiar of these is the flexor reflex, which is activated, for example, when one steps on a sharp object, bangs one's shin against a bench, or touches a hot stove. The response is complex, depending on the location of the offending stimulus, but has two consistent features: (1) Movement of the affected limb is always primarily flexion, and is directed away from the offending stimulus; and (2) if necessary, weight is transferred to the contralateral limb. The synaptic input is from nociceptive and other sensory receptors of the skin; the pattern of movement is determined by a complex interplay of excitation of flexors and reciprocal inhibition of extensors on the side of the stimulus, and by simultaneous extensor excitation and flexor inhibition on the contralateral side (Figure 4C). This synaptic activity is organized within the spinal cord at the segmental level and is supplemented by inputs from higher centers that serve to maintain balance and mediate the appropriate continuation or cessation of movement.

[14]Kirkwood, P. A. and Sears, T. A. 1974. *Nature* 252: 243–244.

[15]Lundberg, A. 1979. *Prog. Brain Res.* 50: 11–28.

[16]Czarkowska, J., Jankowska, E. and Syrbirska, E. 1981. *J. Physiol.* 310: 367–380.

[17]Crago, P. E., Houk, J. C. and Rymer, W. Z. 1982. *J. Neurophysiol.* 47: 1069–1083.

[18]Fukami, Y. 1982. *J. Neurophysiol.* 47: 810–826.

[19]Matthews, P. B. C. 1972. *Mammalian Muscle Receptors and Their Central Action.* Edward Arnold, London.

SUPRASPINAL CONTROL OF MOTONEURONS

Before more complicated movements are discussed, it is useful to out-line briefly the organization of motoneurons in the spinal cord and the arrangement of descending pathways that act upon them. Figure 5 shows a cross section of the cervical spinal cord with the localization of motoneurons supplying a portion of the upper limbs. Within the segment there is an orderly arrangement of motoneurons. Extensor motoneurons tend to be ventral to flexor motoneurons. More importantly, motoneurons supplying proximal muscles are the most medial and ventral, those supplying distal muscles the most lateral and dorsal. This type of organization extends throughout the spinal cord and is important for understanding how motor activities are organized. Medially located motoneurons innervate the trunk and proximal muscles concerned mainly with sustained activites such as stance, while lateral motoneurons innervate distal muscles that tend to be concerned more directly with phasic activities such as manipulation.[20]

Medial–lateral organization of motoneurons

The major pathways descending onto motoneurons arise in the cerebral cortex and the brainstem (Figure 6; see also Appendix C). In accordance with their sites of termination within the ventral horn of the spinal gray matter, these pathways can be classified into two groups,

[20]Crosby, E. C., Humphrey, T. and Lauer, E. W. 1966. *Correlative Anatomy of the Nervous System.* Macmillan, New York.

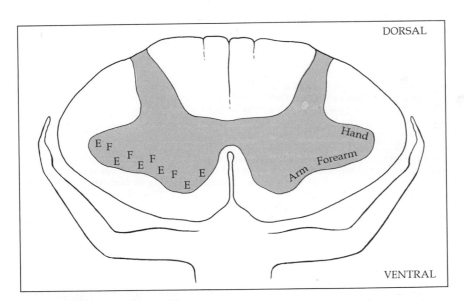

5   ORGANIZATION OF MOTONEURONS supplying the upper extremities, shown in a transverse section of the spinal cord in the cervical region. Muscles of the shoulder and arm are represented most medially, hand most laterally. Extensor motoneurons (E) are located nearest the margin of the gray matter; flexor motoneurons (F) are more central.

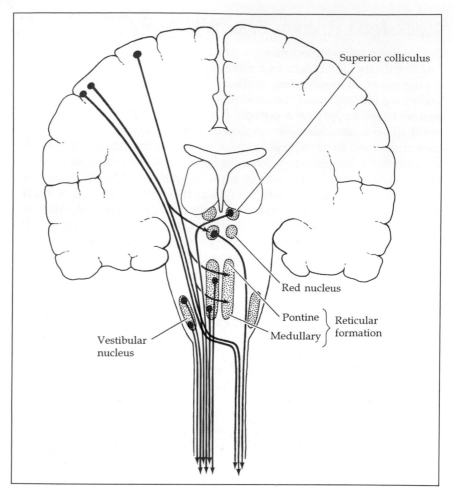

6   MAJOR MOTOR PATHWAYS in the vertebrate central nervous system supplying lateral motoneurons (black) and supplying medial motoneurons (colored), shown schematically on a coronal section of the cerebral hemispheres, continuing to a longitudinal section of the brainstem and spinal cord. Cells in the primary motor area of the cerebral cortex send axons to the contralateral spinal cord to form the *lateral corticospinal tract*, with collateral connections to the red nucleus. Axons from cells in the red nucleus cross the midline and descend in the *rubrospinal tract*. These tracts supply monosynaptic and polysynaptic innervation largely to lateral motoneurons (i.e., those supplying distal musculature). Some cortical fibers descend without crossing to form the *ventral corticospinal tract*, supplying collaterals to brainstem nuclei. The axial musculature is supplied predominantly by the motor regions of the brainstem through the *reticulospinal tracts*, originating in the pontine and medullary reticular formations; the *vestibulospinal tracts*, originating in the vestibular nuclei; and the *tectospinal tract*, originating in the superior colliculus.

lateral and medial.[21] The two major lateral pathways are the LATERAL CORTICOSPINAL TRACT, originating in the cerebral cortex, and the RUBRO-

[21]Kuypers, H. G. J. M. 1981. *In* J. M. Brookhart and V. B. Mountcastle (eds.). *Handbook of Physiology: The Nervous System, Section I*, Vol. II, Pt.2. American Physiological Society, Bethesda, MD.

SPINAL TRACT, originating in the red nucleus of the midbrain. The medial pathways include the VENTRAL CORTICOSPINAL TRACT, the lateral and medial VESTIBULOSPINAL TRACTS, the pontine and medullary RETICULO- SPINAL TRACTS, and the TECTOSPINAL TRACT.

For the reader not familiar with the anatomy of the central nervous system, these terms may seem as bewildering as those used by high-energy physicists. However, one must be able to recognize the names of structures and have some idea of their approximate locations to understand the principles underlying motor performance. Fortunately, anatomists have maintained some degree of rigor in naming fiber tracts. A given pathway is always named according to its origin first, then its termination; in addition, if two or more pathways run in the spinal cord, their individual locations are specified. Thus (as discussed in the previous chapter) the lateral spinothalamic pathway ascends in the lateral portion of the spinal cord to the thalamus; the medullary retic-ulospinal tract originates in the reticular formation of the medulla and ends at various levels in the spinal cord. Terms indicating direction with respect to main axes of the spinal cord and brain, such as "rostral," "anterior," and "ventral," are illustrated in Appendix C.

The lateral corticospinal tract originates in the motor and premotor areas of the cerebral cortex in front of the central sulcus (areas 4 and 6, Figure 7A), as well as from a small strip of the postcentral region (area 3) of the cerebral cortex. Fibers pass downward through the internal capsule and cerebral peduncles to the medullary pyramids, after which most decussate (cross the midline) and continue their descent laterally in the spinal cord, terminating on interneurons and motoneurons at the appropriate levels. The cortical cells from which the pathway originates are arranged in an orderly manner to form a somatotopic pattern of muscle representation (Figure 7B), similar to the somatosensory pattern on the postcentral gyrus. This organization was first demonstrated by Fritsch and Hitzig in 1870 by stimulation of the cerebral cortex of animals.[22] The somatotopic representation in humans was later mapped in detail on the brains of neurosurgical patients by Wilder Penfield and his colleagues.[23] Localized stimulation of the cortical surface with brief electrical shocks produced movements of restricted regions of the body, depending upon the position of the stimulating electrode. As in the sensory system, the face and limbs (particularly, in humans, the hands) receive disproportionate representation.

In the spinal cord, fibers in the lateral corticospinal tracts terminate most prominently on interneurons and motoneurons in the lateral gray matter. Their main influence, then, is on the lateral motoneurons that control the distal muscles concerned with movement and fine manipulation. An important feature of the tract is that many of the fibers originating in the hand area of the motor cortex have terminal branches

Lateral motor pathways

[22]Fritsch, G. and Hitzig, E. 1870. *Arch. Anat. Physiol. Wiss. Med.* 37: 300–332.

[23]Penfield, W. and Rasmussen, T. 1950. *The Cerebral Cortex of Man.* Macmillan, New York.

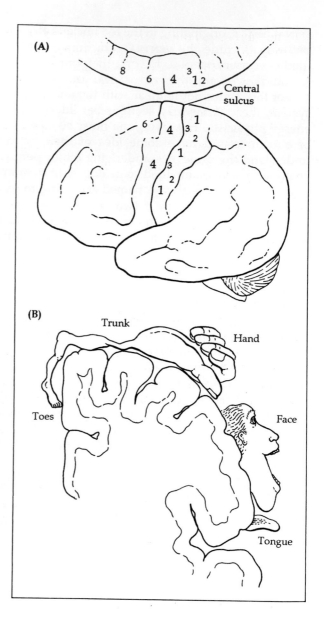

**MOTOR REPRESENTATION ON THE CERE-
BRAL CORTEX. (A)** Lateral view of the surface
of the cerebral cortex. Motor activity is associated
with activation of cells in area 4 of the cerebral
cortex, including the cells of origin of the corticospinal tract. This is the *primary motor area.*
The motor system also includes area 6 (*premotor
area*), extending onto the medial surface of the
hemisphere (upper figure). **(B)** Coronal section
through the cerebral hemisphere anterior to the
central sulcus. The musculature of the human
body is represented in an orderly, but distorted,
fashion, with the leg and foot on the medial
surface of the hemisphere (*cingulate gyrus*) and
the head most lateral. The very large area devoted
to the hand is indicative of the number of neurons involved in control of manipulations by the
digits.

that end directly on motoneurons controlling muscles of the digits.[24] In
humans and other primates, interruption of the lateral corticospinal
tract results primarily in a loss of the ability to move the fingers independently and a consequent deficit in the ability to perform fine, precise
tactile movements.[25,26]

[24]Cheney, D. P. and Fetz, E. E. 1980. *J. Neurophysiol.* 44: 773–791.
[25]Lawrence, D. G. and Kuypers, H. G. J. M. 1968. *Brain* 91: 1–14.
[26]Hepp-Reymond, M.-C. 1988. *In* H. D. Seklis and J. Erwin (eds.). *Comparative Primate
Biology*, Vol. 4. Alan R. Liss, New York, pp. 501–624.

The rubrospinal tracts originate in the red nuclei (Figure 6) and cross the midline before descending in the spinal cord to end on interneurons associated with the lateral motor system. The cells of origin are arranged somatotopically and receive excitatory inputs from the motor cortex and from the cerebellum. The precise functional role of the rubrospinal tract is unclear. It is thought to duplicate many of the functions of the corticospinal tract and therefore constitutes a parallel pathway from the cortex. In primates, lesions of the tract have little obvious effect, but after interruption of both the rubrospinal and corticospinal tracts (thereby removing all direct cortical connections to the affected moto-neurons), coordinated positioning of the hands and feet is severely impaired.[27]

Except for a small component from the ventral corticospinal tract, the medial pathways originate primarily in the brainstem (Figure 6) and distribute their inputs to medial motoneurons innervating proximal muscle groups. The cells of origin of the lateral vestibulospinal tract lie (as the name indicates) in the lateral vestibular nucleus. Each nucleus receives input from the ipsilateral vestibular apparatus, in particular from the utricles of the labyrinth (Chapter 14). The tract descends uncrossed in the spinal cord to supply input to the medial motoneurons supplying postural muscles, with monosynaptic excitatory inputs to extensor muscles and disynaptic inhibitory inputs to flexors. The tract is involved in the maintenance of posture and the regulation of extensor tone. The pontine reticulospinal tract descends ipsilaterally and ends on segmental interneurons that, in turn, provide bilateral excitation to medial extensor motoneurons. The medullary reticulospinal tract descends bilaterally to provide inhibitory inputs to motoneurons supplying the proximal limbs.

Medial motor pathways

Two other brainstem pathways end in the cervical and upper thoracic levels and are concerned with upper body and limb posture, and most particularly with the position of the head. The medial vestibulospinal tract arises from cells in the medial vestibular nucleus that, in turn, receive inputs both from the semicircular canals and from stretch receptors in neck muscles.[28] The tract descends ipsilaterally to mid-thoracic levels and is concerned with postural adjustments of the neck and upper limbs during angular acceleration. The tectospinal tract originates in the superior colliculus and decussates before descending to upper cervical levels. This pathway mediates orientation of the head and eyes to visual and auditory targets.

In summary, descending pathways arising from the cerebral cortex and the red nucleus supply lateral motoneurons and are important for organized movement of small groups of muscles, especially the distal musculature, the corticospinal tract being particularly important for fine movements of the digits. Motor pathways descending from the brain-stem, in contrast, are directed toward large groups of muscles, espe-

[27]Kennedy, P. R. 1990. *Trends Neurosci.* 13: 474–479.
[28]Kaspar, J., Schor, R. H. and Wilson, V. J. 1988. *J. Neurophysiol.* 60: 1765–1768.

cially proximal musculature, concerned with the regulation of position and posture of the lower and upper body and the head. These pathways have prominent inputs from the vestibular apparatus.

THE CEREBELLUM AND BASAL GANGLIA

Two additional structures play a major role in the organization of motor activity: the cerebellum and the basal ganglia. Different regions of the cerebellum are connected to the lateral and medial descending pathways and are closely involved in both the maintenance of balance and posture and the organization of limb movements. The basal ganglia are interconnected with the corticospinal system and hence play a major role in the regulation of limb movements. Less prominent connections descend to the brainstem by way of the pons and thereby influence postural adjustments as well.

Outputs from the cerebellum
 The cerebellum is an outgrowth of the pons and consists of a three-layered cortex overlying deep nuclear groups of cells. Its anatomical features are summarized in Appendix C. Connections from the cortical regions to the deep nuclei, and the nuclear projections out of the cerebellum, have been summarized in a review by Thach and his colleagues.[29] Output from the cerebellar cortex is through the axons of Purkinje cells. The effects of Purkinje cell axons on their target cells are uniformly inhibitory. The projections are organized in an orderly fashion (Figure 8): Cortical cells in the most medial zones (the medial segment of the anterior lobe and the vermis) project to the most medial nucleus—the fastigial nucleus; in addition, some project directly to the brainstem, ending in the ipsilateral vestibular nucleus. Cells in the lateral cerebellar hemisphere project to the most lateral deep nucleus—the dentate nucleus. In the zone intermediate between the vermis and lateral hemispheres, Purkinje cells project to the interposed nuclei. This mediolateral organization of the nuclei is reflected, in turn, in their projections to the medial and lateral descending motor pathways. The dentate nucleus projects to the ventrolateral nucleus of the thalamus and hence to the motor cortex. The interposed nuclei project to the red nucleus and to the thalamic nuclei. Thus, both these cerebellar nuclei have direct inputs to the lateral motor system. Somatotopic order is maintained throughout the projections, beginning with multiple representations of the body musculature on the cerebellar cortex, one for each nucleus, and then in the nuclei themselves. The projections overlap in the thalamus, suggesting multiple control of thalamic targets, and continue through to the cerebral cortex. The fastigial nucleus projects to the vestibular nucleus and the pontine reticular formation, hence influencing the lateral vestibulospinal and pontine reticulospinal tracts (the medial motor system).

[29]Thach, W. T., Goodkin, H. G. and Keating, J. G. 1992. *Annu. Rev. Neurosci.* 15: 403–442.

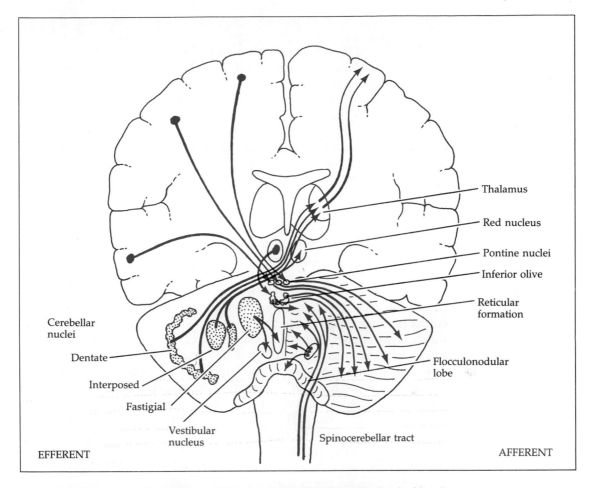

Thalamus

Red nucleus

Pontine nuclei

Inferior olive

Reticular formation

Cerebellar nuclei

Dentate

Interposed

Fastigial

Vestibular nucleus

Flocculonodular lobe

Spinocerebellar tract

EFFERENT

AFFERENT

8 **EFFERENT AND AFFERENT PATHWAYS IN THE CEREBELLUM. Cerebral hemispheres, brainstem, and spinal cord represented as in Figure 6, together with a view of the superior surface of the cerebellum (right) and the underlying cerebellar nuclei (left). Outputs from the cerebellum (left side, black) are through the dentate nucleus, the interposed nuclei, and the fastigial nucleus. Fibers from the dentate nucleus supply the contralateral motor cortex through the ventrolateral and parts of the ventro-posterolateral nuclei of the thalamus. The interposed nuclei project to the contralateral red nucleus as well. Both these pathways, therefore, are associated with the lateral motor system. The fastigial nucleus projects to the vestibular nucleus and the pontine and medullary reticular formations. Inputs (right side, colored) to the lateral hemispheres of the cerebellum are from wide areas of the cerebral cortex, through the pontine nuclei. Afferent input from the red nucleus is relayed through the inferior olive. More medially, the cerebellum receives extensive input from the spinocerebellar tracts. The flocculonodular lobe is supplied by the vestibular nucleus.**

Inputs to the cerebellum (Figure 8) are similarly segregated. The flocculonodular lobe (phylogenetically the oldest part of the cerebellum, the archicerebellum) receives input from the vestibular nucleus. Sherrington defined the cerebellum as "the head ganglion of the proprio-

Cerebellar inputs

ceptive system,"[30] and, indeed, the medial anterior lobe and the posterior vermis (paleocerebellum) receive proprioceptive and cutaneous inputs from all levels of the spinal cord. The most prominent of these incoming pathways are the dorsal and ventral spinocerebellar tracts, from the lumbar and thoracic regions of the cord, and the cuneocerebellar and rostral spinocerebellar pathways from cervical levels. These form multiple somatotopic sensory representations on the cortex of the cerebellum that are connected through to the motor representations discussed above. Input from the head includes auditory and visual information. The lateral hemispheres (neocerebellum) have no direct sensory representation but receive information from a wide area of the cerebral cortex, relayed through cells in the pons, and from the red nucleus via the inferior olive. In summary, the cerebellum receives a wealth of sensory information from the periphery, particularly from peripheral proprioceptors and, in addition, receives input from the cerebral cortex and red nucleus associated with actual or intended movement.

Organization of the cerebellar cortex

The neuronal organization within the cerebellar cortex has been studied in great detail, both morphologically and physiologically.[31] The cortex is composed of three layers, as shown in Figure 9. The inner, or third, layer is packed with GRANULE CELLS; these small cells are estimated to number between 10^{10} and 10^{11}—approximating the sum of all other cells in the nervous system! They send axons outward to the first (molecular) layer to form a system of PARALLEL FIBERS, each several millimeters in length. Also in the granule cell layer are larger and less numerous GOLGI CELLS. The second cortical layer is occupied by PURKINJE CELLS, whose axons constitute the sole output from the cerebellar cortex. The Purkinje cell dendrites extend into the outer molecular layer of the cortex, with their arborizations oriented at right angles to the streams of parallel fibers with which they make synaptic contacts. One can imagine the Purkinje cells as stacked in a row along a folium, like coins on edge, with the parallel fibers extending through them. It is estimated that each Purkinje cell receives inputs from more than 200,000 parallel fibers. Each parallel fiber engages a "beam" of Purkinje cells extending along the folium and projecting in an orderly manner to the underlying cerebellar nucleus. The significance of this arrangement is that such a beam of Purkinje cells may span several joints in a somatotopic region, for example the shoulder, elbow, and wrist joints of the arm, thereby providing a possible mechanism for coordinating multijoint movements.[29]

The length of the parallel fibers is also sufficient to connect cells projecting to adjacent deep nuclei, possibly functioning to provide internuclear coordination. The first cortical layer also contains STELLATE and BASKET CELLS that provide inhibitory inputs to Purkinje cells from

[30]Sherrington, C. S. 1933. *The Brain and Its Mechanism*. Cambridge University Press, London.

[31]Ito, M. 1984. *The Cerebellum and Neural Control*. Raven, New York.

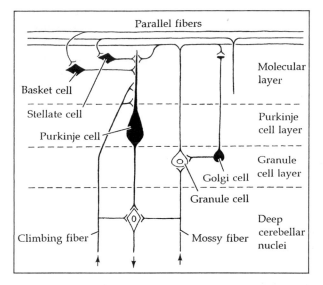

9

**INPUTS AND OUTPUTS WITHIN THE CERE-
BELLUM.** The cerebellar cortex is subdivided
into the granule cell layer, Purkinje cell layer,
and molecular layer. Open boutons indicate excitatory synaptic connections, filled boutons inhibitory connections. The sole output from the
cortical layer to cells of the deep cerebellar nuclei
is by inhibitory axons from Purkinje cells. Mossy
fiber inputs end on granule cells, whose axons
ascend to the molecular layer to form a parallel
fiber network. Parallel fibers form excitatory synapses on Purkinje cells, stellate cells, basket
cells, and dendrites of Golgi cells. Climbing fibers form excitatory synapses on Golgi cells.
Both climbing fibers and mossy fibers make excitatory connections with cells in the deep cerebellar nuclei.

remote parallel fibers—an arrangement equivalent to lateral inhibition
in sensory systems. The major input to the cerebellum is by MOSSY
FIBERS that arise in the spinal cord and brainstem nuclei and form
excitatory synaptic connections with granule and Golgi cells. A second
input is by way of CLIMBING FIBERS that arise in the inferior olive and
make excitatory synaptic contact directly with Purkinje cell dendrites as
they ascend into the molecular layer.

The cerebellar cortex, then, receives input from two sources: mossy
fibers and climbing fibers. Both inputs also make excitatory contacts
with cells in the corresponding cerebellar nuclei. The Purkinje cell axons
that form the sole output to the deep nuclei, being inhibitory, can only
modulate ongoing activity. The regular morphological arrangement of
the fibers and cells in the cerebellar cortex appears to provide a mechanism for restricting and coordinating multijoint movements. These
functions will be considered further when we discuss the activity of
single cells in the cerebellar nuclei seen during the execution of trained
movements.

Beneath the outer cortical layers of the cerebral hemispheres are
groups of neurons collected into nuclear masses known collectively
as the basal ganglia. These consist of the CAUDATE NUCLEUS and the
PUTAMEN (known together as the NEOSTRIATUM), and the external and
internal divisions of the GLOBUS PALLIDUS (Figure 10). Two midbrain
structures, the SUBSTANTIA NIGRA and the SUBTHALAMIC NUCLEUS, have
both afferent and efferent connections with the basal ganglia and are
therefore a functional part of the overall system. In the substantia nigra,
degeneration of dopaminergic neurons that project to the striatum (the
nigrostriatal pathway) is associated with Parkinson's disease (Chapter
10). The basal ganglia receive widespread inputs from the cerebral

The basal ganglia

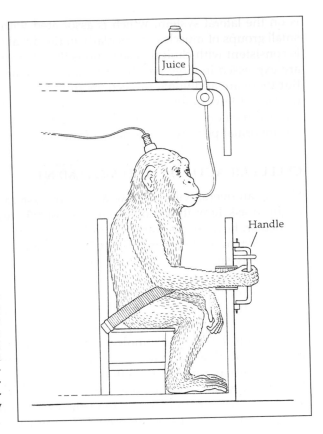

11

EXPERIMENTAL ARRANGEMENT FOR RE-
CORDING CELLULAR ACTIVITY related to
wrist movement. A monkey, previously trained
to perform the required movements, is seated in
the chair with its forearm placed in a cuff. The
wrist is moved to deflect a handle to the left or
right between stops, by flexion or extension of
the wrist. A system of weights, or a torque motor
(not shown), is used to load the handle to oppose
either flexion or extension. For visually guided
movements, the handle position is indicated on
a display screen. Correct movements are re-
warded by a drink of fruit juice. Single-unit activity is recorded from with a microelectrode positioned in an appropriate area of the brain, by
means of a microdrive fixed to the skull.

discharge frequency was related to the force required to execute the
movement.[34] This behavior of the cortical cells was not unlike the behavior of the spinal motoneurons to which they projected. Subsequent
experiments showed that this particular kind of behavior is characteristic
of corticospinal cells that end directly on spinal motoneurons.[24] Other
classes of cells exhibit more complex behavior. For example, Humphrey
and Reed recorded discharges from pyramidal tract neurons in experiments in which monkeys were trained to keep their wrists in a fixed
position when alternating loads were applied against flexion and extension.[35] At low frequencies (less than about 0.6 Hz) the task was accomplished by alternating flexor and extensor contractions to oppose the
load. Individual cortical cells in the wrist area discharged phasically
with the load, in relation to either flexor or extensor activity. At higher
frequencies, flexors and extensors were co-contracted so that changes
in load were opposed by increasing joint stiffness. At the same time a
different, previously silent, class of cortical cells was detected, whose
discharge was continuous and whose discharge rates increased with
increasing frequency of the load perturbations.

[34]Evarts, E. V. 1968. *J. Neurophysiol.* 31: 14–27.
[35]Humphrey, D. R. and Reed, D. J. 1983. *Adv. Neurol.* 39: 347–372.

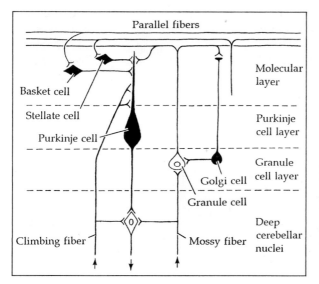

9

INPUTS AND OUTPUTS WITHIN THE CERE-BELLUM. The cerebellar cortex is subdivided into the granule cell layer, Purkinje cell layer, and molecular layer. Open boutons indicate excitatory synaptic connections, filled boutons inhibitory connections. The sole output from the cortical layer to cells of the deep cerebellar nuclei is by inhibitory axons from Purkinje cells. Mossy fiber inputs end on granule cells, whose axons ascend to the molecular layer to form a parallel fiber network. Parallel fibers form excitatory synapses on Purkinje cells, stellate cells, basket cells, and dendrites of Golgi cells. Climbing fibers form excitatory synapses on Golgi cells. Both climbing fibers and mossy fibers make excitatory connections with cells in the deep cerebellar nuclei.

remote parallel fibers—an arrangement equivalent to lateral inhibition in sensory systems. The major input to the cerebellum is by MOSSY FIBERS that arise in the spinal cord and brainstem nuclei and form excitatory synaptic connections with granule and Golgi cells. A second input is by way of CLIMBING FIBERS that arise in the inferior olive and make excitatory synaptic contact directly with Purkinje cell dendrites as they ascend into the molecular layer.

The cerebellar cortex, then, receives input from two sources: mossy fibers and climbing fibers. Both inputs also make excitatory contacts with cells in the corresponding cerebellar nuclei. The Purkinje cell axons that form the sole output to the deep nuclei, being inhibitory, can only modulate ongoing activity. The regular morphological arrangement of the fibers and cells in the cerebellar cortex appears to provide a mechanism for restricting and coordinating multijoint movements. These functions will be considered further when we discuss the activity of single cells in the cerebellar nuclei seen during the execution of trained movements.

Beneath the outer cortical layers of the cerebral hemispheres are groups of neurons collected into nuclear masses known collectively as the basal ganglia. These consist of the CAUDATE NUCLEUS and the PUTAMEN (known together as the NEOSTRIATUM), and the external and internal divisions of the GLOBUS PALLIDUS (Figure 10). Two midbrain structures, the SUBSTANTIA NIGRA and the SUBTHALAMIC NUCLEUS, have both afferent and efferent connections with the basal ganglia and are therefore a functional part of the overall system. In the substantia nigra, degeneration of dopaminergic neurons that project to the striatum (the nigrostriatal pathway) is associated with Parkinson's disease (Chapter 10). The basal ganglia receive widespread inputs from the cerebral

The basal ganglia

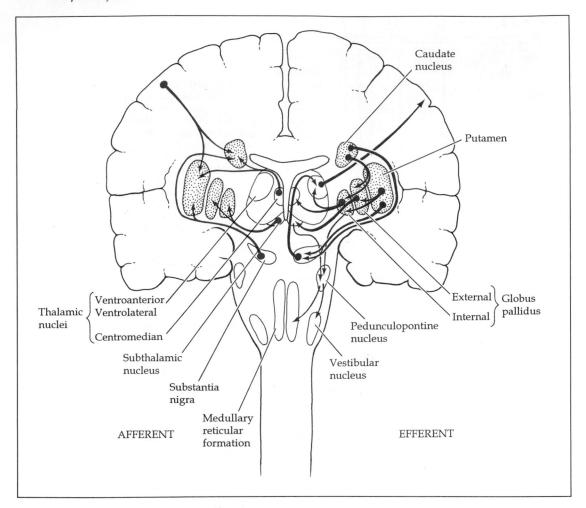

10   MOTOR PATHWAYS IN THE BASAL GANGLIA. Coronal section through the cerebral hemispheres, continuing to a longitudinal section of the brainstem and spinal cord, as in Figure 6. Basal ganglia are the caudate nucleus, putamen, and external and internal divisions of the globus pallidus. The basal ganglia receive inputs (left side, colored) from the motor cortex, the substantia nigra, the subthalamic nucleus, and the centromedian nuclei of the thalamus. Outputs (right side, black) are back to the same nuclei and to the motor cortex through the ventroanterior and ventrolateral nuclei of the thalamus. Additional output pathways project to the vestibular nucleus and medullary reticular formation through the pedunculopontine nucleus.

cortex, particularly from the precentral gyrus. Their major outputs are directed to the ventrolateral and ventroanterior nuclei of the thalamus (overlapping with regions receiving input from the cerebellum) and hence to the cortex.

Although the interconnections of the basal ganglia are complex, some ideas about their function can be deduced from the fact that much, if not most, of their output is to the motor cortex. Thus their influence is on the lateral system, which is associated with organized activity of small groups of muscles, particularly in the distal limbs. This conclusion is consistent with clinical observations that diseases of the basal ganglia are expressed in large part by abnormal movements of the extremities. But there are additional projections from the striatum and subthalamic nuclei to the vestibular and reticular nuclei of the brainstem by way of the pedunculopontine nuclei. This projection, when diseased, accounts for abnormal postures, especially of the trunk and axial musculature.

CELLULAR ACTIVITY AND MOVEMENT

Having an overall view of how the motor system is organized, we can begin to ask how its various elements contribute to motor activity. In particular, how are discharges of individual cells at various levels in the motor system related to movement? At the spinal cord level, the answer is relatively simple: Discharges of α motoneurons cause contractions of their particular motor units. The overall strength of contraction is adjusted by recruitment of more or fewer motor units and by the frequency at which the motoneurons fire. There are, of course, additional factors at work; for example, coactivation of γ motoneurons may be required to maintain the muscle spindles at their optimum length for detecting external disturbances that might interfere with the planned movement (Chapter 14) and providing reflex compensation for such disturbances. Nevertheless, in spite of such complexities, if we were to record the activity of an α motoneuron during the execution of a movement, we would find that its discharge provided an indication of the strength of contraction of its motor unit and (less precisely) of the muscle containing that motor unit. Techniques for recording from single cells in the brains of awake monkeys during the performance of trained tasks have provided similar information about the relation to movement of cellular activity in the cerebral cortex, cerebellum, and basal ganglia.

What descending influences control the spinal motoneurons, and how are these influences coordinated to produce appropriate movements? Are there cells in the cerebral or cerebellar cortex, or in the basal ganglia, whose activities are related in some specific way to movements? And, if so, to what aspect of a movement? Strength of contraction of specific muscle groups? Magnitude of displacement around a joint? Direction in space of the movement? These kinds of questions were first asked by Evarts,[32,33] who recorded the activity of pyramidal tract cells in the motor cortex during the performance of trained wrist movements by awake monkeys (Figure 11). By loading the wrist to oppose either flexion or extension, Evarts could dissociate force required for a movement from its direction. Early results were straightforward: The cells tended to be associated with either extension or flexion, and their

Activity of cortical cells during trained movements

[32]Evarts, E. V. 1965. *J. Neurophysiol.* 28: 216–228.
[33]Evarts, E. V. 1966. *J. Neurophysiol.* 29: 1011–1027.

11

EXPERIMENTAL ARRANGEMENT FOR RE-
CORDING CELLULAR ACTIVITY related to
wrist movement. A monkey, previously trained
to perform the required movements, is seated in
the chair with its forearm placed in a cuff. The
wrist is moved to deflect a handle to the left or
right between stops, by flexion or extension of
the wrist. A system of weights, or a torque motor
(not shown), is used to load the handle to oppose
either flexion or extension. For visually guided
movements, the handle position is indicated on
a display screen. Correct movements are re-
warded by a drink of fruit juice. Single-unit activity is recorded from with a microelectrode positioned in an appropriate area of the brain, by
means of a microdrive fixed to the skull.

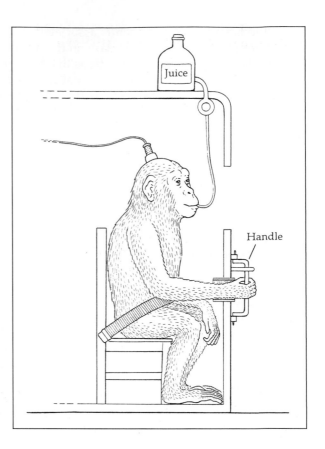

discharge frequency was related to the force required to execute the
movement.[34] This behavior of the cortical cells was not unlike the behavior of the spinal motoneurons to which they projected. Subsequent
experiments showed that this particular kind of behavior is characteristic
of corticospinal cells that end directly on spinal motoneurons.[24] Other
classes of cells exhibit more complex behavior. For example, Humphrey
and Reed recorded discharges from pyramidal tract neurons in experiments in which monkeys were trained to keep their wrists in a fixed
position when alternating loads were applied against flexion and extension.[35] At low frequencies (less than about 0.6 Hz) the task was accomplished by alternating flexor and extensor contractions to oppose the
load. Individual cortical cells in the wrist area discharged phasically
with the load, in relation to either flexor or extensor activity. At higher
frequencies, flexors and extensors were co-contracted so that changes
in load were opposed by increasing joint stiffness. At the same time a
different, previously silent, class of cortical cells was detected, whose
discharge was continuous and whose discharge rates increased with
increasing frequency of the load perturbations.

[34]Evarts, E. V. 1968. *J. Neurophysiol.* 31: 14–27.
[35]Humphrey, D. R. and Reed, D. J. 1983. *Adv. Neurol.* 39: 347–372.

Experiments with visually guided whole-arm movements have indicated that some neurons in the arm area of the cerebral cortex discharge at maximum rates when the movement is in a particular direction.[36,37] Preferred directions were not absolute; discharges of such neurons fell off as the angle of reach was altered, decreasing roughly with the cosine of the angle between the actual and preferred trajectories. Georgopoulos and his colleagues[38] have suggested that the observed relations are causal; that is, the movement trajectory in extrapersonal space is determined by the relative activity within an ensemble of such neurons, the movement being in the preferred direction of those firing most vigorously. Similar observations have been reported on cells in cerebellar nuclei.[39]

Cortical cell activity related to direction of arm movements

The techniques developed by Evarts were used by Thach and his colleagues to examine the relations between trained movements and cellular activity in both the motor cortex and the deep cerebellar nuclei.[28,40,41] Monkeys were trained to perform a sequence of flexion and extension movements with flexor and extensor loading. As in previous experiments, varying the load served to dissociate muscle activity from joint position and direction of movement. Cells in the motor cortex and in the cerebellar nuclei discharged in one of three distinct patterns: some in relation to work against a load (i.e., muscle activity), some in relation to joint position, and some in relation to direction of intended movement.

Cellular activity in cerebellar nuclei

During the trained movements, the *sequence* of cellular activity was in the following order: dentate → motor cortex → interposed nuclei → muscle. This sequence is consistent with the idea that information about planned movements is relayed from the cerebral cortex to the lateral lobe of the cerebellum, where it is processed and sent back to the motor cortex through the dentate nucleus. Signals from the motor cortex are then relayed to the appropriate spinal motoneurons and back through the cerebellar cortex to the interposed nuclei. The idea that the dentate nucleus plays a role in the initiation of planned movements is supported by other experiments in which cooling of the nucleus slowed the onset of both volitional movement and related cellular activity in the motor cortex[42] but had no other effect on execution of the movement.

Purkinje cells in the intermediate zone of the cerebellum that project to the interposed nuclei receive inputs from segmental levels of the spinal cord by way of the spinocerebellar pathways. As a result they are in a position to obtain proprioceptive information from muscles participating in a movement and to modulate cells in the interposed

[36]Schwartz, A. B., Kettner, R. E. and Georgopoulos, A. P. 1988. *J. Neurosci.* 8: 2913–2927.

[37]Caminiti, R., Johnson, P. B. and Urbano, A. 1990. *J. Neurosci.* 10: 2039–2058.

[38]Georgopolous, A. P., Kettner, R. E. and Schwartz, A. B. 1988. *J. Neurosci.* 8: 2928–2937.

[39]Fortier, P. A., Kalaska, J. F. and Smith, A. W. 1989. *J. Neurophysiol.* 62: 198–211.

[40]Thach, W. T. 1978. *J. Neurophysiol.* 41: 654–676.

[41]Schreiber, M. H. and Thach, W. T., Jr. 1985. *J. Neurophysiol.* 54: 1228–1270.

[42]Meyer-Lohmann, J., Hore, J. and Brooks, V. B. 1977. *J. Neurophysiol.* 40: 1038–1050.

12

DISCHARGE PATTERN OF A CELL IN THE INTERPOSED NUCLEUS of the cerebellum (Figure 8) during a visually guided ramp movement in an experiment similar to that shown in Figure 11. The monkey was trained to track a cursor on a visual display with the handle position, by flexing or extending the wrist. The cell is bidirectional, increasing its firing rate on both flexion and extension. Discharge is phasic, rapid near the beginning of the movement and declining gradually in frequency as movement is completed. Bidirectionality and firing pattern mimic discharges from group Ia afferent fibers during the same movements. (After Thach, 1978.)

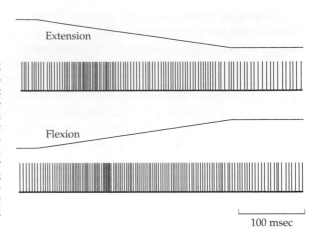

Extension

Flexion

100 msec

nuclei accordingly. During ramp movements many cells in the interposed nucleus discharge with patterns similar to those of Ia afferent fibers (Figure 12). The discharges are bidirectional, with vigorous activity at the beginning of both flexion and extension. Further, tremors during the ramp movements are reflected in discharges of both Ia afferents and cells of the interposed nuclei. It has been suggested from these and other experiments that another function of the interposed nuclei may be to monitor and control, through the γ motor system, the muscle spindle reflex. Such regulation may serve specifically to damp reflex oscillations that might otherwise occur during relatively slow changes in length.

In summary, it appears that the dentate nucleus may be concerned with the organization and execution of planned movements by regulating α and γ motoneurons, and the interposed nuclei, with fine tuning of reflexes and movements, particularly with respect to keeping oscillations under control. The afferent and efferent connections of the fastigial nucleus suggest a role in postural control. Experiments in which synaptic transmission in one or other of the nuclei was blocked either temporarily or permanently by local chemical injections support this view.[29] In the dentate nuclei, such injections introduced a delay in the execution of trained movements around a single joint, and block in the interposed nucleus introduced persistent tremor. In no case, however, did the injections *abolish* the movements. Effects were much more severe on multijoint movements. Block of the fastigial nucleus impaired sitting, standing, and walking; block of the interposed nucleus produced severe tremor during reaching; and block of the dentate nucleus was followed by overreaching targets and incoordination of compound finger movements.

While more recent experiments have provided valuable information about the nature and timing of interactions between the cerebellum and other parts of the motor system, and about the specific roles of the individual cerebellar nuclei in control of movement and posture, the lucid summary of the overall function of the cerebellum given by Adrian almost 50 years ago is still remarkably accurate:[43]

> In spite of its resemblance to the cerebrum, the cerebellum has nothing to do with mental activity. . . . The cerebellum has the more immediate and quite unconscious task of keeping the body balanced whatever the limbs are doing and of insuring that the limbs do whatever is required of them. Its actions show what complex things can be done by the mechanism of the nervous system in carrying out the decisions of the mind. If I decide to raise my arm, a message is dispatched from the motor area of one cerebral hemisphere to the spinal cord and a duplicate of that message goes to the cerebellum. There, as a result of interactions with other sensory impulses, supplementary orders are sent out to the spinal cord so that the right muscles come in at the exact moment when they are needed, both to raise my arm and keep my body from falling over. The cerebellum has access to all the information from the muscle spindles and pressure organs and so can put in the staff work needed to prevent traffic jams and bad coordination. If it is injured the timing breaks down, muscles come in too early or too late and with the wrong force. The staff work needs to be elaborate, particularly when the body has to be balanced on two legs and uses its arms for all manner of movement, but it is done by the machinery of the nervous system after the mind has given its orders. The cerebellum has nothing to do with formulating the general plan of the campaign. Its removal would not affect what we feel or think, apart from the fact that we should be aware that our limbs were not under full control and so should have to plan our activities accordingly.

Single-cell discharges related to movement have also been studied in the basal ganglia. For example, Mink and Thach have examined the behavior of cells in the external and internal globus pallidus of monkeys during a variety of visually guided and self-paced movements.[44,45] The movements consisted of wrist flexion or extension against a load, often involving only a single muscle group. In the visually guided movements, monkeys moved a lever to keep a light spot in register with a moving cursor on a display screen. Of particular interest with respect to basal ganglion function were the relations of cell discharges to gradual ramp movements that involved tracking (CLOSED LOOP movements), as opposed to sudden step movements that were preplanned and did not involve visual feedback (OPEN LOOP). The idea that step movements do not involve immediate feedback is supported by other experiments in which sudden removal of the target after a step movement was begun did not prevent termination of the movement at the final target position.

Cellular activity in basal ganglia

Neurons in the internal globus pallidus usually had relatively high tonic rates of discharge, whereas the activity of those in the external globus pallidus could be either high or low; firing rates could either increase or decrease in relation to movement. In neither nucleus did

[43]Adrian, E. D. 1946. *The Physical Background of Perception.* Clarendon Press, Oxford.
[44]Mink, J. W. and Thach, W. T. 1991. *J. Neurophysiol.* 65: 273–300.
[45]Mink, J. W. and Thach, W. T. 1991. *J. Neurophysiol.* 65: 301–329.

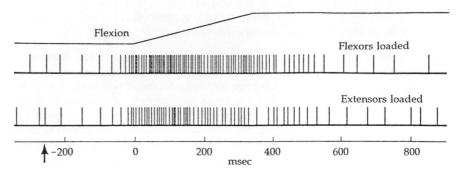

13   **DISCHARGE OF A BASAL GANGLION CELL during a rapid, visually guided step flexion of the wrist. The upper trace indicates handle position; the lower traces show the cell discharge in two separate trials, one with flexors loaded, the other with extensors loaded. The cell discharged with flexion of the wrist, whether or not the movement was against or with the load; that is, discharge of the cell was related to position, not to force of contraction. Time is indicated in relation to beginning of movement, which was about 250 msec after displacement of the cursor (arrow). Movement was complete after about 350 msec. (After Mink and Thach, 1991a.)**

discharges correlate best with initial wrist position, movement velocity, or load. Changes in discharge were related best to visually guided step (open loop) movements (Figure 13). Such changes occurred well after activity in the dentate nucleus and, depending upon load conditions, after the onset of electrical activity in the muscles themselves. Mink and Thach proposed that pallidal discharge is associated with release of holding mechanisms responsible for joint fixation, thereby allowing the programmed movement to occur. The analogy given for the delayed firing was that of starting a car on a hill: The hand-brake is released only after power has been applied to the wheels. Consistent with this idea was the observation that blocking the activity of neurons by injecting muscimol (an agonist of GABA; see Chapter 9) into the internal globus pallidus resulted in increased tone in the wrist due to co-contraction of flexor and extensor muscles.[46] Trained movements were slowed, with no reduction in time to movement onset after displacement of the cursor.

To summarize, some relations have been determined between single-cell discharges in the major motor centers of the brain of primates and the performance of motor tasks. In the simplest experiments, involving movement around a single joint, various neurons in the cortex and cerebellum discharge in relation not only to muscle contraction, but also to more abstract aspects of movement, such as position and direction, and joint fixation. The relative timing of such discharges has provided information about the sequence in which trained movements are processed. Further, detailed analysis of firing patterns in cerebellar neurons, and the effects of block of cerebellar nuclei, have provided

[46]Mink, J. W. and Thach, W. T. 1991. *J. Neurophysiol.* 65: 330–351.

important clues about the role of the cerebellum in motor programming. Finally, the activity of neurons in the globus pallidus suggests that a major function of the basal ganglia is to permit the maintenance of tone when needed to stabilize joint position, and to inhibit the tone during programmed movements.

The precise location of cells used in the actual planning of movements, and the nature of their activity is, at the moment, not accessible to us. However, localized increases in blood flow in the cerebral cortex have been used to identify regions of neuronal activity during movement (Chapter 6).[47] Clearance of radioactive xenon from blood vessels in localized cortical areas was used as a measure of regional blood flow. During the execution of a complex patterned finger movement, blood flow increased bilaterally in the supplementary motor areas and contralateral sensorimotor cortex. When subjects were instructed to think about doing the same movements, but not execute them, blood flow increased in the supplementary motor areas only. Simple movements, such as maintaining isometric contraction of flexors of the finger, resulted in increased blood flow only in the sensorimotor cortex.

We have already alluded to closed loop and open loop motor activity in relation to the presence or absence of sensory feedback in the control of visually guided movements. It is evident that tracking a slowly moving object requires visual feedback, simply because the location of the object must be confirmed repeatedly in order to track it. Apart from visual information, intracortical connections between primary somatosensory and motor cortex,[48,49] and proprioceptive information from the thalamus[50,51] provide ample opportunity for feedback of information about modalities such as touch, joint position, and muscle tension, and for participation of long reflexes through the cortex in control of movement.[52,53]

The role of sensory feedback in movements

On the other hand, in some motor activities sensory feedback is clearly absent. For example, during bird song the movements of the muscles follow each other in a rapid, orderly sequence without sufficient time for a feedback loop to be completed: The next instructions are sent out from the central nervous system before the first movement or sound could be detected by the central nervous system. In fact, most movements, after they are planned, can be completed in the absence of sensory feedback from the limb involved in the task, provided that there are no external perturbations. Both lower primates and humans

[47]Roland, P. E. et al. 1980. *J. Neurophysiol.* 43: 118–136.

[48]Jones, E. G., Coulter, J. D. and Hendry, S. H. C. 1978. *J. Comp. Neurol.* 181: 291–348.

[49]Porter, L. L., Sakamoto, T. and Asanuma, H. 1990. *Exp. Brain Res.* 80: 209–212.

[50]Asanuma, H., Larsen, K. D. and Yumiya, H. 1980. *Exp. Brain Res.* 38: 349–355.

[51]Kosar, E. et al. 1985. *Brain Res.* 345: 68–78.

[52]Favorov, O., Sakamoto, T. and Asanuma, H. 1988. *J. Neurosci.* 8: 3266–3277.

[53]Wannier, T. M. J., Maier, M. A. and Hepp-Reymond, M.-C. 1991. *J. Neurophysiol.* 65: 572–589.

can perform relatively fine, previously learned movements with deaf-ferented limbs.[54] However, in the absence of sensory feedback, the maintenance of fine control tends to deteriorate as the task progresses, possibly because of accumulation of small errors. In humans, for example, writing tends to lose legibility. Deafferentation severely impairs the learning of new movements, as does the removal of somatosensory cortex. After unilateral ablation of areas 1, 2, and 3 in cats, learning an unfamiliar task (retrieval of small pieces of food by lifting them across a gap) was very much prolonged on the side contralateral to the lesion.[55] A similar lesion made after the task had been learned had no such deleterious effect.

With respect to rhythmic activity, such as walking or swimming, both invertebrate and vertebrate nervous systems contain internal circuits that are able to develop programmed sequences of muscular contraction that can unfold and continue in the absence of sensory feedback. For example, after the nervous system of cockroaches has been disconnected from all sensory input, motoneurons continue to generate impulses in patterns that would normally result in walking.[56] Similarly, in isolated spinal cord segments of the lamprey, maintained rhythmic discharges of motoneurons can be elicited that are appropriate for the production of swimming movements in the intact animal.[57] Bursts of activity occur sequentially in the ventral roots, traveling caudally and alternating between sides at rates consistent with the normal swim cycle.

The presence of autonomous internal pattern generators does not mean that during behavior in the intact animal feedback from the periphery is ignored. For example, if a small twig touches the dorsum of the foot of a walking cat during the swing phase (see later), the foot will be lifted elegantly over the twig. Similarly, patterns of respiration are altered by activation of pulmonary stretch receptors.[58]

Two examples—respiration and walking—will be used to illustrate how specific groups of neurons within the mammalian central nervous system produce coordinated movements.

Neural control of respiration

Two antagonistic sets of muscles are responsible for drawing air into the lungs and expelling it. During inspiration, the rib cage is raised by the external intercostal muscles, and the diaphragm contracts. As a result, the volume of the chest is increased, the lungs expand, and air enters. Expiration is not due simply to recoil but is an active process accompanied by contraction of the internal intercostal muscles. Other muscles of the thorax and abdomen also contribute to a variable extent, depending on the posture of the animal and the rate and depth of respiration.[59] An example of the respiratory rhythm is shown in Figure

[54]Marsden, C. D., Rothwell, J. C. and Day, B. L. 1984. *Trends Neurosci.* 7: 253–257.
[55]Sakamoto, T., Arissan, K. and Asanuma, H. 1989. *Brain Res.* 503: 258–264.
[56]Pearson, K. G. and Iles, J. F. 1970. *J. Exp. Biol.* 52: 139–165.
[57]Grillner, S., Wallen, P. and Brodin, L. 1991. *Annu. Rev. Neurosci.* 14: 169–199.
[58]Feldman, J. L. and Grillner, S. 1983. *Physiologist* 26: 310–316.
[59]Da Silva, K. M. C. et al. 1977. *J. Physiol.* 266: 499–521.

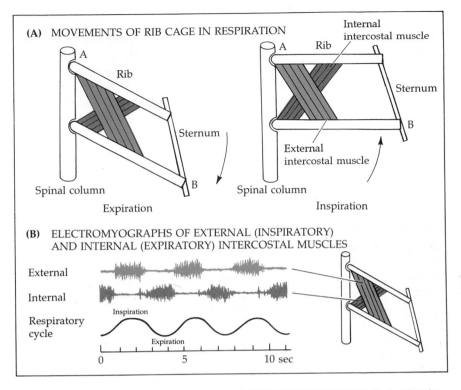

(A) MOVEMENTS OF RIB CAGE IN RESPIRATION

Internal intercostal muscle

Rib

Sternum

Rib

Sternum

A

B

External intercostal muscle

Spinal column

Spinal column

Expiration

Inspiration

(B) ELECTROMYOGRAPHS OF EXTERNAL (INSPIRATORY) AND INTERNAL (EXPIRATORY) INTERCOSTAL MUSCLES

External

Internal

Respiratory cycle

Inspiration

Expiration

0   5   10 sec

14 **MOVEMENTS OF RIB CAGE AND RESPIRATORY MUSCLES during inspiration and expiration. (A) Actions of the external intercostal muscles (raising ribs during inspiration) and internal intercostal muscles (depressing the ribs during expiration). (B) Activity of respiratory muscles in the cat, recorded with needle electrodes. Discharges of the external and internal intercostal muscles are out of phase.**

14. Activity of each muscle can be registered by strain gauges or by recording its electrical activity with wire electrodes embedded in the body of the muscle—electromyography (EMG). Figure 14 shows that inspiratory and expiratory muscle contractions are accompanied by bursts of potentials, indicating motor unit discharges; it is apparent that the two sets of muscles contract out of phase.

As with limb muscles, the stretch reflex contributes to the inspiratory and expiratory movements by maintaining the excitability of the motoneurons. During inspiration and expiration, the internal and external intercostal muscles are stretched alternately; throughout the cycle, as shown earlier, the afferent discharge from muscle spindles is maintained at a high frequency, even when the muscles are actively contracting, owing to the efferent output through the γ system[60,61] (Figure 10 in Chapter 14). Each Ia afferent fiber firing at approximately 100 per second contributes excitatory synaptic potentials to the homonymous motor neurons (i.e., motor neurons supplying the same muscle) and, through

[60]Sears, T. A. 1964. *J. Physiol.* 174: 295–315.
[61]Critchlow, V. and von Euler, C. 1963 *J. Physiol.* 168: 820–847.

neurons (i.e., motor neurons supplying the same muscle) and, through interneurons, inhibition to the motoneurons of antagonist muscles. On the other hand, the diaphragm of the cat has few if any muscle spindles and no true stretch reflex.

How is the respiratory rhythm generated? In cats, after the dorsal roots of the thoracic cord have been severed (a procedure depriving the central nervous system of all sensory feedback from the chest), rhythmic respiration continues. Similarly, if the animal is curarized, a procedure paralyzing the respiratory muscles, a rhythmical motor outflow persists.[58,62] In contrast, section at the level of the medulla abolishes breathing. Within the pons and medulla are situated pools of neurons that fire during inspiration or expiration and that produce excitation and inhibition on the appropriate respiratory motoneurons. For example, during inspiration the motoneurons supplying external intercostal (inspiratory) muscles are depolarized by a barrage of excitatory postsynaptic potentials arising from neurons in higher centers of the medulla or pons, causing bursts of action potentials. The inspiratory phase is terminated by a burst of inhibitory postsynaptic potentials to the inspiratory neurons from other central neurons in closely adjacent regions that are associated with expiration.[63,64]

Examples of such rhythmic activity, recorded from the isolated central nervous system of a neonatal opossum,[65] are shown in Figure 15. The upper trace is an extracellular record from two neurons in the medulla, showing short bursts of impulses with a period of about 2 seconds. In the lower trace, recordings from a thoracic ventral root show corresponding rhythmic discharges of motoneurons supplying respiratory muscles, occurring at the same frequency and with a slight delay. The properties and interconnections of the medullary neurons that generate the respiratory rhythm are not fully understood. One possibility is that central neurons possess inherent rhythmicity (as in heart muscle and certain neurons in invertebrates) and are therefore able to act as pacemakers. Alternatively, reciprocal inhibitory interactions could occur between pools of neurons that turn each other on and off in an alternating rhythm. Both types of mechanism—endogenous pacemaker cells and rhythm-generating circuits—have been observed in invertebrate preparations such as the leech (Chapter 13).

The preceding considerations show that in respiration the role of feedback from the periphery is not to give rise to rhythmicity. Rather, stretch reflexes provide a tonic excitatory input to the motoneurons. Although the average synaptic potential provided by each individual Ia afferent fiber is very small, combined spatial and temporal summation of epsp's from a large number of muscle spindles produces clear effects

[62]Eldridge, F. L. 1977. *Fed. Proc.* 36: 2400–2404.
[63]Sears, T. A. 1964. *J. Physiol.* 175: 404–424.
[64]Hilaire, G. G., Nicholls, J. G. and Sears, T. A. 1983. *J. Physiol.* 342: 527–548.
[65]Nicholls, J. G. et al. 1990. *J. Exp. Biol.* 152: 1–15.

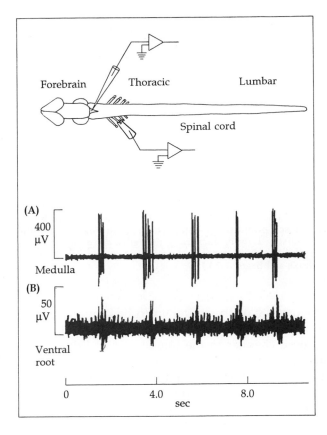

15

RESPIRATORY RHYTHM RECORDED FROM BRAINSTEM NEURONS in an isolated central nervous system from a neonatal opossum. (A) Two brainstem cells discharge regularly with an interburst interval of about 2 seconds. (B) Record from a thoracic ventral root shows corresponding discharge of respiratory motoneurons. (Courtesy of D. J. Zou and J. G. Nicholls, unpublished.)

on motoneurons supplying the respiratory muscles. Figure 16 shows the effect on muscle fiber discharge activity produced by lengthening or shortening an inspiratory muscle (the levator costae) by pulling on its tendon. With each inspiration of the animal, the electromyogram shows a burst of spikes. The activity is enhanced reflexly when the muscle is lengthened and is inhibited when the muscle is allowed to shorten. In addition, stretch reflexes from the lungs as well as from the muscles modulate the rate and depth of breathing; section of the vagus nerve carrying stretch receptor fibers from the lungs leads to prolongation of inspiration.[66]

Of key importance for rhythmicity of breathing is the level of carbon dioxide in the arterial blood.[67] Under conditions of reduced CO_2 concentration, the rate and depth of respiration are reduced; conversely, respiration is increased by raised levels of CO_2. This effect depends upon inputs from chemoreceptors in the carotid arteries and aorta, and from neurons in the medulla that are sensitive to the CO_2 levels. The firing patterns of individual medullary neurons that control expiratory

[66]von Euler, C. 1977. *Fed. Proc.* 36: 2375–2380.
[67]Bainton, C. R., Kirkwood, P. A. and Sears, T. A. 1978. *J. Physiol.* 280: 249–272.

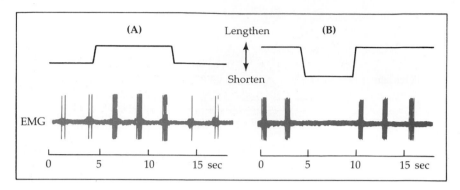

16　**STRETCH REFLEX OF AN INSPIRATORY MUSCLE. During each inspiration the electromyogram (EMG) from a small muscle, the levator costae, shows bursts of action potentials, and the muscle contracts. (A) Lengthening of the muscle by pulling on its tendon increases the burst activity by increasing the reflex drive on the motor unit. (B) Shortening of the muscle, which removes the reflex drive, reduces the EMG activity. (From Hilaire, Nicholls and Sears, 1983.)**

and inspiratory motoneurons have been shown to be influenced critically by the CO_2 concentrations: Changes in the steady level of CO_2 are translated into pronounced changes in the frequencies of firing of the interneurons and, therefore, of the motoneurons.

Locomotion: Walking, trotting, and galloping in cats

A striking feature of locomotion is that in vertebrates there is a consistent, highly stereotyped pattern of limb movements.[68] In a walking cat, the left hindlimb is lifted off the ground first, then the left forelimb, the right hindlimb, and the right forelimb (Figure 17). This sequence provides stabilization by the forelimbs while the hindlimbs propel the animal: The tendency to turn produced by the hindlimb is counteracted by the forelimbs, thereby preventing rotation and enabling movement straight forward. Such a sequence is common to crocodiles, rabbits, cats, and elephants. (Even in invertebrates with six legs, such as cockroaches, a similar sequential pattern is observed.[69]) During locomotion each leg executes an elementary stepping movement that consists of two phases: (1) a *swing* phase during which the leg, having been extended to the rear, is flexed, raised off the ground, swung forward, and extended again to contact the ground; and (2) a *stance* phase during which the leg is in contact with the ground, moving backward in relation to the direction taken by the body.

That the gait of a cat undergoes striking changes as its speed increases is shown in Figure 18. While the cat is walking, a single leg is raised off the ground at any one time. As the speed increases to a trot, two legs are raised off the ground at once—one front and one back on opposite sides of the animal. Still faster, at a gallop, the two front legs and then the two back legs alternate in leaving the ground. The increase in speed is accomplished by shortening the time that each leg stays on

[68]Grillner, S. 1975. *Physiol. Rev.* 55: 274–304.
[69]Pearson, K. 1976. *Sci. Am.* 235: 72–86.

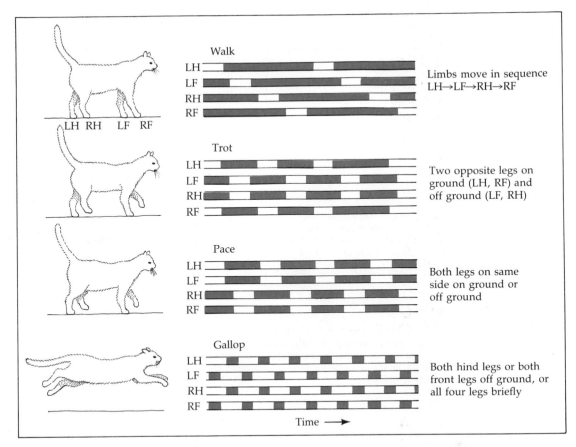

17   **STEPPING PATTERN OF A CAT during walk, trot, pace, and gallop. The white bars show the time that a foot is off the ground, the blue bars on the ground. While the animal walks, the legs are moved in sequence, first on one side, then on the other. In a trot, diagonally opposite limbs are raised together. In a pace the rhythm changes again, the limbs on the same side being raised together. Faster still, the hind limbs, then the front limbs, leave the ground. (After Pearson, 1976.)**

the ground— the stance phase. Thus, as the cat moves faster, each leg is extended for a briefer period before being bent, raised, and moved forward. At all speeds from a slow walk to a gallop, the time spent off the ground by each leg as it swings forward is little altered.

As early as 1911 Graham Brown showed that the elementary circuits required for walking movements in cats appeared to possess semiautomatic properties.[70] The raising and placing of two hind feet in alternation could be achieved in a cat after its spinal cord had been transected. Moreover, other experiments have shown that after certain drugs, such as DOPA (a precursor to biogenic amines; Chapter 9), are given, the hind legs of a spinal cat will pace appropriately on a moving treadmill.[68]

[70]Brown, T. G. 1911. *Proc. R. Soc. Lond. B* 84: 308–319.

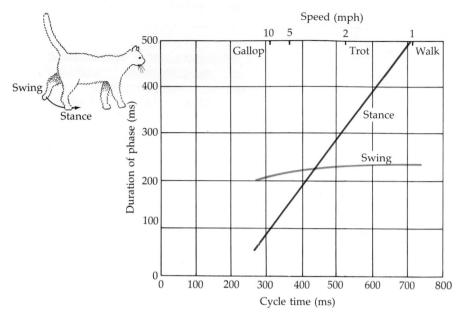

18 **CONSTANCY OF SWING PHASE DURING LOCOMOTION.** As the animal moves more and more rapidly (abscissa), the time spent by each foot on the ground (stance phase) becomes progressively shorter (ordinate). The time that each foot spends in the air (swing phase) is almost the same in a walk and a gallop. (After Pearson, 1976.)

Experiments made in the then U.S.S.R. by Shik, Orlovsky, and Severin[71,72] and in Sweden by Lundberg, Grillner, and their colleagues[68] have provided evidence for the role of central mechanisms in producing coordinated walking movements. When the upper brainstem of a cat is transected, the animal can still stand but does not walk or run spontaneously. In experiments made by Shik and his colleagues, a cat with such a transection was held with its feet touching a moving treadmill. When continuous electrical stimulation at 30 to 60 per second was applied through electrodes placed in the cuneiform nucleus (the MES-ENCEPHALIC LOCOMOTOR REGION), the cat walked (Figure 19). Displacement of the stimulating electrodes by as little as 0.3 mm abolished the walking response. The stance and swing of the forelegs and the electromyograms recorded during walking appeared normal. Stronger stimulation of the mesencephalic locomotor region by larger currents at the same frequency caused the propulsive forces of the leg muscles to be increased. But while the strength of electrical stimulation could influence the force of the walking movements, the frequency of stepping was not altered, provided that the speed of the treadmill remained constant.

In response to acceleration of the treadmill with constant stimulation

[71]Shik, M. L. and Orlovsky, G. N. 1976. *Physiol. Rev.* 56: 465–501.
[72]Shik, M. L., Severin, F. V. and Orlovsky, G. N. 1966. *Biofizika* 11: 756–765.

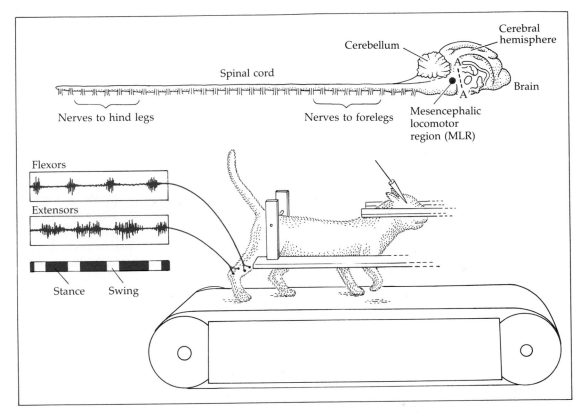

19 **LOCOMOTION BY A CAT ON A TREADMILL after section of the brainstem (A–A' in the diagram). Such an animal does not walk spontaneously. Stimulation of the mesencephalic locomotor region (MLR in diagram) by electrical currents causes the animal to walk on the treadmill. EMGs associated with the locomotor activity can be recorded by electrodes in the limb muscles. The speed of walking or galloping depends on the rate of the treadmill. Increasing the strength or frequency of stimulation increases the strength of limb movements (as though the animal were walking uphill) but not their speed. (After Pearson, 1976.)**

of the mesencephalic locomotor region, the animal changed from walking to trotting, and then to galloping. Sensory feedback plainly controls the rate of stepping with its shortened stance phase. As the leg moving to the rear on the treadmill becomes extended more rapidly it takes less time to reach the point at which afferent signals initiate the swing phase in which the leg is lifted and swung forward. The organization within the central nervous system that controls the switching from walk to trot to gallop is not yet understood. As expected, cutting dorsal roots abolishes the response to different treadmill speeds, but not the walking evoked by electrical stimulation.[73]

The general conclusion from these kinds of experiments is that within the central nervous system there exists a hierarchically ordered

[73]Grillner, S. and Zanger, P. 1974. *Acta Physiol. Scand.* 91: 38A–39A.

series of connections that can initiate and control a programmed series of movements. Within the spinal cord, interconnections between appropriate pools of motor neurons are acted upon by descending influences. In the absence of phasic sensory input from the limbs, each limb can be programmed by the central nervous system to be raised, swung forward, and lowered by contraction of the appropriate muscle groups acting on the joints. The role of feedback is to modulate these centrally initiated responses in accord with different needs or loads imposed by the external world.

From these considerations, certain striking parallels emerge in the automaticity of breathing and of walking. For both types of movement, a central program orders the contractions of appropriate groups of muscles in a predestined sequence. The drive for the alternation of leg movements or of inspiration and expiration depends on a descending stimulus—for respiration the influence of carbon dioxide on the medullary neurons and for walking the activity of neurons in the mesencephalic locomotor region. In both cases sensory feedback modulates the frequency and extent of the rhythmic activity to be consistent with the moment-to-moment requirements of the animal.

SUGGESTED READINGS

General

Asanuma, H. 1989. *The Motor Cortex*. Raven, New York.

Bernardi, G., Carpenter, M., DiChiaria, G., Morelli, M. and Stanzioni, P. (eds.). 1991. *The Basal Ganglia. III. Proceedings of the Third Triennial Symposium of the International Basal Ganglion Society*. Plenum, New York.

Binder, M. D. and Mendell, L. M. (eds.). 1990. *The Segmental Motor System*. Oxford University Press, New York.

Ito, M. 1984. *The Cerebellum and Neural Control*. Raven, New York.

Thach, W. T., Goodkin, H. G. and Keating, J. G. 1992. Cerebellum and the adaptive coordination of movement. *Annu. Rev. Neurosci.* 15: 403–442.

Original Papers

Evarts, E. V. 1968. Relation of pyramidal tract activity to force exerted during voluntary movement. *J. Neurophysiol.* 31: 14–27.

Henneman, E. Somjen, G. and Carpenter, D. O. 1965. Functional significance of cell size in spinal motoneurons. *J. Neurophysiol.* 28: 560–580.

Meyer-Lohmann, J., Hore, J. and Brooks, V. B. 1977. Cerebellar participation in generation of prompt arm movements. *J. Neurophysiol.* 40: 1038–1050.

Mink, J. W. and Thach, W. T. 1991. Basal ganglia motor control. I. Nonexclusive relation of pallidal discharge to five moment modes. *J. Neurophysiol.* 65: 273–300.

Thach, W. T. 1978. Correlation of neural discharge with pattern and force of muscular activity, joint position, and direction of next intended movement in motor cortex and cerebellum. *J. Neurophysiol.* 41: 654–676.

PART FIVE

The Visual System

CHAPTER SIXTEEN

RETINA AND LATERAL GENICULATE NUCLEUS

The image of the outside world that falls on the retina provides the eye with raw information that sets in motion signals leading to perception. The sequence of steps by which the neuronal signals are evoked, transmitted, and combined to produce a scene with objects and background, light and shade, and color can now be followed step by step through the retina and the first relay station on the way to the cortex, the lateral geniculate nucleus. In both structures the nerve cells are arranged in layers; individual types of neurons can be identified by their anatomical, physiological, and molecular characteristics.

The neuronal responses to light start at receptors known as rods and cones. Already at this earliest stage correlations between molecular mechanisms and perception are evident. Rods contain molecules of the visual pigment rhodopsin that absorb light most effectively in the blue-green region of the spectrum. Absorption of light leads to a conformational change in rhodopsin and the activation of a G protein. As a result, a cascade of reactions is initiated, cyclic guanosine monophosphate (cGMP) is hydrolyzed, sodium channels in the rod membrane close, and the membrane potential increases in magnitude. This hyperpolarization reduces ongoing transmitter release by rods onto postsynaptic cells. Excellent agreement has been found between measurements of the absorption of light by rhodopsin and psychophysical measurements of visual experience in dim light. The sensitivity is such that a few single quanta of light trapped by rods can produce signals that reach consciousness. In bright light the photopigment of rods is unresponsive to illumination. Color and daylight vision depend on the cones. Three types of cones, known as red, green, and blue, contain pigments that differ in the molecular composition of the protein. Each of the three pigments absorbs best in one region of the spectrum. As with rods, light that is absorbed leads to a hyperpolarizing signal and reduction of transmitter release. Together, the properties of the three pigments in the three types of cones can account quantitatively and qualitatively for many aspects of color perception. Detailed information at the molecular level is now available to explain color-blindness.

The transformation or integration that occurs at successive levels in the visual system is best analyzed in terms of the *receptive fields* of individual neurons. This term refers to the restricted area of the retinal surface that influences, upon illumination, the signaling of a neuron in the visual system. The influence may be excitatory or inhibitory. Receptive fields are the building blocks for the synthesis and perception of the complex visual world.

The output of the eye is produced by ganglion cells, each of which sends an axon along the optic nerve to the lateral geniculate nucleus. The receptive field of each ganglion cell is a small circular area on the retina. The best responses are evoked in *"on" cells* by small spots of light shone onto the center of the field surrounded by darkness, or in *"off" cells* by small dark spots surrounded by light. Certain other ganglion cells respond best to moving spots or edges and brisk changes in intensity; others to spots of one color in the center of a larger background field of contrasting color. These cells demonstrate a general rule in vision: The system is designed to respond to differences in intensity or wavelength (that is, contrast) rather than absolute intensities or colors of light. All ganglion cells share the common property of responding best to small spots and relatively poorly to diffuse illumination.

The connections between receptors and ganglion cells involve bipolar, horizontal, and amacrine cells. Like rods and cones, bipolar and horizontal cells produce graded local potentials, not action potentials. Detailed information is available about synaptic mechanisms, transmitters, and patterns of connections used by these cells to synthesize the receptive fields of ganglion cells.

In the lateral geniculate nucleus, neurons are arranged in well-defined layers. Each layer is supplied by ganglion cell axons from one eye only, forming a map of that retina. The spatial maps in the various layers are in precise register. The receptive fields of lateral geniculate nucleus cells resemble those of ganglion cells in the retina, being concentric with "on-" or "off-" centers. Lateral geniculate cells respond best to small light or dark spots and poorly or not at all to diffuse illumination. Cells having similar characteristics are grouped together in each layer. Thus, in the monkey, four layers of smaller cells, the parvocellular division, receive inputs from ganglion cells with fine spatial discrimination that are sensitive to color and relatively insensitive to small changes in contrast. The two deeper layers, the magnocellular division, contain larger cells that are not color-sensitive but respond well to moving stimuli and to small changes in contrast. The magnocellular and parvocellular divisions of the lateral geniculate nucleus send their outputs to different groups of cortical cells; the two modalities maintain their identities as distinct pathways through successive levels of the brain.

To convey the relevance of a molecular and cellular approach to higher functions, this chapter deals principally with the performance of nerve cells at initial stages or relays of the visual system. A comprehensive critical treatment of the retina and lateral geniculate nucleus is not possible within the scope of this book; the past few years have produced an overwhelming body of work on the structure and function of the visual system. Psychophysics, color vision, dark adaptation, retinal pigments, transduction, transmitters, and the organization of the retina could each form the basis of a self-contained monograph (see references at the end of the chapter). The same applies to comparative aspects of the visual system in invertebrates (such as the horseshoe crab *Limulus*, or the barnacle) and in lower vertebrates (fish, frog, mud puppy, and turtle) as well as in mammals (rabbits and squirrels). Instead, we describe selected experiments made in the cat and the monkey that provide a clear, continuous thread extending from molecular mechanisms and signaling to perception. Such information is a prerequisite for the analysis of structure and function in the cerebral cortex discussed in Chapter 17.

The eye acts as a self-contained outpost of the brain. It collects information, analyzes it, and hands it on for further processing by the brain through a well-defined pathway, the optic nerve. The first step is the formation on each retina of a sharp image of the outside world. Essential for clear vision are correct focus of the image by accommodation of the lens, which changes its thickness; regulation of light entering the eye by adjustment of the pupil; and convergence of the two eyes to ensure that appropriate images fall on corresponding points of both retinas. Our acuity of vision depends critically upon the region of the visual field that is being inspected. We can read small print only at the center of gaze, not in the peripheral field of vision. This loss of acuity arises from the way in which visual information is processed by the retina; it is not the result of blurred images outside the central region. Before dealing with such problems it is necessary to describe the principal anatomical features of the visual pathway and then the stepwise transformation as light becomes trapped by visual pigments to generate electrical currents.

The eye

The pathways from the eye to the cerebral cortex are illustrated in Figure 1, which depicts some of the major landmarks of the human brain that are useful in the context of the following discussion. The optic nerve fibers arise from ganglion cells in the retina and end on layers of cells in a relay station (the lateral geniculate nucleus), whose axons in turn project through the optic radiation to the cerebral cortex. From here on the progression becomes ever more complex, with no end point in sight.

Anatomical pathways in the visual system

Figure 1A shows how the output from each retina divides in two at the optic chiasm to supply the lateral geniculate nucleus and cortex on both sides of the animal. As a result, the right side of each retina projects to the right cerebral hemisphere. Figure 1 also shows that the right side of each retina receives the image of the visual world on the left side of the animal. Each cerebral hemisphere, therefore, "sees" the visual field of the opposite side of the world. Accordingly, people with damage

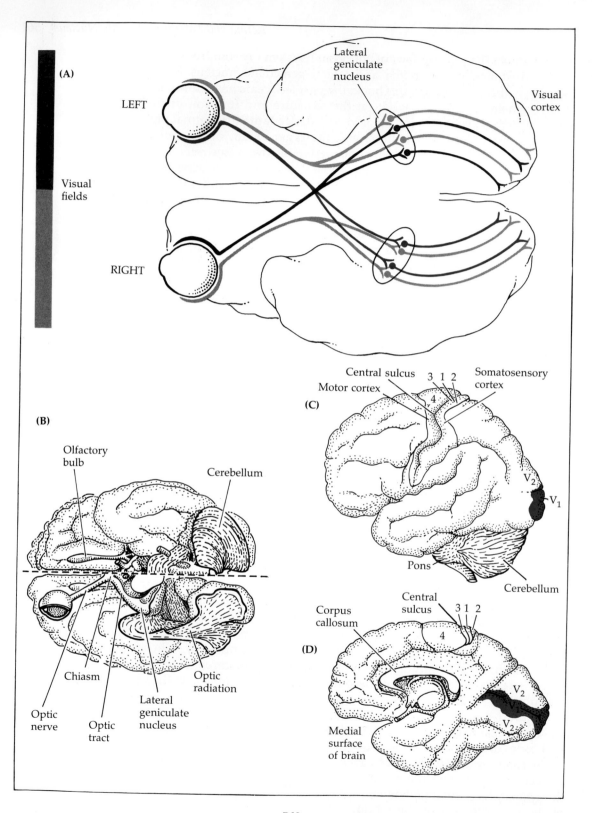

(A)

Visual fields

LEFT

RIGHT

Lateral geniculate nucleus

Visual cortex

(B)

Olfactory bulb

Cerebellum

Chiasm

Optic radiation

Optic nerve

Optic tract

Lateral geniculate nucleus

(C)

Central sulcus

Motor cortex

3 1 2

Somatosensory cortex

4

V_2

V_1

Pons

Cerebellum

(D)

Corpus callosum

Central sulcus

3 1 2

4

V_2

V_1

V_2

Medial surface of brain

◄ **1** VISUAL PATHWAYS. (A) Outline of the visual pathways seen from below (base of the brain) in primates. The right side of each retina, shown in blue, projects to the right lateral geniculate nucleus, and the right visual cortex receives information exclusively from the left half of the visual field. Note that inputs from each eye end in separate layers of the lateral geniculate nucleus. (B) Visual pathways in a partially dissected human brain seen from below. (C,D) Lateral and medial views of the cortical surface. Visual area 1 (V_1) is also known as the striate cortex, or area 17; V_2 corresponds to area 18. Areas V_3, V_4, and V_5 (not shown) are described in Chapter 17. (C) and (D) also show area 4 (the motor cortex) and areas 3, 1, and 2 of the sensory cortex.

to the right side of the cortex caused by trauma or disease become blind to the left side of the world, and vice versa.

Other pathways that branch off to the midbrain are not described here. In higher vertebrates they are primarily concerned with regulating eye movements and pupillary responses and are not directly relevant for pattern recognition.

By examining the cellular anatomy of the various structures in the visual pathway, one can exclude the possibility that information is handed on unchanged from level to level. The neurons *converge* and *diverge* extensively at every stage; that is, each cell receives and makes connections with a number of other cells. For example, the human eye contains over 100 million primary receptors, the rods and cones, but only about 1 million optic nerve fibers are sent by ganglion cells into the brain. In the monkey and the cat the same principle holds: a step-down in neuronal numbers from receptors to ganglion cells. Therefore, within the eye as a whole there occurs a funneling of information. An individual neuron that receives impulses from several incoming nerve fibers cannot reflect separately the signals of any one of them. Instead, converging impulses of different origin are integrated at each stage into an entirely new message that takes account of all the inputs.

THE RETINA

What makes the retina so especially inviting for physiological research is the neat layering and orderly repetition of the relatively few cell types—there are only five main classes. The arrangement and typical positions of various cells are illustrated in Figure 2B, which shows a cross section of a human retina.[1] On the deep surface, farthest from the incoming light, lie the RODS and CONES, which are concerned with night and daytime vision, respectively. They are connected to the BI-POLAR CELLS, which in turn connect to the GANGLION CELLS and so to the optic nerve fibers. Apart from this through-line, there are other cells that make predominantly lateral, or side-to-side, connections. These are the HORIZONTAL CELLS and the AMACRINE CELLS. Only ama-crine and ganglion cells give propagated action potentials; photorecep-tors, horizontal, and bipolar cells give local graded signals. Within each

[1]Boycott, B. B. and Dowling, J. E. 1969. *Philos. Trans. R. Soc. Lond.* B 255: 109–184.

2

PATHWAYS FOR LIGHT AND ARRANGE-
MENT OF CELLS in the retina. (A) A cross section through the eye. Light must pass through
the lens and the layers of cells in order to reach
the rod and cone photoreceptors. The fovea is a
specialized area, containing only cones, that is
used for fine discrimination. At this point the
superficial layers of cells spread apart, permitting
more direct access of light to the photoreceptors.
(B) Section through a human retina showing the
five principal cell types arranged in layers. Light
enters the retina and reaches the rods and cones,
where it is absorbed, starting excitation in the
outer segments. Synaptic connections are made
in the outer and the inner plexiform layers. (From
Boycott and Dowling, 1969.)

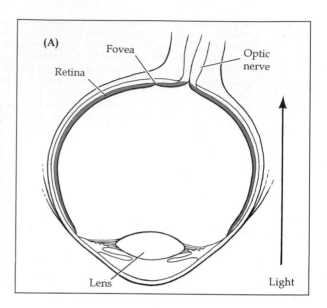

of these major classes, there are subgroups exhibiting important differences in structure and function; the role played by these cells and their
interconnections will be discussed later. We shall see that the aggregation of comparable cells into layers and well-defined groups is a feature
of the central nervous system and is particularly apparent in the lateral
geniculate nucleus and the visual cortex.

Photoreceptors   The rods and cones set the stage for vision and define the limits for
how the outside world can be perceived. Unlike us, snakes have specialized receptors in their eyes to detect radiation in the infrared, and
ants can use blue sky to navigate by polarized light. Since cats lack
appropriate receptors, they are color-blind—as we are at night (when
all cats are gray). The sensitivity of our rods in darkness is such that a
single quantum of light can give rise to a measurable signal; and only
seven or so rods need to be activated by single quanta for a conscious
sensation.[2] Yet with different sets of cone photoreceptors, we can still
detect subtle tints and differences in contrast or color on a bright day
when the light intensity is eight log units greater. Mechanisms of transduction and signaling are now better understood in rods and cones
than in other sensory end organs (Chapter 14). Much information is
available about the molecular and biochemical events that occur when
photons are trapped and ion channels are closed to generate electrical
signals in photoreceptors. What is particularly appealing is that molecular, biophysical, and psychophysical approaches have been combined
to show how the properties of visual pigments and photoreceptors can
account for perceptual phenomena.

Morphology and   The rods and cones constitute a densely packed array of photode-
arrangement of   tectors in the layer of retina adjacent to pigment epithelium, farthest
photoreceptors   from the cornea and the incoming light. Light must traverse dense

[2]Hecht, S., Shlaer, S. and Pirenne, M. H. 1942. *J. Gen. Physiol.* 25: 819–840.

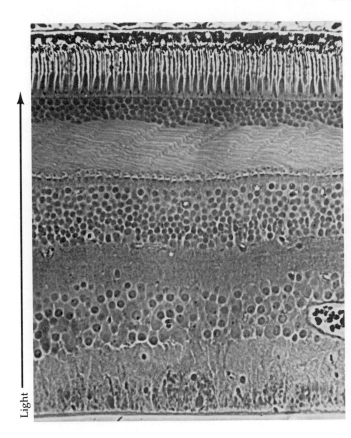

(B)

Photoreceptor cells:
Rods and cones

Outer plexiform layer

Horizontal, bipolar,
and amacrine cells

Inner plexiform layer

Ganglion cells

Optic nerve fibers

Light

layers of cells and fibers before reaching the outer segments of rods and cones where photons are absorbed. The arrangement in the retina is not uniform. As Helmholtz described it in 1867:[3]

> There is in the retina a remarkable spot which is placed near its center . . . and which . . . is called the fovea or pit. . . . (It) is of great importance for vision since it is the spot where the most exact discrimination is made. The cones are here packed most closely together and receive light which has not been impeded by other semi-transparent parts of the retina. We may assume that a single nervous . . . connection . . . runs from each of these cones through the trunk of the optic nerve to the brain . . . and there produces its special impression so that the excitation of each individual cone will produce a distinct and separate effect upon the sense.

It is worth remembering that Helmoltz formulated these concepts before the word *synapse* or even the cell doctrine existed.

Figure 3 shows three principal regions of the photoreceptors: (1) An outer segment at which light is trapped by visual pigment; (2) an inner segment containing the nucleus, ribosomes, mitochondria and endo-

[3]Helmholtz, H. 1962/1927. *Helmholtz's Treatise on Physiological Optics.* J. P. C. Southhall (ed). Dover, New York.

plasmic reticulum; and (3) the synaptic terminal, which releases transmitter onto second-order cells and which also receives synaptic inputs.

The visual pigments described below are concentrated in membranes of the outer segments. Rods contain approximately 10^8 pigment mole-

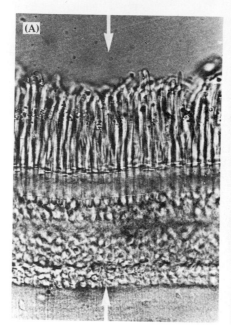

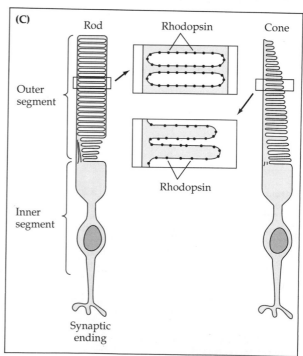

3

PHOTORECEPTORS IN RETINA. (A,B) Rod in the retina of toad injected with a fluorescent dye, Lucifer Yellow, as seen in normal and ultraviolet light. Arrows mark identical points on retina. **(C)** Diagram of a rod and a cone. In the rod, the pigment rhodopsin (black dots) is embedded in membranes arranged in the form of disks, not continuous with the outer membrane of the cell. In the cone, the pigment molecules are on infolded membranes that are continuous with the surface membrane. The outer segment is connected to the inner segment by a narrow stalk. The synaptic endings continually release transmitter in the dark. (A and B courtesy of the late B. Nunn, unpublished. C after Baylor, 1987.)

cules. They are aggregated on several hundred discrete disks (approximately 750 in a monkey rod) that do not make contact with the outer membrane. By contrast, the disklike infoldings in cones are continuous with the cell membrane. The visual pigment is so densely packed on outer segment membranes that the distance between two visual pigment molecules in a rod is about 20 nm;[4] it makes up about 80 percent of the total protein. This dense packing of sensitive molecules in serial layers of membranes traversed by light enhances the probability that a photon will be trapped on its way through the outer segment. The question arises: How are electrical signals generated when light is absorbed by visual pigments?

The events that occur when light is absorbed by the rod pigment rhodopsin have been studied over the years by psychophysical, biochemical, physiological, and most recently molecular techniques. Rhodopsin molecules consist of two moieties: a protein known as opsin, combined with a chromophore, 11 *cis* vitamin A aldehyde, known as RETINAL (Figure 4). (A chromophore is a chemical group producing color in a compound.) The absorption characteristics of the visual pigments have been quantitatively measured by spectrophotometry.[5,6] When different wavelengths of light are shone through a rhodopsin solution, blue-green light at wavelengths of about 500 nm is most effectively absorbed. The absorption spectrum is similar when small spots of light of different wavelengths are shone onto single rods under the microscope. An elegant correspondence has been demonstrated between the absorption characteristics of rhodopsin and our perception

Visual pigments

[4]Dowling, J. E. 1987. *The Retina*. Harvard University Press, Cambridge, MA.
[5]Brown, P. K. and Wald, G. 1963. *Nature* 200: 37–43.
[6]Marks, W. B., Dobelle, W. H. and MacNichol, E. F. 1964. *Science* 143: 1181–1183.

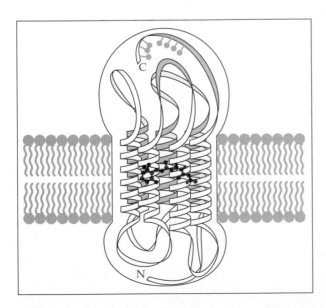

4

STRUCTURE OF VERTEBRATE RHODOPSIN in the membrane. The helix is partly opened to show the position of retinal (color). (After Stryer and Bourne, 1986.)

in dim light. Quantitative psychophysical measurements made in human subjects show that blue-greenish lights at about 500 nm are optimal for perception of a weak stimulus in the dark. In daylight, when rods are inactive and cones are used, we are most sensitive to redder lights, at longer wavelengths in the spectrum, corresponding to the absorption spectra of cones in the fovea (see below).[4]

Once a photon has been absorbed by rhodopsin, the retinal chromophore undergoes photoisomerization and changes from the 11-*cis* to the all-*trans* configuration as the terminal chain connected to opsin rotates. This transition is extremely rapid; it takes place in approximately 10^{-12} seconds. The protein then undergoes a series of transformations through various intermediaries.[7] One of these, metarhodopsin II, is of crucial importance for transduction (see below). Figure 5 shows the sequence of biochemical changes occurring in bleaching and in regeneration of active rhodopsin. Metarhodopsin II appears after about 1 msec. Once rhodopsin has been bleached and transformed to all-*trans* retinal and opsin, it no longer responds to light and loses its color.

[7]Matthews, R. G. et al. 1963. *J. Gen. Physiol.* 47: 215–240.

Wilhelm Kühne (1837–1900). Under the portrait on the left is shown the view from his room that was presented to the retina of a rabbit, causing bleaching and a clearly discernible image of the window arrangement. Kühne isolated visual purple (shown on the right) for the first time. (Photomontage courtesy of Dr. Rolf Boch.)

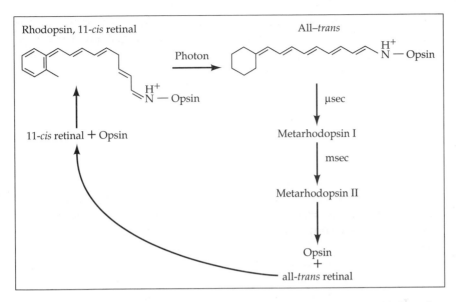

Rhodopsin, 11-*cis* retinal

All–*trans*

Photon

11-*cis* retinal + Opsin

μsec

Metarhodopsin I

msec

Metarhodopsin II

Opsin
+
all-*trans* retinal

5   **BLEACHING OF RHODOPSIN BY LIGHT. In the dark, 11-*cis* retinal is bound to the protein opsin. Capture of a photon causes the transformation of the 11-*cis* retinal to all-*trans* retinal. This in turn is rapidly converted to metarhodopsin II, then to opsin and all-*trans* retinal, which become regenerated to rhodopsin. Metarhodopsin II is the trigger that sets in motion activation of the second-messenger system. (After Dowling, 1987.)**

Regeneration of pigments is slow, taking many minutes. The stability of rhodopsin in the dark is extraordinary. Baylor has calculated that spontaneous thermal isomerization of a single rhodopsin molecule occurs about once every 3000 years, or 10^{23} times more slowly than photoisomerization.[8]

At the molecular level, the opsin protein of rhodopsin has been sequenced and the gene has been cloned.[9] Rhodopsin consists of 348 amino acid residues with seven hydrophobic regions, each containing 20 to 25 amino acids. The putative structure involves seven transmembrane helices. The amino terminus is located in the extracellular space (i.e., within the disk in rods) and the carboxy terminus in the cytoplasm. The lysine site at which retinal is attached to the opsin is in the seventh transmembrane-spanning segment. Nathans has cloned the genes for the red, green, and blue pigments of cones responsible for color vision and has shown them to be similar to the gene for rhodopsin.[10,11] Two other biologically important molecules discussed elsewhere bear a high resemblance to visual pigments in their sequences and structure. Both the β_1-adrenergic receptor and the muscarinic acetylcholine receptor have similar transmembrane-spanning regions with highly conserved

[8]Baylor, D. A. 1987. *Invest. Ophthalmol. Vis. Sci.* 28: 34–49.
[9]Nathans, J. and Hogness, D. S. 1984. *Proc. Natl. Acad. Sci. USA* 81: 4851–4855.
[10]Nathans, J. 1987. *Annu. Rev. Neurosci.* 10: 163–194.
[11]Nathans, J. 1989. *Sci. Am.* 260(2): 42–49.

structures. Both these receptors bind with transmitters and exert their effects through second messengers by activating G proteins (see Chapters 2 and 8).

Transduction   How does photoisomerization of rhodopsin give rise to a change in membrane potential? For many years it was clear that some sort of internal transmitter was required for the generation of electrical signals in rods and cones. One reason was the enormous amplification. In 1970 Baylor and Fuortes,[12] working on turtle photoreceptors, showed that a membrane conductance change of 0.1 percent was produced when a single photon was absorbed and activated one pigment molecule out of about 10^8. Information about the capture of photons in a rod must be conveyed by an intracellular messenger from the disk across the cytoplasm to the outer membrane. Calcium, which had originally been postulated as the coupling agent, has now been ruled out by decisive experiments.[13]

The sequence of events through which pigment molecules activate a cascade of steps to influence channels has been elucidated. By patch clamp recordings from rod and cone outer segments and by molecular

[12]Baylor, D. A. and Fuortes, M. G. F. 1970. *J. Physiol.* 207: 77–92.
[13]Yau, K. W. and Baylor, D. A. 1989. *Annu. Rev. Neurosci.* 12: 289–327.

6

DARK CURRENT IN A ROD. (A) In darkness, sodium ions flow through specialized channels of the rod outer segment, causing a depolarization. The current loop is completed through the neck of the rod, with the outward movement of potassium through the inner segment membrane. When the outer segment is illuminated, as in (B), the channels close (due to the decrease in intracellular cyclic GMP) and the rod then becomes hyperpolarized. This hyperpolarization prevents transmitter release. Sodium and potassium concentrations of the rod are maintained by the sodium–potassium pump in the inner segment (colored circle). In the dark, calcium ions also enter the rod through the cation channels. (After Baylor, 1987.)

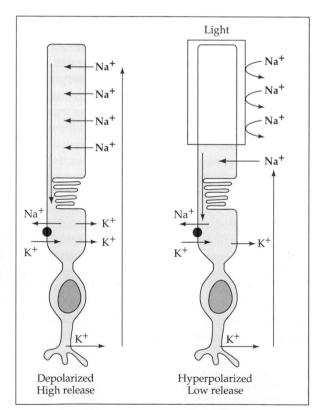

Depolarized
High release

Hyperpolarized
Low release

techniques, Fesenko, Yau, Baylor, Stryer, and their colleagues[14-16] have shown how cyclic GMP acts as an internal transmitter from disk to surface membrane and fulfills the requirements of appropriate kinetics and great amplification. The scheme is shown in Figure 6. In darkness rods and cones have membrane potentials of approximately −40 mV, far from the potassium equilibrium potential E_K (−80 mV). This is because a continuous "dark" current flows into the outer segment as sodium ions move down their electrochemical gradient through the open channels. Light causes a hyperpolarization; this is brought about by closure of cation channels in the cell membrane of the outer segment. (The properties of these channels and the electrical signals are described below.) Cytoplasmic cyclic GMP keeps these channels in the open state. In the absence of cGMP the channels close and the electrical resistance of rod outer segment membrane increases and approaches that of a lipid bilayer. The role of cGMP has been demonstrated directly by measuring currents in patches of rod outer segment membrane. Application of cGMP to the cytoplasmic surface increased the sodium current, whereas removal of cGMP abolished it (Figure 7). Increased internal calcium concentration, on the other hand, did not open these sodium channels nor close them when they were open.

How do these sodium channels in the outer segment differ from those described earlier in conventional neurons? Baylor, Yau, and their colleagues have defined the following properties:[8,13]

1. They are ligand- but not voltage-activated; it is cGMP that opens the pore.
2. They do not inactivate with steady depolarization or with maintained application of cGMP.
3. They are not blocked by tetrodotoxin.
4. Their selectivity is low: Ions such as potassium can pass through the channels almost as readily as sodium (the permeability ratio of potassium to sodium is 0.7). Moreover, the permeability for calcium is even higher than for sodium (12.5:1). Nevertheless, with normal concentrations of ions in the extracellular fluid, the sodium conductance predominates so that the bulk of the inward current is carried by sodium.
5. The single-channel conductance of approximately 0.1 pS is increased to approximately 20 pS if divalent cations are removed.

A cDNA for the rod channel has been isolated and the complete amino acid sequence deduced. The protein contains several transmembrane segments and a region similar to the cGMP-binding domains found in cGMP protein kinase.[17] After mRNA injection into oocytes, the channels that appear in the membrane resemble those found in photoreceptors: They are cGMP-activated; they have a conductance of approximately 20 pS in the absence of divalent cations; and their cation

[14]Fesenko, E. E., Kolesnikov, S. S. and Lyubarsky, A. L. 1985. *Nature* 313: 310–313.
[15]Yau, K. W. and Nakatani, K. 1985. *Nature* 317: 252–255.
[16]Stryer, L. and Bourne, H. R. 1986. *Annu. Rev. Cell Biol.* 2: 391–419.
[17]Kaupp, U. B. et al. 1989. *Nature* 342: 762–766.

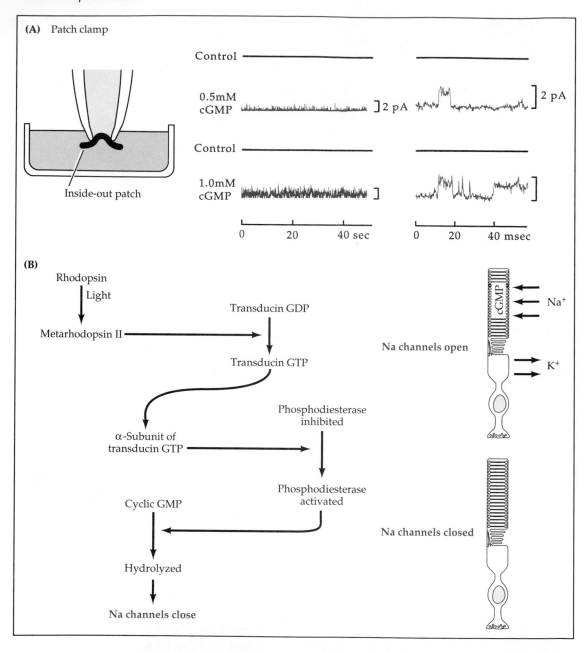

7   ROLE OF CYCLIC GMP IN OPENING SODIUM CHANNELS in rod outer segment membranes. **(A)** Single-channel recordings made from inside-out patches bathed in various concentrations of cyclic GMP. Channel openings cause deflections in the upward direction. The frequency of channel opening is extremely low in control recordings. Addition of cGMP causes single-channel openings, the frequency of which increases with increased concentrations. **(B)** Coupling of photopigment activation to G-protein activation. The G protein transducin binds GTP in the presence of metarhodopsin II, leading to activation of phosphodiesterase, which in turn hydrolyzes cGMP. With the reduced concentration of cGMP, sodium channels close. (After Baylor, 1987.)

selectivity is also similar to that of the photoreceptor sodium dark-current channels.

The decrease in internal cyclic GMP concentration triggered by light is brought about by metarhodopsin II, an intermediary in the bleaching process. It is enzymatically active and acts on a membrane G protein.[18] This protein, known as TRANSDUCIN, consists of three polypeptide chains, α, β, and γ. Through metarhodopsin II, the GDP bound to the α subunit of transducin is exchanged for GTP (see Chapter 8). The activated α subunit becomes separated from the β and γ subunits and in turn activates a phosphodiesterase, the enzyme that hydrolyzes cGMP. The cGMP concentration falls, sodium channels close, and the rods hyperpolarize. The cascade is terminated by phosphorylation of the carboxy-terminal region of the active metarhodopsin II. The key role of cyclic GMP in controlling the channels during the response is supported by the biochemical observation that internal cGMP concentration decreases by about 20 percent upon illumination.[8,13]

The cyclic GMP cascade provides great amplification of responses to light. A single molecule of active metarhodopsin II can catalyze the exchange of many molecules of GDP for GTP and thereby liberate hundreds of G-protein α subunits. Each of these can activate phosphodiesterase molecules that can hydrolyze large numbers of cGMP molecules and close many channels. Calcium, which had originally been proposed as the intracellular signal for transduction, plays a key role in influencing the sensitivity of rods to light as well as dark adaptation. Intracellular calcium ions inhibit guanylate cyclase, the enzyme responsible for the synthesis of cGMP. Illumination of the outer segment and closure of the cation channels causes the intracellular calcium concentration to drop. As a result, the activity of guanylate cyclase increases, allowing channels to reopen and current to flow.[19]

As shown in Chapter 14, sensory receptors of various modalities respond to appropriate stimuli by graded local depolarizations. In "long" receptors, action potentials are initiated, which in turn cause voltage-activated calcium channels in the presynaptic terminals to open; calcium ions enter and trigger the release of transmitter. Thus activation of a conventional sensory receptor causes transmitter liberation that excites or inhibits the next cell in line. Invertebrate photoreceptors have long been known to conform to this scheme. Figure 8 shows the depolarizing receptor potential and impulses in a photoreceptor cell in the eye of the horseshoe crab *Limulus*.[20] The very different hyperpolarizing responses to light recorded with an intracellular microelectrode in a vertebrate photoreceptor are shown in Figure 8B.[21] "At rest" in the dark, with a continuous inward current flowing through the outer segment, the synaptic ending is depolarized and continually releases

Electrical responses to light

[18]Stryer, L. 1987. *Sci. Am.* 257(7): 42–50.
[19]Tamura, T., Nakatani, K. and Yau, K. W. 1991. *J. Gen. Physiol.* 98: 95–130.
[20]Fuortes, M. G. F. and Poggio, G. F. 1963. *J. Gen. Physiol.* 46: 435–452.
[21]Baylor, D. A., Fuortes, M. G. F. and O'Bryan, P. M. 1971. *J. Physiol.* 214: 265–294.

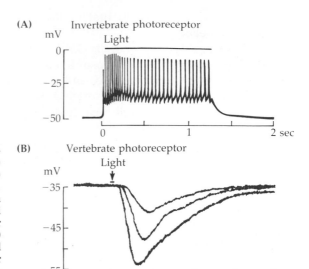

(A)

mV   Invertebrate photoreceptor

Light

0

−25

−50

0   1   2 sec

(B)   Vertebrate photoreceptor

mV   Light

−35

−45

−55

0   0.2   0. 4 sec

8

**RESPONSES OF PHOTORECEPTORS. (A) Pho-
toreceptors of an invertebrate (a horseshoe crab)
respond to light (indicated by bar above record)
with a depolarization that gives rise to impulses.
This is the usual type of response elicited from
sensory receptors activated by various stimuli
such as touch, pressure, or stretch (see Chapter
14). (B) Photoreceptors of a vertebrate (a turtle)
respond by a hyperpolarization that is graded
according to the intensity of the flash. (A after
Fuortes and Poggio, 1963; B after Baylor, Fuortes
and O'Bryan, 1971.)**

transmitter. Light, by causing a hyperpolarization of the photoreceptor,
turns off the release onto the second-order cells.

Quantal responses   Records such as those shown in Figure 8 were produced by bright
flashes of light. The finding that a single quantum of light can give rise
to a conscious sensation raises a number of tantalizing problems. What
is the unitary response produced by a single quantum of light in a
vertebrate photoreceptor and how likely is it to occur? To approach such
problems, Baylor and his colleagues have made recordings using the
experimental arrangement shown in Figure 9 for directly measuring the
currents generated by an individual rod in a toad, a monkey, or a
human.[22] A piece of retina is isolated from the animal or from a cadaver
and maintained in a chamber. Part of the outer segment of the rod is
sucked into a fine pipette (a snug fit is necessary). At rest in the dark,
as expected, a "dark" current flows continuously into the outer segment. Flashes of light close channels in the outer segment, causing a
decrease in this steady dark current. Typical responses to light flashes
are shown in Figure 10. Figure 10A shows the results of very dim
flashes, corresponding to 1 or 2 quanta of light. The currents are small
and quantal in nature. That is, sometimes a dim flash evokes a unitary
response, sometimes a doublet, and sometimes nothing at all. In monkey rods the unit current caused by a single photon is about 0.5×10^{-12}
amperes. Because of the large amplification through the cyclic GMP
cascade, a single photon can close approximately 300 channels—about
3 to 5 percent of the rod channels that are open in the dark.[8,13] Moreover,
because of the extreme stability of visual pigments mentioned earlier,
random isomerizations and spurious channel closings are rare events.
This allows the effects of single light quanta to stand out against an

Denis Baylor, 1991

[22]Schnapf, J. L. and Baylor, D. A. 1987. *Sci. Am.* 256(4): 40–47.

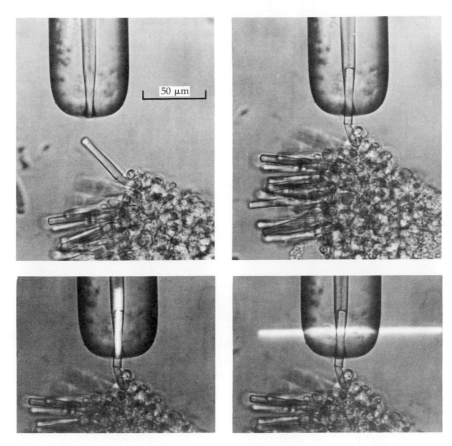

9 METHOD FOR RECORDING MEMBRANE CURRENTS of rod outer segment. A suction electrode with a fine tip is used to suck up the outer segment of a rod that protrudes from a piece of toad retina. Slits of light illuminate the receptor with precision. Since the electrode fits tightly around the photoreceptor, current flowing into it or out of it is recorded. Similar recordings have also been made with photoreceptors from mammalian retina. (From Baylor, Lamb and Yau, 1979.)

extremely quiet background. These experiments provide a rare example of the way in which a process as complex as seeing the dimmest possible flashes of light can be correlated with the events that occur in single molecules.

Extraordinary insights and experiments by Young and Helmholtz in the nineteenth century defined crucial questions for color vision and at the same time provided clear, unequivocal explanations. Their conclusion that there must be three types of sensory photoreceptors for color has stood the test of time and been confirmed even at the molecular level. To set the stage we quote again from Helmholtz, who compares the perception of light and sound, color and tone.[3] One envies the clarity, force, and timeless beauty of his thinking, especially in view of the confusing, vitalistic concepts that were current in the nineteenth century.

Cones and color vision

(A)

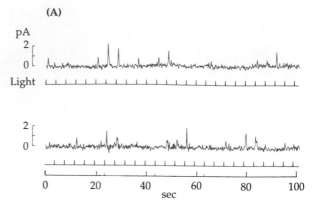

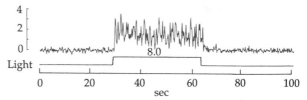

(B)

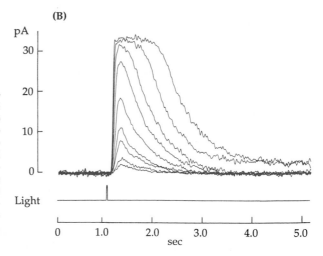

10

**RECORDINGS MADE BY SUCTION ELEC-
TRODE from monkey rod outer segment. (A)
Responses to dim flashes are shown in upper two
traces. The currents fluctuate in a quantal manner. Smaller deflections are the currents generated by single photons interacting with visual
pigments. Often photoisomerizations fail to occur. Steady, more intense illumination gives rise
to a burst of signals (lower trace). (B) Records
from a rod in monkey retina with flashes of increasing intensity. These currents are the counterpart of voltage traces shown in Figure 8B.
(From Baylor, Nunn and Schnapf, 1984.)**

All differences of hue depend upon combinations in different proportions
of the three primary colors . . . red, green and violet. . . . Just as the difference of sensation of light and warmth depends . . . upon whether the
rays of the sun fall upon nerves of sight or nerves of feeling, so it is supposed
in Young's hypothesis that the difference of sensation of colors depends
simply upon whether one or the other kind of nervous fibers are more
strongly affected. When all three kinds are equally excited, the result is the
sensation of white light. . . . If we allow two different colored lights to fall
at the same time upon a white screen . . . we see only a single compound,
more or less different from the two original ones. We shall better understand

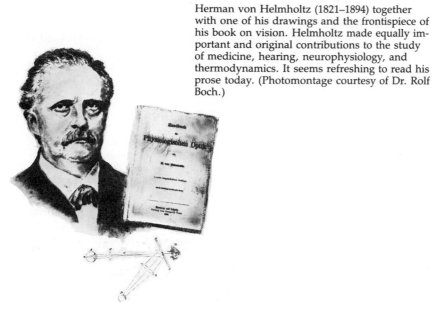

Herman von Helmholtz (1821–1894) together with one of his drawings and the frontispiece of his book on vision. Helmholtz made equally important and original contributions to the study of medicine, hearing, neurophysiology, and thermodynamics. It seems refreshing to read his prose today. (Photomontage courtesy of Dr. Rolf Boch.)

the remarkable fact that we are able to refer all the varieties in the composition of external light to mixtures of three primitive colors if we compare the eye with the ear. Sound is, like light, an undulating movement spreading by waves. In the case of sound we have to distinguish waves of various length which produce upon our ear impressions of different quality. We recognize the long waves as low notes, the short as high-pitched, and the ear may receive at once many waves of sound, that is to say many notes. But here these do not melt into compound notes in the same way that colors when perceived at the same time and place melt into compound colors. The eye cannot tell the difference if we substitute orange for red and yellow; but if we hear the notes C and E sounded at the same time, we cannot put D instead of them without entirely changing the impression upon the ear. The most complicated harmony of a full orchestra becomes changed to our perception if we alter any one of its notes. No accord is completely like another, composed of different tones; whereas if the ear perceived musical tones as the eye colors, every accord might be completely represented by combining only three constant notes, one very low, one very high and one intermediate, simply changing the relative strength of these three primary notes to produce all possible musical effects. . . . If we wish to describe an accord exactly and completely the strength of each of its component tones must be exactly stated. In the same way, the physical nature of a particular kind of light can only be fully ascertained by measuring and noting the amount of light of each of the simple colors which it contains. But in sunlight and in flames we find a continuous transition of colors into one another through numberless intermediate gradations. Accordingly, we must ascertain the amount of light of an infinite number of compound rays . . . to arrive at an exact physical knowledge of sun or starlight. In the sensations of the eye we need to distinguish . . . only the varying intensities of three components. The . . . musician is able to catch the separate notes of the various instruments among the complicated harmonies of an entire orchestra, but the optician cannot directly ascertain the composition of light by means of the eye.

We must not be led astray by confounding the notions of a *phenomenon* and an *appearance*. The colors of objects are phenomena caused by certain real differences in their constitution. They are, according to the scientific as well as to the uninstructed view, no mere appearance, even though the way in which they appear depends chiefly upon the constitution of our nervous system . . . It must be confessed that both in men and in quadrupeds we have at present no anatomical basis for this theory of colors.

These farsighted predictions were first validated by spectrophotometry. Wald, Brown, MacNichol, Dartnall and their colleagues showed the existence of three types of cones with different pigments in human retina.[5,6,23] Second, Baylor and his colleagues recorded from monkey and human cones by the suction pipette technique[24] (Figure 9). The results are shown in Figure 11. Three populations of cones were found with distinct but overlapping sensitivities in the blue, green, or red part of the spectrum. The wavelengths of light optimal for initiating electrical signals coincided precisely with those demonstrated by psychophysical and spectrophotometric measurements.

Nathans has cloned the genes for the red, green, and blue cone opsin pigments, as well as the gene for rhodopsin, and compared the sequences.[9–11] Slight differences in the opsin molecules, all of which combine with 11-*cis*-retinal, account for their ability to trap different wavelengths of light preferentially. The very close homology in the sequences suggest that all four genes evolved from a common ancestor (Figure 12). In males with normal color vision, the gene encoding the green pigment has been found to be present in multiple copies. The visual pigment genes are aligned in a head-to-tail arrangement on the X chromosome.

How do the red, green, and blue cones interact to produce color mixing? First, it is evident that a single type of photoreceptor cannot on its own provide information about color. Thus rhodopsin is bleached equally well by weak lights at the peak of its wavelength sensitivity (500 nm) or by stronger illumination with wavelengths on either side. With three types of cone, however, each absorbing better in a different region of the spectrum, a comparison of the output provides information about color. For example, a pure monochromatic light at 575 nm will activate both green and red cones to produce a sensation of yellow. It is the ratio of activation of the two cones that determines which color is seen: A shift of the illuminating light to shorter wavelengths gives rise to a stronger green-cone output, whereas a shift to longer wavelengths influences red cones more strongly. It is evident that the red–green cone output produced by monochromatic yellow light can be

[23]Dartnall, H. J. A., Bowmaker, J. K. and Molino, J. D. 1983. *Proc. R. Soc. Lond. B.* 220: 115–130.

[24]Schnapf, J. L. et al. 1988. *Vis. Neurosci.* 1: 255–261.

(A)

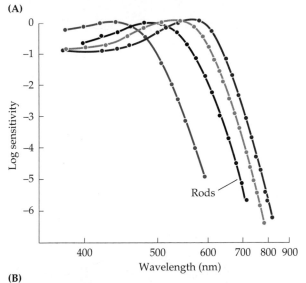

Rods

(B)

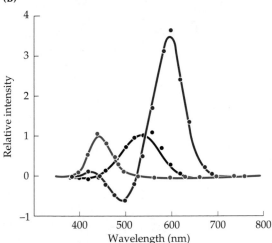

(C)

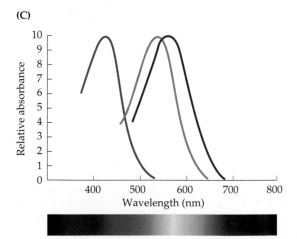

11

**SPECTRAL SENSITIVITY OF PHOTORECEP-
TORS** of human subjects and of visual pigments.
(A) The spectral sensitivities of blue, green, and
red cones (as colored) and rods (black) from macaque monkeys. The responses were recorded by
suction electrodes, averaged, and normalized.
Smooth curves through the cone spectra were
drawn by eye; the curve through the rod spectrum is obtained from visual pigments in human
subjects. **(B)** Comparison of spectral sensitivity
of monkey cones with those obtained by human
color matching. The continuous curves represent
color-matching experiments in which the sensitivity at various wavelengths was determined in
human subjects (see Baylor, 1987). The symbols
show results predicted from electrical measurements made by recording currents from single
cones, after correcting for absorption in the lens
and by pigments on the path to the outer segment. The correspondence between the results
obtained on single cells and by color matching
is extraordinarily good. **(C)** Spectral sensitivity
curves of the three colored pigments showing
absorbance peaks for blue, green, and red pigments. **(A** after Schnapf and Baylor, 1987; **B** after
Baylor, 1987; **C** after Dowling, 1987.)

duplicated by mixing monochromatic red and green lights of appropriate wavelength and intensity (see Figure 13).

Similarly, white can be produced by a uniform spread of wavelengths or by mixtures of red, green, and blue. As Helmholtz pointed

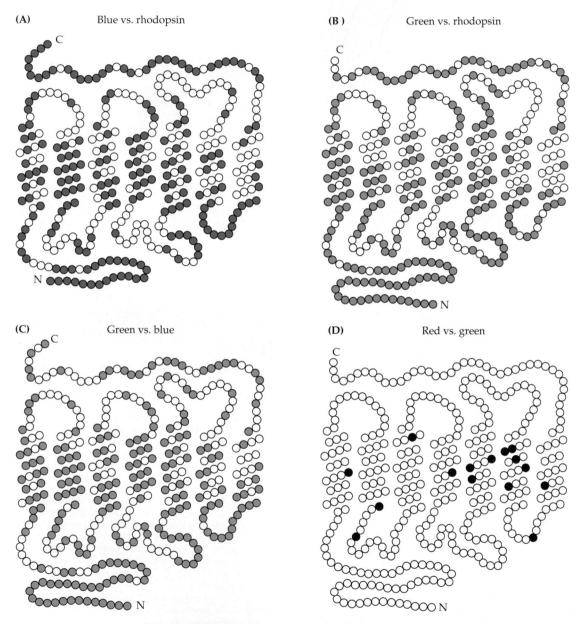

(A)   Blue vs. rhodopsin

(B)   Green vs. rhodopsin

(C)   Green vs. blue

(D)   Red vs. green

12 **COMPARISONS OF NUCLEOTIDE SEQUENCES of red, green, and blue pigments with each other and with rhodopsin. Each colored dot represents an amino acid difference. (A,B) Blue and green pigments compared with rhodopsin. (C,D) Green pigment compared with blue and red pigments. Note that the sequences of red and green pigments are highly similar. (From Nathans, 1989.)**

out, colors can be distinguished and matched, but the visual system cannot determine the wavelength composition of light falling on the retina. In principle, two types of cones with different pigments would be sufficient to recognize colors; but many different mixtures of wavelengths would then appear identical. This is the situation in color blindness.[3]

> In this condition the differences of color are reduced to combinations of only two primary colors. Persons so affected confound . . . certain hues which appear very different to ordinary eyes. At the same time they distinguish other colors . . . accurately. They are usually "red-blind"; that is to say, there is no red in their system of colors and they see no difference which is produced by the addition of red. All tints are for them varieties of blue and green, or, as they call it, yellow. . . . The scarlet flowers of the geranium have for them exactly the same color as its leaves. They cannot distinguish between the red and green signals of trains. Very full scarlet appears to them almost black, so that a red-blind Scotch clergyman went to buy scarlet cloth for his gown, thinking it was black.

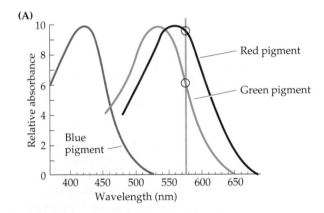

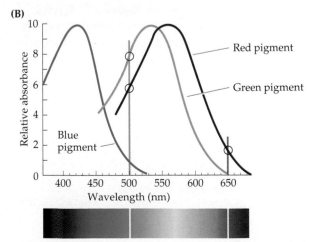

13

COLOR PRODUCED BY MONOCHROMATIC YELLOW LIGHT duplicated by mixing red and green lights of appropriate wavelength and intensity. (A) Spectral sensitivity of red, green, and blue cones. Pure monochromatic yellow light at a wavelength of 575 nm activates both red and green cones in the proportions shown. (B) Exactly the same effect on the red and the green cones can be produced by monochromatic wavelengths of light farther in the red and green ranges. Hence green together with red can produce a sensation of yellow.

In his detailed studies of color blindness, Nathans has confirmed that the genetic defect in such people consists of an absence of the red pigment.[10] This work represents a major advance and a triumph of molecular neurobiology. From our present perspective one can only marvel at explanations at this level that confirm so beautifully the brilliant but rigorous speculations of Young and Helmholtz: Their idea that major attributes of color vision and color blindness are to be found within the receptors themselves has now been confirmed in terms of genes and protein structure.

Bipolar, horizontal, and amacrine cells

From the morphological descriptions of Ramón y Cajal mentioned in Chapter 1, a general scheme for the wiring diagram of the retina emerges: a through-line of transmission from photoreceptors to ganglion cells by way of bipolar cells, with interactions mediated through horizontal and amacrine cells. From diagrams of connections in primate retina,[25,26] it is clear that the output of the eye must be the result of highly complex integrative processes. For example, the horizontal cells shown in Figure 14 receive synapses from many receptors and in turn feed back onto them as well as ending on bipolar cells. Similarly, certain amacrine cells, which receive inputs from bipolar cells, send synapses back to them as well as to the ganglion cells. One can conclude that horizontal cells and amacrine cells transmit and modify signals traveling through the retina. An additional source of complexity is that each of the major classes of neurons shown in Figure 14 is known to have various morphological and pharmacological subtypes, often with clear implications for function. By physiological, biochemical, and anatomical criteria at least 4 types of bipolar cell, 2 types of horizontal cell, and 10 to 20 types of amacrine cell have been described.[26,27]

Numerous questions arise about the transmission of signals. For example, what signals do bipolar cells give? How are they influenced by rods and cones? And how is signaling to ganglion cells modified by horizontal and amacrine cells? The crucial problem is to find out what is happening to each type of cell in the two intraretinal synaptic stations, in the outer and inner plexiform layers. The pioneering work of Svaetichin,[28] Tomita,[29] Dowling,[4] Baylor,[8] Fuortes,[12] Kaneko,[30] Raviola,[31] Wässle,[26] and Daw[32] and their colleagues in which microelectrode recordings were combined with dye injection and morphological studies gave rise to a major revolution in retinal physiology that was further augmented by the use of cellular neurochemistry and the identification

[25]Dowling, J. E. and Boycott, B. B. 1966. *Proc. R. Soc. Lond. B* 166: 80–111.
[26]Wässle, H. and Boycott, B. B. 1991. *Physiol. Rev.* 71: 447–480.
[27]Masland, R. H. 1988. *Trends Neurosci.* 11: 405–410.
[28]Svaetichin, G. 1953. *Acta Physiol. Scand.* 29: 565–600.
[29]Tomita, T. 1965. *Cold Spring Harbor Symp. Quant. Biol.* 30: 559–566.
[30]Kaneko, A. 1979. *Annu. Rev. Neurosci.* 2: 169–191.
[31]Raviola, E. and Dacheux, R. F. 1990. *J. Neurocytol.* 19: 731–736.
[32]Daw, N. W., Brunken, W. J. and Parkinson, D. 1989. *Annu. Rev. Neurosci.* 12: 205–225.

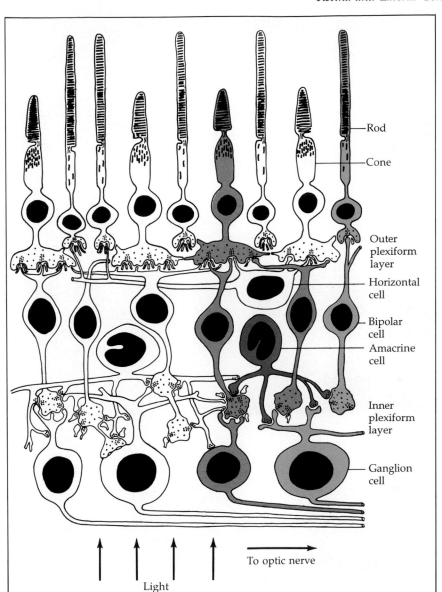

Rod

Cone

Outer plexiform layer

Horizontal cell

Bipolar cell

Amacrine cell

Inner plexiform layer

Ganglion cell

To optic nerve

Light

14 **PRINCIPAL CELL TYPES AND CONNECTIONS of primate retina to illustrate rod and cone pathways to ganglion cells. (After Dowling and Boycott, 1966; Daw, Jensen and Brunken, 1990.)**

of transmitters. Examples of various types of cells are shown in Figure 15.

The technique of illuminating selected areas of the retina introduced the important concept of the receptive field, a concept that provided a key for understanding the significance of the signals not only in the retina but at successive stages in the cortex. The term *receptive field* was

The concept of receptive fields

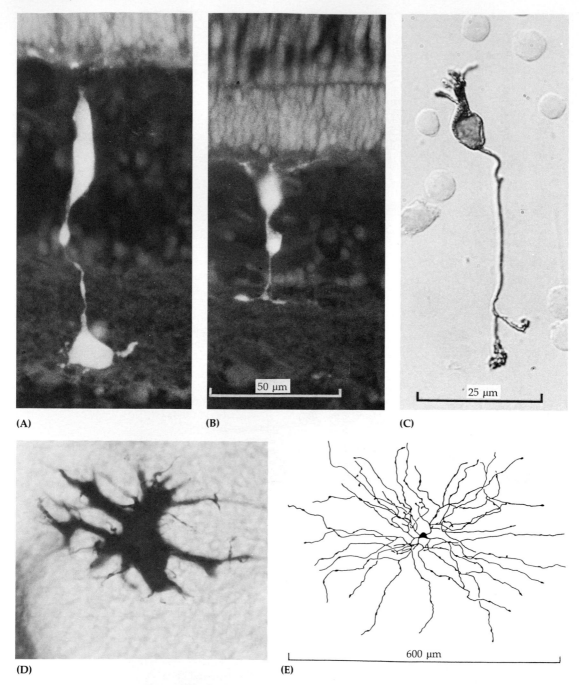

(A)   (B)   (C)

(D)   (E)

15   BIPOLAR, HORIZONTAL, AND AMACRINE CELLS. (A,B) Bipolar cells of
goldfish injected with fluorescent dye. (A) A depolarizing "on"-center cell. (B) A
hyperpolarizing "off"-center bipolar cell. (C) A bipolar cell isolated from rat retina
stained for protein kinase C. (D) A horizontal cell in dogfish retina injected with
horseradish peroxidase. (E) A starburst amacrine cell from rabbit retina stained intracellularly with Lucifer Yellow. (A, B, and D courtesy of A. Kaneko, unpublished; C
from Yamashita and Wässle, 1991; E from Masland, 1988.)

coined originally by Sherrington in relation to reflex actions and was reintroduced by Hartline.[33] Applied to the visual system, the RECEPTIVE FIELD OF A NEURON can be defined as *the area on the retina from which the activity of a neuron can be influenced by light* (see also Chapter 14). For example, a record of the activity of one particular neuron in the optic nerve or cortex of a cat shows that the rate of firing increases or decreases only when a defined area of retina is illuminated (Figures 18 and 22). This area is its receptive field. By definition, illumination outside a receptive field produces no effect at all. The area itself can be subdivided into distinct regions, some of which increase activity and others of which suppress impulses in the cell. This description of receptive fields also applies to neurons such as bipolar or horizontal cells in which graded local potentials are influenced by the activity of photoreceptors (see below). It will be shown that diffuse flashes of light are of little or no use for assessing function in the visual system.

Each bipolar cell receives its input from rods or cones. Rod bipolar cells are typically supplied by 15 to 45 receptors. One type of cone bipolar, the midget bipolar, receives its input from a single cone. As one might expect, midget bipolar cells are found in the through-line from the fovea, where acuity is highest. Other bipolar cells are supplied by a convergent input from 5 to 20 adjacent cones.[34] Bipolar cells and horizontal cells do not generate impulses but respond with graded, sustained depolarizations or hyperpolarizations to illumination of appropriate receptors. The responses and receptive fields of bipolar cells at first seem somewhat bewildering. One reason is that light causes photoreceptors to become hyperpolarized and reduces transmitter release onto the bipolar cells; a second reason is that the continuous release of transmitter in the dark depolarizes some bipolar cells and hyperpolarizes others, depending on the types of receptors at the postsynaptic sites. There is good evidence that the transmitter liberated by photoreceptors that produces these different effects is glutamate.[32,35,36]

In the dark, with ongoing bombardment of glutamate receptors, some bipolar cells are maintained in an excited depolarized state. Light will therefore decrease tonic release by the photoreceptors and give rise to a hyperpolarization in such hyperpolarizing (H) bipolar cells. Conversely, the membrane potential of a different bipolar cell (D) that is continually hyperpolarized by glutamate in the dark will, with light, be deprived of the hyperpolarizing input and will become depolarized. Characteristics of hyperpolarizing and depolarizing bipolar cells are of key importance in shaping the response characteristics of subsequent relays in the visual system; D cells respond with depolarization to lights being turned "on"; H cells respond with depolarization to light being turned "off". The properties of the glutamate-activated channels in

Responses of bipolar and horizontal cells

[33]Hartline, H. K. 1940. *Am. J. Physiol.* 130: 690–699.
[34]Sterling, P., Freed, M. A. and Smith, R. G. 1988. *J. Neurosci.* 8: 623–642.
[35]Miller, A. E. and Schwartz, E. A. 1983. *J. Physiol.* 334: 325–349.
[36]Sarthy, P. V., Hendrickson, A. E. and Wu, J. Y. 1986. *J. Neurosci.* 6: 637–643.

bipolar cells have been analyzed by the use of pharmacological agonists and antagonists;[32] they have not been characterized at the molecular level.

The receptive field of a hyperpolarizing bipolar cell is shown in Figure 16. A small spot of light shone onto the central part of the field causes a sustained hyperpolarization. The explanation is shown in the diagram: When light shines on the central photoreceptors that are directly connected to the bipolar cell, they hyperpolarize and stop releasing depolarizing transmitter. An important feature of the receptive fields of all bipolar cells is that they are concentric in shape. Thus the central area driven directly by photoreceptors is enveloped by an antagonistic SURROUND. In the H, or "off"-center, bipolar cell of Figure 16 illumination of this surround with light shining in the form of an annulus leaves the center in the dark and gives rise to a depolarization.

How is this antagonistic effect of the surround produced? Figure 16 shows that the pathway is indirect and is mediated by horizontal cells.[37] Since sign inversions, such as reduction of inhibition, occur, the sequence tends to be confusing and difficult to follow. Like bipolar cells, horizontal cells do not give impulses; they respond to illumination of photoreceptors by hyperpolarization (i.e., again, removal of a tonic depolarizing influence). Each horizontal cell is supplied by a large number of photoreceptors. Moreover, horizontal cells are electrically coupled to each other. Lucifer Yellow injected into one horizontal cell will spread through gap junctions to others.[38] This feature increases the area of retina that can influence the membrane potential of a horizontal cell. The output of horizontal cells is fed back onto photoreceptors and presumably onto bipolar cells with endings close to the sites of receptor–bipolar cell synapses. In darkness, horizontal cells, in their depolarized state, continually release the transmitter GABA, which tends to hyperpolarize the photoreceptor terminals. When light causes the horizontal cells to become hyperpolarized, the photoreceptors supplied by them become depolarized (owing to removal of hyperpolarizing GABA influence). Thus, in the example shown in Figure 16, diffuse illumination antagonizes the effect of light shining on the central group of photoreceptors.

Depolarizing (D) "on"-center bipolar cells have similarly shaped concentric fields except that illumination of the center causes depolarization by removal of the continuous inhibition occurring in darkness; conversely, illumination of the surround acts antagonistically to hyperpolarize. Again, the surround effect is mediated indirectly through horizontal cells that stop releasing GABA and thereby depolarize receptors.

What are the physiological implications of such receptive fields? For a horizontal cell, the best stimulus is illumination of a large area of retina containing all the receptors that feed onto it. D and H bipolar

[37]Daw, N. W., Jensen, R. J. and Brunken, W. J. 1990. *Trends Neurosci.* 13: 110–115.
[38]Kaneko, A. 1971. *J. Physiol.* 213: 95–105.

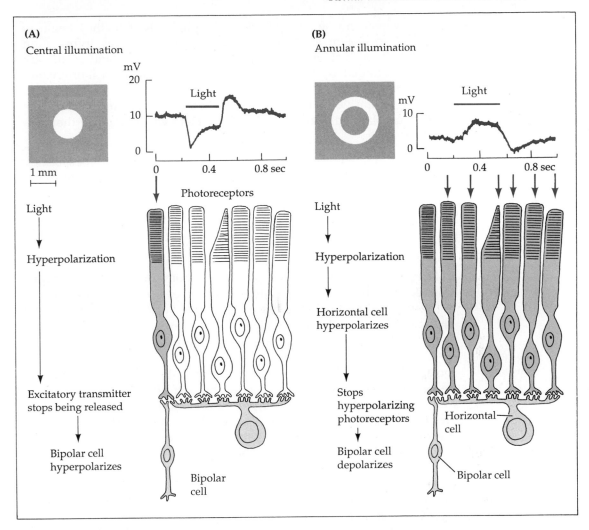

(A)

Central illumination

1 mm

Light

↓

Hyperpolarization

↓

Excitatory transmitter
stops being released

↓

Bipolar cell
hyperpolarizes

Photoreceptors

Bipolar
cell

(B)

Annular illumination

Light

↓

Hyperpolarization

↓

Horizontal cell
hyperpolarizes

↓

Stops
hyperpolarizing
photoreceptors

↓

Bipolar cell
depolarizes

Horizontal
cell

Bipolar cell

16 **RESPONSES AND CONNECTIONS OF BIPOLAR CELL. Records show the
receptive field of a bipolar cell in a goldfish retina responding by hyperpolarization (H) to illumination of its center and by depolarization to a ring of light. Other
bipolar cells (D) respond in the opposite way (depolarization with central illumination);
neither type of bipolar cell generates impulses. The diagrams illustrate connections
required to elicit these responses. This is a difficult series of interactions to understand
because in one part of the circuit the "stimulus" (light) stops transmitter release, while
at another synapse it indirectly causes an increase in transmitter release. (A) Light falling
on the photoreceptor causes hyperpolarization. As a result, excitatory transmitter stops
being released and the bipolar cell becomes hyperpolarized. (B) Light falling on the
surrounding area again prevents transmitter from being released by photoreceptors; as
a result, horizontal cells become hyperpolarized. This hyperpolarization prevents the
horizontal cell from releasing its inhibitory transmitter onto the photoreceptor. The
photoreceptor therefore becomes depolarized and starts to release its excitatory transmitter once again. The bipolar cell becomes depolarized. (Bipolar cell records from
Kaneko, 1970.)**

cells, however, do not simply respond to light. Rather, they begin to analyze information about objects. Their signals convey information about small spots of lights surrounded by darkness or about small dark spots surrounded by light. They respond to contrast in illumination over a small area of retina.

Comprehensive reviews and papers describe the properties of the various bipolar and horizontal cells.[4,26,39] Here we summarize the principal subtypes.

1. D and H cone bipolar cells respond best to small light or dark spots.
2. The centers of D and H midget bipolar cells are supplied by single cones.
3. Rod bipolar cells are D, or "on"-center, and respond best to small bright spots.

Detailed descriptions of the complex electron microscopical appearance of photoreceptor terminals, bipolar cells, and the feedback synapses of horizontal cells are provided by Dowling,[4] Raviola,[40] and their colleagues. An important principle derives from the pattern of synaptic connections in Figure 16. A single photoreceptor contributes to the receptive field centers of numerous "on" and "off" bipolar cells and to the surrounds of others. The effects of red, green, and blue cones on responses of horizontal and bipolar cells are mentioned below.

Receptive fields of ganglion cells   Many years before the electrical responses of photoreceptors or bipolar cells could be measured, important information was obtained by recording from ganglion cells. Thus, the first analysis of signaling in the retina was made at the output stage. The first advantage of this approach was that ganglion cells gave all-or-nothing action potentials: Records could be made with extracellular electrodes at a time before intracellular microelectrodes had been perfected or dye injections developed. Second, going straight to the output was an essential simplification and shortcut. The finding of concentric "on" and "off" receptive fields in ganglion cells made it possible later to understand the responses in bipolar and horizontal cells.

It was Kuffler who pioneered the experimental analysis of the mammalian visual system by first concentrating on receptive field organization and the meaning of signals in the cat.[41] The end results of synaptic interactions rather than the synaptic mechanisms themselves were the objective. Hubel has succinctly put the achievement in perspective:[42]

> What is especially interesting to me is the unexpectedness of the results, as reflected in the failure of anyone before Kuffler to guess that something like center–surround receptive fields could exist or that the optic nerve would virtually ignore anything so boring as diffuse light levels.

[39]Cohen, E. and Sterling, P. 1990. *Philos. Trans. R. Soc. Lond. B.* 330: 323–328.
[40]Raviola, E. and Dacheux, R. F. 1987. *Proc. Natl. Acad. Sci. USA* 84: 7324–7328.
[41]Kuffler, S. W. 1953. *J. Neurophysiol.* 16: 37–68.
[42]Hubel, D. H. 1988. *Eye, Brain, and Vision.* Scientific American Library. New York.

17

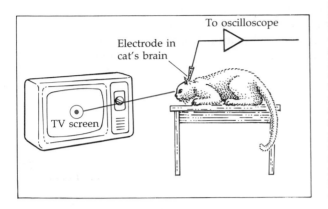

STIMULATION OF RETINA WITH PATTERNS OF LIGHT. The eyes of an anesthetized, light-adapted cat or monkey focus on a television screen with various patterns of light generated by a computer. (Alternatively, a projecter can be used to project light patterns onto a screen.) An electrode records the responses from a single cell in the visual pathway. Light or shadow falling onto a restricted area of the screen may accelerate (excite) or slow (inhibit) the signals given by a neuron. By determining the areas on the screen from which a neuron's firing is influenced, one can delineate the receptive field of the cell. The positions of cells in the brain and the tracks of electrode penetration are reconstructed histologically after the experiment. In his original experiments Kuffler shone light directly into the eye by means of a specially constructed ophthalmoscope.

The principal new approach was not so much a matter of technique; rather it consisted of formulating the following question: What is the best way to stimulate individual ganglion cells whose axons carry information to the higher centers through the optic nerve? This question led logically to the use of discrete circumscribed spots and patterns of light for stimulation of selected areas of the cat retina, instead of diffuse uniform illumination, which produces a confusing array of responses. Such procedures had been foreshadowed by Hartline's pioneering work on the eye of a simple invertebrate, the horseshoe crab *Limulus*,[33] and on the retina of the frog, the study of which was later pursued by Barlow,[43] Maturana, Lettvin, and their colleagues.[44] Kuffler's initial choice of the cat was a lucky one; in the rabbit, for example, the situation would have been more complicated. The ganglion cells in the rabbit have more elaborate receptive fields and can respond specifically to such complex features as edges or to movement in one direction rather than another.[45] Equally complex are lower vertebrates, such as frogs. A general law seems to emerge: The dumber the animal, the smarter its retina (D. A. Baylor, personal communication).

A methodological feature of Kuffler's early experiments that became essential was the use of the intact undissected eye, the normal refracting channels of which served as pathways for stimulation. A convenient way of illuminating particular portions of the retina is to anesthetize the animal lightly and place it facing a screen or a television monitor at a distance for which its eyes are properly refracted. When one then shines patterns of light onto the screen or displays computer-generated television images, these will be well focused on the retinal surface (Figure 17).

[43]Barlow, H. B. 1953. *J. Physiol.* 119: 69–88.
[44]Maturana, H. R. et al. 1960. *J. Gen. Physiol.* 43: 129–175.
[45]Barlow, H. B., Hill, R. M. and Levick, W. R. 1964. *J. Physiol.* 173: 377–407.

When one records from a particular ganglion cell, the first task is to find the location of its receptive field. Characteristically, most ganglion cells and neurons throughout the visual system show continued discharges at rest even in the absence of illumination. Appropriate stimuli do not necessarily initiate activity but may modulate the background firing; responses of ganglion cells can consist of either an increase or a decrease of ongoing discharges.

Figure 18, adapted from a paper by Kuffler,[41] shows that for a ganglion cell a small spot of light, 0.2 mm in diameter, shone onto a part of the receptive field is far more effective than diffuse illumination in producing excitation. Furthermore, the same spot of light can have opposite effects, depending on the exact position of the stimulus within the receptive field. For example, in one area the spot of light excites a ganglion cell for the duration of illumination. Such an "on" response can be converted into an inhibitory "off" response by simply shifting the spot by 1 mm or less across the retinal surface. As was the case for bipolar cells, which were studied many years later, there are two basic receptive field types: "on"-center and "off"-center ganglion cells. The receptive fields are roughly concentric, with a ganglion cell in the geometrical central region of any field.

In an "on"-center receptive field, light produces the most vigorous response if it completely fills the center, whereas for most effective inhibition of firing, the light must cover the entire ring-shaped surround (annular illumination in Figure 18). When the inhibitory annular light is turned off, the ganglion cell gives an exuberant "off" discharge. An "off"-center field has a converse organization, with inhibition arising in the circular center. The spotlike center and its surround are antagonistic; therefore, if they are illuminated simultaneously they tend to cancel each other's contribution.

Sizes of receptive fields

Neighboring ganglion cells collect information from very similar, but not quite identical, areas of the retina. Even a small (0.1 mm) spot of light on the retina covers the receptive fields of many ganglion cells that have diverse responses, some ganglion cells being inhibited, others excited. (In practice it is inconvenient to describe the dimensions of fields as millimeters on the retina; instead, degrees of arc are a more useful measure. In our eyes 1 mm on the retina corresponds to about 4°; for reference, the image of the moon, which subtends 0.5°, has a diameter of ⅛ mm on our retina.) This characteristic organization, with neighboring groups of receptors projecting indirectly onto neighboring ganglion cells in the retina, is retained at all levels in visual pathways. The systematic analysis of receptive fields demonstrates the general principle that *neurons processing related information are clustered together.* In sensory systems this means that the central neurons dealing with a particular area of the surface can communicate with each other over short distances. This appears to be an economical arrangement, as it saves long lines of communication and simplifies the making of connections.

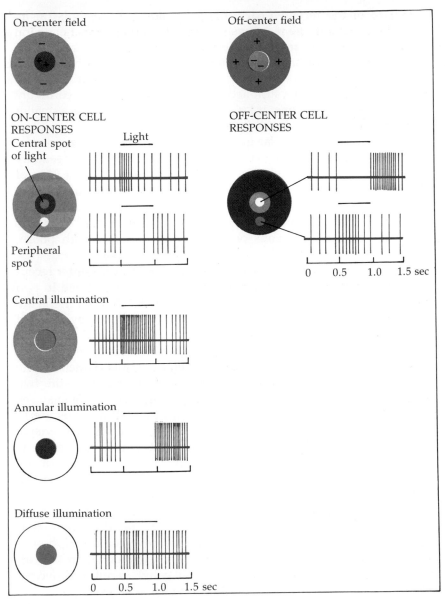

18 **RECEPTIVE FIELDS OF GANGLION CELLS in the retinas of cats and monkeys** are grouped into two main classes: "on"-center (+) and "off"-center (−) fields. "On"-center cells respond best to a spot of light shone onto the central part of a receptive field. Illumination (indicated by bar above records) of the surrounding area with a spot or a ring of light reduces or suppresses the discharges and causes responses when the light is turned off. Illumination of the entire receptive field elicits relatively weak discharges because center and surround oppose each other's effects. "Off"-center cells slow down or stop signaling when the central area of their field is illuminated and accelerate when the light is turned off. Light shone onto the surround of an "off"-center receptive field excites the corresponding cell. (After Kuffler, 1953.)

The size of the receptive field of a ganglion cell depends on its location in the retina. The receptive fields of cells situated in the central areas of the retina have much smaller centers than those at the periphery; receptive fields are smallest in the fovea, where the acuity of vision is highest. The central "on" or "off" region of such a "midget" ganglion cell's receptive field can be supplied by a single cone and is accordingly only about 2.5 μm in diameter subtending 0.5 *minutes* of arc—smaller than the period at the end of this sentence.

There is a strikingly similar gradation of receptive field size in relation to fine resolution by the eye and by touch. A higher order sensory neuron in the brain responding to a fine touch applied to the skin of the fingertip has a receptive field that is very small compared with that of a neuron having a field on the skin of the upper arm (Chapter 14). To discern the form of an object, we use our fingertips and foveas, not the less discriminating regions on the receptor surfaces with poorer resolution.

Classification of ganglion cell receptive fields

Superimposed on the general scheme of "on"- or "off"-center receptive fields, ganglion cells in the monkey retina can be grouped in two main categories, which will be denoted as M and P.[46,47] The criteria are both anatomical and physiological. The M and P terminology is based on the anatomical projections of these neurons to the lateral geniculate nucleus and from there to the cortex. P ganglion cells project to the four dorsal layers of smaller cells in the lateral geniculate nucleus (the parvocellular division), M ganglion cells to the larger cells in the two ventral layers (the magnocellular division) (see Figure 19). We shall see that the characteristic response properties of M and P neurons are in general maintained through separate specialized pathways in the visual system. In brief, P ganglion cells have small receptive field centers, have high spatial resolution, and are sensitive to color. P cells provide information about fine detail at high contrast. M cells have larger receptive fields and show little or no sensitivity to color. M cells are more sensitive to small differences in contrast than P cells, give higher-frequency discharges, and conduct impulses more rapidly along their larger-diameter axons.[46,47] For convenience we defer the description of color coding by monkey P ganglion cells until Chapter 17. In the cat, which has no color vision, the classification of ganglion cells is different, with X, Y, and W groups.[48,49] X and Y are in some respects parallel to P and M in their properties, but there are major differences and the two classifications are not interchangeable. In the cat, X ganglion cells are smaller than Y and have smaller receptive fields; Y cells are more sensitive to moving stimuli than X. Key procedures for determining differences between X and Y cells involve the assessment of spatial summation by the use of sinusoidal gratings, a procedure that does not

[46]Kaplan, E. and Shapley, R. M. 1986. *Proc. Natl. Acad. Sci. USA* 83: 2755–2757.

[47]Shapley, R. and Perry, V. H. 1986. *Trends Neurosci.* 9: 229–235.

[48]Wässle, H., Peichl, L. and Boycott, B. B. 1981. *Proc. R. Soc. Lond. B* 212: 157–175.

[49]Enroth-Cugell, C. and Robson, J. G. 1966. *J. Physiol.* 187: 517–552.

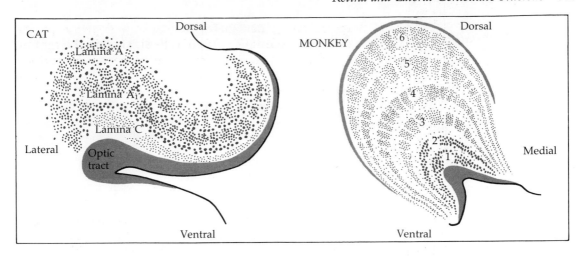

19 **LATERAL GENICULATE NUCLEUS. (A)** In the lateral geniculate nucleus of the cat, there are three layers of cells: A, A₁, and C. **(B)** In the monkey, the lateral geniculate nucleus has six layers. In the four dorsal layers (3, 4, 5, 6; parvocellular), the cells are smaller than in layers 1 and 2 (magnocellular). In both animals, each layer is supplied by only one eye and contains cells with specialized properties. (From Szentágothai, 1973.)

reliably differentiate between P and M cells in the monkey. The magno-parvo classification in the visual system provides a convenient and useful framework for studying pathways and properties, but mixing and exceptions do occur, particularly at higher levels.[50]

The complex interactions between bipolar, amacrine, and ganglion cells occur in the inner plexiform layer (see Figure 14). The figure in Box 1 illustrates essential features of the pathways to "on"- and "off"-center ganglion cells. We have seen that many of the key features of receptive fields of ganglion cells are already present in bipolar cells. As expected, depolarizing "on"-center and hyperpolarizing "off"-center cone bipolar cells make chemical synapses with appropriate "on" or "off" ganglion cells. These synapses are excitatory; consequently the change in membrane potential of a bipolar cell causes the ganglion cell to which it is connected to change its membrane potential in the same direction. The complex pathway from rods to ganglion cells is described in Box 1. Rods and cones supply the same ganglion cell but by way of different interposed cells.

An elegant demonstration of the through-line from photoreceptor to ganglion cell was provided by Baylor and Fettiplace.[51] They changed the membrane potential of a single photoreceptor by passing current through an intracellular microelectrode while recording from a ganglion cell to which a connection was made through bipolar cells. Hyperpo-

How are bipolar and amacrine cells connected to ganglion cells?

[50]Schiller, P. H. and Logothetis, N. K. 1990. *Trends Neurosci.* 13: 392–398.
[51]Baylor, D. A. and Fettiplace, R. 1977. *J. Physiol.* 271: 391–424.

BOX 1   CONNECTIONS OF RODS TO GANGLION CELLS VIA BIPOLAR AND AMACRINE CELLS

As shown in the figure, bipolar cells that are influenced by rods do not connect directly to ganglion cells.[37,26,52] Instead, a specialized amacrine cell (A2) is interposed. In response to illumination the rod bipolar depolarizes and secretes excitatory transmitter, again presumably glutamate. It rapidly depolarizes the A2 amacrine cell which, like other amacrine cells, can give action potentials. Excitation from the A2 amacrine to the depolarizing cone bipolar cell is mediated by an electrical synapse through gap junctions. The "on"-center ganglion cell in the figure is then excited by way of its normal depolarizing bipolar cell input. The same A2 amacrine cell also makes inhibitory hyperpolarizing synapses (using glycine as a transmitter) on the hyperpolarizing cone bipolar cell that supplies the "off"-center ganglion cell in the figure. This second pathway shows several reversals of sign in the transmission of signals across synapses: (1) Illumination of the rod causes hyperpolarization and a reduction in the secretion of transmitter; (2) the rod bipolar cell becomes less inhibited, depolarizes, and secretes transmitter; (3) the A2 amacrine cell becomes depolarized and secretes glycine; and (4) glycine inhibits the "off"-center ganglion cell, hyperpolarizing it and suppressing impulses. This same ganglion cell also gives "off" responses to signals from cones. Thus an elegant and precise convergence of rod and cone systems is apparent in ganglion cell receptive fields. The same center ganglion cell that responds to illumination of cones in daylight responds to similar faint light patterns shone at the same spot of the retina in dim ambient light.

In spite of the alarming complexity of this scheme, it represents an oversimplification. The main outline has been established by careful experiments but many open questions remain. Approximately 20 different types of amacrine cells have been described.[27] Some secrete dopamine, others indolamines, and there is evidence that these transmitters, as well as GABA and ACh, regulate overall sensitivity and contribute to the center–surround antagonism. In addition, peptides such as VIP may play a more subtle trophic role in the development of the eye.[53]

[52]Dacheux, R. F. and Raviola, E. 1986. *J. Neurosci.* 6: 331–345.

[53]Stone, R. A. et al. 1988. *Proc. Natl. Acad. Sci. USA* 85: 257–60.

larization of the single receptor, a cone sensitive to red light, initiated firing in the ganglion cell. Membrane potential changes produced by light and by artificially applied currents had identical effects on the firing of the ganglion cell. As expected, appropriate currents could quantitatively counteract or enhance the effects of light. This experiment showed unequivocally that the hyperpolarization or depolarization per se of a receptor is the event that signals information to other cells in the brain about light or darkness in the outside world.

What information do ganglion cells convey?

The most striking feature of ganglion cell signals is that they tell a different story from that of primary sensory receptors. They do not convey information about absolute levels of illumination because they behave in a similar fashion at different background levels of light. They ignore much of the information of the photoreceptors, which work more like a photographic plate or a light meter. Rather, they measure differences within their receptive fields by comparing the degree of illumi-

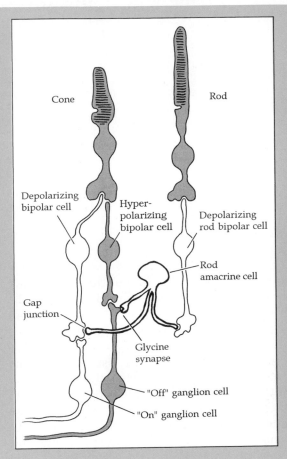

ROD AND CONE PATHWAYS IN THE MAMMAL-IAN RETINA. The pathway from rods to ganglion cells is highly complex, as shown. Cells that give depolarizing responses to light are shown as white, while those that give "off" responses and hyperpolarize are shaded in blue. Rods connect to specialized rod bipolar cells. Hyperpolarization of the rod by illumination reduces its transmitter output. The rod bipolar ceases to be inhibited and becomes depolarized. The depolarized rod bipolar excites the rod amacrine cell. Through gap junctions this produces excitation of the depolarizing cone bipolar cell shown on the left. The excitation in turn causes increased transmitter release and excitation of the "on"-center ganglion cell. At the same time depolarization of the rod amacrine cell causes it to release inhibitory transmitter (glycine) onto the "off"-center ganglion cell shown in blue. The hyperpolarization of the "off"-center ganglion cell contributes to its "off" response. Hence the rod amacrine cell can excite one ganglion cell and inhibit another, depending upon the junctions that it makes. A nice feature of the arrangement is that ganglion cells receive appropriate and matching "on" or "off" inputs from rods and cones. (After Daw, Jensen and Brunken, 1990.)

nation between the center and the surround. Apparently they are designed to notice simultaneous contrast and ignore gradual changes in overall illumination. They are exquisitely tuned to detect such contrast as the edge of an image or a bar crossing the opposing regions of a receptive field. It is evident, therefore, that the three retinal layers extract and analyze a great deal of information about the outside world. By choosing some aspects of the information collected by the primary receptors, and not others, a start has been made in selecting features that are important for form vision while jettisoning the level of background illumination.

LATERAL GENICULATE NUCLEUS

The optic nerve fibers running from each eye terminate on cells of the right and left LATERAL GENICULATE NUCLEUS, a distinctively layered

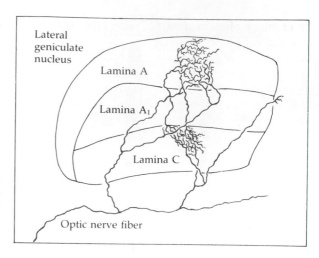

20

TERMINATION OF OPTIC NERVE FIBER in
lateral geniculate nucleus of cat. A single "on"-
center Y axon from the contralateral eye was injected with horseradish peroxidase. Branches end
in layers A and C but not in A₁. (Modified from
Bowling and Michael, 1980.)

structure (*geniculate* means "bent like a knee"). In the lateral geniculate
nucleus of the cat, there are three obvious, well-defined layers of cells
(A, A₁, C), one of which (C) has a complex structure that has been
further subdivided.[54] In the monkey the lateral geniculate nucleus has
six layers of cells (Figure 19). The cells in the deeper layers 1 and 2 are
larger than those in layers 3, 4, 5, and 6, giving rise to the terms
MAGNOCELLULAR and PARVOCELLULAR layers. In both cat and monkey,
each layer is predominantly supplied by one or the other eye. In the
cat the inputs to layers A and C originate in ganglion cells in the eye
on the other side of the animal, the fibers having crossed at the chiasm;
A₁ is supplied with fibers that come from the eye on the same side and
do not cross. In the monkey, layers 6, 4, and 1 are supplied by the
opposite eye; layers 5, 3, and 2 by the eye on the same side. The
segregation of endings from each eye into separate layers has been
shown by recording electrically and also by a variety of anatomical
techniques.[55–57] Particularly striking is the arborization of a single optic
nerve fiber that has been injected with the enzyme horseradish peroxidase (Figure 20). The terminals are all confined to the layers supplied
by that eye, with no spillover across the border. Because of the orderly
and systematic separation of fibers at the chiasm, the receptive fields
of the cells in the lateral geniculate nucleus are all situated in the visual
field on the opposite side of the animal (see Figure 1).

An important topographical feature is the highly ordered arrangement of receptive fields within individual layers of the geniculate.
Neighboring regions of the retina make connections with neighboring
geniculate cells, so that the receptive fields of adjacent neurons overlap

[54]Guillery, R. W. 1970. *J. Comp. Neurol.* 138: 339–368.
[55]Hubel, D. H. and Wiesel, T. N. 1972. *J. Comp. Neurol.* 146: 421–450.
[56]Hubel, D. H. and Wiesel, T. N. 1961. *J. Physiol.* 155: 385–398.
[57]Bowling, D. B. and Michael, C. R. 1980. *Nature* 286: 899–902.

over most of their area.[55,56] The *area centralis* in the cat (the region of the cat retina with small receptive field centers) and the fovea in the monkey project onto the greater portion of each geniculate layer. There are relatively few cells devoted to the peripheral retina. This heavy representation (the same as in the cortex) reflects the use of these regions for high-acuity vision and the need for fine-grained resolution. Although there are probably equal numbers of optic nerve fibers and geniculate cells, each geniculate cell receives its input from several optic fibers, which in turn split up to make synapses with several geniculate neurons.

The topographically ordered sequence of connections is not confined to one individual layer; even cells in the different layers are in register. Thus if a penetration is made perpendicularly through the geniculate, as the microelectrode passes from one layer to the next, records are obtained from successive cells driven by first one eye and then the other. The positions of the receptive fields remain in corresponding positions on the two retinas, representing the same region in the visual field.[55,56] For cells in the lateral geniculate nucleus, no extensive mixing of information or interaction between the eyes occurs, and binocularly excited cells (neurons with receptive fields in both eyes) are rare. However, inhibitory interneurons exist that receive inputs from both eyes.[58]

Surprisingly, perhaps, the responses from geniculate cells do not differ drastically from those of retinal ganglion cells (Figure 21). Geniculate neurons also have concentrically arranged antagonistic receptive fields, of either the "off"-center or the "on"-center type, but the contrast mechanism is more finely tuned by more equal matching of the inhibitory and excitatory areas. Thus the geniculate neurons, like retinal ganglion cells, require contrast for optimal stimulation but give even weaker responses to diffuse illumination.

Studies of receptive fields of lateral geniculate neurons are still incomplete. For example, there are interneurons whose contribution has not been established and pathways that descend from the cortex to end in the lateral geniculate nucleus.[59] Knowledge of the synaptic organization is likely to be increased through analyses made by intracellular electrodes.

Why is there more than one layer of the lateral geniculate devoted to each eye? It is now becoming apparent that different functional properties are represented in the different layers. For example, cells in the four dorsal layers of the monkey lateral geniculate nucleus (the parvocellular layers) resemble in their properties P ganglion cells and respond to lights of different colors with fine discrimination. In contrast, layers 1 and 2 (the magnocellular layers) contain M-like cells that give brisk responses and are not selective for wavelength. In the cat, X and Y fibers end in different sublayers of A, C, and A_1. "On"- and "off"-center cells are also segregated into different layers of the lateral genic-

Functional significance of layering

[58]Dubin, M. W. and Cleland, B. G. 1977. *J. Neurophysiol.* 40: 410–427.
[59]Gilbert, C. D. 1983. *Annu. Rev. Neurosci.* 6: 217–247.

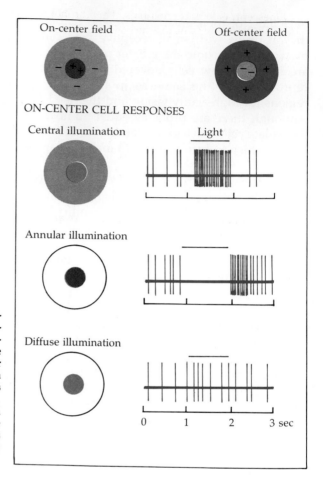

21

**RECEPTIVE FIELDS OF LATERAL GENICU-
LATE NUCLEUS CELLS.** The concentric receptive fields of cells in the lateral geniculate nucleus resemble those of ganglion cells in the
retina, consisting of "on"-center and "off"-center
types. The responses illustrated are from an
"on"-center cell in the lateral geniculate nucleus
of a cat. Bar above records indicates illumination.
The central and surround areas antagonize each
other's effects, so that diffuse illumination of the
entire receptive field gives only weak responses
(bottom record), less pronounced than in retinal
ganglion cells. (After Hubel and Wiesel, 1961.)

ulate nucleus of mink and ferret and, to some extent, monkey.[60–62] In
summary the lateral geniculate nucleus is a way station in which ganglion cell axons are sorted so that nearby cells receive inputs from the
same region of the visual field; the principle that neurons processing
like information are clustered together is retained.

It is appropriate to close this chapter with the following quotation
from Sherrington, written long before receptive fields were mapped for
single cells:[63]

> The chief wonder of all we have not touched on yet. Wonder of wonders,
> though familiar even to boredom. So much with us that we forget it all our
> time. The eye sends, as we saw, into the cell-and-fibre forest of the brain
> throughout the waking day continual rhythmic streams of tiny, individual

[60]LeVay, S. and McConnell, S. K. 1982. *Nature* 300: 350–351.
[61]Stryker, M. P. and Zahs, K. R. 1983. *J. Neurosci.* 10: 1943–1951.
[62]Schiller, P. H. and Malpeli, J. G. 1978. *J. Neurophysiol.* 41: 788–797.
[63]Sherrington, C. S. 1951. *Man on His Nature*. Cambridge University Press, Cambridge.

evanescent, electrical potentials. This throbbing streaming crowd of electrified shifting points in the spongework of the brain bears no obvious semblance in space-pattern, and even in temporal relation resembles but a little remotely the tiny two-dimensional upside-down picture of the outside world which the eyeball paints on the beginnings of its nerve-fibres to electrical storm. And the electrical storm so set up is one which affects a whole population of brain-cells. Electrical charges having in themselves not the faintest elements of the visual—having, for instance, nothing of "distance", "right-side-upness", no "vertical", nor "horizontal", nor "color", nor "brightness", nor "shadow", nor "roundness", nor "squareness", nor "contour", nor "transparency", nor "opacity", nor "near", nor "far", nor visual anything—yet conjour up all these. A shower of little electrical leaks conjures up for me, when I look, the landscape; the castle on the height, or when I look at him, my friend's face and how distant he is from me they tell me. Taking their word for it, I go forward and my other senses confirm that he is there.

SUGGESTED READINGS

General Reviews

Baylor, D. A. 1987. Photoreceptor signals and vision. Proctor Lecture. *Invest. Ophthalmol. Vis. Sci.* 28: 34–49.

Daw, N. W., Brunken, W. J. and Parkinson, D. 1989. The function of synaptic transmitters in the retina. *Annu. Rev. Neurosci.* 12: 205–225.

Daw, N. W., Jensen, R. J. and Brunken, W. J. 1990. Rod pathways in mammalian retinae. *Trends Neurosci.* 13: 110–115.

Dowling, J. E. 1987. *The Retina: An Approachable Part of the Brain.* Harvard University Press, Cambridge, MA.

Masland, R. H. 1988. Amacrine cells. *Trends Neurosci.* 11: 405–410.

McNaughton, P. A. 1990. Light response of vertebrate photoreceptors. *Physiol. Rev.* 70: 847–883.

Nathans, J. 1989. The genes for color vision. *Sci. Am.* 260(2): 42–49.

Schnapf, J. L. and Baylor, D. A. 1987. How photoreceptor cells respond to light. *Sci. Am.* 256(4): 40–47.

Shapley, R. and Perry, V. H. 1986. Cat and monkey retinal ganglion cells and their visual functional roles. *Trends Neurosci.* 9: 229–235.

Stryer, L. and Bourne, H. R. 1986. G proteins: A family of signal transducers. *Annu. Rev. Cell Biol.* 2: 391–419.

Wässle, H. and Boycott, B. B. 1991. Functional architecture of the mammalian retina. *Physiol. Rev.* 71: 447–480.

Yau, K. W. and Baylor, D. A. 1989. Cyclic GMP-activated conductance of retinal photoreceptor cells. *Annu. Rev. Neurosci.* 12: 289–327.

Original Papers

Baylor, D. A., Nunn, B. J. and Schnapf, J. L. 1987. Spectral sensitivity of cones of the monkey *Macaca fascicularis. J. Physiol.* 390: 145–160.

Baylor, D. A., Nunn, B. J. and Schnapf, J. L. 1984. The photocurrent, noise and spectral sensitivity of rods of the monkey *Macaca fascicularis. J. Physiol.* 357: 575–607.

Dacheux, R. F. and Raviola, E. 1986. The rod pathway in the rabbit retina: A depolarizing bipolar and amacrine cell. *J. Neurosci.* 6: 331–345.

Dowling, J. E. and Boycott, B. B. 1966. Organization of the primate retina: Electron microscopy. *Proc. R. Soc. Lond. B* 166: 80–111.

Fesenko, E. E., Kolesnikov, S. S. and Lyubarsky, A. L. 1985. Induction by cyclic GMP of cationic conductance in plasma membrane of retinal rod outer segment. *Nature* 313: 310–313.

Hecht, S., Shlaer, S. and Pirenne, M. H. 1942. Energy, quanta, and vision. *J. Gen. Physiol.* 25: 819–840.

Hubel, D. H. and Wiesel, T. N. 1961. Integrative action in the cat's lateral geniculate body. *J. Physiol.* 155: 385–398.

Kaneko, A. 1971. Electrical connexions between horizontal cells in the dogfish retina. *J. Physiol.* 213: 95–105.

Kaplan, E. and Shapley, R. M. 1986. The primate retina contains two types of ganglion cells, with high and low contrast sensitivity. *Proc. Natl. Acad. Sci. USA* 83: 2755–2757.

Kaupp, U.B., Niidome, T., Tanabe, T., Terada, S., Bonigk, W., Stühmer, W., Cook, N. J., Kangawa, K., Matsuo, H. and Hirose, T. 1989. Primary structure and functional expression from complementary DNA of the rod photoreceptor cyclic GMP-gated channel. *Nature* 342: 762–766.

Kuffler, S. W. 1953. Discharge patterns and functional organization of the mammalian retina. *J. Neurophysiol.* 16: 37–68.

Nathans, J. 1990. Determinants of visual pigment absorbance: Identification of the retinylidene Schiff's base counterion in bovine rhodopsin. *Biochemistry* 29: 9746–9752.

Schnapf, J. L., Nunn, B. J., Meister, M. and Baylor, D. A. 1990. Visual transduction in cones of the monkey *Macaca fascicularis*. *J. Physiol.* 427: 681–713.

Sterling, P., Freed, M.A. and Smith, R.G. 1988. Architecture of rod and cone circuits to the on-β ganglion cell. *J. Neurosci.* 8: 623–642.

Tamura, T., Nakatani, K. and Yau, K.-W. 1991. Calcium feedback and sensitivity regulation in primate rods. *J. Gen. Physiol.* 98: 95–130.

Yamashita, M. and Wässle, H. 1991. Responses of rod bipolar cells isolated from the rat retina to the glutamate agonist 2-amino-4-phosphonobutyric acid (APB). *J. Neurosci.* 11: 2372–2382.

CHAPTER SEVENTEEN

THE VISUAL CORTEX

Groups of neurons in the visual cortex process information about form, contrast, location, distance, movement, and color of objects in the external world. In the primary visual area (known as V_1, area 17, or striate cortex) neurons ignore uniform illumination. Those involved in the initial stages of pattern recognition require highly specific shapes or forms—in particular, lines or edges with a certain orientation or position on the retina. Some categories of neurons are specialized to respond to angles or corners or to movements in one direction but not in another. According to the type of information they carry, cortical neurons have been classified as simple and complex. Individual cortical neurons receive inputs from corresponding areas of retina in both eyes, with similar receptive field organization. Neighboring simple and complex cells share common functional properties; such cells are stacked in the form of columns that run at right angles to the cortical surface. For example, some columns of cells are best stimulated by vertical bars shone onto a small region of a particular area of the retina. Other columns of neurons in nearby areas of the visual cortex respond preferentially to a horizontal bar or to bars at other angles. A comparable segregation into columns has been shown for cells that are influenced preferentially by one or the other eye. Orientation and ocular dominance columns have been demonstrated in living animals by optical recording techniques. Within these columns, cells that deal with color are grouped in separate clusters known as blobs. Most blob cells' receptive fields are circular, with center–surround organization for different wavelengths of light. Aggregation of cortical neurons with related receptive field positions and functions makes it easier for them to interconnect so that they can perform the type of analysis required of them. Maps in visual areas of the cortex are not simple representations of the retina. Far larger areas of cortex are devoted to the fovea than to the peripheral retina.

Several lines of evidence indicate that specialized lines of transmission project from relay to relay through the primary visual cortex and then on to higher visual areas known as V_2, V_3, V_4, and V_5. These pathways, each of which is primarily (but not exclusively) concerned with information about either depth, movement, color, or form, are supplied by inputs originating from one of the two divisions of the lateral geniculate nucleus. (1) Neurons in the parvocellular pathway are concerned primarily with form and color and have small receptive fields. They are relatively insensitive to small differences in contrast so that bright patterns are required to activate them optimally. (2) Neurons in the magnocellular pathway have larger receptive fields, are insensitive to differences in wavelength, respond well to moving stimuli, and detect small changes in contrast. Magnocellular and parvocellular geniculate axons end in different layers of the primary visual cortex and supply segregated groups of neurons with only occasional mixing. Further projections from visual area 1 continue to supply groups of neurons with magnocellular or parvocellular properties arranged in orderly stripes in the second visual area. Still higher regions of the cortex such as V_4 and V_5 contain cells responding mainly to color or to movement. The separation of major modalities is reflected in human perceptual experiments in which contrast, color, movement, and depth are varied independently. Schemes that assume ordered series of connections can be derived to explain many features of how neurons respond selectively to specific stimuli such as a bar of light, a corner, or a colored square. Cells with relatively simple properties combine to form receptive fields of progressively greater complexity and visual content.

General problems and questions of numbers

Proceeding from the retina and lateral geniculate nucleus to the cerebral cortex raises questions that go beyond simple matters of technique. It has long been acknowledged that understanding the workings of any part of the nervous system requires knowledge of the cellular properties of its neurons—how they conduct and carry information and how they transmit that information from one cell to the next at synapses. Yet the monitoring of activity in single neurons might seem an unprofitable way to study higher functions in which large numbers of cells take part. The argument usually took (and still does, at times) the following form: The brain contains some 10^{10} or more cells. Even the simplest task or event, like a movement or looking at a line or square, engages hundreds of thousands of nerve cells in various parts of the nervous system. What chance do physiologists have of gaining insight into complex actions within the brain when they sample only one or a few of these units, a hopelessly small fraction of the total number?

On closer scrutiny, the logic of the argument about basic difficulties introduced by large numbers and complex higher functions is not so impeccable as it seems. As so frequently happens, some simplifying principle turns up, opening a new and clarifying view. In the case of the retina, after all, well over 100 million cells deal with the infinite variety of the world. What simplifies the situation is the major cell types that are laid out in an apparently well ordered manner as repeating units. Receptive field studies now indicate that this maze of 100 million cells and many hundred million cross-connections results in certain stereotyped kinds of responses within the cortex, even though the layering and the cytoarchitecture there are far more complicated.

A clear, continuous thread extending from signaling to perception is provided by the pioneering experiments begun by Kuffler[1] in the retina and followed into the visual cortex by Hubel and Wiesel. Hubel has given a vivid description of the early experiments on visual cortex in Stephen Kuffler's laboratory at Johns Hopkins University in 1958.[2] Since then our understanding of the physiology and anatomy of the cerebral cortex has blossomed through the experiments of Hubel and Wiesel and also through the large body of work for which it supplied the starting point or inspiration. As a result only in a specialized book devoted to the mammalian visual system is it now possible to do justice to copious papers dealing with complex problems such as color vision, depth perception, or the interconnections of neurons in various visual areas. In this chapter our aim is to give a brief, narrative description of signaling and cortical architecture in relation to perception, based on the classical work of Hubel and Wiesel and on recent experiments made by them together with their colleagues and by Zeki, Ferster, Daw, Schiller, Wong-Riley, Grinvald, Gilbert, Land, Lund, Van Sluyters, Van Essen, Tootell, and their colleagues.

The problem faced by Hubel and Wiesel in 1958 was to find out how signals denoting small, bright, dark, or colored spots in the retina could be transmuted into signals that conveyed information about the shape, size, color, movement, and depth of objects. Techniques that are routine now—such as optical recording, horseradish peroxidase injection, or brain scanning—had not yet been thought of. At the outset Hubel and Wiesel faced completely unanswered questions, which they tackled by assuming that visual centers in the cortex would perform their processing according to principles similar to those in the retina but at a more advanced level.

Strategies for exploring the cortex

One crucial strategy in their analysis was the use of stimuli that mimic those occurring under natural conditions. For example, edges, contours, and simple patterns presented to the eye revealed features of the organization that could never have been detected by using bright flashes without form. Another key to the success of Hubel and Wiesel's approach lay in asking not simply what stimulus evokes a response in a particular neuron, but what is the most effective stimulus that produces the highest-frequency discharge. Pursuit of this question through the various stages of the visual system has elicited many surprising and remarkable results. Early papers demonstrated that the receptive fields of simple and complex cells in the primary visual cortex could serve as building blocks for the initial stages of pattern recognition. In addition analysis of receptive fields clearly revealed the helpful, simplifying principle that neurons performing similar tasks are stacked in well-defined COLUMNS. Hubel has written:[3]

I think that the most important advance was the strategy of making long

[1]Kuffler, S. W. 1953. *J. Neurophysiol.* 16: 37–68.
[2]Hubel, D. H. 1982. *Nature* 299: 515–524.
[3]Hubel, D. H. 1982. *Annu. Rev. Neurosci.* 5: 363–370.

microelectrode penetrations through the cortex, recording from cell after cell, comparing responses to natural stimuli, with the object not only of finding the optimal stimulus for particular cells, but also of learning what the cells had in common and how they were grouped.

A major development has been to follow functionally defined connections from one area in the brain to another by sophisticated anatomical techniques. This has revealed the presence of separate pathways for processing information about specific attributes of visual stimuli. Inputs derived from the magnocellular and parvocellular divisions of the lateral geniculate nucleus (Chapter 16) project from relay to relay in groupings that are largely separate. Moreover, the distinctive attributes of magnocellular and parvocellular systems for the analysis of form, color, movement, and depth persist right through to consciousness. Before we describe the properties of cortical neurons, it is useful to summarize briefly the structure of the primary visual cortex.

Cytoarchitecture of the cortex

Visual information passes to the cortex from the lateral geniculate nucleus through the optic radiation. In the monkey, the part of the visual cortex where the optic radiation ends consists of a folded plate of cells about 2 mm thick (Figure 1). This region of the brain, primary visual cortex, visual area 1, or V_1, is also called the striate cortex or area 17, older terms that were based on anatomical criteria developed at the beginning of the twentieth century. V_1 lies posteriorly in the occipital lobe and can be recognized in cross section by its characteristic appearance. Incoming bundles of fibers form in this area a clear stripe that can be seen by the naked eye—hence the name *striate*. Adjacent extrastriate regions of cortex are also concerned with vision. The area that immediately surrounds V_1 is called V_2 (or area 18) and receives inputs from it (Figure 1). The exact boundaries of V_2, V_3, V_4, and V_5 cannot be defined by simple inspection of the brain, but a number of criteria exist.[4-9] For example, in V_2 the striate appearance is lost, large cells are found superficially, and coarse, obliquely running myelinated fibers are seen in the deeper layers. It will be shown that the different visual areas perform different types of analysis. Each area contains its own representation of the visual field projected in an orderly manner. Projection maps were made before the era of single-cell analysis, by shining light onto small parts of the retina and recording with gross electrodes. These maps, like those made by positron emission tomography, demonstrated that much more cortical area is devoted to representation of the fovea than to representation of the rest of the retina—as expected,

[4]Hubel, D. H. and Wiesel, T. N. 1965. *J. Neurophysiol.* 28: 229–289.
[5]Shipp, S. and Zeki, S. 1985. *Nature* 315: 322–325.
[6]Zeki, S. and Shipp, S. 1988. *Nature* 335: 311–317.
[7]DeYoe, E. A. and Van Essen, D. C. 1988. *Trends Neurosci.* 11: 219–226.
[8]DeYoe, E. A., Hockfield, S., Garren, H. and Van Essen, D. C. 1990. *Vis. Neurosci.* 5: 67–81.
[9]Maunsell, J. H. and Newsome, W. T. 1987. *Annu. Rev. Neurosci.* 10:363–401.

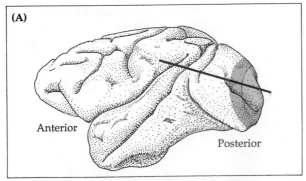

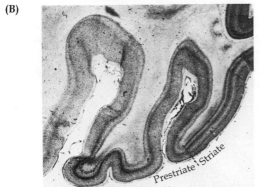

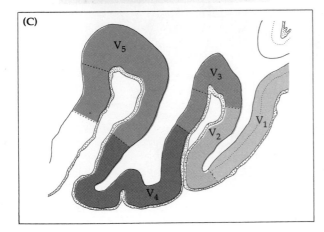

1

RELATION OF PRIMARY VISUAL CORTEX V₁ to V₂, V₃, V₄, and V₅ in the monkey. **(A)** The cortex and the plane of section passing through V₁ and V₂. The boundary between V₁ and V₂ is unambiguous. **(B,C)** A section through the occipital cortex. In **(B)**, the boundary between V₁ and V₂ occurs at the arrow, where the striped appearance is lost. The boundaries between V₂, V₃, V₄, and V₅ are revealed by a combination of physiological and anatomical studies; V₄ is divided into subregions. (After Zeki, 1990; micrograph courtesy of S. Zeki and M. Rayan.)

since form vision is principally confined to foveal and parafoveal areas.[10–12]

In sections of the cortex, neurons can be classified according to their shapes. The two principal groups of neurons are stellate cells and pyramidal cells. Examples of these cells are shown in Figure 2. The

[10]Talbot, S. A. and Marshall, W. H. 1941. *Am. J. Ophthalmol.* 24:1255–1264.
[11]Daniel, P. M. and Whitteridge, D. 1961. *J. Physiol.* 159: 203–221.
[12]Fox, P. T. et al. 1987. *J. Neurosci.* 7: 913–922.

(A)

(B)

(C)

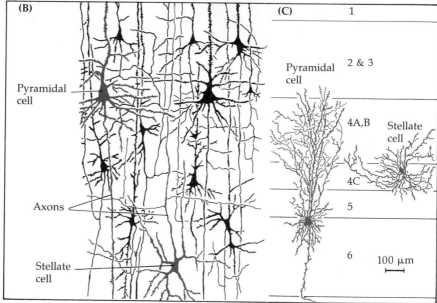

2   **ARCHITECTURE OF VISUAL CORTEX. (A)** Distinct layering of cells in a section
of striate cortex of the macaque monkey, stained to show cell bodies (Nissl stain).
Fibers arriving from the lateral geniculate nucleus end in layers 4A, 4B, and 4C. **(B)**
Drawing of pyramidal and stellate cells in the visual cortex of the cat. The connections
stained with Golgi technique for the most part run radially through the thickness of the
cortex and extend for relatively short distances laterally. **(C)** Drawing from photographs
of a pyramidal cell and a spiny stellate cell in the cat cortex that had been injected with
horseradish peroxidase after their activity had been recorded. Both were simple cells.
(A from Hubel and Wiesel, 1972; B after Ramón y Cajal; C from Gilbert and Wiesel,
1979.)

main differences are in the lengths of the axons and the shapes of the cell bodies. The axons of pyramidal cells are longer, dip down into white matter, and leave the cortex; those of stellate cells tend to terminate locally. These two groups of cortical cells exhibit variations, such as the presence or absence of spines on the dendrites, that bear on their functional properties.[13] Thus smooth stellate cells (without spines) are inhibitory and secrete GABA. There are other fancifully named neurons (double bouquet cells, chandelier cells, basket cells, and crescent cells) as well as neuroglial cells.

Characteristically, the processes of cells run for the most part in a radial direction, up and down through the thickness of the cortex (at right angles to the surface). In contrast, many (but not all; see Figure 16) of their lateral processes are short. Connections between areas V_1, V_2, V_3, V_4, and V_5 are made by axons that dip down and run in bundles through the white matter to surface again elsewhere (see Figure 4).

An important development has been the discovery of regional specializations in the cytoarchitecture of the visual cortex of the monkey. Discrete, circumscribed clusters of neurons have been found mainly in layers 2 and 3, but also in layers 5 and 6. These patches, or "blobs," form a highly regular array ("a polka-dot pattern . . . as if the animal's brain had the measles. . . . we call them 'blobs' because the word is graphic, legitimate and seems to annoy our competitors."[2,14]) (Figure 3). Blobs were first seen in cortical tissue stained for cytochrome oxidase,

Cytochrome oxidase-stained "blobs"

[13]Lund, J. S. 1988. *Annu. Rev. Neurosci.* 11: 253–288.
[14]Hubel, D. H. 1988. *Eye, Brain and Vision*. Scientific American Library, New York.

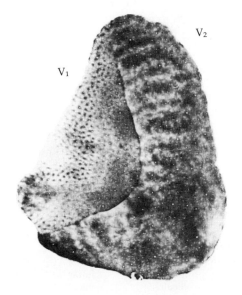

V_2

V_1

3

SECTION OF MONKEY VISUAL CORTEX stained by cytochrome oxidase to show blobs in V_1 and stripes in V_2. The blobs are arranged in polka-dot pattern. A clear boundary can be discerned between V_1 and V_2. At this line the blobs change to stripes, thick and thin, running at right angles to the border. The border with area V_3, adjacent to V_2, is less well defined. (From Livingstone and Hubel, 1988.)

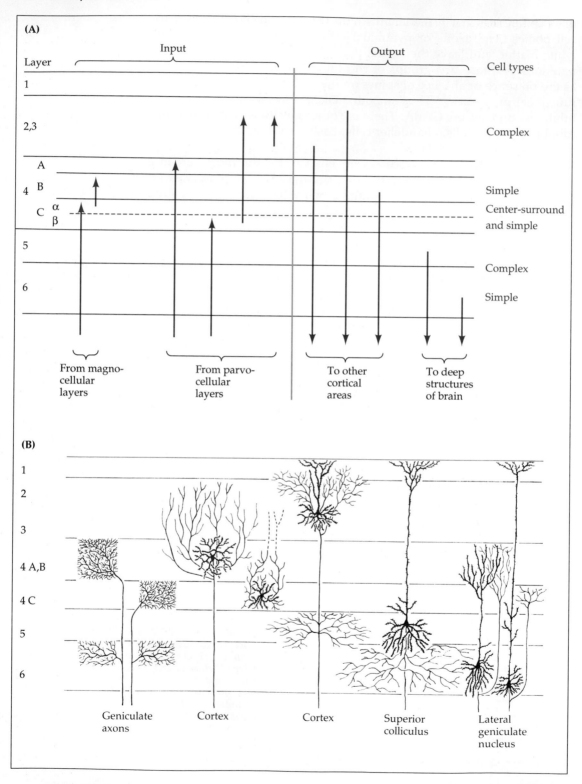

(A)

Input

Output

Layer

Cell types

1

2,3   Complex

4   A
    B   Simple
    C   α
        β   Center-surround and simple

5   Complex

6   Simple

From magno-cellular layers

From parvo-cellular layers

To other cortical areas

To deep structures of brain

(B)

1
2
3
4 A,B
4 C
5
6

Geniculate axons

Cortex

Cortex

Superior colliculus

Lateral geniculate nucleus

◄ **4** CONNECTIONS OF VISUAL CORTEX. (A) The layers with their various inputs, outputs, and types of cells. Note that inputs from the lateral geniculate nucleus end mainly in layer 4. Those arising from magnocellular layers end principally in 4Cα and 4B, while those from parvocellular layers end in 4A and 4Cβ. Simple cells are found mainly in layers 4 and 6, complex cells in layers 2, 3, 5, and 6. (B) Principal arborizations of processes in layers 1 to 6 of cat cortex. In addition to these vertical connections, many cells have long horizontal connections running within a layer to distant regions of the cortex. (A after Hubel, 1988; B after Gilbert and Wiesel, 1981.)

an enzyme indicative of high metabolic activity.[15,16] In macaque monkeys, the patches of cytochrome oxidase stain are precisely arranged in parallel rows about 0.5 mm apart. At the border of V_1 and V_2 the pattern of cytochrome oxidase stain changes abruptly. In V_2 are seen alternating thick, thin, dark, and pale stripes that run perpendicular to the V_1–V_2 border.[17,18] The properties and connections of blob neurons are described below in relation to higher visual areas and color vision.

A general feature of the mammalian cortex is that the cells are arranged in six layers within the gray matter (Figure 4). These layers vary in appearance from area to area in the cortex depending on the density of cell packing and the thickness. Incoming geniculate fibers end for the most part, but not exclusively, in layer 4. The inputs are shown in Figure 4. Geniculate afferents also supply cells in layer 6; superficial layers receive inputs from the pulvinar, a region of the thalamus.[13,19] Numerous cortical cells, especially those in layers 2, upper 3, and 5 receive inputs from neurons within the cortex. The specialized projection to and from blobs in layer 3 of the macaque monkey constitutes a quite separate system.

Inputs, outputs, and layering of cortex

The outputs from layers 6, 5, 4, 3, and 2 are shown in Figure 4 and are as follows:[13,19] Axons of cells in layer 6 project back to the lateral geniculate nucleus and to another deep structure, the claustrum. Cells in layer 5 also project down, principally to the superior colliculus, a structure in the midbrain concerned with eye movements. Cells in layer 4 project to layers 3 and 2; they also send axons out to end in other areas of cortex. Cells in layers 2 and 3 provide a major output to other cortical areas; they also project down to layer 5. A single cell that carries efferent signals out of the cortex can also mediate intracortical connections from one layer to another. For example, axons of a cell in layer 6, in addition to supplying the geniculate, end in layer 4. From this anatomy a general pattern emerges: Information from the retina is transmitted to cells (mainly in layer 4) by geniculate axons, handed on from neuron to neuron through the thickness of the cortex and then sent out to other regions of the brain by fibers looping down through white matter.

[15]Wong-Riley, M. 1979. *Brain Res.* 171: 11–28.
[16]Wong-Riley, M. T. 1989. *Trends Neurosci.* 12: 94–101.
[17]Hendrickson, A. E. 1985. *Trends Neurosci.* 8: 406–410.
[18]Livingstone, M. S. and Hubel, D. H. 1984. *J. Neurosci.* 4: 309–356.
[19]Gilbert, C. D. 1983. *Annu. Rev. Neurosci.* 6: 217–247.

It is important to note that layer 4, the region in which many of the geniculate fibers end, is further subdivided into A, B, and C. Layer 4C is itself subdivided into two sublayers: 4Cα above and 4Cβ below.[13] Fibers of the lateral geniculate originating in parvocellular (P) or magnocellular (M) divisions end in discrete sublaminae. In monkey visual cortex, parvocellular inputs supply cells in layers 4A and 4Cβ and the upper part of layer 6. Magnocellular inputs supply cells in 4Cα and the deeper part of 6. Cells in layer 4Cβ receiving parvocellular input supply layers 2 and 3; cells in layer 4Cα receiving magnocellular input in turn supply 4B. Both M and P systems supply the blobs in layer 3. It will be shown below that M and P projections retain distinct identities with characteristic properties in visual areas V_2, V_3, V_4, and V_5. Moreover, a monoclonal antibody preferentially labels the distinct magnocellular pathways in the visual areas of the cortex.[8] Thus, separation into M and P divisions first observed in the retina and then in the lateral geniculate nucleus is in general maintained in the cortex.

Also preserved in layer 4 is the segregation of inputs derived from the two eyes. In adult cats and monkeys, the cells in one layer of the geniculate nucleus receiving their inputs from one eye, project to aggregates of target cells in layer 4C separate from those supplied by the other eye. These aggregates are grouped as alternating stripes or bands of cortical cells that are supplied exclusively by one eye or the other.[20] Above and below layer 4 most cells are driven by both eyes, although one eye usually dominates. The way in which inputs from the two eyes are combined at the cellular level in the deeper and more superficial cortical layers is described below. The original demonstration of eye segregation and ocular dominance in the primary visual cortex was made by Hubel and Wiesel with electrical recording techniques.[21] They used the term COLUMN, a term introduced by Mountcastle for somatosensory cortex,[22] to describe the organization of cells sharing similar properties stacked on top of one another, through the thickness of the cortex. Although the aggregates of neurons are shaped more like slabs, the term column has become established and is generally retained.

A variety of experimental procedures has been developed for demonstrating the alternating groups of cells in layer 4 supplied by the left or the right eye. One of the first was to make a small lesion in one layer of the lateral geniculate nucleus; degenerating terminals subsequently appear in layer 4 in a characteristic pattern of alternating stripes (or slabs).[23] These correspond to areas driven by the eye in whose line of connection the lesion is made. Later a striking demonstration of the ocular dominance pattern was provided by the transport of radioactive amino acids from one eye. The principle is to inject an amino acid such as proline or leucine into the vitreous of one eye, from which it is taken up by nerve cell bodies of the retina and incorporated into protein. The

[20]Hubel, D. H. and Wiesel, T. N. 1977. *Proc. R. Soc. Lond. B* 198: 1–59.
[21]Hubel, D. H. and Wiesel, T. N. 1962. *J. Physiol.* 160: 106–154.
[22]Mountcastle, V. B. 1957. *J. Neurophysiol.* 20: 408–434.
[23]Hubel, D. H. and Wiesel, T. N. 1972. *J. Comp. Neurol.* 146: 421–450.

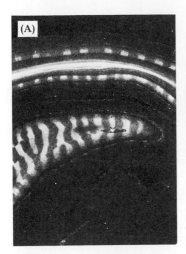

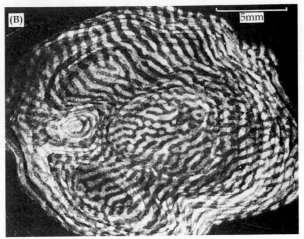

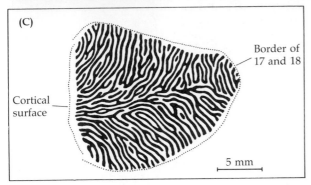

5 OCULAR DOMINANCE COLUMNS IN MONKEY CORTEX demonstrated by injection of radioactive proline into one eye. (A,B) Autoradiographs photographed with dark-field illumination in which the silver grains appear white. (A) At the top of the picture the section passes through layer 4 of the visual cortex at right angles to the surface, displaying columns cut perpendicularly. In the center, layer 4 has been cut horizontally, showing that the columns consist of longer slabs. (B) Reconstruction made from numerous horizontal sections of layer 4C in another monkey in which the ipsilateral eye had been injected. (No single horizontal section can encompass more than a part of layer 4 because of the curvature of the cortex.) Dorsal is above, medial to the right. In both (A) and (B), the ocular dominance columns appear as stripes of equal width supplied by one eye or the other. (C) Reconstruction of the pattern of ocular dominance columns over the entire exposed part of layer 4C. Scale, 5 mm. (A and B courtesy of S. LeVay; C from LeVay, Hubel and Wiesel, 1975.)

labeled protein is then transported from ganglion cells through the optic nerve fibers to their terminals within the lateral geniculate nucleus. An extraordinary feature is that the label is also transferred from neuron to neuron across synapses.[24–26] Thus the endings of geniculate fibers are labeled in layer 4. Figure 5 clearly shows the radioactivity around the terminals of geniculate axons supplied by the injected eye. Zones that

[24]Specht, S. and Grafstein, B. 1973. *Exp. Neurol.* 41: 705–722.

[25]LeVay, S., Hubel, D. H. and Wiesel, T. N. 1975. *J. Comp. Neurol.* 159: 559–576.

[26]LeVay, S. et al. 1985. *J. Neurosci.* 5: 486–501.

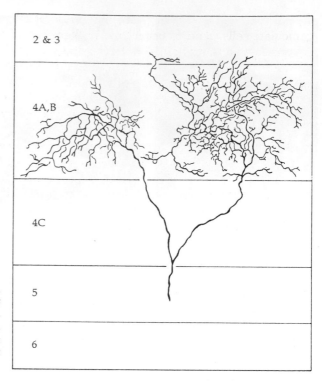

2 & 3	
4A,B	
4C	
5	
6	

6

"OFF"-CENTER Y AXON FROM LATERAL GE-NICULATE NUCLEUS terminating in layer 4 of cat. The axon was injected with horseradish peroxidase through a microelectrode. The terminals are grouped in two clusters separated by a vacant zone supplied by the other eye. (After Gilbert and Wiesel, 1979.)

get their inputs from the other eye remain clear. The striped arrangement demonstrated in this way is the same as that produced by lesions of the lateral geniculate nucleus, but now the pattern is seen in all parts of the primary visual cortex. The pattern of eye segregation is also defined by the blobs; blobs stained by cytochrome oxidase lie in the center of each ocular dominance column.[13,27,28] It is also possible to observe ocular dominance columns and blobs in live monkeys by optical recording techniques (see Figure 20 below). At the cellular level a similar pattern has been revealed in layer 4 by retrograde uptake or by injection of horseradish peroxidase into individual axons of the lateral geniculate nucleus as they approach the cortex.[13,29,30] The axon shown in Figure 6 is an "off"-center afferent that gives transient responses to dark moving spots. It ends in two distinct clumps of processes in layer 4. The processes are separated by a blank area of a size appropriate for the territory supplied by the other eye. Together these morphological studies have borne out and added depth to the original description of ocular dominance columns presented by Hubel and Wiesel in 1962.

[27]Hendrickson, A. E., Hunt, S. P. and Wu, J.-Y. 1981. *Nature* 292: 605–607.
[28]Horton, J. C. and Hubel, D. H. 1981. *Nature* 292: 762–764.
[29]Gilbert, C. D. and Wiesel, T. N. 1979. *Nature* 280: 120–125.
[30]Blasdel, G. G. and Lund, J. S. 1983. *J. Neurosci.* 3:1389–1413.

Responses of cortical neurons, like those of the retinal ganglion and geniculate cells, tend to occur on a background of maintained activity. A consistent observation is that discharges of cortical neurons are not significantly influenced by diffuse illumination of the retina. Almost complete insensitivity to diffuse light is an intensification of the process already noted in the retina and the lateral geniculate nucleus; it results from an equally matched antagonistic action of the inhibitory and excitatory regions in the receptive fields of cortical cells. The neuronal firing rate is altered only when certain demands about the position and form of the stimulus on the retina are met. The receptive fields of most cortical neurons have configurations that differ from those of retinal or geniculate cells, so that spots of light often have little or no effect. In his Nobel address, Hubel described the experiment in which Wiesel and he first recognized this essential property:[2]

> Our first real discovery came about as a surprise . . for three or four hours we got absolutely nowhere. Then gradually we began to elicit some vague and inconsistent responses by stimulating somewhere in the midperiphery of the retina. We were inserting the glass slide with its black spot into the slot of the ophthalmoscope when suddenly, over the audio monitor, the cell went off like a machine gun. After some fussing and fiddling, we found out what was happening. The response had nothing to do with the black dot. As the glass slide was inserted its edge was casting onto the retina a faint but sharp shadow, a straight dark line on a light background. That was what the cell wanted, and it wanted it, moreover, in just one narrow range of orientations. This was unheard of. It is hard now to think back and realize just how free we were from any idea of what cortical cells might be doing in an animal's daily life.

By following a progression of clues, Hubel and Wiesel worked out the appropriate light stimuli for various cortical cells; they classified the receptive fields on the basis of relative simplicity of construction as SIMPLE and COMPLEX. Each of these categories includes a number of subgroups and important variables that bear on perceptual mechanisms.

David H. Hubel (left) and Torsten N. Wiesel during an experiment, about 1969. The cat, not shown, also faces the screen.

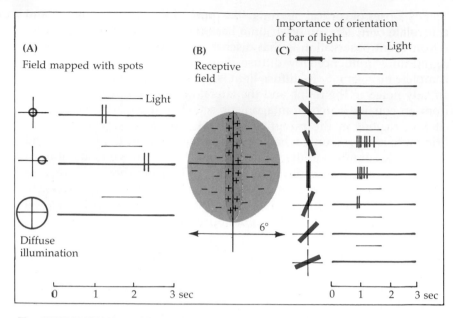

7 **RESPONSES OF A SIMPLE CELL in cat striate cortex to spots of light (A) and bars (C).** The receptive field (B) has a narrow central "on" area (+) flanked by symmetrical antagonistic "off" areas (−). The best stimulus for this cell is a vertically oriented light bar (1° × 8°) in the center of its receptive field (fifth record from top in C). Other orientations are less effective or ineffective. Diffuse light (third record from top in A) does not stimulate. Illumination is indicated by bar. (After Hubel and Wiesel, 1959.)

Functional properties of simple cells

The receptive fields of simple cells exhibit several varieties.[14,21,31,32] In the monkey, one class of cortical cells with properties even more elementary than those of simple cells has a concentric center-surround receptive field, much like that of a geniculate cell or a ganglion cell. Such neurons are found exclusively in layer 4C, the region where most geniculate fibers end. Little or no obvious transformation has occurred in the properties or the meaning of the signal. Most simple cells, however, have quite different receptive fields. They are found in layers 4 and 6 and deep in layer 3. All these regions also receive direct input from the geniculate. One type of simple cell has a receptive field that consists of an extended narrow central portion flanked by two antagonistic areas. The center may be either excitatory or inhibitory. Figure 7 shows a receptive field of a simple cell in the striate cortex mapped out with spots of light that excited weakly in the center (because the spots covered only a small part of the central area (marked with +) and inhibited in the surround (−).

The requirements of such a simple cell are exacting. For optimal activation it needs a bar of light that is not more than a certain width, that entirely fills the central area, and that is oriented at a certain angle.

[31]Hubel, D. H. and Wiesel, T. N. 1959. *J. Physiol.* 148: 574–591.
[32]Hubel, D. H. and Wiesel, T. N. 1968. *J. Physiol.* 195: 215–243.

This is illustrated by the records in Figure 7. As one might expect for this cell, a vertically oriented slit of light (fifth record from top) is most effective. With a small deviation from this requirement, the response is diminished.

Different cells have receptive fields requiring a wide range of different orientations and positions. A new population of simple cells is therefore activated by rotating the stimulus or by shifting its position in the visual field. The distribution of inhibitory–excitatory flanks in various simple receptive fields may not be symmetrical, or the field may consist of two longitudinal regions facing each other—one excitatory, the other inhibitory. Figure 8 shows examples of four receptive fields, all with a common axis of orientation but with differences in the distribution of areas within the field. For the receptive field in Figure 8A, a narrow light slit with appropriate orientation elicits the best response. A dark bar in the same place with light flanks suppresses ongoing spontaneous activity. Cells with the field shapes shown in Figure 8B and 8C produce the opposite responses. For the field shown in Figure 8D, an edge with light on the left and darkness on the right is most effective for "on" responses, whereas reversing the dark and light areas is best for "off" discharges.

Another type of simple cell has been found. Once again the orientation and the position of the stimulus are critical and the field is made of antagonistic "on" and "off" areas. But in addition the length of the bar or edge is important: Stretching the bar beyond an optimal length reduces its effectiveness as a stimulus.[33] It is as though there is an additional "off" area that exists at the top end or the bottom end of the fields shown in Figure 8 and that tends to suppress firing. Hence, for such simple cells, the best stimulus is an appropriately oriented bar or edge that stops in a particular place (END-INHIBITION or END-STOPPING).

The common properties of all simple cells are (1) that they respond best to a properly oriented stimulus positioned so as not to encroach

[33]Gilbert, C. D. 1977. *J. Physiol.* 268: 391–421.

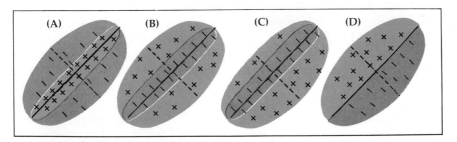

8 RECEPTIVE FIELDS OF SIMPLE CELLS in cat striate cortex. In practice all possible orientations are observed for each type of field. The optimal stimuli are for (A), a narrow slit (or bar) of light in the center; for (B) and (C), a dark bar; and for (D), an edge with dark on the right. Considerable asymmetry can be present, as in (C). (After Hubel and Wiesel, 1962.)

on antagonistic zones, and (2) that stationary slits or spots can be used to define "on" and "off" areas. Another constant and remarkable feature is that in spite of all the different proportions of inhibitory and excitatory areas, the two contributions match exactly and cancel each other's effectiveness, so that diffuse illumination of the entire receptive field produces a feeble response at best. The "off" areas in cortical fields are not always able to initiate impulses in response to dark bars. Frequently (particularly in end-inhibition and in the more elaborate fields to be described shortly) illumination of the "off" area can only be detected as a reduction in the discharge evoked from the "on" area. Moving edges or bars of the appropriate orientation are highly effective in initiating impulses.[14,21]

In simple cells the optimal width of the narrow light or dark bar is comparable to the diameters of the "on"- or "off"-center regions in the doughnut-shaped receptive fields of ganglion or lateral geniculate cells (Chapter 16). Once again there is a specialization for detecting differences, but the spotlike contrast representation of ganglion cells has been transformed and extended into a line or an edge. Resolution has not been lost but has been incorporated into a more complex pattern. Cortical cells that have fields derived from the fovea are best excited by narrower bars than those that excite cells with fields in the peripheral area of the eye, corresponding to the smaller receptive fields of foveal ganglion cells.

Properties of complex cells

In recordings made from individual neurons in the visual cortex, one finds, in addition to simple cells, other neurons that behave quite differently. These complex cells, which are abundant in layers 2, 3, and 5, have two important properties in common with simple cells: They require specific field-axis orientation of a dark–light boundary, and illumination of the entire field is ineffective.[21] The demand, however, for precise positioning of the stimulus, observed in simple cells, is relaxed in complex cells. In addition, there are no longer distinct "on" and "off" areas that can be mapped with small spots of light. As long as a properly oriented stimulus falls within the boundary of the receptive field, most complex cells will respond, as in the examples illustrated in Figure 9. The meaning of the signals arising from complex cells, therefore, differs significantly from that of simple cells. The simple cell localizes an oriented bar of light (Figure 8) to a particular position within the receptive field, while the complex cell signals the abstract concept of *orientation without strict reference to position.* Although the receptive field has been enlarged, resolution is not lost but incorporated into a more elaborate design.

Two main classes of complex cells can be distinguished; both respond best to moving edges or slits of fixed width and precise orientation. One type of cell gives the responses shown in Figure 9. For such cells the response improves as the edge or slit becomes longer, up to a point; making the stimulus even longer gives rise to no additional effect. Other complex cells, like end-stopped simple cells, require slits or edges

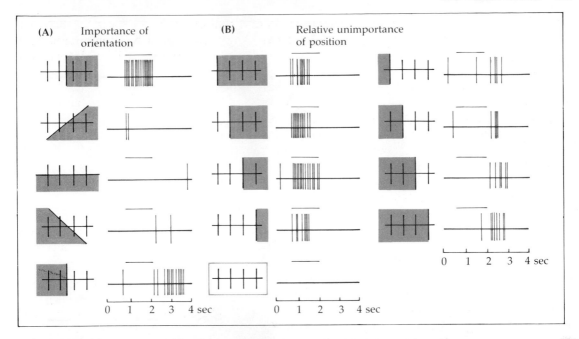

9 RESPONSES OF A COMPLEX CELL in the striate cortex of the cat. The cell responds best to a vertical edge. (A) With light on the left and dark on the right (first record), there is an "on" response. With light on the right (fifth record), there is an "off" response. Orientation other than vertical is less effective. (B) The position of the border within the field is not important. Illumination of the entire receptive field (bottom record) gives no response. (After Hubel and Wiesel, 1962.)

that stop.[4,34] (In an earlier nomenclature, these would have been called hypercomplex cells, a term that is now no longer used.) The best stimulus for such cells, therefore, requires not only a certain orientation but in addition some discontinuity, such as a line that stops, an angle, or a corner. Figure 10 shows the responses of a cell that responds best to an edge with dark below and light above at an angle of about 45°. Diffuse illumination, other axis orientations, or spots are without effect. At first glance one might classify this as a complex cell similar to that giving the responses shown in Figure 9. The difference, however, becomes clear in the fourth and fifth records from the top in Figure 10, when the dark edge is extended; this elongation to the right depresses the response in this complex cell. Interestingly, diffuse illumination of the right-hand field does not diminish the response (last record). It is therefore not a simple "off" area. One description of the best stimulus for this cell is a corner; moreover, the stimulus must move in one direction. Such directional sensitivity is a feature commonly found in complex cells.

Other variants of complex cells may be mentioned. For example, the

[34]Palmer, L. A. and Rosenquist, A. C. 1974. *Brain Res.* 67: 27–42.

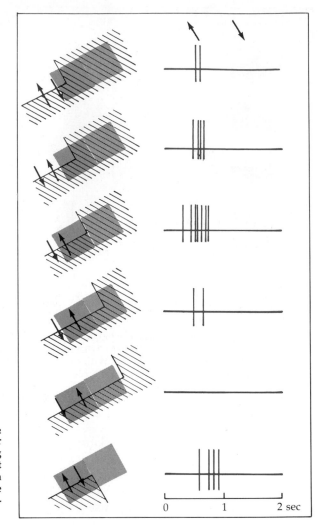

10

END-INHIBITION OF A COMPLEX CELL in V₂ (area 18) of the cat cortex. The best stimulus for this cell (third record from the top) is a moving (arrows), oriented edge (a corner) that does not encroach on the antagonistic right-hand portion of the receptive field. The records also show the selective sensitivity of the cell to upward movement. (After Hubel and Wiesel, 1965.)

cell giving the responses shown in Figure 11 responds to a narrow bar or slit; unlike the cell just discussed, it behaves as though an extra end-inhibitory effect were added on the left side. The field can be considered as made up of three components, two inhibitory and one excitatory. The best stimulus for this complex cell is a moving, oriented bar or slit that covers the middle region. The stimulus, however, must not extend into either of the two inhibitory lateral portions of the receptive field. When widened or displaced in either direction, the stimulus is weakened or ineffective. This cell, therefore, is even more specific in its requirements. It signals that a narrow dark-oriented line is stopping in this part of the retina. It does not tell exactly where the line is in terms of position in the plane of movement (up or down), but it does indicate that (1) the line is oriented; (2) the line is not wider than the amount

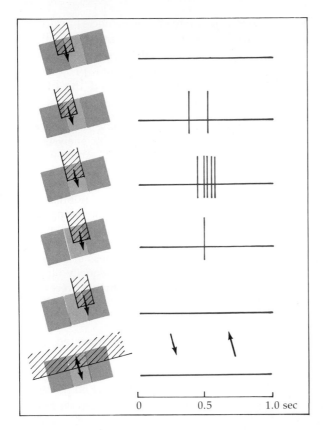

11

RESPONSES OF A COMPLEX CELL with end-inhibition in cat cortex. The best stimulus is a narrow dark tongue in the middle moving downward. Shifting the dark edge sideways or widening it encroaches on antagonistic flanking areas and diminishes the response. (After Hubel and Wiesel, 1965.)

specified; and (3) the line stops within the central part of the field. Such complex cells are found largely in layers 3 and 5. Often it takes several hours of trying out different patterns on the screen in front of the animal before the most effective stimulus can be identified.

At this stage the following question is frequently asked: How can you be sure that the best stimuli for the simple and complex cells are in fact straight lines rather than curves, the letter E, or perhaps the shape of a mouse or a banana? In an experiment, of course, it is not possible to try all combinations on each cell. In general, however, it has been found in many laboratories that curves, circles, and complex patterns do not stimulate cells in V_1 as effectively as straight bars or edges.[14] Naturally, "straight" can only be considered in relation to the size of the receptive fields, so that a segment of a large circle on the retina is virtually a straight line. Just as a line can be constructed out of closely spaced dots, any desired shape can be formed out of straight lines of the proper length. Cells that respond selectively to more complex shapes, such as faces, have been found in higher visual centers.[35,36]

[35]Damasio, A. R., Tranel, D. and Damasio, H. 1990. *Annu. Rev. Neurosci.* 13: 89–109.
[36]Perett, D. I., Mistlin, A. J. and Chitty, A. J. 1987. *Trends Neurosci.* 10: 358–364.

Receptive fields from
both eyes converging
on cortical neurons

Binocular interaction is introduced here because it provides another example of part of the design the brain uses to end up with perception of form. When we look at an object with one or both eyes, we see only one image, even if the size and the position of the object's projection are slightly different on the two retinas. Sherrington posed the problem:[37]

> How habitually and unwittingly the self regards itself as one is instanced by binocular vision. Our binocular visual field is shown by analysis to presuppose outlook from the body by a single eye centered at a point in the midvertical of the forehead at the level of the root of the nose. It, unconsciously, takes for granted that its seeing is done by a cyclopean eye having a center of rotation at the point of intersection just mentioned. In the visual field it obtains visual depth by unknowingly combining . . . crossed images of not too great lateral disparation . . . . Oneness is obtained by compromise between differences, if not too great, offered to the perceiving "self." There are other perceptual instances. The brightness of a binocular field differs hardly sensibly from that of either of two equally illuminated uniocular fields composing it. But the quantity of stimulus received by the eyes is roughly double in the binocular observation than that which it is in the uniocular.

Interestingly, well over 100 years ago Johannes Müller suggested that individual nerve fibers from the two eyes might fuse or become connected to the same cells in the brain. Thereby, he almost exactly anticipated Hubel and Wiesel's results.[21,31] They found that about 80 percent of all cortical neurons in the visual areas of the cat can be driven from both eyes. Since the neurons in the various layers in the lateral geniculate nucleus are predominantly innervated from one eye or the other, the first opportunity for significant interaction between the eyes must occur in the cortex. As mentioned above, the separation is maintained in the fourth layer of area 17, where each simple cell is driven by only one eye, the other being without effect. Mixing between the two eyes occurs in the subsequent relay stations, that is, in layers deeper toward the white matter and in more superficial cortical layers (toward the roots of the hair).

Examination of the receptive fields of a binocularly driven cell shows that (1) they are usually in exactly equivalent positions in the visual field of the two eyes, (2) their preferred orientation is the same, and (3) the corresponding areas in the receptive fields add to each other's effect. An example of synergistic action between the two eyes is shown for a simple cell in Figure 12. Shining light onto an "on" region in the left eye sums with illumination onto the "on" area of the right eye. Simultaneous illumination of antagonistic areas in the two eyes reduces ongoing responses and increases "off" discharges. Such cells would be useful for unifying images from the two eyes.

[37]Sherrington, C. S. 1906. *The Integrative Action of the Nervous System.* (1961 Edition). Yale University Press, New Haven.

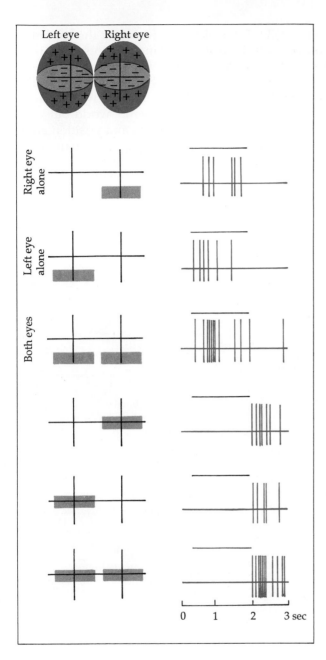

12

BINOCULAR ACTIVATION of a simple cortical neuron that has identical receptive fields in both eyes. Simultaneous illumination of corresponding "on" areas (+) of right and left receptive fields is more effective than stimulation of one alone (upper three records). In the same way, stimulation of "off" areas (−) in the two eyes reinforces each other's "off" discharges (lower records). Cells used for depth perception have receptive fields in the two eyes in disparate regions of the visual field. Such cells require that the bar is farther from or closer to the eye than the plane of focus (see text). (After Hubel and Wiesel, 1959.)

For depth perception there exists another binocular specialization of receptive fields. An object out of the plane of focus casts images on disparate parts of the two retinas. Neurons with properties that fit them for depth perception have been found in monkey cortex, particularly in areas V_2 and V_5. For such cells the best stimulus is an appropriately oriented bar in front of the plane of focus (for certain cells), or beyond it (for others).[38-40] Impulses fail to be evoked by presenting the bar only to one eye or to the other, or to both in the plane of focus. It is the disparity of the position on the two retinas that these cells require.

A separate problem concerns the way in which the two cortices are knit together to produce a single image of the body and the world. Each hemicortex is wired to perceive one half of the external world but not the other. This is equally true for sensations of touch and position and constitutes a general situation in relation to perception. It is natural to wonder what happens in the midline. How do the two sides of our brain mesh the right world and the left world together with no hint of a seam or discontinuity?

The obvious way to preserve continuity is to join the left and the right visual fields together at the midline in register. To achieve this a

[38]Fischer, B. and Poggio, G. F. 1979. *Proc. R. Soc. Lond. B* 204: 409–414.
[39]Ferster, D. 1981. *J. Physiol.* 311: 623–655.
[40]Hubel, D. H. and Livingstone, M. S. 1987. *J. Neurosci.* 7: 3378–3415.

BOX 1   CORPUS CALLOSUM

The general question of transfer of information between the hemispheres has been studied in a most rewarding manner in humans and in monkeys by Sperry, Myers, Gazzaniga, and their colleagues.[42-44] Concentrating on the coordinating role of the corpus callosum, a bundle of fibers that runs between the two hemispheres, they have shown that the fibers are actually involved in the transfer of information and learning from one hemisphere to the other. To cite one example, a normal person can name an object, such as a coin or a key, when it is placed in either hand (stereognosis). After section of the corpus callosum, however, a right-handed person can name the object only when it is placed in the right hand; the information from the right hand crosses before reaching the cortex and projects to the left hemisphere. It is in the *left* hemisphere that the main area responsible for language lies. What happens when the object is placed in the left hand, which projects to the right hemisphere? Information still reaches consciousness when the key is placed in the left hand. There is, however, no way in which the concept *key*

[42]Sperry, R. W. 1970. *Proc. Res. Assoc. Nerv. Ment. Dis.* 48: 123–138.
[43]Gazzaniga, M. S. 1967. *Sci. Am.* 217(8): 24–29.
[44]Gazzaniga, M. S. 1989. *Science* 245: 947–952.

can be verbally expressed because the center for language, which is situated on the left side of the brain, cannot be reached without the corpus callosum. Thus a person without a corpus callosum can recognize an object with the left hand and may be able to use it, but cannot say the word *key* in relation to the object. The figure expresses some of these ideas. Other experiments on the corpus callosum provide surprising insight into higher functions. For example, when deprived of cross connections, the two hemispheres can lead virtually separate existences.

The fusing, or knitting together, of the two fields of vision is mediated by fibers in the corpus callosum. Certain cells have receptive fields that straddle the midline and receive information about both sides of the visual world. These cells lie at the boundary of V_1 and V_2, and they combine inputs from both hemispheres by way of the corpus callosum. Interestingly, these cells lie in layer 3, which contains neurons known to send their axons to other regions of cortex. The organization and orientation preference of the two receptive fields that are brought together in this way are similar. Such projections from nerve cells in one hemisphere to the other have been shown anatomically by injecting horseradish peroxidase into the cortex at the boundary of V_1

cell in the right cortex that responds to a horizontal bar in the middle of the field of vision should be somehow connected to its counterpart in the left cortex that responds to the continuation of the same bar. Such interactions would allow a complete picture to be formed with a minimum number of connections between the two hemispheres. On the other hand, there would be little purpose in linking fields seen out of the corners of the two eyes that look on quite different parts of the world and observe, for example, a cathedral and a dolphin. Highly specific connections between neurons with receptive fields exactly at the midline have been found experimentally to run from cortex to cortex through the corpus callosum.[41] They are described in Box 1.

A scheme of organization was proposed early on by Hubel and Wiesel.[4,14,21] This scheme had the advantage of using known mechanisms to explain how a nerve cell can respond so selectively to a visual

Schemes for elaborating receptive fields

[41]Hubel, D. H. and Wiesel, T. N. 1967. *J. Neurophysiol.* 30: 1561–1573.

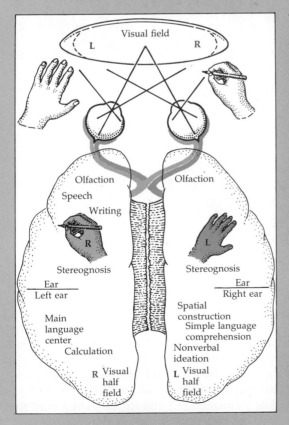

and V$_2$.[45] Enzyme taken up by terminals is transported back to and fills neuronal cell bodies that are situated at exactly corresponding sites in opposite hemispheres. Cutting or cooling (which blocks conduction) the corpus callosum in an animal causes the receptive field to shrink and become confined to just one side of the midline (the usual arrangement with cortical cells). Furthermore, recordings from the callosum show that single fibers have receptive fields close to the midline, not in the periphery. The role of callosal fibers is clearly demonstrated in an experiment of Berlucchi and Rizzolatti.[46] They made a longitudinal cut through the optic chiasm, thereby severing all direct connections to the cortex from the contralateral eye. Yet, provided the corpus callosum was intact, some cells in the cortex with fields close to the midline could still respond to appropriate illumination of the contralateral eye.

[45]Shatz, C. J. 1977. *J. Comp. Neurol.* 173: 497–518.
[46]Berlucchi, G. and Rizzolatti, G. 1968. *Science* 159: 308–310.

THE CORPUS CALLOSUM. Transection of the corpus callosum interrupts connections between the cerebral hemispheres, including those of receptive fields that straddle the midline. (After Sperry, 1970.)

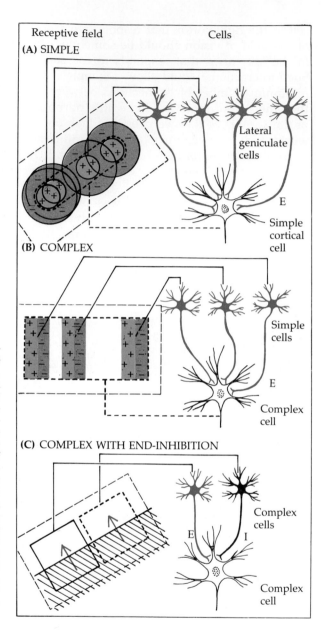

13

SYNTHESIS OF RECEPTIVE FIELDS. Hypothesis devised by Hubel and Wiesel to explain the synthesis of simple and complex receptive fields. In each case, lower-order cells converge to form receptive fields of higher-order neurons. (A) Fields of simple cells are elaborated by the convergence of many geniculate neurons with displaced concentric fields (only four appear in the sketch). They must be arranged in a straight line on the retina according to the axis orientation of simple receptive fields. (B) Simple cells responding best to a vertically oriented edge at slightly different positions could bring about the behavior of a complex cell that responds well to a vertically oriented edge situated anywhere within its field. (C) Each of the two complex cells responds best to an obliquely oriented edge. But one cell is excitatory and the other is inhibitory to the next cell, producing end-inhibition. Hence an edge that covers both fields, as in the sketch, is ineffective, whereas a corner restricted to the left field would excite. (After Hubel and Wiesel, 1962, 1965.)

pattern—say, only to a horizontal line of a certain length that is moving upward through a small part of the visual field. The key idea is that more elaborate receptive fields are constructed by an ordered synthesis of simpler ones originating at a lower level. In the cortex, fields of simple cells behave as if they were built up of large numbers of geniculate fields. This idea is illustrated in Figure 13A, where the centers of concentric fields of geniculate neurons are lined up in such a way that a bar of light through their centers would excite them strongly and a parallel shift of the bar into the inhibitory surround would reduce or stop the excitatory output of the cells. This scheme results in a central excitatory area and its opposing surround. In the same way, one can tentatively construct complex receptive fields by lining up appropriate rows of simple fields. Figure 13B satisfies the requirements of a complex cell that is excited by a vertical edge stimulus that falls anywhere within the area of the receptive field. This is so because wherever the edge falls, one of the simple fields is traversed at its vertical inhibitory–excitatory boundary. None of the other components respond because they are uniformly covered by light or darkness. Diffuse illumination of the entire field covers all component fields equally, and therefore none fires. One can postulate that only one or a few of the simple cells fire at any one position of the stimulus to evoke a response in a complex cell.

The connections shown in Figure 13 were postulated by Hubel and Wiesel as the simplest that could account for orientation selectivity. Ferster[47–49] has provided experimental evidence to show how the receptive fields of simple cells are synthesized in the cat cortex. Microelectrodes were used to make intracellular recordings from neurons in layer 4 of the cat visual cortex that had simple receptive fields. Electrical stimulation of the lateral geniculate nucleus evoked direct monosynaptic excitatory synaptic potentials and indirect disynaptic inhibitory synaptic potentials. Figure 14 shows excitatory synaptic potentials in a simple cell in response to an appropriately oriented dark bar. The excitatory potentials were shown to be elicited by stimuli appropriate for driving geniculate afferents. Ferster's results suggest that lateral geniculate neurons combine as predicted by Hubel and Wiesel to provide inputs to the central "on" region and the flanking "off" regions. Although inhibitory connections between cortical cells have also been observed, they seem not to be required for orientation selectivity. Unlike simple cells, complex cells showed long latency, disynaptic excitatory synaptic potentials after electrical stimulation of the lateral geniculate nucleus. The excitatory potentials were evoked by moving bars of the appropriate orientation and direction rather than small spots, suggesting that, as in the scheme of Figure 13, complex cells receive input mainly from simple cells.

The analysis of cortical circuitry has been further advanced by pharmacological techniques. A localized group of neurons in one layer of the cortex can be blocked reversibly by applying GABA, the predominant inhibitory transmitter in cortex, through a pipette.[50,51] One by one, cells become inhibited by GABA and unresponsive to visual stimuli. As

[47]Ferster, D. 1987. *J. Neurosci.* 7: 1780–1791.
[48]Ferster, D. 1988. *J. Neurosci.* 8: 1172–1180.
[49]Ferster, D. and Koch, C. 1987. *Trends Neurosci.* 10: 487–492.
[50]Bolz, J. and Gilbert, C. D. 1986. *Nature* 320: 362–365.
[51]Bolz, J., Gilbert, C. D. and Wiesel, T. N. 1989. *Trends Neurosci.* 12: 292–296.

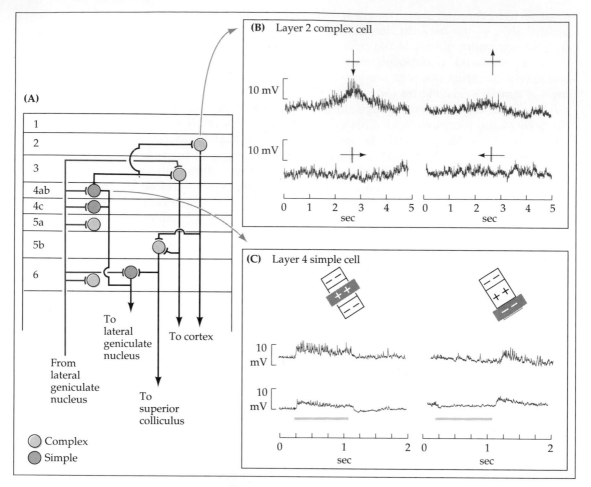

14   CONNECTIONS AND RESPONSES OF CELLS in the visual cortex of the cat, from intracellular recordings made with microelectrodes. **(A)** Direct connections of fibers from the lateral geniculate nucleus to cells in the various layers. These connections were shown to be monosynaptic by intracellular recording. Simple cells are shown as gray circles, complex cells as blue circles. Disynaptic connections are inferred from latencies and properties of intracellular recordings. **(B)** Intracellular recordings from a layer 2 complex cell that responded with excitatory synaptic potentials to a horizontally oriented bar of light moving downward. **(C)** Intracellular recordings from a layer 4 simple cell. The receptive field organization, shown above, is an "on" region surrounded by inhibitory flanks. The recordings clearly show increased excitation with the appropriately oriented bar over the "on" region of the field. The bar shone onto the "off" area removes excitatory input, to be followed by a burst of excitatory potentials after removal of the light from the inhibitory flank. The lower records of each pair represent averages from 10 consecutive traces. These are the responses that would be expected from the receptive field arrangements suggested by Hubel and Wiesel and illustrated in Figure 13. Inhibitory synaptic potentials have also been revealed. They are not monosynaptic, and they are not apparent in these recordings, which were made at the inhibitory reversal potential. (After Ferster, 1988).

they drop out, the receptive fields of neurons that they supply become modified. Figure 15 shows the contribution of simple cells in layer 6 to the end-inhibition of a simple cell in layer 4. Under normal conditions, a bar of the appropriate orientation causes a brisk discharge of the layer 4 simple cell. When the bar is lengthened, however, the response is end-inhibited. After cells in layer 6 have been inactivated by local application of GABA, the long bar acts as effectively as the short bar in initiating impulses: The inhibitory side fields normally provided by the layer 6 cells have been eliminated. The cells in layer 6 have unusually long receptive fields and receive inputs from layer 5. When a cell in layer 5 contributing to the field of a layer 6 simple cell is blocked by GABA, the receptive field is reduced in length as shown in Figure 15.

Together these results lend support to the idea of hierarchical organization. This does not mean that each receptive field of succeeding complexity is generated by combining inputs derived solely from the immediately preceding level. Thus, complex cells can receive inputs from lateral geniculate nucleus cells.[19] Rather, increasing complexity in receptive field organization is produced by convergence of appropriate inputs in an orderly manner. The main point to be emphasized here is that the various schemes serve as a conceptual framework, many details of which still need to be worked out. It is also worth emphasizing that the original scheme proposed by Hubel and Wiesel in 1962 as a working hypothesis still seems clear, elegant, and reasonable.

Table 1 summarizes some of the characteristics of receptive fields at successive levels of the visual system. Each eye conveys to the brain information collected from regions of various sizes on the retinal surface. The emphasis is not on diffuse illumination or the energy absorbed by the photoreceptors but on contrast. What types of signals are generated if a square patch of light such as that in Figure 16 is presented to the retina?

Starting with the optic nerve fibers, the following types of signaling occur: the "on"-center ganglion cells within the square increase their discharge (at least for the initial period) while the "off"-center ganglion cells are suppressed. The best-stimulated ganglion cells, however, are those subjected to the maximum of contrast, that is, those having centers lying immediately adjacent to the boundary between the light and the dark areas. Similarly, the geniculate cells that fire best have their centers close to the border. This process is still further enhanced in the cortex. Cortical cells having receptive fields lying either completely within the square or outside it send no signals because diffuse illumination is not an effective stimulus. Only those simple cells with receptive fields having axis orientation coinciding with the horizontal or vertical boundaries of the square will be stimulated.

Similar considerations apply to the stimulation of complex cells, which also requi properly oriented bars or edges. There is an important difference, however, which depends upon the fact that, at rest,

Receptive fields: Units for form perception

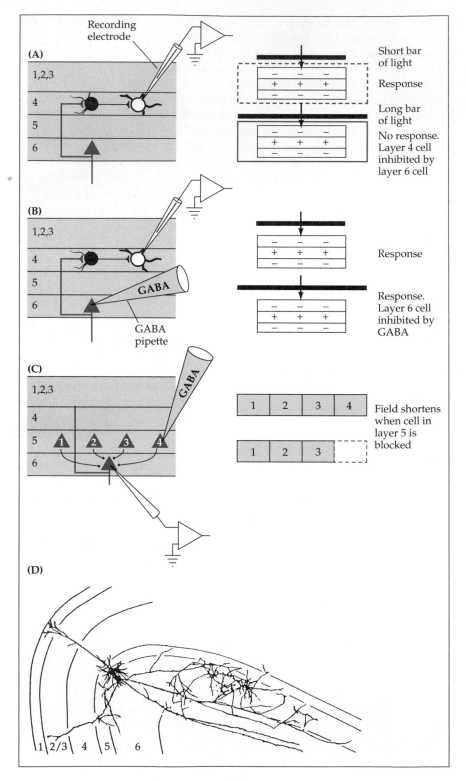

◄**15**   HOW CELLS IN LAYER 6 CONTRIBUTE TO END-STOPPING of receptive fields of layer 4 simple cells. (A) The simple cell in layer 6 has a long receptive field (blue rectangle shown on right). Four cells in layer 5 (numbered 1, 2, 3, 4 in C) converge on the simple cell in layer 6, each contributing to the total length of its receptive field. The layer 6 cell projects to an inhibitory interneuron in layer 4 (filled circle). Consequently a long bar of light activates the layer 6 cell, which in turn activates the inhibitory neuron. As a result, the simple cell in layer 4 penetrated by the microelectrode fails to respond to a long bar of light. Its receptive field (indicated by the dashed line) consists of the smaller area indicated by antagonistic "on" (+) and "off" (−) symbols with end-stopping beyond the field. (B) After the simple cell in layer 6 has been prevented from firing by applying GABA to it through a pipette, the receptive field of the layer 4 simple cell retains the same organization but now without end-stopping. A long bar is as effective as a short one. (C) The contribution of layer 5 cells to the receptive field of a layer 6 cell is shown by applying GABA. Localized application of GABA to layer 5 shortens the receptive field length of the layer 6 cell by inhibiting cells that contribute to its receptive field. (D) Injection of layer 5 cell with horseradish peroxidase reveals its long arborization in layers 5 and 6. (After Bolz, Gilbert and Wiesel, 1989.)

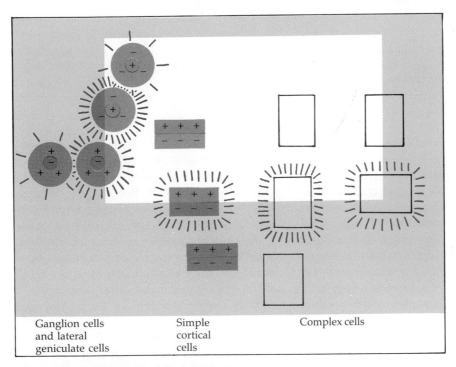

| Ganglion cells and lateral geniculate cells | Simple cortical cells | Complex cells |

16   RESPONSES OF NEURONS TO A PATTERN. When a square patch of light is presented to the retina, signals arise predominantly from receptive fields close to the borders of the square. The ganglion cells and lateral geniculate cells whose receptive fields are situated close to the border fire better than those subjected to uniform light or darkness. The only simple and complex cells that fire are those with appropriate receptive fields situated along a border or corner with the correct orientation.

TABLE 1
CHARACTERISTICS OF RECEPTIVE FIELDS
AT SUCCESSIVE LEVELS OF THE VISUAL SYSTEM

Type of cell	Shape of field	What is best stimulus?	How good is diffuse light as a stimulus?
Photoreceptor	⊕	Light	Good
Ganglion	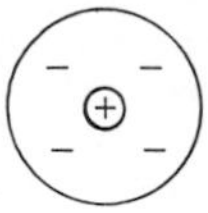	Small spot or narrow bar over center	Moderate
Geniculate	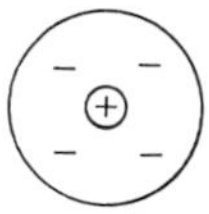	Small spot or narrow bar over center	Poor
Simple (layers 4 and 6 only)	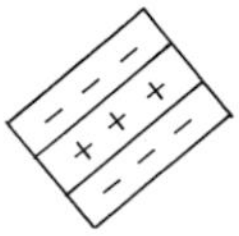	Narrow bar or edge (some end-inhibited)	Ineffective
Complex (not layer 4)	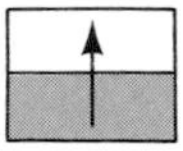	Bar or edge	Ineffective
End-inhibited complex (not layer 4)	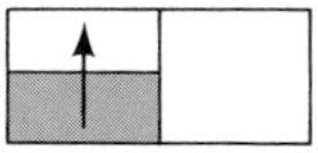	Line or edge that stops; corner or angle	Ineffective

small rapid saccadic movements are made continually by the eyes. They are essential for vision to be preserved while the eyes fixate, but they are not perceived as motion. Each movement causes a new population of simple cells with exactly the same orientation but slightly different receptive field location to be thrown into action. For those complex cells that "see" the square, however, a boundary of appropriate orientation anywhere within the field is the only requirement. Many of the same complex cells therefore continue to fire even during eye movements, as long as the displacement is small and the pattern does not pass outside the receptive field of the complex cell. For such cells the position of the square on the retina does not appear to change. In the lower right corner of the square is shown the field of a complex cell with end-inhibition that responds best to a right angle—a corner.

If the preceding considerations are valid, the surprising conclusion is that the primary visual cortex receives little information about the absolute level of uniform illumination within the square. Signals arrive only from the cells with receptive fields situated close to the border. This hypothesis is supported by a well-known psychophysical experiment. A square that appears white when surrounded by a black border

Is orientation of stimulus important?	Are there distinct "on" and "off" areas within receptor fields?	Are cells driven by both eyes?	Can cells respond selectively to movement in one direction?
No	No	No	No
No	Yes	No	No
No	Yes	No	No
Yes	Yes	Yes (except in layer 4)	Some can
Yes	No	Yes	Some can
Yes	No	Yes	Some can

can be made to appear dark merely by increasing the brightness of the surround. In other words, we perceive the difference or contrast at the boundary, and it is by that standard that the brightness in the uniformly illuminated central area is judged. The eye does, however, have an index of the level of brightness that is expressed by the pupillary size, which varies according to the absolute strength of ambient light over a wide range. Pupillary size is adjusted by a feedback mechanism, the incoming loop of which leaves the eye through the optic nerve.

Hubel and Wiesel's early experiments showed cortical cells with similar properties to be aggregated together in a vertical, columnar organization.[20,21,32,52,53] In any one penetration through the cortex as the electrode moved from the surface through layer after layer to the white matter, all cells were found to have the same field axis orientation, ocular dominance, and position in the visual field. The columns for eye preference have already been mentioned. As shown in Figures 5 and 6, the inputs from the two eyes are segregated in layer 4, in which cortical neurons are monocularly driven. Outside layer 4 both eyes are

Ocular dominance and orientation columns

[52]Hubel, D. H. and Wiesel, T. N. 1963. *J. Physiol.* 165: 559–568.
[53]Hubel, D. H. and Wiesel, T. N. 1974. *J. Comp. Neurol.* 158: 267–294.

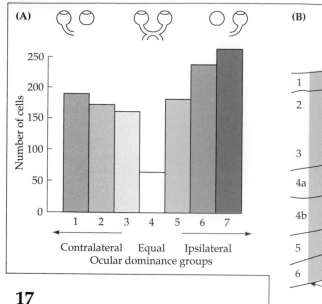

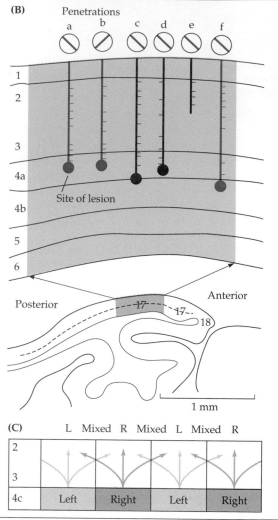

17

PHYSIOLOGICAL DEMONSTRATION OF OC-ULAR DOMINANCE COLUMNS. (A) Eye preference of 1116 cells in 28 rhesus monkeys. Most cells (groups 2 through 6) are driven by both eyes. **(B)** When penetrations a, b, c, d, e, and f are made through the cortex at right angles to the surface, the cells encountered in one track have similar eye preference (black or blue corresponding to contralateral or ipsilateral), as well as similar orientation specificity, indicated by lines within open circles. Small bars indicate units recorded from. **(C)** Diagram to show how inputs from the two eyes arriving in layer 4 are combined in more superficial layers through horizontal or diagonal connections to create cells with binocular fields. (After Hubel and Wiesel, 1968; Wiesel and Hubel, 1974; Hubel, 1988.)

effective but one usually dominates. For example, when a microelectrode is driven through the thickness of the cortex at right angles to its surface, all the cells encountered in deeper and more superficial layers exhibit the same eye preference. Inputs from the two eyes are blended in deeper and more superficial cortical layers to generate binocularly driven receptive fields. Figure 17 illustrates the ocular dominance characteristics of neurons in the striate cortex in the monkey. The cells (total 1116) are subdivided into seven groups. Groups 1 and 7 are driven exclusively by one of the two eyes and are found in layer 4 of the cortex. In groups 2, 3 and 5, 6 the effect of one eye is stronger than that of the other, and the cells in the middle group, 4, are equally influenced. It is clear from the histogram that the majority of cells respond to both eyes.

The cortical cells that are grouped closely together show not only the same eye preference but also the same orientation specificity and receptive field position. A sample experiment is shown in Figure 18. A microelectrode is inserted normal to the surface of the cortex in area V_1 of the cat. Each bar indicates one cell and its preferred receptive field orientation in the progression through the cortex. The circle at the end marks the site of a lesion at the final position of the electrode tip. From this end point and the electrode track observed in the fixed brain after the end of the experiment, the following sequence is seen: At first, all the cells are optimally driven by bars or edges at about 90° to the vertical, at one position in the visual field. After a penetration of about 0.6 mm, the axis of the receptive field orientation changes to about 45°. A second track to the right shows other cells, with slightly different receptive field positions and field axis orientations. With this more oblique penetration, the field axis changes repeatedly with only small movements of the electrode tip, as though a series of columns with different axis orientations is being traversed. The orientation columns receive their input from cells with largely overlapping receptive fields on the retinal surface.

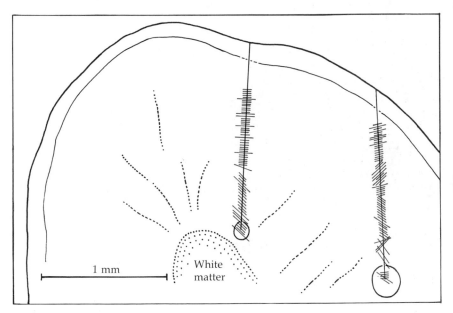

18   AXIS ORIENTATION OF RECEPTIVE FIELDS of neurons encountered as an electrode traverses the cortex normal to its surface. Cell after cell tends to have the same axis orientation, indicated by the angle of the bar to the electrode track. The penetration to the right is more oblique; consequently, the track crosses several columns and the axis orientations change frequently. The electrode track is reconstructed by making a lesion at the end of the penetration (circle) and cutting serial sections through the brain. Such experiments have established that cat and monkey cells with similar axis orientation are stacked in columns running at right angles to the cortical surface. (After Hubel and Wiesel, 1962.)

Information about the arrangement of orientation columns in the visual cortex of monkeys and cats was first obtained by making oblique rather than vertical electrode penetrations through the cortex.[53] An example is shown in Figure 19, which reveals once again the orderliness

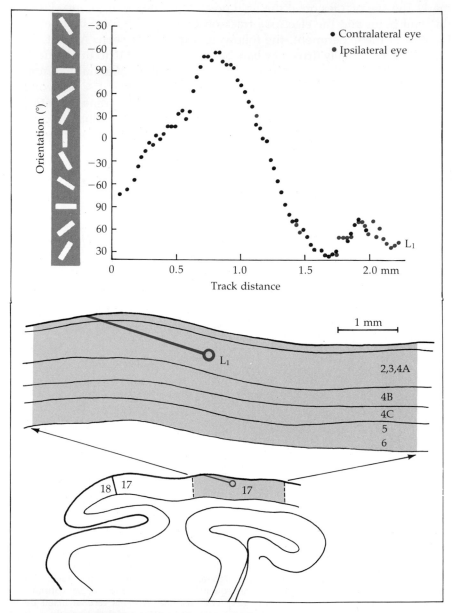

19   **ARRANGEMENT OF ORIENTATION COLUMNS. As the electrode moves obliquely through the cortex, the orientation specificity of the cells encountered shifts systematically. The shift is about 10° for each 50 μm, as though a series of columns were being traversed in a regular sequence. (After Hubel and Wiesel, 1974.)**

of the arrangement of cells (in a spider monkey named George). Each 50-μm shift of the electrode along the cortex is accompanied by a change in field axis orientation of about 10° in a regular sequence through 180°. The field axis orientation columns are narrower than those for ocular dominance—20 to 50 μm compared with 0.25 to 0.5 mm. The columns for orientation have been demonstrated anatomically by the use of 2-deoxyglucose, a technique pioneered by Sokoloff.[54] The principle is that active cells take up radioactive deoxyglucose as though it were glucose. This molecule, however, cannot be degraded or transported out; as a result, metabolically active cells become labeled with radioactivity and their distribution can be seen by autoradiography. In monkeys and cats, after the eyes were exposed to vertical stripes, bands of radioactivity appeared in the cortex, corresponding to orientation columns that had been demonstrated physiologically. Labeling extends from the surface of the cortex to the white matter and also highlights the "blobs."[55]

The arrangement of orientation and ocular dominance columns as well as blobs has now been revealed in living cortex by recording optically.[56–58] The results are shown in Figure 20. A major advantage of this technique is that large areas of cortex can be surveyed simultaneously during the experiment, and a population of columns can be visualized. In brief, photosensitive dyes are applied to the cortex, or alternatively changes in intrinsic absorbance are measured as patterned visual stimuli of various orientations (Figure 20) are presented in different parts of the visual field of each eye.[59] The ocular dominance columns have the slablike shape expected from anatomical studies. The orientation columns appear as beaded bands that are patchy and discontinuous, unlike the ocular dominance columns.

Figure 20 also shows experiments of Bonhoeffer and Grinvald in which areas of cortex lit up after presentation of differently oriented visual stimuli. What is particularly striking is the arrangement of these orientation columns to each other. At a first inspection the organization appears disorderly. However, the close-up of Figure 20D reveals the presence of ORIENTATION CENTERS, focal points at which all the orientations come together. From there the cells responsive to various orientations radiate in an extraordinarily regular manner. Some "pinwheels" are systematically organized with clockwork progression, others anticlockwise. Thus orientation is represented in a radial rather than a linear fashion with each orientation appearing only once in the cycle.[60] One or two such centers, evenly spaced, occur in each square

[54]Sokoloff, L. 1977. *J. Neurochem.* 29:13–26.

[55]Tootell, R. B. et al. 1988. *J. Neurosci.* 8: 1500–1530.

[56]Grinvald, A. et al. 1986. *Nature* 324: 361–364.

[57]Ts'o, D. Y. et al. 1990. *Science* 249: 417–420.

[58]Blasdel, G. G. 1989. *Annu. Rev. Physiol.* 51: 561–581.

[59]Bonhoeffer, T. and Grinvald, A. 1991. *Nature* 353: 429–431.

[60]Swindale, N. V., Matsubara, J. A. and Cynader, M. S. 1987. *J. Neurosci.* 7: 1414–1427.

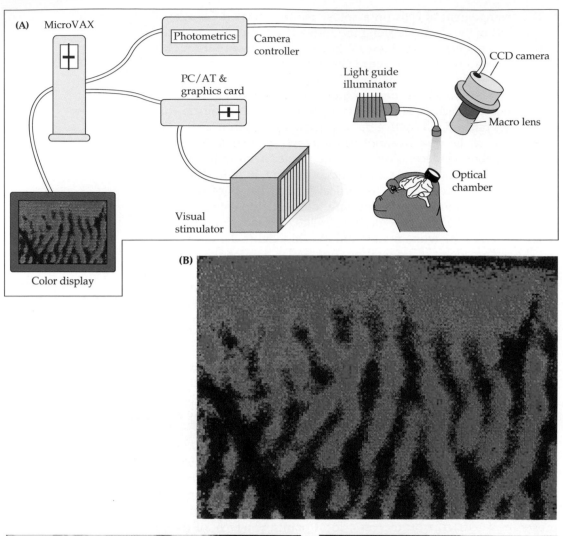

(A) MicroVAX — Photometrics — Camera controller — CCD camera — Macro lens — Light guide illuminator — Optical chamber — PC/AT & graphics card — Visual stimulator — Color display

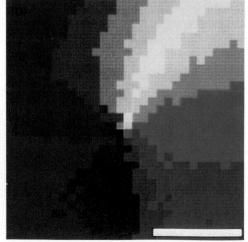

◄ **20**   DISPLAY OF OCULAR DOMINANCE AND ORIENTATION COLUMNS by optical imaging. (A) Schematic diagram of method. The camera records changes in light reflected from cortex (without dyes) following activity induced by patterned lights shone into one eye or the other with various orientations. The skull is opened but records are taken through the intact dura. (B) Ocular dominance columns in living monkey displayed by shining lights into one eye. (C,D) Orientation columns displayed by shining lights with various orientations indicated by colors. (C) The overall pattern of bean-shaped areas responding to selective orientation (indicated by color). Although this pattern seems at first disorderly, close inspection reveals centers at which all orientations come together. In (D) a typical "pinwheel" is shown. Note that each orientation is represented only once and that the sequence is beautifully precise. Such pinwheel foci occur at regular distances from each other. Scale is 300 μm. (After Ts'o et al., 1990; Bonhoeffer and Grinvald, 1991.)

millimeter of cortex. The pattern is similar to models proposed earlier on theoretical grounds.[61] Colored stimuli also reveal the presence of the blobs in the living brain. Recordings such as those of Figure 20 represent a major advance in scanning activity in an area of cortex layer by layer without electrodes or fixation.

A convenient conceptual scheme for thinking about the way in which orientation and ocular dominance columns might interact in the cortex was proposed by Hubel and Wiesel.[23] Before optical recordings had been made, they suggested a roughly cuboidal aggregate of cells in which all the possible orientations could be represented for a receptive field area in the same place in each of the two eyes. This they called the HYPERCOLUMN. In this unit the orientation and ocular dominance columns would combine inputs from the two eyes to produce coherent binocular fields for the synthesis of simple or complex fields positioned in the same parts of the visual field. An adjacent group of cortical cells would analyze information in the same way for an adjacent but overlapping part of the visual field and so on. In regions of cortex that deal with peripheral fields of vision, the receptive fields of individual cells become larger. A hypercolumn here would deal in the same way with a relatively large area of peripheral retina. Moving from one small region of cortex to the next is accompanied by a much larger shift in the position of the receptive field than for columns concerned with the fovea.[62] But the basic organization remains similar throughout the cortex; the orientation and ocular dominance columns have constant anatomical dimensions in cortical areas dealing with the periphery and the fovea. The hypercolumn remains useful as a concept for explaining how cells combine their inputs and how the cortex is parceled into functional units.

An extra dimension of horizontal connections has been added to the modular concept of cortical organization. Classical staining techniques, such as Golgi impregnation, revealed a preponderance of neuronal processes that ran perpendicularly from layer to layer; intracellular injections of single cells have demonstrated that cortical neurons also

Horizontal connections between columns

[61]Linsker, R. 1989. *Proc. Natl. Acad. Sci. USA* 83: 8779–8783.
[62]Tootell, R. B. et al. 1988. *J. Neurosci.* 8: 1531–1568.

have long horizontal processes that extend laterally from column to column.[29,63–65] Such connections were mentioned earlier as important for synthesizing elongated receptive fields of simple cells in layer 6: The receptive fields of layer 5 cells with long horizontal axons were combined and added end to end on the layer 6 simple cell. Many simple and complex cells are found with horizontal projections that extend over 8 mm, corresponding to several hypercolumns. An individual neuron can therefore integrate information over an area of retina several times larger than the receptive field as understood by conventional criteria.[66] Such horizontal connections could generate subthreshold modulatory influences. Of particular interest is that connections are made only between columns that have similar orientation specificity. Evidence for these specific interconnections has been obtained by two additional methods. First, when tiny labeled microbeads are injected into one column, they are transported to another hypercolumn at a distance; both sets of labeled neurons have the same orientation preference. Second, by cross correlation of the firing patterns of neurons with the same orientation preference in two widely separated columns, one can show that they are interconnected.[66,67] Moreover after a lesion has been made in the retina, cortical cells that have been deprived of input can show responses to distant stimuli that would be outside their "normal" receptive fields.[68] Together these results suggest that many aspects of receptive field organization remain unsolved and that eventually the concept may need to be refined.

Relation between blobs and columns

Separate from, yet related to, the columns of the primary visual cortex are the blobs, the clusters of cells that stain with cytochrome oxidase and other enzymes. For many years these cells had eluded the electrodes of Hubel and Wiesel. Although "the historically minded reader may have wondered how so prominent a group of cells could have been missed by so prominent a pair of investigators,"[18] there are many reasons why substructures cannot be discerned before the major architectural features have been defined.

The blobs (Figure 3) are precisely arranged in parallel rows when horizontal sections through layer 3 are viewed from above. One blob is situated in the center of each ocular dominance column.[17,55] Their projections and role in color vision are described below. In brief, most blob cells have receptive fields that are concentric, with "on" and "off" regions.[40,69] Unlike other neurons outside layer 4, they do not require

[63]Gilbert, C. D. and Wiesel, T. N. 1983. *J. Neurosci.* 3: 1116–1133.

[64]Gilbert, C. D. and Wiesel, T. N. 1989. *J. Neurosci.* 9: 2432–2442.

[65]Katz, L. C., Gilbert, C. D. and Wiesel, T. N. 1989. *J. Neurosci.* 9: 1389–1399.

[66]Ts'o, D. Y., Gilbert, C. D. and Wiesel, T. N. 1986. *J. Neurosci.* 6: 1160–1170.

[67]Gray, C. M. et al. 1989. *Nature* 338: 334–337.

[68]Gilbert, C. D., Hirsch, J. A. and Wiesel, T. N. 1990. Cold Spring Harbor Symp. *Quant. Biol.* 55, 663–677.

[69]Ts'o, D. Y. and Gilbert, C. D. 1988. *J. Neurosci.* 8: 1712–1727.

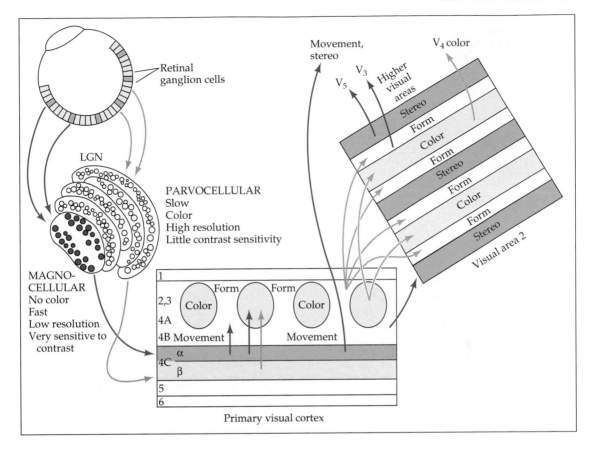

21   **SCHEME OF CONNECTIONS of magnocellular and parvocellular divisions of the lateral geniculate nucleus to V₁, V₂, V₃, V₄, and V₅. The magnocellular pathway—which is sensitive to contrast, with low resolution, fast responses, and no color sensitivity—projects primarily to layer 4Cα. The parvocellular division—slower, responsive to color, with high resolution and low sensitivity to contrast—projects mainly to layer 4Cβ. Both divisions project to blobs, which in turn project to thin stripes in V₂ and from there to V₄. Interblob regions project principally to unstained areas in V₂, then to V₃. Magnocellular pathways project principally to V₅. Back-projections to V₁ and interconnections of the various areas are not shown. (See also Table 2.) (After Livingstone and Hubel, 1987b.)**

bars or edges with precise orientation. Instead, it is the wavelength of light falling on center and surround that determines firing for most blob cells. The relations between receptive field properties of such cells and color vision are discussed below.

An extraordinarily regular and well-defined series of projections has been found from the columnar and the blob systems of the primary visual cortex to the second visual area, V_2, and to higher visual areas, V_3, V_4, and V_5. The segregation is particularly clear in V_2, which surrounds V_1. Figure 21 and Table 2 show the main pathways that have

Connections from V_1 to higher visual areas

TABLE 2
MAJOR SUBDIVISIONS AND CONNECTIONS OF THE PRIMATE VISUAL SYSTEM

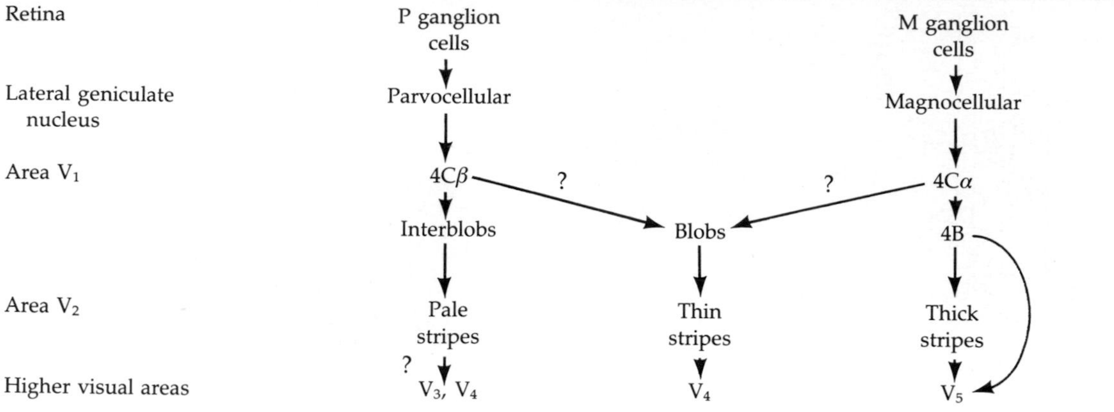

Retina	P ganglion cells				M ganglion cells
Lateral geniculate nucleus	Parvocellular				Magnocellular
Area V_1	4Cβ ———	?	→ Blobs ←	?	——— 4Cα
	Interblobs		Blobs		4B
Area V_2	Pale stripes		Thin stripes		Thick stripes
Higher visual areas	? → V_3, V_4		V_4		V_5
Property					
Color-coded	Yes/no		Yes		No
Contrast sensitivity	Low		High		High
Spatial resolution	High		Low		Low
Orientation selectivity	Yes		No		Yes
Movement sensitivity	Yes		No		Yes
Directionality	No		No		Yes
Stereopsis	No		No		Yes

After Livingstone and Hubel, 1987b.

been traced by combined physiological and anatomical techniques.[5,6,40,70–74]

When visual area 2 is stained for cytochrome oxidase, a picture of dark stripes becomes apparent: Thick and thin dark stripes alternate with paler areas having less enzyme activity. These parallel stripes run at right angles to the border between V_1 and V_2. After horseradish peroxidase has been injected into blobs of V_1, it selectively marks the thin stripes in V_2; the connections are reciprocal—injections into thin stripes label blobs in V_1.[75] Similarly, the interblob regions, where the simple and complex cells are located, project to the pale stripes, except for cells of layer 4B, which project to the thick stripes. These projections from V_1 to V_2 therefore continue the segregation of magnocellular and parvocellular divisions (Figure 21; Table 2).[40,70] As expected, neurons in the thin stripes of V_2 that receive inputs from blobs in V_1 are sensitive to color or brightness but not to orientation. Neurons in the pale stripes that receive inputs from columnar interblob regions in V_1 are orienta-

[70]Livingstone, M. and Hubel, D. 1988. *Science* 240: 740–749.
[71]Van Essen, D. C. 1979. *Annu. Rev. Neurosci.* 2: 227–263.
[72]Zeki, S. M. 1978. *J. Physiol.* 277: 245–272.
[73]Zeki, S. 1990. *Disc. Neurosci.* 6: 1–64.
[74]Tootell, R. B., Hamilton, S. L. and Switkes, E. 1988. *J. Neurosci.* 8: 1594–1609.
[75]Livingstone, M. S. and Hubel, D. H. 1987. *J. Neurosci.* 7: 3371–3377.

tion-selective, often with end-stopping but without color sensitivity or selectivity for direction of movement. These two inputs are derived mainly from parvocellular relays and are concerned with the analysis of color and form. Neurons in thick stripes of V_2 that receive magnocelluar inputs from layer 4B of V_1 (via $4C\alpha$) are orientation-selective, color-insensitive and only rarely end-stopped. Thick stripe cells characteristically require binocular stimuli and respond best when a bar, edge, or slit of the correct orientation lies just in front of or behind the plane of focus. Such cells seem adapted for depth perception.[40]

Further projections to V_3, V_4, and V_5 are shown in Figure 21. Each of these areas contains a representation of the contralateral visual field, usually coarser than that in V_1. Not shown in the figure are the various subdivisions of these areas.[76] According to this simplified scheme, areas V_3 and V_5 receive mainly magnocellular projections from layer 4B of V_1 and thick stripes of V_2. Neurons in V_3, which are as expected mainly color-insensitive, are specialized for processing information relating to form, while those in V_5 are specialized for the detection of movement, often in one direction, as well as textures and binocular disparity. V_5 is also known as MT (middle temporal). The inputs to V_4 via parvocellular pathways through the thin stripes of V_2 result in an area whose principal function is concerned with color.

A cautionary note is appropriate here. The various areas of visual cortex can be defined in principle by a number of criteria, including representation of the visual field, distinctive cytoarchitectural features, anatomical connections, receptive field properties, and behavioral deficits resulting from localized lesions.[6,7,9,77] For many areas the identification and precise boundaries are not complete. Receptive field properties may be nonuniform, the representation of the visual field may be disorderly, and the input and output connections may be mixed with numerous back-projections. The identification of areas V_1 and V_2 seems satisfactory and convincing. For V_3, V_4, and V_5 a general pattern has emerged, but it is not unlikely that subdivision of higher visual areas may well become modified in time. Zeki, who with his colleagues pioneered the exploration of visual areas outside V_1, has used position emission tomography (PET scanning) to analyze functional specialization in the human brain. By this technique it has been possible to identify areas comparable to V_1, V_2, V_4, and V_5 after presenting appropriate visual stimuli to conscious subjects, and also to observe the sequence in which they become activated.[78] V_1 parcels out information in parallel to the various areas.

At the level of the cones we have seen a clear correlation between neuronal signals and the wavelength of light falling on the retina: Red, green, and blue cones preferentially absorb light in long-, medium-, and short-wavelength ranges. In principle, therefore, by comparing the activity of each type of cone, the nervous system could compute wave-

Color vision

[76]Felleman, D. J. and Van Essen, D. C. 1987. *J. Neurophysiol.* 57: 889–920.
[77]Schiller, P. H. and Logothetis, N. K. 1990. *Trends Neurosci.* 13: 392–398.
[78]Zeki, S. et al. 1991. *J. Neurosci.* 11: 641–649.

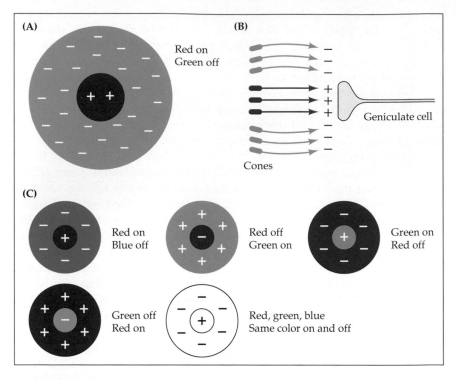

22   **RECEPTIVE FIELD ORGANIZATION OF CELLS in monkey lateral geniculate nucleus, parvocellular layers, responding to color. (A) Receptive field organization with red "on"-center, green "off"-surround. (B) Postulated connections for synthesizing a receptive field of the type shown in (A). A cell of this sort would fire best in response to a small or large red spot on a neutral background. A red spot on a green background would be relatively ineffective. (C) Examples of receptive field organizations with various spectral sensitivities. (After Wiesel and Hubel 1966; Hubel 1988.)**

length. How is this achieved by neurons? The convergence of inputs from cones begins with color-coded horizontal cells and continues with ganglion cells and lateral geniculate cells of the parvocellular division. The properties of such color-coded ganglion and geniculate cells are shown in Figure 22. No major transformation of receptive field properties has been found between optic nerve fibers and geniculate cells.[79] The receptive field of the geniculate cell in Figure 22 is concentric, with a red "on"-center and a green "off"-surround. A small red spot illuminating the center causes a brisk discharge; a larger green light over the surround inhibits it. Such a cell responds best to a small or large red light, on a neutral or blue, but not on a green, background. It responds to small and large spots of white light like a normal cat ganglion cell (which is insensitive to color). Other cells show blue–yellow antagonistic areas (yellow representing a mixed red and green cone contribution). Representative types of center–surround organiza-

[79]Wiesel, T. N. and Hubel, D. H. 1966. *J. Neurophysiol.* 29: 1115–1156.

tions that have been observed in the parvocellular layers of the monkey geniculate nucleus are summarized in Figure 22. Such neurons are known as COLOR-OPPONENT CELLS and analyze wavelength by comparing their cone inputs in just the way that Young or Helmholtz would have enjoyed finding out about. A red, green, blue, yellow, black, or white ball on a billiard table evokes in an array of color-opponent cells unambiguous signals that are then presented to the brain.

A coherent and somewhat surprising picture of subsequent steps in cortical analysis of color and color perception has been provided by experiments of Zeki, Hubel, Daw, Land, and their colleagues. It has been mentioned already that the pathways for color tend to be segregated from those for analysis of other properties such as depth, movement, contrast and form. Zeki has demonstrated parvocellular pathways that lead from V_1, through V_2, to V_4 in which color-coded cells are abundant.[73] Evidence for the key role of V_4 in color vision is provided by PET scan studies in normal individuals. Increased activity is seen in an area homologous to V_4 when colored patterns of light are shone into the eyes.[78]

Pathways for color

The separation of color from form perception is convincingly demonstrated in rare cases of patients suffering from localized brain damage. For example, an unusual adult patient has been described who suffered bilateral lesions to cortex anterior to V_1, probably corresponding to V_4.[80] Before the lesion he had normal color vision but afterwards he became unable to see colors. He knew that a strawberry was red and a banana yellow, but after the lesion the world appeared like a black and white movie. Other functions such as memory and form vision were little affected and he was able to continue his job as a customs inspector. When a familiar object was pointed out he could describe what color it ought to have, but he could not match it with an appropriate colored felt-tip marker. His deficit was not in speech or in object recognition but in color perception per se.

Psychophysical tests in normal subjects confirm the separation of color pathways right up to perception. Full descriptions are provided in papers and reviews by Zeki,[73] Hubel,[14] and Livingstone.[81] For example, it is hard if not impossible to discern patterns or forms in a scene unless the magnocellular pathways are activated by the contrast that is normally present as different degrees of brightness and shadow. The parvocellular system with its emphasis on color and high spatial resolution has only limited ability on its own to provide information about the form of objects. Hence, a colored picture of a complex structure with all its components reflecting equal quantities of light appears formless. This is because the magnocellular pathway is not activated. Similarly, our sensations of depth and movement tend to be lost when black and white contrast is insufficient to bring in magnocellular pathways. An impressive demonstration is to move a pattern of green and

[80]Pearlman, A. L., Birch, J. and Meadows, J. C. 1979. *Ann. Neurol.* 5: 253–261.
[81]Livingstone, M. S. and Hubel, D. H. 1987. *J. Neurosci.* 7: 3416–3468.

red stripes across a television screen. The intensity of each color can then be adjusted to be equiluminant (i.e., each red or green stripe emits the same effective amount of light as its neighbor, although at a different wavelength). One still sees the colored stripes, but they seem to stop moving. (Full accounts of these and related phenomena showing the separate analysis of color information in the brain are provided by reviews cited at the end of this chapter.)

Color constancy
 A major problem in our understanding of color vision has been to understand how the cortex decides upon the color of an object in the visual scene. Our brains program these calculations so successfully that we do not intuitively realize that there is indeed a problem. Surely illustrations in this book appear blue because they reflect light at short wavelength. From all that has been said so far, one might imagine that the colors we see are determined directly and simply by the wavelength of light. That this is not so was apparent to Helmholtz (again!).[82] He pointed out that an apple seen in daylight, in sunset, or by candlelight appears red. Yet the light reflected from its surface contains far more red light at sunset and far more yellow in candlelight. Somehow the brain has assigned a red color to the apple that does not shift under very different conditions; it "discounts the illuminant." A familiar example is the tint of two correctly exposed photographs taken on the same film in daylight and in the artificial room light provided by electric lamp bulbs. The daylight colors seem realistic, whereas those taken indoors appear too yellow. Yet we are not at all aware of this yellowness in an artificially lit room. (This phenomenon until recently was a commonplace observation; the flash present on nearly every camera nowadays emits wavelengths of light balanced to resemble daylight.) The biological advantages of color constancy are clear: Green berries should not turn red at sunset; pink lips should not turn yellow by candlelight.

 Spectacular demonstrations of color constancy designed by Land[83,84] have acted as a powerful stimulus for neurobiological research into color vision. What his demonstrations showed was that the way we see the color of an object depends critically upon the light reflected from the entire scene, not just from the object itself. We cannot assign a color— red, yellow, green, blue, or white—to an area just by determining the wavelength of light reflected from it. We need also to know the composition of light reflected from surrounding areas. This uncomfortable conclusion seems counterintuitive and contrary to our experience. As with black and white, however, the brain assigns the sensation of color by comparing light falling on different parts of the retina, instead of measuring absolute brightness or wavelengths at one spot on the retina. Rather, it is as though overall comparisons were made in the cortex of contrasts at boundaries, within three separate pictures of a complex

[82]Helmholtz, H. 1962/1927. *Helmholtz's Treatise on Physiological Optics.* J. P. C. Southhall (eds.). Dover, New York.
[83]Land, E. H. 1986. *Vision Res.* 26: 7–21.
[84]Land, E. H. 1986. *Proc. Natl. Acad. Sci. USA* 83: 3078–3080.

scene viewed through short-, medium-, and long-wavelength filters. Over a broad range the exact wavelength of each of these blue, green, and red lines of analysis does not determine the color that is assigned to an object within the field of view (Box 2).

It has not been possible to provide a comprehensive and satisfactory explanation of the Land phenomena in terms of the receptive field properties of color-coded cells in V_1, V_2, and V_4. One type of cell, known as DOUBLE-OPPONENT, has properties that could participate in color constancy. Originally described in the goldfish retina by Daw,[85,86] double-opponent cells have been found in primate cortex parvocellular blobs and stripes, but not in the lateral geniculate nucleus or retina.[40,69,75,87,88] They therefore appear to represent a higher stage of color processing. In brief, such cells have roughly concentric center–surround receptive fields with red–green or blue–yellow antagonism (Figure 23). But, unlike color opponent cells in the geniculate, each color produces antagonistic effects both in center and surround of the double-opponent cell. Thus shining red light in the center evokes an "on"-discharge, red in the surround evokes an "off." Green in the surround evokes an "on"-discharge, in the center an "off." Suppose one shines a small red spot on the center of such a cell's receptive field with a neutral surround in white ambient light to evoke a discharge. If one now increases the proportion of red in the ambient light, the discharge will change only little: Increased excitation of the central area by red will be antagonized by increased inhibition from the periphery by red. Presumably, the long horizontal connections from blob to blob play an important role in the spatial interactions required to explain the Land phenomena.

It has now become possible to approach experimentally many of the questions posed by Helmholtz, Hering, and, in our own time, Land about how the brain analyzes visual scenes falling on the retina. Anatomical, physiological, and psychophysical experiments have each revealed designs in the cortex for perception with remarkable consistency. One important principle derived from functional anatomy is that separate pathways beginning at the retina extend through to consciousness.[67,73,89] You can fool the systems used for assessing depth and detecting movement by using light conditions that bring in only the parvo system; and patients with lesions to specific areas of cortex can lose color vision with little impairment of their ability to recognize patterns. Nevertheless, magno and parvo divisions are not entirely separate, either anatomically or physiologically: Both combine for recognition of patterns or forms. Presumably fish snapping at flies, cormorants diving for fish, and cats pouncing on birds all use movement–depth detecting fast systems; this system would, however, be relatively

Where do we go from here?

[85]Daw, N. W. 1984. *Trends Neurosci.* 7: 330–335.

[86]Daw, N. W. 1968. *J. Physiol.* 197: 567–592.

[87]Hubel, D. H. and Livingstone, M. S. 1990. *J. Neurosci.* 10: 2223–2237.

[88]Tootell, R. B. et al. 1988. *J. Neurosci.* 8: 1569–1593.

[89]Schiller, P. H. and Lee, K. 1991. *Science* 251: 1251–1253.

BOX 2   COLOR CONSTANCY

We show a color figure, taken from one of Land's extraordinary demonstrations, in which a complex pattern of about 100 rectangles and squares of colored paper is illuminated by lights shining from three separate projectors.[90] The picture is abstract, nonrepresentational, and reminiscent of a painting by Mondrian. The slide projectors contain no slides. The brightness of each projector is controlled by a rheostat and the wavelength by an interference filter. As expected, when the three projectors shine lights of equal intensity at long, medium, and short wavelengths, corresponding to the red, green, and blue of our cone systems, we see the colors of the figure. The first surprising result occurs when one changes the relative intensities of red, green, and blue by increasing the current through the red projector while reducing it through the green and even more through the blue. The adjustments are made so that the overall light reflected back has the same intensity as before. Although the wavelength composition of the light has changed, the colors of the squares do not change when the pattern is viewed as a whole. The yellow, green, red, blue, and white squares retain their colors even when the red or the green or the blue component of the light is increased or decreased over a wide range. Land further describes the following experiment: A photometer equally sensitive to all wavelengths is pointed at a yellow square. With the green and blue projectors turned off, the red projector is adjusted until a reading of 1 (in arbitrary units) is obtained. It is then turned off and the green projector is adjusted until it too gives a reading of 1; the procedure is repeated with the blue projector. From this piece of yellow paper, then, equal amounts of red, green, and blue light are reflected when all three projectors are turned on. It looks yellow when we view the picture. Now exactly the same is done with a piece of green paper, adjusting the red, green, and blue lights so that the same intensity at each wavelength is reflected back to the photometer. By definition the light coming back to our eyes from the green square has exactly the same wavelength composition as that previously reflected from the yellow. What color will the green square appear to have? The intuitive answer is "yellow." Yet the square still appears green. This result is always true provided that the picture is viewed as a whole. If, however, all that can be seen by the observer is a small patch of uniform color, surrounded by darkness, its color constancy is not preserved: Green does turn to yellow. Thus the color of a green square viewed on its own with the rest of the picture masked does depend critically on the relative intensities of the three wavelengths. In the context of the whole pattern, as soon as the mask is removed the square becomes green in the same ambient illumination. This remarkable effect does not depend on judgment, adaptation, or afterimages. From these and other sophisticated experiments Land concluded:

> If you contemplate the Mondrian while I make these sweeping changes in relative illumination on the separate wavebands, you cannot possibly understand why there are not concomitant changes in color—as long as you think in terms of color-mixing at each point. If, however, you can imagine the Mondrian as being a composite of three independent images, one carried by long waves, one carried by middle waves, and one carried by short waves, I can then introduce the proposition that, first, each of these images is unaltered by a change in the flux by which it is carried and second, that it is the comparing of these three complete images, rather than the merging of their fluxes, which produces the array of colors.

His *retinex theory* (*retinex* is retina and cortex combined) provides a framework for computing the color seen at a particular part of the retina on the basis of the relative intensities of three wavelengths and their spatial interactions.

[90]Land, E. H. 1983. *Proc. Natl. Acad. Sci. USA* 80: 5163–5169.

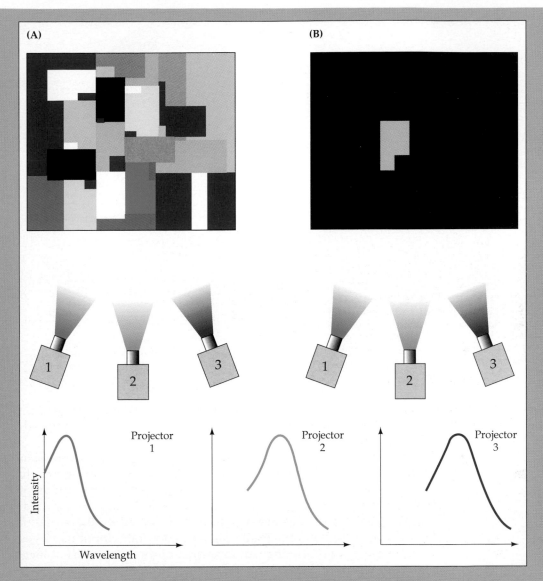

MONDRIAN-LIKE ARRANGEMENT OF COLORED RECTANGLES to display Land phenomena. The diagram below shows three projectors, each of which shines monochromatic red, green, or blue light onto the picture. The amount of light projected can be adjusted. (A) When the whole picture is viewed with equal intensities of red, green, and blue light, the colors appear normal. Land's surprising demonstration of color constancy shows that these colors are maintained for an observer even though the relative contributions of the red, green, and blue projectors are drastically altered. (B) If, however, a square is viewed on its own, surrounded by total darkness, the color of the square appears commensurate with the mixtures of the three wavelengths of light falling on it. (After Land, 1963; Zeki, 1990.)

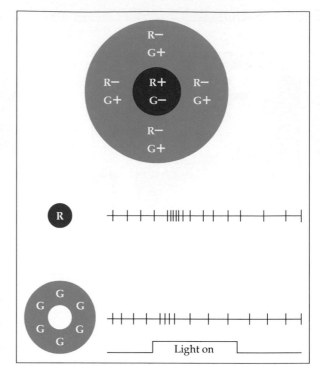

23

THE RECEPTIVE FIELD OF A DOUBLE-OP-
PONENT CELL, the properties of which could
help to explain color constancy. A cell of this
type responds best to a small red spot shining on
a green background or to green annular illumination. (After Daw, 1984.)

useless for matching bedspreads, cushions, carpets, and curtains in a
boudoir. Although contributions from the two systems can be confused,
they usually work together.

How is the whole picture put together so that we see the expression
on a tennis player's face and at the same time see the ball so precisely
as it comes toward us? Partial answers are provided by organization of
the first steps in visual signaling described here. A major conceptual
discovery was that the abstract significance conveyed by the signal of
a neuron can be highly complex; the signal incorporates input messages
and gives them new meaning. An extension of the hypothesis of hierarchical organization predicts that cells should be discovered that bring
together larger and larger parts of the information that appears in the
field of vision. Indeed, in higher visual areas "face" neurons have been
described.[35] But how far can this concentration of information go? Will
there be a small group of cells or one pontifical cell that synthesizes
and combines into a whole picture all the features perceived? The
neurophysiologist obviously faces a dilemma. On the one hand we
know that visual information is synthesized into diverse meaningful
units in terms of receptive fields. These are distributed among an in-
calculable number of cells; in this sense perception is scattered and
diffuse. On the other hand, there should be cells that see the "big
picture," because without them we end up stating that each group of
cells looks at the next group, and vice versa. In discussions of this type

at this stage, the dreaded homunculus makes his appearance—the cell or the little man in the brain who actually sees what we see. To ridicule this concept is both fashionable and a sure sign of sophistication. Nevertheless, the homunculus does have a useful function: He represents and continually reminds us of our ignorance about higher cortical function. As soon as answers are found, he will die a natural death like phlogiston. We have no way as yet to replace him by a computer. What is so encouraging is that perception can now be thought of in terms of known cell properties after the signals have traversed only seven or eight synapses, and there are many more synapses to come before the cells run out. In no other system have mind and cell come so closely together.

SUGGESTED READING

General Reviews

Bolz, J., Gilbert, C. D. and Wiesel, T. N. 1989. Pharmacological analysis of cortical circuitry. *Trends Neurosci.* 12: 292–296.

Daw, N. W. 1984. The psychology and physiology of colour vision. *Trends Neurosci.* 7: 330–335.

Gilbert, C. D., Hirsch, J. A. and Wiesel, T. N. 1990. Lateral interactions in visual cortex. *Cold Spring Harbor Symp. Quant. Biol.* 55: 663–677.

Helmholtz, H. 1962/1927. *Helmholtz's Treatise on Physiological Optics.* J. P. C. Southhall (ed.). Dover, New York.

Hubel, D. H. 1982. Exploration of the primary visual cortex. *Nature* 299: 515–524.

Hubel, D. H. 1988. *Eye, Brain and Vision.* Scientific American Library, New York.

Hubel, D. H. and Wiesel, T. N. 1977. Ferrier Lecture: Functional architecture of macaque monkey visual cortex. *Proc. R. Soc. Lond. B* 198: 1–59.

Lund, J. S. 1988. Anatomical organization of macaque monkey striate visual cortex. *Annu. Rev. Neurosci.* 11: 253–288.

Maunsell, J. H. and Newsome, W. T. 1987. Visual processing in monkey extrastriate cortex. *Annu. Rev. Neurosci.* 10: 363–401.

Zeki, S. 1990. Colour vision and functional specialisation in the visual cortex. *Disc. Neurosci.* 6: 1–64.

Original papers

Bonhoeffer, T. and Grinvald, A. 1991. Iso-orientation domains in cat visual cortex are arranged in pinwheel-like patterns. *Nature* 353: 429–431.

Ferster, D. 1988. Spatially opponent excitation and inhibition in simple cells of the cat visual cortex. *J. Neurosci.* 8: 1172–1180.

Gilbert, C. D. and Wiesel, T. N. 1989. Columnar specificity of intrinsic horizontal and corticocortical connections in cat visual cortex. *J. Neurosci.* 9: 2432–2442.

Hubel, D. H. and Livingstone, M. S. 1987. Segregation of form, color, and stereopsis in primate area 18. *J. Neurosci.* 7: 3378–3415.

Hubel, D. H. and Wiesel, T. N. 1959. Receptive fields of single neurones in the cat's striate cortex. *J. Physiol.* 148: 574–591.

Hubel, D. H. and Wiesel, T. N. 1962. Receptive fields, binocular interaction, and functional architecture in the cat's visual cortex. *J. Physiol.* 160: 106–154.

Hubel, D. H. and Wiesel, T. N. 1968. Receptive fields and functional architecture of monkey striate cortex. *J. Physiol.* 195: 215–243.

Land, E. H. 1983. Recent advances in retinex theory and some implications for cortical computations: color vision and the natural image. *Proc. Natl. Acad. Sci. USA* 80: 5163–5169.

Livingstone, M. S. and Hubel, D. H. 1984. Anatomy and physiology of a color system in the primate visual cortex. *J. Neurosci.* 4: 309–356.

Livingstone, M. S. and Hubel, D. H. 1987a. Connections between layer 4B of area 17 and the thick cytochrome oxidase stripes of area 18 in the squirrel monkey. *J. Neurosci.* 7: 3371–3377.

Livingstone, M. S. and Hubel, D. H. 1987b. Psychophysical evidence for separate channels for the perception of form, color, movement, and depth. *J. Neurosci.* 7: 3416–3468.

Livingstone, M. S. and Hubel, D. H. 1988. Segregation of form, color, movement, and depth: Anatomy, physiology, and perception. *Science* 240: 740–749.

Shipp, S. and Zeki, S. 1985. Segregation of pathways leading from area V_2 to areas V_4 and V_5 of macaque monkey visual cortex. *Nature* 315: 322–325.

Tootell, R. B., Switkes, E., Silverman, M. S. and Hamilton, S. L. 1988. Functional anatomy of macaque striate cortex. II. Retinotopic organization. *J. Neurosci.* 8: 1531–1568.

Ts'o, D. Y., Frostig, R. D., Lieke, E. E. and Grinvald, A. 1990. Functional organization of primate visual cortex revealed by high-resolution optical imaging. *Science* 249: 417–420.

Wiesel, T. N. and Hubel, D. H. 1966. Spatial and chromatic interactions in the lateral geniculate body of the rhesus monkey. *J. Neurophysiol.* 29: 1115–1156.

Zeki, S. and Shipp, S. 1988. The functional logic of cortical connections. *Nature* 335: 311–317.

Zeki, S., Watson, J. D., Lueck, C. J., Friston, K. J., Kennard, C. and Frackowiak, R. S. 1991. A direct demonstration of functional specialization in human visual cortex. *J. Neurosci.* 11: 641–649.

GENETIC AND ENVIRONMENTAL INFLUENCES IN THE MAMMALIAN VISUAL SYSTEM

CHAPTER EIGHTEEN

Experiments have been made in the mammalian visual system to assess the relative importance of genetic factors and of the environment for the establishment and proper performance of synaptic interactions. In newborn, visually naive kittens and monkeys, many features of the neuronal organization are already present. Cells in the retina and the lateral geniculate nucleus respond to stimuli in much the same way as those in adult animals. In the visual cortex, the neurons have characteristic receptive fields and require oriented bars or edges. Clear differences are apparent, however, particularly in layer 4, where geniculate fibers end. At the time of birth in kittens and monkeys, the ocular dominance columns are not fully formed: Cells in layer 4 are driven by both eyes. The adult pattern—in which cells in layer 4 are supplied by one or the other, but not both, eyes—is established in the first 6 weeks after birth. During this time geniculate fibers retract to form columns with clear boundaries.

In early life the connections of neurons in the visual cortex are susceptible to change and can be irreversibly affected by inappropriate use. For example, closure of the lids of one eye during the first 3 months of life leads to blindness in that eye. The abnormality occurs chiefly at the level of the cortex; although geniculate cells are still driven by the eye that had been closed, the great majority of cortical cells are not. The other eye functions normally. The period of greatest susceptibility in cats and monkeys occurs during the first 6 weeks of life. Lid closure in adult animals has no effect.

Deprivation during the first 6 weeks leads to a shrinkage of the cortical dominance columns supplied by that eye. A corresponding increase is seen in the width of columns supplied by the normal undeprived eye. Thus when retraction of geniculate inputs to layer 4 occurs, columns that are normally of equal width become unequal: Fibers supplied by the deprived eye retract more than usual, whereas those supplied by the normal eye hardly retract at all. During this early critical period, these effects can be reversed by opening the sutured eye and closing the undeprived eye. That competition between the two eyes for cortical territory occurs is further suggested by the effects of bilateral deprivation. In monkeys with both eyes closed, neither eye has an advantage and the columnar structure develops. But each cell in the cortex is driven by only one or the other eye and not both.

Abnormal patterns of sensory input without deprivation also lead to similar effects attributable to competition. When a squint, or strabismus, is produced by cutting extraocular muscles, each eye is exposed to the normal amount of visual input and only the fixation of the two eyes upon objects is altered. Yet the way cortical cells are driven by the two eyes is changed in kittens or immature monkeys that have been made to squint. The cells have normal receptive field properties, but only a few are driven by both eyes. Instead, one eye or the other eye is effective on its own. Since there is no disuse, it again appears that impulse traffic in convergent pathways must continue in an appropriately balanced manner for the normal functional organization to be maintained. This idea is reinforced by experiments in which all impulses in the optic nerves, including spontaneous activity, were blocked in kittens by the use of tetrodotoxin injected into both eyes. Columns then failed to develop altogether, the areas in layer 4 supplied by each eye remaining coextensive as at birth. The cellular and molecular mechanisms through which coherent and incoherent impulse activity affect structure are not known. A major problem concerns how some neurons are induced to sprout and form new synaptic connections, while others retract to lose those that had been present, and still others remain unaffected.

These results have a wide significance for considering a variety of aspects of development in the central nervous system. Other sensory systems and higher functions may also have critical periods during which their performance can be sharpened by appropriate use or irreversibly damaged by disuse or inappropriate use. Well-defined genetically determined errors of connections also occur. In the Siamese cat, some optic nerve fibers consistently take the wrong pathway during development and become connected to inappropriate cells in the lateral geniculate nucleus and cortex. Developmental studies of the mammalian visual system provide a first step for understanding how experience in early life influences the structure and function of our brains in later life.

We have emphasized repeatedly the specificity of the wiring that is necessary for the nervous system to function properly. It is also clear that development continues after birth for various periods in different animals. For example, kittens are born with their eyes closed. If the lids are opened and light is shone into an eye, the pupil constricts, although the animal had not previously been exposed to light and appears to be completely blind.[1] By 10 days, the kitten shows evidence of vision and thereafter begins to recognize objects and patterns. When kittens are

[1]Riesen, A. H. and Aarons, L. 1959. *J. Comp. Physiol. Psychol.* 52: 142–149.

brought up in darkness instead of in their normal environment, the pupillary reflex continues to function but they remain blind. It is as though there were a hierarchy of susceptibility, with "hard" and "soft" wiring in different parts of the brain.

Changes in the performance of the nervous system during development raise a number of questions. What are the relative contributions of genetic factors and experience (summed up by the phrase *nature and nurture*)? In this chapter we describe one system without presenting a full review of the field of sensory deprivation. To what extent are the neuronal circuits required for vision already present and ready to work at birth? What effect on their development has light falling into the eyes? Does a kitten or a monkey brought up in darkness become blind because the connections fail to form, or because some of the connections that had originally been there have withered away? Georg von Lichtenberg (1742–1799), who seemed to have a keen appreciation of such developmental problems, wrote pithily: "What astonished him was that cats should have two holes cut in their skins at exactly the same places where their eyes were." The visual system offers great advantages for approaching directly questions relating to development because the relay stations are accessible and the amount of light and natural stimulation can readily be altered. Within the visual system we again emphasize the pathways from retina to cortex in cat and monkey. For our purposes, it is convenient to focus largely on work that follows logically from the material presented Chapters 16 and 17.

A good deal is known about the organization of the visual connections that underlie perception in the adult cat and monkey. A simple cell in the cortex selectively "recognizes" one well-defined type of visual stimulus, such as a narrow bar of light oriented vertically, in a particular region of the visual field of either eye. It is natural to wonder whether cells of this type are already present in the newborn animal or whether visual experience and learning are required so that a random set of preexisting connections is reformed or modified for such a specific task. For technical reasons, it is difficult to record from cells in newborn animals. Most experiments in kittens have been made during the first 3 weeks after birth. To prevent form vision, the lids were sutured or the cornea was covered by a translucent occluder. Similarly, visually naive monkeys were produced by suturing the animals' lids immediately or several days after birth; in some instances, monkeys were delivered by Caesarean section for later examination, care being taken to avoid exposure to light.[2,3]

> The visual system in newborn kittens and monkeys

A newborn monkey appears visually alert and is able to fixate. In contrast, a newborn kitten whose lids have been opened by surgery is behaviorally blind. Nevertheless, many of the features seen in adults are already present in the performance of cortical neurons in both animals. For example, recordings made from individual cells in the

[2]Hubel, D. H. and Wiesel, T. N. 1963. *J. Neurophysiol.* 26: 994–1002.
[3]Wiesel, T. N. and Hubel, D. H. 1974. *J. Comp. Neurol.* 158: 307–318.

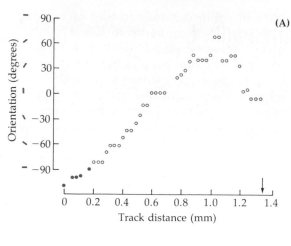

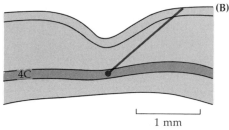

1

ORIENTATION COLUMNS in the absence of visual experience. **(A)** Axis orientation of receptive fields encountered by an electrode during an oblique penetration through the right cortex of a 17-day-old baby monkey whose eyes had been sutured closed on the second day after birth. **(B)** The circle marks a lesion made at the end of the electrode track in layer 4. The orientation of receptive fields in (A) changes progressively as orientation columns are traversed, indicating that normally organized orientation columns are present in the visually naive animal. Closed circles are from the ipsilateral eye; open circles are from the other eye. (From Wiesel and Hubel, 1974.)

primary visual cortex V_1 show that the cells are not driven by diffuse illumination of the eyes. As in a mature animal, they fire best when light or dark bars with a particular orientation are shone onto a particular region of the retina of either eye. In many cells the range of orientations recorded from animals lacking prior visual experience cannot be distinguished from that in adults. The responses of the receptive fields are also organized into antagonistic "on" and "off" areas that are similar in the two eyes. Moreover, with oblique penetrations the preferred orientation changes in a regular sequence as the electrode moves through the cortex[3] (Figure 1). In newborn animals, however, the discharges of cortical cells tend to be weaker than those of the adult and some cells are unresponsive.

A major difference, which has turned out to be of key importance in assessing the changes that occur during normal and abnormal development, is apparent in layer 4 of the visual cortex. Unlike the pattern in the adult, where the ocular dominance columns corresponding to the inputs from the two eyes can be clearly distinguished, in the newly born kitten or monkey extensive overlap occurs in the arborizations of geniculate fibers ending in layer 4.[4,5] This is shown in Figure 2A and schematically in Figure 3. Only a hint of ocular dominance can be

[4]LeVay, S., Wiesel, T. N. and Hubel, D. H. 1980. *J. Comp. Neurol.* 191: 1–51.
[5]LeVay, S., Stryker, M. P. and Shatz, C. J. 1978. *J. Comp. Neurol.* 179: 223–244.

2   DEVELOPMENT OF OCULAR DOMINANCE COLUMNS in layer 4 of cat cortex. Autoradiographs of sections of the visual cortex of animals in which one eye had been previously injected with tritium-labeled proline. Photos taken in dark-field illumination show silver grains as white dots. **(A)** Brain of a 15-day-old kitten. Layer 4 is labeled continuously and uniformly with no evidence of columns. At this stage and for the next few weeks, fibers ending in the cortex from the geniculate overlap in layer 4. **(B)** Similar section through the visual cortex of an adult cat (92 days). Note the patchy distribution of label in layer 4, corresponding to ocular dominance columns for the injected eye. (After LeVay, Stryker, and Shatz, 1978.)

resolved; the reason is that the individual fibers of the geniculate nucleus spread over a wide area in layer 4. During the first 6 weeks of the animal's life, axons retract, establishing more restricted and separate domains of cortex that become supplied exclusively by one eye or the other. In parallel, physiological changes develop: Initially, simple cortical cells of layer 4 are driven by both eyes. By 6 weeks, as in the adult, each cell can be driven by only one eye. The postnatal development of ocular dominance columns proceeds in animals reared in total darkness.

That sorting out of columns actually begins before birth and without visual experience was demonstrated by Rakic.[6–9] He injected the eyes of monkeys in utero with radioactive amino acids at different stages; this procedure made it possible to observe the outgrowth of fibers and their distribution in the lateral geniculate nucleus and layer 4 of the cortex. At early stages the overlap of the territories supplied by the two eyes is virtually complete; by a few days before birth, hints of columnar organization are discernible. Similarly, in fetal cats it has been shown by Shatz and her colleagues that the optic nerve fiber terminals overlap as they reach the lateral geniculate nucleus and that the sorting out into separate layers (A, A₁, and C) begins during the last third of gestation

[6]Rakic, P. 1977. *J. Comp. Neurol.* 176: 23–52.

[7]Rakic, P. 1977. *Philos. Trans. R. Soc. Lond. B* 278: 245–260.

[8]Rakic, P. 1986. *Trends Neurosci.* 9: 11–15.

[9]Rakic, P. 1988. *Science* 241: 170–176.

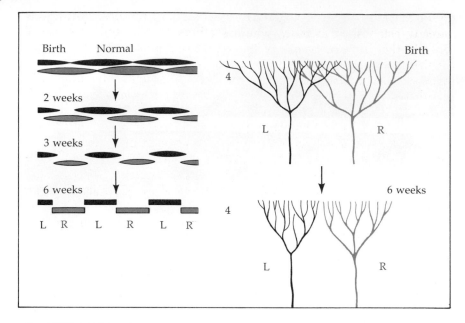

3 RETRACTION OF LATERAL GENICULATE NUCLEUS AXONS ending in layer 4 of the cortex during the first 6 weeks of life. Diagrams on left and right show schematically the overlap of inputs from the right (R) and left (L) eyes present at birth and the subsequent segregation into separate clusters corresponding to ocular dominance columns. (Modified from Hubel and Wiesel, 1977.)

and forms the full adult pattern about 2 weeks after birth.[10] Ocular dominance columns in visual cortex only become recognizable at about 30 days after the birth of the cat. The kitten is born at a more immature stage of development than the monkey.[5]

In newly born kittens and monkeys, as in adults, cortical cells outside layer 4 tend to be driven by both eyes, some better by one eye, some by the other, and some equally well by both.[2,3] About 20 percent of all the 1116 cells in Figure 4 from visually normal monkeys were driven solely by one eye and about the same percentage by the other. The degree of dominance can be conveniently expressed in a histogram by grouping neurons into seven categories according to the discharge frequency with which they respond to stimulation of one or the other eye (Chapter 17). In the visually inexperienced kitten and the 2-day-old monkey, the histogram appears rather normal, the majority of cells responding to appropriate illumination of either eye.

These findings in immature animals come as no great surprise. Although the development of the cortex is susceptible to environmental changes, one would be rather surprised if the basic outline of neural organization with its orderly and intricate visual connections were totally dependent on the vagaries of the visual environment. The principal

[10]Shatz, C. J. and Sretavan, D. W. 1986. *Annu. Rev. Neurosci.* 9: 171–207.

point to be made here is that certain features of the basic wiring are already established at birth, whereas others become fully developed only in the first few weeks of life. This is reminiscent of the events occurring during the development of nerve–muscle synapses in neonatal rats: At birth, each motor end plate is supplied by numerous motoneurons, but in a few weeks, most retract, leaving each muscle fiber supplied by just one axon[11] (Chapter 11).

The remainder of this chapter is devoted mainly to an account of the ways in which abnormal sensory experience in early life can drastically affect the anatomy and physiology of the brain. However, systematic abnormalities also occur as a result of genetic defects. Color blindness is one familiar example, and others are provided by mutations occurring in Siamese cats and in albino animals.

In Siamese cats, certain optic nerve fibers fail to grow along their usual pathways during development. In addition the animals are frequently cross-eyed.[12] The defect is accompanied by abnormal crossing of optic nerve fibers at the chiasm. Some of the fibers that should remain

Abnormal connections in the visual system of the Siamese cat

[11]Redfern, P. A. 1970. *J. Physiol.* 209: 701–709.
[12]Hubel, D. H. and Wiesel, T. N. 1971. *J. Physiol.* 218: 33–62.

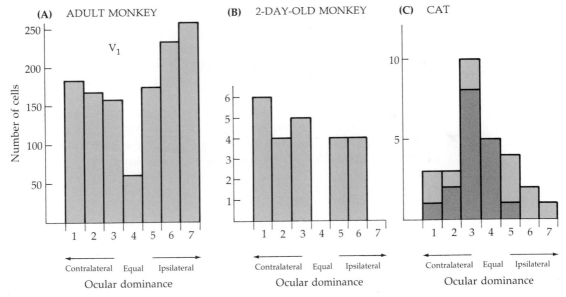

4 OCULAR DOMINANCE DISTRIBUTION in the visual cortex. Cells in groups 1 and 7 of the histogram are driven by one eye only (ipsilateral or contralateral). All other cells have input from both eyes. In groups 2, 3 and 5,6 one eye predominates; in group 4 both have a roughly equal influence. (A) Normal adult monkey. (B) A similar ocular dominance distribution in a normal 2-day-old monkey. (C) Histogram of ocular dominance distribution in a 20-day-old kitten (light blue) with normal visual experience and two kittens (8 and 16 days old) without previous visual exposure (dark blue). More monocularly driven cells are present in monkey than in cat visual cortex. (A and B from Wiesel and Hubel, 1974; C from Hubel and Wiesel, 1963.)

on one side of the animal, instead of doing so, cross to the other side. As a result, the lateral geniculate nucleus receives a disproportionately large input from the contralateral eye and a correspondingly diminished input from the ipsilateral eye. Interestingly, the optic nerve fibers that have taken a wrong course terminate in the layer of the lateral geniculate nucleus normally reserved for the ipsilateral eye. The cells there receive information from unaccustomed regions of the visual field and from the wrong eye. These abnormal fibers are accommodated in the "incorrect" layer of the lateral geniculate by those cells left "vacant" because their expected axons did not arrive. Farther on in the pathways, specific regions of the visual cortex receive an orderly projection from the lateral geniculate nucleus. Complete scrambling of connections does not occur because the ground rules for normal cell connections are still followed. As a result, the abnormal input is segregated from the normal.[12]

Albinism

The genetic aspects of the defect in visual development shared by albinos of many species are of considerable interest. These aspects have been explored in detailed systematic studies by Guillery and his colleagues, who examined a large variety of animals, including rats, minks, mice, guinea pigs, Himalayan rabbits, ferrets, monkeys, and even one white tiger.[13] In all these species, a direct correlation can be made between the absence of pigment in the eye and the abnormal course taken by the optic nerve fibers of certain ganglion cells. In view of this, it has been suggested that melanin itself or another related gene product of the pigment epithelial cell may influence the fate of retinal ganglion cell axons as they cross at the chiasm. Here, then, is an example of an intriguing genetic defect. It involves the albino gene that determines the color of an animal and also causes errors in connections and a modification of specificity in the visual system.[14]

Another example of a genetic effect on cortical organization is one in which a strain of mice was bred selectively with extra face whiskers. These animals also displayed appropriately placed extra barrels (Chapter 14) in their cortex, representing the extra whiskers.[15]

EFFECTS OF ABNORMAL EXPERIENCE

This section describes three types of experiments, mostly by Hubel and Wiesel, in which animals were deprived of normal visual stimuli. They studied the effects on the physiological responses of the nerve cells in the visual system after (1) closing the lids of one or both eyes; (2) preventing form vision, but not access of light to the eye; and (3) leaving light and form vision intact, but producing an artificial strabismus (squint) in one eye. These procedures cause abnormalities in function and in some instances profound anatomical changes.

When the lids of one eye were sutured during the first 2 weeks of life, kittens and monkeys still developed normally and used their un-

[13]Guillery, R. W. 1974. *Sci. Am.* 230(5): 44–54.
[14]Guillery, R. W., Jeffery, G. and Cattanach, B. M. 1987. *Development* 101: 857–867.
[15]Welker, E. and Van der Loos, H. 1986. *J. Neurosci.* 6: 3355–3373.

operated eye. However, at the end of 1 to 3 months, when the operated eye was opened and the normal one closed, it was clear that the animals were practically blind in the operated eyes. For example, kittens would bump into objects and fall off tables.[16,17] There was no gross evidence of a physiological defect within such eyes; pupillary reflexes appeared normal and so did the electroretinogram, which serves as an index of the average electrical activity of the eye. Records made from retinal ganglion cells in deprived animals showed no obvious changes in their responses, and their receptive fields appeared normal.

Although responses of cells in the lateral geniculate nucleus appeared relatively unchanged after monocular deprivation,[18] there were major changes in the responses of cortical cells.[4,19] When electrical recordings were made in the visual cortex, only a few of the cells could be driven by the eye that had been closed. The majority of those that did respond had abnormal receptive fields. Figure 5 shows the ocular dominance histograms obtained from the cells examined in kittens and monkeys raised with closure of one eye during the first weeks of life.

Cortical cells after monocular deprivation

The results described so far indicate that if one eye is not used normally in the first weeks of life, its power wanes and it ceases to be effective. These far-reaching changes are produced by the relatively minor procedure of sewing the lids, without cutting any nerves. What is the important condition for maintaining and developing proper visual responses? Is diffuse light adequate? Lid closure reduces light that reaches the retina but does not exclude it. One would suspect, therefore, that diffuse light alone would not keep an eye functioning normally. To test this idea, a series of experiments were made in which a plastic occluder (like frosted glass or a ping-pong ball) was placed over the cornea of a newborn kitten instead of closing its eyelids; the occluder prevented form vision but still admitted light.[17] All these cats were functionally blind in the deprived eye. Furthermore, cortical cells were no longer driven by the deprived eye. Neither retinal nor geniculate responses were noticeably changed under such conditions. Thus, *form vision*, rather than the presence of light, is an important stimulus required to prevent abnormal development of cortical connections. Surprisingly, however, form vision alone may not be adequate to maintain complete normality (see later).

Relative importance of diffuse light and form for maintaining normal responses

Cells in the lateral geniculate nucleus of cat and of monkey are arranged in layers, each predominantly supplied by one or the other eye (Chapter 16). In the same animals that showed marked abnormalities in the cortex after lid closure, the geniculate cells seemed to behave normally. The cells in the appropriate layers responded with "on" or "off" discharges to small spots of light shone into the deprived or the normal eye, and no clear-cut differences from normal firing patterns could be observed. Nevertheless, it was shown that marked changes in

Morphological changes in the lateral geniculate nucleus after visual deprivation

[16]Wiesel, T. N. 1982. *Nature* 299: 583–591.
[17]Wiesel, T. N. and Hubel, D. H. 1963. *J. Neurophysiol.* 26:1003–1017.
[18]Wiesel, T. N. and Hubel, D. H. 1963. *J. Neurophysiol.* 26: 978–993.
[19]Wiesel, T. N. and Hubel, D. H. 1965. *J. Neurophysiol.* 28: 1029–1040.

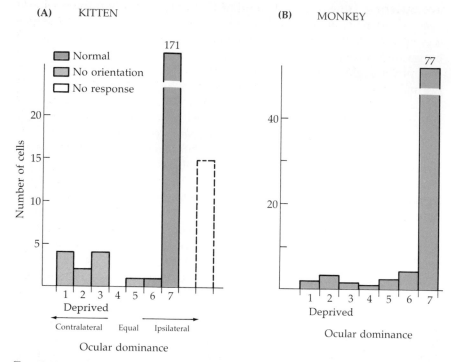

(A)   KITTEN

(B)   MONKEY

5   DAMAGE PRODUCED BY CLOSURE OF ONE EYE. Ocular dominance distribution in kittens and in a monkey. (A) In five kittens 8 to 14 weeks old deprived of vision in the right eye, only 13 out of 199 cells responded to stimulation of the deprived eye. All but one of those cells had abnormal receptive fields. The column with dashed lines represents spontaneously active cells that did not respond to either eye. (B) Ocular dominance distribution in a monkey whose right eye was closed from 21 to 30 days of age. In spite of a subsequent 4 years of binocular vision, most cortical neurons were unresponsive to stimulation of the right eye. (A from Hubel and Wiesel, 1965; B from LeVay, Wiesel and Hubel, 1980.)

morphology occurred after lid closure of one eye: The cells were noticeably smaller in the layers supplied by the deprived eye. The cell bodies were only about one-half as large as those in the normal layers, the reduction in size depending on the duration of lid closure.[18] It seems surprising that cells in the lateral geniculate nucleus showed obvious morphological changes but little significant physiological deficit. Several lines of evidence suggest that the size of cells in the lateral geniculate nucleus may depend on the extent of their arborization in the cortex.[20]

The morphological consequences of eye closure are particularly conspicuous in layer 4 of the visual cortex V$_1$ where geniculate fibers terminate.[4,21] Changes in ocular dominance columns develop in monkeys after closure of one eye at birth. These have been revealed by the technique of autoradiography as in Figure 2, after injection of radioac-

Morphological
changes in the cortex
after visual
deprivation

[20]Guillery, R. W. and Stelzner, D. J. 1970. *J. Comp. Neurol.* 139: 413–422.
[21]Hubel, D. H., Wiesel, T. N. and LeVay, S. 1977. *Philos. Trans. R. Soc. Lond. B* 278: 377–409.

tive materials into one eye (Chapter 17). After deprivation there occurs a marked reduction in width of the ocular dominance columns that receive their input from the eye that had been occluded. At the same time, the columns with input from the normal eye show a corresponding increase in width compared with that seen in a normal adult monkey. The shrinkage of ocular dominance columns is evident in Figure 6, in which the normal columns can be compared with columns in animals in which one eye had been closed at 2 weeks and left closed for 18 months. The changes indicate that geniculate axons activated by the normal eye retained the territory in the cortex lost by their weaker, visually deprived neighbors. These results have also been confirmed physiologically by recording from layer 4 where the geniculate fibers terminate. Almost all cells were driven by the eye that had not been deprived.

When the lids of one eye are closed in an adult cat or monkey, no abnormal consequences are seen.[4] For example, in an adult animal, even if an eye is closed for over a year the cells in the cortex continue to be driven normally by both eyes and display the normal ocular

Critical period of susceptibility to lid closure

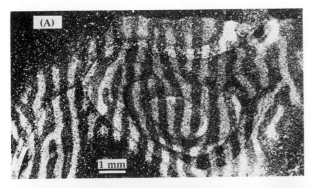

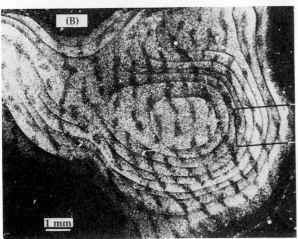

6

OCULAR DOMINANCE COLUMNS AFTER CLOSURE OF ONE EYE. (A) Normal adult rhesus monkey. The right eye had been injected with a radioactive proline-fucose mixture 10 days before sections were made tangential to the exposed dome-shaped primary visual cortex of the right hemisphere. Layer 4 displays fingerlike alternating dark and light processes. With dark-field illumination, the radioactivity in the geniculate axon terminals in layer 4 appears as fine white granules, forming the light stripes (columns) that correspond to the injected eye. The dark intervening bands respond to the other eye. This picture was made as a photomontage reconstruction from eight parallel sections through layer 4. (B) A similar reconstruction of layer 4 in an 18-month-old monkey whose right eye had been closed at the age of 2 weeks. Proline-fucose was injected into the normal left eye. The white label demonstrates the columns in layer 4 whose input is derived from the nondeprived eye. The columns are larger than normal and alternate with narrowed columns, seen as dark gaps, supplied by the eye whose lids had been closed. (From Hubel, Wiesel, and LeVay, 1977.)

dominance histogram. Moreover, even if one eye is completely removed in an adult monkey, the structure of layer 4 remains normal when observed with autoradiography or other staining methods. This finding indicates a remarkable resistance to change of layer 4 in the adult animal when compared with the changes seen in the immature animal. Such resistance to change occurs even though surgical removal of an eye in an adult leads to considerable atrophy in the lateral geniculate nucleus.

The period during which susceptibility to lid closure in kittens is highest has been narrowed down to the fourth and fifth weeks after birth. During the first 3 weeks or so of life, eye closure has little effect. This is not surprising since the kittens' eyes are normally closed for the first 10 days. But abruptly, during weeks 4 and 5, sensitivity increases. Closure at that age for as little as 3 to 4 days leads to a sharp decline in the number of cells that can be driven by the deprived eye.[22,23] An experiment in which littermates are compared is shown in Figure 7. In this example, 6- and 8-day closures starting at the age of 23 and 30 days (Figure 7A and 7B) caused about as great an effect as 3 months of monocular deprivation from birth. The susceptibility to lid closure declines after the critical period has passed and eventually disappears by about 3 months of age (Figure 7C and 7D). The critical period can, however, be prolonged to more than 6 months by rearing kittens in the dark. In the absence of visual experience, the susceptibility to monocular closure can still be demonstrated at these late times. There is evidence that even a brief exposure of the kitten to light for a few hours may be sufficient to prevent such extension of the critical period.[24,25]

In monkeys the greatest sensitivity to lid closure is during the first 6 weeks.[4,16,21] Before six weeks, substantial changes in eye preference and columnar architecture develop if one eye is closed for a few days. During the subsequent months (up to about 12 to 18 months), several weeks of closure are required to produce obvious changes in ocular dominance histograms or the width of columns in layer 4. At later times, no changes can be produced even by surgical removal of one eye.

The susceptibility of kittens and monkeys during early life is reminiscent of some clinical observations made in humans. It has long been known that removal of a clouded or opaque lens (cataract) can lead to a restoration of vision, even though the patient has been blind for many years. In contrast, a cataract that develops in a baby can lead to blindness without the possibility of recovery unless the operation is performed very early in the critical period. A familiar clinical procedure used in the past for the treatment of children with strabismus (or squint) was to patch the good eye for prolonged periods in order to encourage the weaker eye to be used. There is evidence that damage in acuity

[22]Hubel, D. H. and Wiesel, T. N. 1970. *J. Physiol.* 206: 419–436.
[23]Malach, R., Ebert, R. and Van Sluyters, R. C. 1984. *J. Neurophysiol.* 51: 538–551.
[24]Cynader, M. and Mitchell, D. E. 1980. *J. Neurophysiol.* 43: 1026–1040.
[25]Mower, G. D., Christen, W. G. and Caplan, C. J. 1983. *Science* 221: 178–180.

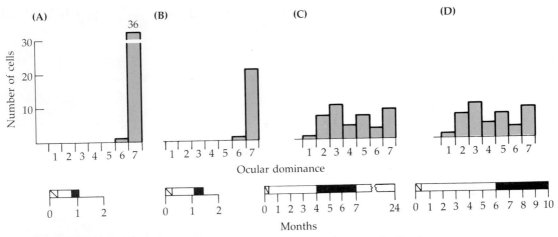

7 **CRITICAL PERIOD IN KITTENS. Histograms of ocular dominance distribution in the visual cortex in kittens that are littermates. (A) The right eye was sutured shut during the critical period for 6 days (age 23–29 days). (B) The right eye was closed from age 30 to 39 days during the critical period. In each animal (A and B) only one cell was weakly influenced by the temporarily deprived eye. The damage was about as great as eye closure for 3 months or longer. (C) The right eye was open for the first 4 months, then sutured for 3 months, then opened again. Recordings were made at the age of 2 years. (D) The eye was open for the first 6 months, then closed for 4 months. Ocular dominance was determined at 10 months of age. In (C) and (D) the ocular dominance distributions appear normal for both eyes. The black segment below the abscissa indicates the closure period. (From Hubel and Wiesel, 1970.)**

may result, depending on the child's age at the time and the duration of the patching. Clinical observations suggest that the greatest sensitivity occurs in babies during the first year but that the critical period may persist for several years.[26]

An interesting change that develops after lid closure in neonatal monkeys is elongation of the eye. Such elongation causes blurred images and short-sightedness (myopia); thus it is known that children develop myopia if the lids interfere with vision or if the transparency of the eye is reduced. Experiments by Wiesel, Raviola, and their colleagues[27] have been made to test whether a reduced level of impulse activity influences transmitter synthesis in a deprived eye. Their hypothesis is that transmitters, particularly peptides, could play a part in regulating the growth of the eye during development as well as transmitting signals from cell to cell. Their results suggest that abnormal levels of transmitters do occur in eyes deprived of normal input. In particular the level of vasoactive intestinal polypeptide of amacrine cells, a peptide that is known to act on blood vessels and smooth muscles, increases. Whether it actually produces the abnormal elongation following lid suture is not known.

[26]Jacobson, S. G., Mohindra, J. and Held, R. 1981. *Br. J. Opthalmol.* 65: 727–735.
[27]Stone, R. A. et al. 1988. *Proc. Natl. Acad. Sci. USA* 85: 257–260.

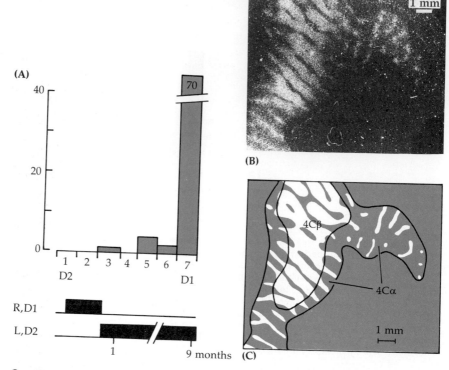

(A)

(B)

(C)

8   **EFFECTS OF REVERSE SUTURE ON OCULAR DOMINANCE** in monkey. At 2 days of age, the right eye was closed. At 3 weeks (19 days later), the right eye was opened and the left closed. The left eye was then kept closed for 9 months, when the right (initially deprived) eye was injected with tritium-labeled proline. (A) Ocular dominance histogram from monkey cortex. Almost all cells are driven exclusively by the right eye (D1), virtually none by the left (D2). Had both eyes been open at 3 weeks, the histogram would be reversed; almost none of the cells would then be driven by the right eye. Accordingly, fibers driven by the right eye had retracted and then recaptured cells they had previously lost. (B) Tangential section of cortex passing through layers 4Cα and 4Cβ (layer 4Cα is superficial to 4Cβ). In dark-field illumination, silver grains appear white. The bands labeled by the right eye are expanded in layer 4Cβ even though it had been deprived of light for 19 days. During those first days, fields supplied by the right eye had shrunk before expanding. (C) Diagram to show expansion of territory supplied by the right, initially deprived eye in layer 4Cβ. In 4Cα, the distribution is different: The territory supplied by the left eye is expanded and that by the right eye is reduced. These results show (1) that ocular dominance columns in layer 4Cβ shrink with disuse and expand after reverse suture during the critical period; and (2) that recovery does not occur equally well in other layers, such as 4Cα. Note that the bands in 4Cα and 4Cβ are still in precise register as in normal monkeys. (After LeVay, Wiesel and Hubel, 1980.)

Recovery during the
critical period

To what extent is recovery possible after lid closure during the critical period? Even if the deprived eye in a cat or monkey is subsequently opened for months or years, the damage remains permanent, with little or no recovery: The animal continues to be blind in that eye, with shrunken columns and skewed ocular dominance histograms.[28] In animals with monocular closure, experiments have been made in which

[28]Wiesel, T. N. and Hubel, D. H. 1965. *J. Neurophysiol.* 28: 1060–1072.

the lids were opened in the deprived eye and closed over the normal eye. This procedure, termed REVERSE SUTURE, leads to a dramatic recovery of vision provided it is carried out during the critical period. Kittens and monkeys not only begin to see again with the initially deprived eye, but they become blind in the other eye.[29,4] Accompanying these changes, the ocular dominance histograms switch, so that the newly opened eye drives most cells, while the eye that had been opened for the first weeks (now closed) cannot. Moreover, the anatomical pattern in layer 4 revealed by autoradiography shows a similar switch in preference: The shrunken regions supplied by the initially closed eye expand at the expense of the other eye. Figure 8 shows recordings and autoradiographs of the cortex in a monkey in which the right eye had been closed at 2 days and left closed for 3 weeks. By this time the deprived eye would no longer be able to drive cortical cells and the columns supplied by it would have retracted. The right eye was then opened and the left eye closed for the next 8 months. Nearly all neurons responded only to the initially deprived right eye, and the areas of cortex supplied by it had expanded.

The remarkable conclusions from these experiments are that (1) during the critical period in a normal animal, geniculate fibers supplying layer 4 of the cortex retract so that each eye supplies areas of comparable extent; (2) lid closure of one eye during the critical period leads to unequal retraction; and (3) reverse suture during the critical period produces *sprouting* of the geniculate axons so that an eye can recapture the cells it had lost (Figure 9). If postponed until adulthood, reverse

[29]Blakemore, C. and Van Sluyters, R. C. 1974. *J. Physiol.* 237:195–216.

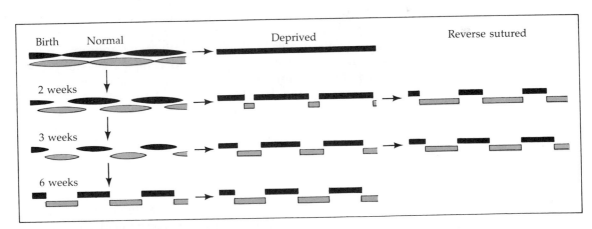

9 **SCHEME ILLUSTRATING EFFECTS OF EYE CLOSURE. As shown in Figure 3,** in the normal cat or monkey by 6 weeks ocular dominance columns have become well defined in layer 4 of the cortex. Lid closure causes excessive retraction of fibers supplied by the deprived eye. Those supplied by the open eye retract less than usual and in the adult supply larger areas of cortex than in normal animals, where the competition is more equal. After reverse suture during the critical period, the initially deprived eye can recapture the territory it had lost in layer 4Cβ. (After Hubel and Wiesel, 1977.)

suture is without effect. For example, in a monkey in which the reverse suture was performed at 1 year of age, the labeled columns for the initially deprived eye remained shrunken. Thus, though it is possible to cause change by lid closure at one year, connections that have already been changed once are not readily regained.

The concept of a well-defined, hard-and-fast critical period may be an oversimplification. Experiments on reverse suture in monkeys suggest that different layers of the striate cortex may develop at different rates; the critical period may be over for one layer while an adjacent layer is still capable of being modified in structure and function (see Figure 8). The molecular mechanisms involved in maintaining and terminating the critical period are as yet unknown.[30]

REQUIREMENTS FOR MAINTENANCE OF FUNCTIONING CONNECTIONS IN THE VISUAL SYSTEM: THE ROLE OF COMPETITION

At this stage one might be tempted to conclude that loss of activity in the visual pathways is the main factor that tends to disrupt normal responses of cortical neurons. After all, cortical cells are driven not by diffuse illumination but by shapes and forms. The following discussion shows that there must be additional causes of a far more subtle nature. In particular there must occur in addition a special interaction between the two eyes, an interaction we cannot yet explain fully.

Binocular lid closure

The first clue that loss of visually evoked activity cannot on its own account for the changed performance of neurons is shown by the following experiments. Both eyes were closed in monkeys, newborn or delivered by Caesarean section.[4] From the preceding discussion one would expect that cells in the cortex would subsequently be driven by neither eye. Surprisingly, however, after binocular closure for 17 days or longer, most cortical cells could still be driven by appropriate illumination; the receptive fields of simple and complex cells appeared largely normal. The columnar organization for orientation was similar to that in controls (Figure 1). The principal abnormality was that a substantial fraction of the cells could not be driven binocularly (Figure 10). In addition, some spontaneously active cells could not be driven at all, and others did not require specifically oriented stimuli. However, the areas of cortex supplied by each eye were equal, and the pattern resembled that seen in normal or adult monkeys: In layer 4, cells were driven by one eye only and the columns were well defined when marked by autoradiography. Binocular closure in kittens leads to similar effects except that more cortical cells continue to be binocularly driven.[19] At the same time cells in the lateral geniculate body do show atrophy (a decrease in size of approximately 40%) in layers supplied by each eye. The conclusion one can draw from these experiments is that some,

---

[30]Daw, N. W. et al. 1983. *J. Neurosci.* 3: 907–914.

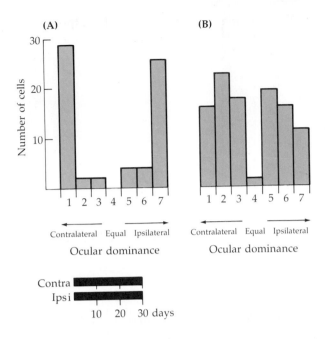

10

OCULAR DOMINANCE HISTOGRAMS after closure of both eyes at birth. (A) A monkey was delivered by Caesarean section and the lids of both eyes were immediately sutured. Recordings were made at 30 days of age. In contrast to the effects of monocular deprivation, each of the two deprived eyes could drive cells in the visual cortex. The receptive fields appeared normal, except that relatively few cells were driven by both eyes. The black area at the bottom of the histogram indicates closure time. (B) Ocular dominance histogram from a normal 21-day-old monkey. (From Wiesel and Hubel, 1974.)

but not all, of the ill effects expected from closing one eye are reduced or averted by closing both eyes. Once more one might theorize that inputs from the two eyes are somehow competing for representation in cortical cells, and with one eye closed the contest becomes unequal.

The abnormal effects described in the preceding discussion were produced by suturing eyelids or by using translucent diffusers, implicating loss of form vision. Following the clue that in children a squint (strabismus) can produce severe loss of vision or blindness in one eye, Hubel and Wiesel produced artificial squint in cats and monkeys by cutting an eye muscle.[31,16] The optical axis of the eye was thereby deflected from normal. Under such conditions, illumination and pattern stimulation for each eye remained unchanged. The experiment at first seemed disappointing because after several months vision in both eyes of the operated kittens appeared normal, and Hubel and Wiesel were about to abandon a laborious set of experiments (personal communication). Nevertheless, they recorded from cortical cells and obtained the following surprising results. Individual cortical cells had normal receptive fields and responded briskly to precisely oriented stimuli. But almost every cell responded only to one eye; some were driven only by the ipsilateral eye and others only by the contralateral, but almost none were driven by both. The cells were, as usual, grouped in columns with respect to eye preference and field axis orientation. As expected, no atrophy occurred in the lateral geniculate body. Similar results were obtained in monkeys in which strabismus had been induced during the

Effects of artificial squint

[31]Hubel, D. H. and Wiesel, T. N. 1965. *J. Neurophysiol.* 28:1041–1059.

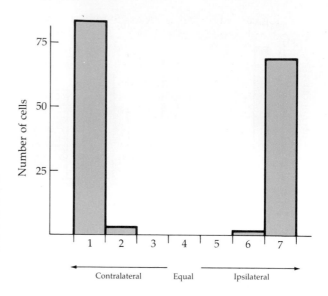

11

EFFECT OF SQUINT ON MONKEY OCULAR DOMINANCE. The histogram shows eye preference of cells in a 3-year-old monkey in which strabismus had been produced by cutting one eye muscle at 3 weeks of age. Cells are driven by one eye or the other eye, but not both. The cells were grouped in a typical columnar fashion. (After Hubel and Wiesel, unpublished, in Wiesel, 1982.)

critical period. The almost complete lack of binocular representation in cortical cells is shown in a histogram (Figure 11) from a monkey with artificial squint. The critical period for squint to produce changes is comparable to that for monocular deprivation.[32]

Squint provides an example in which all the usual parameters of light are normal—the amount of illumination and form and pattern stimuli. The only apparent change consists of a failure of the images on the two retinas to fall on appropriate areas. Because the cortex in such an animal is rich in responsive cells and columns, it seems unlikely that the large percentage of cells that had originally been driven by both eyes could simply have dropped out.

We have no detailed explanation for the loss of binocular connections with squint. The factor that seems important for maintaining the normal effectiveness of connections from the lateral geniculate body to the cortex is some form of congruity of input from the eyes. It is as though the homologous receptive fields in both eyes must be in register with, and superimposable on, each other, so that excitation will be simultaneous. The following experiments further support this idea. During the first 3 months or longer, the eyes of a kitten were occluded with a plastic occluder that was switched on alternate days from one eye to the other, so that the two eyes received the same total experience, but at different times.[31] Once again the result was the same as in the squint experiments: Cells were driven predominantly by either one eye or the other, but not by both. The maintenance of normal binocularity, therefore, depends not only on the amount of impulse traffic but also on the appropriate spatial and temporal overlap of activity in the different incoming fibers.

[32] Baker, F. H., Grigg, P. and van Noorden, G. K. 1974. *Brain. Res.* 66: 185–208.

Orientation
preferences of
cortical cells

A logical extension is to ask whether the orientation preference of cortical cells can be changed by raising kittens in an environment in which they see only one orientation. The results of such experiments are somewhat difficult to interpret. However, clear changes in orientation preference have been found in kittens in which one eye was presented with bars or stripes of only one orientation while the other eye saw a normal spectrum of orientation.[33,34] Using a slightly different experimental approach, Carlson, Hubel, and Wiesel sutured the lids of one eye in a newly born monkey.[35] The animal was kept in darkness except when it placed its head in a holder. Then, with the head held vertically, it would see vertical stripes with the unsutured eye. Since the monkey received orange juice each time it placed its head in the holder correctly, it performed this maneuver frequently. Thus, during the critical period, one eye received no visual input while the other saw only vertical stripes. After 57 hours of experience between 12 and 54 days after birth, normal levels of cortical activity were found, with cells of all orientations arranged as usual in columns. As expected, the open eye tended to dominate. When tests were made for orientation preference, the results shown in Figure 12 were obtained. Both eyes could drive cells equally well when horizontal lines were the stimulus. However, the right eye (the open eye) was considerably more effective for vertical stripes. The probable explanation for this result is that neither eye saw horizontal bars or edges during the critical period. Hence, the stimulation for horizontality is analogous to binocular closure with equal competition; the cortical cells respond to horizontal bars in one eye or the other but not both. For the vertical input, however, the open eye had an enriched experience and "captured" cells in vertical orientation columns that had previously been supplied by the deprived eye.

Effects of impulse
activity on structure

Two separate but related problems are raised by the experiments on sensory deprivation in early life. First, what are the effects of impulse activity on the outgrowth patterns of neurons? Second, how do coherent and incoherent activity in two pathways determine how they compete for territory and define their fields?

To test directly how different levels of activity promote neurite outgrowth or retraction, experiments have been made on various vertebrate and invertebrate neurons in culture.[36–38] For example, in leech and snail neurons (Chapter 13), trains of action potentials evoked by electrical stimulation at different frequencies have been shown to give rise to pronounced neurite retraction followed by regrowth. Moreover, this effect depends not only on the frequency and duration of the trains but on the molecular environment in which the leech neuron is growing and on the stage of outgrowth. That action potentials can modulate structure is clear; how they do so is still uncertain.

[33]Cynader, M. and Mitchell, D. E. 1977. *Nature* 270: 177–178.

[34]Rauschecker, J. P. and Singer, W. 1980. *Nature* 280: 58–60.

[35]Carlson, M., Hubel, D. H. and Wiesel, T. N. 1986. *Brain Res.* 390: 71–81.

[36]Cohan, C. S. and Kater, S. B. 1986. *Science* 232: 1638–1640.

[37]Fields, R. D., Neale, E. A. and Nelson, P. G. 1990. *J. Neurosci.* 10: 2950–2964.

[38]Grumbacher-Reinert, S. and Nicholls, J. G. 1992. *J. Exp. Biol.* 167: 1–14.

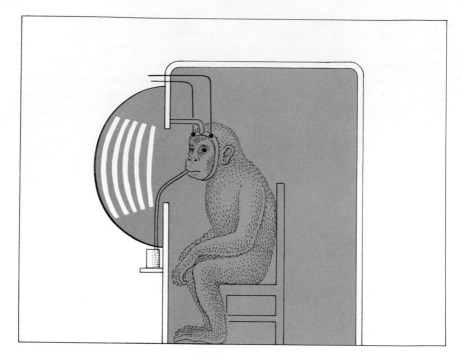

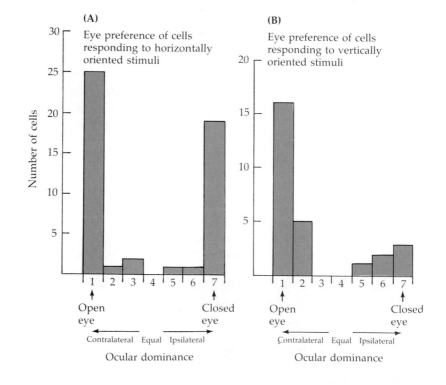

(A) Eye preference of cells responding to horizontally oriented stimuli

(B) Eye preference of cells responding to vertically oriented stimuli

◄ **12** ORIENTATION PREFERENCES OF CORTICAL CELLS in a monkey with altered visual experience. The monkey was kept in a darkened room. At 12 days the right eye was closed. Whenever the monkey placed its head in a holder, it received orange juice. At that time it also saw vertical stripes with its left eye. (The head holder ensured that the head was not tilted.) Thus, for a total of 57 hours of exposure, one eye saw vertical lines, the other nothing. (A,B) Ocular dominance histograms. When horizontally oriented light stimuli were shone onto the screen, cortical cells responded equally well to the left eye and the right eye. Thus, no deprivation is apparent for such orientation except for the lack of binocular cells. With vertically oriented stimuli, the left (open) eye was much more effective in driving cortical cells. The histogram resembles that seen after monocular deprivation. The results suggest competition that was equal for horizontal (which neither eye had ever seen) and unequal for vertical (favored by the left, open eye). (After Carlson, Hubel and Wiesel, 1986.)

For neurons in the visual system, the role of the action potentials themselves in shaping the fields of arborization has been shown in experiments made in kittens. When the lids are closed or an animal is brought up in complete darkness, impulse traffic in the visual pathways does not stop entirely. Neurons continue to fire spontaneously, and ocular dominance columns develop as segregated areas in layer 4.[16] Stryker, Shatz, and their colleagues have shown that this presumably equal low level of activity from the two eyes is important for normal development.[39,40] Tetrodotoxin (TTX), which blocks action potentials (Chapter 4), was injected into both eyes of newly born kittens. Several days later, after removal of the toxin, the visual pathways from retina through the geniculate to the cortex conducted once again. The interesting result was that in the lateral geniculate nucleus inputs from the two eyes failed to segregate into separate layers. After injection of TTX into the eyes, optic nerve axons remained in layers from which they would have retracted.[39,41,42] Moreover, cells in layer 4 of the visual cortex were still driven by both eyes as in the newly born animal, and the ocular dominance columns revealed by autoradiography resembled the neonatal pattern, with extensive overlap and no clear boundaries. Thus, in the absence of *all* firing, geniculate fibers failed to retract normally in layer 4 of the cortex. In amphibians, as in mammals, TTX prevents the development of orderly structure in the visual system.[43] (In embryonic fish, however, TTX does not influence the formation of retinotectal connections.[44])

In other experiments on kittens, it has been shown that the effects of lid closure on ocular dominance shifts can be dramatically modified by blocking the impulse activity in the cortex.[45] For example, during prolonged, continous local infusion of TTX into the visual cortex of a

[39]Shatz, C. J. 1990. *Neuron* 5: 745–756.

[40]Stryker, M. P. and Harris, W. A. 1986. *J. Neurosci.* 6: 2117–2133.

[41]Shatz, C. J. and Stryker, M. P. 1988. *Science* 242: 87–89.

[42]Shatz, C. J. 1990. *J. Neurobiol.* 21: 197–211.

[43]Cline, H. T. 1991. *Trends Neurosci.* 14: 104–111.

[44]Stuermer, C. A., Rohrer, B. and Munz, H. 1990. *J. Neurosci.* 10: 3615–3626.

[45]Reiter, H. O., Waitzman, D. M. and Stryker, M. P. 1986. *Exp. Brain Res.* 65: 182–188.

kitten, geniculate axons and cortical neurons become inexcitable. Yet after the TTX has been removed, cells remain responsive to stimuli in both eyes even though one had been deprived of form and light for several days during the critical period.

An area of intense research is whether the postsynaptic target cells play a role in determining which of their inputs shall maintain connections, depending on the relative balance of activity.[39] In amphibian tectum and kitten cortex, there is evidence to suggest that chronic blockage of synaptic transmission from geniculate axons to cortical cells by NMDA (glutamate) antagonists (see Chapter 10) prevents retraction.[46,47] The mechanism by which a postsynaptic cell could act this way on its input fibers has not yet been determined.

Role of synchronized activity

From the squint experiments it is clear that rough synchrony of inputs from the two eyes is necessary for binocular connections to develop normally in the cortex. Yet, as we have seen, much of the development has already proceeded by the time of birth. In the darkness of the womb, before a kitten or a monkey has seen anything, and before photoreceptors have become functional, the layers of the lateral geniculate nucleus and the cortical columns are becoming recognizable. Is the traffic of action potentials along the optic nerve already a shaping factor at this early stage? And how is synchrony achieved?[48–50]

That patterned waves of activity are produced by retinal ganglion cells of immature ferrets or fetal kittens was shown by Meister, Wong, Baylor, and Shatz.[51] Retinas were isolated and placed on an array of 61 electrodes built into a dish. Each electrode was connected to an amplifier and measured spikes from four or so ganglion cells, the signals of which could be individually resolved. The ferret was chosen because at birth its lateral geniculate nucleus is highly immature and unlayered and the photoreceptors are unresponsive to light. The initial hypothesis was that the firing of the ganglion cells in an eye might be correlated and that this firing could somehow direct the segregation of the terminals into one layer of the lateral geniculate nucleus. (A neuron would somehow have affinity for a neighboring cell firing in synchrony with it.) With the multielectrode array recording from a large number of neurons at once, an extraordinary pattern of activity became apparent.

Figure 13 shows the results of an experiment in which discharges from 82 ganglion cells in a neonatal ferret retina were recorded and analyzed. The small dots represent the positions of the neurons from which activity was recorded. The larger dots represent active neurons, the spot diameter being proportional to the average firing rate of the

[46]Cline, H. T. and Constantine-Paton, M. 1990. *J. Neurosci.* 10: 1197–1216.

[47]Bear, M. F. et al. 1990. *J. Neurosci.* 10: 909–925.

[48]Kuljis, R. O. and Rakic, P. 1990. *Proc. Natl. Acad. Sci. USA* 87: 5303–5306.

[49]Maffei, L. and Galli-Resta, L. 1990. *Proc. Natl. Acad. Sci. USA* 87: 2861–2964.

[50]Constantine-Paton, M., Cline, H. T. and Debski, E. 1990. *Annu. Rev. Neurosci.* 13: 129–154.

[51]Meister, M. et al. 1991. *Science* 252: 939–943.

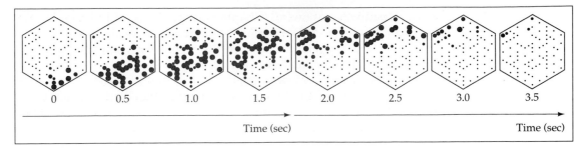

13 A wave of impulse activity spreading across the isolated retina of a neonatal ferret. Within the hexagonally shaped dish are imbedded recording electrodes in a regular array. The location of each of 82 retinal neurons is represented by a small spot. Electrically active neurons are marked by a larger spot, the size of which is proportional to its firing rate. Each frame represents the activity averaged over successive 0.5-second intervals. During the 3.5 seconds represented by the eight frames, action potentials begin with one small group of cells and spread slowly across the retina. Shortly thereafter a new wave begins, and then another, flowing in different directions. (After Meister et al., 1991.)

ganglion cell during the period of 0.5 second represented by the frame. A wave of activity is seen to spread across the retina from the beginning to the end of this series of records. Typically, ganglion cells fire bursts lasting for a few seconds separated by about 2 minutes of silence. Wave after wave of excitation sweeps across the retina as a concerted firing pattern. It is tempting to suggest that the patterns of activity in an eye might be appropriate for allowing neurons that are next to one another to form coherent fields of innervation. That the two eyes could fire in synchrony seems inconceivable. At the same time it is not clear how synchronized activity in one eye could lead to segregation of its optic nerve fiber inputs to the geniculate and prevention of overlap into adjacent geniculate layers. Also tantalizing is the mechanism by which bursts of activity are generated and how they propagate as a wave from neighbor to neighbor across the retina.[49]

SENSORY DEPRIVATION IN EARLY LIFE

The effects produced by altered sensory inputs in kittens and immature monkeys have a number of important implications for our understanding of the nervous system. Critical periods, retraction, and susceptibility to damage during the development of the nervous system have been observed in other animals and other sensory systems, as for example in the projections of the corpus callosum,[52,53] in the somatosensory cortex of mammals,[54] especially in barrel fields of mice,[55] and in the

[52]Innocenti, G. M. 1981. *Science* 212: 824–827.

[53]Olavarria, J. and Van Sluyters, R. C. 1985. *J. Comp. Neurol.* 239: 1–26.

[54]Kaas, J. H., Merzenich, M. M. and Killackey, H. P. 1983. *Annu. Rev. Neurosci.* 6: 325–356.

[55]Jeanmonod, D., Rice, F. L. and Van der Loos, H. 1981. *Neuroscience* 6: 1503–1535.

auditory pathways of owls[56] (Chapter 14). At the level of behavior, the demonstration of a critical period of vulnerability to deprivation or to abnormal experience is not new. In the recent experiments, however, abnormalities in signaling have been pinned down to relays in the cortex and not significantly to lower levels.

There exists a wealth of literature reporting other complex behavioral processes in a variety of animals that show periods of susceptibility. Imprinting is one example. Lorenz[57] has shown that birds will follow any moving object presented during the first day after hatching, as if it were their mother. In higher animals—for example, in dogs—behavioral studies indicate that if they are handled by humans during a critical period of 4 to 8 weeks after birth, they are far more tractable and tame than animals that have been isolated from human contact.[58] The critical period in an animal's development seems to represent a time during which a significant sharpening of senses or faculties occurs.

Why should vulnerability in early life be so pronounced? The developing brain must not only form itself but must be able to represent the outside world, the body, and its movements. The eye, for example, must grow to be the right size for distant objects to be in focus on the retina through the relaxed lens. And the two eyes, separated by different distances in different newborn babies, must act together. As if this were not enough, gross changes occur in limb length, skull diameter and therefore of body image in the first few months and years of life. At all times the maps in the brain for different functions must be in register. An example of how adaptable the brain is during the critical period is provided by the experiments of Knudsen and his colleagues[56,59] (see also Chapter 14). In adult barn owls the neural maps for visual and auditory space are precisely aligned in the optic tectum: The owl turns its head to the exact spot at which a sound arises (important for catching squeaking mice). If baby owls are raised with visual fields displaced by placing prisms over their eyes, the maps in the tectum change. The auditory space map comes to fit in exact register with the new visual map, which early experience has validated. Plasticity in the critical period therefore enables fine tuning of brain function to be carried out in response to knowledge of the world. The critical period corresponds to the time for maximal growth of the skull and the facial ruff.[60]

In adult life too there is evidence for plasticity. Merzenich,[61] Kaas,[62] Pons,[63] Wall,[64] and their colleagues have shown major changes in the

[56]Knudsen, E. I., Esterly, S. D. and Du Lac, S. 1991. *J. Neurosci.* 11: 1727–1747.

[57]Lorenz, K. 1970. *Studies in Animal and Human Behavior*, Vols. 1 and 2. Harvard University Press, Cambridge.

[58]Fuller, J. L. 1967. *Science* 158: 1645–1652.

[59]Knudsen, E. I. and Knudsen, P. F. 1989. *J. Neurosci.* 9: 3306–3313.

[60]Knudsen, E. I., Esterly, S. D. and Knudsen, P. F. 1984. *J. Neurosci.* 4: 1001–1011.

[61]Jenkins, W. M. et al. 1990. *J. Neurophysiol.* 63: 82–104.

[62]Kaas, J. H. 1991. *Annu. Rev. Neurosci.* 14: 137–167.

[63]Pons, T. P. et al. 1991. *Science* 252: 1857–1860.

[64]Wall, J. T. 1988. *Trends Neurosci.* 11: 549–557.

representation of the body in the somatosensory cortex of adult monkey. After lesions and deafferentation, cells in the cortex that are normally driven by inputs from one part of the body can start to respond to different inputs from a different position. At a functional level we can rapidly adapt to prisms over our eyes that displace the visual fields.

Since the flexibility of connections and structure in early life make the brain vulnerable to sensory deprivation, it is natural to wonder whether enrichment and a fuller early life during the critical period could enhance cortical function. Such tests are in practice difficult to devise, first, because newly born animals need to be with their mothers for much of the time and, second, because it is hard to know what would constitute an extra-rich and pleasant set of stimuli for, say, baby rats, mice, or monkeys.[65] In one set of experiments[66] rats and mice were presented with a variety of mild, nonaversive visual and sensory stimuli for the first weeks of life during the preweaning period. Subsequently, detailed and systematic measurements of neurons in the visual cortex revealed a more profuse dendritic arborization than that measured in controls. What is hard is to compare the level of sensory enrichment in the laboratory to that occurring in real life, with its sounds, smells, patterns, and variety.

Mechanisms responsible for the delicate balance between intrinsic and extrinsic influences in shaping our brain remain unknown. It is still not possible to provide a full and comprehensive cellular or molecular explanation for describing how columns are assembled and why neurons in the visual cortex should be so much more vulnerable than those in the retina. Local differences in the molecular environment as well as the inherent properties of the neurons themselves presumably contribute to variability of structure and function. A remarkable example of the role of local determinants is provided by experiments in which embryonic visual cortex was transposed and grafted into somatosensory cortex of neonatal rats.[67] A well-organized array of barrels corresponding to the vibrissae subsequently developed in what would have been visual cortex. Each barrel was delineated, as in normal cortex, by its characteristic group of thalamic afferent fibers and by a rim of glycoproteins. This experiment shows that factors extrinsic to the cortex itself trigger its specialized development. It is known from studies made in tissue culture that large protein molecules, such as laminin (Chapter 11), can influence the outgrowth, the branching patterns, and the physiological properties of nerve cells. It remains to be determined which specific cells and molecules provide the cues for forming elaborate structures in the cortex and endowing them with plasticity.

It is tempting to speculate about the effects of deprivation on higher functions in humans. One can imagine, as Hubel has said,[68]

[65]Turner, A. M. and Greenough, W. T. 1985. *Brain. Res.* 329: 195–203.

[66]Venable, N. et al. 1989. *Brain Res. Dev. Brain Res.* 49: 140–144.

[67]Schlaggar, B. L. and O'Leary, D. D. M. 1991. *Science* 252: 1556–1560.

[68]Hubel, D. H. 1967. *Physiologist* 10: 17–45.

Perhaps the most exciting possibility for the future is the extension of this type of work to other systems besides sensory. Experimental psychologists and psychiatrists both emphasize the importance of early experience on subsequent behavior patterns—could it be that deprivation of social contacts or the existence of other abnormal emotional situations early in life may lead to a deterioration or distortion of connections in some yet unexplored parts of the brain?

To find a physiological basis for such behavioral problems seems a distant, but not impossible, goal.

SUGGESTED READING

General Reviews

Hubel, D. H. 1988. *Eye, Brain and Vision*. Scientific American Library, New York.

Rakic, P. 1986. Mechanism of ocular dominance segregation in the lateral geniculate nucleus: Competitive elimination hypothesis. *Trends Neurosci.* 9: 11–15.

Shatz, C. J. 1990. Impulse activity and the patterning of connections during CNS development. *Neuron* 5: 745–756.

Singer, W. 1990. The formation of cooperative cell assemblies in the visual cortex. *J. Exp. Biol.* 153: 177–197.

Wiesel, T. N. 1982. The postnatal development of the visual cortex and the influence of environment. *Nature* 299: 583–591.

Original Papers

Carlson, M., Hubel, D. H. and Wiesel, T. N. 1986. Effects of monocular exposure to oriented lines on monkey striate cortex. *Brain Res.* 390: 71–81.

Hubel, D. H. and Wiesel, T. N. 1965. Binocular interaction in striate cortex of kittens reared with artificial squint. *J. Neurophysiol.* 28: 1041–1059.

Hubel, D. H., Wiesel, T. N. and LeVay, S. 1977. Plasticity of ocular dominance columns in monkey striate cortex. *Philos. Trans. R. Soc. Lond. B* 278: 377–409.

Knudsen, E. I., Esterly, S. D. and Du Lac, S. 1991. Stretched and upside-down maps of auditory space in the optic tectum of blind-reared owls; acoustic basis and behavioral correlates. *J. Neurosci.* 11: 1727–1747.

LeVay, S., Wiesel, T. N. and Hubel, D. H. 1980. The development of ocular dominance columns in normal and visually deprived monkeys. *J. Comp. Neurol.* 191: 1–51.

Meister, M., Wong, R. O., Baylor, D. A. and Shatz, C. J. 1991. Synchronous bursts of action potentials in ganglion cells of the developing mammalian retina. *Science* 252: 939–943.

Rakic, P. 1977. Prenatal development of the visual system in rhesus monkey. *Philos. Trans. R. Soc. Lond. B* 278: 245–260.

Stryker, M. P. and Harris, W. A. 1986. Binocular impulse blockade prevents the formation of ocular dominance columns in cat visual cortex. *J. Neurosci.* 6: 2117–2133.

Wiesel, T. N. and Hubel, D. H. 1963. Single-cell responses in striate cortex of kittens deprived of vision in one eye. *J. Neurophysiol.* 26: 1003–1017.

Wiesel, T. N. and Hubel, D. H. 1974. Ordered arrangement of orientation columns in monkeys lacking visual experience. *J. Comp. Neurol.* 158: 307–318.

PART SIX

Conclusion

CHAPTER NINETEEN

PERSPECTIVES

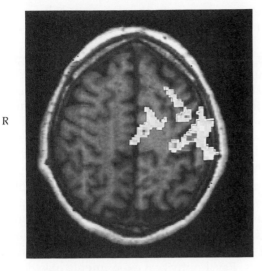

R

L

MAGNETIC RESONANCE IMAGING AND PET SCANS integrated to show area of cortical activity during tactile exploration of an object with the right hand. The image shows activation of the left supplementary motor area, motor cortex, and somatosensory cortex. (Photograph courtesy of R. J. Seitz.)

Since the previous edition of *From Neuron to Brain*, our understanding of how nerve cells produce their electrical signals, how they communicate with one another, and how they become connected during development has become deeper and richer as the result of the use of molecular biological techniques. At the same time, important noninvasive techniques have been developed to study the conscious brain as it sees, hears, thinks, and initiates voluntary movements. A lesson from the past persists: Under exceptional circumstances brilliant scientists with a firm grasp of reality can predict with amazing accuracy from their observations new concepts and directions that will emerge. Examples are provided by the links between Helmholtz and isolation of the visual pigments, between Sherrington and the integrative events at synapses, between Hodgkin, Huxley, and Katz and the electrical properties of membrane channel molecules. By contrast, there seems no way for the practical scientist to guess what *techniques* will become available, such as site-directed mutagenesis, expression of messenger RNA in oocytes, or use of positron emission tomography to scan the living brain. The solution of many problems, now seemingly insoluble, will await techniques and approaches yet to be devised. For example, we know only little about sleep or about higher functions such as consciousness.

What can reasonably be predicted about advances that would be incorporated into the next edition of this book? One reasonable guess is that once again, as in the past, collaboration between basic scientists and neurologists will be an important factor for understanding integrative brain functions, and that this understanding will lead to the prevention and alleviation of many diseases of the nervous system, some of whose origins are still unknown and whose symptoms cannot yet be treated effectively.

Higher functions of the brain

In spite of the conspicuous advances in our understanding of higher functions, we are still woefully ignorant about how the complete scene of the world around us is fabricated by the brain. It is now possible to find cortical cells that respond to a highly complex pattern, and hints are emerging about how small parts of the picture are put together. Yet we cannot even begin to discuss how the sight of a laughing child will make us smile, or how the *Eroica* symphony conjures up emotions not to be expressed in words. To deal with the topic of higher brain function, as throughout the rest of the book, we select a few examples. From the preceding chapters it is evident that certain modes of experimentation and thinking have proved essential for approaching the problem of how higher functions are performed. These include electrophysiology, cellular neurobiology, biochemistry, pharmacology, clinical neurology, and psychophysics. Psychophysical experiments have led to precise quantitative descriptions of sensory phenomena such as visual sensitivity, dark adaptation, color constancy, and analysis of contrast, pattern, movement, and color by the visual system. Such experiments set the limits for the properties required of individual cells, as well as for the way aggregates of neurons hand on information in conscious subjects. A wonderful example is the precise correspondence between the absorbance of visual pigments, the corresponding electrical responses of photoreceptors, and, in matching experiments, the conscious perception of color. Psychophysics, by dealing directly with the end-stage of sensory perception, can provide a comprehensive scheme of how sensory signals are processed when one sees, or hears, or feels a touch.

Integration in the visual system

Nowhere have the mechanisms involved in analysis of sensory signals been more clearly demonstrated than in the cellular studies of the visual system described in Chapters 17 and 18. Hubel and Wiesel's observations on the receptive fields of visual cortical cells explain various optical illusions and clinical disorders of vision. One example of interest is the proposed explanation of the COMPLETION PHENOMENON found in patients suffering from migraine or with a small retinal or cortical lesion from another cause, causing a blind area (scotoma). In this condition forms or shapes projected onto the retina appear to contain an empty area corresponding to the lesion. Yet the subject who looks at striped wallpaper, a zebra, or anything with a continuous striped pattern sees the pattern continue through the blind area. An interesting observation was made by Lashley[1], a perceptive reporter of psychophysical phenomena, during a migraine attack:

[1]Lashley, K. S. 1941. *Arch. Neurol. Psychiat.* 46, 331–339.

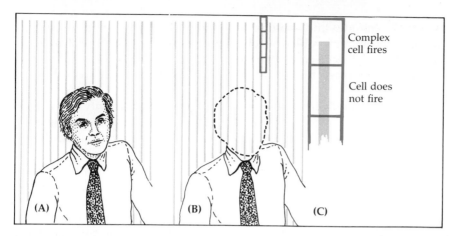

1   THE COMPLETION PHENOMENON. (A,B) During Lashley's migraine attack, a small area of complete blindness occurs in the visual field occupied by his friend's head. Yet the stripes on the wallpaper continue through the blind area. A possible explanation is that complex cells of the type described in Chapter 17 respond best to corners or lines that stop. Hence a stripe on the wallpaper acts as a good stimulus for a complex cell only at its end. (C) A complex cell with its receptive field within the blind area would no longer be activated by the lines of the face; however, its silence could be interpreted by higher centers as an indication that the gray stripe had not stopped.

> Talking with a friend I glanced just to the right of his face whereupon his head disappeared. His shoulders and necktie were still visible but the vertical stripes on the wallpaper behind him seemed to extend right down to his necktie. Quick mapping revealed an area of total blindness covering about 30° just off the macula. It was quite impossible to see this as a blank area when projected on the striped wallpaper of uniformly patterned surface although any intervening object failed to be seen.

The events described by Lashley can be interpreted in terms of complex cells that are end-stopped and signal information only about the terminations of lines (Figure 1). With an array of such cells, silence of members in the array during an episode of temporary (in Lashley's case) blindness produces little change in signaling since they are normally silent anyway unless the line stops or changes direction (Chapter 17). This type of reasoning provides a conceptual scheme that uses known properties of cells to explain perceptual phenomena. Some of the weaknesses and strengths of this approach have already been discussed. It is ironic, incidentally, that this interpretation rests on cortical cells that have specific connections. This is contrary to the principles enunciated by Lashley himself, who did not believe in specificity or localization in the cortex.

One major strength of Hubel and Wiesel's work on cortical organization and sensory deprivation has been to point toward fundamental problems requiring quite different experimental approaches, in this case biophysics and molecular neurobiology. Their experiments, made with external electrodes and anatomical techniques, clearly posed for the first time an intriguing problem: How do coherent and incoherent trains of impulses traveling in convergent pathways compete to influence neurite outgrowth, synapse formation, and synaptic stability in layer 4 of the visual cortex?

Motor integration

Processes involved in the generation of motor activity have been more difficult to analyze. Here the end stage is in the periphery—a coordinated pattern of muscle contractions. Rather than converging systematically on a small group of cells with stereotyped behavior, as in the visual system, the paths backward from the muscles through the motoneurons and descending pathways diverge to a number of structures involved in the movement: the brainstem, cerebellum, basal ganglia, and motor cortex. We are only beginning to understand how these structures are regulated to produce coordinated muscle activity. At the moment it seems quite impossible to describe how one can throw a stone accurately to hit (or even come close to) an object 10 meters distant. To add to this complexity, many learned motor tasks can be transferred to other muscle groups. Adrian pointed out that once you have learned to write your name, you can do it at once by holding the pencil between your toes![2] In the brain must be lodged a scheme for writing one's name irrespective of the muscles to be used. This phenomenon illustrates the ability of the nervous system to abstract concepts and encode them in such a way that they can be expressed more generally.

Psychopharmacology

A different approach to exploring higher functions is provided by psychopharmacology. An example is the benzodiazepines—drugs that have proved to be of great benefit for mankind. They are the remedy of choice for various forms of epilepsy and insomnia, for premedication before operations, for mild surgical interventions, for the treatment of spasticity, and specifically for the relief of crippling anxiety. It is this aspect that is of interest for higher functions. Psychologists have devised elegant tests to distinguish between drugs that influence anxiety or alertness or memory or motor activity or pain thresholds. Benzodiazepines have been shown to target anxiety selectively, with minimal side effects. The molecule that they influence, described in Chapters 2 and 10, is the $GABA_A$ receptor. Benzodiazepines enable GABA released by presynaptic endings in the brain to give rise to larger conductance changes and larger inhibitory currents.[3] The sites on the subunits to which the benzodiazepines bind have been identified in cloned channels expressed in oocytes. What is not yet known is the site of those particular GABA synapses in the brain that are responsible for the alleviation of anxiety and fear; nor can one begin to guess how the potentiation of those inhibitory synapses acts on succeeding relays. In this case the molecular action and effects are well understood but the overall mech-

[2]Adrian, E. D. 1946. *The Physical Background of Perception*. Clarendon Press, Oxford.
[3]Richards, G., Schoch, P. and Haefely, W. 1991. *Sem. Neurosci.* 3: 191–203.

anism is not. Other important pharmacological approaches to higher functions mentioned already include links between drugs, transmitters, and schizophrenia.

Inevitably, throughout the previous discussions of higher functions, diseases of the brain and lesions have been recurrent themes. For many years neurology was not only inseparable from neurobiology but provided the only method for studying higher functions in relation to structure. The story of Phineas Gage is still dramatic, evocative, and instructive. In 1848, at the age of 25, Phineas Gage suffered a massive lesion to the brain while working as a construction foreman on the railways in Vermont. He had been using a tamping iron to place a charge of dynamite into a rock that was to be blasted so as to make way for the tracks. As he pushed on the dynamite it exploded and blew the iron rod clean through his skull. The large rod, which weighed 13 pounds, was 3 feet long and more than 1 inch in diameter, entered near the angle of the left jaw, passed behind the left eye, and flew out through the midline at the top of his head. It fractured the skull, made a large communicating lesion, and destroyed much of the left frontal lobe. The facts about this incident and Gage's amazing recovery were recorded in detail by observers and by the local doctor, John Harlow.[4] Gage lost consciousness only briefly and could sit up and speak on his way to the doctor—whom he was able to talk to about the accident and about the extent of his injury. What astounded the doctor and all who had seen the accident was that he soon recovered and was able to lead a relatively normal life for more than 12 years. Persistent infection of the wound was the main overt problem, coupled with blindness of the left eye and paralysis on the left side of his face. Gage's personality, however, underwent a major change. In Harlow's words, "He was no longer the same Gage!" From being a well-liked, quiet, sober, industrious, and careful worker, he changed to a loud-mouthed, boastful, impatient, and restless braggart. When he applied for reinstatement to his old job he was rejected because of the change in his character. Unable to focus his mind, work towards a goal, or keep friends, he wandered about the United States and Chile accompanied by his favorite memento, the tamping iron, telling his tale.

The importance of neurology

The story of Phineas Gage emphasizes the advantages and pitfalls of lesions and deficits as a means of analyzing brain function. At a time when nothing was known of sensory, motor, visual, or auditory cortex, it became clear that the prefrontal area was somehow associated with the very highest functions of human conduct and personality. But Gage's accident was unlike most neurological cases. The lesion, its timing, and its extent were known and obvious as the major symptoms developed. Until recently, however, neurologists could in general know only which area of the brain had been affected after the patient had died.

[4]Harlow, J. M. 1868. *Publ. Mass. Med. Soc.* 2: 328–334.

Other examples of neurological observations made in the last century that defined specific brain areas involved in higher functions were those of Broca and Wernicke on disorders of language. A clear picture emerged through the correlation of defects in speech with the areas of cortex that had been damaged by vascular accidents or tumors. Language almost always depends on one hemisphere, the dominant hemisphere corresponding to handedness; in 95 percent of the population this is the left hemisphere. A dramatic way to demonstrate this in conscious subjects is to inject a short-acting barbiturate drug into the carotid artery supplying one hemisphere. Injection into the left side puts the left hemisphere to sleep for a minute or two and abolishes speech but not singing; injection into the right artery produces the opposite effect—the subject can talk but not sing.[5] Lesions occurring in the area described by Broca in 1861, an area that lies in close association with the motor cortex (Figure 2), lead to difficulty in enunciating words and stringing them together in a sentence. The patients, whose intelligence is unaffected, can read and understand language but speak with difficulty or not at all. Typically, the patient may say haltingly, "Went . . . doctor . . . ill," or in severe cases mumble only "yes" or "no." The deficit is to the output or motor aspect of brain function even though the muscles for speech and their direct control are not affected.

Lesions to the areas described by Wernicke in 1874 (Figure 2), a region closely applied to sensory cortex, also give rise to disorders of language, but of a different nature. The problem here is sensory rather than motor. Speech is profuse, rapid, garbled, inaccurate, and incoherent. Although the patient can talk fluently, comprehension of spoken language is defective as though an unknown foreign tongue were being heard. Patients string together real and fabricated words: "Trees are greel when went gone home long life to the bean." What seems to have meaning is clouded by confusion.

Comparable failures of function occur in the visual system after lesions to the preoccipital cortex. Objects can no longer be recognized by sight. Shown a key, the patient will be unable to identify it. Once put into the hand, it ceases to be a meaningless object and is recognized at once as a key. Another fascinating, related disability following lesions of the parietal lobe is the loss of programmed sequences of movement in performing purposeful actions. When the patient tries to put on a shirt, the buttons may be done up first, then one arm pushed through the collar, then one through a sleeve, and so on.

Highly counterintuitive and difficult to comprehend at first sight are other effects of lesions to the parietal lobe on one side, usually the right. Patients with such lesions may lose the concept of having two sides to their body and their world. When the patient with a lesion of the right parietal lobe is asked to draw a daisy, all the petals are on the right; in a drawing of a clock the numbers are crowded on the right, as are all the spokes of a bicycle wheel; lipstick is applied only to the right side

[5]Bogen, J. E. and Gordon, H. W. 1971. *Nature* 230: 524–525.

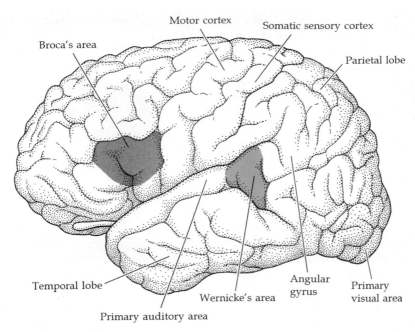

Motor cortex

Somatic sensory cortex

Broca's area

Parietal lobe

Temporal lobe

Angular gyrus

Primary visual area

Wernicke's area

Primary auditory area

2 **AREAS DESCRIBED BY BROCA AND WERNICKE, associated with speech in the left dominant hemisphere.**

of the mouth. If the patient's own right hand is held up by the doctor, it is immediately described as "my hand." If the left hand is held up the reply is, "I don't know what that thing is. It's not mine." When asked whether it is a hand, the patient may say, "I don't know. You're the doctor." The left side of the body gives reflex responses and can on occasion move, but it has been banished from the conscious life of the subject. It is important to emphasize that these are true neurological defects, not hysterical reactions of the patient.

Together these clinical observations show that our inner world, which seems so complete, so unitary, and so perfect, is composed of elementary components welded together to form a continuum. The seamless match is so flawless that no clue can be obtained by introspection. A triumph of the early neurologists was to use nature's own experiments to describe functions of various brain areas from careful correlation of symptoms with lesions; neither of these parameters was quantitatively measurable. Their achievements are all the more remarkable because the use of lesions to assess function is fraught with pitfalls. The symptoms may be weak or ill-defined. In Chapter 17 we described effects observed in patients who had the entire corpus callosum transected. Sophisticated tests were required to show the deficits, since the patients were often not aware that the right and left worlds had been completely separated (just as a normal subject has no idea that activity jumps from one cerebral hemisphere to the other as a visual or a tactile stimulus crosses the midline). Worse still, while some massive lesions

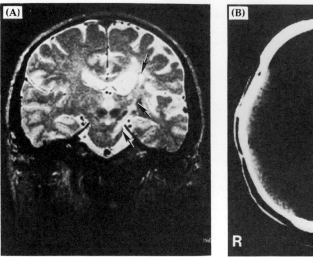

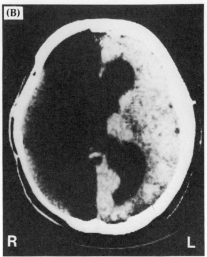

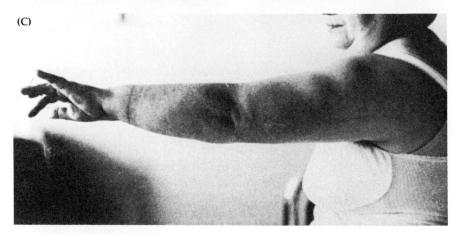

3   MAGNETIC RESONANCE IMAGING SCANS of two patients. (A) This patient had a small lesion or infarct at the site marked by the arrow. The degenerating pyramidal tract is marked by arrowheads. As a result, there was severe persistent hemiplegia (paralysis) on the opposite side of the body and the patient was severely disabled by this restricted lesion. (B,C) This patient had had a complete hemispherectomy at age 18 for intractable epilepsy due to early brain damage with infantile hemiplegia. The scan in (B) shows the absence of brain tissue on the right side. (C) The same patient voluntarily raising her left arm. (Photographs courtesy of H.-J. Freund.)

fail to produce obvious deficits, others give rise to symptoms corresponding to hyperactivity of other areas.

With the newer techniques now available, such as magnetic resonance imaging (MRI) and positron emission tomography (PET) scanning, the neurologist is now for the first time in a position to locate and observe lesions directly and follow their progress in the living brain. The major problem that remains, however, is that fixed, reproducible, cut-and-dried correlations between lesions and functions can usually be established only for small lesions.[6] Large lesions and gross pathology illustrate our ignorance about compensatory mechanisms that may be operative, producing unpredictable symptoms. Figure 3 shows scans of two brains of living patients. The small lesion in Figure 3A produced a permanent severe paralysis on the opposite side of the body. The patient shown in Figure 3B and 3C had a far more extensive and drastic lesion. The entire right cerebral hemisphere had been removed in an operation. Yet she could walk and could move her left arm. It is at present not understood how this type of operation (performed in only the most serious of cases) can be followed by recovery of coordinated movement of the body, speech, sensation, and a relatively normal life. Such recovery can occur only if the brain damage occurs early in life. The time of surgery seems less critical. In the patient shown in Figure 3 hemispherectomy was performed at age 18. Even more remarkable perhaps is the scan shown in Figure 4, which was taken from a 63-year-old woman who led what seemed to be a perfectly normal life. With what appears to be virtually no brain at all, she was, for example, able to play the organ regularly at her local church. Her hydrocephalus had developed as a result of meningitis at the age of 4 when the shape of her head became typically hydrocephalic. In both of these patients the damage occurred early in life. There are, however, examples of recovery after lesions occurring in later life, emphasizing the functional plasticity of the brain.

Whatever the difficulties in interpretation of clinical observations, there is no doubt that neurology will continue to play a key role in exploring the flow of information and integration at higher levels as more information becomes available from brain scans of patients and normal subjects. At the same time there clearly exists a two-way street between basic and applied neuroscience. Molecular biological and genetic techniques are already beginning to play a role in diagnoses of such conditions as retinoblastoma, Huntington's disease, and perhaps Alzheimer's disease; the possibility of treatment with genetically engi-

[6]Müller, F. et al. 1991. *Neuropsychologia* 29: 125–145.

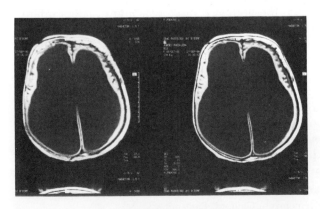

4

BRAIN SCAN of a 63-year-old woman without neurological deficits until the current disorder that brought her to the hospital. In spite of the extensive hydrocephalus, which had developed early in life (producing a characteristically shaped, enlarged head), this patient led an apparently normal life and had far less disability than the patients shown in Figure 3. (Photographs courtesy of H.-J. Freund.)

neered cells is now being intensively investigated in a type of muscular dystrophy known as Duchenne's. Sophisticated electrophysiological techniques are used by neurosurgeons for recording from individual neurons, for implanting electrodes (as for control of the bladder), and for devising prostheses to replace lost functions.[7] Without being unduly optimistic, one can begin to hope for the development of strategies to understand the origin and eventually the treatment of such devastating diseases as multiple sclerosis, degenerative diseases, and mental disorders.

Perspectives   The perspectives from which we have assessed what is important and interesting in our field, and what its future directions might be, are colored by our experience. Two of the authors were graduate students in the 1950s. At that time exciting new experiments and concepts were opening up hitherto unapproachable fields. We knew that sensory signals were initiated by receptor potentials, that voltage- and receptor-activated ion channels were responsible for signaling, and that release of transmitters from nerve terminals was quantal; fine structural studies by electron microscopy promised to tell us much about neuronal function. Intracellular recording techniques had been extended into the mammalian spinal cord to study the behavior of individual motoneurons, and a neuroinhibitory substance had been found in extracts of beef brain. However, apart from anatomy, regions of the central nervous system higher than the spinal cord were still largely a *terra incognita*, and there was no hint of advances to be made in transmitter neurochemistry at the cellular level. Through the pioneering work of H. Berger, H. H. Jasper, and E. D. Adrian, the EEG had provided a way to study overall activity of the brain and to make important clinical diagnoses, such as the localization of epileptogenic foci. Such gross electrical techniques also provided a means for localization of function in the brain but were of little help in understanding mechanism, since they could not detect the underlying fine grain of neuronal activity. Nevertheless, experimenters had begun to study the behavior of individual neurons in the cerebral cortex with microelectrodes, and how that behavior could be influenced by activity of cells in other regions of the brain.

By the late 1960s, when the youngest of us was a graduate student, a major revolution had occurred. Inhibitory and excitatory transmitters could be identified, and the chemistry of their synthesis and regulation could be analyzed in single cells. Moreover, through the work of Stephen Kuffler, followed by that of Hubel and Wiesel, mechanisms of perception became approachable in terms of orderly arrangements of connections between cells. A far-reaching innovation, again introduced by Kuffler, was the concept of bringing every possible tool to bear on problems of the nervous system. Electrophysiology was no longer alone but inextricably linked to anatomy, fine structure, and biochemistry. An extension to molecular biology was on the horizon.

[7]Young, R. R. and Delwaide, P. J. (eds.). 1992. *The Principles and Practice of Restorative Neurosurgery*. Butterworth-Heinemann, Guilford, England.

One feature of the development of neuroscience is how quantitatively accurate, imaginative, and far-reaching in their importance were predictions made by a few outstanding scientists. One inspiring example from the psychophysical studies of Helmholtz is his prediction of the three photoreceptors and their pigments in the retina. Another is the Golgi studies of Ramón y Cajal that revealed the neural pattern for flow of information through circuits in the retina and in the cerebellum. One reads with amazement Sherrington's account of integration of excitatory and inhibitory inputs in the spinal cord, expounded long before electrical recordings were made from motoneurons by J. C. Eccles, his onetime pupil. Perhaps the greatest influence was from the experiments of Hodgkin, Huxley, and Katz that predicted many of the properties of membrane channels long before the means were at hand for direct measurements.

The ability to record the membrane potential of a neuron with an intracellular electrode provided a major and, prior to the development of the technique, unpredictable direction for advance over the following 25 years. During that period, the future use of site-directed mutagenesis to alter the function of single channel proteins, the infection of cells with mRNA to produce the altered channels, and the use of patch clamp techniques to detect and analyze their behavior were equally unpredictable developments before the techniques were at hand. Future developments can again be expected to follow the emergence of new techniques.

Many problems remain unsolved at the level of subcellular organization, in relation to development and growth, and at the level of intercellular circuitry. A key problem at the cellular level is how various molecules are directed to the appropriate places in the cell. The various channels for sodium, potassium, and calcium, receptors for various ligands, and proteins associated with transmitter release must be synthesized in the cell body and shipped selectively to the soma membrane, nodes of Ranvier, dendrites, and axon terminals. An important new tool is the resurgence of microscopy, coupled to the use of a variety of vital dyes, that allows noninvasive monitoring of membrane potential changes, intracellular ion concentrations, and membrane cycling associated with transmitter release. One can be confident that time and hard work with tools already available will eventually lead to clear answers. Similar considerations apply to the task of cataloging and describing the molecular properties of receptors and channels. Here we may expect some surprises, and with them unanticipated benefits. An example is the unexpected occurrence of a wide diversity of channel isotypes. There are a number of sodium and potassium channel types in skeletal muscle, heart, and brain, and ligand-activated channels show similar diversity. In the future, it may be possible to design therapeutic drugs to target one particular channel isotype specifically. Thus the sodium channel blocker lidocaine, which is sometimes used to control cardiac arrhythmias but has side effects on the CNS, could be replaced by a drug directed only at cardiac sodium channels.

Unsolved problems

Our understanding of nervous system development and regeneration, in relation to phenomena such as axon guidance, synapse formation and elimination, and receptor aggregation again can be expected to be advanced with currently available tools. Although much is known at the cellular and molecular level about how neurons and glial cells originate, migrate, send out processes, and form connections, open questions still abound. Here biochemical techniques can be expected to provide new information about such problems as the mechanisms whereby trophic substances affect cell viability, how specific molecules attract or repel growing ends of nerves, and how a substance secreted from a nerve terminal can cause receptor aggregation. Although the prospects for effecting repair in the central nervous system after injury seem brighter than before, restoration of function still seems a distant goal, requiring tracts to reestablish their connections with appropriate targets and form appropriately tuned synapses.

Harder to foresee are approaches and answers to major.questions relating to the behavior of aggregates of neurons and neuronal circuitry. Here the problems of numbers appears overwhelming. What, for example, is the role of the hundred thousand parallel fibers connecting to a single Purkinje cell in the cerebellum? In this context, previous work on the visual system is important: Hubel and Weisel showed that one need not necessarily be intimidated by numbers when looking for principles of cellular organization. The particular fascination of neurobiology is the combination of exciting intellectual problems with the long-range hope of alleviating the suffering and anguish caused by diseases of the nervous system.

SUGGESTED READING

Adams, R. D. and Victor, M. 1989. *Principles of Neurology*, 4th Ed. McGraw-Hill, New York.

Damasio, H. and Damasio, A. R. 1989. *Lesion Analysis in Neurophysiology*. Oxford University Press, Oxford.

Freund, H.-J. 1991. What is the evidence for multiple motor areas in the human brain? *In* D. R. Humphrey and H.-J. Freund (eds.). *Motor Control: Concepts and Issues*. Wiley, New York.

Kandel, E. R., Schwartz, J. H. and Jessell, T. M. 1991. *Principles of Neural Science*, 3rd Ed. Elsevier, New York

Young, R. R. and Delwaide, P. J. (eds.) 1992. *The Principles and Practice of Restorative Neurosurgery*. Butterworth-Heinemann, Guilford, England.

CURRENT FLOW IN ELECTRICAL CIRCUITS

A few basic concepts are required to understand the electrical circuits used in this presentation. An especially clear and lively treatment is found in a book by Rogers.[1] For our purposes it is sufficient to describe the properties of a few circuit elements and explain how they work when connected together in ways that correspond to the circuits described for nerves. The difficulties sometimes encountered on first reading accounts of electrical circuits often stem from the apparently abstract nature of the forces and movements involved. It is reassuring, therefore, to realize that many of the original pioneers in the field must have been faced with similar problems, since the terms devised in the last century are mainly related to the movement of fluids. Thus, the words "current," "flow," "potential," "resistance," and "capacitance" apply equally well to both electricity and hydraulics. The analogy between the two systems is illustrated by the fact that complex problems in hydraulics may be solved by using solutions to equivalent electrical circuits.

The analogy between a simple electrical circuit and its hydraulic equivalent is illustrated in Figure 1. The first point to be made is that a source of energy is required to keep the current flowing. In the hydraulic circuit, it is a pump; in the electrical circuit, a battery. The second point is that neither water nor electrical charge is created or lost within such a system. Thus, the flow rate of water is the same at points a, b, and c in the hydraulic circuit, since no water is added or removed between them. Similarly the electrical current in the equivalent circuit is the same at the three corresponding points. In both circuits, there are a number of *resistances* to current flow. In the hydraulic circuit, such resistance is offered by narrow tubes; similarly thinner wires offer greater resistance to electrical current flow.

Terms and units describing electric currents

The unit used to express rate of flow is to some extent a matter of choice; one can measure flow of water through a pipe in cubic feet per minute, for example, although in some other situation milliliters per hour might be more suitable. Electrical current flow is conventionally measured in COULOMBS/SEC or AMPERES (abbreviated A). One coulomb is equal to the charge carried by 6.24×10^{18} electrons. In electrical circuits and equations, current is usually designated by I or i. As with flow of water, flow of current is a vector quantity, which is just a way of saying that it has a specified direction. The direction of flow is often indicated by arrows, as in Figure 1, current always being assumed to flow from the positive to the negative pole of a battery.

What do *positive* and *negative* mean with regard to current flow?

[1]Rogers, E. M. 1960. *Physics for the Enquiring Mind*. Princeton University Press, Princeton.

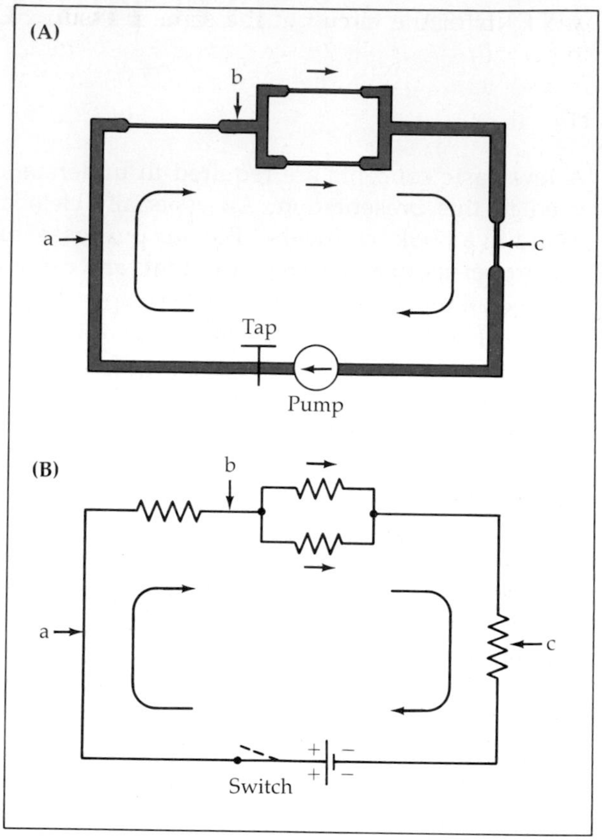

1

HYDRAULIC AND ELECTRICAL CIRCUITS.
(A), (B) Analogous circuits for the flow of water
and of electrical current. A battery is analogous
to a pump which operates at constant pressure,
the switch to a tap in the hydraulic line, and
resistors to constrictions in the tubes.

Here the hydraulic analogy does not help. It is useful instead to consider
the effects of passing current through a chemical solution. For example,
suppose two copper wires are dipped into a solution of copper sulfate
and connected to the positive and negative poles of a battery. Copper
ions in solution are repelled from the positive wire, move through the
solution, and are deposited from the solution onto the negative wire.
In short, positive ions move in the direction conventionally designated
for current: from positive to negative in the circuit. At the same time,
sulfate ions move in the opposite direction and are deposited onto the
positive wire. The direction specified for current, then, is the direction
in which positive charges move in the circuit; negative charges move
in the reverse direction. Students are usually told that the direction
specified for current in a wire is opposite to the direction in which the
electrons move. Such a statement is true, but irrelevant.

To explain the energy source for current flow and the meaning of
electrical POTENTIAL, the hydraulic analogy is again useful. The flow of
fluid depicted in Figure 1 depends on a pressure difference, the direction of flow being from high to low. No net movement occurs between

two parts of the circuit at the same pressure. The overall pressure in the circuit is supplied by expenditure of energy in driving the pump. In the electrical circuit shown here, the electrical "pressure" or POTEN-TIAL is provided by a BATTERY, in which chemical energy is stored. Hydraulic pressure is measured in dynes/cm^2; electrical potential is measured in VOLTS.

Symbols used in electrical circuit diagrams and arrangements of circuit elements in series and in parallel are illustrated in Figure 2. As the names imply, a VOLTMETER measures electrical potential and is equivalent to a pressure gauge in hydraulics; an AMMETER measures current flowing in a circuit and is equivalent to a flowmeter.

In hydraulic systems, at least under ideal circumstances, the amount of current flowing through the system increases with pressure. The factor that determines the relation between pressure and flow rate is an inherent characteristic of the pipes, their RESISTANCE. Small-diameter, long pipes have greater resistances than large-diameter, short ones. Similarly, current flow in electrical circuits depends on the resistance in the circuit. Again, small, long wires have larger resistances than large, short ones. If current is being passed through an ionic solution, the resistance of the solution will increase as the solution is made more dilute. This is because there are fewer ions available to carry the current. In conductors such as wires, the relation between current and potential difference is described by Ohm's law, formulated by Ohm in the 1820s. The law says that the amount of current (I) flowing in a conductor is related to the potential difference (V) applied to it, $I = V/R$. The constant R is the resistance of the wire. If I is in amperes and V is in volts, then

Ohm's law and electrical resistance

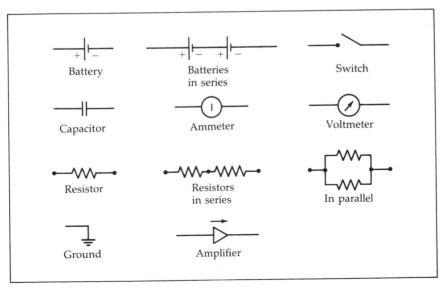

2   **SYMBOLS IN ELECTRIC CIRCUIT DIAGRAMS.**

R is in units of OHMS (Ω). The reciprocal of resistance is a measure of the ease with which current flows through a conductor and is called CONDUCTANCE and is indicated by $g = 1/R$; units of conductance are SIEMENS (S). Thus, Ohm's law may also be written $I = gV$.

Use of Ohm's law in understanding circuits

Ohm's law holds whenever the graph of current against potential is a straight line. In any circuit or part of a circuit for which this is true, any one of the three variables in the equation may be calculated if the other two are known. For example,

1. We can pass a known current through a nerve membrane, measure the change in potential, and then calculate the membrane resistance ($R = V/I$).
2. If we measure the potential difference produced by an unknown current and know the membrane resistance, we can calculate the applied current ($I = V/R$).
3. If we pass a known current through the membrane and know its resistance, then we can calculate the change in potential ($V = IR$).

Two additional simple, but important, rules (Kirchoff's laws) should be mentioned:

1. The algebraic sum of all the currents flowing toward any juncti.n is zero. For example, at point a in Figure 4, $I_{total} + I_{R1} + I_{R3} = 0$, which means that I_{total} (arriving) $= -I_{R1} - I_{R3}$ (leaving) (this is merely a statement that charge is neither created nor destroyed anywhere in the circuit).
2. The algebraic sum of all the battery voltages is equal to the algebraic sum of all the IR voltage drops in a loop. An example of this is shown in Figure 3B: $V = IR_1 + IR_2$ (this is a statement of the conservation of energy).

We can now examine in more detail the circuits of Figures 3 and 4,

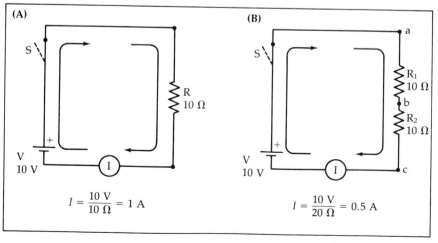

$$(A)\quad I = \frac{10\ V}{10\ \Omega} = 1\ A$$

$$(B)\quad I = \frac{10\ V}{20\ \Omega} = 0.5\ A$$

3   **OHMS'S LAW** applied to simple circuits. **(A)** Current $I = 10\ V/10\ \Omega = 1\ A$. **(B)** Current $= 10\ V/20\ \Omega = 0.5\ A$, and the voltage across each resistor is 5 V.

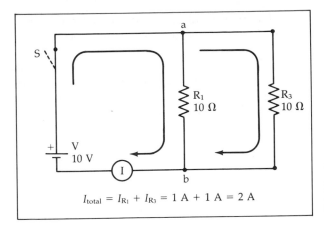

$$I_{total} = I_{R_1} + I_{R_3} = 1\text{ A} + 1\text{ A} = 2\text{ A}$$

4

PARALLEL RESISTORS. When R_1 and R_3 are in parallel, the voltage drop across each resistor is 10 V and the total current is 2 A.

which are needed to construct a model of the membrane. Figure 3A shows a battery (V) of 10 V connected to a resistance (R) of 10 Ω. The switch S can be opened or closed, thereby interrupting or establishing current flow. The voltage applied to R is 10 V; therefore the current measured by the ammeter, I, is, by Ohm's law, 1.0 A. In Figure 3B, the resistor is replaced by two resistors, R_1 and R_2, IN SERIES. By the first of Kirchoff's laws, the current flowing into point b must be equal to that leaving. Therefore, the same current, I, must flow through both the resistors. By the second of Kirchoff's laws, then, $IR_1 + IR_2 = V$ (10 V). It follows that the current, $I = V/(R_1 + R_2) = 0.5$ A. The voltage at b, then, is 5 V positive to that at c and a is 5 V positive to b. Note that because there is only one path for the current, the total resistance, R_{total}, seen by the battery is simply the sum of the two resistors; that is, $R_{total} = R_1 + R_2$.

What happens if, as shown in Figure 4, we add a second resistor, also of 10 Ω, IN PARALLEL, rather than in series? In the circuit, the two resistors R_1 and R_3 provide two separate pathways for current. Both have a voltage V (10 V) across them, so the respective currents will be

$$I_{R_1} = V/R_1 = 1\text{ A}$$

$$I_{R_3} = V/R_3 = 1\text{ A}$$

Therefore, to satisfy the first of Kirchoff's laws, there must be 2 A arriving at point a and 2 A leaving point b. The ammeter, then, will read 2 A. Now the combined resistance of R_1 and R_3 is $R_{total} = V/I = $ 10 V/2 A $= 5$ Ω, or half that of the individual resistors. This makes sense if one thinks of the hydraulic analogy: Two pipes in parallel will offer less resistance to flow than one pipe alone. In the electrical circuit, the *conductances* add: $g_{total} = g_1 + g_3$, or $1/R_{total} = 1/R_1 + 1/R_3$.

If we now generalize to any number (n) of resistors, resistances in series add simply:

$$R_{total} = R_1 + R_2 + R_3 + \ldots + R_n$$

and in parallel:

$$1/R_{total} = 1/R_1 + 1/R_2 + 1/R_3 + \ldots + 1/R_n$$

Applying circuit
analysis to the
membrane model

Figure 5A shows a circuit similar to that used to represent nerve membranes. Notice that the two batteries drive current around the circuit in the same direction and that the resistances R_1 and R_2 are in series. What is the potential difference between points b and d (which represent the outside and inside of the membrane)? The total potential across the two resistors between a and c is 150 mV, a being positive to c. Therefore, the current flowing between a and c through the resistors is 150 mV/100,000 Ω = 1.5 μA. When 1.5 μA flows across 10,000 Ω, as between a and b, a potential drop of 15 mV is produced, a being positive with respect to b. The potential difference between outside and inside is therefore 100mV $-$ 15 mV = 85 mV. We can obtain the same result by considering the voltage drop across R_2 (1.5 μA $\times$ 90,000 Ω = 135 mV) and adding it to V_2 (135 mV $-$ 50 mV = 85 mV). This *must* be so, as the potential between b and d must have a unique value.

In Figure 5B, R_1 and R_2 have been exchanged. As the total resistance in the circuit is the same, the current must be the same, 1.5 μA. Now the potential drop across R_2, between a and b, is 90,000 Ω $\times$ 1.5 μA = 135 mV, a being positive. Now the potential across the membrane is 100 mV $-$ 135 mV = $-$35 mV, OUTSIDE NEGATIVE; the same result can, of course, be obtained from the current through R_1. This simple circuit

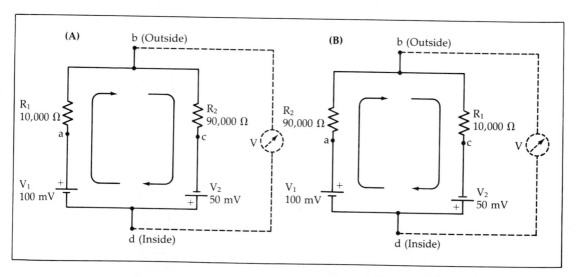

5   **ANALOGUE CIRCUITS FOR NERVE MEMBRANES.** In (A) and (B) the resistors R_1 and R_2 are reversed; otherwise the circuits are the same. The batteries V_1 and V_2 are in series. In (A), b (the "outside" of the membrane) is positive with respect to d (the "inside") by 85 mV; in (B) it is negative by 35 mV (see text). These circuits illustrate how changes in resistance can give rise to membrane potential changes even though the batteries (which represent ionic equilibrium potentials) remain constant.

illustrates an important point about membrane physiology: *The potential across a membrane can change as a result of resistance changes while the batteries remain unchanged.* A general expression for the membrane potential in the circuit shown in Figure 5A can be derived simply, as follows:

$$V_m = V_1 - IR_1$$

As $I = (V_1 + V_2)/(R_1 + R_2)$:

$$V_m = V_1 - \frac{(V_1 + V_2)R_1}{R_1 + R_2}$$

On rearranging:

$$V_m = \frac{V_1 R_2/R_1 - V_2}{1 + R_2/R_1}$$

In the circuits described in Figures 3 and 4, closing or opening the switch produces instantaneous and simultaneous changes in current and potential. Capacitors introduce a time element into the consideration of current flow. They accumulate and store electrical charge and, when they are present in a circuit, current and voltage changes are no longer simultaneous. A capacitor consists of two conducting plates (usually of metal) separated by an insulator (air, mica, oil, or plastic). When voltage is applied between the plates (Figure 6A), there is an instantaneous displacement of charge from one plate to the other through the external circuit. Once the capacitor is fully charged, however, there is no further current, as none can flow across the insulator. The CAPACITANCE (C) of a capacitor is defined by how much charge (q) it can store for each volt applied to it:

$$C = q/V$$

The units of capacitance are coulombs/volt or FARADS (F). The larger the plates of a capacitor and the closer together they are, the greater its capacitance. A one-farad capacitor is very large; capacitances in common use are in the range of microfarads (μF) or smaller.

When the switch in Figure 6A is closed, then, there is an instantaneous charge separation at the plates. The amount of charge stored in the capacitor is proportional to its capacitance and to the magnitude of the applied voltage (V_A). When the switch is opened, as in B, the charge on the capacitor remains, as does the voltage (V) between the plates. (One can sometimes get a surprising shock from electronic apparatus after it has been turned off because some of the capacitors in the circuits may remain charged.) The capacitor can be discharged by shorting it with a second switch, as in Figure 6C. Again, the current flow is instantaneous, returning the charge and the voltage on the capacitor to zero. If, instead, the capacitor is discharged through a resistor (R; Figure 6D), the discharge is no longer instantaneous. This is because the resistor limits the current flow. If the voltage on the capacitor is V, then

Electrical capacitance and time constant

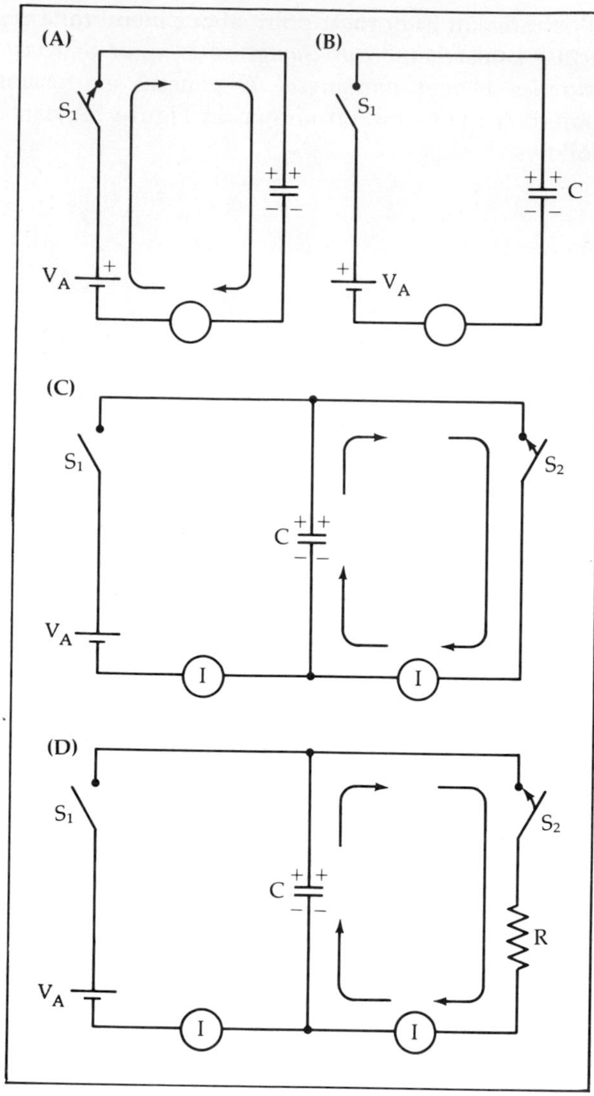

6

CAPACITORS in electrical circuits (A), (B) and (C) are idealized circuits having no resistance. When S_1 is closed in (A), the capacitor is charged instantaneously to voltage V_A. If S_1 is then opened (B), the potential remains on the capacitor. Closing switch S_2 in (C) discharges the capacitor instantaneously. In (D) the capacitor is discharged through resistor R. The maximum discharge current is $I = V_A/R$.

by Ohm's law the maximum current is $I = V/R$. With no resistor in the circuit, the current becomes infinitely large and the capacitor is discharged in an infinitesimal time period; if R is very large, the discharge process takes a very long time. The rate of discharge at any given time, dq/dt, is simply equal in magnitude to the current. In other words, $dq/dt = -V/R$ (negative because the charge is decreasing with time), where V initially is equal to the battery voltage and decreases as the capacitor is discharged. As $q = CV$, $dq/dt = CdV/dt$, and we can then write $CdV/dt = -V/R$, or

$$dV/dt = -V/RC$$

The equation says that the rate of loss of voltage from the capacitor is proportional to the voltage remaining. Thus, as the voltage decreases, the rate of discharge decreases. The constant of proportionality, $1/RC$, is the RATE CONSTANT for the process: RC is its TIME CONSTANT. This kind of process arises over and over again in nature. For example, the rate at which water drains from a bathtub decreases as the depth, and hence the pressure at the drain, decreases. In this kind of situation, the discharge process is described by an exponential function:

$$V = V_0\, e^{-t/\tau}$$

where V_0 is the initial charge on the capacitor and the time constant $\tau = RC$. Similarly, when the capacitor is charged through a resistor, as in Figure 7, the charging process takes a finite time. The voltage between the plates increases with time until the battery voltage is reached and no further current flows. The charging process is now a rising exponential, with a time constant $\tau = RC$:

$$V = V_A(1 - e^{-t/\tau})$$

These examples illustrate another property of a capacitor. Current flows into and out of the capacitor only when the potential is changing:

$$I_C = dQ/dt = C dV/dt$$

When the voltage across the capacitor is steady ($dV/dt = 0$), the capacitive current, I_C, is zero. In other words, the capacitance has an "infinite resistance" for a steady potential difference and a "low resistance" for a rapidly changing potential. Figure 7B shows a circuit in which current flows through a resistor and capacitor in parallel and Figure 7E the time courses of the capacitative current and voltage.

The properties of a capacitor in a circuit can be illustrated by the slightly more elaborate hydraulic analogy shown in Figure 7C. The capacitor is represented by an elastic diaphragm that forms a partition in a fluid-filled chamber. When the tap is opened, the pressure generated by the pump causes fluid flow into the chamber, bulging the diaphragm until, because of its elasticity, it provides an equal and opposite pressure; then there is no more fluid flow and the chamber is fully charged. If a tube is placed alongside, as in Figure 7D, some fluid flows through the tube and some is used to expand the diaphragm. If the tube is of high resistance, the pressure difference between its two ends is larger for a given flow than for a lower resistance tube. In this case, the distention of the diaphragm is greater and takes longer to achieve. Similarly, if the capacity of the chamber is larger, more fluid is diverted during the filling (or "charging") process and a longer time is required to reach a steady state. Thus, the characteristic time constant of the system is determined by the product of resistance and capacitance.

When capacitors are arranged in parallel, as in Figure 8A, the total capacitance is increased. The total charge stored is the sum of that

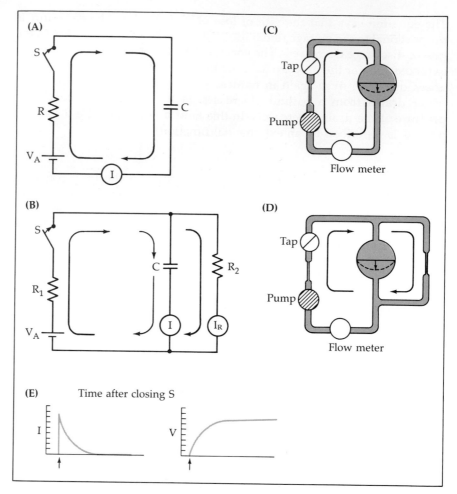

7   CHARGING OF A CAPACITOR. In (A) the capacitor is charged at a rate limited by the resistor, the initial rate being $I = V_A/R$. In (B) the charging rate depends on both resistors in the circuit. The capacitative current and the voltage across the capacitor are shown as functions of time in (E). The voltage reaches its final value only when the capacitor is fully charged ($I_c = 0$). (C) and (D) are hydraulic analogues of the circuits in (A) and (B).

stored in each: $q_1 + q_2 = C_1V_A + C_2V_A$, or $q_{total} = C_{total}V_A$, where $C_{total} = C_1 + C_2$. In contrast, capacitance *decreases* when capacitors are arranged in series (Figure 8B). It turns out that the relation is the same as for resistors in parallel: their reciprocals sum. In summary, then, for a number (N) of capacitors in parallel,

$$C_{total} = C_1 + C_2 + C_3 + \ldots + C_n$$

and in series,

$$1/C_{total} = /C_1 + 1/C_3 + \ldots + 1/C_n$$

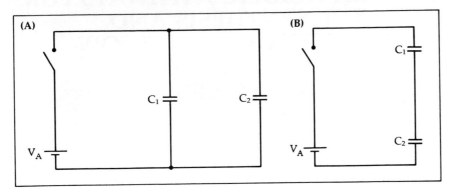

8 CAPACITORS IN PARALLEL (A) AND IN SERIES (B).

METABOLIC PATHWAYS FOR THE SYNTHESIS AND INACTIVATION OF LOW-MOLECULAR-WEIGHT TRANSMITTERS

The diagrams on the following pages summarize the predominant metabolic pathways for the low-molecular-weight transmitters acetylcholine, GABA, dopamine, norepinephrine, epinephrine, 5-HT, and histamine. Glutamate, glycine, and ATP are not included; there appear to be no special neuronal pathways for their synthesis or degradation. For each metabolic step the portion of the molecule being modified is printed in color. Further information can be found in several comprehensive books:

Gilman, A. G., Rall, T. W., Nies, A. S. and Taylor, P. (eds.). 1990. *Goodman and Gilman's The Pharmacological Basis of Therapeutics*, 8th Ed. Pergamon Press, New York.

Siegel, G., Agranoff, B., Albers, R. W. and Molinoff, P. (eds.). 1989. *Basic Neurochemistry*, 4th Ed. Raven Press, New York.

Stryer, L. 1988. *Biochemistry*, 3rd Ed. W. H. Freeman, New York.

Acetylcholine (ACh)

SYNTHESIS

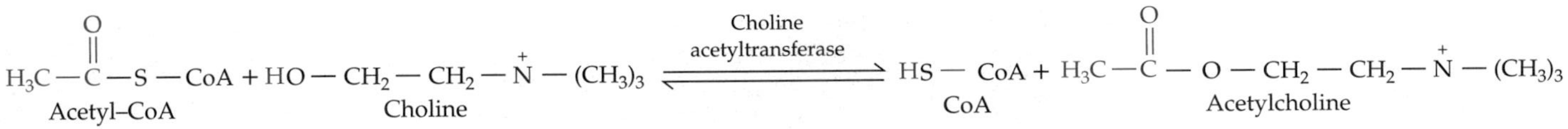

$$H_3C-C(=O)-S-CoA + HO-CH_2-CH_2-\overset{+}{N}-(CH_3)_3 \underset{\text{acetyltransferase}}{\overset{\text{Choline}}{\rightleftharpoons}} HS-CoA + H_3C-C(=O)-O-CH_2-CH_2-\overset{+}{N}-(CH_3)_3$$

Acetyl–CoA   Choline   CoA   Acetylcholine

INACTIVATION

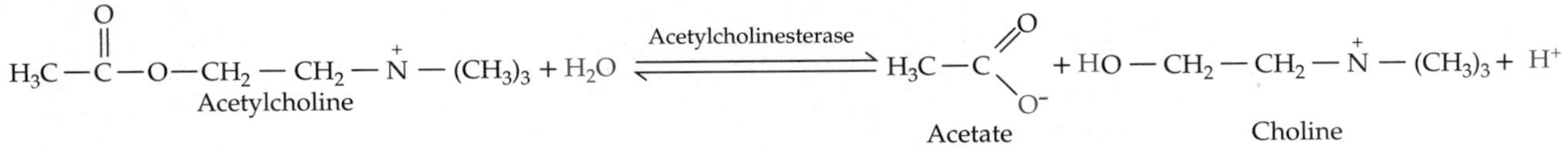

$$H_3C-C(=O)-O-CH_2-CH_2-\overset{+}{N}-(CH_3)_3 + H_2O \overset{\text{Acetylcholinesterase}}{\rightleftharpoons} H_3C-C(=O)(O^-) + HO-CH_2-CH_2-\overset{+}{N}-(CH_3)_3 + H^+$$

Acetylcholine   Acetate   Choline

γ–Aminobutyric acid (GABA)

SYNTHESIS

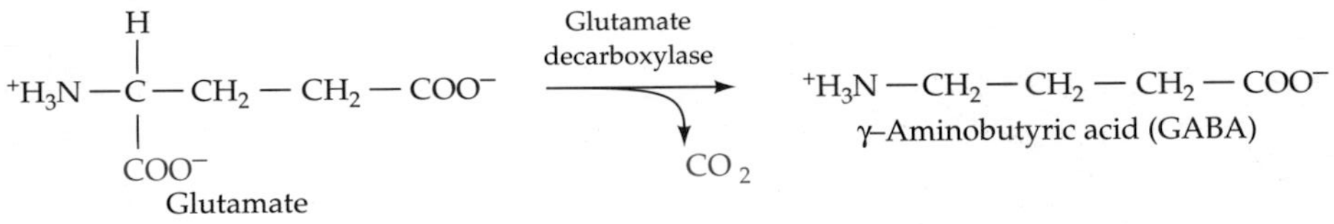

$$^+H_3N-\overset{H}{\underset{COO^-}{C}}-CH_2-CH_2-COO^- \underset{\text{decarboxylase}}{\overset{\text{Glutamate}}{\longrightarrow}} {}^+H_3N-CH_2-CH_2-CH_2-COO^-$$

Glutamate CO_2 γ-Aminobutyric acid (GABA)

INACTIVATION

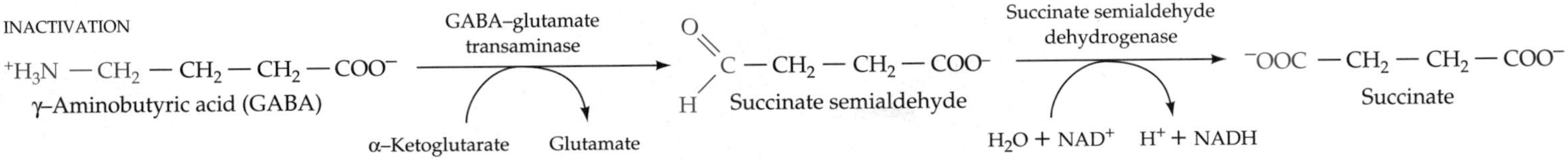

$$^+H_3N-CH_2-CH_2-CH_2-COO^- \underset{\text{transaminase}}{\overset{\text{GABA–glutamate}}{\longrightarrow}} O=\overset{}{\underset{H}{C}}-CH_2-CH_2-COO^- \underset{\text{dehydrogenase}}{\overset{\text{Succinate semialdehyde}}{\longrightarrow}} {}^-OOC-CH_2-CH_2-COO^-$$

γ-Aminobutyric acid (GABA)   α-Ketoglutarate   Glutamate   Succinate semialdehyde $H_2O + NAD^+$ $H^+ + NADH$ Succinate

Catecholamines: Dopamine

SYNTHESIS

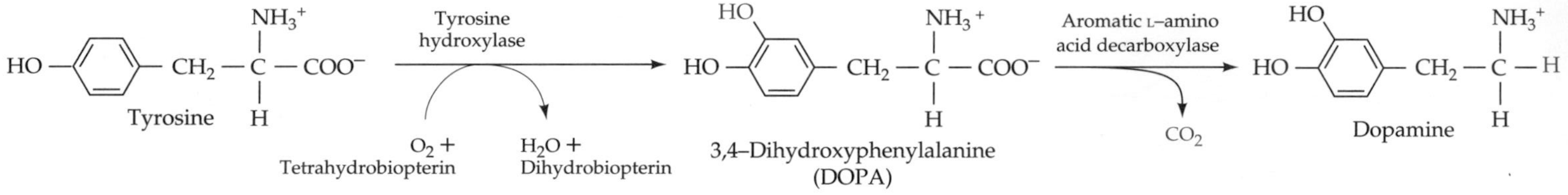

INACTIVATION

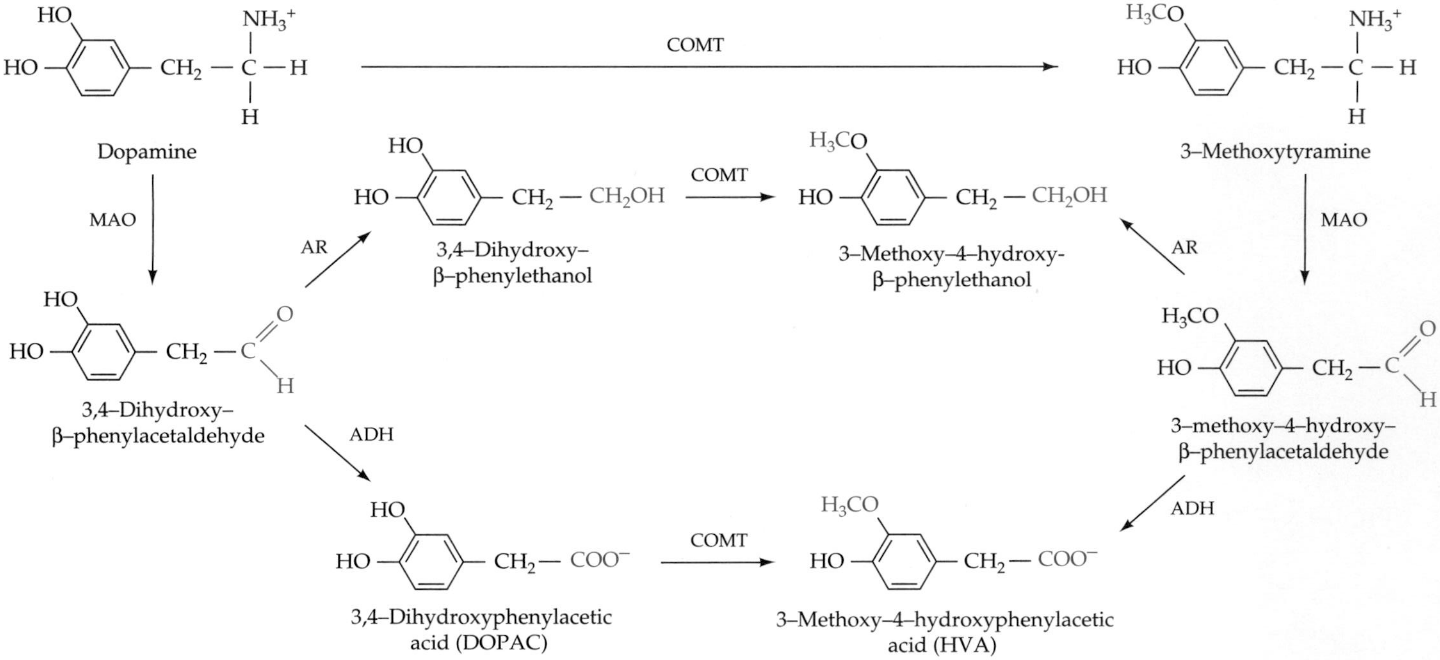

Catecholamines: Norepinephrine and epinephrine

SYNTHESIS

INACTIVATION OF NOREPINEPHRINE

5–Hydroxytryptamine (5–HT; Serotonin)

SYNTHESIS

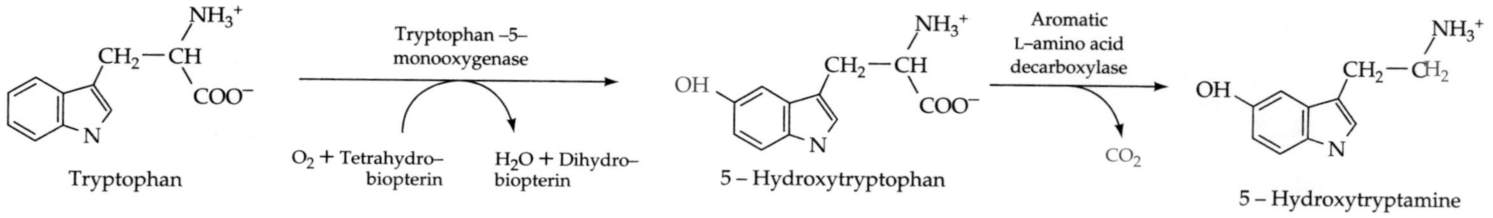

Tryptophan

Tryptophan –5–
monooxygenase

O_2 + Tetrahydro–
biopterin

H_2O + Dihydro–
biopterin

5 – Hydroxytryptophan

Aromatic
L–amino acid
decarboxylase

CO_2

5 – Hydroxytryptamine

INACTIVATION

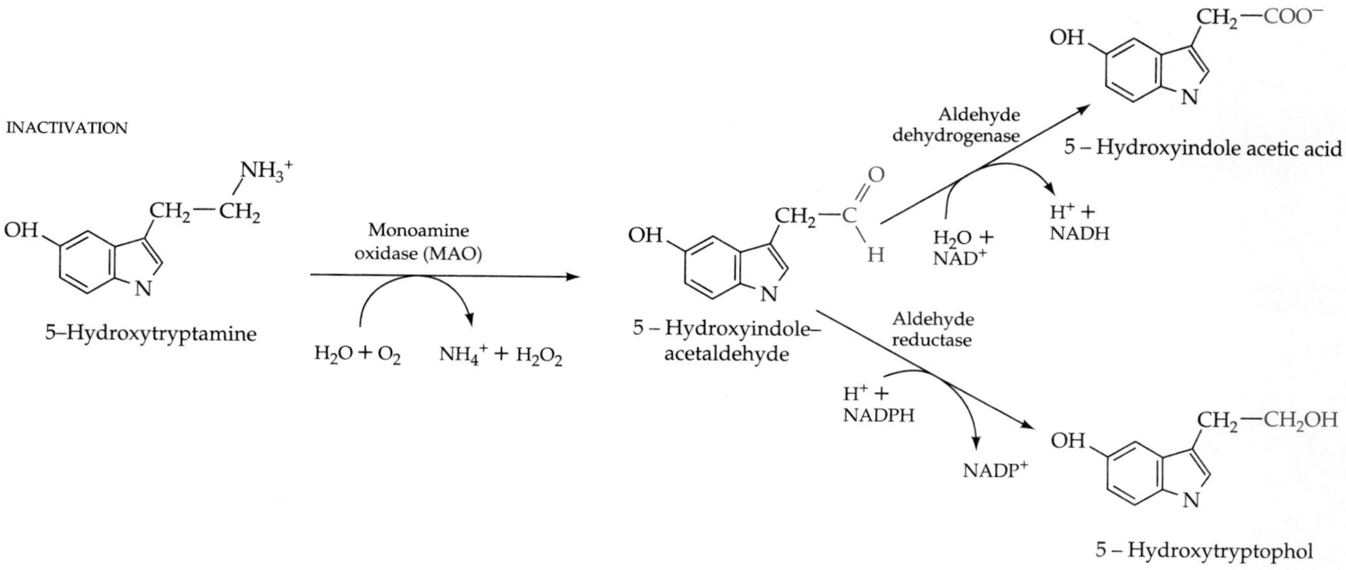

5-Hydroxytryptamine

Monoamine
oxidase (MAO)

$H_2O + O_2$ $NH_4^+ + H_2O_2$

5 – Hydroxyindole–
acetaldehyde

Aldehyde
dehydrogenase

$H_2O +$
NAD^+

$H^+ +$
NADH

5 – Hydroxyindole acetic acid

Aldehyde
reductase

$H^+ +$
NADPH

$NADP^+$

5 – Hydroxytryptophol

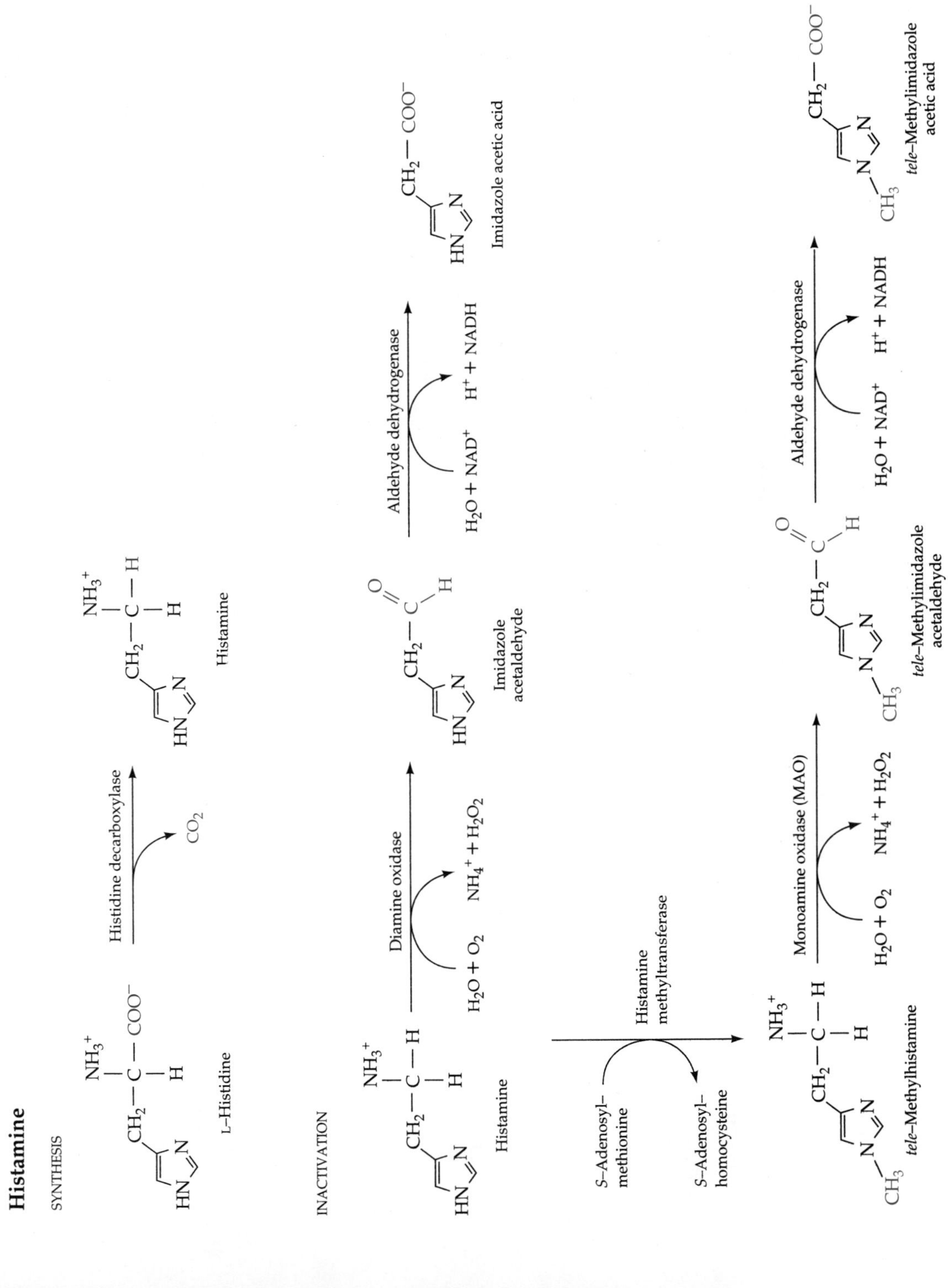

Histamine

SYNTHESIS

L–Histidine

Histidine decarboxylase

CO_2

Histamine

INACTIVATION

Histamine

Diamine oxidase

$H_2O + O_2$ $NH_4^+ + H_2O_2$

Imidazole acetaldehyde

Aldehyde dehydrogenase

$H_2O + NAD^+$ $H^+ + NADH$

Imidazole acetic acid

Histamine methyltransferase

S–Adenosyl-methionine

S–Adenosyl-homocysteine

tele–Methylhistamine

Monoamine oxidase (MAO)

$H_2O + O_2$ $NH_4^+ + H_2O_2$

tele–Methylimidazole acetaldehyde

Aldehyde dehydrogenase

$H_2O + NAD^+$ $H^+ + NADH$

tele–Methylimidazole acetic acid

Inactivation of biogenic amines

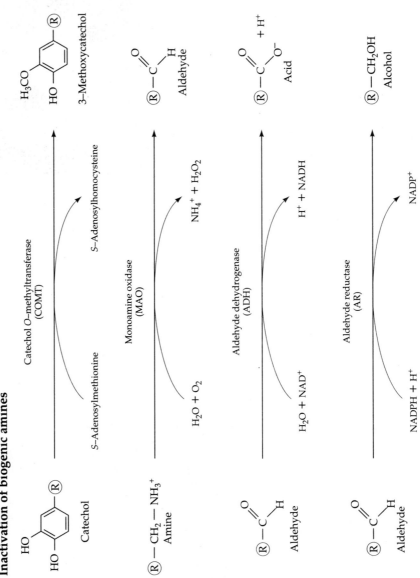

STRUCTURES AND PATHWAYS OF THE BRAIN

The following drawings show the brain viewed from different aspects and cut in different sections. The aim is to provide the visual equivalent of a glossary relating to material in the text rather than to present a full atlas. Consequently, only key landmarks and structures are illustrated. Further anatomical information can be found in a number of comprehensive books, including:

Carpenter, M. B. 1991. *Core Text of Neuroanatomy*, 4th Ed. Williams and Wilkins, Baltimore.

Martin, J. H. 1989. *Neuroanatomy*. Elsevier, New York.

Nolte, J. 1992. *The Human Brain*, 3rd Ed. Mosby, St. Louis.

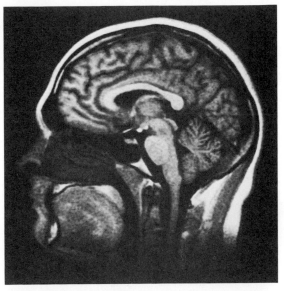

Magnetic resonance image of a living human brain (sagittal section). Copyright 1984 by the General Electric Co. Reproduced with permission.

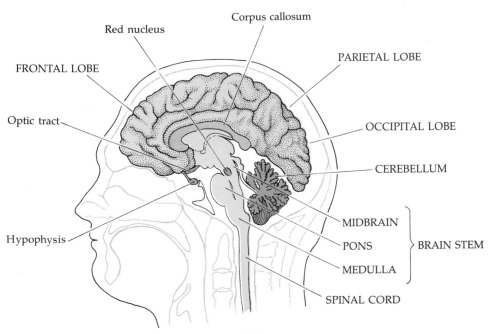

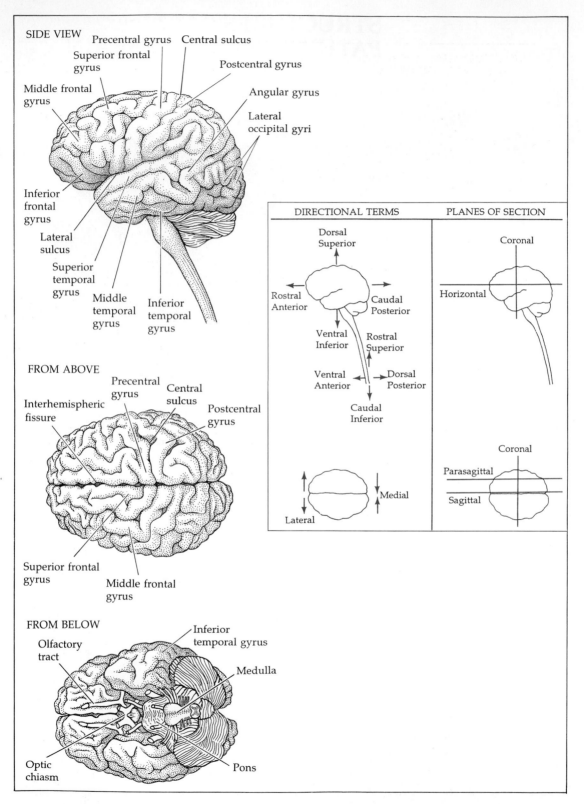

SIDE VIEW

Precentral gyrus Central sulcus
Superior frontal gyrus
Postcentral gyrus
Middle frontal gyrus
Angular gyrus
Lateral occipital gyri
Inferior frontal gyrus
Lateral sulcus
Superior temporal gyrus
Middle temporal gyrus
Inferior temporal gyrus

FROM ABOVE

Precentral gyrus Central sulcus
Interhemispheric fissure
Postcentral gyrus
Superior frontal gyrus
Middle frontal gyrus

FROM BELOW

Olfactory tract
Inferior temporal gyrus
Medulla
Optic chiasm
Pons

DIRECTIONAL TERMS

Dorsal Superior
Rostral Anterior
Caudal Posterior
Ventral Inferior
Rostral Superior
Ventral Anterior
Dorsal Posterior
Caudal Inferior

Lateral
Medial

PLANES OF SECTION

Coronal
Horizontal

Coronal
Parasagittal
Sagittal

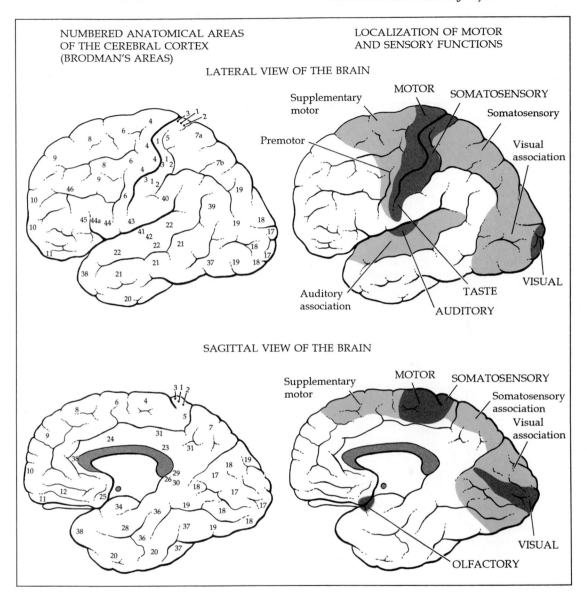

NUMBERED ANATOMICAL AREAS
OF THE CEREBRAL CORTEX
(BRODMAN'S AREAS)

LOCALIZATION OF MOTOR
AND SENSORY FUNCTIONS

LATERAL VIEW OF THE BRAIN

SAGITTAL VIEW OF THE BRAIN

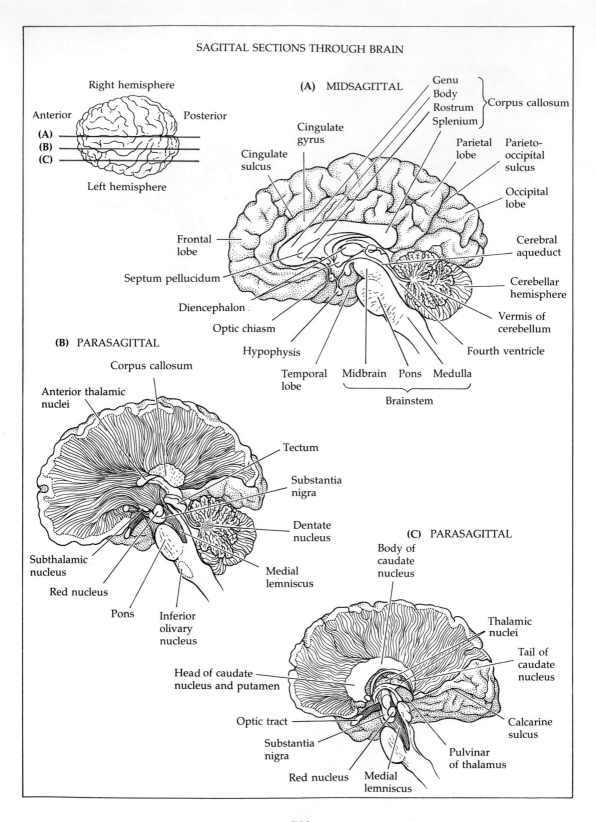

SAGITTAL SECTIONS THROUGH BRAIN

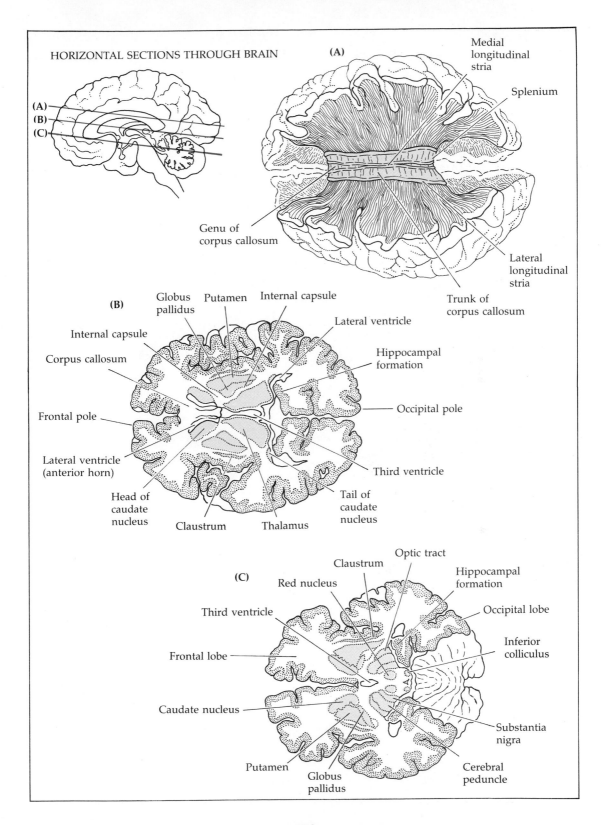

HORIZONTAL SECTIONS THROUGH BRAIN

(A)

Medial longitudinal stria

Splenium

Genu of corpus callosum

Lateral longitudinal stria

Trunk of corpus callosum

(B)

Globus pallidus

Putamen

Internal capsule

Lateral ventricle

Internal capsule

Hippocampal formation

Corpus callosum

Occipital pole

Frontal pole

Lateral ventricle (anterior horn)

Third ventricle

Head of caudate nucleus

Tail of caudate nucleus

Claustrum

Thalamus

(C)

Optic tract

Claustrum

Hippocampal formation

Red nucleus

Occipital lobe

Third ventricle

Inferior colliculus

Frontal lobe

Caudate nucleus

Substantia nigra

Putamen

Globus pallidus

Cerebral peduncle

711

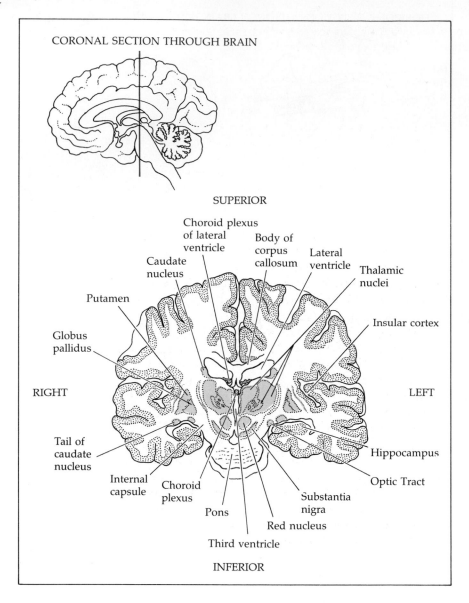

CORONAL SECTION THROUGH BRAIN

SUPERIOR

Choroid plexus
of lateral
ventricle

Body of
corpus
callosum

Lateral
ventricle

Thalamic
nuclei

Caudate
nucleus

Putamen

Insular cortex

Globus
pallidus

RIGHT

LEFT

Tail of
caudate
nucleus

Hippocampus

Optic Tract

Internal
capsule

Choroid
plexus

Substantia
nigra

Pons

Red nucleus

Third ventricle

INFERIOR

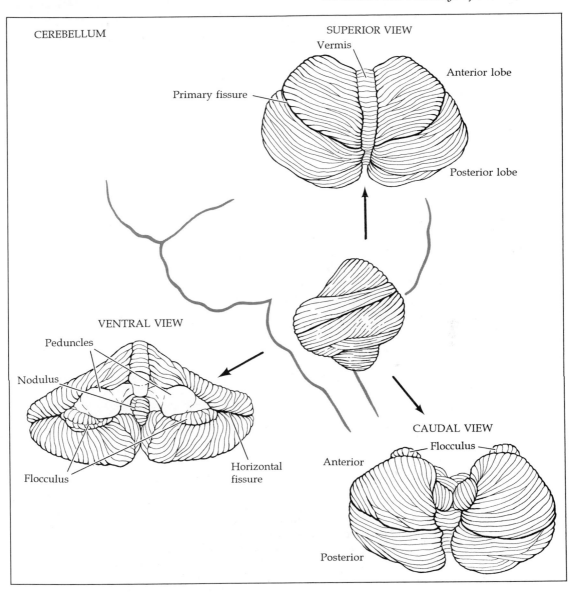

CEREBELLUM

SUPERIOR VIEW

Vermis

Anterior lobe

Primary fissure

Posterior lobe

VENTRAL VIEW

Peduncles

Nodulus

Flocculus

Horizontal
fissure

CAUDAL VIEW

Flocculus

Anterior

Posterior

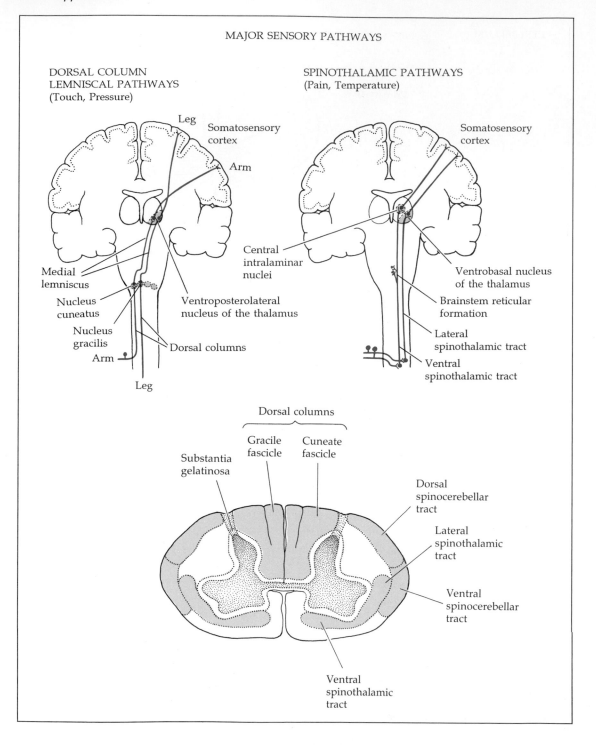

MAJOR SENSORY PATHWAYS

DORSAL COLUMN
LEMNISCAL PATHWAYS
(Touch, Pressure)

SPINOTHALAMIC PATHWAYS
(Pain, Temperature)

Leg

Somatosensory
cortex

Arm

Somatosensory
cortex

Central
intralaminar
nuclei

Ventrobasal nucleus
of the thalamus

Medial
lemniscus

Brainstem reticular
formation

Nucleus
cuneatus

Ventroposterolateral
nucleus of the thalamus

Lateral
spinothalamic tract

Nucleus
gracilis

Dorsal columns

Ventral
spinothalamic tract

Arm

Leg

Dorsal columns

Gracile
fascicle

Cuneate
fascicle

Substantia
gelatinosa

Dorsal
spinocerebellar
tract

Lateral
spinothalamic
tract

Ventral
spinocerebellar
tract

Ventral
spinothalamic
tract

MAJOR MOTOR PATHWAYS

TRACTS DESCENDING TO THE SPINAL CORD

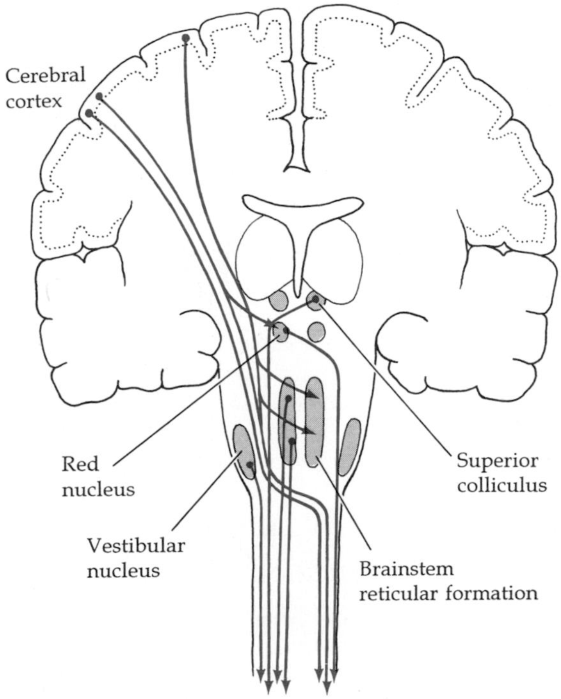

Cerebral cortex

Red nucleus

Vestibular nucleus

Superior colliculus

Brainstem reticular formation

CROSS SECTION OF CERVICAL SPINAL CORD

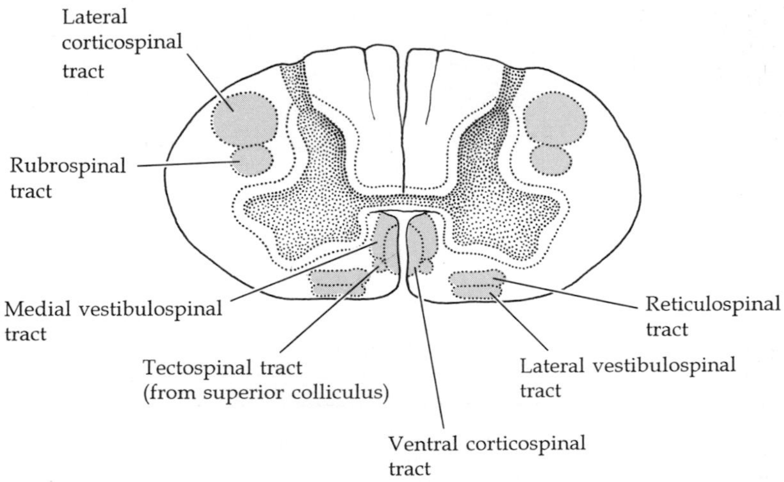

Lateral corticospinal tract

Rubrospinal tract

Medial vestibulospinal tract

Tectospinal tract (from superior colliculus)

Reticulospinal tract

Lateral vestibulospinal tract

Ventral corticospinal tract

CENTRAL PATHWAYS FOR NOREPINEPHRINE, DOPAMINE,
5-HYDROXYTRYPTAMINE, AND HISTAMINE

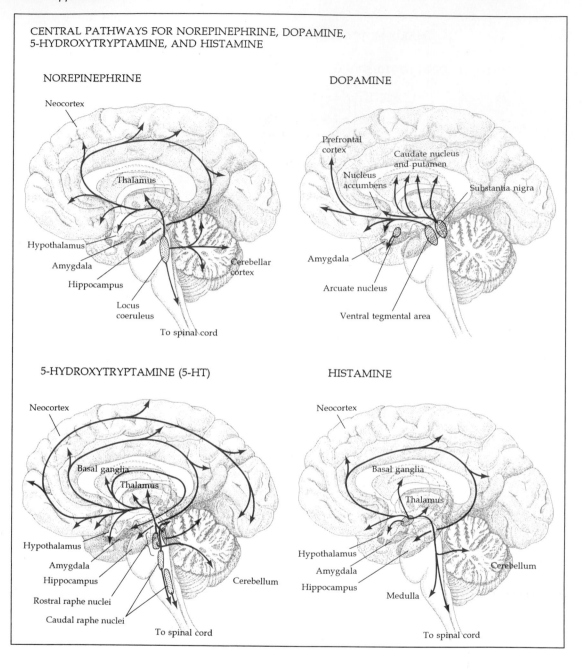

GLOSSARY

The definitions below apply to the terms as they are used in the context of this book. Words such as **excitation, adaptation,** and **inhibition** all have additional meanings that are not included.

For structural formulae of transmitters see Appendix B.

For anatomical terms see Appendix C.

acetylcholine (ACh)   Transmitter liberated by vertebrate motoneurons, preganglionic autonomic neurons, and in various central nervous system pathways.

acetylcholine receptor (ACh receptor)   Membrane protein that binds ACh. There are two different varieties.

   nicotinic ACh receptor   Activated by nicotine; consists of five polypeptide subunits that form a cation channel when activated.

   muscarinic ACh receptor   Activated by muscarine; consists of a single polypeptide chain and is coupled by a G protein to one or more intracellular second messenger systems.

action potential   Brief, regenerative, all-or-nothing electrical potential that propagates along an axon or muscle fiber. Also called an **impulse.**

active transport   Movement of ions or molecules against an electrochemical gradient by utilization of metabolic energy.

active zone   Region in a presynaptic nerve terminal characterized by densely staining material on the cytoplasmic surface of the presynaptic membrane and a cluster of synaptic vesicles; believed to be the site of transmitter release.

adaptation   The decline in response of a sensory neuron to a maintained stimulus.

adenosine 5'-triphosphate (ATP)   A common metabolite; hydrolysis of the terminal phosphoester linkage provides energy for many cellular reactions. ATP serves as the phosphate donor in phosphorylation reactions. Also found in adrenergic and cholinergic synaptic vesicles; acts as a transmitter at synapses made by vertebrate sympathetic neurons.

adenylyl cyclase   Enzyme catalyzing the synthesis of cyclic AMP from ATP.

adrenaline   See **epinephrine.**

adrenergic   Refers to neurons that release norepinephrine as a transmitter.

afferent   An axon conducting impulses toward the central nervous system; also called **primary afferent.** See **group I afferents; group II afferents.**

afterhyperpolarization   Slow hyperpolarizing potential seen in many neurons following a train of action potentials.

agonist   A molecule that activates a receptor.

γ-aminobutyric acid (GABA)   An inhibitory neurotransmitter.

amphipathic   Containing separate regions of hydrophilic and hydrophobic groups.

ampulla   The sensory region of a semicircular canal in the vestibular apparatus.

anion   A negatively charged ion.

antagonist   A molecule that prevents activation of a receptor.

anterograde   In the direction from the neuronal cell body toward the axon terminal. Compare with **retrograde**.

antibody   An immunoglobulin molecule.

anticholinesterase   Any cholinesterase inhibitor (e.g., neostigmine, eserine); such agents prevent the hydrolysis of ACh and thereby allow its action to be prolonged.

astrocyte   A class of glial cells found in the vertebrate central nervous system.

area centralis   In the cat, the area of retina with highest acuity; contains cones.

autonomic nervous system   Part of the vertebrate nervous system innervating viscera, skin, smooth muscle, glands, and heart. The autonomic nervous system consists of two distinct divisions, the **parasympathetic** and the **sympathetic**.

axon   The process or processes of a neuron specialized to conduct impulses, usually over long distances.

axon hillock   The region of the cell body from which the axon originates; often the site of impulse initiation. See **initial segment**.

axonal transport   Movement of material along an axon.

axoplasm   The intracellular constituents of an axon.

axotomy   Severing of an axon.

basal lamina   An extracellular glycoprotein- and proteoglycan-containing matrix that ensheaths many tissues in the body, including nerves and muscle fibers.

basilar membrane   The membrane in the cochlea upon which the hair cells sit; separates the scala tympani from the scala media.

biogenic amine   A general term referring to any of several bioactive amines, including norepinephrine, epinephrine, dopamine, 5-hydroxytryptamine (serotonin), and histamine.

bipolar cell   A neuron with two major processes arising from the cell body; in the vertebrate retina, bipolar cells are interposed between receptors and ganglion cells.

blobs   Small, regularly spaced assemblies of neurons in the visual cortex of the monkey; the neurons are stained with cytochrome oxidase and respond mainly to color stimuli.

blood–brain barrier   Refers to the restricted access of substances to neurons and glial cells within the brain.

botulinum toxin   A bacterial toxin that blocks release of transmitter from vertebrate motor nerve terminals.

bouton   The small terminal expansion of the presynaptic nerve fiber at a synapse; site of transmitter release.

α-bungarotoxin   Toxin from venom of the snake *Bungarus multicinctus*; binds to the nicotinic ACh receptor with high affinity.

capacitance of the membrane (C_m)   Property of the cell membrane enabling electrical charge to be stored and separated and introducing distortion in the time course of passively conducted signals; usually measured in microfarads (μF).

carrier molecule   A molecule involved in transporting ions or other molecules across cell membranes.

catecholamine   A general term referring to molecules having both a catechol ring and an amino group; typically dopamine, norepinephrine, and epinephrine.

cation   A positively charged ion.

cDNA (complementary DNA)   DNA synthesized by reverse transcriptase using mRNA as a template.

centrifugal control   Regulation of performance of peripheral sense organs by axons coming from the central nervous system.

cerebrospinal fluid (CSF)   The clear liquid filling the ventricles of the brain and spaces between the meninges. See **subarachnoid space; ventricles**.

channel   A pathway through a membrane allowing the passage of ions or molecules. All channels that have been characterized are aqueous pores formed by a single large protein or by an assembly of polypeptide subunits.

charybdotoxin (CTX)   A toxin that blocks potassium A channels; obtained from scorpions.

chimera   An experimentally derived embryo or organ comprising cells arising from two or more genetically distinct sources.

cholinergic   Refers to a neuron that releases acetylcholine as a transmitter.

cholinesterase   An enzyme that hydrolyzes acetylcholine into acetic acid and choline.

choroid plexus   Folded processes that are rich in blood vessels and project into ventricles of the brain and secrete cerebrospinal fluid.

chromaffin granules   Large vesicles found in cells of the adrenal medulla, containing epinephrine (often norepinephrine as well), ATP, dopamine β-hydroxylase, and chromogranins.

clone   All the progeny of a single cell. Also, to obtain a clone of cells containing a molecule of interest, such as a particular cDNA.

cochlea   The bony canal containing the sensory apparatus for hearing.

conductance (*g*)   The reciprocal of electrical resistance and thus a measure of the ability to conduct electricity; in cell membranes or ion channels, a measure of permeability to one or more ion species.

connexin   The subunit from which a connexon is formed.

connexon   A membrane channel bridging the space between two adjacent cells, connecting the cytoplasm of one to that of the other. See **gap junction**.

contralateral   On the opposite side of the body. Compare with **ipsilateral**.

convergence   The coming together of presynaptic nerve fibers to make synapses on one postsynaptic neuron.

cortical barrel   Barrel-shaped aggregate of cortical neurons related to a specific peripheral sensory structure (e.g., a facial whisker).

cortical column   An aggregate of cortical neurons, extending inward from the pial surface, that share common properties (e.g., sensory modality, receptive field position, eye dominance, orientation, movement sensitivity).

coulomb   A unit of electrical charge.

coupling potential   A potential change in one cell produced by current spread from another through an electrical synapse.

crustacean   Any member of the class Crustacea, which includes arthropods with hard shells, such as lobsters, crabs, barnacles, and shrimp.

curare   A plant extract that blocks nicotinic ACh receptors.

cytosol   The part of the cell cytoplasm that occupies the space between membrane-bounded organelles.

dalton (Da)   A unit of molecular mass, numerically equal to one-twelfth the mass of a carbon atom; often expressed in thousands, or **kilodaltons (kDa)**.

dendrite   Process of a neuron specialized to act as a postsynaptic receptor region of a neuron.

depolarization Reduction in magnitude of the resting membrane potential toward zero.

depression Reduction of transmitter release from presynaptic terminals due to previous synaptic activity.

desensitization Reduction of the response of a receptor to a ligand after prolonged or repeated exposure.

diacylglycerol (DAG) An intracellular second messenger produced by phospholipase C-catalyzed hydrolysis of phospholipids. DAG activates protein kinase C.

divalent Having an electric charge of $+2$ or -2.

divergence Branching of an axon to form synapses with several other neurons.

domain A particular region of a polypeptide; for example, one of the four repeating regions of the voltage-activated sodium channel.

dopamine Transmitter liberated by neurons in some autonomic ganglia and in the central nervous system.

driving potential The difference between the membrane potential and the equilibrium potential for movement of an ion species through a membrane channel.

efferent An axon conducting impulses outward from the central nervous system. Compare with **afferent**.

γ-efferent fiber Small, myelinated motor axon supplying an intrafusal muscle fiber. See **fusimotor**.

EGTA Ethylene glycol bis-(β-aminoethyl ether) N,N,N′N′-tetraacetic acid. A high-affinity calcium-binding compound.

electrochemical gradient The transmembrane difference in potential energy of an ion arising from the combined electrical and diffusional forces acting on it.

electroencephalogram (EEG) A recording of the electrical activity of the brain by external electrodes on the scalp.

electrogenic pump Active transport of ions across a cell membrane in which a net transfer of electrical charges contributes directly to the membrane potential.

electromyogram (EMG) A recording of muscular activity by external electrodes.

electroretinogram (ERG) Potential changes in response to light recorded by external electrodes on the eye.

electrotonic potentials Localized, graded, subthreshold potentials, produced by artificially applied currents and characterized by the passive electrical properties of cells.

endocytosis The process whereby membrane, together with some extracellular fluid, is internalized by the invagination and pinching off of a vesicle from the plasma membrane.

endothelial cells Layer of cells lining the blood vessels.

end plate Postsynaptic area of vertebrate skeletal muscle fiber.

end plate potential (epp) Postsynaptic potential change in a skeletal muscle fiber produced by ACh liberated from presynaptic terminals.

ependyma Layer of cells lining the cerebral ventricles and central canal of spinal cord.

epinephrine (adrenaline) Hormone secreted by the adrenal medulla; certain of its actions resemble those of sympathetic nerves.

equilibrium potential Membrane potential at which there is no net passive movement of a permeant ion species into or out of a cell.

eserine An anticholinesterase (see above); also known as physostigmine.

excitation A process tending to produce action potentials.

excitatory postsynaptic potential (epsp) Depolarization of the postsynaptic membrane of a neuron produced by an excitatory transmitter released from presynaptic terminals.

exocytosis The process whereby synaptic vesicles fuse with presynaptic terminal membrane and empty transmitter molecules into the synaptic cleft.

explant A piece of tissue placed in culture.

extracellular matrix A scaffolding of large glycoproteins and proteoglycans that surrounds and separates cells or tissues.

extrafusal Muscle fibers making up the mass of a skeletal muscle (i.e., not within the sensory muscle spindles).

facilitation Increased release of transmitter from nerve terminals due to previous synaptic activity.

family A group of gene products with closely similar structure and function (e.g., nicotinic ACh receptors). See **isotype; superfamily**.

farad (F) Unit of capacitance; the **microfarad** ($\mu F = 10^{-6}$ F) is more commonly used.

faraday (*F*) The number of coulombs carried by one mole of a univalent ion (96,500).

fasciculation 1. The aggregation of neuronal processes to form a bundle. 2. The spontaneous contraction of muscle fibers in groups.

fetus A relatively late-stage mammalian embryo.

field axis orientation For simple cortical neurons in the visual system, the angle of the long axis of the receptive field (e.g., horizontal, vertical, oblique).

formants Frequency components of basic speech elements.

fovea Central depression in the retina composed of slender cones; the area of greatest visual resolution.

fusimotor Motoneurons supplying muscle fibers in a muscle spindle.

G proteins Receptor-coupled proteins that bind guanine nucleotides and activate intracellular messenger systems.

GABA (γ-aminobutyric acid) An inhibitory neurotransmitter.

ganglion A discrete collection of nerve cells.

gap junction The region of contact between two cells in which the space between adjacent membranes is reduced to about 2 nm and bridged by connecting channels. See **connexon**.

gate The mechanism whereby a channel is opened and closed.

gene expression The transcription of DNA into mRNA and the translation of mRNA into protein.

glia See **neuroglia**.

glioblast A dividing cell, the progeny of which will develop into glial cells.

glutamate A transmitter liberated at many excitatory synapses in the vertebrate central nervous system.

glycine A transmitter liberated at many inhibitory synapses in the spinal cord and brain stem.

glycoprotein A protein containing carbohydrate residues.

Golgi tendon organ Sensory element in muscle tendons activated by muscle stretch or contraction.

gray matter Part of the central nervous system composed predominantly of the cell bodies of neurons and fine terminals, as opposed to major axon tracts (**white matter**).

group I afferents   Sensory fibers from muscle with conduction velocities in the range of 80 to 120 meters per second.

    group Ia   Arise from muscle spindles.

    group Ib   Arise from Golgi tendon organs.

group II afferents   Sensory fibers from muscle spindles with conduction velocities in the range of 30 to 80 meters per second.

growth cone   The expanded tip of a growing axon.

hair cells   Sensory cells in which bending of stereocilia ("hairs") causes a change in membrane potential; responsible for auditory transduction, transduction of vestibular stimuli, and vibratory transduction in lateral line organs of fish.

histamine   A transmitter liberated by a small number of neurons in the vertebrate central nervous system.

homology   The occurrence of identical bases at corresponding positions in two nucleotide sequences, or of amino acids in two polypeptide sequences.

horseradish peroxidase (HRP)   An enzyme used as a histochemical marker for tracing processes of neurons or spaces between cells.

hydropathy index   A measure of the insolubility of an amino acid, or amino acid sequence, in water; hence, a measure of preference for a lipid environment.

hydrophilic   Having a relatively high water solubility; polar.

hydrophobic   Having a relatively low water solubility; nonpolar.

5-hydroxytryptamine (5-HT)   A neurotransmitter; also known as serotonin.

hyperpolarization   An increase in magnitude of the membrane potential from the resting value, tending to reduce excitability.

impulse   See **action potential**.

inactivation   Reduction in conductance of a voltage-activated channel even though the activating voltage change is maintained.

inhibition   A process tending to prevent a nerve or muscle cell from initiating action potentials.

    postsynaptic inhibition   Mediated through a permeability change in the postsynaptic cell, holding the membrane potential away from threshold.

    presynaptic inhibition   Mediated by an inhibitory fiber ending on an excitatory terminal, reducing the release of transmitter.

inhibitory postsynaptic potential (ipsp)   The potential change (usually hyperpolarizing) in a nerve or muscle cell produced by an inhibitory transmitter released from presynaptic terminals.

initial segment   The region of an axon close to the cell body; often the site of impulse initiation. See **axon hillock**.

inositol-1,4,5-triphosphate (IP$_3$)   An intracellular second messenger liberated by phospholipase C-catalyzed hydrolysis of phosphatidyl inositol. IP$_3$ triggers the release of calcium from intracellular stores.

input resistance (r_{input})   The resistance measured by injecting current into a cell or fiber; in a cylindrical fiber, $r_{input} = 0.5\sqrt{r_m r_i}$.

in situ hybridization   A technique for visualizing the distribution of mRNA for a particular protein by labeling the tissue with an antisense oligonucleotide probe.

integration   The process whereby a neuron sums the various excitatory and inhibitory influences converging upon it and synthesizes a new output signal.

intercellular clefts   Narrow, fluid-filled spaces between membranes of adjacent cells; usually about 20 nm wide.

interneuron   A neuron that is neither purely sensory nor purely motor but connects other neurons.

internode   The myelinated portion of an axon lying between two nodes of Ranvier.

intrafusal fiber   Muscle fiber within a muscle spindle; its contraction initiates or modulates sensory discharge.

invertebrate   Any member of the animal kingdom that does not have vertebrae (a "backbone").

in vitro   Literally, "in glass;" refers to a biological process studied outside an intact living organism.

in vivo   Literally, "in life;" refers to a biological process studied within an intact living organism.

ionophoresis   Ejection of ions by passing current through a micropipette; used for applying charged molecules with a high degree of temporal and spatial resolution. Can also be spelled **iontophoresis**.

ipsilateral   On the same side of the body. Compare with **contralateral**.

isotypes   Gene products of the same family, but with variations in amino acid sequence (e.g., voltage-activated sodium channels from brain and from muscle).

laminin   A prominent extracellular matrix glycoprotein; promotes neurite outgrowth in vitro.

lateral geniculate nucleus   Small, knee-shaped nucleus acting as a relay in the visual pathway; part of posteroinferior aspect of the thalamus.

length constant (λ)   The distance (usually in millimeters) over which a localized graded potential decreases to $1/e$ of its original size in an axon or a muscle fiber. $\lambda = \sqrt{r_m/r_i}$.

long-term potentiation (LTP)   An increase in size of a synaptic potential lasting hours or more; produced by previous synaptic activation.

magnetic resonance imaging (MRI)   A visualization technique that provides high-resolution pictures of brain structures. Formerly known as nuclear magnetic resonance imaging.

magnocellular pathways   In the visual system, large retinal ganglion cells and lateral geniculate cells that project to discrete cortical areas; particularly sensitive to movement and small changes in contrast.

Mauthner cell   A large nerve cell—up to 1 mm in length—in the mesencephalon of fishes and amphibians.

Meissner's corpuscle   A rapidly adapting mechanoreceptor in superficial skin.

Merkel's disc   A slowly adapting mechanoreceptor in superficial skin.

microglia   Wandering, macrophage-like cells in the central nervous system that accumulate at sites of injury and scavenge debris.

microtubule   A component of the cytoskeleton prominent in axons; formed by polymerization of tubulin monomers.

miniature end plate potential (mepp)   A small postsynaptic depolarization of a muscle fiber, caused by the spontaneous release of a single quantum of transmitter from the presynaptic nerve terminal. See **quantal release**.

modality   Class of sensation (e.g., touch, vision, olfaction).

monoclonal antibody   An antibody molecule raised from a clone of transformed lymphocytes.

monovalent   Having an electric charge of +1 or −1.

monosynaptic   A direct pathway from one neuron to the next, involving only one synapse.

motoneuron (motor neuron)   A neuron that innervates muscle fibers.

α-motoneuron   Supplies extrafusal muscle fibers.

γ-motoneuron   Supplies intrafusal muscle fibers.

motor unit   A single motoneuron and the muscle fibers it innervates.

MRI   See **magnetic resonance imaging**.

mRNA (messenger RNA)   Polymer of ribonucleic acids transcribed from DNA that serves as a template for protein synthesis.

multimeric   Composed of more than one polypeptide subunit; e.g., the pentameric acetylcholine receptor (five polypeptides).

   homomultimeric   Composed of identical subunits.

   heteromultimeric   Composed of nonidentical subunits.

muscarinic   See **acetylcholine receptor**.

muscle spindle   Fusiform (spindle-shaped) structure in skeletal muscles containing small muscle fibers and sensory receptors activated by stretch.

mutagenesis   Alteration of a gene to produce a product different from that of the standard, or "wild-type," gene.

myelin   Fused membranes of Schwann cells or glial cells forming a high-resistance sheath surrounding an axon.

myoblast   A dividing cell, the progeny of which develop into muscle cells.

myotube   A developing muscle fiber formed by the fusion of myoblasts.

naloxone   An agent that blocks opiate receptors specifically.

neostigmine   An anticholinesterase; also known as prostigmine.

neurite   Any neuronal process (axon or dendrite); typically used to refer to the processes of neurons in cell culture.

neuroblast   A dividing cell, the progeny of which develop into neurons.

neuroglia   Non-neuronal satellite cells associated with neurons. In the mammalian central nervous system the main groupings are astrocytes and oligodendrocytes; in peripheral nerves the satellite cells are called Schwann cells.

neuromodulator   A substance liberated from a neuron that modifies the efficacy of synaptic transmission.

neuropil   A network of axons, dendrites, and synapses.

neurotransmitter   See **transmitter**.

nicotinic   See **acetylcholine receptor**.

nociceptive   Responding to noxious (tissue-damaging or painful) stimulation.

node of Ranvier   Localized area devoid of myelin occurring at intervals along a myelinated axon.

noise   Fluctuations in membrane potential or current due to random opening and closing of ion channels.

norepinephrine (noradrenaline)   Transmitter liberated by most sympathetic nerve terminals.

ocular dominance   The greater effectiveness of one eye over the other for driving simple or complex cells in the visual cortex.

Ohm's law   Relates current (I) to voltage (V) and resistance (R); $I = V/R$.

oligodendroglia   Class of vertebrate CNS glial cells that form myelin.

opiate   Term denoting products derived from the juice of opium poppy seed capsules.

opioid   Any directly acting compound whose actions are similar to those of opiates and are specifically antagonized by naloxone.

optic chiasm   The point of crossing, or decussation, of the optic nerves. In cats and primates, fibers arising from the medial part of the retina cross to supply the opposite lateral geniculate nucleus.

ouabain G-strophanthidin, a glycoside that specifically blocks sodium–potassium ATPase.

oval window The membraneous partition between the scala vestibuli and the middle ear that receives auditory vibrations from the eardrum through bony interconnections. Compare with **round window**.

overshoot Reversal of the membrane potential during the peak of the action potential.

Pacinian corpuscle Rapidly adapting mechanoreceptor sensitive to vibration; found in deep skin and other tissues.

parasympathetic nervous system The cranial and sacral divisions of the autonomic nervous system.

parvocellular pathways In the visual system, small retinal ganglion cells and lateral geniculate cells that project to discrete areas of the visual cortex; concerned with color detection and fine discrimination.

patch clamp A technique whereby a small patch of membrane is sealed to the tip of a micropipette, enabling currents through single membrane channels to be recorded.

permeability Property of a membrane or channel allowing substances to pass into or out of the cell.

PET scan See **positron emission tomography**.

phagocytosis Endocytosis and the degradation of foreign or degenerating material.

phenotype The physical characteristics of an animal or cell.

phoneme A basic sound element of speech.

phospholipase C; phospholipase A2 Enzymes that hydrolyze phospholipids.

phosphorylation The covalent addition of one or more phosphate ions to a molecule, for example to a channel protein.

polar A molecule with separate positively charged and negatively charged regions. See **hydrophilic**.

polysynaptic A pathway involving a series of synaptic connections.

positron emission tomography (PET) A technique for mapping active areas of the brain. Glucose is labeled with isotopes that emit positrons; sites of glucose uptake are then located by detecting positron emissions.

postmitotic A cell that is no longer capable of undergoing cell division.

posttetanic potentiation An increase in transmitter release from nerve terminals following a train of repetitive stimuli.

protease An enzyme that hydrolyzes protein molecules. Can also be spelled **proteinase**.

protein kinase An enzyme that phosphorylates proteins.

protein phosphatase An enzyme that cleaves phosphate residues from proteins.

pump An active transport mechanism.

pyramidal cell Any neuron with a long apical dendrite and shorter basal dendrites; this morphology is characteristic of many cortical neurons.

quantal release Secretion of multimolecular packets (quanta) of transmitter by the presynaptic nerve terminal.

quantal size The number of molecules of neurotransmitter in a quantum.

quantum content The number of quanta in a synaptic response.

receptive field The area of the periphery whose stimulation influences the firing of a neuron. For cells in the visual pathway, the receptive field refers to an area on the retina whose illumination influences the activity of a neuron.

receptor   1. A nerve terminal or accessory cell, associated with sensory transduction. 2. A molecule in the cell membrane that combines with a specific chemical substance.

directly coupled receptors   Molecules that, when activated, form ion channels that span the membrane.

indirectly coupled receptors   Molecules that activate G proteins, which, in turn, modify the activity of channels or pumps either directly or through a second messenger pathway.

receptor potential   Graded, localized potential change in a sensory receptor initiated by the appropriate stimulus; the electrical manifestation of the transduction process.

reciprocal innervation   Interconnections of neurons arranged so that pathways that excite one group of muscles inhibit the antagonistic motoneurons.

rectification   The property of a membrane, or membrane channel, that allows it to conduct ionic current more readily in one direction than in the other.

reflex   Involuntary movement or other response elicited by activation of sensory receptors and involving conduction through one or more synapses in the central nervous system.

refractory period   The time following an impulse during which a stimulus cannot elicit a second impulse.

resistance of the membrane (R_m)   Property of the cell membrane reflecting the difficulty ions encounter in moving across it. Inverse of conductance.

resting potential   The steady electrical potential across a membrane in its quiescent state.

retinotectal   Refers to the projection of retinal ganglion cells to neurons in the optic tectum.

retrograde   In the direction from the axon terminal toward the cell body. Compare with **anterograde**.

reversal potential   The value of the membrane potential at which a chemical transmitter produces no change in potential.

Ringer's fluid   A saline solution containing sodium chloride, potassium chloride, and calcium chloride; named after Sidney Ringer.

round window   The membranous partition between the scala tympani and the middle ear. Compare with **oval window**.

Ruffini's capsule   Slowly adapting mechanoreceptor found deep in skin.

saccule   The part of the vestibular apparatus that responds to vertical acceleration of the head.

saltatory conduction   Conduction along a myelinated axon whereby the leading edge of the propagating action potential leaps from node to node.

scala tympani; scala media; scala vestibuli   Fluid-filled compartments of the cochlea.

Schwann cell   Satellite cell in the peripheral nervous system; responsible for making the myelin sheath.

second messenger   A molecule forming part of a second messenger system.

second messenger system   A series of molecular reactions inside a cell initiated by occupation of extracellular receptor sites and leading to a functional response, such as opening or closing of membrane channels.

semicircular canal   Fluid-filled loop in the vestibular apparatus associated with detection of head rotation.

serotonin   Another name for 5-hydroxytryptamine or 5-HT; a neurotransmitter.

siemens (S) Unit of conductance; the reciprocal of ohm.

size principle The orderly recruitment of motor units of increasing size as the strength of a muscle contraction increases.

soma The cell body.

somatotopic Organized in an orderly manner in relation to an outline of the body, or a part of the body.

stereocilia Specialized microvilli of graded length projecting from the apical surface of a hair cell.

striate cortex Also known as area 17 or visual I; primary visual region of occipital lobe marked by striation of Gennari, visible with the naked eye.

subarachnoid space Space filled by cerebrospinal fluid between the arachnoid and pia, two layers of connective tissue (meninges) surrounding the brain.

subunit The basic structural building block of a multimeric protein such as a membrane channel; usually a subunit is a single polypeptide.

summation The additive action of synaptic potentials.

> **temporal summation** Addition in sequence, with each potential adding to the one preceding it.

> **spatial summation** Addition of potentials arising in different parts of a cell, for example, of potentials spreading to the axon hillock from different branches of a dendritic tree.

superfamily A group of gene product families with similar structure and function (e.g., ligand-activated ion channels). See **family**.

supersensitivity An increase in responsiveness of target cells to chemical transmitters seen, for example, after denervation.

sympathetic nervous system The thoracic and lumbar divisions of the autonomic nervous system.

synapse Site at which neurons make functional contact; a term coined by Sherrington.

synaptic cleft The space between the membranes of the pre- and postsynaptic cells at a chemical synapse across which transmitter diffuses.

synaptic delay The time between a presynaptic nerve impulse and a postsynaptic response.

synaptic vesicles Small, membrane-bounded sacs contained in presynaptic nerve terminals. Those that appear in electron micrographs to have dense cores contain catecholamines and serotonin; clear vesicles are the storage sites for other transmitters.

tectorial membrane Membrane in the cochlea in which hair cell cilia are imbedded.

tetanus A train of action potentials; also a sustained muscular contraction produced by such a train.

tetraethylammonium (TEA) Quarternary ammonium compound that selectively blocks certain voltage-activated potassium channels in neurons and muscle fibers.

tetrodotoxin (TTX) Toxin from puffer fish that selectively blocks the voltage-activated sodium channel in neurons and muscle fibers.

threshold 1. The critical value of membrane potential or depolarization at which an impulse is initiated. 2. The minimal stimulus required for a sensation.

tight junction Site at which fusion occurs between the outer leaflets of membranes of adjacent cells, resulting in a five-layered junction. It is called a

macula occludens if the area is a spot and a **zonula occludens** if the junction is a circumferential ring. Such complete junctions prevent the movement of substances through the extracellular space between the cells.

time constant (τ)   A measure of the rate of buildup or decay of a localized graded potential; equal numerically to the product of the resistance and the capacitance of the membrane.

transcription   Synthesis of mRNA using DNA as a template.

transducer   Device for converting one form of energy into another (e.g., a microphone, photoelectric cell, loudspeaker, or light bulb).

translation   Synthesis of protein using mRNA as a template.

transmitter   Chemical substance liberated by a presynaptic nerve terminal causing an effect on the membrane of the postsynaptic cell, usually an increase in permeability to one or more ions.

tuning   The property of a receptor (e.g., a cochlear hair cell) that restricts its response to a specified frequency range.

undershoot   Transient hyperpolarization following an action potential; caused by increased potassium conductance.

utricle   Part of the vestibular apparatus that responds to horizontal acceleration of the head.

varicosity   Swelling along an axon from which transmitter is liberated.

ventricles   Cavities within the brain containing cerebrospinal fluid and lined by ependymal cells.

ventricular zone   Region adjacent to the lumen of the neural tube (future ventricle) in the developing vertebrate neuroepithelium where cell proliferation occurs.

vertebrate   An animal with a backbone (vertebrae).

voltage clamp   Technique for displacing membrane potential abruptly to a desired value and keeping the potential constant while measuring currents across the cell membrane; devised by Cole and Marmont.

white matter   Part of the central nervous system appearing white; consisting of myelinated fiber tracts. Compare with **gray matter**.

whole-cell recording   Recording of membrane currents in an intact cell with a patch clamp electrode, through an opening in the cell membrane.

BIBLIOGRAPHY

THE NUMBERS THAT FOLLOW EACH ENTRY IDENTIFY THE CHAPTER(S) IN WHICH THE REFERENCE IS CITED.

Abbott, N. J., Lane, N. J. and Bundgaard, M. 1986. The blood–brain interface in invertebrates. *Ann. NY Acad. Sci.* 481: 20–42. [6]

Abbott, N. J., Liebe, E. M. and Raff, M. (eds.). 1992. *Glial–Neuronal Interaction. Ann. NY Acad. Sci.* 633. [6]

Acklin, S. E. 1988. Electrical properties and anion permeability of doubly rectifying junctions in the leech central nervous system. *J. Exp. Biol.* 137: 1–11. [7]

Adams, P. R. and Brown, D. A. 1980. Luteinizing hormone-releasing factor and muscarinic agonists act on the same voltage-sensitive K^+-current in bullfrog sympathetic neurones. *Br. J. Pharmacol.* 68: 353–355. [8]

Adams, P. R., Brown, D. A. and Constanti, A. 1982a. Pharmacological inhibition of the M-current. *J. Physiol.* 332: 223–262. [8]

Adams, P. R., Brown, D. A. and Constanti, A. 1982b. M-currents and other potassium currents in bullfrog sympathetic neurones. *J. Physiol.* 330: 537–572. [8]

Adams, R. D. and Victor, M. 1989. *Principles of Neurology*, 4th Ed. McGraw-Hill, New York. [19]

Adams, W. B. and Levitan, I. B. 1985. Voltage and ion dependences of the slow currents which mediate bursting in *Aplysia* neurone R15. *J. Physiol.* 360: 69–93. [13]

Adler, E. M., Augustine, G. J., Duffy, S. N. and Charlton, M. P. 1991. Alien intracellular calcium chelators attenuate neurotransmitter release at the squid giant synapse. *J. Neurosci.* 11: 1496–1507. [7]

Adrian, E. D. 1946. *The Physical Background of Perception*. Clarendon Press, Oxford. [1, 14, 15, 19]

Adrian, E. D. 1959. *The Mechanism of Nervous Action*. University of Pennsylvania Press, Philadelphia. [15]

Adrian, E. D. and Zotterman, Y. 1926. The impulses produced by sensory nerve endings. Part II. The response of a single end organ. *J. Physiol.* 61: 151–171. [14]

Aguayo, A. J., Bray, G. M. and Perkins, S. C. 1979. Axon–Schwann cell relationships in neuropathies of mutant mice. *Ann. NY Acad. Sci.* 317: 512–531. [6]

Aguayo, A. J., Charron, L. and Bray, G. M. 1976. Potential of Schwann cells from unmyelinated nerves to produce myelin: A quantitative ultrastructural and radiographic study. *J. Neurocytol.* 5: 565–573. [6]

Aguayo, A. J., Dickson, R., Trecarten, J. Attiwell, M., Bray, G. M. and Richardson, P. 1978. Ensheathment and myelination of regenerating PNS fibres by transplanted optic nerve glia. *Neurosci. Lett.* 9: 97–104. [12]

Aguayo, A. J., Bray, G. M., Rasminsky, M., Zwimpfer, T., Carter, D. and Vidal-Sanz, M. 1990. Synaptic connections made by axons regenerating in the central nervous system of adult mammals. *J. Exp. Biol.* 153: 199–224. [6, 12]

Aguayo, A. J., Rasminsky, M., Bray, G. M., Carbonetto, S., McKerracher, L., Villegas-Perez, M. P., Vidal-Sanz, M. and Carter, D. A. 1991. Degenerative and regenerative responses of injured neurons in the central nervous system of adult mammals. *Philos. Trans. R. Soc. Lond. B* 331: 337–343. [12]

Akam, M. 1989. Hox and HOM: Homologous gene clusters in insects and vertebrates. *Cell* 57: 347–349. [11]

Akil, H., Watson, S. J., Young, E., Lewis, M. E., Khachaturian, H. and Walker, J. M. 1984. Endogenous opioids: Biology and function. *Annu. Rev. Neurosci.* 7: 223–255. [10]

Aldrich, R. W. and Stevens, C. F. 1983. Inactivation of open and closed sodium channels determined separately. *Cold Spring Harbor Symp. Quant. Biol.* 48: 147–153. [4]

Aldrich, R. W. and Stevens, C. F. 1987. Voltage-dependent gating of single sodium channels from mammalian neuroblastoma. *J. Neurosci.* 7: 418–431. [4]

Alexandrowicz, J. S. 1951. Muscle receptor organs in the abdomen of *Homarus vulgaris* and *Parlinurus vulgaris*. *Q. J. Microsc. Sci.* 92: 163–199. [14]

Alkon, D. 1987. *Memory Traces in the Brain*. Cambridge University Press, Cambridge. [13]

Alkon, D. L. 1989. Memory storage and neural systems. *Sci. Am.* 261(1): 42–50. [13]

Allen, R. D., Allen, N. S. and Travis, J. L. 1981. Video-enhanced differential interference contrast (AVEC-DIC) microscopy: A new method capable of analyzing microtubule-related movement in the reticulopodial network of *Allogromia laticollaris*. *Cell Motil.* 1: 291–302. [9]

Almers, W. and Tse, F. W. 1990. Transmitter release from synapses: Does a preassembled fusion pore initiate exocytosis? *Neuron* 4: 813–818. [9]

Almers, W., Stanfield, P. and Stühmer, W. 1983. Lateral distribution of sodium and potassium channels in frog skeletal muscle: Measurements with a patch clamp technique. *J. Physiol.* 336: 261–284. [4]

Altamirano, A. A. and Russell, J. M. 1987. Coupled Na/K/Cl efflux. "Reverse" unidirectional fluxes in squid giant axons. *J. Gen. Physiol.* 89: 669–686. [3]

Anderson, C. R. and Stevens, C. F. 1973. Voltage clamp analysis of acetylcholine-produced end-plate current fluctuations at frog neuromuscular junction. *J. Physiol.* 235: 655–691. [2,7]

Anderson, H., Edwards, J. S. and Palka, J. 1980. Developmental neurobiology of invertebrates. *Annu. Rev. Neurosci.* 3: 97–139. [12]

Anderson, M. J. and Cohen, M. W. 1977. Nerve-induced and spontaneous redistribution of acetylcholine receptors on cultured muscle cells. *J. Physiol.* 268: 757–773. [11]

Angeletti, R. H., Hermodson, M. A. and Bradshaw, R. A. 1973. Amino acid sequences of mouse 2.5S nerve growth factor. II. Isolation and characterization of the thermolytic and peptic peptides and the complete covalent structure. *Biochemistry* 12: 100–115. [11]

Angeletti, R. H., Mercanti, D. and Bradshaw, R. A. 1973. Amino acid sequences of mouse 2.5S nerve growth factor. I. Isolation and characterization of the soluble tryptic and chymotryptic peptides. *Biochemistry* 12: 90–100. [11]

Angevine, J. B. and Sidman, R. L. 1961. Autoradiographic study of cell migration during histogenesis of cerebral cortex in the mouse. *Nature* 192: 766–768. [11]

Anglister, L. and McMahan, U. J. 1985. Basal lamina directs acetylcholinesterase accumulation at synaptic sites in regenerating muscle. *J. Cell Biol.* 101: 735–743. [12]

Anglister, L., Faber, I. C., Sharar, C. A. and Grinvald, A. 1982. Location of voltage-sensitive calcium channels along developing neurites: Their possible role in regulation of neurite elongation. *Dev. Biol.* 94: 351–365. [11]

Arbas, E. A. and Calabrese, R. L. 1987. Slow oscillations of membrane potential in interneurons that control heartbeat in the medicinal leech. *J. Neurosci.* 7: 3953–3960. [13]

Arbuthnott, E. R., Boyd, I. A. and Kalu, K. U. 1980. Ultrastructural dimensions of myelinated peripheral nerve fibres in the cat and their relation to conduction velocity. *J. Physiol.* 308: 125–157. [5]

Arbuthnott, E. R., Gladden, M. H. and Sutherland, F. I. 1989. The selectivity of fusimotor innervation in muscle spindles of the rat studied by light microscopy. *J. Anat.* 163: 183–190. [14]

Arbuthnott, E. R., Ballard, K. J., Boyd, I. A., Gladden, M. H. and Sutherland, F. I. 1982. The ultrastructure of cat fusimotor endings and their relationship to foci of sarcomere convergence in intrafusal fibres. *J. Physiol.* 331: 285–309. [14]

Arch, S. and Berry, R. W. 1989. Molecular and cellular regulation of neuropeptide expression: The bag cell model system. *Brain Res. Rev.* 14: 181–201. [13]

Aréchiga, H., Cortés, J. L., Garcia, U. and Rodríguez-Sosa, L. 1985. Neuroendocrine correlates of circadian rhythmicity in crustaceans. *Am. Zool.* 25: 265–274. [13]

Armstrong, C. M. 1981. Sodium channels and gating currents. *Physiol. Rev.* 61: 644–683. [4]

Armstrong, C. M. and Bezanilla, F. 1974. Charge movement associated with the opening and closing of the activation gates of Na channels. *J. Gen. Physiol.* 63: 533–552. [4]

Armstrong, C. M. and Bezanilla, F. 1977. Inactivation of the sodium channel II. Gating current experiments. *J. Gen. Physiol.* 70: 567–590. [4]

Armstrong, C. M. and Gilly, W. F. 1979. Fast and slow steps in the activation of sodium channels. *J. Gen. Physiol.* 74: 691–711. [4]

Armstrong, C. M. and Hille, B. 1972. The inner quaternary ammonium ion receptor in potassium channels of the node of Ranvier. *J. Gen. Physiol.* 59: 388–400. [4]

Armstrong, C. M., Bezanilla, F. and Rojas, E. 1973. Destruction of sodium conductance inactivation in squid axons perfused with pronase. *J. Gen. Physiol.* 62: 375–391. [4]

Art, J. J. and Fettiplace, R. 1987. Variation of membrane properties in hair cells isolated from the turtle cochlea. *J. Physiol.* 385: 207–242. [14]

Art, J. J., Fettiplace, R. and Fuchs, P. A. 1984. Synaptic hyperpolarization and inhibition of turtle cochlear hair cells. *J. Physiol.* 356: 525–550. [14]

Art, J. J., Crawford, A. C., Fettiplace, R. and Fuchs, P. A. 1985. Efferent modulation of hair cell tuning in the cochlea of the turtle. *J. Physiol.* 360: 397–421. [14]

Arthur, R. M., Pfeiffer, R. R. and Suga, N. 1971. Properties of 'two-tone inhibition' in primary auditory neurones. *J. Physiol.* 212: 593–609. [14]

Arvanitaki, A. and Cardot, H. 1941. Les caracteristiques de l'activité rythmique ganglionnaire "spontanée" chez l'Aplysie. *C. R. Soc. Biol.* 135: 1207–1211. [13]

Asanuma, H. 1989. *The Motor Cortex.* Raven, New York. [15]

Asanuma, H., Larsen, K. D. and Yumiya, H. 1980. Peripheral input pathways to the monkey motor cortex. *Exp. Brain Res.* 38: 349–355. [15]

Ascher, P., Nowak, L. and Kehoe, J. 1986. Glutamate-activated channels in molluscan and vertebrate neurons. *In* J. M. Ritchie, R. D. Keynes and L. Bolis (eds.). *Ion Channels in Neural Membranes.* Alan R. Liss, New York, pp. 283–295. [7]

Ashmore, J. F. 1987. A fast motile response in guinea-pig outer hair cells: The cellular basis of the cochlear amplifier. *J. Physiol.* 388: 323–347. [14]

Astion, M. L., Obaid, A. L. and Orkand, R. K. 1989. Effects of barium and bicarbonate on glial cells of *Necturus* optic nerve: Studies with microelectrodes and voltage-sensitive dyes. *J. Gen. Physiol.* 93: 731–744. [6]

Atwood, H. L. and Morin, W. A. 1970. Neuromuscular and axoaxonal synapses of the crayfish opener muscle. *J. Ultrastruct. Res.* 32: 351–369. [7]

Atwood, H. L. and Sandeman, D. C. (eds.). 1982. Synapses and neurotransmitters. *In* H. L. Atwood (ed.), *Biology of Crustacea*, Vol. 3. Academic Press, New York. [13]

Atwood, H. L., Dudel, J., Feinstein, N. and Parnas, I. 1989. Long–term survival of decentralized axons and incorporation of satellite cells in motor neurons of rock lobsters. *Neurosci. Lett.* 101: 121–126. [6]

Ausoni, S., Gorza, L., Schiaffino, S., Gundersen, K. and Lømo, T. 1990. Expression of myosin heavy-chain isoforms in stimulated fast and slow rat muscles. *J. Neurosci.* 10: 153–160. [12]

Avenet, P. and Lindemann, B. 1987a. Action potentials in epithelial taste receptors induced by mucosal calcium. *J. Memb. Biol.* 95: 265–269. [14]

Avenet, P. and Lindemann, B. 1987b. Patch-clamp study of isolated taste receptor cells of the frog. *J. Membr. Biol.* 97: 223–240. [14]

Avenet, P. and Lindemann, B. 1988. Amiloride-blockable sodium currents in isolated taste receptor cells. *J. Memb. Biol.* 105: 245–255. [14]

Avenet, P., Hofmann, F. and Lindemann, B. 1988. K-channels in taste receptor cells closed by cAMP-dependent protein kinase. *Nature* 331: 351–354. [14]

Axelrod, J. 1971. Noradrenaline: Fate and control of its biosynthesis. *Science* 173: 598–606. [9]

Axelsson, J. and Thesleff, S. 1959. A study of supersensitivity in denervated mammalian skeletal mucle. *J. Physiol.* 147: 178–193. [12]

Bailey, C. H. and Chen, M. 1988. Morphological basis of short-term habituation in *Aplysia*. *J. Neurosci.* 8: 2452–2459. [13]

Bailey, C. H. and Chen, M. 1989. Structural plasticity at identified synapses during long-term memory in *Aplysia*. *J. Neurobiol.* 20: 356–372. [13]

Bainton, C. R., Kirkwood, P. A. and Sears, T. A. 1978. On the transmission of the stimulating effects of carbon dioxide to the muscles of respiration. *J. Physiol.* 280: 249–272. [15]

Baker, F. H., Grigg, P. and van Noorden, G. K. 1974. Effects of visual deprivation and strabismus on the response of neurons in the visual cortex of the monkey, including studies on the striate and prestriate cortex in the normal animal. *Brain Res.* 66: 185–208. [18]

Baker, P. F., Hodgkin, A. L. and Ridgeway, E. B. 1971. Depolarization and calcium entry in squid giant axons. *J. Physiol.* 218: 709–755. [3,4]

Baker, P. F., Hodgkin, A. L. and Shaw, T. I. 1962. Replacement of the axoplasm of giant nerve fibres with artificial solutions. *J. Physiol.* 164: 330–354. [3]

Baker, P. F., Blaustein, M. P., Keynes, R. D., Manil, J., Shaw, T. I. and Steinhardt, R. A., 1969. The ouabain-sensitive fluxes of sodium and potassium in squid giant axons. *J. Physiol.* 200: 459–496. [3]

Baldessarini, R. J. and Tarsy, D. 1980. Dopamine and the pathophysiology of dyskinesias induced by antipsychotic drugs. *Annu. Rev. Neurosci.* 3: 23–41. [10]

Balice-Gordon, R. J. and Lichtman, J. W. 1990a. In vivo visualization of the growth of pre- and postsynaptic elements of mouse neuromuscular junctions. *J. Neurosci.* 10: 894–908. [11]

Balice-Gordon, R. J. and Lichtman, J. W. 1990b. Loss of synaptic sites during competitive synapse elimination is both rapid and saltatory. *Soc. Neurosci. Abstr.* 16: 456. [11]

Baptista, C. A. and Macagno, E. R. 1988. The role of the sexual organs in the generation of postembryonic neurons in the leech *Hirudo medicinalis*. *J. Neurobiol.* 19: 707–726. [13]

Barchi, R. L. 1983. Protein components of the purified sodium channel from rat skeletal muscle sarcolemma. *J. Neurochem.* 40: 1377–1385. [2]

Barde, Y.-A. 1989. Trophic factors and neuronal survival. *Neuron* 2: 1525–1534. [11]

Barker, D. 1962. *Muscle Receptors*. Hong Kong University Press, Hong Kong. [14]

Barker, D., Emonet-Dénand, F., Laporte, Y. and Stacey, M. J. 1980. Identification of the intrafusal endings of skeletofusimotor axons in the cat. *Brain Res.* 185: 227–237. [14]

Barlow, H. B. 1953. Summation and inhibition in the frog's retina. *J. Physiol.* 119: 69–88. [16]

Barlow, H. B., Hill, R. M. and Levick, W. R. 1964. Retinal ganglion cells responding selectively to direction and speed of image motion in the rabbit. *J. Physiol.* 173: 377–407. [16]

Barres, B. A., Chun, L. L. and Corey, D. P. 1990. Ion channels in vertebrate glia. *Annu. Rev. Neurosci.* 13: 441–471. [6]

Barrionuevo, G. and Brown, T. H. 1983. Associative long-term potentiation in hippocampal slices. *Proc. Natl. Acad. Sci. USA* 80: 7347–7351. [10]

Basbaum, A. I. 1985. Functional analysis of the cytochemistry of the spinal dorsal horn. *In* H. L. Fields, R. Dubner and F. Cervero (eds.). *Advances in Pain Research and Therapy*, Vol.9. Raven Press, New York. [14]

Basbaum, A. I. and Fields, H. L. 1979. The origin of descending pathways in the dorsolateral funiculus of the spinal cord of the cat and rat: Further studies on the anatomy of pain modulation. *J. Comp. Neurol.* 187: 513–522. [14]

Baux, G., Fossier, P. and Tauc, L. 1990. Histamine and FLRFamide regulate acetylcholine release at an identified synapse in *Aplysia* in opposite ways. *J. Physiol.* 429: 147–168. [13]

Baxter, D. A. and Byrne, J. H. 1990. Differential effects of cAMP and serotonin on membrane current, action potential duration, and excitability of pleural sensory neurons of *Aplysia*. *J. Neurophysiol.* 64: 978–990.

Bayliss, W. M. and Starling, E. H. 1902. The mechanism of pancreatic secretion. *J. Physiol.* 28: 325–353. [10]

Baylor, D. A. 1987. Photoreceptor signals and vision. Proctor lecture. *Invest. Ophthalmol. Vis. Sci.* 28: 34–49. [16]

Baylor, D. A. and Fettiplace, R. 1977. Transmission from photoreceptors to ganglion cells in turtle retina. *J. Physiol.* 271: 391–424. [16]

Baylor, D. A. and Fuortes, M. G. F. 1970. Electrical responses of single cones in the retina of the turtle. *J. Physiol.* 207: 77–92. [16]

Baylor, D. A. and Nicholls, J. G. 1969a. Changes in extracellular potassium concentration produced by neuronal activity in the central nervous system of the leech. *J. Physiol.* 203: 555–569. [6]

Baylor, D. A. and Nicholls, J. G. 1969b. Chemical and electrical synaptic connexions between cutaneous mechanoreceptor neurones in the central nervous system of the leech. *J. Physiol.* 203: 591–609. [7]

Baylor, D. A., Fuortes, M. G. F. and O'Bryan, P. M. 1971. Receptive fields of cones in the retina of the turtle. *J. Physiol.* 214: 265–294. [16]

Baylor, D. A., Lamb, T. D. and Yau, K.-W. 1979. The membrane current of single rod outer segments. *J. Physiol.* 288: 589–611. [16]

Baylor, D. A., Nunn, B. J. and Schnapf, J. L. 1984. The photocurrent, noise and spectral sensitivity of rods of the monkey *Macaca fascicularis*. *J. Physiol.* 357: 575–607. [16]

Baylor, D. A., Nunn, B. J. and Schnapf, J. L. 1987. Spectral sensitivity of cones of the monkey *Macaca fascicularis*. *J. Physiol.* 390: 145–160. [16]

Beadle, D. J., Lees, G. and Kater, S. B. 1988. *Cell Culture Approaches to Invertebrate Neuroscience*. Academic Press, London. [13]

Beam, K. G., Caldwell, J. H. and Campbell, D. T. 1985. Na channels in skeletal muscle concentrated near the neuromuscular junction. *Nature* 313: 588–590. [4]

Bean, B. P. 1981. Sodium channel inactivation in the crayfish giant axon. Must a channel open before inactivating? *Biophys. J.* 35: 595–614. [4]

Bear, M. F., Kleinschmidt, A., Gu, Q. A. and Singer, W. 1990. Disruption of experience-dependent synaptic modifications in striate cortex by infusion of an NMDA receptor antagonist. *J. Neurosci.* 10: 909–925. [18]

Bekkers, J. M. and Stevens, C. F. 1990. Presynaptic mechanism for long-term potentiation in the hippocampus. *Nature* 346: 724–729. [10]

Belardetti, F. and Siegelbaum, S. A. 1988. Up- and down-modulation of single K^+ channel function by distinct second messengers. *Trends Neurosci.* 11: 232–238. [13]

Belardetti, F., Kandel, E. R. and Siegelbaum, S. A. 1987. Neuronal inhibition by the peptide FMRFamide involves opening of S K^+ channels. *Nature* 325: 153–156. [8]

Belmar, J. and Eyzaguirre, C. 1966. Pacemaker site of fibrillation potentials in denervated mammalian muscle. *J. Neurophysiol.* 29: 425–441. [12]

Bennett, M. V. L. 1973. Function of electrotonic junctions in embryonic and adult tissues. *Fed. Proc.* 32: 65–75. [7]

Bennett, M. V. L. 1974. Flexibility and rigidity in electrotonically coupled systems. *In* M. V. L. Bennett (ed.). *Synaptic Transmission and Neuronal Interactions*. Raven Press, New York, pp. 153–158. [7]

Bennett, M. V. L., Barrio, L. C., Bargiello, T. A., Spray, D. C., Herzberg, E. and Saez, J. C. 1991. Gap junctions: New tools, new answers, new questions. *Neuron* 6: 305–320. [5]

Bentley, D. and Caudy, M. 1983. Navigational substrates for peripheral pioneer growth cones: Limb–axis polarity cues, limb segment boundaries, and guidepost neurons. *Cold Spring Harbor Symp. Quant. Biol.* 48: 573–585. [11]

Bentley, D. and Toroian-Raymond, A. 1989. Pre-axonogenesis migration of afferent pioneer cells in the grasshopper embryo. *J. Exp. Zool.* 251: 217–223. [13]

Bentley, D., Guthrie, P. B. and Kater, S. B. 1991. Calcium ion distribution in nascent pioneer axons and coupled preaxonogenesis neurons in situ. *J. Neurosci.* 11: 1300–1308. [11]

Benzer, S. 1971. From the gene to behavior. *J.A.M.A.* 218: 1015–1022. [13]

Berg, D. K. and Hall, Z. W. 1975. Increased extrajunctional acetylcholine sensitivity produced by chronic postsynaptic neuromuscular blockade. *J. Physiol.* 244: 659–676. [12]

Berlucchi, G. and Rizzolatti, G. 1968. Binocularly driven neurons in visual cortex of split-chiasm cats. *Science* 159: 308–310. [17]

Bernardi, G., Carpenter, M., Di Chiara, G., Morelli, M. and Stanzioni, P. (eds.). 1991. *The Basal Ganglia III. Proceedings of the Third Triennial Symposium of the International Basal Ganglion Society*. Plenum, New York. [15]

Berne, R. M. and Levy, M. N. (eds.). 1988. *Physiology*, 2nd ed. Mosby, St. Louis. [14]

Bernstein, J. 1902. Untersuchungen zur Thermodynamik der bioelektrischen Ströme. *Pflügers Arch.* 92: 521–562. [3]

Berridge, M. J. 1988. Inositol lipids and calcium signalling. *Proc. R. Soc. Lond. B* 234: 359–378. [8]

Betz, H. 1990. Ligand-gated ion channels in the brain: The amino acid receptor superfamily. *Neuron* 5: 383–392. [2,10]

Betz, W. J. 1987. Motoneuron death and synapse elimination in vertebrates. *In* M. M. Salpeter (ed.). *The Vertebrate Neuromuscular Junction.* Alan R. Liss, New York, pp. 117–162 [11]

Betz, W. J. and Sakmann, B. 1973. Effects of proteolytic enzyme on function and structure of frog neuromuscular junctions. *J. Physiol.* 230: 673–688. [9]

Betz, W. J., Caldwell, J. H. and Ribchester, R. R. 1980. The effects of partial denervation at birth on the development of muscle fibers and motor units in rat lumbrical muscle. *J. Physiol.* 303: 265–279. [11]

Betz, W. J., Mao, F. and Bewick, G. S. 1992. Activity-dependent fluorescent staining and destaining of living vertebrate motor nerve terminals. *J. Neurosci.* 12: 363–375. [9]

Bevan, S. and Steinbach, J. H. 1977. The distribution of α-bungarotoxin binding sites on mammalian skeletal muscle developing in vivo. *J. Physiol.* 267: 195–213. [12]

Bevan, S., Chiu, S. Y., Gray, P. T. and Ritchie, J. M. 1985. The presence of voltage-gated sodium, potassium, and chloride channels in rat cultured astrocytes. *Proc. R. Soc. Lond. B* 225: 299–313. [6]

Beyer, E. C., Paul, D. L. and Goodenough, D. A. 1987. Connexon43: A protein from rat heart homologous to a gap junction from liver. *J. Cell Biol.* 105: 2621–2629. [5]

Bialeck, W. 1987. Physical limits to sensation. *Annu. Rev. Biophys. Biophys. Chem.* 16: 455–478. [14]

Bignami, A. and Dahl, D. 1974. Astrocyte-specific protein and neuroglial differentiation: An immunofluorescence study with antibodies to the glial fibrillary acidic protein. *J. Comp. Neurol.* 153: 27–38. [6]

Binder, M. D. and Mendell, L. M. (eds.). 1990. *The Segmental Motor System.* Oxford University Press, New York. [15]

Birks, R., Katz, B. and Miledi, R. 1960. Physiological and structural changes at the amphibian myoneural junction in the course of nerve degeneration. *J. Physiol.* 150: 145–168. [7]

Birks, R. I. and MacIntosh, F. C. 1961. Acetylcholine metabolism of a sympathetic ganglion. *J. Biochem. Physiol.* 39: 787–827. [9]

Birnbaumer, L., Codina, J., Mattera, R., Yatani, A., Scherer, N., Toro, M.-J. and Brown, A. M. 1987. Signal transduction by G proteins. *Kidney Int.* 32 (Suppl. 23): S14–S37. [8]

Bittner, G. D. 1991. Long-term survival of anucleate axons and its implications for nerve regeneration. *Trends Neurosci.* 14: 188–193. [12]

Bixby, J. L. and Van Essen, D. C. 1979. Competition between foreign and original nerves in adult mammalian skeletal muscle. *Nature* 282: 726–728. [12]

Bixby, J. L., Lilien, J. and Reichardt, L. F. 1988. Identification of the major proteins that promote neuronal process outgrowth on Schwann cells in vitro. *J. Cell Biol.* 107: 353–361. [11]

Björklund, A. 1991. Neural transplantation--an experimental tool with clinical possibilities. *Trends Neurosci.* 14: 319–322. [12]

Björklund, A., Dunnett, S. B., Stenevi, U., Lewis, N. E. and Iversen, S. D. 1980. Reinnervation of the denervated striatum by substantia nigra transplants: Functional consequences as revealed by pharmacological and sensorimotor testing. *Brain Res.* 199: 307–333. [12]

Black, I. B. 1978. Regulation of autonomic development. *Annu. Rev. Neurosci.* 1: 183–214. [11]

Black, J. A. and Waxman, S. G. 1988. The perinodal astrocyte. *Glia* 1: 169–183. [6]

Black, J. A., Kocsis, J. D. and Waxman, S. G. 1990. Ion channel organization of the myelinated fiber. *Trends Neurosci.* 13: 48–54. [5,6]

Blackman, J. G. and Purves, R. D. 1969. Intracellular recordings from the ganglia of the thoracic sympathetic chain of the guinea-pig. *J. Physiol.* 203: 173–198. [7]

Blackman, J. G., Ginsborg, B. L. and Ray, C. 1963. Synaptic transmission in the sympathetic ganglion of the frog. *J. Physiol.* 167: 355–373. [7]

Blackshaw, S. E. 1981a. Sensory cells and motor neurons. *In* K. J. Muller, J. G. Nicholls and G. S. Stent (eds.). *Neurobiology of the Leech.* Cold Spring Harbor Laboratory, Cold Spring Harbor, NY, pp. 51–78. [13]

Blackshaw, S. E. 1981b. Morphology and distribution of touch cell terminals in the skin of the leech. *J. Physiol.* 320: 219–228. [13]

Blackshaw, S. E. and Thompson, S. W. 1988. Hyperpolarizing responses to stretch in sensory neurones innervating leech body wall muscle. *J. Physiol.* 396: 121–137. [13, 14]

Blackshaw, S. E., Nicholls, J. G. and Parnas, I. 1982. Expanded receptive fields of cutaneous mechanoreceptor cells after single neurone deletion in leech central nervous system. *J. Physiol.* 326: 261–268. [12, 13]

Blair, L. A. C., Levitan, E. S., Marshall, J., Dionne, V. E. and Barnard, E. A. 1988. Single subunits of the GABA$_A$ receptor form ion channels with properties of the native receptor. *Science* 242: 577–579. [10]

Blakemore, C. and Van Sluyters, R. C. 1974. Reversal of the physiological effects of monocular deprivation in kittens: Further evidence for a sensitive period. *J. Physiol.* 237: 195–216. [18]

Blasdel, G. G. 1989. Visualization of neuronal activity in monkey striate cortex. *Annu. Rev. Physiol.* 51: 561–581. [17]

Blasdel, G. G. and Lund, J. S. 1983. Termination of afferent axons in macaque striate cortex. *J. Neurosci.* 3: 1389–1413. [17]

Blauert, J. 1983. *Spatial Hearing: The Psychophysics of Human Sound Localization.* MIT Press, Cambridge. [14]

Blaustein, M. P. 1988. Calcium transport and buffering in neurons. *Trends Neurosci.* 11: 438–443. [3]

Bliss, T. V. P. and Lømo, T. 1973. Long-lasting potentiation of synaptic transmission in the dentate of the anesthetized rabbit following stimulation of the perforant path. *J. Physiol.* 232: 331–356. [10]

Bloom, F. E. and Iversen, L. L. 1971. Localizing [^{3}H]-GABA in nerve terminals of rat cerebral cortex by electron microscopic autoradiography. *Nature* 229: 628–630. [9]

Blumenfeld, H., Spira, M. E., Kandel, E. R. and Siegelbaum, S. A. 1990. Facilitatory and inhibitory transmitters modulate calcium influx during action potentials in *Aplysia* sensory neurons. *Neuron* 5: 487–499. [8]

Bogen, J. E. and Gordon, H. W. 1971. Musical tests for functional lateralization with intracarotid amobarbital. *Nature* 230: 524–525. [19]

Bolz, J. and Gilbert, C. D. 1986. Generation of end-inhibition in the visual cortex via interlaminar connections. *Nature* 320: 362–365. [17]

Bolz, J., Gilbert, C. D. and Wiesel, T. N. 1989. Pharmacological analysis of cortical circuitry. *Trends Neurosci.* 12: 292–296. [17]

Bonhoeffer, T. and Grinvald, A. 1991. Iso-orientation domains in cat visual cortex are arranged in pinwheel-like patterns. *Nature* 353: 429–431. [17]

Bonner, T. I. 1989. The molecular basis of muscarinic receptor diversity. *Trends Neurosci.* 12: 148–151. [8]

Born, D. E. and Rubel, E. W. 1988. Afferent influences on brain stem auditory nuclei of the chicken: Presynaptic action potentials regulate protein synthesis in nucleus magnocellularis neurons. *J. Neurosci.* 8: 901–919. [12]

Bostock, H. and Sears, T. A. 1978. The internodal axon membrane: Electrical excitability and continuous conduction in segmental demyelination. *J. Physiol.* 280: 273–301. [5,6]

Bostock, H., Sears, T. A. and Sherratt, R. M. 1981. The effects of 4-aminopyridine and tetraethylammonium on normal and demyelinated mammalian nerve fibres. *J. Physiol.* 313: 301–315. [5]

Bowling, D. B. and Michael, C. R. 1980. Projection patterns of single physiologically characterized optic tract fibres in cat. *Nature* 286: 899–902. [16]

Bowling, D., Nicholls, J. G. and Parnas, I. 1978. Destruction of a single cell in the central nervous system of the leech as a means of analyzing its connections and functional role. *J. Physiol.* 282: 169–180. [12, 13]

Boyd, I. A. and Martin, A. R. 1956. The end-plate potential in mammalian muscle. *J. Physiol.* 132: 74–91. [7]

Boycott, B. B. and Dowling, J. E. 1969. Organization of primate retina: Light microscopy. *Philos. Trans. R. Soc. Lond. B* 255: 109–184. [16]

Bradbury, M. 1979. *The Concept of a Blood–Brain Barrier.* Wiley, Chichester. [6]

Brady, S. T., Lasek, R. J. and Allen, R. D. 1982. Fast axonal transport in extruded axoplasm from squid giant axon. *Science* 218: 1129–1131. [9]

Brand, J. G., Teeter, J. H., Cagan, R. H. and Kare, M. R. (eds.). 1989. *Receptor Events in Transduction and Olfaction.* Volume 1 of *Chemical Senses.* Marcel Dekker, New York. [14]

Bray, D. and Hollenbeck, P. J. 1988. Growth cone motility and guidance. *Annu. Rev. Cell Biol.* 4: 43–61. [11]

Bray, G. M., Rasminsky, M. and Aguayo, A. J. 1981. Interactions between axons and their sheath cells. *Annu. Rev. Neurosci.* 4: 127–162. [6]

Bredt, D. S., Hwang, P. M., Glatt, C. E., Lowenstein, C., Reed, R. R. and Snyder, S. H. 1991. Cloned and expressed nitric oxide synthase structurally resembles cytochrome P-450 reductase. *Nature* 351: 714–718. [8]

Breitwieser, G. E. and Szabo, G. 1985. Uncoupling of cardiac muscarinic and β-adrenergic receptors from ion channels by a guanine nucleotide analogue. *Nature* 317: 538–540. [8]

Brenner, H. R. and Johnson, E. W. 1976. Physiological and morphological effects of post-ganglionic axotomy on presynaptic nerve terminals. *J. Physiol.* 260: 143–158. [12]

Brenner, H. R. and Martin, A. R. 1976. Reduction in acetylcholine sensitivity of axotomized ciliary ganglion cells. *J. Physiol.* 260: 159–175. [12]

Brew, H., Gray, P. T., Mobbs, P. and Attwell, D. 1986. Endfeet of retina glial cells have higher densities of ion channels that mediate K^+ buffering. *Nature* 324: 466–468. [6]

Bridge, J. H. B., Smolley, J. R. and Spitzer, N. W. 1990. The relationship between charge movements associated with I_{Ca} and I_{Na-Ca} in cardiac myocytes. *Science* 248: 376–378. [3]

Brigant, J. L. and Mallart, A. 1982. Presynaptic currents in mouse motor endings. *J. Physiol.* 333: 619–636. [7]

Brightman, M. W. and Reese, T. S. 1969. Junctions between intimately apposed cell membranes in the vertebrate brain. *J. Cell Biol.* 40: 668–677. [6]

Brightman, M. W., Reese, T. S. and Feder, N. 1970. Assessment with the electronmicroscope of the permeability to peroxidase of cerebral endothelium in mice and sharks. *In* C. Crone and N. A. Lassen (eds.). *Capillary Permeability. Alfred Benzon Symposium II.* Munskgaard, Copenhagen, pp. 468–476. [6]

Brodfuehrer, P. D. and Friesen, W. O. 1986. From stimulation to undulation: A neuronal pathway for the control of swimming activity in the leech. *Science* 234: 1002–1004. [13]

Bronner-Fraser, M. 1985. Alterations in neural crest migration by a monoclonal antibody that affects cell adhesion. *J. Cell Biol.* 101: 610–617. [11]

Bronner-Fraser, M. and Fraser, S. E. 1988. Cell lineage analysis reveals multipotency of some avian neural crest cells. *Nature* 335: 161–164. [11]

Brooks, V. B. 1956. An intracellular study of the action of repetitive nerve volleys and of botulinum toxin on miniature end-plate potentials. *J. Physiol.* 134: 264–277. [7]

Brown, A. G. and Fyffe, R. E. W. 1981. Direct observations on the contacts made between Ia afferent fibers and α-motoneurones in the cat's lumbosacral spinal cord. *J. Physiol.* 313: 121–140. [15]

Brown, A. M. and Birnbaumer, L. 1990. Ionic channels and their regulation by G protein subunits. *Annu. Rev. Physiol.* 52: 197–213. [8]

Brown, D. A. 1990. G-proteins and potassium currents in neurons. *Annu. Rev. Physiol.* 52: 215–242. [8]

Brown, G. L. 1937. The actions of acetylcholine on denervated mammalian and frog's muscle. *J. Physiol.* 89: 438–461. [12]

Brown, H. M., Ottoson, D. and Rydqvist, B. 1978. Crayfish stretch receptor: An investigation with voltage-clamp and ion-sensitive electrodes. *J. Physiol.* 284: 155–179. [14]

Brown, M. C. and Ironton, R. 1977. Motor neurone sprouting induced by prolonged tetrodotoxin block of nerve action potentials. *Nature* 265: 459–461. [12]

Brown, M. C., Holland, R. L. and Hopkins, W. G. 1981. Motor nerve sprouting. *Annu. Rev. Neurosci.* 4: 17–42. [12]

Brown, M. C., Hopkins, W. G. and Keynes, R. J. 1982. Short- and long-term effects of paralysis on the motor innervation of two different neonatal mouse muscles. *J. Physiol.* 329: 439–450. [11]

Brown, M. C., Jansen, J. K. S. and Van Essen, D. 1976. Polyneuronal innervation of skeletal muscle in newborn rats and its elimination during maturation. *J. Physiol.* 261: 387–422. [11]

Brown, P. K. and Wald, G. 1963. Visual pigments in human and monkey retinas. *Nature* 200: 37–43. [16]

Brown, R. O., Pulst, S. M. and Mayeri, E. 1989. Neuroendocrine bag cells of *Aplysia* are activated by bag cell peptide-containing neurons in the pleural ganglion. *J. Neurophysiol.* 61: 1142–1152. [13]

Brown, T. G. 1911. The intrinsic factor in the act of progression in the mammal. *Proc. R. Soc. Lond. B* 84: 308–319. [15]

Brown, T. H., Kairiss, E. W. and Keenan, C. L. 1990. Hebbian synapses: Biophysical mechanisms and algorithms. *Annu. Rev. Neurosci.* 13: 475–511. [10]

Buck, L. and Axel, R. 1991. A novel multigene family may encode odorant receptors: A molecular basis for odor recognition. *Cell* 65: 175–187. [14]

Bullock, T. H. and Hagiwara, S. 1957. Intracellular recording from the giant synapse of the squid. *J. Gen. Physiol.* 40: 565–577. [7]

Bundgaard, M. and Cserr, H. F. 1981. A glial blood–brain barrier in elasmobranchs. *Brain Res.* 226: 61–73. [6]

Bunge, R. P. 1968. Glial cells and the central myelin sheath. *Physiol. Rev.* 48: 197–251. [6]

Bunge, R. P., Bunge, M. B. and Bates, M. 1989. Movements of the Schwann cell nucleus implicate progression of the inner (axon-related) Schwann cell process during myelination. *J. Cell Biol.* 109: 273–284. [6]

Burden, S. J., Sargent, P. B. and McMahan, U. J. 1979. Acetylcholine receptors in regenerating muscle accumulate at original synaptic sites in the absence of the nerve. *J. Cell Biol.* 82: 412–425. [9, 12]

Burger, P. M., Mehl, E., Cameron, P. L., Maycox, P. R., Baumert, M., Lottspeich, F., De Camilli, P. and Jahn, R. 1989. Synaptic vesicles immunoisolated from rat cerebral cortex contain high levels of glutamate. *Neuron* 3: 715–720. [9]

Burger, P. M., Hell, J., Mehl, E., Krasel, C., Lottspeich, F. and Jahn, R. 1991. GABA and glycine in synaptic vesicles: Storage and transport characteristics. *Neuron* 7: 287–293. [9]

Burke, R. E. 1978. Motor units: Physiological histochemical profiles, neural connectivity and functional specialization. *Am Zool.* 18: 127–134. [15]

Burke, R. E., Walmsley, B. and Hodgson, J. A. 1979. HRP anatomy of group Ia afferent contacts on alpha motoneurones. *Brain Res.* 160: 347–352. [15]

Burnstock, G. 1990. Overview: Purinergic mechanisms. *Ann. NY Acad. Sci.* 603: 1–17. [9]

Burrows, M. and Siegler, M. V. 1985. Organization of receptive fields of spiking local interneurons in the locust with inputs from hair afferents. *J. Neurophysiol.* 53: 1147–1157. [13]

Bursztajn, S., Berman, S. A. and Gilbert, W. 1989. Differential expression of acetylcholine receptor mRNA in nuclei of cultured muscle cells. *Proc. Natl. Acad. Sci. USA* 86: 2928–2932. [12]

Buttner, N., Siegelbaum, S. A. and Volterra, A. 1989. Direct modulation of *Aplysia* S–K^+ channels by a 12-lipoxygenase metabolite of arachidonic acid. *Nature* 342: 553–555. [8]

Byrne, J. H. 1987. Cellular analysis of associative learning. *Physiol. Rev.* 67: 329–439. [13]

Byrne, J., Castellucci, V. F. and Kandel, E. R. 1974. Receptive fields and response properties of mechanoreceptor neurons innervating siphon skin and mantle shelf in *Aplysia*. *J. Neurophysiol.* 37: 1041–1064. [13]

Caldwell, J. H. and Daw, N. W. 1978. Effects of picro-toxin and strychnine on rabbit retinal ganglion cells: Changes in centre surround receptive fields. *J. Physiol.* 276: 299–310. [10]

Caminiti, R., Johnson, P. B. and Urbano, A. 1990. Making arm movements within different parts of space: Dynamic aspects in the primate motor cortex. *J. Neurosci.* 10: 2039–2058. [15]

Campbell, K. P., Leung, A. T. and Sharp, A. H. 1988. The biochemistry and molecular biology of the di-hydropyridine-sensitive calcium channel. *Trends Neurosci.* 11: 425–430. [4]

Campenot, R. B. 1977. Local control of neurite development by nerve growth factor. *Proc. Natl. Acad. Sci. USA* 74: 4516–4519. [11]

Campenot, R. B. 1982. Development of sympathetic neurons in compartmentalized cultures. II. Local control of neurite survival by nerve growth factor. *Dev. Biol.* 93: 13–21. [11]

Cannon, W. B. and Rosenblueth, A. 1949. *The Supersensitivity of Denervated Structures: Law of Denervation.* Macmillan, New York. [12]

Cantino, D. and Mugnani, E. 1975. The structural basis for electrotonic coupling in the avian ciliary ganglion: A study with thin sectioning and freeze-fracturing. *J. Neurocytol.* 4: 505–536. [5]

Caprio, J. 1978. Olfactory and taste in the channel catfish: An electrophysiological study of responses to amino acids and their derivatives. *J. Comp. Physiol.* 132: 357–371. [14]

Caputo, C., Bezanilla, F. and DiPolo, R. 1989. Currents related to the sodium–calcium exchange in squid giant axon. *Biochim. Biophys. Acta* 986: 250–256. [3]

Carafoli, E. 1988. Membrane transport of calcium: an overview. *Meth. Enzymol.* 157: 3–11 [3]

Carew, T. J. 1989. Developmental assembly of learning in *Aplysia. Trends Neurosci.* 12: 389–394. [13]

Carew, T. J., Castellucci, V. F. and Kandel, E. R. 1979. Sensitization in *Aplysia*: Rapid restoration of transmission in synapses inactivated by long-term habituation. *Science* 205: 417–419. [13]

Carlson, M., Hubel, D. H. and Wiesel, T. N. 1986. Effects of monocular exposure to oriented lines on monkey striate cortex. *Brain Res.* 390: 71–81. [18]

Caroni, P. and Grandes, P. 1990. Nerve sprouting in innervated adult muscle induced by exposure to elevated levels of insulin-like growth factors. *J. Cell Biol.* 110: 1307–1317. [11]

Caroni, P. and Schwab, M. E. 1988. Two membrane protein fractions from rat central myelin with inhibitory properties for neurite growth and fibroblast spreading. *J. Cell Biol.* 106: 1281–1288. [6]

Casey, P. J., Graziano, M. P., Freissmuth, M. and Gilman, A. G. 1988. Role of G proteins in transmembrane signaling. *Cold Spring Harbor Symp. Quant. Biol.* 53: 203–208. [8]

Caspar, D. L. D., Goodenough, D. A., Makowski, L.

and Phillips, W. C. 1977. Gap junction structures. I. Correlated electron microscopy and X-ray diffraction. *J. Cell Biol.* 74: 605–628. [5]

Castellucci, V. F. and Kandel, E. R. 1974. A quantal analysis of the synaptic depression underlying habituation of the gill-withdrawal reflex in *Aplysia. Proc. Natl. Acad. Sci. USA* 71: 5004–5008. [13]

Castle, N. A., Haylett, D. G. and Jenkinson, D. H. 1989. Toxins in the characterization of potassium channels. *Trends Neurosci.* 12: 59–65 [4]

Catterall, W. A. 1980. Neurotoxins that act on voltage-sensitive sodium channels in excitable membranes. *Annu. Rev. Pharmacol. Toxicol.* 20: 15–43. [4]

Catterall, W. A. 1988. Structure and function of voltage-sensitive ion channels. *Science* 242: 50–61. [2]

Caviness, V. S., Jr. 1982. Neocortical histogenesis in normal and reeler mice: A developmental study based upon [^{3}H]-thymidine autoradiography. *Dev. Brain Res.* 4: 293–302. [11]

Ceccarelli, B. and Hurlbut, W. P. 1980. Vesicle hypothesis of the release of quanta of acetylcholine. *Physiol. Rev.* 60: 396–441. [9]

Changeux, J. P. 1991. Compartmentalized transcription of acetylcholine receptor genes during motor end-plate epigenesis. *New Biol.* 3: 413–429. [12]

Cheney, D. P. and Fetz, E. E. 1980. Functional classes of primate corticomotoneuronal cells and their relation to active force. *J. Neurophysiol.* 44: 773–791. [15]

Chiba, A. and Murphey, R. K. 1991. Connectivity of identified central synapses in the cricket is normal following regeneration and blockade of presynaptic activity. *J. Neurobiol.* 22: 130–142. [12]

Chitnis, A. B. and Kuwada, J. Y. 1991. Elimination of a brain tract increases errors in pathfinding by follower growth cones in the zebrafish embryo. *Neuron* 7: 277–285. [11]

Chiu, S. Y. and Ritchie, J. M. 1981. Evidence for the presence of potassium channels in the paranodal region of acutely demyelinated mammalian nerve fibres. *J. Physiol.* 313: 415–437. [5]

Chiu, S. Y., Ritchie, J. M., Rogart, R. B. and Stagg, D. 1979. A quantitative description of potassium currents in the paranodal region of acutely demyelinated mammalian nerve fibres. *J. Physiol.* 292: 149–166. [4]

Choi, D. W., Koh, J.-Y. and Peters, S. 1988. Pharmacology of glutamate neurotoxicity in cortical cell culture: attenuation by NMDA antagonists. *J. Neurosci.* 8: 185–196. [10]

Chua, M. and Betz, W. J. 1991. Characteristics of a nonselective cation channel in the surface membrane of adult rat skeletal muscle. *Biophys. J.* 59: 1251–1260. [3]

Clausen, T. and Flatman, J. A. 1977. The effect of catecholamines on Na–K transport and membrane potential in rat soleus muscle. *J. Physiol.* 270: 383–414. [8]

Cline, H. T. 1991. Activity-dependent plasticity in the visual systems of frogs and fish. *Trends Neurosci.* 14: 104–111. [18]

Cline, H. T. and Constantine-Paton, M. 1990. NMDA receptor agonist and antagonists alter retinal ganglion cell arbor structure in the developing frog retinotectal projection. *J. Neurosci.* 10: 1197–1216. [18]

Close, R. I. 1972. Dynamic properties of mammalian skeletal muscles. *Physiol. Rev.* 52: 129–197. [12]

Codina, J., Yatani, A., Grenet, D., Brown, A. M. and Birnbaumer, L. 1987. The α subunit of the GTP binding protein G_k opens atrial potassium channels. *Science* 236: 442–445. [8]

Coggeshall, R. E. and Fawcett, D. W. 1964. The fine structure of the central nervous system of the leech, *Hirudo medicinalis. J. Neurophysiol.* 27: 229–289. [13]

Cohan, C. S. and Kater, S. B. 1986. Suppression of neurite elongation and growth cone motility by electrical activity. *Science* 232: 1638–1640. [18]

Cohen, A. H., Mackler, S. A. and Selzer, M. 1988. Behavioral recovery following spinal transection: Functional regeneration in the lamprey CNS. *Trends Neurosci.* 11: 227–231. [12]

Cohen, E. and Sterling, P. 1990. Convergence and divergence of cones onto bipolar cells in the central area of cat retina. *Philos. Trans. R. Soc. Lond. B* 330: 323–328. [16]

Cohen, M. W. 1970. The contribution by glial cells to surface recordings from the optic nerve of an amphibian. *J. Physiol.* 210: 565–580. [6]

Cohen, M. W. 1972. The development of neuromuscular connexions in the presence of D-tubocurarine. *Brain Res.* 41: 457–463. [12]

Cohen, M. W., Jones, O. T. and Angelides, K. J. 1991. Distribution of Ca^{2+} channels on frog motor nerve terminals revealed by fluorescent Ω-conotoxin. *J. Neurosci.* 11: 1032–1038. [7]

Cohen, S. 1959. Purification and metabolic effects of a nerve growth-promoting protein from snake venom. *J. Biol. Chem.* 234: 1129–1137. [11]

Cohen, S. 1960. Purification of a nerve growth-promoting protein from the mouse salivary gland and its neuro-cytotoxic antiserum. *Proc. Natl. Acad. Sci. USA* 46: 302–311. [11]

Cold Spring Harbor Symposium on Quantitative Biology. 1983. 48: 1–146. [2]

Cole, K. S. 1968. *Membranes, Ions and Impulses.* University of California Press, Berkeley. [4]

Collingridge, G. L. and Lester, R. A. J. 1989. Excitatory amino acid receptors in the vertebrate central nervous system. *Pharmacol. Rev.* 40: 143–219. [7]

Collingridge, G. L., Kehl, S. J. and McLennan, H. 1983. Excitatory amino acids in synaptic transmission in the Schaffer collateral–commissural pathway of the rat hippocampus. *J. Physiol.* 334: 33–46. [10]

Collins, W. F., Jr., Nulsen, F. E. and Randt, C. T. 1960. Relation of peripheral nerve fiber size and sensation in man. *Arch. Neurol.* 3: 381–385. [14]

Colquhoun, D. and Sakmann, B. 1981. Fluctuations in the microsecond time range of current through single acetylcholine receptor ion channels. *Nature* 294: 464–466. [7]

Comb, M., Hyman, S. E. and Goodman, H. M. 1987. Mechanisms of trans-synaptic regulation of gene expression. *Trends Neurosci.* 10: 473–478. [9]

Connor, J. A. and Stevens, C. F. 1971. Voltage clamp studies of a transient outward membrane current in gastropod neural somata. *J. Physiol.* 213: 21–30. [2,4]

Consiglio, M. 1900. Sul decorso delle fibre irido-costritricci neglu ucelli: Nota sperimentale. *Arch. Farmacol. Terap.* 8: 268–275. [7]

Constantine-Paton, M., Cline, H. T. and Debski, E. 1990. Patterned activity, synaptic convergence, and the NMDA receptor in developing visual pathways. *Annu. Rev. Neurosci.* 13: 129–154. [11, 18]

Constantine-Paton, M., Blum, A. S., Mendez-Otero, R. and Barnstable, C. J. 1986. A cell surface molecule distributed in a dorsoventral gradient in the perinatal rat retina. *Nature* 324: 459–462. [11]

Conte, F., DeFelice, L. J. and Wanke, E. 1975. Potassium and sodium ion current noise in the membrane of the squid giant axon. *J. Physiol.* 248: 45–82. [4]

Conti, F. and Neher, E. 1980. Single channel recordings of K^+ currents in squid axons. *Nature* 285: 140–143. [4]

Conti, F. and Stühmer, W. 1989. Quantal charge redistributions accompanying structural transitions of sodium channels. *Eur. Biophys. J.* 17: 53–59. [4]

Coombs, J. S., Eccles, J. C. and Fatt, P. 1955. The specific ion conductances and the ionic movements across the motoneuronal membrane that produce the inhibitory post-synaptic potential. *J. Physiol.* 130: 326–373. [7]

Cooper, E., Couturier, S. and Ballivet, M. 1991. Pentameric structure and subunit stoichiometry of a neuronal nicotinic acetylcholine receptor. *Nature* 350: 235–238. [2]

Cooper, J. R., Bloom, F. E. and Roth, R. H. 1991. *The Biochemical Basis of Pharmacology,* 6th Ed. Oxford University Press, New York. [10]

Cooper, N. G. F. and Steindler, D. A. 1986. Monoclonal antibody to glial fibrillary acidic protein reveals a parcellation of individual barrels in the early postnatal mouse somatosensory cortex. *Brain Res.* 380: 341–348. [6]

Costa, E. 1991. The allosteric modulation of $GABA_A$ receptors. *Neuropsychopharmacology* 4: 225–235. [10]

Costanzo, R. M. and Gardiner, E. P. 1980. A quantitative analysis of responses of direction-sensitive neurons in somatosensory cortex of awake monkeys. *J. Neurophysiol.* 43: 1319–1341. [14]

Cota, G. and Armstrong, C. M. 1989. Sodium channel gating in clonal pituitary cells. The inactivation state is not voltage dependent. *J. Gen. Physiol.* 94: 213–232. [4]

Cotman, C. W., Monaghan, D. T. and Ganong, A. H. 1988. Excitatory amino acid neurotransmission: NMDA receptors and Hebb-type synaptic plasticity. *Annu. Rev. Neurosci.* 11: 61–80. [10]

Couteaux, R. and Pecot-Déchavassine, M. 1970. Vésicules synaptiques et poches au niveau des zones actives de la jonction neuromusculaire. *C. R. Acad. Sci.* (Paris) 271: 2346–2349. [7]

Couturier, S., Bertrand, D., Matter, J.-M., Hernandez, M.-C., Bertrand, S., Millar, N., Valera, S., Barkas, Y. and Ballivet, M. 1990. A neuronal nicotinic acetylcholine receptor subunit (α7) is developmentally regulated and forms a homo-oligomeric channel blocked by α-BTX. *Neuron* 5: 847–856. [2]

Cowan, W. M., Fawcett, J. W., O'Leary, D. D. M. and Stanfield, B. B. 1984. Regressive events in neurogenesis. *Science* 225: 1258–1265. [11]

Cox, E. C., Müller, B. and Bonhoeffer, F. 1990. Axonal guidance in the chick visual system: Posterior tectal membranes induce collapse of growth cones from the temporal retina. *Neuron* 4: 31–47. [11]

Crago, P. E., Houk, J. C. and Rymer, W. Z. 1982. Sampling of total muscle force by tendon organs. *J. Neurophysiol.* 47: 1069–1083. [15]

Crawford, A. C. and Fettiplace, R. 1981. An electrical tuning mechanism in turtle cochlear hair cells. *J. Physiol.* 312: 377–412. [14]

Crawford, A. C. and Fettiplace, R. 1985. The mechanical properties of ciliary bundles of turtle cochlear hair cells. *J. Physiol.* 364: 359–379. [14]

Critchlow, V. and von Euler, C. 1963. Intercostal muscle spindle activity and its γ-motor control. *J. Physiol.* 168: 820–847. [14, 15]

Crosby, E. C., Humphrey, T. and Lauer, E. W. 1966. *Correlative Anatomy of the Nervous System.* Macmillan, New York. [15]

Crowe, A. and Matthews, P. B. C. 1964. The effects of stimulation of static and dynamic fusimotor fibres on the response to stretching of the primary endings of muscle spindles. *J. Physiol.* 174: 109–131. [14]

Cserr, H. F. 1988. Role of secretion and bulk flow of brain interstitial fluid in brain volume regulation. *Ann. NY Acad. Sci.* 529: 9–20. [6]

Cull-Candy, S. G., Miledi, R. and Parker, I. 1980. Single glutamate-activated channels recorded from locust muscle fibres with perfused patch-clamp electrodes. *J. Physiol.* 321: 195–210. [2]

Curtis, B. M. and Catterall, W. A. 1986. Reconstitution of the voltage-sensitive calcium channel purified from skeletal muscle transverse tubules. *Biochemistry* 25: 3077–3083. [8]

Curtis, D. R. and Eccles, J. C. 1959. Repetitive synaptic activation. *J. Physiol.* 148: 43P–44P. [15]

Curtis, H. J. and Cole, K. S. 1940. Membrane action potentials from squid giant axon. *J. Cell. Comp. Physiol.* 15: 147–157. [4]

Cynader, M. and Mitchell, D. E. 1977. Monocular astigmatism effects on kitten visual cortex development. *Nature* 270: 177–178. [18]

Cynader, M. and Mitchell, D. E. 1980. Prolonged sensitivity to monocular deprivation in dark-reared cats. *J. Neurophysiol.* 43: 1026–1040. [18]

Czarkowska, J., Jankowska, E. and Syrbirska, E. 1981. Common interneurones in reflex pathways from group Ia and Ib afferents of knee flexors and extensors in the cat. *J. Physiol.* 310: 367–380. [15]

Dacheux, R. F. and Raviola, E. 1986. The rod pathway in the rabbit retina: A depolarizing bipolar and amacrine cell. *J. Neurosci.* 6: 331–345. [16]

Dahlström, A. 1971. Axoplasmic transport (with particular respect to adrenergic neurones). *Philos. Trans. R. Soc. Lond. B* 261: 325–358. [9]

Dale, H. H. 1953. *Adventures in Physiology.* Pergamon Press, London. [7]

Dale, H. H., Feldberg, W. and Vogt, M. 1936. Release of acetylcholine at voluntary motor nerve endings. *J. Physiol.* 86: 353–380. [7]

Damasio, A. R., Tranel, D. and Damasio, H. 1990. Face agnosia and the neural substrates of memory. *Annu. Rev. Neurosci.* 13: 89–109. [17]

Damasio, H. and Damasio, A. R. 1989. *Lesion Analysis in Neurophysiology.* Oxford University Press, Oxford. [19]

Dani, J. A. 1989. Site-directed mutagenesis and single-channel currents define the ionic channel of the nicotinic acetylcholine receptor. *Trends Neurosci.* 12: 127–128. [2]

Daniel, P. M. and Whitteridge, D. 1961. The representation of the visual field on the cerebral cortex in monkeys. *J. Physiol.* 159: 203–221. [17]

Dartnall, H. J. A., Bowmaker, J. K. and Molino, J. D. 1983. Human visual pigments: Microspectrophotometric results from the eyes of seven persons. *Proc. R. Soc. Lond. B* 220: 115–130. [16]

Dash, P. K., Hochner, B. and Kandel, E. R. 1990. Injection of the cAMP-responsive element into the nucleus of *Aplysia* sensory neurons blocks long-term facilitation. *Nature* 345: 718–721. [13]

Da Silva, K. M. C., Sayers, B. M., Sears, T. A. and Stagg, D. T. 1977. The changes in configuration of the rib cage and abdomen during breathing in the anaesthetized cat. *J. Physiol.* 266: 499–521. [15]

David, S. and Aguayo, A. J. 1981. Axonal elongation into peripheral nervous system "bridges" after central nervous system injury in adult rats. *Science* 214: 931–933. [12]

Davies, A. M. and Lumsden, A. 1990. Ontogeny of the somatosensory system: Origins and early development of primary sensory neurons. *Annu. Rev. Neurosci.* 13: 61–73. [11]

Davies, J. A., Cook, G. M. W., Stern, C. D. and Keynes, R. J. 1990. Isolation from chick somites of a glycoprotein fraction that causes collapse of dorsal root ganglion growth cones. *Neuron* 2: 11–20. [11]

Daw, N. W. 1968. Colour-coded ganglion cells in the goldfish retina: Extension of their receptive fields by means of new stimuli. *J. Physiol.* 197: 567–592. [17]

Daw, N. W. 1984. The psychology and physiology of colour vision. *Trends Neurosci.* 7: 330–335. [17]

Daw, N. W., Brunken, W. J. and Parkinson, D. 1989. The function of synaptic transmitters in the retina. *Annu. Rev. Neurosci.* 12: 205–225. [16]

Daw, N. W., Jensen, R. J. and Brunken, W. J. 1990. Rod pathways in mammalian retinae. *Trends Neurosci.* 13: 110–115. [16]

Daw, N. W., Rader, R. K., Robertson, T. W. and Ariel, M. 1983. Effects of 6-hydroxy-dopamine on visual deprivation in the kitten striate cortex. *J. Neurosci.* 3: 907–914. [18]

de Boer, E. 1983. No sharpening? A challenge for cochlear mechanics. *J. Acoust. Soc. Am.* 73: 567–573. [14]

De Boysson-Bardies, B., Halle, P., Sagart, L. and Durand, C. 1989. A crosslinguistic investigation of vowel formants in babbling. *J. Child Lang.* 16: 1–17. [14]

De Camilli, P. and Greengard, P. 1986. Synapsin I: A synaptic vesicle-associated neuronal phosphoprotein. *Biochem. Pharmacol.* 35: 4349–4357. [9]

DeRobertis, E. M., Oliver, G. and Wright, C. V. E. 1990. Homeobox genes and the vertebrate body plan. *Sci. Am.* 263(1): 46–52. [11]

Decker, E. R. and Dani, J. A. 1990. Calcium permeability of the nicotinic acetylcholine receptor: The single channel calcium influx is significant. *J. Neurosci.* 10: 3413–3420. [7]

Deitmer, J. W. and Schlue, W. R. 1989. An inwardly directed electrogenic sodium-bicarbonate co-transport in leech glial cells. *J. Physiol.* 411: 179–194. [6]

de la Torre, J. C. 1980. An improved approach to histofluorescence using the SPG method for tissue monoamines. *J. Neurosci. Methods* 3: 1–5. [10]

del Castillo, J. and Katz, B. 1954a. Quantal components of the end-plate potential. *J. Physiol.* 124: 560–573. [7]

del Castillo, J. and Katz, B. 1954b. Statistical factors involved in neuromuscular facilitation and depression. *J. Physiol.* 124: 574–585. [7]

del Castillo, J. and Katz, B. 1954c. Changes in end-plate activity produced by presynaptic polarization. *J. Physiol.* 124: 586–604. [7]

del Castillo, J. and Katz, B. 1955. On the localization of end-plate receptors. *J. Physiol.* 128: 157–181. [7]

del Castillo, J. and Stark, L. 1952. The effect of calcium ions on the motor end-plate potentials. *J. Physiol.* 116: 507–515. [7]

Del Rio-Hortega, P. 1920. La microglía y su transforma-

ción células en basoncito y cuerpos gránulo-adiposos. *Trab. Lab. Invest. Biol. Madrid* 18: 37–82. [6]

Dennis, M. J. 1981. Development of the neuromuscular junction: Inductive interactions between cells. *Annu. Rev. Neurosci.* 4: 43–68. [12]

Dennis, M. and Miledi, R. 1974a. Electrically induced release of acetylcholine from denervated Schwann cells. *J. Physiol.* 237: 431–452. [6]

Dennis, M. J. and Miledi, R. 1974b. Characteristics of transmitter release at regenerating frog neuromuscular junctions. *J. Physiol.* 239: 571–594. [7]

Dennis, M. J. and Sargent, P. B. 1979. Loss of extrasynaptic acetylcholine sensitivity upon reinnervation of parasympathetic ganglion cells. *J. Physiol.* 289: 263–275. [12]

Dennis, M. J. and Yip, J. W. 1978. Formation and elimination of foreign synapses on adult salamander muscle. *J. Physiol.* 274: 299–310. [12]

Dennis, M. J., Harris, A. J. and Kuffler, S. W. 1971. Synaptic transmission and its duplication by focally applied acetylcholine in parasympathetic neurones of the heart of the frog. *Proc. R. Soc. Lond. B* 177: 509–539. [7]

De Potter, W. P., Smith, A. D. and De Schaepdryver, A. F. 1970. Subcellular fractionation of splenic nerve: ATP, chromogranin A, and dopamine β-hydroxylase in noradrenergic vesicles. *Tissue Cell* 2: 529–546. [9]

DeRiemer, S. A., Elliott, E. J., Macagno, E. R. and Muller, K. J. 1983. Morphological evidence that regenerating axons can fuse with severed axon segments. *Brain Res.* 272: 157–161. [12]

DeYoe, E. A. and Van Essen, D. C. 1988. Concurrent processing streams in monkey visual cortex. *Trends Neurosci.* 11: 219–226. [17]

DeYoe, E. A., Hockfield, S., Garren, H. and Van Essen, D. C. 1990. Antibody labeling of functional subdivisions in visual cortex: Cat-301 immunoreactivity in striate and extrastriate cortex of the macaque monkey. *Vis. Neurosci.* 5: 67–81. [17]

Diamond, J. and Miledi, R. 1962. A study of foetal and new–born muscle fibres. *J. Physiol.* 162: 393–408. [12]

Dickinson-Nelson, A. and Reese, T. S. 1983. Structural changes during transmitter release at synapses in the frog sympathetic ganglion. *J. Neurosci.* 3: 42–52. [9]

Dietzel, I., Heinemann, U. and Lux, H. D. 1989. Relations between slow extracellular potential changes, glial potassium buffering, and electrolyte and cellular volume changes during neuronal hyperactivity in cat brain. *Glia* 2: 25–44. [6]

Dionne, V. E. and Leibowitz, M. D. 1982. Acetylcholine receptor kinetics. A description from single-channel currents at the snake neuromuscular junction. *Biophys. J.* 39: 253–261. [7]

DiPaola, M., Czajkowski, C. and Karlin, A. 1989. The sidedness of the COOH terminus of the acetylcholine receptor δ subunit. *J. Biol. Chem.* 264: 15457–15463. [2]

Dodd, J. and Jessell, T. M. 1988. Axon guidance and the patterning of neuronal projections in vertebrates. *Science* 242: 692–699. [11]

Dolkart-Gorin, P. and Johnson, E. M. 1979. Experimental auto-immune model of nerve growth factor deprivation: effects on developing peripheral sympathetic and sensory neurons. *Proc. Natl. Acad. Sci. USA* 76: 5382–5386. [11]

Dolphin, A. C. 1990. G protein modulation of calcium currents in neurons. *Annu. Rev. Physiol.* 52: 243–255. [8]

Douglas, W. W. 1978. Stimulus–secretion coupling: Variations on the theme of calcium-activated exocytosis involving cellular and extracellular sources of calcium. *Ciba Found. Symp.* 54: 61–90. [7]

Douglas, W. W. 1980. Autocoids. *In* A. G. Gilman, L. S. Goodman and A. Gilman (eds.). *Goodman and Gilman's The Pharmacological Basis of Therapeutics.* Macmillan, New York, pp. 608–618. [10]

Dowdall, M. J., Boyne, A. F. and Whittaker, V. P. 1974. Adenosine triphosphate: A constituent of cholinergic synaptic vesicles. *Biochem. J.* 140: 1–12. [9]

Dowling, J. E. 1987. *The Retina: An Approachable Part of the Brain.* Harvard University Press, Cambridge, MA. [1, 16]

Dowling, J. E. and Boycott, B. B. 1966. Organization of the primate retina: Electron microscopy. *Proc. R. Soc. Lond. B* 166: 80–111. [16]

Drapeau, P. 1990. Loss of channel modulation by transmitter and protein kinase C during innervation of an identified leech neuron. *Neuron* 4: 875–882. [13]

Droz, B. and Leblond, C. P. 1963. Axonal migration of proteins in the central nervous system and peripheral nerves as shown by radioautography. *J. Comp. Neurol.* 121: 325–346. [9]

Dubin, M. W. and Cleland, B. G. 1977. Organization of visual inputs to interneurons of the lateral geniculate nucleus of the cat. *J. Neurophysiol.* 40: 410–427. [16]

Du Bois-Reymond, E. 1848. *Untersuchungen über thierische Electricität.* Reimer, Berlin. [7]

Duchen, L. W. and Strich, S. J. 1968. The effects of botulinum toxin on the pattern of innervation of skeletal muscle in the mouse. *Q. J. Exp. Physiol.* 53: 84–89. [12]

Dudai, Y. 1989. *The Neurobiology of Memory.* Oxford University Press, New York. [13]

Dudel, J. and Kuffler, S. W. 1961. Presynaptic inhibition at the crayfish neuromuscular junction. *J. Physiol.* 155: 543–562. [7]

Dudel, J. and Parnas, I. 1987. Augmented synaptic release by one excitatory axon in regions in which a synergistic axon was removed in lobster muscle. *J. Physiol.* 390: 189–199. [12]

Dudel, J., Franke, C. and Hatt, H. 1990. Rapid activation, desensitization, and resensitization of synaptic channels of crayfish muscle after glutamate pulses. *Biophys. J.* 57: 533–545. [9]

Dunant, Y. and Israel, M. 1985. The release of acetylcholine. *Sci. Am.* 252(4): 58–66. [9]

Dunlap, K. and Fischbach, G. D. 1978. Neurotransmitters decrease the calcium component of sensory neurone action potentials. *Nature* 276: 837–839. [8]

Dunlap, K. and Fischbach, G. D. 1981. Neurotransmitters decrease the calcium conductance activated by depolarization of embryonic chick sensory neurones. *J. Physiol.* 317: 519–535. [8]

Dunlap, K., Holz, G. G. and Rane, S. G. 1987. G proteins as regulators of ion channel function. *Trends Neurosci.* 10: 244–247. [8]

Dunn, P. M. and Marshall, L. M. 1985. Lack of nicotinic supersensitivity in frog sympathetic neurones following denervation. *J. Physiol.* 363: 211–225. [12]

Dunnett, S. 1991. Cholinergic grafts, memory and aging. *Trends Neurosci.* 14: 371–376. [10]

Easter, S. E., Jr. and Stuermer, C. A. O. 1984. An evaluation of the hypothesis of shifting terminals in goldfish optic tectum. *J. Neurosci.* 4: 1052–1063. [11]

Ebihara, L., Beyer, E. C., Swensen, K. I., Paul, D. L. and Goodenough, D. A. 1989. Cloning expression of a *Xenopus* embryonic gap junction protein. *Science* 243: 1194–1195. [5]

Eccles, J. C. 1964. *The Physiology of Synapses.* Springer-Verlag, Berlin. [15]

Eccles, J. C. and O'Connor, W. J. 1939. Responses which nerve impulses evoke in mammalian striated muscles. *J. Physiol.* 97: 44–102. [7]

Eccles, J. C. and Sherrington, C. S. 1930. Numbers and contraction-values of individual motor-units examined in some muscles of the limb. *Proc. R. Soc. Lond. B* 106: 326–357. [14]

Eccles, J. C., Eccles, R. M. and Fatt, P. 1956. Pharmacological investigations on a central synapse operated by acetylcholine. *J. Physiol.* 131: 154–169. [10]

Eccles, J. C., Eccles, R. M. and Magni, F. 1961. Central inhibitory action attributable to presynaptic depolarization produced by muscle afferent volleys. *J. Physiol.* 159: 147–166. [7]

Eccles, J. C., Katz, B. and Kuffler, S. W. 1942. Effects of eserine on neuromuscular transmission. *J. Neurophysiol.* 5: 211–230. [7]

Edmondson, J. C., Liem, R. K. H., Kuster, J. E. and Hatten, M. E. 1988. Astrotactin: A novel neuronal surface antigen that mediates neuron–astroglial interactions in cerebellar microcultures. *J. Cell Biol.* 106: 505–517. [11]

Edwards, C. 1982. The selectivity of ion channels in nerve and muscle. *Neuroscience* 7: 1335–1366. [3,7]

Edwards, C. and Ottoson, D. 1958. The site of impulse initiation in a nerve cell of a crustacean stretch receptor. *J. Physiol.* 143: 138–148. [14]

Edwards, C., Ottoson, D., Rydqvist, B. and Swerup, C. 1981. The permeability of the transducer membrane of the crayfish stretch receptor to calcium and other divalent cations. *Neuroscience* 6: 1455–1460. [14]

Edwards, F. A., Konnerth, A., Sakmann, B. and Takahashi, T. 1989. A thin slice preparation for patch clamp recordings from neurones of the mammalian central nervous system. *Pflügers Arch.* 414: 600–612. [2]

Edwards, J. S. and Palka, J. 1976. Neural generation and regeneration in insects. *In* J. Fentress (ed.), *Simpler Networks and Behavior*. Sinauer, Sunderland, MA, pp. 167–185. [12]

Edwards, J. S., Reddy, G. R. and Rani, M. U. 1989. Central projections of a homoeotic regenerate, antennapedia, in a stick insect, *Carausius morosus* (Phasmida). *J. Neurobiol.* 20: 101–114. [13]

Eldridge, F. L. 1977. Maintenance of respiration by central neural feedback mechanisms. *Fed. Proc.* 36: 2400–2404. [15]

Elliott, C. J. H. and Benjamin, P. R. 1989. Esophageal mechanoreceptors in the feeding system of the pond snail, *Lymnaea stagnalis*. *J. Neurophysiol.* 61: 727–736. [13]

Elliott, E. J. and Muller, K. J. 1983. Sprouting and regeneration of sensory axons after destruction of ensheathing glial cells in the leech central nervous system. *J. Neurosci.* 3: 1994–2006. [6, 13]

Elliott, T. R. 1904. On the action of adrenalin. *J. Physiol.* 31: (Proc.) xx–xxi. [9]

Emonet-Dénand, F., Jami, L. and Laporte, Y. 1980. Histophysiological observation on the skeleto–fusimotor innervation of mammalian spindles. *Prog. Clin. Neurophysiol.* 8: 1–11. [14]

England, J., Gamboni, F., Levinson, S. R. and Finger, T. E. 1990. Formations of new distributions of sodium channels along demyelinated axons. *Proc. Natl. Acad. Sci. USA* 87: 6777–6780. [5]

Enroth-Cugell, C. and Robson, J. G. 1966. The contrast sensitivity of retinal ganglion cells of the cat. *J. Physiol.* 187: 517–552. [16]

Erulkar, S. D. and Weight, F. F. 1977. Extracellular potassium and transmitter release at the giant synapse of squid. *J. Physiol.* 266: 209–218. [6]

Erxleben, C. 1989. Stretch-activated current through single ion channels in the abdominal stretch receptor organ of the crayfish. *J. Gen. Physiol.* 94: 1071–1083. [14]

Evans, P. D., Reale, V., Merzon, R. M. and Villegas, J. 1992. Role of glutamate in axon-Schwann cell signalling in the squid nerve fiber. *Ann. NY Acad. Sci.* 633: 434–447. [6]

Evarts, E. V. 1965. Relation of discharge frequency to conduction velocity in pyramidal neurons. *J. Neurophysiol.* 28: 216–228. [15]

Evarts, E. V. 1966. Pyramidal tract activity associated with a conditioned hand movement in the monkey. *J. Neurophysiol.* 29: 1011–1027. [15]

Evarts, E. V. 1968. Relation of pyramidal tract activity to force exerted during voluntary movement. *J. Neurophysiol.* 31: 14–27. [15]

Eyzaguirre, C. and Kuffler, S. W. 1955. Processes of excitation in the dendrites and in the soma of single isolated sensory nerve cells of the lobster and crayfish. *J. Gen. Physiol.* 39: 87–119. [14]

Falck, B., Hillarp, N.-A., Thieme, G. and Thorp, A. 1962. Fluorescence of catecholamines and related compounds condensed with formaldehyde. *J. Histochem. Cytochem.* 10: 348–354. [10]

Falls, D. L., Harris, D. A., Johnson, F. A., Morgan, M. M., Corfas, G. and Fischbach, G. D. 1990. M_r 42,000 ARIA: A protein that may regulate the accumulation of acetylcholine receptors at developing chick neuromuscular junctions. *Cold Spring Harbor Symp. Quant. Biol.* 55: 397–406. [11]

Fambrough, D. M. 1979. Control of acetylcholine receptors in skeletal muscle. *Physiol. Rev.* 59: 165–227. [12]

Farmer, L., Sommer, J. and Monard, D. 1990. Glia-derived nexin potentiates neurite extension in hippocampal pyramidal cells in vitro. *Dev. Neurosci.* 12: 73–80. [11]

Fatt, P. and Ginsborg, B. L. 1958. The ionic requirements for the production of action potentials in crustacean muscle fibres. *J. Physiol.* 142: 516–543. [4]

Fatt, P. and Katz, B. 1951. An analysis of the end-plate potential recorded with an intra-cellular electrode. *J. Physiol.* 115: 320–370. [7,9]

Fatt, P. and Katz, B. 1952. Spontaneous subthreshold potentials at motor nerve endings. *J. Physiol.* 117: 109–128. [7]

Fatt, P. and Katz, B. 1953. The effect of inhibitory nerve impulses on a crustacean muscle fibre. *J. Physiol.* 121: 374–389. [7]

Favorov, O., Sakamoto, T. and Asanuma, H. 1988. Functional role of corticoperipheral loop circuits during voluntary movements in the monkey: a preferential bias theory. *J. Neurosci.* 8: 3266–3277. [15]

Fawcett, J. W. and Keynes, R. J. 1990. Peripheral nerve regeneration. *Annu. Rev. Neurosci.* 13: 43–60. [11, 12]

Feldberg, W. 1945. Present views of the mode of action of acetylcholine in the central nervous system. *Physiol. Rev.* 25: 596–642. [7]

Feldman, J. L. and Grillner, S. 1983. Control of vertebrate respiration and locomotion: A brief account. *Physiologist* 26: 310–316. [15]

Felleman, D. J. and Van Essen, D. C. 1987. Receptive field properties of neurons in area V3 of macaque monkey extrastriate cortex. *J. Neurophysiol.* 57: 889–920. [17]

Fernandez, J. and Stent, G. 1982. Embryonic development of the hirudinid leech *Hirudo medicinalis*: Structure, development and segmentation of the germinal plate. *J. Embryol. Exp. Morphol.* 72: 71–96. [13]

Fernandez, J. M., Neher, E. and Gomperts, B. D. 1984. Capacitance measurements reveal stepwise fusion events in degranulating mast cells. *Nature* 312: 453–455. [9]

Fernandez, J., Tellez, V. and Olea, N. 1992. Hirudinea. In F. W. Harrison (ed.). *Microscopic Anatomy of Invertebrates*, Vol. 7, *Annelida*. Wiley-Liss, New York, pp. 323-394. [13]

Ferster, D. 1981. A comparison of binocular depth mechanisms in areas 17 and 18 of the cat visual cortex. *J. Physiol.* 311: 623–655. [17]

Ferster, D. 1987. Origin of orientation-selective epsps in simple cells of cat visual cortex. *J. Neurosci.* 7: 1780–1791. [17]

Ferster, D. 1988. Spatially opponent excitation and inhibition in simple cells of the cat visual cortex. *J. Neurosci.* 8: 1172–1180. [17]

Ferster, D. and Koch, C. 1987. Neuronal connections underlying orientation selectivity in cat visual cortex. *Trends Neurosci.* 10: 487–492. [17]

Fertuck, H. C. and Salpeter, M. M. 1974. Localization of acetylcholine receptor by ^{125}I-labeled α-bungarotoxin binding at mouse motor endplates. *Proc. Natl. Acad. Sci. USA* 71: 1376–1378. [9]

Fesenko, E. E., Kolesnikov, S. S. and Lyubarsky, A. L. 1985. Induction by cyclic GMP of cationic conductance in plasma membrane of retinal rod outer segment. *Nature* 313: 310–313. [16]

Fettiplace, R. 1987. Electrical tuning of hair cells in the inner ear. *Trends Neurosci.* 10: 421–425. [14]

Fex, S. Sonessin, B., Thesleff, S. and Zelena, J. 1966. Nerve implants in botulinum poisoned mammalian muscle. *J. Physiol.* 184: 872–882. [12]

ffrench-Constant, C., Miller, R. H., Kruse, J., Schachner, M. and Raff, M. C. 1986. Molecular specialization of astrocyte processes at nodes of Ranvier in rat optic nerve. *J. Cell Biol.* 102: 844–852. [6]

Fibiger, H. C. 1991. Cholinergic mechanisms in learning, memory and dementia: a review of recent evidence. *Trends Neurosci.* 14: : 220–223. [10]

Fields, H. L. 1987. *Pain.* McGraw-Hill, New York. [14]

Fields, H. L. and Basbaum, A. I. 1978. Brain stem control of spinal pain transmission neurons. *Annu. Rev. Physiol.* 40: 217–248. [10, 14]

Fields, H. L. and Besson, J.-M. (eds.). 1988. *Pain Modulation. Progress in Brain Research* Vol. 77. Elsevier, New York. [14]

Fields, R. D., Neale, E. A. and Nelson, P. G. 1990. Effects of patterned electrical activity on neurite outgrowth from mouse sensory neurons. *J. Neurosci.* 10: 2950–2964. [18]

Fillenz, M. 1990. *Noradrenergic Neurons.* Cambridge University Press, Cambridge. [10]

Finer-Moore, J. and Stroud, R. M. 1984. Amphipathic analysis and possible formation of the ion channel in an acetylcholine receptor. *Proc. Natl. Acad. Sci. USA* 81: 155–159. [2]

Firestein, S., Shepherd, G. M. and Werblin, F. 1990. Time course of the membrane current underlying sensory transduction in salamander olfactory receptor neurones. *J. Physiol.* 430: 135–158. [14]

Fischer, B. and Poggio, G. F. 1979. Depth sensitivity of binocular cortical neurones of behaving monkeys. *Proc. R. Soc. Lond. B* 204: 409–414. [17]

Fischer, W., Gage, F. H. and Björklund, A. 1989. Degenerative changes in forebrain cholinergic nuclei correlate with cognitive impairments in aged rats. *Eur. J. Neurosci.* 1: 34–45. [11]

Fischer, W., Björklund, A., Chen, K. and Gage, F. H. 1991. NGF improves spatial memory in aged rodents as a function of age. *J. Neurosci.* 11: 1889–1906. [11]

Flock, A. 1964. Transducing mechanisms in the lateral line canal organ receptors. *Cold Spring Harbor Symp. Quant. Biol.* 30: 133–145. [14]

Flockerzi, V., Oeken, H.-J., Hofman, F., Pelzer, D., Cavalie, A. and Trautwein, W. 1986. Purified dihydropyridine-binding site from skeletal muscle t-tubules is a functional calcium channel. *Nature* 323: 66–68. [4,8]

Fonnum, F. 1984. Glutamate: A neurotransmitter in mammalian brain. *J. Neurochem.* 42: 1–11. [10]

Fontaine, B. and Changeux, J.-P. 1989. Localization of nicotinic acetylcholine receptor α-subunit transcripts during myogenesis and motor endplate development in the chick. *J. Cell Biol.* 108: 1025–1037. [12]

Foote, S. L., Bloom, F. E. and Aston-Jones, G. 1983. Nucleus locus ceruleus: New evidence of anatomical and physiological specificity. *Physiol. Rev.* 63: 844–914. [10]

Fortier, P. A., Kalaska, J. F. and Smith, A. W. 1989. Cerebellar neuronal activity related to whole arm reaching movements in the monkey. *J. Neurophysiol.* 62: 198–211. [15]

Fox, P. T., Miezin, F. M., Allman, J. M., Van Essen, D. C. and Raichle, M. E. 1987. Retinotopic organization of human visual cortex mapped with positron-emission tomography. *J. Neurosci.* 7: 913–922. [17]

Frank, E. and Fischbach, G. D. 1979. Early events in neuromuscular junction formation in vitro: Induction of acetylcholine receptor clusters in the postsynaptic membrane and morphology of newly formed synapses. *J. Cell Biol.* 83: 143–158. [11, 12]

Frank, E., Jansen, J. K. S., Lømo, T. and Westgaard, R. H. 1975. The interaction between foreign and original nerves innervating the soleus muscle of rats. *J. Physiol.* 247: 725–743. [12]

Frank, K. and Fuortes, M.G.F. 1957. Presynaptic and postsynaptic inhibition of monosynaptic reflexes. *Fed. Proc.* 16: 39–40. [7]

Frankenhaeuser, B. and Hodgkin, A. L. 1956. The aftereffects of impulses in the giant nerve fibres of *Loligo*. *J. Physiol.* 131: 341–376. [6]

Frankenhaeuser, B. and Hodgkin, A. L. 1957. The actions of calcium on the electrical properties of squid axons. *J. Physiol.* 137: 218–244. [4]

Fraser, S., Keynes, R. and Lumsden, A. 1990. Segmentation in the chick embryo hindbrain is defined by cell lineage restrictions. *Nature* 344: 431–435. [11]

Fraser, S. E., Murray, B. A., Chuong, C.-M. and Edelman, G. M. 1984. Alteration of the retinotectal map in *Xenopus* by antibodies to neural cell adhesion molecules. *Proc. Natl. Acad. Sci. USA* 81: 4222–4226. [11]

French, K. A. and Muller, K. J. 1986. Regeneration of a distinctive set of axosomatic contacts in the leech central nervous system. *J. Neurosci.* 6: 318–324. [13]

Freund, H.-J. 1991. What is the evidence for multiple motor areas in the human brain? *In* D. R. Humphrey and H.-J. Freund (eds.), *Motor Control: Concepts and Issues.* Wiley, New York. [19]

Fritsch, G. and Hitzig, E. 1870. Ueber die electrische Erregbarkeit des Grosshirns. *Arch. Anat. Physiol. Wiss. Med.* 37: 300–332. [15]

Fuchs, P. A. and Evans, M. G. 1990. Potassium currents in hair cells isolated from the cochlea of the chick. *J. Physiol.* 429: 529–551. [14]

Fuchs, P. A. and Murrow, B. W. 1992. Cholinergic inhibition of short (outer) hair cells of the chick's cochlea. *J. Neurosci.* 12: 2460–2467. [7]

Fuchs, P. A. and Sokolowski, B. H. A. 1990. The acquisition during development of Ca-activated potassium currents by cochlear hair cells of the chick. *Proc. R. Soc. Lond. B* 241: 122–126. [14]

Fuchs, P. A., Evans, M. G. and Murrow, B. W. 1990. Calcium currents in hair cells isolated from the cochlea of the chick. *J. Physiol.* 429: 553–568. [14]

Fuchs, P. A., Henderson, L. P. and Nicholls, J. G. 1982. Chemical transmission between individual Retzius and sensory neurones of the leech in culture. *J. Physiol.* 323: 195–210. [6]

Fuchs, P. A., Nagai, T. and Evans, M. G. 1988. Electrical tuning in hair cells isolated from the chick cochlea. *J. Neurosci.* 8: 2460–2467. [14]

Fuortes, M. G. F. and Poggio, G. F. 1963. Transient responses to sudden illumination in cells of the eye of *Limulus. J. Gen. Physiol.* 46: 435–452. [16]

Fujisawa, H. 1987. Mode of growth of retinal axons within the tectum of *Xenopus* tadpoles, and implications in the ordered neuronal connections between the retina and the tectum. *J. Comp. Neurol.* 260: 127–139. [11]

Fukami, Y. 1982. Further morphological and electrophysiological studies on snake muscle spindles. *J. Neurophysiol.* 47: 810–826. [15]

Fukami, Y. and Hunt, C. C. 1977. Structures in sensory region of snake spindles and their displacement during stretch. *J. Neurophysiol.* 40: 1121–1131. [14]

Fuller, J. L. 1967. Experimental deprivation and later behavior. *Science* 158: 1645–1652. [18]

Fumagalli, G., Balbi, S., Cangiano, A. and Lømo, T. 1990. Regulation of turnover and number of acetylcholine receptors at neuromuscular junctions. *Neuron* 4: 563–569. [12]

Furley, A. J., Morton, S. B., Manalo, D., Karagogeos, D., Dodd, J. and Jessell, T. M. 1990. The axonal glycoprotein TAG-1 is an immunoglobulin superfamily member with neurite outgrowth-promoting activity. *Cell* 61: 157–170. [11]

Furshpan, E. J. and Potter, D. D. 1959. Transmission at the giant motor synapses of the crayfish. *J. Physiol.* 145: 289–325. [7]

Furshpan, E. J., MacLeish, P. R., O'Lague, P. H. and Potter, D. D. 1976. Chemical transmission between rat sympathetic neurons and cardiac myocytes developing in microcultures: Evidence for cholinergic, adrenergic, and dual function neurons. *Proc. Natl. Acad. Sci. USA* 73: 4225–4229. [11]

Gage, F. H., Armstrong, D. M., Williams, L. R. and Varon, S. 1988. Morphologic response of axotomized septal neurons to nerve growth factor. *J. Comp. Neurol.* 269: 147–155. [11]

Gage, P. W. and Armstrong, C. M. 1968. Miniature endplate currents in voltage clamped muscle fibres. *Nature* 218: 363–365. [7]

Gainer, H., Sarne, Y. and Brownstein, M. J. 1977. Biosynthesis and axonal transport of rat neurohypophysial proteins and peptides. *J. Cell Biol.* 73: 366–381. [9]

Galambos, R. 1956. Suppression of auditory nerve activity by stimulation of efferent fibers to the cochlea. *J. Neurophysiol.* 19: 424–437. [14]

Gao, W. Q. and Macagno, E. R. 1987. Extension and retraction of axonal projections by some developing neurons in the leech depends upon the existence of neighboring homologues. II. The AP and AE neurons. *J. Neurobiol.* 18: 295–313. [13]

Gazzaniga, M. S. 1967. The split brain in man. *Sci. Am.* 217(8): 24–29. [17]

Gazzaniga, M. S. 1989. Organization of the human brain. *Science* 245: 947–952. [17]

Georgopolous, A. P., Kettner, R. E. and Schwartz, A. B. 1988. Primate motor cortex and free arm movements to visual targets in three-dimensional space. II. Coding of directional movement by a neuronal population. *J. Neurosci.* 8: 2928–2937. [15]

Ghosh, A., Antonini, A., McConnell, S. K. and Shatz, C. J. 1990. Requirement for subplate neurons in the formation of thalamocortical connections. *Nature* 347: 179–181. [11]

Gilbert, C. D. 1977. Laminar differences in receptive field properties of cells in cat primary visual cortex. *J. Physiol.* 268: 391–421. [17]

Gilbert, C. D. 1983. Microcircuitry of the visual cortex. *Annu. Rev. Neurosci.* 6: 217–247. [10, 16, 17]

Gilbert, C. D. and Wiesel, T. N. 1979. Morphology and intracortical projections of functionally characterised neurones in the cat visual cortex. *Nature* 280: 120–125. [17]

Gilbert, C. D. and Wiesel, T. N. 1983. Clustered intrinsic connections in cat visual cortex. *J. Neurosci.* 3: 1116–1133. [17]

Gilbert, C. D. and Wiesel, T. N. 1989. Columnar specificity of intrinsic horizontal and corticocortical connections in cat visual cortex. *J. Neurosci.* 9: 2432–2442. [17]

Gilbert, C. D., Hirsch, J. A. and Wiesel, T. N. 1990. Lateral interactions in visual cortex. *Cold Spring Harbor Symp. Quant. Biol.* 55: 663–677. [17]

Gilbert, S. F. 1991. *Developmental Biology*, 3rd ed. Sinauer, Sunderland, MA. [11]

Gilman, A. G. 1987. G proteins: Transducers of receptor-generated signals. *Annu. Rev. Biochem.* 56: 615–649. [8]

Gimlich, R. L. and Braun, J. 1986. Improved fluorescent compounds for tracing cell lineage. *Dev. Biol.* 109: 509–514. [11]

Gloor, S., Odnik, K., Guenther, J., Nick, H. and Monard, D. 1986. A glia-derived neurite promoting factor with protease inhibitory activity belongs to the protease nexins. *Cell* 47: 687–693. [6]

Glover, J. C. and Kramer, A. P. 1982. Serotonin analog selectively ablates identified neurons in the leech embryo. *Science* 216: 317–319. [13]

Gold, M. R. and Martin, A. R. 1983a. Characteristics of inhibitory post-synaptic currents in brain-stem neurones of the lamprey. *J. Physiol.* 342: 85–98. [7]

Gold, M. R. and Martin, A. R. 1983b. Analysis of glycine-activated inhibitory post-synaptic channels in brainstem neurones of the lamprey. *J. Physiol.* 342: 99–117. [2, 3, 7]

Goldman, D. E. 1943. Potential, impedance and rectification in membranes. *J. Gen. Physiol.* 27: 37–60. [3]

Goldowitz, D. 1989. The *weaver* phenotype is due to intrinsic action of the mutant locus in granule cells: Evidence from homozygous *weaver* chimeras. *Neuron* 2: 1565–1575. [11]

Goldowitz, D. and Mullen, R. J. 1982. Granule cell as a site of gene action in the *weaver* mouse cerebellum: Evidence from heterozygous mutant chimeras. *J. Neurosci.* 2: 1474–1485. [11]

Golgi, C. 1903. *Opera Omnia*, Vols. I and II. U. Hoepli, Milan. [6]

Goodman, C. S. and Spitzer, N. C. 1979. Embryonic development of identified neurones: Differentiation from neuroblast to neurone. *Nature* 280: 208–214. [11]

Göpfert, H. and Schaefer, H. 1938. Uber den direkt und indirekt erregten Aktionsstrom und die Funktion der motorischen Endplate. *Pflügers Arch.* 239: 597–619. [7]

Grafstein, B. 1983. Chromatolysis reconsidered: A new view of the reaction of the nerve cell body to axon injury. In F. J. Seil (ed.). *Nerve, Organ, and Tissue Regeneration: Research Perspectives.* Academic Press, New York, pp. 37–50. [12]

Grafstein, B. and Forman, D. S. 1980. Intracellular transport in neurons. *Physiol. Rev.* 60: 1167–1283. [9]

Gray, C. M., Konig, P., Engel, A. K. and Singer, W. 1989. Oscillatory responses in cat visual cortex exhibit intercolumnar synchronization which reflects global stimulus properties. *Nature* 338: 334–337. [17]

Gray, P. T. and Ritchie, J. M. 1986. A voltage-gated chloride conductance in rat cultured astrocytes. *Proc. R. Soc. Lond. B* 228: 267–288. [6]

Gray, R. and Johnston, D. 1985. Rectification of single GABA-gated chloride channels in adult hippocampal neurons. *J. Neurophysiol.* 54: 134–142. [2]

Greene, L. A., Varon, S., Piltch, A. and Shooter, E. M. 1971. Substructure of the β subunit of mouse 7S nerve growth factor. *Neurobiology* 1: 37–48. [11]

Greer, J. J. and Stein, R. B. 1990. Fusimotor control of muscle spindle sensitivity during respiration in the cat. *J. Physiol.* 422: 245–264. [14]

Grenningloh, G., Rienitz, A., Schmitt, B., Methfessel, C., Zensen, M., Beyreuther, K., Gundelfinger, E. D. and Betz, H. 1987. A strychnine-binding subunit of the glycine receptor shows homology with nicotinic acetylcholine receptors. *Nature* 328: 215–220 [2]

Griffin, D. R. 1958. *Listening in the Dark.* Yale University Press, New Haven. [14]

Grillner, S. 1975. Locomotion in vertebrates: Central mechanisms and reflex interaction. *Physiol. Rev.* 55: 274–304. [15]

Grillner, S. and Zanger, P. 1974. Locomotor movements generated by the deafferented spinal cord. *Acta Physiol. Scand.* 91: 38A–39A. [15]

Grillner, S., Wallen, P. and Brodin, L. 1991. Neuronal network generating locomotor behavior in lamprey: Circuitry, transmitters, membrane properties and simulation. *Annu. Rev. Neurosci.* 14: 169–199. [15]

Grinnell, A. D. 1970. Electrical interaction between antidromically stimulated frog motoneurones and dorsal root afferents: Enhancement by gallamine and TEA. *J. Physiol.* 210: 17–43. [7]

Grinvald, A., Lieke, E., Frostig, R. D., Gilbert, C. D. and Wiesel, T. N. 1986. Functional architecture of cortex revealed by optical imaging of intrinsic signals. *Nature* 324: 361–364. [17]

Grossman, Y., Parnas, I. and Spira, M. E. 1979. Ionic mechanisms involved in differential conduction of action potentials at high frequency in a branching axon. *J. Physiol.* 295: 307–322. [5, 13]

Grumbacher-Reinert, S. 1989. Local influence of substrate molecules in determining distinctive growth patterns of identified neurons in culture. *Proc. Natl. Acad. Sci. USA* 86: 7270–7274. [11]

Grumbacher-Reinert, S. and Nicholls, J. G. 1992. Influence of substrate on retraction of neurites following electrical activity of leech Retzius cells in culture. *J. Exp. Biol.* 167: 1–14. [18]

Gu, X. 1991. Effect of conduction block at axon bifurcations on synaptic transmission to different postsynaptic neurones in the leech. *J. Physiol.* 441, 755–778. [13]

Gu, X. N., Macagno, E. R. and Muller, K. J. 1989. Laser microbeam axotomy and conduction block show that electrical transmission at a central synapse is distributed at multiple contacts. *J. Neurobiol.* 20: 422–434. [5, 13]

Guastella, J., Nelson, N., Nelson, H., Czyzyk, L., Keynan, S., Miedel, M. C., Davidson, N., Lester, H. A. and Kanner, B. I. 1990. Cloning and expression of a rat brain GABA transporter. *Science* 249: 1303–1306. [9]

Guharay, R. and Sachs, F. 1984. Stretch-activated single ion channel currents in tissue-cultured embryonic chick skeletal muscle. *J. Physiol.* 352: 685–701. [14]

Guillery, R. W. 1970. The laminar distribution of retinal fibers in the dorsal lateral geniculate nucleus of the rat: A new interpretation. *J. Comp. Neurol.* 138: 339–368. [16]

Guillery, R. W. 1974. Visual pathways in albinos. *Sci. Am.* 230(5): 44–54. [18]

Guillery, R. W. and Stelzner, D. J. 1970. The differential effects of unilateral lid closure upon the monocular and binocular segments of the dorsal lateral geniculate nucleus in the cat. *J. Comp. Neurol.* 139: 413–422. [18]

Guillery, R. W., Jeffery, G. and Cattanach, B. M. 1987. Abnormally high variability in the uncrossed retinofugal pathway of mice with albino mosaicism. *Development* 101: 857–867. [18]

Gundersen, R. W. and Barrett, J. N. 1980. Characterization of the turning response of dorsal root neurites toward nerve growth factor. *J. Cell Biol.* 87: 546–554. [11]

Guth, L. 1968. "Trophic" influences of nerve. *Physiol. Rev.* 48: 645–687. [12]

Guthrie, S. C. and Gilula, N. B. 1990. Gap junction communication and development. *Trends Neurosci.* 12: 12–16. [5]

Gutmann, E. 1976. Neurotrophic relations. *Annu. Rev. Physiol.* 38: 177–216. [12]

Gutnick, M. J., Connors, B. W. and Ransom, B. R. 1981. Dye-coupling between glial cells in the guinea pig neocortical slice. *Brain Res.* 213: 486–492. [6]

Guy, H. R. and Conti, F. 1990. Pursuing the structure and function of voltage-gated channels. *Trends Neurosci.* 13: 201–206. [2]

Hagiwara, S. and Byerly, L. 1981. Calcium channel. *Annu. Rev. Neurosci.* 4: 69–125. [4]

Hagiwara, S. and Harunori, O. 1983. Studies of single calcium channel currents in rat clonal pituitary cells. *J. Physiol.* 336: 649–661. [4]

Hall, Z. W. 1992. *An Introduction to Molecular Neurobiology.* Sinauer, Sunderland, MA. [8]

Hall, Z. W., Bownds, M. D. and Kravitz, E. A. 1970. The metabolism of γ-aminobutyric acid in the lobster nervous system. *J. Cell Biol.* 46: 290–299. [9]

Hall, Z. W., Hildebrand, J. G. and Kravitz, E. A. 1974. *Chemistry of Synaptic Transmission.* Chiron Press, Newton, MA. [9]

Hamburger, V. 1939. Motor and sensory hyperplasia following limb-bud transplantations in chick embryos. *Physiol. Zool.* 12: 268–284. [11]

Hamill, O. P. and Sakmann, B. 1981. Multiple conductance states of single acetylcholine receptor channels in embryonic muscle cells. *Nature* 294: 462–464. [2, 7]

Hamill, O. P., Marty, A., Neher, E., Sakmann, B. and Sigworth, J. 1981. Improved patch-clamp techniques for high-resolution current recording from cells and cell-free membrane patches. *Pflügers Arch.* 391: 85–100. [2]

Hamon, M. Bourgoin, S., Artaud, F. and El Mestikawy, S. 1981. The respective roles of tryptophan uptake and tryptophan hydroxylase in the regulation of serotonin synthesis in the central nervous system. *J. Physiol.* (Paris) 77: 269–279. [9]

Harik, S. I. 1984. Locus ceruleus lesion by local 6-hydroxydopamine infusion causes marked and specific destruction of noradrenergic neurons, long-term depletion of norepinephrine and the enzymes that synthesize it, and enhanced dopaminergic mechanisms in the ipsilateral cerebral cortex. *J. Neurosci.* 4: 699–707. [10]

Harlow, J. M. 1868. Recovery from passage of an iron bar through the head. *Publ. Mass. Med. Soc.* 2: 328–334. [19]

Harrelson, A. L. and Goodman, C. S. 1988. Growth cone guidance in insects: fasciclin II is a member of the immunoglobulin superfamily. *Science* 242: 700–708. [11]

Harris, G. W., Reed, M. and Fawcett, C. P. 1966. Hypothalamic releasing factors and the control of anterior pituitary function. *Br. Med. Bull.* 22: 266–272. [10]

Hartline, H. K. 1940. The receptive fields of optic nerve fibers. *Am. J. Physiol.* 130: 690–699. [16]

Hartshorn, R. P. and Catterall, W. A. 1984. The sodium channel from rat brain: Purification and subunit composition. *J. Biol. Chem.* 259: 1667–1675. [2]

Hartzell, H. C., Kuffler, S. W. and Yoshikami, D. 1975. Postsynaptic potentiation: Interaction between quanta of acetylcholine at the skeletal neuromuscular synapse. *J. Physiol.* 251: 427–463. [9]

Hatten, M. E. 1990. Riding the glial monorail: A common mechanism for glial-guided neuronal migration in different regions of the developing mammalian brain. *Trends Neurosci.* 13: 179–184. [6]

Hatten, M. E., Liem, R. K. H. and Mason, C. A. 1986. Weaver mouse cerebellar granule neurons fail to migrate on wild-type astroglial processes in vitro. *J. Neurosci.* 6: 2675–2683. [11]

Hausdorff, W. P., Caron, M. G. and Lefkowitz, R. J. 1990. Turning off the signal: Desensitization of beta-adrenergic receptor function. *FASEB J.* 4: 2881–2889. [9]

Hayashi, J. H. and Hildebrand, J. G. 1990. Insect olfactory neurons in vitro: Morphological and physiological characterization of cells from the developing antennal lobes of *Manduca sexta*. *J. Neurosci.* 10: 848–859. [13]

Hecht, S., Shlaer, S. and Pirenne, M. H. 1942. Energy, quanta and vision. *J. Gen. Physiol.* 25: 819–840. [16]

Heffner, C. D., Lumsden, A. G. S. and O'Leary, D. D. M. 1990. Target control of collateral extension and directional axon growth in the mammalian brain. *Science* 247: 217–220. [11]

Heiligenberg, W. 1989. Coding and processing of electrosensory information in gymnotiform fish. *J. Exp. Biol.* 146: 255–275. [14]

Helmholtz, H. 1889. *Popular Scientific Lectures.* Longmans, London. [1]

Helmholtz, H. 1962/1927. *Helmholtz's Treatise on Physiological Optics.* J. P. C. Southhall (ed.). Dover, New York. [1, 16, 17]

Hendrickson, A. E. 1985. Dots, stripes and columns in monkey visual cortex. *Trends Neurosci.* 8: 406–410. [17]

Hendrickson, A. E., Hunt, S. P. and Wu, J.-Y. 1981. Immunocyotchemical localization of glutamic acid decarboxylase in monkey striate cortex. *Nature* 292: 605–607. [17]

Hendrickson, A. E., Ogren, M. P., Vaughn, J. E., Barber, R. P. and Wu, J.-Y. 1983. Light and electron microscope immunocytochemical localization of glutamic acid decarboxylase in monkey geniculate complex: Evidence for GABAergic neurons and synapses. *J. Neurosci.* 3: 1245–1262. [10]

Hendry, I. A., Stockel, K., Thoenen, H. and Iversen, L. L. 1974. The retrograde axonal transport of nerve growth factor. *Brain Res.* 68: 103–121. [11]

Hendry, S. H. C. and Jones, E. G. 1981. Sizes and distributions of intrinsic neurons incorporating tritiated GABA in monkey sensory-motor cortex. *J. Neurosci.* 1: 390–408. [10]

Hendry, S. H. C., Fuchs, J., deBlas, A. L. and Jones, E. G. 1990. Distribution and plasticity of immunocytochemically localized GABA$_A$ receptors in adult monkey visual cortex. *J. Neurosci.* 10: 2438–2450. [10]

Henneman, E., Somjen, G. and Carpenter, D. O. 1965. Functional significance of cell size in spinal motoneurons. *J. Neurophysiol.* 28: 560–580. [15]

Hepp-Reymond, M. C. 1988. Functional organization of motor cortex and participation in voluntary movements. *In* H. D. Seklis and J. Erwin (eds.), *Comparative Primate Biology*, Vol. 4. Alan R. Liss, New York, pp. 501–624. [15]

Heuman, R. 1987. Regulation of the synthesis of nerve growth factor. *J. Exp. Biol.* 132: 133–150. [6]

Heumann, R., Korsching, S., Brandtlow, C. and Thoenen, H. 1987. Changes of nerve growth factor synthesis in nonneuronal cells in response to sciatic nerve transection. *J. Cell Biol.* 104: 1623–1631. [12]

Heuser, J. E. and Reese, T. S. 1973. Evidence for recycling of synaptic vesicle membrane during transmitter release at the frog neuromuscular junction. *J. Cell Biol.* 57: 315–344. [9]

Heuser, J. E. and Reese, T. S. 1981. Structural changes after transmitter release at the frog neuromuscular junction. *J. Cell Biol.* 88: 564–580. [9]

Heuser, J. E., Reese, T. S. and Landis, D. M. D. 1974. Functional changes in frog neuromuscular junction studied with freeze-fracture. *J. Neurocytol.* 3: 109–131. [7]

Heuser, J. E., Reese, T. S., Dennis, M. J., Jan, Y., Jan, L. and Evans, L. 1979. Synaptic vesicle exocytosis captured by quick freezing and correlated with quantal transmitter release. *J. Cell Biol.* 81: 275–300. [9]

Hilaire, G. G., Nicholls, J. G. and Sears, T. A. 1983. Central and proprioceptive influences on the activity of levator costae motoneurones in the cat. *J. Physiol.* 342: 527–548. [15]

Hilakivi, I. 1987. Biogenic amines in the regulation of wakefulness and sleep. *Med. Biol.* 65: 97–104. [10]

Hill, D. R. and Bowery, N. G. 1981. [^{3}H]-baclofen and [^{3}H]-GABA bind to bicuculline-insensitive GABA$_B$ sites in rat brain. *Nature* 290: 149–152. [10]

Hille, B. 1970. Ionic channels in nerve membranes. *Prog. Biophys. Mol. Biol.* 21: 1–32. [4]

Hille, B. 1992. *Ionic Channels of Excitable Membranes*, 2nd Ed. Sinauer, Sunderland, MA. [2, 3, 4]

Hirning, L. D., Fox, A. P., McCleskey, E. W., Olivera, B. M., Thayer, S. A., Miller, R. J. and Tsien, R. W. 1988. Dominant role of N-type Ca^{2+} channels in evoked release of norepinephrine from sympathetic neurons. *Science* 239: 57–61. [8]

Hishinuma, A., Hockfield, S., McKay, R. and Hildebrand, J. G. 1988. Monoclonal antibodies reveal cell-type-specific antigens in the sexually dimorphic olfactory system of *Manduca sexta*. I. Generation of monoclonal antibodies and partial characterization of the antigens. *J. Neurosci.* 8: 296–307. [11]

Ho, R. K. and Weisblat, D. A. 1987. A provisional epithelium in leech embryo: Cellular origins and influence on a developmental equivalence group. *Dev. Biol.* 120: 520–534. [13]

Hochner, B., Parnas, H. and Parnas, I. 1989. Membrane depolarization evoked neurotransmitter release in the absence of calcium entry. *Nature* 342: 433–435. [7]

Hochner, B., Klein, M., Schacher, S. and Kandel, E. R. 1986. Additional component in the cellular mechanism of presynaptic facilitation contributes to behavioral dishabituation in *Aplysia. Proc. Natl. Acad. Sci. USA* 83: 8794–8798. [13]

Hodgkin, A. L. 1937. Evidence for electrical transmission in nerve. I, II *J. Physiol.* 90: 183–210, 211–232. [5]

Hodgkin, A. L. 1954. J. A note on conduction velocity. *J. Physiol.* 125: 221–224. [5]

Hodgkin, A. L. 1964. *The Conduction of the Nervous Impulse.* Liverpool University Press, Liverpool. [1, 3]

Hodgkin, A. L. 1973. Presidential address. *Proc. R. Soc. Lond. B.* 183: 1–19. [3]

Hodgkin, A. L. 1977. Obituary: Lord Adrian, 1889–1977. *Nature* 269: 543–544. [1]

Hodgkin, A. L. and Horowitz, P. 1959. The influence of potassium and chloride ions on the membrane potential of single muscle fibres. *J. Physiol.* 148: 127–160. [3]

Hodgkin, A. L. and Huxley, A. F. 1939. Action potentials recorded from inside a nerve fibre. *Nature* 144: 710–711. [4]

Hodgkin, A. L. and Huxley, A. F. 1952a. Currents carried by sodium and potassium ion through the membrane of the giant axon of *Loligo. J. Physiol.* 116: 449–472. [4]

Hodgkin, A. L. and Huxley, A. F. 1952b. The components of the membrane conductance in the giant axon of *Loligo. J. Physiol.* 116: 473–496. [4]

Hodgkin, A. L. and Huxley, A. F. 1952c. The dual effect of membrane potential on sodium conductance in the giant axon of *Loligo. J. Physiol.* 116: 497–506. [4]

Hodgkin, A. L. and Huxley, A. F. 1952d. A quantitative description of membrane current and its application to conduction and excitation in nerve. *J. Physiol.* 117: 500–544. [2, 4]

Hodgkin, A. L. and Katz, B. 1949. The effect of sodium ions on the electrical activity of the giant axon of the squid. *J. Physiol.* 108: 37–77. [3, 4]

Hodgkin, A. L. and Keynes, R. D. 1955a. Active transport of cations in giant axons from *Sepia* and *Loligo. J. Physiol.* 128: 28–60. [3]

Hodgkin, A. L. and Keynes, R. D. 1955b. The potassium permeability of a giant nerve fibre. *J. Physiol.* 128: 253–281. [4]

Hodgkin, A. L. and Keynes, R. D. 1956. Experiments on the injection of substances into squid giant axons by means of a microsyringe. *J. Physiol.* 131: 592–617. [3]

Hodgkin, A. L. and Rushton, W. A. H. 1946. The electrical constants of a crustacean nerve fibre. *Proc. R. Soc. Lond. B* 133: 444–479. [5]

Hodgkin, A. L., Huxley, A. F. and Katz, B. 1952. Measurement of current–voltage relations in the membrane of the giant axon of *Loligo. J. Physiol.* 116. 424–448. [4]

Holland, R. L. and Brown, M. C. 1980. Postsynaptic transmission block can cause terminal sprouting of a motor nerve. *Science* 207: 649–651. [12]

Hollyday, M. and Hamburger, V. 1976. Reduction of the naturally occurring motor neuron loss by enlargement of the periphery. *J. Comp. Neurol.* 170: 311–320. [11]

Honig, M. C., Collins, W. F. and Mendell, L. M. 1983. Alpha-motoneuron epsps exhibit different frequency sensitivities to single Ia-afferent fiber stimulation. *J. Neurophysiol.* 49: 886–901. [15]

Horton, J. C. and Hubel, D. H. 1981. A regular patchy distribution of cytochrome-oxidase staining in primary visual cortex of the macaque monkey. *Nature* 292: 762–764. [17]

Horvitz, H. R. 1982. Factors that influence neural development in nematodes. *In* J. G. Nicholls (ed.). *Repair and Regeneration of the Nervous System.* Springer-Verlag, New York, pp. 41–55. [11]

Hoshi, T., Zagotta, W. N. and Aldrich, R. W. 1990. Biophysical and molecular mechanisms of *Shaker* potassium channel inactivation. *Science* 250: 533–538. [4]

Howard, J. and Hudspeth, A. J. 1988. Compliance of the hair bundle associated with gating of mechanoelectrical transduction channels in the bullfrog's sacular hair cell. *Neuron* 1: 189–199. [14]

Howard, J., Hudspeth, A. J. and Vale, R. D. 1989. Movement of microtubules by single kinesin molecules. *Nature* 342: 154–158. [9]

Howe, J. R. and Ritchie, J. M. 1988. Two types of potassium current in rabbit cultured Schwann cells. *Proc. R. Soc. Lond. B* 235: 19–27. [6]

Hoy, R. R. 1989. Startle, categorical response, and attention in acoustic behavior of insects. *Annu. Rev. Neurosci.* 12: 355–375. [13]

Hoy, R., Bittner, G. D. and Kennedy, D. 1967. Regeneration in crustacean motoneurons: Evidence for axonal fusion. *Science* 156: 251–252. [12]

Hubel, D. H. 1967. Effects of distortion of sensory input on the visual system of kittens. *Physiologist* 10: 17–45. [18]

Hubel, D. H. 1982a. Exploration of the primary visual cortex. *Nature* 299: 515–524. [17]

Hubel, D. H. 1982b. Cortical neurobiology: A slanted historical perspective. *Annu. Rev. Neurosci.* 5: 363–370. [17]

Hubel, D. H. 1988. *Eye, Brain and Vision.* Scientific American Library, New York. [6, 16, 17, 18]

Hubel, D. H. and Livingstone, M. S. 1987. Segregation of form, color, and stereopsis in primate area 18. *J. Neurosci.* 7: 3378–3415. [17]

Hubel, D. H. and Livingstone, M. S. 1990. Color and contrast sensitivity in the lateral geniculate body and primary visual cortex of the macaque monkey. *J. Neurosci.* 10: 2223–2237. [17]

Hubel, D. H. and Wiesel, T. N. 1959. Receptive fields of single neurones in the cat's striate cortex. *J. Physiol.* 148: 574–591. [17]

Hubel, D. H. and Wiesel, T. N. 1961. Integrative action in the cat's lateral geniculate body. *J. Physiol.* 155: 385–398. [16]

Hubel, D. H. and Wiesel, T. N. 1962. Receptive fields, binocular interaction and functional architecture in the cat's visual cortex. *J. Physiol.* 160: 106–154. [17]

Hubel, D. H. and Wiesel, T. N. 1963a. Shape and arrangement of columns in cat's striate cortex. *J. Physiol.* 165: 559–568. [17]

Hubel, D. H. and Wiesel, T. N. 1963b. Receptive fields of cells in striate cortex of very young, visually inexperienced kittens. *J. Neurophysiol.* 26: 994–1002. [18]

Hubel, D. H. and Wiesel, T. N. 1965a. Receptive fields and functional architecture in two non-striate visual areas (18 and 19) of the cat. *J. Neurophysiol.* 28: 229–289. [17]

Hubel, D. H. and Wiesel, T. N. 1965b. Binocular interaction in striate cortex of kittens reared with artificial squint. *J. Neurophysiol.* 28: 1041–1059. [18]

Hubel, D. H. and Wiesel, T. N. 1967. Cortical and callosal connections concerned with the vertical meridian of visual field in the cat. *J. Neurophysiol.* 30: 1561–1573. [17]

Hubel, D. H. and Wiesel, T. N. 1968. Receptive fields and functional architecture of monkey striate cortex. *J. Physiol.* 195: 215–243. [17]

Hubel, D. H. and Wiesel, T. N. 1970. The period of susceptibility to the physiological effects of unilateral eye closure in kittens. *J. Physiol.* 206: 419–436. [18]

Hubel, D. H. and Wiesel, T. N. 1971. Aberrant visual projections in the Siamese cat. *J. Physiol.* 218: 33–62. [18]

Hubel, D. H. and Wiesel, T. N. 1972. Laminar and columnar distribution of geniculo-cortical fibers in the macaque monkey. *J. Comp. Neurol.* 146: 421–450. [16, 17]

Hubel, D. H. and Wiesel, T. N. 1974. Sequence regularity and geometry of orientation columns in the monkey striate cortex. *J. Comp. Neurol.* 158: 267–294. [17]

Hubel, D. H. and Wiesel, T. N. 1977. Ferrier Lecture. Functional architecture of macaque monkey visual cortex. *Proc. R. Soc. Lond. B* 198:1–59. [17, 18]

Hubel, D. H., Wiesel, T. N. and LeVay, S. 1977. Plasticity of ocular dominance columns in monkey striate cortex. *Philos. Trans. R. Soc. Lond. B* 278: 377–409. [18]

Huganir, R. L. and Greengard, P. 1990. Regulation of neurotransmitter receptor desensitization by protein phosphorylation. *Neuron* 5: 555–567. [9]

Hudspeth, A. J. 1989. How the ear's works work. *Nature* 341: 397–404. [14]

Hudspeth, A. J. and Corey, D. P. 1977. Sensitivity, polarity and conductance change in the response of vertebrate hair cells to controlled mechanical stimuli. *Proc. Natl. Acad. Sci. USA* 74: 2407–2411. [14]

Hudspeth, A. J. and Lewis, R. S. 1988. Kinetic analysis of voltage- and ion-dependent conductances in saccular hair cells of the bull-frog, *Rana catesbeiana. J. Physiol.* 400: 237–274. [14]

Hudspeth, A. J., Poo, M. M. and Stuart, A. E. 1977. Passive signal propagation and membrane properties in median photoreceptors of the giant barnacle. *J. Physiol.* 272: 25–43. [14]

Hughes, J. (ed.). 1983. Opioid peptides. *Br. Med. Bull.* 39: 1–106. [14]

Hughes, J., Smith, T. W., Kosterlitz, H. W., Fothergill, L. A., Morgan, B. A. and Morris, H. R. 1975. Identification of two related pentapeptides from the brain with potent opiate agonist activity. *Nature* 258: 577–579. [10]

Humphrey, D. R. and Reed, D. J. 1983. Separate cortical systems for the control of joint movement and joint stiffness: reciprocal activation and coactivation of antagonist muscles. *Adv. Neurol.* 39: 347–372. [15]

Hunt, C. C. 1990. Mammalian muscle spindle: Peripheral mechanisms. *Physiol. Rev.* 70: 643–663. [14]

Hunt, C. C., Wilkerson, R. S. and Fukami, Y. 1978. Ionic basis of the receptor potential in primary endings of mammalian muscle spindles. *J. Gen. Physiol.* 71: 683–698. [14]

Huxley, A. 1928. *Point Counter Point,* Chapter 3. Harper Collins, New York. [14]

Huxley, A. F. and Stämpfli, R. 1949. Evidence for saltatory conduction in peripheral myelinated nerve fibers. *J. Physiol.* 108: 315–339. [5]

Hyvärinen, J. and Poranen, A. 1978. Movement-sensitive and direction and orientation-selective cutaneous receptive fields in the hand area of the postcentral gyrus in monkeys. *J. Physiol.* 283: 523–537. [14]

Ignarro, L. J. 1990. Haem-dependent activation of guanylate cyclase and cyclic GMP formation by endogenous nitric oxide: a unique transduction mechanism for transcellular signaling. *Pharmacol. Toxicol.* 67: 1–7. [8]

Innocenti, G. M. 1981. Growth and reshaping of axons in the establishment of visual callosal connections. *Science* 212: 824–827. [11, 18]

Inoué, S. 1981. Video image processing greatly enhances contrast, quality, and speed in polarization-based microscopy. *J. Cell Biol.* 89: 346–356. [9]

Isacoff, E. Y., Jan, N. J. and Jan, L. Y. 1990. Evidence for the formation of heteromultimeric potassium channels in *Xenopus* oocytes. *Nature* 345: 530–534. [2]

Ito, M. 1984. *The Cerebellum and Neural Control*. Raven Press, New York. [15]

Iversen, L. L. 1967. *The Uptake and Storage of Noradrenaline in Sympathetic Nerves*. Cambridge Unversity Press, London. [9]

Iversen, L. L., Lee, C. M., Gilbert, R. F., Hunt, S. and Emson, P. C. 1980. Regulation of neuropeptide release. *Proc. R. Soc. Lond. B* 210: 91–111. [10]

Ivy, G. O. and Killackey, H. P. 1982. Ontogenetic changes in the projections of neocortical neurons. *J. Neurosci.* 2: 735–743. [11]

Jack, J. J. B., Redman, S. J. and Wong, K. 1981. The components of synaptic potentials evoked in cat spinal motoneurones by impulses in single group Ia afferents. *J. Physiol.* 321: 65–96. [15]

Jacobson, M. and Hirose, G. 1981. Clonal organization of the central nervous system of the frog. II. Clones stemming from individual blastomeres of the 32- and 64-cell stages. *J. Neurosci.* 1: 271–284. [11]

Jacobson, S. G., Mohindra, J. and Held, R. 1981. Development of visual acuity in infants with congenital cataracts. *Br. J. Opthalmol.* 65: 727–735. [18]

Jan, Y. N. and Jan, L. Y. 1990. Genes required for specifying cell fates in *Drosophila* embryonic sensory nervous system. *Trends Neurosci.* 13: 493–498. [13]

Jan, Y. N., Jan, L. Y. and Kuffler, S. W. 1979. A peptide as a possible transmitter in sympathetic ganglia of the frog. *Proc. Natl. Acad. Sci. USA* 76: 1501–1505. [8]

Jan, Y. N., Jan, L. Y. and Kuffler, S. W. 1980. Further evidence for peptidergic transmission in sympathetic ganglia. *Proc. Natl. Acad. Sci. USA* 77: 5008–5012. [8]

Jan, Y. N., Bowers, C. W., Branton, D., Evans, L. and Jan, L. Y. 1983. Peptides in neuronal function: Studies using frog autonomic ganglia. *Cold Spring Harbor Symp. Quant. Biol.* 48: 363–374. [8]

Jansen, J. K. S. and Matthews, P. B. C. 1962. The central control of the dynamic response of muscle spindle receptors. *J. Physiol.* 161: 357–378. [14]

Jansen, J. K. S., Njå, A., Ormstad, K. and Walloe, L. 1971. On the innervation of the slowly adapting stretch receptor of the crayfish abdomen: An electrophysiological approach. *Acta. Physiol. Scand.* 81: 273–285. [14]

Jansen, J. K. S., Lømo, T., Nicholaysen, K. and Westgaard, R. H. 1973. Hyperinnervation of skeletal muscle fibers: Dependence on muscle activity. *Science* 181: 559–561. [12]

Jeanmonod, D., Rice, F. L. and Van der Loos, H. 1981. Mouse somatosensory cortex: Alterations in the barrelfield following receptor injury at different early postnatal ages. *Neuroscience* 6: 1503–1535. [18]

Jellies, J. and Kristan, W. B., Jr. 1988. Embryonic assembly of a complex muscle is directed by a single identified cell in the medicinal leech. *J. Neurosci.* 8: 3317–3326. [13]

Jellies, J., Loer, C. M. and Kristan, W. B., Jr. 1987. Morphological changes in leech Retzius neurons after target contact during embryogenesis. *J. Neurosci.* 7: 2618–2629. [13]

Jendelová, P. and Syková, E. 1991. Role of glia in K^+ and pH homeostasis in the neonatal rat spinal cord. *Glia* 4: 56–63. [6]

Jenkins, W. M., Merzenich, M. M., Ochs, M. T., Allard, T. T. and Guic-Robles, E. 1990. Functional reorganization of primary somatosensory cortex in adult owl monkeys after behaviorally controlled tactile stimulation. *J. Neurophysiol.* 63: 82–104. [14, 18]

Jessell, T. M. and Iversen, L. L. 1977. Opiate analgesics inhibit substance P release from rat trigeminal nucleus. *Nature* 268: 549–551. [10]

Joh, T. H., Park, D. H. and Reis, D. J. 1978. Direct phosphorylation of brain tyrosine hydroxylase by cyclic AMP-dependent protein kinase: Mechanism of enzyme activation. *Proc. Natl. Acad. Sci. USA* 75: 4744–4748. [9]

Johansson, R. S. and Vallbo, Å. B. 1983. Tactile sensory coding in the glabrous skin of the human hand. *Trends Neurosci.* 6: 27–32. [14]

Johnson, E. M., Jr., Taniuchi, M. and DiStefano, P. S. 1988. Expression and possible function of nerve growth factor receptors on Schwann cells. *Trends Neurosci.* 11: 299–304. [12]

Johnson, E. W. and Wernig, A. 1971. The binomial nature of transmitter release at the crayfish neuromuscular junction. *J. Physiol.* 218: 757–767. [7]

Johnson, F. H., Eyring, H. and Polissar, M. J. 1954. *The Kinetic Basis of Molecular Biology*. Wiley, New York. [2]

Johnson, J. W. and Ascher, P. 1984. Glycine potentiates the NMDA response in cultured mouse brain neurones. *Nature* 325: 529–531. [7]

Johnson, R. G., Jr. 1988. Accumulation of biological amines into chromaffin granules: A model of hormone and neurotransmitter transport. *Physiol. Rev.* 68: 232–307. [9]

Jones, E. G., Coulter, J. D. and Hendry, S. H. 1978. Intracortical connectivity of architectonic fields in the somatic sensory, motor and parietal cortex of monkeys. *J. Comp. Neurol.* 181: 291–348. [15]

Jones, S. W. and Adams, P. R. 1987. The M-current and other potassium currents of vertebrate neurons. *In* L. K. Kaczmarek and I. B. Levitan (eds.). *Neuromodulation: The Biochemical Control of Neuronal Excitability*. Oxford University Press, New York, pp. 159–186. [8]

Jope, R. 1979. High-affinity choline uptake and acetylcholine production in the brain. Role of regulation of ACh synthesis. *Brain Res. Rev.* 1: 313–344. [9]

Jouvet, M. 1972. The role of monoamines and acetylcho-line-containing neurons in the regulation of the sleep–waking cycle. *Ergeb. Physiol.* 64: 166–307. [10]

Julius, D. 1991. Molecular biology of serotonin receptors. *Annu. Rev. Neurosci.* 14: 335–360. [8]

Junge, D. 1992. *Nerve and Muscle Excitation*, 3rd Ed. Sinauer, Sunderland, MA. [3]

Kaas, J. H. 1983. What if anything is SI? Organization of first somatosensory area of cortex. *Physiol. Rev.* 63: 206–231. [14]

Kaas, J. H. 1991. Plasticity of sensory and motor maps in adult mammals. *Annu. Rev. Neurosci.* 14: 137–167. [18]

Kaas, J. H., Merzenich, M. M. and Killackey, H. P. 1983. The reorganization of somatosensory cortex following peripheral nerve damage in adult and developing mammals. *Annu. Rev. Neurosci.* 6: 325–356. [18]

Kaczmarek, L. K. and Levitan, I. B. (eds.). 1987. *Neuromodulation: The Biochemical Control of Neuronal Excitability*. Oxford University Press, New York. [8]

Kalinoski, D. L., Huque, T., LaMorte, V. J., and Brand, J. G. 1989. Second messenger events in taste. *In* J. G. Brand, J. H. Teeter, R. H. Cagan and M. R. Kare (eds.), *Receptor Events in Transduction and Olfaction*. Volume 1 of *Chemical Senses*. Marcel Dekker, New York, pp. 85–101. [14]

Kallen, R. G., Sheng, Z.-H., Yang, J., Chen, L, Rogart, R. B. and Barchi, R. L. 1990. Primary structure and expression of a sodium channel characteristic of denervated and immature rat skeletal muscle. *Neuron* 4: 233–242. [2, 12]

Kalmidjn, A. J. 1982. Electric and magnetic field detection in elasmobranch fishes. *Science* 218: 916–918. [14]

Kandel, E. R. 1976. *Cellular Basis of Behavior.* W. H. Freeman, San Francisco. [13]

Kandel, E. R. 1979. *Behavioral Biology of Aplysia.* W. H. Freeman, San Francisco. [13]

Kandel, E. R. 1989. Genes, nerve cells, and the remembrance of things past. *J. Neuropsychiatr.* 1: 103–125. [13]

Kandel, E. R., Schwartz, J. H. and Jessell, T. M. (eds.). 1991. *Principles of Neural Science*, 3rd Ed. Elsevier, New York. [10, 19]

Kaneko, A. 1971. Electrical connexions between horizontal cells in the dogfish retina. *J. Physiol.* 213: 95–105. [16]

Kaneko, A. 1979. Physiology of the retina. *Annu. Rev. Neurosci.* 2: 169–191. [16]

Kao, C. T. 1966. Tetrodotoxin, saxotoxin and their significance in the study of excitation phenomena. *Pharmacol. Rev.* 18: 997–1049. [4]

Kao, C. T. and Levinson, S. R. (eds.) 1983. *Tetrodotoxin, Saxotoxin, and the Molecular Biology of the Sodium Channel. Ann. N.Y. Acad. Sci.* 479. [2]

Kao, P. N. and Karlin, A. 1986. Acetylcholine receptor binding site contains a disulfide cross-link between adjacent half-cystinyl residues. *J. Biol. Chem.* 261: 8085–8088. [2]

Kaplan, E. and Shapley, R. M. 1986. The primate retina contains two types of ganglion cells, with high and low contrast sensitivity. *Proc. Natl. Acad. Sci. USA* 83: 2755–2757. [16]

Kaspar, J., Schor, R. H. and Wilson, V. J. 1988. Response of vestibular neurons to head rotations in vertical planes. II. Response to neck stimulation and vestibular-neck interactions. *J. Neurophysiol.* 60: 1765–1768. [15]

Kass, J. H. 1991. Plasticity of sensory and motor maps in adult mammals. *Annu. Rev. Neurosci.* 14: 137–167. [12]

Kater, S. B., Mattson, M. P., Cohan, C. and Connor, J. 1988. Calcium regulation of the neuronal growth cone. *Trends Neurosci.* 11: 315–321. [11]

Katsuki, Y. 1961. Neural mechanisms of auditory sensation in cats. *In* W. A. Rosenblith (ed.). *Sensory Communication.* MIT Press, Cambridge, MA., pp. 561–583. [14]

Katsumaru, H., Murakami, F., Wu, J.-Y. and Tsukahara, N. 1986. Sprouting of GABAergic synapses in the red nucleus after lesions of the nucleus interpositus in the cat. *J. Neurosci.* 6: 2864–2874. [12]

Katz, B. 1949. The efferent regulation of the muscle spindle in the frog. *J. Exp. Biol.* 26: 201–217. [14]

Katz, B. 1950. Depolarization of sensory nerve terminals and the initiation of impulses in the muscle spindle. *J. Physiol.* 111: 261–282. [14]

Katz, B. 1966. *Nerve, Muscle, and Synapse.* McGraw-Hill, New York. [8]

Katz, B. and Miledi, R. 1964. The development of acetylcholine sensitivity in nerve-free segments of skeletal muscle. *J. Physiol.* 170: 389–396. [12]

Katz, B. and Miledi, R. 1965. The measurement of synaptic delay, and the time course of acetylcholine release at the neuromuscular junction. *Proc. R. Soc. Lond. B* 161: 483–495. [7]

Katz, B. and Miledi, R. 1967a. The timing of calcium action during neuromuscular transmission. *J. Physiol.* 189: 535–544. [7]

Katz, B. and Miledi, R. 1967b. A study of synaptic transmission in the absence of nerve impulses. *J. Physiol.* 192: 407–436. [7]

Katz, B. and Miledi, R. 1968a. The role of calcium in neuromuscular facilitation. *J. Physiol.* 195: 481–492. [7]

Katz, B. and Miledi, R. 1968b. The effect of local blockage of motor nerve terminals. *J. Physiol.* 199: 729–741. [7]

Katz, B. and Miledi, R. 1972. The statistical nature of the acetylcholine potential and its molecular components. *J. Physiol.* 244: 665–699. [2, 7]

Katz, B. and Miledi, R. 1973. The binding of acetylcholine to receptors and its removal from the synaptic cleft. *J. Physiol.* 231: 549–574. [9]

Katz, B. and Miledi, R. 1977. Transmitter leakage from motor nerve terminals. *Proc. R. Soc. Lond. B* 196: 59–72. [7]

Katz, B. and Thesleff, S. 1957. A study of "desensitization" produced by acetylcholine at the motor endplate. *J. Physiol.* 138: 63–80. [9]

Katz, L. C., Gilbert, C. D. and Wiesel, T. N. 1989. Local circuits and ocular dominance columns in monkey striate cortex. *J. Neurosci.* 9: 1389–1399. [17]

Kaupp, U. B., Niidome, T., Tanabe, T., Terada, S., Bonigk, W., Stühmer, W., Cook, N. J., Kangawa, K., Matsuo, H. and Hirose, T. 1989. Primary structure and functional expression from complementary DNA of the rod photoreceptor cyclic GMP-gated channel. *Nature* 342: 762–766. [16]

Kawai, N., Yamagishi, S., Saito, M. and Furuya, K. 1983. Blockade of synaptic transmission in the squid giant synapse by a spider toxin (JSTX). *Brain Res.* 278: 346–349. [10]

Kawakami, K., Noguchi, S., Noda, M., Takahashi, H., Ohta, T., Kawamura, M., Nojima, H., Nagano, K., Hirose, T., Inayama, S., Hayashida, H., Miyata, T. and Numa, S. 1985. Structure of α-subunit of *Torpedo californica* (Na$^+$-K$^+$)ATPase deduced from cDNA sequence. *Nature* 316: 733–736. [3]

Kawamura, Y. and Kare, M. R. 1987. *Umami: A Basic Taste.* Marcel Dekker, New York. [14]

Kehoe, J. 1985. Synaptic block of a calcium–activated potassium conductance in *Aplysia* neurones. *J. Physiol.* 369: 439–474. [13]

Keinanen, K. Wisden, W., Sommer, B., Werner, P., Herb, A., Verdoorn, T. A., Sakmann, B. and Seeburg, P. H. 1990. A family of AMPA-selective glutamate receptors. *Science* 249: 556–560. [2]

Kelly, J. P. and Van Essen, D. C. 1974. Cell structure and function in the visual cortex of the cat. *J. Physiol.* 238: 515–547. [6]

Kennedy, M. B. 1989. Regulation of neuronal function by calcium. *Trends Neurosci.* 12: 417–420. [8]

Kennedy, P. R. 1990. Corticospinal, rubrospinal and rubro-olivary projections: A unifying hypothesis. *Trends Neurosci.* 13: 474–479. [15]

Kettenmann, H. and Schachner, M. 1985. Pharmacological properties of γ-aminobutyric acid-, glutamate-, and aspartate-induced depolarizations in cultured astrocytes. *J. Neurosci.* 5: 3295–3301. [6]

Keynes, R. D. 1990. A series-parallel model of the voltage-gated sodium channel. *Proc. R. Soc. Lond. B* 240: 425–432. [4]

Keynes, R. D. and Lumsden, A. 1990. Segmentation and the origin of regional diversity in the vertebrate central nervous system. *Neuron* 2: 1–9. [11]

Keynes, R. D. and Rojas, E. 1974. Kinetics and steady-state properties of the charged system controlling sodium conductance in the squid giant axon. *J. Physiol.* 239: 393–434. [4]

Kim, D., Lewis, D. L., Graziadei, L., Neer, E. J., Bar-Sagi, D. and Clapham, D. E. 1989. G-protein βγ subunits activate the cardiac muscarinic K$^+$ channel via phospholipase A$_2$. *Nature* 337: 557–560. [8]

Kimmel, C. B. and Warga, R. M. 1988. Cell lineage and developmental potential of cells in the zebrafish embryo. *Trends Genet.* 4: 68–74. [11]

Kinnamon, S. C. 1988. Taste transduction: A diversity of mechanisms. *Trends Neurosci.* 11: 491–496. [14]

Kinnamon, S. C. and Roper, S. D. 1987. Passive and active membrane properties of mudpuppy taste receptor cells. *J. Physiol.* 383: 601–614. [14]

Kinnamon, S. C. and Roper, S. D. 1988. Evidence for a role of voltage-sensitive apical K$^+$ channels in sour and salt taste transduction. *Chem. Senses* 13: 115–121. [14]

Kinnamon, S. C., Dionne, V. E. and Beam, K. G. 1988. Apical localization of K$^+$ channels in taste cells provide the basis for sour taste transduction. *Proc. Natl. Acad. Sci. USA* 85: 7023–7027. [14]

Kirkwood, P. A. 1979. On the use and interpretation of cross–correlation measurements in the mammalian central nervous system. *J. Neurosci. Methods* 1: 107–132. [15]

Kirkwood, P. A. and Sears, T. A. 1974. Monosynaptic excitation of motoneurones from secondary endings of muscle spindles. *Nature* 252: 243–244. [15]

Kirkwood, P. A. and Sears, T. A. 1982. Excitatory postsynaptic potentials from single muscle spindle afferents in external intercostal motoneurones of the cat. *J. Physiol.* 322: 287–314. [15]

Kirn, J. R., Alvarez-Buylla, A. and Nottebohm, F. 1991. Production and survival of projection neurons in a forebrain vocal center of adult male canaries. *J. Neurosci.* 11: 1756–1762. [11]

Kirshner, N. 1969. Storage and secretion of adrenal catecholamines. *Adv. Biochem. Psychopharm.* 1: 71–89. [9]

Kistler, J., Stroud, R. M., Klymkowski, M. W., Lalancette, R. A. and Fairclough, R. H. 1982. Structure and function of an acetylcholine receptor. *Biophys. J.* 37: 371–383. [2]

Klassen, H. and Lund, R. D. 1990. Retinal graft-mediated pupillary responses in rats: Restoration of a reflex function in the mature mammalian brain. *J. Neurosci.* 10: 578–587. [12]

Klein, M., Shapiro, E. and Kandel, E. R. 1980. Synaptic plasticity and the modulation of the Ca^{2+} current. *J. Exp. Biol.* 89: 117–157. [13]

Kleinfeld, D., Parsons, T. D., Raccuia-Behling, F., Salzberg, B. M. and Obaid, A. L. 1990. Foreign connections are formed in vitro by *Aplysia californica* interneuron L10 and its in vivo followers and nonfollowers. *J. Exp. Biol.* 154: 237–255. [13]

Klinke, R. 1986. Neurotransmission in the inner ear. *Hearing Res.* 22: 235–243. [14]

Knudsen, E. I. and Konishi, M. 1979. Mechanisms of sound location in the barn owl (*Tyto alba*). *J. Comp. Physiol.* 133: 13–21. [14]

Knudsen, E. I. and Knudsen, P. F. 1989. Vision calibrates sound localization in developing barn owls. *J. Neurosci.* 9: 3306–3313. [14, 18]

Knudsen, E. I., Esterly, S. D. and Du Lac, S. 1991. Stretched and upside-down maps of auditory space in the optic tectum of blind-reared owls: Acoustic basis and behavioral correlates. *J. Neurosci.* 11: 1727–1747. [18]

Knudsen, E. I., Esterly, S. D. and Knudsen, P. F. 1984. Monaural occlusion alters sound localization during a sensitive period in the barn owl. *J. Neurosci.* 4: 1001–1011. [18]

Kohler, C., Swanson, L. W., Haglund, L. and Wu, J.-Y. 1985. The cytoarchitecture, histochemistry and projections of the tuberomammillary nucleus in the rat. *Neuroscience* 16: 85–110. [10]

Koike, H., Kandel, E. R. and Schwartz, J. H. 1974. Synaptic release of radioactivity after intrasomatic injection of choline-H^3 into an identified cholinergic interneuron in abdominal ganglion of *Aplysia californica*. *J. Neurophysiol.* 37: 815–827. [9]

Konnerth, A., Orkand, P. M. and Orkand, R. K. 1988. Optical recording of electrical activity from axons and glia of frog optic nerve: Potentiometric dye responses and morphometrics. *Glia* 1: 225–232. [6]

Kopin, I. J. 1968. False adrenergic transmitters. *Annu. Rev. Pharmacol.* 8: 377–394. [9]

Kopin, I. J., Breese, G. R., Krauss, K. R. and Weise, V. K. 1968. Selective release of newly synthesized norepinephrine from the cat spleen during sympathetic nerve stimulation. *J. Pharmacol. Exp. Therap.* 161: 271–278. [9]

Kosar, E., Waters, S., Tsukahara, N. and Asanuma, H. 1985. Anatomical and physiological properties of the projection from the sensory cortex to the motor cortex in normal cats: The difference between corticocortical and thalamocortical projections. *Brain Res.* 345: 68–78. [15]

Kramer, A. P. and Weisblat, D. A. 1985. Developmental neural kinship groups in the leech. *J. Neurosci.* 5: 388–407. [13]

Krämer, H., Cagan, R. L. and Zipursky, S. L. 1991. Interaction of *bride of sevenless* membrane-bound ligand and the *sevenless* tyrosine-kinase receptor. *Nature* 352: 207–212. [11]

Krasne, F. B. 1987. Silencing normal input permits regenerating foreign afferents to innervate an identified crayfish sensory interneuron. *J. Neurobiol.* 18: 61–73. [13]

Kreutzberg, G. W., Graeber, M. B. and Streit, W. J. 1989. Neuron–glial relationship during regeneration of motorneurons. *Metab. Brain Dis.* 4: 81–85. [6]

Kriebel, M. E., Vautrin, J. and Holsapple, J. 1990. Transmitter release: prepackaging and random mechanism or dynamic and deterministic process. *Brain Res. Rev.* 15: 167–78. [7]

Krieger, D. T. 1983. Brain peptides: What, where and why? *Science* 222: 975–985. [10]

Kristan, W. B. 1983. The neurobiology of swimming in the leech. *Trends Neurosci.* 6: 84–88. [13]

Krouse, M. E., Schneider, G. T. and Gage, P. W. 1986. A large anion-selective channel has seven conductance levels. *Nature* 319: 58–60. [3]

Kruger, L., Perl, E. R. and Sedivic, M. J. 1981. Fine structure of myelinated mechanical nociceptor endings in cat hairy skin. *J. Comp. Neurol.* 198: 137–154. [14]

Kuba, K. 1970. Effects of catecholamines on the neuromuscular junction in the rat diaphragm. *J. Physiol.* 211: 551–570. [8]

Kuffler, D. P. 1987. Long-distance regulation of regenerating frog axons. *J. Exp. Biol.* 132: 151–160. [6]

Kuffler, D. P., Thompson, W. and Jansen, J. K. S. 1980. The fate of foreign endplates in cross-innervated rat soleus muscle. *Proc. R. Soc. Lond. B* 208: 189–222. [12]

Kuffler, S. W. 1943. Specific excitability of the endplate region in normal and denervated muscle. *J. Neurophysiol.* 6: 99–110. [12]

Kuffler, S. W. 1953. Discharge patterns and functional organization of the mammalian retina. *J. Neurophysiol.* 16: 37–68. [16, 17]

Kuffler, S. W. 1954. Mechanisms of activation and motor control of stretch receptors in lobster and crayfish. *J. Neurophysiol.* 17: 558–574. [14]

Kuffler, S. W. 1967. Neuroglial cells: Physiological properties and a potassium mediated effect of neuronal activity on the glial membrane potential. *Proc. R. Soc. Lond. B* 168: 1–21. [6]

Kuffler, S. W. 1980. Slow synaptic responses in autonomic ganglia and the pursuit of a peptidergic transmitter. *J. Exp. Biol.* 89: 257–286. [8, 9]

Kuffler, S. W. and Eyzaguirre, C. 1955. Synaptic inhibition in an isolated nerve cell. *J. Gen Physiol.* 39: 155–184. [7, 14]

Kuffler, S. W. and Nicholls, J. G. 1966. The physiology of neuroglial cells. *Ergeb. Physiol.* 57: 1–90. [6]

Kuffler, S. W. and Potter, D. D. 1964. Glia in the leech central nervous system: Physiological properties and neuron–glia relationship. *J. Neurophysiol.* 27: 290–320. [6, 13]

Kuffler, S. W. and Sejnowski, T. J. 1983. Peptidergic and muscarinic excitation at amphibian sympathetic synapses. *J. Physiol.* 341: 257–278. [8]

Kuffler, S. W. and Vaughan Williams, E. M. 1953. Small-nerve junctional potentials: The distribution of small motor nerves to frog skeletal muscle, and the membrane characteristics of the fibres they innervate. *J. Physiol.* 121: 289–317. [12]

Kuffler, S. W. and Yoshikami, D. 1975a. The distribution of acetylcholine sensitivity at the post-synaptic membrane of vertebrate skeletal twitch muscles: Iontophoretic mapping in the micron range. *J. Physiol.* 244: 703–730. [7, 9]

Kuffler, S. W. and Yoshikami, D. 1975b. The number of transmitter molecules in a quantum: An estimate from iontophoretic application of acetylcholine at the neuromuscular junction. *J. Physiol.* 251: 465–482. [7, 9]

Kuffler, S. W., Dennis, M. J. and Harris, A. J. 1971. The development of chemosensitivity in extrasynaptic areas of the neuronal surface after denervation of parasympathetic ganglion cells in the heart of the frog. *Proc. R. Soc. Lond. B* 177: 555–563. [12]

Kuffler, S. W., Hunt, C. C. and Quilliam, J. P. 1951. Function of medullated small-nerve fibers in mammalian ventral roots: Efferent muscle spindle innervation. *J. Neurophysiol.* 14: 29–54. [14]

Kuffler, S. W., Nicholls, J. G. and Orkand, R. K. 1966. Physiological properties of glial cells in the central nervous system of amphibia. *J. Neurophysiol.* 29: 768–787. [6]

Kuhar, M. J., De Souza, E. B. and Unnerstall, J. R. 1986. Neurotransmitter receptor mapping by autoradiography and other methods. *Annu. Rev. Neurosci.* 9: 27–59. [10]

Kuhlman, J. R., Li, C. and Calabrese, R. L. 1985. FMRFamide-like substances in the leech. II. Bioactivity on the heartbeat system. *J. Neurosci.* 5: 2310–2317. [13]

Kuljis, R. O. and Rakic, P. 1990. Hypercolumns in primate visual cortex can develop in the absence of cues from photoreceptors. *Proc. Natl. Acad. Sci. USA* 87: 5303–5306. [18]

Kuno, M. 1964a. Quantal components of excitatory synaptic potentials in spinal motoneurones. *J. Physiol.* 175: 81–99. [7]

Kuno, M. 1964b. Mechanism of facilitation and depression of the excitatory synaptic potential in spinal motoneurones. *J. Physiol.* 175: 100–112. [7]

Kuno, M. 1971. Quantum aspects of central and ganglionic synaptic transmission in vertebrates. *Physiol. Rev.* 51: 647–678. [15]

Kuno, M. and Weakly, J. N. 1972. Quantal components of the inhibitory synaptic potentials in spinal motoneurones of the cat. *J. Physiol.* 224: 287–303. [7]

Kupfermann, I. 1967. Stimulation of egg laying: Possible neuroendocrine function of bag cells of abdominal ganglion of *Aplysia californica*. *Nature* 216: 814–815. [13]

Kupfermann, I. 1991. Functional studies of cotransmission. *Physiol. Rev.* 71: 683–732. [9]

Kuwada, J. Y. 1986. Cell recognition by neuronal growth cones in a simple vertebrate embryo. *Science* 233: 740–746. [11]

Kuwada, J. Y. and Kramer, A. P. 1983. Embryonic development of the leech nervous system: Primary axon outgrowth of identified neurons. *J. Neurosci.* 3: 2098–2111. [13]

Kuypers, H. G. J. M. 1981. Anatomy of the descending pathways. *In* J. M. Brookhart and V. B. Mountcastle (eds.). *Handbook of Physiology: The Nervous System*, Section 1, Vol. 2, Part 1. The American Physiological Society, Bethesda, MD., pp. 597–666. [15]

Kuypers, H. G. J. M. and Ugolini, G. 1990. Viruses as transneuronal tracers. *Trends Neurosci.* 13: 71–75. [1]

Kyte, J. and Doolittle, R. F. 1982. A simple method for displaying the hydrophobic character of a protein. *J. Molec. Biol.* 157: 105–132. [2]

Lam, D. M.-K. and Ayoub, G. S. 1983. Biochemical and biophysical studies of isolated horizontal cells from the teleost retina. *Vision Res.* 23: 433–444. [10]

Lampson, L. A. 1987. Molecular bases of the immune response to neural antigens. *Trends Neurosci.* 10: 211–216. [6]

Lance-Jones, C. and Landmesser, L. 1980. Motoneurone projection patterns in the chick hind limb following early partial reversals of the spinal cord. *J. Physiol.* 302: 581–602. [11]

Lance-Jones, C. and Landmesser, L. 1981a. Pathway selection by chick lumbosacral motoneurons during normal development. *Proc. R. Soc. Lond. B* 214: 1–18. [11]

Lance-Jones, C. and Landmesser, L. 1981b. Pathway selection by embryonic chick motoneurons in an experimentally altered environment. *Proc. R. Soc. Lond. B* 214: 19–52. [11]

Lancet, D. 1986. Vertebrate olfactory reception. *Annu. Rev. Neurosci.* 9: 329–355. [14]

Lancet, D. 1988. Molecular components of olfactory reception and transduction. *In* F. L. Margolis and T. V. Getchell, T. V. (eds.). *Molecular Biology of the Olfactory System*. Plenum, New York, pp. 25–50. [14]

Land, E. H. 1983. Recent advances in retinex theory and some implications for cortical computations: color vision and the natural image. *Proc. Natl. Acad. Sci. USA* 80: 5163–5169. [17]

Land, E. H. 1986a. Recent advances in retinex theory. *Vision Res.* 26: 7–21. [17]

Land, E. H. 1986b. An alternative technique for the computation of the designator in the retinex theory of color vision. *Proc. Natl. Acad. Sci. USA* 83: 3078–3080. [17]

Lander, A. D. 1989. Understanding the molecules of neural cell contacts: Emerging patterns of structure and function. *Trends Neurosci.* 12: 189–195. [11]

Landis, S. C. 1990. The regulation of transmitter phenotype. *Trends Neurosci.* 13: 344–350. [11]

Langer, G., Bronke, D. and Scheich, H. 1981. Neuronal discrimination of natural and synthetic vowels in field L of trained mynah birds. *Exp. Brain Res.* 43: 11–24. [14]

Landmesser, L. 1971. Contractile and electrical responses of vagus-innervated frog sartorius muscles. *J. Physiol.* 213: 707–725. [12]

Landmesser, L. 1972. Pharmacological properties, cholinesterase activity and anatomy of nerve–muscle junctions in vagus-innervated frog sartorius. *J. Physiol.* 220: 243–256. [12]

Lane, N. J. 1981. Invertebrate neuroglia–junctional structure and development. *J. Exp. Biol.* 95: 7–33. [6]

Langasch, D., Thomas, L. and Betz, H. 1988. Conserved quaternary structure of ligand-gated ion channels: The postsynaptic glycine receptor is a pentamer. *Proc. Natl. Acad. Sci. USA* 85: 7394–7398. [2]

Langley, J. N. 1895. Note on regeneration of pre-ganglionic fibres of the sympathetic. *J. Physiol.* 22: 215–230. [11]

Langley, J. N. 1907. On the contraction of muscle, chiefly in relation to the presence of "receptive" substances. *J. Physiol.* 36: 347–384. [9]

Langley, J. N. and Anderson, H. K. 1892. The action of nicotin on the ciliary ganglion and on the endings of the third cranial nerve. *J. Physiol.* 13: 460–468. [7]

Langley, J. N. and Anderson, H. K. 1904. The union of different kinds of nerve fibres. *J. Physiol.* 31: 365–391. [12]

Lankford, K. L and Letourneau, P. 1989. Evidence that calcium may control neurite outgrowth by regulating the stability of actin filaments. *J. Cell Biol.* 109: 1229–1243. [11]

Lankford, K. L., Cypher, C. and Letourneau, P. 1990. Nerve growth motility. *Curr. Opin. Cell Biol.* 2: 80–85. [11]

Lanno, I., Webb, C. K. and Bezanilla, F. 1988. Potassium conductance of the squid giant axon. Single channels studies. *J. Gen. Physiol.* 92: 179–196. [4]

Lashley, K. S. 1941. Pattern of cerebral integration indicated by the scotomas of migraine. *Arch. Neurol. Psychiat.* 46: 331–339. [19]

Laskowski, M. B. and Sanes, J. R. 1988. Topographically selective reinnervation of adult mammalian skeletal muscles. *J. Neurosci.* 8: 3094–3099. [12]

Lassignal, N. and Martin, A. R. 1977. Effect of acetylcholine on postjunctional membrane permeability in eel electroplaque. *J. Gen. Physiol.* 70: 23–36. [7]

Latorre, R. and Miller, C. 1983. Conduction and selectivity in potassium channels. *J. Memb. Biol.* 71: 11–30. [4]

La Vail, J. H. and La Vail, M. M. 1974. The retrograde intraaxonal transport of horseradish peroxidase in the chick visual system: A light and electron microscopic study. *J. Comp. Neurol.* 157: 303–358. [9]

Lawrence, P. A. and Tomlinson, A. 1991. A marriage is consummated. *Nature* 352: 193. [11]

LeBlanc, N. and Hume, J. R. 1990. Sodium-current induced release of calcium from cardiac sarcoplasmic reticulum. *Science* 248: 372–376. [3]

Le Douarin, N. M. 1980. The ontogeny of the neural crest in avian embryo chimeras. *Nature* 286:663–669. [11]

Le Douarin, N. M. 1982. *The Neural Crest.* Cambridge University Press, Cambridge, England. [11]

Le Douarin, N. M. 1986. Cell line segregation during peripheral nervous system ontogeny. *Science* 231: 1515–1522. [11]

Lefkowitz, R. J., Hoffman, B. B. and Taylor, P. 1990. Neurohumoral transmission: The autonomic and somatic motor nervous systems. *In* A. G. Gilman, T. W. Rall, A. S. Nies and P. Taylor (eds.). *Goodman and Gilman's the Pharmacolgical Basis of Therapeutics,* 8th Ed. Pergamon, New York, pp. 84–121. [8]

Leksell, L. 1945. The action potential and excitatory effects of the small ventral root fibres to skeletal muscle. *Acta. Physiol. Scand.* 10(Suppl. 31): 1–84. [14]

Lent, C. M., Zundel, D., Freedman, E. and Groome, J. R. 1991. Serotonin in the leech central nervous system: Anatomical correlates and behavioral effects. *J. Comp. Physiol. A* 168: 191–200. [13]

Leonard, J. P. and Wickelgren, W. O. 1986. Prolongation of calcium potentials by γ-aminobutyric acid in primary sensory neurones of the lamprey. *J. Physiol.* 375: 481–497. [8]

Leonard, R. J., Labarca, C. G., Charnet, P., Davidson, N. and Lester, H. A. 1988. Evidence that the M2 membrane-spanning region lines the ion channel pore of the nicotinic receptor. *Science* 242: 1578–1581. [2]

Letourneau, P. C. 1975. Cell-to-substratum adhesion and guidance of axonal elongation. *Dev. Biol.* 44: 92–101. [11]

Leutje, C. W. and Patrick, J. 1991. Both α- and β-subunits contribute to agonist sensitivity of neuronal nicotinic acetylcholine receptors. *J. Neurosci.* 11: 837–845. [2]

Leutje, C. W., Patrick, J. and Seguela, P. 1990. Nicotinic receptors in mammalian brain. *FASEB J.* 4: 2753–2760. [2]

LeVay, S. and McConnell, S. K. 1982. On and off layers in the lateral geniculate nucleus of the mink. *Nature* 300: 350–351. [16]

LeVay, S., Hubel, D. H. and Wiesel, T. N. 1975. The pattern of ocular dominance columns in macaque visual cortex revealed by a reduced silver stain. *J. Comp. Neurol.* 159: 559–576. [17]

LeVay, S., Stryker, M. P. and Shatz, C. J. 1978. Ocular dominance columns and their development in layer IV of the cat's visual cortex: A quantitative study. *J. Comp. Neurol.* 179: 223–244. [18]

LeVay, S., Wiesel, T. N. and Hubel, D. H. 1980. The development of ocular dominance columns in normal and visually deprived monkeys. *J. Comp. Neurol.* 191: 1–51. [18]

LeVay, S., Connolly, M., Houde, J. and Van Essen, D. C. 1985. The complete pattern of ocular dominance stripes in the striate cortex and visual field of the macaque monkey. *J. Neurosci.* 5: 486–501. [17]

Levi-Montalcini, R. 1982. Developmental neurobiology and the natural history of nerve growth factor. *Annu. Rev. Neurosci.* 5: 341–362. [11]

Levi-Montalcini, R. and Angeletti, P. U. 1968. Nerve growth factor. *Physiol. Rev.* 48: 534–569. [11]

Levi-Montalcini, R. and Cohen, S. 1960. Effects of the extract of the mouse submaxillary salivary glands on the sympathetic system of mammals. *Ann. NY Acad. Sci.* 85: 324–341. [11]

Levinson, S. R. and Meves, H. 1975. The binding of tritiated tetrodotoxin to squid giant axon. *Philos. Trans. R. Soc. Lond. B* 270: 349–352. [4]

Levitan, I. B. and Kaczmarek, L. K. 1991. *The Neuron: Cell and Molecular Biology.* Oxford University Press, New York. [8]

Lev-Ram, V. and Grinvald, A. 1986. Ca^{2+}- and K^{+}-dependent communication between central nervous system myelinated axons and oligodendrocytes revealed by voltage-sensitive dyes. *Proc. Natl. Acad. Sci. USA* 83: 6651–6655. [6]

Lev-Tov, A., Miller, J. P., Burke, R. E. and Rall, W. 1983. Factors that control amplitude of EPSPs in dendritic neurons. *J. Neurophysiol.* 50: 399–412. [5]

Lewis, B. (ed.). *Bioacoustics: A Comparative Approach.* Academic Press, New York. [14]

Lichtman, J. W., Wilkinson, R. S. and Rich, M. M. 1985. Multiple innervation of tonic endplates revealed by activity-dependent uptake of fluorescent probes. *Nature* 314: 357–359. [9]

Lieberman, E. M., Abbott, N. J. and Hassan, S. 1989. Evidence that glutamate mediates axon-to-Schwann cell signaling in the squid. *Glia* 2: 94–102. [6]

Liley, A. W. 1956. The quantal components of the mammalian end-plate potential. *J. Physiol.* 132: 650–666. [7]

Lim, N. F., Nowycky, M. C. and Bookman, R. J. 1990. Direct measurement of exocytosis and calcium currents in single vertebrate nerve terminals. *Nature* 344: 449–451. [9]

Lin, J. S., Sakai, K., Vanni-Mercier, G., Arrang, J. M., Garbarg, M., Schwartz, J. C. and Jouvet, M. 1990. Involvement of histaminergic neurons in arousal mechanisms demonstrated with [^{3}H]-receptor ligands in the cat. *Brain Res.* 523: 325–330. [10]

Lindvall, O. 1991. Prospects of transplantation in human neurodegenerative diseases. *Trends Neurosci.* 14: 376–384. [10]

Ling, G. and Gerard, R. W. 1949. The normal membrane potential of frog sartorius fibers. *J. Cell. Comp. Physiol.* 34: 383–396. [7]

Linsker, R. 1989. From basic network principles to neural

architecture: Emergence of orientation columns. *Proc. Natl. Acad. Sci. USA* 83: 8779–8783. [17]

Lipscombe, D., Kongsamut, S. and Tsien, R. W. 1989. α-Adrenergic inhibition of sympathetic neurotransmitter release mediated by modulation of N-type calcium-channel gating. *Nature* 340: 639–642. [8]

Lipscombe, D., Madison, D. V., Poenie, M., Reuter, H., Tsien, R. W. and Tsien, R. Y. 1988. Imaging of cytosolic Ca^{2+} transients arising from Ca^{2+} stores and Ca^{2+} channels in sympathetic neurons. *Neuron* 1: 355–365. [8]

Lipscombe, D., Madison, D. V., Poenie, M., Reuter, H., Tsien, R. Y. and Tsien, R. W. 1989. Spatial distribution of calcium channels and cytosolic calcium transients in growth cones and cell bodies of sympathetic neurons. *Proc. Natl. Acad. Sci. USA* 85: 2398–2402. [8]

Liu, Y. and Nicholls, J. G. 1989. Steps in the development of chemical and electrical synapses by pairs of identified leech neurones in culture. *Proc. R. Soc. Lond. B* 236: 253–268. [13]

Livett, B. G., Geffen, L. B. and Austin, L. 1968. Proximo-distal transport of [^{14}C]noradrenaline and protein in sympathetic nerves. *J. Neurochem.* 15: 931–939. [9]

Livingstone, M. S. and Hubel, D. H. 1984. Anatomy and physiology of a color system in the primate visual cortex. *J. Neurosci.* 4: 309–356. [17]

Livingstone, M. S. and Hubel, D. H. 1987a. Connections between layer 4B of area 17 and the thick cytochrome oxidase stripes of area 18 in the squirrel monkey. *J. Neurosci.* 7: 3371–3377. [17]

Livingstone, M. S. and Hubel, D. H. 1987b. Psychophysical evidence for separate channels for the perception of form, color, movement, and depth. *J. Neurosci.* 7: 3416–3468. [17]

Livingstone, M. S. and Hubel, D. H. 1988. Segregation of form, color, movement, and depth: anatomy, physiology, and perception. *Science* 240: 740–749. [17]

Llinás, R. 1982. Calcium in synaptic transmission. *Sci. Am.* 247(4): 56–65. [7]

Llinás, R. and Sugimori, M. 1980. Electrophysiological properties of in vitro Purkinje cell dendrites in mammalian cerebellar slices. *J. Physiol.* 305: 197–213. [4]

Llinás, R., Baker, R. and Sotelo, C. 1974. Electrotonic coupling between neurons in cat inferior olive. *J. Neurophysiol.* 37: 560–571. [7]

Llinás, R., McGuinness, T. L., Leonard, C. S., Sugimori, M. and Greengard, P. 1985. Intraterminal injection of synapsin I or calcium/calmodulin-dependent protein kinase II alters neurotransmitter release at the squid giant synapse. *Proc. Natl. Acad. Sci. USA* 82: 3035–3039. [9]

Lloyd, D. P. C. 1943. Conduction and synaptic transmission of the reflex response to stretch in spinal cats. *J. Neurophysiol.* 6: 317–326. [15]

Lockery, S. R. and Kristan, W. B., Jr. 1990. Distributed processing of sensory information in the leech. II. Identification of interneurons contributing to the local bending reflex. *J. Neurosci.* 10: 1816–1829. [13]

Loewenstein, W. 1981. Junctional intercellular communication: The cell-to-cell membrane channel. *Physiol. Rev.* 61: 829–913. [5, 6]

Loewenstein, W. R. and Mendelson, M. 1965. Components of adaptation in a Pacinian corpuscle. *J. Physiol.* 177: 377–397. [14]

Loewi, O. 1921. Über humorale Übertragbarkeit der Herznervenwirkung. *Pflügers Arch.* 189: 239–242. [7]

Loh, Y. P. and Parish, D. C. 1987. Processing of neuropeptide precursors. *In* A. J. Turner (ed.). *Neuropeptides and Their Peptidases.* Horwood, New York, pp. 65–84. [9]

Lømo, T. and Rosenthal, J. 1972. Control of ACh sensitivity by muscle activity in the rat. *J. Physiol.* 221: 493–513. [12]

Lømo, T., Westgaard, R. H. and Dahl, H. A. 1974. Contractile properties of muscle: Control by pattern of muscle activity in the rat. *Proc. R. Soc. Lond. B* 187: 99–103. [12]

Lorenz, K. 1970. *Studies in Animal and Human Behavior,* Vols. 1 and 2. Harvard University Press, Cambridge. [18]

Loring, R. H. and Zigmond, R. E 1987. Ultrastructural distribution of $[^{125}I]$-toxin F binding sites on chick ciliary neurons: Synaptic localization of a toxin that blocks ganglionic nicotinic receptors. *J. Neurosci.* 7: 2153–2162. [9, 12]

Luddens, H. and Wisden, W. 1991. Function and pharmacology of multiple GABA$_A$ receptor subunits. *Trends Pharmacol. Sci.* 12: 49–51. [10]

Lumsden, A. G. S. 1991. Cell lineage restrictions in the chick embryo hindbrain. *Philos. Trans. R. Soc. Lond. B* 331: 281–286. [11]

Lumsden, A. G. S. and Davies, A. M. 1986. Chemotropic effect of specific target epithelium in the developing mammalian nervous system. *Nature* 323: 538–539. [11]

Lund, J. S. 1988. Anatomical organization of macaque monkey striate visual cortex. *Annu. Rev. Neurosci.* 11: 253–288. [17]

Lundberg, A. 1979. Multisensory control of spinal reflex pathways. *Prog. Brain. Res.* 50: 11–28. [15]

Lundberg, J. M. and Hökfelt, T. 1983. Coexistence of peptides and classical neurotransmitters. *Trends Neurosci.* 6: 325–333.

Lundberg, J. M., Änggård, A., Fahrenkrug, J., Hökfelt, T. and Mutt, V. 1980. Vasoactive intestinal polypeptide in cholinergic neurons of exocrine glands: Functional significance of coexisting transmitters for vasodilation and secretion. *Proc. Natl. Acad. Sci. USA* 77: 1651–1655. [9]

Luqmani, Y. A., Sudlow, G. and Whittaker, V. P. 1980. Homocholine and acetylhomocholine: False transmitters in the cholinergic electromotor system of *Torpedo.* *Neuroscience* 5: 153–160. [9]

Luskin, M. B. and Shatz, C. J. 1985a. Neurogenesis of the cat's primary visual cortex. *J. Comp. Neurol.* 242: 611–631. [11]

Luskin, M. B. and Shatz, C. J. 1985b. Studies of the earliest generated cells of the cat's visual cortex: Cogeneration of subplate and marginal zones. *J. Neurosci.* 5: 1062–1075. [11]

Luskin, M. B., Pearlman, A. L. and Sanes, J. R. 1988. Cell lineage in the cerebral cortex of the mouse studied in vivo and in vitro with a recombinant retrovirus. *Neuron* 1: 635–647. [6, 11]

Lynch, D. R. and Snyder, S. H. 1986. Neuropeptides: Multiple molecular forms, metabolic pathways, and receptors. *Annu. Rev. Biochem.* 55: 773–799. [10]

Lynch, G., Larson, J., Kelso, S., Barrionuevo, G. and Schottler, F. 1983. Intracellular injections of EGTA block induction of hippocampal long-term potentiation. *Nature* 304: 719–721. [10]

Lynn, B. and Hunt, S. P. 1984. Afferent C-fibres: Physiological and biochemical correlations. *Trends Neurosci.* 7: 186–188. [14]

Macagno, E. R. 1980. Number and distribution of neurons in the leech segmental ganglion. *J. Comp. Neurol.* 190: 283–302. [13]

Macagno, E. R. and Stewart, R. R. 1987. Cell death during gangliogenesis in the leech: Competition leading to the death of PMS neurons has both random and nonrandom components. *J. Neurosci.* 7: 1911–1918. [13]

Macagno, E. R., Muller, K. J. and DeRiemer, S. A. 1985. Regeneration of axons and synaptic connections by touch sensory neurons in the leech central nervous system. *J. Neurosci.* 5: 2510–2521. [12, 13]

Macagno, E. R., Muller, K. J. and Pitman, R. M. 1987. Conduction block silences parts of a chemical synapse in the leech central nervous system. *J. Physiol.* 387: 649–664. [13]

Macdonald, R. L. and Werz, M. A. 1986. Dynorphin A decreases voltage-dependent calcium conductance of mouse dorsal root ganglion neurones. *J. Physiol.* 377: 237–249. [10]

MacKinnon, R. 1991. Determination of the subunit stoichiometry of a voltage-activated potassium channel. *Nature* 350: 232–238. [2]

Madison, D. V. and Nicoll, R. A. 1986. Actions of noradrenalin recorded intracellularly in rat hippocampal CA1 pyramidal neurones, in vitro. *J. Physiol.* 372: 221–244. [8]

Madison, D. V., Malenka, R. A. and Nicoll, R. A. 1991. Mechanisms underlying long-term potentiation of synaptic transmission. *Annu. Rev. Neurosci.* 14: 379–397. [10]

Maeno, T., Edwards, C. and Anraku, M. 1977. Permeability of the endplate membrane activated by acetylcholine to some organic cations. *J. Neurobiol.* 8: 173–184. [2]

Maffei, L. and Galli-Resta, L. 1990. Correlation in the discharges of neighboring rat retinal ganglion cells during prenatal life. *Proc. Natl. Acad. Sci. USA* 87: 2861–2964. [18]

Maggio, J. E. 1988. Tachykinins. *Annu. Rev. Neurosci.* 11: 13–28. [10]

Magleby, K. L. and Stevens, C. F. 1972. The effect of voltage on the time course of end-plate currents. *J. Physiol.* 223: 151–171. [7]

Magleby, K. L. and Zengel, J. E. 1982. A quantitative description of stimulation-induced changes in transmitter release at the frog neuromuscular junction. *J. Gen. Physiol.* 80: 613–638. [7]

Mains, R. E. and Patterson, P. H. 1973. Primary cultures of dissociated sympathetic neurons. I. Establishment of long-term growth in culture and studies of differentiated properties. *J. Cell Biol.* 59: 329–345. [11]

Mains, R. E., Eipper, B. A. and Ling, N. 1977. Common precursor to corticotropins and endorphins. *Proc. Natl. Acad. Sci. USA* 74: 3014–3018. [10]

Makowski, L., Caspar, D. L., Phillips, W. C. and Goodenough, D. A. 1977. Gap junction structure. II. Analysis of the X-ray diffraction data. *J. Cell Biol.* 74: 629–645. [5]

Malach, R., Ebert, R. and Van Sluyters, R. C. 1984. Recovery from effects of brief monocular deprivation in the kitten. *J. Neurophysiol.* 51: 538–551. [18]

Malenka, R. C., Kauer, J. A., Zucker, R. S. and Nicoll, R. A. 1988. Postsynaptic calcium is sufficient for potentiation of hippocampal synaptic transmission. *Science* 242: 81–84. [10]

Malenka, R. C., Kauer, J. A., Perkel, D. J., Kelly, P. T., Nicoll, R. A. and Waxham, M. N. 1989. An essential role for post-synaptic calmodulin and protein kinase activity in long-term potentiation. *Nature* 340: 554–557. [10]

Malinow, R. and Tsien, R. W. 1990. Presynaptic enhancement shown by whole-cell recordings of long-term potentiation in hippocampal slices. *Nature* 346: 177–180. [10]

Malinow, R., Schulman, H. and Tsien, R. W. 1989. Inhibition of postsynaptic PKC or CaMKII blocks induction but not expression of LTP. *Science* 245: 862–866. [10]

Mallart, A. and Martin, A. R. 1967. Analysis of facilitation of transmitter release at the neuromuscular junction of the frog. *J. Physiol.* 193: 679–697. [7]

Mallart, A. and Martin, A. R. 1968. The relation between quantum content and facilitation at the neuromuscular junction of the frog. *J. Physiol.* 196: 593–604. [7]

Manabe, T., Renner, P. and Nicoll, R. A. 1992. Postsynaptic contribution to long-term potentiation revealed by the analysis of miniature synaptic currents. *Nature* 355: 50–55. [10]

Marcel, D., Pollard, H., Verroust, P., Schwartz, J. C. and Beaudet, A. 1990. Electron microscopic localization of immunoreactive enkephalinase (EC 3.4.24.11) in the neostriatum of the rat. *J. Neurosci.* 10: 2804–2817. [9]

Marks, W. B., Dobelle, W. H. and MacNichol, E. F. 1964. Visual pigments of single primate cones. *Science* 143: 1181–1183. [16]

Marmont, G. 1949. Studies on the axon membrane. *J. Cell. Comp. Physiol.* 34: 351–382. [4]

Marsden, C. D., Rothwell, J. C. and Day, B. L. 1984. The use of peripheral feedback in the control of movement. *Trends Neuroci.* 7: 253–257. [15]

Marshall, I. G. and Parsons, S. M. 1987. The vesicular acetylcholine transport system. *Trends Neurosci.* 10: 174–177. [9]

Martin, A. R. 1977. Junctional transmission II. Presynaptic mechanisms. *In* E. Kandel (ed.). *Handbook of the Nervous System*, Vol. 1. American Physiological Society, Baltimore, pp. 329–355. [7]

Martin, A. R. 1990. Glycine- and GABA-activated chloride conductances in lamprey neurons. *In* O. P. Otterson and J. Storm-Mathisen (eds.). *Glycine Neurotransmission*. Wiley, New York, pp. 171–191. [7, 9, 10]

Martin, A. R. and Dryer, S. E. 1989. Potassium channels activated by sodium. *Q. J. Exp. Physiol.* 74: 1033–1041. [4]

Martin, A. R. and Levinson, S. R. 1985. Contribution of the Na^+–K^+ pump to membrane potential in familial periodic paralysis. *Muscle Nerve* 8: 359–362. [3]

Martin, A. R. and Pilar, G. 1963. Dual mode of synaptic transmission in the avian ciliary ganglion. *J. Physiol.* 168: 443–463. [7]

Martin, A. R. and Pilar, G. 1964a. Quantal components of the synaptic potential in the ciliary ganglion of the chick. *J. Physiol.* 175: 1–16. [7]

Martin, A. R. and Pilar, G. 1964b. Presynaptic and postsynaptic events during post-tetanic potentiation and facilitation in the avian ciliary ganglion. *J. Physiol.* 175: 16–30. [7]

Martin, A. R. and Ringham, G. L. 1975. Synaptic transfer at a vertebrate central nervous system synapse. *J. Physiol.* 251: 409–426. [7]

Martin, A. R., Patel, V. V., Faille, L. and Mallart, A. 1989. Presynaptic calcium currents recorded from calyciform nerve terminals in the lizard ciliary ganglion. *Neurosci. Lett.* 105: 14–18. [7]

Marty, A. 1989. The physiological role of calcium-dependent channels. *Trends Neurosci.* 12: 420–424. [8]

Masland, R. H. 1988. Amacrine cells. *Trends Neurosci.* 11: 405–410. [16]

Massoulié, J. and Bon, S. 1982. The molecular forms of cholinesterase and acetylcholinesterase in vertebrates. *Annu. Rev. Neurosci.* 5: 57–106. [9]

Masuda-Nakagawa, L. M. and Nicholls, J. G. 1991. Extracellular matrix molecules in development and regeneration of the leech CNS. *Philos. Trans. R. Soc. Lond. B* 331: 323–335. [11]

Masuda-Nakagawa, L. M., Muller, K. J. and Nicholls, J. G. 1990. Accumulation of laminin and microglial cells at sites of injury and regeneration in the central nervous system of the leech. *Proc. R. Soc. Lond. B* 241: 201–206. [6, 13]

Mathie, A., Cull-Candy, S. G. and Colquhoun, D. 1991. Conductance and kinetic properties of single nicotinic acetylcholine receptor channels in rat sympathetic neurons. *J. Physiol.* 439: 717–750. [7]

Matsuda, T., Wu, J.-Y. and Roberts, E. 1973. Immunochemical studies on glutamic acid decarboxylase (EC 4.1.1.15) from mouse brain. *J. Neurochem.* 21: 159–166, 167–172. [10]

Matthews, B. H. C. 1931a. The response of a single end organ. *J. Physiol.* 71: 64–110. [14]

Matthews, B. H. C. 1931b. The response of a muscle spindle during active contraction of a muscle. *J. Physiol.* 72: 153–174. [14]

Matthews, B. H. C. 1933. Nerve endings in mammalian muscle. *J. Physiol.* 78: 1–53. [14]

Matthews, G. and Wickelgren, W. O. 1979. Glycine, GABA and synaptic inhibition of reticulospinal neurones of the lamprey. *J. Physiol.* 293: 393–415 [3, 10]

Matthews, P. B. 1988. Proprioceptors and their contribution to somatosensory mapping: Complex messages require complex processing. *Can. J. Physiol. Pharmacol.* 66: 430–438. [14]

Matthews, P. B. C. 1964. Muscle spindles and their motor control. *Physiol. Rev.* 44: 219–288. [14]

Matthews, P. B. C. 1972. *Mammalian Muscle Receptors and Their Central Action.* Edward Arnold, London. [15]

Matthews, P. B. C. 1981. Evolving views on the internal operation and functional role of the muscle spindle. *J. Physiol.* 320: 1–30. [14]

Matthews, P. B. C. 1982. Where does Sherrington's "muscular sense" originate? Muscles, joints, corollary discharges? *Annu. Rev. Neurosci.* 5: 189–218. [14]

Matthews, R. G., Hubbard, R., Brown, P. K. and Wald. G. 1963. Tautomeric forms of metarhodopsin. *J. Gen. Physiol.* 47: 215–240. [16]

Matthey, R. 1925. Récupération de la vue après résection des nerfs optiques chez le triton. *C. R. Soc. Biol.* 93: 904–906. [12]

Maturana, H. R., Lettvin, J. Y., McCulloch, W. S. and Pitts, W. H. 1960. Anatomy and physiology of vision in the frog (*Rana pipiens*). *J. Gen. Physiol.* 43: 129–175. [16]

Maunsell, J. H. and Newsome, W. T. 1987. Visual processing in monkey extrastriate cortex. *Annu. Rev. Neurosci.* 10: 363–401. [17]

Maycox, P. R., Hell, J. W. and Jahn, R. 1990. Amino acid neurotransmission: Spotlight on synaptic vesicles. *Trends Neurosci.* 13: 83–87. [9]

Mayer, M. L. and Westbrook, G. L. 1987.The physiology of excitatory amino acids in the vertebrate central nervous system. *Prog. Neurobiol.* 28: 198–276. [7, 10]

Mayer, M. L., Westbrook, G. L. and Guthrie, P. B. 1984. Voltage-dependent block by Mg^{2+} of NMDA responses in spinal cord neurones. *Nature* 309: 261–263. [7]

McArdle, J. J. 1983. Molecular aspects of the trophic influence of nerve on muscle. *Prog. Neurobiol.* 21: 135–198. [12]

McCloskey, D. I. 1978. Kinesthetic sensibility. *Physiol. Rev.* 58: 763–820. [14]

McConnell, S. K. 1989. The determination of neuronal fate in the cerebral cortex. *Trends Neurosci.* 12: 342–349. [11]

McConnell, S. K. 1991. The generation of neuronal diversity in the central nervous system. *Annu. Rev. Neurosci.* 14: 269–300. [11]

McCrea, P.D., Popot, J.-L. and Engleman, D. M. 1987. Transmembrane topography of the nicotinic acetylcholine receptor subunit. *EMBO J.* 6: 3619–3626. [2]

McEachern, A. E., Jacob, M. H. and Berg, D. K. 1989. Differential effects of nerve transection on the ACh and GABA receptors of chick ciliary ganglion neurons. *J. Neurosci.* 9: 3899–3907. [12]

McEwen, B. S. and Grafstein, B. 1968. Fast and slow components in axonal transport of protein. *J. Cell Biol.* 38: 494–508. [9]

McGlade-McCulloh, E., Morrissey, A. M., Norona, F. and Muller, K. J. 1989. Individual microglia move rapidly and directly to nerve lesions in the leech central nervous system. *Proc. Natl. Acad. Sci. USA* 86: 1093–1097. [6, 13]

McIntyre, A. K. 1980. Biological seismography. *Trends Neurosci.* 3: 202–205. [14]

McLachlan, E. M. 1978. The statistics of transmitter release at chemical synapses. *Int. Rev. Physiol. Neurophysiol. III* 17: 49–117. [7]

McMahan, U. J. 1990. The agrin hypothesis. *Cold Spring Harbor Symp. Quant. Biol.* 50: 407–418. [12]

McMahan, U. J. and Slater, C. R. 1984. The influence of basal lamina on the accumulation of acetylcholine receptors at synaptic sites in regenerating muscle. *J. Cell Biol.* 98: 1452–1473. [12]

McMahan, U. J. and Wallace, B. G. 1989. Molecules in basal lamina that direct the formation of synaptic specializations at neuromuscular junctions. *Dev. Neurosci.* 11: 227–247. [11, 12]

McMahan, U. J., Sanes, J. R. and Marshall, L. M. 1978. Cholinesterase is associated with the basal lamina at the neuromuscular junction. *Nature* 271: 172–174. [9]

McMahan, U. J., Spitzer, N. C. and Peper, K. 1972. Visual identification of nerve terminals in living isolated skeletal muscle. *Proc. R. Soc. Lond. B* 181: 421–430. [9]

McManaman, J. L., Haverkamp, L. J. and Oppenheim, R. W. 1991. Skeletal muscle proteins rescue motor neurons from cell death in vivo. *Adv. Neurol.* 56: 81–88. [11]

McNaughton, P. A. 1990. Light response of vertebrate photoreceptors. *Physiol. Rev.* 70: 847–883. [16]

Meech, R. W. 1974. The sensitivity of *Helix aspersa* neurones to injected calcium ions. *J. Physiol.* 237: 259–277. [4]

Meister, M., Wong, R. O., Baylor, D. A. and Shatz, C. J. 1991. Synchronous bursts of action potentials in ganglion cells of the developing mammalian retina. *Science* 252: 939–943. [18]

Melloni, E. and Pontremoli, S. 1989. The calpains. *Trends Neurosci.* 12: 438–444. [8]

Mendell, L. N. and Henneman, E. 1971. Terminals of single Ia fibers: Location, density, and distribution within a pool of 300 homonymous motoneurons. *J. Neurophysiol.* 34: 171–187. [15]

Menesini-Chen, M. G., Chen, J. S. and Levi-Montalcini, R. 1978. Sympathetic nerve fibers ingrowth in the central nervous system of neonatal rodents upon intracerebral NGF injections. *Arch. Natl. Biol.* 116: 53–84. [11]

Mengod, G., Charli, J. L. and Palacios, J. M. 1990. The use of in situ hybridization histochemistry for the study of neuropeptide gene expression in the human brain. *Cell. Mol. Neurobiol.* 10: 113–126. [10]

Merlie, J. P. and Sanes, J. R. 1985. Concentration of acetylcholine receptor mRNA in synaptic regions of adult muscle fibers. *Nature* 317: 66–68. [12]

Merzenich, M. M., Kaas, J. H., Sur, M. and Lin, C.-S. 1978. Double representation of the body surface within cytoarchitecture areas 3B and 1 in "SI" in the owl monkey (*Aotus trivirgatus*). *J. Comp Neurol.* 181: 41–74. [14]

Meyer, M. R., Reddy, G. R. and Edwards, J. S. 1987. Immunological probes reveal spatial and developmental diversity in insect neuroglia. *J. Neurosci.* 7: 512–521. [6]

Meyer-Lohmann, J., Hore, J. and Brooks, V. B. 1977. Cerebellar participation in generation of prompt arm movements. *J. Neurophysiol.* 40: 1038–1050. [15]

Miachon, S., Berod, A., Leger, L., Chat, M., Hartman, B. and Pujol, J. F. 1984. Identification of catecholamine cell bodies in the pons and pons–mesencephalon junction of the cat brain, using tyrosine hydroxylase and dopamine-β-hydroxylase immunohistochemistry. *Brain Res.* 305: 369–374. [10]

Miledi, R. 1960a. Junctional and extra-junctional acetylcholine receptors in skeletal muscle fibres. *J. Physiol.* 151: 24–30. [7]

Miledi, R. 1960b. The acetylcholine sensitivity of frog muscle fibers after complete or partial denervation. *J. Physiol.* 151: 1–23. [12]

Miledi, R., Parker, I. and Sumikawa, K. 1983. Recording single γ-aminobutyrate- and acetylcholine-activated receptor channels translated by exogenous mRNA in *Xenopus* oocytes. *Proc. R. Soc. Lond.* B 218: 481–484. [2]

Miledi, R., Stefani, E. and Steinbach, A. B. 1971. Induction of the action potential mechanism in slow muscle fibres of the frog. *J. Physiol.* 217: 737–754. [12]

Miller, A. E. and Schwartz, E. A. 1983. Evidence for the identification of synaptic transmitters released by photoreceptors of the toad retina. *J. Physiol.* 334: 325–349. [16]

Miller, J., Agnew, W. S. and Levinson, S. R. 1983. Principle glycopeptide of the tetrodotoxin/saxitoxin binding protein from *Electrophorus electricus*: Isolation and partial chemical and physical characterization. *Biochemistry* 22: 462–470. [2, 4]

Miller, R. H., ffrench-Constant, C. and Raff, M. C. 1989. The macroglial cells of the rat optic nerve. *Annu. Rev. Neurosci.* 12: 517–534. [6]

Miller, T. M. and Heuser, J. E. 1984. Endocytosis of synaptic vesicle membrane at the frog neuromuscular junction. *J. Cell Biol.* 98: 685–698. [9]

Mink, J. W. and Thach, W. T. 1991a. Basal ganglia motor control. I. Nonexclusive relation of pallidal discharge to five moment modes. *J. Neurophysiol.* 65: 273–300. [15]

Mink, J. W. and Thach, W. T. 1991b. Basal ganglia motor control. II. Late pallidal timing relative to movement onset and inconsistent pallidal coding of movement parameters. *J. Neurophysiol.* 65: 301–329. [15]

Mink, J. W. and Thach, W. T. 1991c. Basal ganglia motor control. III. Pallidal ablation: Normal reaction time, muscle co-contraction, and slow movement. *J. Neurophysiol.* 65: 330–351. [15]

Mishina, M., Takai, T., Imoto, K., Noda, M., Takahashi, T., Numa, S., Methfessel, C. and Sakmann, B. 1986. Molecular distinction between fetal and adult forms of muscle acetylcholine receptor. *Nature* 321: 406–411. [2, 12]

Moiseff, M. 1989. Bi-coordinate sound localization by the barn owl. *J. Comp. Physiol.* 164: 637–644. [14]

Møllgård, K., Dziegielewska, K. M, Saunders, N. R., Zakut, H. and Soreq, H. 1988. Synthesis and localization of plasma proteins in the developing human brain. Integrity of the fetal blood–brain barrier to endogenous proteins of hepatic origin. *Dev. Biol.* 128: 207–221. [6]

Monaghan, D. T., Bridges, R. J. and Cotman, C. W. 1989. The excitatory amino acid receptors: Their classes, pharmacology and distinct properties in the function of the central nervous system. *Annu. Rev. Pharmacol. Toxicol.* 29: 365–402. [7]

Monard, D. 1988. Cell-derived proteases and protease inhibitors as regulators of neurite outgrowth. *Trends Neurosci.* 11: 541–544. [6]

Moody, W. J. 1981. The ionic mechanisms of intracellular pH regulation in crayfish neurones. *J. Physiol.* 316: 293–308. [3]

Moore, R. Y. and Bloom, F. E. 1978. Central catecholamine neuron systems: Anatomy and physiology of the dopamine systems. *Annu. Rev. Neurosci.* 1: 129–169. [10]

Moore, R. Y. and Bloom, F. E. 1979. Central catecholamine neuron systems: Anatomy and physiology of the norepinephrine and epinephrine systems. *Annu. Rev. Neurosci.* 2: 113–168. [10]

Morgan, J. I. and Curran, T. 1991. Stimulus–transcription coupling in the nervous system: Involvement of the inducible proto-oncogenes *fos* and *jun*. *Annu. Rev. Neurosci.* 14: 421–451. [8]

Morita, K. and Barrett, E. F. 1990. Evidence for two calcium-dependent potassium conductances in lizard motor nerve terminals. *J. Neurosci.* 10: 2614–2625. [7]

Mountcastle, V. B. 1957. Modality and topographic properties of single neurons of cat's somatic sensory cortex. *J. Neurophysiol.* 20: 408–434. [17]

Mountcastle, V. B. and Powell, T. P. S. 1959. Neural mechanisms subserving cutaneous sensibility with special reference to the role of afferent inhibition in sensory perception and discrimination. *Bull. Johns Hopkins Hosp.* 105: 201–232. [14]

Mower, G. D., Christen, W. G. and Caplan, C. J. 1983. Very brief visual experience eliminates plasticity in the cat visual cortex. *Science* 221: 178–180. [18]

Mudge, A. W., Leeman, S. E. and Fischbach, G. D. 1979. Enkephalin inhibits release of substance P from sensory neurons in culture and decreases action potential duration. *Proc. Natl. Acad. Sci. USA* 76: 526–530. [8, 10]

Mugnaini, E. 1982. Membrane specializations in neuroglial cells and at neuron–glial contacts. *In* T. A. Sears (ed.). *Neuronal–Glial Cell Interrelationships*. Springer-Verlag, New York, pp. 39–56. [6]

Mulkey, R. M. and Zucker, R. S. 1991. Action potentials must admit calcium to evoke transmitter release. *Nature* 350: 153–155. [7]

Muller, D., Joly, M. and Lynch, G. 1988. Contributions of quisqualate and NMDA receptors to the induction and expression of LTP. *Science* 242: 1694–1697. [10]

Müller, F., Junesch, E., Binkofski, F. and Freund, H.-J. 1991. Residual sensorimotor functions in a patient after right-sided hemispherectomy. *Neuropsychologia* 29: 125–145. [19]

Muller, K. J. 1981. Synapses and synaptic transmission *In* K. J. Muller, J. G. Nicholls and G. S. Stent (eds.). *Neurobiology of the Leech*. Cold Spring Harbor Laboratory, Cold Spring Harbor, NY, pp. 79–111. [13]

Muller, K. J. and McMahan, U. J. 1976. The shapes of sensory and motor neurones and the distribution of their synapses in ganglia of the leech: A study using intracellular injection of horseradish peroxidase. *Proc. R. Soc. Lond. B* 194: 481–499. [13]

Muller, K. J. and Nicholls, J. G. 1974. Different properties of synapses between a single sensory neurone and two different motor cells in the leech CNS. *J. Physiol.* 238: 357–369. [13]

Muller, K. J. and Nicholls, J. G. 1981. Regeneration and plasticity. *In* K. J. Muller, J. G. Nicholls and G. S. Stent (eds.). *Neurobiology of the Leech*. Cold Spring Harbor Laboratory, Cold Spring Harbor, N. Y. [12]

Muller, K. J. and Scott, S. A. 1981. Transmission at a "direct" electrical connexion mediated by an interneurone in the leech. *J. Physiol.* 311: 565–583. [13]

Muller, K. J., Nicholls, J. G. and Stent, G. S. (eds.). 1981. *Neurobiology of the Leech*. Cold Spring Harbor Laboratory, Cold Spring Harbor, NY, pp. 197–226. [13]

Mullins, L. J. and Noda, K. 1963. The influence of sodium-free solutions on membrane potential of frog muscle fibers. *J. Gen. Physiol.* 47: 117–132. [3]

Murphey, R. K., Johnson, S. E. and Sakaguchi, D. S. 1983. Anatomy and physiology of supernumerary cercal afferents in crickets: Implication for pattern formation. *J. Neurosci.* 3: 312–325. [12]

Murrell, J., Farlow, M., Ghetti, B. and Benson, M. D. 1991. A mutation in the amyloid precursor protein associated with hereditary Alzheimer's disease. *Science* 254: 97–99. [10]

Naegele, J. R. and Barnstable, C. J. 1989. Molecular determinants of GABAergic local-circuit neurons in the visual cortex. *Trends Neurosci.* 12: 28–34. [10]

Nagai, T., McGeer, P. L., Araki, M. and McGeer, E. G. 1984. GABA-T intensive neurons in the rat brain. *In* A. Björklund, T. Hökfelt and M. J. Kuhar (eds.). *Classical Transmitters and Transmitter Receptors in the CNS*, Part II, *Handbook of Chemical Neuroanatomy*, Vol. 3. Elsevier, Amsterdam pp. 247–272. [10]

Nagle, G. T., de Jong-Brink, M., Painter, S. D., Bergamin-Sassen, M. M., Blankenship, J. E. and Kurosky, A. 1990. Delta-bag cell peptide from the egg-laying hormone precursor of *Aplysia*. Processing, primary structure, and biological activity. *J. Biol. Chem.* 265: 22329–22335. [13]

Nakajima, S. and Onodera, K. 1969a. Membrane properties of the stretch receptor neurones of crayfish with particular reference to mechanisms of sensory adaptation. *J. Physiol.* 200: 161–185. [14]

Nakajima, S. and Onodera, K. 1969b. Adaptation of the generator potential in the crayfish stretch receptors under constant length and constant tension. *J. Physiol.* 200: 187–204. [14]

Nakajima, S. and Takahashi, K. 1966. Post-tetanic hyperpolarization and electrogenic Na pump in stretch receptor neurone of crayfish. *J. Physiol.* 187: 105–127. [14]

Nakajima, Y., Tisdale, A. D. and Henkart, M. P. 1973. Presynaptic inhibition at inhibitory nerve terminals: A new synapse in the crayfish stretch receptor. *Proc. Natl. Acad. Sci. USA* 70: 2462–2466. [7]

Nakamura, H. and O'Leary, D. D. M. 1989. Inaccuracies in initial growth and arborization of chick retinotectal axons followed by course corrections and axon remodeling to develop topographic order. *J. Neurosci.* 9: 3776–3795. [11]

Nakanishi, N. Schneider, N. A. and Axel, R. 1990. A family of glutamate receptor genes: Evidence for the formation of heteromultimeric receptors with distinct channel properties. *Neuron* 5: 569–581. [2]

Nakanishi, S. 1991. Mammalian tachykinin receptors. *Annu. Rev. Neurosci.* 14: 123–136. [8]

Narahashi, T., Moore, J. W. and Scott, W. R. 1964. Tetrodotoxin blockage of sodium conductance increase in lobster giant axons. *J. Gen. Physiol.* 47: 965–974. [4]

Nastuk, W. L. 1953. Membrane potential changes at a single muscle end-plate produced by transitory application of acetylcholine with an electrically controlled microjet. *Fed. Proc.* 12: 102. [7]

Nathans, J. 1987. Molecular biology of visual pigments. *Annu. Rev. Neurosci.* 10: 163–194. [16]

Nathans, J. 1989. The genes for color vision. *Sci. Am.* 260(2): 42–49. [16]

Nathans, J. 1990. Determinants of visual pigment absorbance: Identification of the retinylidene Schiff's base counterion in bovine rhodopsin. *Biochemistry* 29: 9746–9752. [16]

Nathans, J. and Hogness, D. S. 1984. Isolation and nucleotide sequence of the gene encoding human rhodopsin. *Proc. Natl. Acad. Sci. USA* 81: 4851–4855. [16]

Nauta, W. J. H. and Feirtag, M. 1986. *Fundamental Neuroanatomy*. Freeman, New York. [1]

Neher, E. and Sakmann, B. 1976. Single channel currents recorded from membrane of denervated frog muscle fibres. *Nature* 260: 799–801. [7]

Neher, E. and Steinbach, J. H. 1978. Local anaesthetics transiently block currents through single acetylcholine receptor channels. *J. Physiol.* 277: 153–176. [7]

Neher, E., Sakmann, B. and Steinbach, J. H. 1978. The extracellular patch clamp: A method for resolving currents through individual open channels in biological membranes. *Pflügers Arch.* 375: 219–228 [2]

Neuweiler, G. 1990. Auditory adaptations for prey capture in echolocating bats. *Physiol. Rev.* 70: 615–641. [14]

Newcomb, R. W. and Scheller, R. H. 1990. Regulated release of multiple peptides from the bag cell neurons of *Aplysia californica*. *Brain Res.* 521: 229–237. [13]

Newman, E. A. 1985. Voltage-dependent calcium and potassium channels in retinal glial cells. *Nature* 317: 809–811. [6]

Newman, E. A. 1986. High potassium conductance in astrocyte endfeet. *Science* 233: 453–454. [6]

Newman, E. A. 1987. Distribution of potassium conductance in mammalian Müller (glial) cells: A comparative study. *J. Neurosci.* 7: 2423–2432. [6]

Neyton, J. and Trautmann, A. 1985. Single-channel currents of an intercellular junction. *Nature* 317: 331–335. [5]

Nicholls, J. G. 1987. *The Search for Connections. Studies of Regeneration in the Nervous System of the Leech*. Sinauer, Sunderland, MA. [13]

Nicholls, J. G. and Baylor, D. A. 1968. Specific modalities and receptive fields of sensory neurons in the CNS of the leech. *J. Neurophysiol.* 31: 740–756. [13]

Nicholls, J. G. and Purves, D. 1972. A comparison of chemical and electrical synaptic transmission between single sensory cells and a motoneurone in the central nervous system of the leech. *J. Physiol.* 225: 637–656. [7, 13]

Nicholls, J. G. and Wallace, B. G. 1978. Modulation of transmission at an inhibitory synapse in the central nervous system of the leech. *J. Physiol.* 281: 157–170. [13]

Nicholls, J. G., Stewart, R. R., Erulkar, S. D. and Saunders, N. R. 1990. Reflexes, fictive respiration, and cell division in the brain and spinal cord of the newborn opossum, *Monodelphis domestica*, isolated and maintained in vitro. *J. Exp. Biol.* 152: 1–15. [15]

Nicoll, R. A. 1988. The coupling of neurotransmitter receptors to ion channels in the brain. *Science* 241: 545–551. [8]

Nicoll, R. A., Malenka, R. C. and Kauer, J. A. 1990. Functional comparison of neurotransmitter receptor subtypes in mammalian central nervous system. *Physiol. Rev.* 70: 513–565. [10]

Nicoll, R. A., Schenker, C. and Leeman, S. E. 1980. Substance P as a transmitter candidate. *Annu. Rev. Neurosci.* 3: 227–268. [10]

Niemierko, S. and Lubinska, L. 1967. Two fractions of axonal acetylcholinesterase exhibiting different behaviour in severed nerves. *J. Neurochem.* 14: 761–769. [9]

Nitkin, R. M., Smith, M. A., Magill, C., Fallon, J. R., Yao, M. Y.-M., Wallace, B. G. and McMahan, U. J. 1987. Identification of agrin, a synapse organizing protein from *Torpedo* electric organ. *J. Cell Biol.* 105: 2471–2478. [12]

Njå, A. and Purves, D. 1978. The effects of nerve growth factor and its antiserum on synapses in the superior cervical ganglion of the guinea-pig. *J. Physiol.* 277: 53–75. [12]

Noda, M., Takahashi, H., Tanabe, T., Toyosato, M., Furutani, Y., Hirose, T., Asai, M., Inayama, S., Miyata, T. and Numa, S. 1982. Primary structure of α-subunit precursor of *Torpedo californica* acetylcholine receptor deduced from a cDNA sequence. *Nature* 299: 793–797. [2]

Noda, M., Takahashi, H., Tanabe, T., Toyosato, M., Kikyotani, S., Hirose, T., Asai, M., Takashima, H., Inayama, S., Miyata, T. and Numa, S. 1983a. Primary structure of β- and δ-subunit precursors of *Torpedo californica* acetylcholine receptor deduced from cDNA sequences. *Nature* 301: 251–255. [2]

Noda, M., Takahashi, H., Tanabe, T., Toyosato, M., Kikyotani, S., Furutani, Y., Hirose, T., Takashima, H., Inayama, S., Miyata, T. and Numa, S. 1983b. Structural homology of *Torpedo californica* acetylcholine receptor subunits. *Nature* 302: 528–32. [2]

Noda, M., Shimizu, S., Tanabe, T., Takai, T., Kayano, T., Ikeda, T., Takahashi, H., Nakayama, H., Kanaoka, Y., Minamino, N., Kangawa, K., Matsuo, H., Raftery, M. A., Hirose, T., Inagama, S., Hayashida, H., Miyata, T. and Numa, S. 1984. Primary structure of *Electrophorus electricus* sodium channel deduced from cDNA sequence. *Nature* 312: 121–127. [2]

Noguchi, S., Noda, M., Takahashi, H., Kawakami, K., Ohta, T., Nagano, K., Hirosi, T., Inayama, S., Kawamura, M. and Numa, S. 1986. Primary structure of the β-subunit of *Torpedo californica* (Na$^+$-K$^+$)ATPase deduced from cDNA sequence. *FEBS Lett.* 196: 315–320. [3]

Nottebohm, F. 1989. From bird song to neurogenesis. *Sci. Am.* 260(2): 74–79. [11]

Numa, S., Noda, M., Takahashi, H., Tanabe, T., Toyosato, M., Furutani, Y. and Kikyotani, S. 1983. Molecular structure of the nicotinic acetylcholine receptor. *Cold Spring Harbor Symp. Quant. Biol.* 48: 57–69. [2]

Nusbaum, M. P., Friesen, W. O., Kristan, W. B., Jr. and Pearce, R. A. 1987. Neural mechanisms generating the leech swimming rhythm: Swim-initiator neurons excite the network of swim oscillator neurons. *J. Comp. Physiol. A* 161: 355–366. [13]

Nussbaumer, J. C. and Van der Loos, H. 1984. An electrophysiological and anatomical study of projections to the mouse cortical barrelfield and its surroundings. *J. Neurophysiol.* 53: 686–698. [14]

Nussinovitch, I. and Rahamimoff, R. 1988. Ionic basis of tetanic and post-tetanic potentiation at a mammalian neuromuscular junction. *J. Physiol.* 396: 435–455. [7]

Obaid, A. L., Socolar, S. J. and Rose, B. 1983. Cell-to-cell channels with two independently-regulated gates in series: Analysis of junctional conductance modulation by membrane potential, calcium and pH. *J. Memb. Biol.* 73: 68–89. [5]

Obata, K. 1969. Gamma-aminobutyric acid in Purkinje cells and motoneurones. *Experientia* 25: 1283. [10]

Obata, K., Takeda, K. and Shinozaki, H. 1970. Further study on pharmacological properties of the cerebellar-induced inhibition of Deiters neurones. *Exp. Brain Res.* 11: 327–342. [10]

O'Brien, R. A. D., Ostberg, A. J. C. and Vrbova, G. 1978. Observations on the elimination of polyneuronal innervation in developing mammalian skeletal muscle. *J. Physiol.* 282: 571–582. [11]

Ochoa, E. L., Chattopadhyay, A. and McNamee, M. G. 1989. Desensitization of the nicotinic acetylcholine receptor: Molecular mechanisms and effect of modulators. *Cell Mol. Neurobiol.* 9: 141–178. [9]

Ocorr, K. A. and Byrne, J. H. 1985. Membrane responses and changes in cAMP levels in *Aplysia* sensory neurons produced by serotonin, tryptamine, FMRFamide and small cardioactive peptide B (SCPB). *Neurosci. Lett.* 55: 113–118. [13]

Odette, L. L. and Newman, E. A. 1988. Model of potassium dynamics in the central nervous system. *Glia* 1: 198–210. [6]

O'Dowd, B. F., Lefkowitz, R. J. and Caron, M. G. 1989. Structure of the adrenergic and related receptors. *Annu. Rev. Neurosci.* 12: 67–83. [8]

Olavarria, J. and Van Sluyters, R. C. 1985. Organization and postnatal development of callosal connections in the visual cortex of the rat. *J. Comp. Neurol.* 239: 1–26. [18]

O'Leary, D. D. M. and Stanfield, B. B. 1989. Selective elimination of axons extended by developing cortical neurons is dependent on regional locale. *J. Neurosci.* 9: 2230–2246. [11]

O'Leary, D. D. M. and Terashima, T. 1988. Cortical axons branch to multiple subcortical targets by interstitial axon budding: Implications for target recognition and "waiting periods". *Neuron* 1: 901–910. [11]

O'Leary, D. D. M., Stanfield, B. B. and Cowan, W. M. 1981. Evidence that the early postnatal restriction of the cells of origin of the callosal projection is due to the elimination of axonal collaterals rather than to the death of neurons. *Dev. Brain Res.* 1: 607–617. [11]

Olsen, R. W. and Leeb-Lundberg, F. 1981. Convulsant and anticonvulsant drug binding sites related to GABA-regulated chloride ion channels. *In* E. Costa, G. DiChiara and G. L. Gessa (eds.). *GABA and Benzodiazepine Receptors*. Raven Press, New York, pp. 93–102. [10]

Olson, R. W. and Tobin, A. J. 1990. Molecular biology of GABA$_A$ receptors. *FASEB J.* 4: 1469–1480. [2]

Oppenheim, R. W. 1991. Cell death during development of the nervous system. *Annu. Rev. Neurosci.* 14: 453–501. [11]

Orbeli, L. A. 1923. Die sympathetische innervation der skelettmuskeln. *Bull. Inst. Sci. Leshaft* 6: 194–197. [8]

Orkand, P. M. and Kravitz, E. A. 1971. Localization of the sites of γ-aminobutyric acid (GABA) uptake in lobster nerve–muscle preparations. *J. Cell Biol.* 49: 75–89. [9]

Orkand, P. M., Bracho, H. and Orkand, R. K. 1973. Glial metabolism: Alteration by potassium levels comparable to those during neural activity. *Brain Res.* 55: 467–471. [6]

Orkand, R. K., Nicholls, J. G. and Kuffler, S. W. 1966. Effect of nerve impulses on the membrane potential of glial cells in the central nervous system of amphibia. *J. Neurophysiol.* 29: 788–806. [6]

O'Rourke, N. A. and Fraser, S. E. 1990. Dynamic changes in optic fiber terminal arbors lead to retinotopic map formation: An in vivo confocal microscopic study. *Neuron* 5: 159–171. [11]

Otsuka, M., Obata, K., Miyata, Y. and Tanaka, Y. 1971. Measurement of γ-aminobutyric acid in isolated nerve cells of cat central nervous system. *J. Neurochem.* 18: 287–295. [10]

Ottoson, D. 1956. Analysis of the electrical activity of the olfactory epithelium. *Acta Physiol. Scand.* 35(Suppl. 122): 1–83. [14]

Ottoson, D. 1983. *Physiology of the Nervous System*, Chapter 31. Oxford University Press, New York. [14]

Overton, E. 1902. Beitrage zur allgemeinen Muskel- und Nervenphysiologie. II. Über die Unentbehrlichkeit von Natrium- (oder Lithium-) Ionen für den Kontraktionsakt des Muskels. *Pflügers Arch.* 92: 346–386. [4]

Pacholczyk, T., Blakely, R. D. and Amara, S. G. 1991. Expression cloning of a cocaine and antidepressant-sensitive human noradenaline transporter. *Nature* 350: 350–354. [9]

Palacios, J. M., Waeber, C., Hoyer, D. and Mengod, G. 1990. Distribution of serotonin receptors. *Ann. NY Acad. Sci.* 600: 36–52. [10]

Palacios, J. M., Mengod, G., Vilaro, M. T., Wiederhold, K. H., Boddeke, H., Alvarez, F. J., Chinaglia, G. and Probst, A. 1990. Cholinergic receptors in the rat and human brain: Microscopic visualization. *Prog. Brain Res.* 84: 243–253. [10]

Palka, J. and Schubiger, M. 1988. Genes for neural differentiation. *Trends Neurosci.* 11: 515–517. [13]

Palmer, L. A. and Rosenquist, A. C. 1974. Visual receptive fields of single striate cortical units projecting to the superior colliculus in the cat. *Brain Res.* 67: 27–42. [17]

Panula, P., Airaksinen, M. S., Pirvola, U. and Kotilainen, E. 1990. A histamine-containing neuronal system in human brain. *Neuroscience* 34: 127–132. [10]

Papazian, D. M., Timpe, L. C., Jan, N. Y. and Jan, L. Y. 1991. Alteration of voltage-dependence of Shaker potassium channel by mutations in the S4 sequence. *Nature* 349: 305–310. [2]

Parnas, H., Parnas, I. and Segel, L. A. 1990. On the contribution of mathematical models to the understanding of neurotransmitter release. *Int. Rev. Neurobiol.* 32: 1–50. [7]

Parnas, I. 1987. Strengthening of synaptic inputs after elimination of a single neurone innervating the same target. *J. Exp. Biol.* 132: 231–247. [12, 13]

Parnas, I. and Strumwasser, F. 1974. Mechanisms of long-lasting inhibition of a bursting pacemaker neuron. *J. Neurophysiol.* 37: 609–620. [13]

Partridge, L. D. and Thomas, R. C. 1976. The effects of lithium and sodium on potassium conductance in snail neurones. *J. Physiol.* 254: 551–563. [4]

Paschal, B. M. and Vallee, R. B. 1987. Retrograde transport by the microtubule associated protein MAP 1C. *Nature* 330: 181–183. [9]

Patterson, P. H. and Chun, L. L. Y. 1974. The influence of non-neuronal cells on catecholamine and acetylcholine synthesis and accumulation in cultures of dissociated sympathetic neurons. *Proc. Natl. Acad. Sci. USA* 71: 3607–3610. [11]

Patterson, P. H. and Chun, L. L. Y. 1977. The induction of acetylcholine synthesis in primary cultures of dissociated rat sympathetic neurons. I. Effects of conditioned medium. *Dev. Biol.* 56: 263–280. [11]

Patterson, P. H. and Purves, D. (eds). 1982. *Readings in Developmental Neurobiology.* Cold Spring Harbor Laboratory, Cold Spring Harbor, NY. [11]

Paul, D. L. 1986. Molecular cloning of cDNA for rat liver gap junction protein. *J. Cell Biol.* 103: 123–134. [5]

Paulson, O. B. and Newman, E. A. 1987. Does the release of potassium from astrocyte endfeet regulate cerebral blood flow? *Science* 237: 896–898. [6]

Payton, W. B. 1981. History of medicinal leeching and early medical references. In K. J. Muller, J. G. Nicholls and G. S. Stent (eds.) *Neurobiology of the Leech.* Cold Spring Harbor Laboratory, Cold Spring Harbor, NY, pp. 27–34. [13]

Pearlman, A. L., Birch, J. and Meadows, J. C. 1979. Cerebral color blindness: an acquired defect in hue discrimination. *Ann. Neurol.* 5: 253–261. [17]

Pearson, K. 1976. The control of walking. *Sci. Am.* 235(6): 72–86. [15]

Pearson, K. G. and Iles, J. F. 1970. Discharge patterns of coxal levator and depressor motoneurons of the cockroach *Periplaneta americana. J. Exp. Biol.* 52: 139–165. [15]

Pelosi, P. and Tirindelli, R. 1989. Structure/activity studies and characterization of an odorant-binding protein. In J. G. Brand, J. H. Teeter, R. H. Cagan and M. R. Kare (eds.), *Receptor Events in Transduction and Olfaction.* Volume 1 of *Chemical Senses.* Marcel Dekker, New York, pp. 207–226. [14]

Penfield, W. 1932. *Cytology and Cellular Pathology of the Nervous System*, Vol. II. Hafner, New York. [6]

Penfield, W. and Rasmussen, T. 1950. *The Cerebral Cortex of Man: A Clinical Study of Localization of Function.* Macmillan, New York. [14, 15]

Penner, R. and Neher, E. 1989. The patch-clamp technique in the study of secretion. *Trends Neurosci.* 12: 159–163. [9]

Peper, K., Dreyer, F., Sandri, C., Akert, K. and Moore, H. 1974. Structure and ultrastructure of the frog motor end-plate: A freeze-etching study. *Cell Tissue Res.* 149: 437–455. [7]

Perett, D. I., Mistlin, A. J. and Chitty, A. J. 1987. Visual neurones responsive to faces. *Trends Neurosci.* 10: 358–364. [17]

Perry, V. H. and Gordon, S. 1988. Macrophages and microglia in the nervous system. *Trends Neurosci.* 11: 273–277. [6]

Perschak, H. and Cuenod, M. 1990. In vivo release of endogenous glutamate and aspartate in the rat striatum during stimulation of the cortex. *Neuroscience* 35: 283–287. [9]

Pert, C. B. and Snyder, S. H. 1973. Opiate receptor: Demonstration in nervous tissue. *Science* 179: 1011–1014. [10]

Peters, A., Palay, S. L. and Webster, H. de F. 1976. *The Fine Structure of the Nervous System*. Saunders, Philadelphia. [1, 6, 15]

Pevsner, J., Sklar, P. B., Hwang, P. M. and Snyder, S. H. 1989. Odorant-binding protein. Sequence analysis and localization suggest an odorant transport function. *In* J. G. Brand, J. H. Teeter, R. H. Cagan and M. R. Kare (eds.), *Receptor Events in Transduction and Olfaction*. Volume 1 of *Chemical Senses*. Marcel Dekker, New York, pp. 227–242. [14]

Pfaffinger, P. J., Martin, J. M., Hunter, D. D., Nathanson, N. M. and Hille, B. 1985. GTP-binding proteins couple cardiac muscarinic receptors to a K channel. *Nature* 317: 536–538. [8]

Phelan, K. A. and Hollyday, M. 1990. Axon guidance in muscleless chick wings: The role of muscle cells in motoneuronal pathway selection and muscle nerve formation. *J. Neurosci.* 10: 2699–2716. [11]

Pickenhagen, W. 1989. Summation of the conference. *In* J. G. Brand, J. H. Teeter, R. H. Cagan and M. R. Kare (eds.), *Receptor Events in Transduction and Olfaction*. Volume 1 of *Chemical Senses*. Marcel Dekker, New York, pp. 505–509. [14]

Piomelli, D., Volterra, A., Dale, N., Siegelbaum, S. A., Kandel, E. R., Schwartz, J. H. and Belardetti, F. 1987. Lipoxygenase metabolites of arachidonic acid as second messengers for presynaptic inhibition of *Aplysia* sensory cells. *Nature* 328: 38–43. [8]

Poggio, G. F. and Mountcastle, V. B. 1960. A study of the functional contributions of the lemniscal and spinothalamic systems to somatic sensibility. *Bull. Johns Hopkins Hosp.* 106: 266–316. [14]

Pons, T. P., Garraghty, P. E., Ommaya, A. K., Kaas, J. H., Taub, E. and Mishkin, M. 1991. Massive cortical reorganization after sensory deafferentation in adult macaques. *Science* 252: 1857–1860. [18]

Poritsky, R. 1969. Two- and three-dimensional ultrastructure of boutons and glial cells on the motoneuronal surface in the cat spinal cord. *J. Comp. Neurol.* 135: 423–452. [1, 15]

Porter, C. W. and Barnard, E. A. 1975. The density of cholinergic receptors at the postsynaptic membrane: Ultrastructural studies in two mammalian species. *J. Membr. Biol.* 20: 31–49. [7]

Porter, L. L. , Sakamoto, T. and Asanuma, H. 1990. Morphological and physiological identification of neurons in the cat motor cortex which receive direct input from the somatic sensory cortex. *Exp. Brain Res.* 80: 209–212. [15]

Porter, M. E. and Johnson, K. A. 1989. Dynein structure and function. *Annu. Rev. Cell Biol.* 5: 119–151. [9]

Potter, L. T. 1970. Synthesis, storage, and release of [^{14}C]-acetylcholine in isolated rat diaphragm muscles. *J. Physiol.* 206: 145–166. [9]

Powell, T. P. S. and Mountcastle, V. B. 1959. Some aspects of the functional organization of the cortex of the postcentral gyrus of the monkey: A correlation of findings obtained in a single unit analysis with cytoarchitecture. *Bull. Johns Hopkins Hosp.* 105: 133–162. [14]

Prell, G. D. and Green, J. P. 1986. Histamine as a neuroregulator. *Annu. Rev. Neurosci.* 9: 209–254. [10]

Pritchett, D. B., Sontheimer, H., Shivers, B. D. S., Ymer, S., Kettenmann, H., Schofield, P. R. and Seeburg, P. H. 1989. Importance of a novel GABA$_A$ receptor subunit for benzodiazepine pharmacology. *Nature* 338: 582–585. [10]

Purves, D. 1975. Functional and structural changes in mammalian sympathetic neurones following interruption of their axons. *J. Physiol.* 252: 429–463. [12]

Purves, D. and Lichtman, J. W. 1983. Specific connections between nerve cells. *Annu. Rev. Physiol.* 45: 553–565. [11]

Purves, D. and Lichtman, J. W. 1985. *Principles of Neural Development*. Sinauer, Sunderland, MA. [11, 12]

Purves, D. and Sakmann, B. 1974a. Membrane properties underlying spontaneous activity of denervated muscle fibers. *J. Physiol.* 239: 125–153. [12]

Purves, D. and Sakmann, B. 1974b. The effect of contractile activity on fibrillation and extrajunctional acetylcholine sensitivity in rat muscle maintained in organ culture. *J. Physiol.* 237: 157–182. [12]

Purves, D., Thompson, W. and Yip, J. W. 1981. Reinnervation of ganglia transplanted to the neck from different levels of the guinea pig sympathetic chain. *J. Physiol.* 313: 49–63. [12]

Quick, D. C., Kennedy, W. R. and Donaldson, L. 1979. Dimensions of myelinated nerve fibers near the motor and sensory terminals in cat tenuissimus muscles. *Neuroscience* 4: 1089–1096. [5]

Quilliam, T. A. and Armstrong, J. 1963. Mechanoreceptors. *Endeavour* 22: 55–60. [14]

Rabacchi, S. A., Neve, R. L. and Dräger, U. C. 1990. A positional marker for the dorsal embryonic retina is homologous to the high-affinity laminin receptor. *Development* 109: 521–531. [11]

Raff, M. C. 1989. Glial cell diversification in the rat optic nerve. *Science* 243: 1450–1455. [6]

Raftery, M. A., Hunkapiller, M. W., Strader, C. D. and Hood, L. E. 1980. Acetylcholine receptor: Complex of homologous subunits. *Science* 208: 1454–1457. [2]

Ragsdale, C. and Woodgett, J. 1991. *trking* neurotrophic receptors. *Nature* 350: 660–661. [11]

Rakic, P. 1971. Neuron–glia relationship during granule cell migration in developing cerebellar cortex. A Golgi and electron-microscopic study in *Macacus rhesus*. *J. Comp. Neurol.* 141: 283–312. [6, 11]

Rakic, P. 1972. Mode of cell migration to the superficial layers of the fetal monkey neocortex. *J. Comp. Neurol.* 145: 61–83. [11]

Rakic, P. 1974. Neurons in rhesus monkey visual cortex: Systematic relationship between time of origin and eventual disposition. *Science* 183: 425–427. [11]

Rakic, P. 1977a. Genesis of the dorsal lateral geniculate nucleus in the rhesus monkey: Site and time of origin, kinetics of proliferation, routes of migration and pattern of distribution of neurons. *J. Comp. Neurol.* 176: 23–52. [18]

Rakic, P. 1977b. Prenatal development of the visual system in rhesus monkey. *Philos. Trans. R. Soc. Lond. B* 278: 245–260. [18]

Rakic, P. 1986. Mechanism of ocular dominance segregation in the lateral geniculate nucleus: Competitive elimination hypothesis. *Trends Neurosci.* 9: 11–15. [18]

Rakic, P. 1988a. Defects of neuronal migration and the pathogenesis of cortical malformations. *Prog. Brain. Res.* 73: 15–37. [6]

Rakic, P. 1988b. Specification of cerebral cortical areas. *Science* 241: 170–176. [6, 18]

Rall, W. 1967. Distinguishing theoretical synaptic potentials computed from different soma-dendritic distributions of synaptic input. *J. Neurophysiol.* 30: 1138–1168. [5]

Ramirez, J. M. and Pearson, K. G. 1988. Generation of motor patterns for walking and flight in motoneurons supplying bifunctional muscles in the locust. *J. Neurobiol.* 19: 257–282. [13]

Ramón y Cajal, S. 1955. *Histologie du Système Nerveux*, Vol. II. C.S.I.C., Madrid. [6]

Rane, S. G. and Dunlap, K. 1986. Kinase C activator, 1,2-oleoylacetylglycerol attenuates voltage-dependent calcium current in sensory neurons. *Proc. Natl. Acad. Sci. USA* 83: 184–188. [10]

Ransom, B. R. and Goldring, S. 1973. Slow depolarization in cells presumed to be glia in cerebral cortex of cat. *J. Neurophysiol.* 36: 869–878. [6]

Raper, J. A. and Kapfhammer, J. P. 1990. The enrichment of a neuronal growth cone collapsing activity from embryonic chick brain. *Neuron* 2: 21–29. [11]

Rasmussen, C. D. and Means, A. R. 1989. Calmodulin, cell growth and gene expression. *Trends Neurosci.* 12: 433–438. [8]

Ratnam, M., Le Nguyen, D., Rivier, J., Sargent, P.B. and Lindstrom, J. 1986. Transmembrane topography of nicotinic acetylcholine receptor: Immunochemical tests contradict theoretical predictions based on hydrophobicity profiles. *Biochemistry* 25: 2633–2643. [2]

Rauschecker, J. P. and Singer, W. 1980. Changes in the circuitry of the kitten visual cortex are gated by postsynaptic activity. *Nature* 280: 58–60. [18]

Ravdin, P. and Axelrod, D. 1977. Fluorescent tetramethyl rhodamine derivatives of α-bungarotoxin: Preparation, separation, and characterization. *Anal. Biochem.* 80: 585–592. [9]

Raviola, E. and Dacheux, R. F. 1987. Excitatory dyad synapse in rabbit retina. *Proc. Natl. Acad. Sci. USA* 84: 7324–7328. [16]

Raviola, E. and Dacheux, R. F. 1990. Axonless horizontal cells of the rabbit retina: Synaptic connections and origin of the rod aftereffect. *J. Neurocytol.* 19: 731–736. [16]

Rayport, S. G. and Schacher, S. 1986. Synaptic plasticity in vitro: Cell culture of *Aplysia* neurons mediating short-term habituation and sensitization. *J. Neurosci.* 6: 759–763. [13]

Ready, D. F., Handson, T. E. and Benzer, S. 1976. Development of the *Drosophila* retina, a neurocrystalline lattice. *Dev. Biol.* 53: 217–240. [11]

Recio-Pinto, E., Thornhill, W. B., Duch, D. S., Levinson, S. R. and Urban, B. W. 1990. Neuraminidase treatment modifies the function of electroplax sodium channels in planar lipid bilayers. *Neuron* 5: 675–684. [2]

Redfern, P. A. 1970. Neuromuscular transmission in new-born rats. *J. Physiol.* 209: 701–709. [11, 18]

Reese, T. S. and Karnovsky, M. J. 1967. Fine structural localization of a blood–brain barrier to exogenous peroxidase. *J. Cell Biol.* 34: 207–217. [6]

Reichert, H. and Rowell, C. H. 1985. Integration of non-phaselocked exteroceptive information in the control of rhythmic flight in the locust. *J. Neurophysiol.* 53: 1201–1218. [13]

Reichardt, L. F. and Kelly, R. B. 1983. A molecular description of nerve terminal function. *Annu. Rev. Biochem.* 52: 871–926. [9]

Reichardt, L. F. and Tomaselli, K. J. 1991. Extracellular matrix molecules and their receptors: Functions in neural development. *Annu. Rev. Neurosci.* 14: 531–570. [11]

Reiter, H. O., Waitzman, D. M. and Stryker, M. P. 1986. Cortical activity blockade prevents ocular dominance plasticity in the kitten visual cortex. *Exp. Brain Res.* 65: 182–188. [18]

Reuter, H. 1974. Localization of β adrenergic receptors, and effects of noradrenaline and cyclic nucleotides on action potentials, ionic currents and tension in mammalian cardiac muscle. *J. Physiol.* 242: 429–451. [8]

Reuter, H., Cachelin, A. B., DePeyer, J. E. and Kokubun, S. 1983. Modulation of calcium channels in cultured cardiac cells by isoproternenol and 8-bromo-cAMP. *Cold Spring Harbor Symp. Quant. Biol.* 48: 193–200. [8]

Ribchester, R. R. and Taxt, T. 1983. Motor unit size and synaptic competition in rat lumbrical muscles reinnervated by active and inactive motor axons. *J. Physiol.* 344: 89–111. [11]

Ribchester, R. R. and Taxt, T. 1984. Repression of inactive motor nerve terminals in partially denervated rat muscle after regeneration of active motor axons. *J. Physiol.* 347: 497–511. [11]

Richards, J. G., Schoch, P. and Haefely, W. 1991. Benzodiazepine receptors: New vistas. *Sem. Neurosci.* 3: 191–203. [19]

Richards, J. G., Schoch, P., Haring, P., Takacs, B. and Mohler, H. 1987. Resolving GABA$_A$/benzodiazepine receptors: Cellular and subcellular localization in the CNS with monoclonal antibodies. *J. Neurosci.* 7: 1866–1886. [9]

Richardson, P. M., McGuinness, U. M. and Aguayo, A. J. 1980. Axons from CNS neurones regenerate into PNS grafts. *Nature* 284: 264–265. [12]

Ridley, R. M. and Baker, H. F. 1991. Can fetal neural transplants restore function in monkeys with lesion-induced behavioural deficits? *Trends Neurosci.* 14: 366–370. [12]

Riesen, A. H. and Aarons, L. 1959. Visual movement and intensity discrimination in cats after early deprivation of pattern vision. *J. Comp. Physiol. Psychol.* 52: 142–149. [18]

Ringham, G. 1975. Localization and electrical characteristics of a giant synapse in the spinal cord of the lamprey. *J. Physiol.* 251: 385–407. [7]

Ripps, H. and Witkovsky P. 1985. Neuron–glia interaction in the brain and retina. *Progr. Retinal Res.* 4: 181–219. [6]

Ritchie, J. M. 1986. Distribution of saxitoxin binding sites in mammalian neural tissue. *In* C. Y. Kao and S. R. Levinson (eds.). *Tetrodotoxin, Saxitoxin and the Molecular Biology of the Sodium Channel. Ann. NY Acad. Sci.* 479: 385–401. [4]

Ritchie, J. M. 1987. Voltage-gated cation and anion channels in mammalian Schwann cells and astrocytes. *J. Physiol.* (Paris) 82: 248–257. [2, 6]

Ritchie, J. M. 1990. Voltage-gated cation and anion channels in the satellite cells of the mammalian nervous system. *Advances in Neural Regeneration Research*, pp. 237–252. [6]

Ritchie, J. M., Black, J. A., Waxman, S. G. and Angelides, K. J. 1990. Sodium channels in the cytoplasm of Schwann cells. *Proc. Natl. Acad. Sci. USA* 87: 9290–9294. [6]

Ritz, M. C., Cone, E. J. and Kuhar, M. J. 1990. Cocaine inhibition of ligand binding at dopamine, norepinephrine, and serotonin transporters: A structure-activity study. *Life Sci.* 46: 635–645. [9]

Roberts, A. and Bush, B. M. H. 1971. Coxal muscle receptors in the crab: The receptor current and some properties of the receptor nerve fibres. *J. Exp. Biol.* 54: 515–524. [14]

Roberts, J. L. and Herbert, E. 1977. Characterization of a common precursor to corticotropin and β-lipotropin: Cell-free synthesis of the precursor and identification of corticotropin peptides in the molecule. *Proc. Natl. Acad. Sci. USA* 77: 4826–4830. [10]

Robitaille, R., Adler, E. M., and Charlton, M. P. 1990. Strategic location of calcium channels at release sites of frog neuromuscular synapses. *Neuron* 5: 773–779. [7]

Rodriguez-Tebar, A., Dechant, G. and Barde, Y.-A. 1991. Neurotrophins: structural relatedness and receptor interactions. *Philos. Trans. R. Soc. Lond. B* 331: 255–258. [11]

Rogart, R. B., Cribbs, L. L., Muglia, L. K., Kephart, D. D. and Kaiser, M. W. 1989. Molecular cloning of a putative tetrodotoxin-resistant heart Na$^+$ channel isoform. *Proc. Natl. Acad. Sci. USA* 86: 8170–8174. [2]

Rogers, J. H. 1989. Two calcium-binding proteins mark many chick sensory neurons. *Neuroscience* 31: 697–709. [1]

Roland, P. E., Larson, B., Lasser, N. A. and Skinhoj, E. 1980. Supplementary motor area and other cortical areas in organization of voluntary movements in man. *J. Neurophysiol.* 43: 118–136. [15]

Roper, S. D. 1983. Regenerative impulses in taste cells. *Science* 220: 1311–1312. [14]

Roper, S. D. 1989. The cell biology of vertebrate taste receptors. *Annu. Rev. Neurosci.* 12: 329–353. [14]

Roper, S. D. and McBride D. W. 1989. Distribution of ion channels of taste cells and its relationship to chemosensory transduction. *J. Membr. Biol.* 109: 29–39. [14]

Rose, S. P. R. 1991. How chicks make memories: The cellular cascade from c-*fos* to dendritic remodelling. *Trends Neurosci.* 14: 390–397. [8]

Rosenbluth, J. 1988. Role of glial cells in the differentiation and function of myelinated axons. *Int. J. Dev. Neurosci.* 6: 3–24. [6]

Rosenthal, J. L. 1969. Post-tetanic potentiation at the neuromuscular junction of the frog. *J. Physiol.* 203: 121–133. [7]

Ross, W. N., Aréchiga, H. and Nicholls, J. G. 1988. Influence of substrate on the distribution of calcium channels in identified leech neurons in culture. *Proc. Natl. Acad. Sci. USA* 85: 4075–4078. [8, 13]

Ross, W. N., Lasser-Ross, N. and Werman, R. 1990. Spatial and temporal analysis of calcium-dependent electrical activity in guinea pig Purkinje cell dendrites. *Proc. R. Soc. Lond. B* 240: 173–185. [4, 8]

Rossant, J. 1985. Interspecific cell markers and lineage in mammals. *Philos. Trans. R. Soc. Lond. B* 312: 91–100. [11]

Rotshenker, S. 1988. Multiple modes and sites for the induction of axonal growth. *Trends Neurosci.* 11: 363–366. [12]

Rotzler, S., Schramek, H. and Brenner, H. R. 1991. Metabolic stabilization of endplate acetylcholine receptors regulated by Ca^{2+} influx associated with muscle activity. *Nature* 349: 337–339. [12]

Rovainen, C. M. 1967. Physiological and anatomical studies on large neurons of the central nervous system of the sea lamprey (*Petromyzon marinus*). II. Dorsal cells and giant interneurons. *J. Neurophysiol.* 30: 1024–1042. [7]

Rubin, G. M. 1989. Development of the *Drosophila* retina: Inductive events studied at single cell resolution. *Cell* 57: 519–520. [11]

Rubin, L. L., Barbu, K., Bard, F., Cannon, C., Hall, D. E., Horner, H., Janatpour, M., Liaw, C., Manning, K., Morales, J., Porter, S., Tanner, L., Tomaselli, K. and Yednock, T. 1992. Differentiation of brain endothelial cells in cell culture. *Ann. NY Acad. Sci.* 633: 420–425. [6]

Rudy, B. 1988. Diversity and ubiquity of K channels. *Neuroscience* 25: 729–749. [4]

Rupp, F., Payan, D. G., Magill-Solc, C., Cowan, D. M. and Scheller, R. H. 1991. Structure and expression of a rat agrin. *Neuron* 6: 811–823. [12]

Ruppersburg, J. P., Schröter, K. H., Sakmann, B., Stocker, M., Sewing, S. and Pongs, O. 1990. Heteromultimeric channels formed by rat brain potassium channel proteins. *Nature* 345: 535–537. [2]

Rushton, W. A. H. 1951. A theory of the effects of fibre size in medullated nerve. *J. Physiol.* 115: 101–122. [5]

Russell, J. M. 1983. Cation-coupled chloride influx in squid axon. Role of potassium and stoichiometry of the transport process. *J. Gen. Physiol.* 81: 909–925. [3]

Saadat, S., Sendtner, M. and Rohrer, H. 1989. Ciliary neurotrophic factor induces cholinergic differentiation of rat sympathetic neurons in culture. *J. Cell Biol.* 108: 1807–1816. [11]

Sachs, F. 1988. Mechanical transduction in biological systems. *CRC Crit. Rev. Biomed. Eng.* 16: 141–169. [14]

Sahley, C. L., Martin, K. A. and Gelperin, A. 1990. Analysis of associative learning in the terrestrial mollusc *Limax maximus*. II. Appetitive learning. *J. Comp. Physiol. A* 167: 339–345. [13]

Sakamoto, T., Arissan, K. and Asanuma, H. 1989. Functional role of sensory cortex in learning motor skills in cats. *Brain Res.* 503: 258–264. [15]

Sakmann, B., Noma, A. and Trautwein, W. 1983. Acetylcholine activation of single muscarinic K$^+$ channels in isolated pacemaker cells of the mammalian heart. *Nature* 303: 250–253. [8]

Sakmann, B., Edwards, F., Konnerth, A. and Takahashi, T. 1989. Patch clamp techniques used for studying synaptic transmission in slices of mammalian brain. *Q. J. Exp. Physiol.* 74: 1107–1118. [7]

Salpeter, M. M. (ed.). 1987. *The Vertebrate Neuromuscular Junction*. Alan R. Liss, New York. [11]

Salpeter, M. M. 1987. Vertebrate neuromuscular junctions: General morphology, molecular organization, and function consequences. *In* M. M. Salpeter (ed.). *The Vertebrate Neuromuscular Junction*. Alan R. Liss, New York, pp. 1–54. [9]

Salpeter, M. M. and Loring, R. H. 1985. Nicotinic acetylcholine receptors in vertebrate muscle: Properties, distribution and neural control. *Prog. Neurobiol.* 25: 297–325. [12]

Salpeter, M. M. and Marchaterre, M. 1992. Acetylcholine receptors in extrajunctional regions of innervated muscle have a slow degradation rate. *J. Neurosci.* 12: 35–38. [12]

Sanes, J. R. 1989. Analyzing cell lineage with a recombinant retrovirus. *Trends Neurosci.* 12: 21–28. [11]

Sanes, J. R., Marshall, L. M. and McMahan, U. J. 1978. Reinnervation of muscle fiber basal lamina after removal of muscle fibers. *J. Cell Biol.* 78: 176–198. [12]

Sargent, P. B. and Pang, D. Z. 1988. Denervation alters the size, number, and distribution of clusters of acetylcholine receptor-like molecules on frog cardiac ganglion neurons. *Neuron* 1: 877–886. [12]

Sargent, P. B. and Pang, D. Z. 1989. Acetylcholine receptor-like molecules are found in both synaptic and extrasynaptic clusters on the surface of neurons in the frog cardiac ganglion. *J. Neurosci.* 9: 1062–1072. [12]

Sargent, P. B., Yau, K.-W. and Nicholls, J. G. 1977. Extrasynaptic receptors on cell bodies of neurons in central nervous system of leech. *J. Neurophysiol.* 40: 446–452. [13]

Sarthy, P. V., Hendrickson, A. E. and Wu, J. Y. 1986. L-glutamate: A neurotransmitter candidate for cone photoreceptors in the monkey retina. *J. Neurosci.* 6: 637–643. [16]

Schacher, S., Montarolo, P. and Kandel, E. R. 1990. Selective short- and long-term effects of serotonin, small cardioactive peptide, and tetanic stimulation on sensorimotor synapses of *Aplysia* in culture. *J. Neurosci.* 10: 3286–3294. [13]

Schacher, S., Glanzman, D., Barzilai, A., Dash, P., Grant, S. G. N., Keller, F., Mayford, M. and Kandel. E. R. 1990. Long-term facilitation in *Aplysia*: Persistent phosporylation and structural changes. *Cold Spring Harbor Symp. Quant. Biol.* 55: 187–202. [13]

Schachner, M. 1982. Cell type-specific surface antigens in the mammalian nervous system. *J. Neurochem.* 39: 1–8. [6]

Schalling, M., Stieg, P. E., Lindquist, C., Goldstein, M. and Hökfelt, T. 1989. Rapid increase in enzyme and peptide mRNA in sympathetic ganglia after electrical stimulation in humans. *Proc. Natl. Acad. Sci. USA* 86: 4302–4305. [9]

Scheutze, S. M. and Role, L. W. 1987. Developmental regulation of nicotinic acetylcholine receptors. *Annu. Rev. Neurosci.* 10: 403–457. [11, 12]

Schiller, P. H. and Lee, K. 1991. The role of the primate extrastriate area V4 in vision. *Science* 251: 1251–1253. [17]

Schiller, P. H. and Logothetis, N. K. 1990. The color-opponent and broad-band channels of the primate visual system. *Trends Neurosci.* 13: 392–398. [16, 17]

Schiller, P. H. and Malpeli, J. G. 1978. Functional specificity of lateral geniculate nucleus laminae of the rhesus monkey. *J. Neurophysiol.* 41: 788–797. [16]

Schlaggar, B. L. and O'Leary, D. D. M. 1991. Potential of visual cortex to develop an array of functional units unique to somatosensory cortex. *Science* 252: 1556–1560. [11, 18]

Schmidt, R. F. 1971. Presynaptic inhibition in the vertebrate central nervous system. *Ergeb. Physiol.* 63: 20–101. [7]

Schnapf, J. L. and Baylor, D. A. 1987. How photoreceptor cells respond to light. *Sci. Am.* 256(4): 40–47. [16]

Schnapf, J. L., Kraft, T. W., Nunn, B. J. and Baylor, D. A. 1988. Spectral sensitivity of primate photoreceptors. *Vis. Neurosci.* 1: 255–261. [16]

Schnapf, J. L., Nunn, B. J., Meister, M. and Baylor, D.A. 1990. Visual transduction in cones of the monkey *Macaca fascicularis. J. Physiol.* 427: 681–713. [16]

Schnapp, B. J. and Reese, T. S. 1989. Dynein is the motor for retrograde axonal transport of organelles. *Proc. Natl. Acad. Sci. USA* 86: 1548–1552. [9]

Schnapp, B. J., Vale, R. D., Sheetz, M. P. and Reese, T. S. 1985. Single microtubules from squid axoplasm support bidirectional movement of organelles. *Cell* 40: 455–462. [9]

Schnell, L. and Schwab, M. E. 1990. Axonal regeneration in the rat spinal cord produced by an antibody against myelin-associated neurite growth inhibitors. *Nature* 343: 269–272. [12]

Schofield, P. R. 1989. GABA$_A$ receptor complexity. *Trends Pharmacol. Sci.* 10: 476–478. [2]

Schofield, P. R., Darlison, M. G., Fujita, N., Burt, D. R., Stephenson, F. A., Rodriguez, H., Rhee, L. M., Ramchandran, J., Reale, V., Glencorse, T. A., Seeburg, P. H. and Barnard, E. A. 1987. Sequence and functional expression of the GABA$_A$ receptor shows a ligand-gated receptor superfamily. *Nature* 328: 221–227. [2]

Schon, F. and Kelly, J. S. 1974. Autoradiographic localization of [³H]GABA and [³H]glutamate over satellite glial cells. *Brain Res.* 66: 275–288. [6]

Schramm, M. and Selinger, Z. 1984. Message transmission: Receptor-controlled adenylate cyclase system. *Science* 225: 1350–1356. [8]

Schrieber, M. H. and Thach, W. T., Jr. 1985. Trained slow tracking. II. Bidirectional discharge patterns of cerebellar nuclear, motor cortex and spindle afferent neurons. *J. Neurophysiol.* 54: 1228–1270. [15]

Schuch, U., Lohse, M. J. and Schachner, M. 1989. Neural cell adhesion molecules influence second messenger systems. *Neuron* 3: 13–20. [11]

Schultzberg, M., Hökfelt, T. and Lundberg, J. M. 1982. Coexistence of classical neurotransmitters and peptides in the central and peripheral nervous system. *Br. Med. Bull.* 38: 309–313. [9]

Schwab, M. E. 1991. Nerve fibre regeneration after traumatic lesions of the CNS: Progress and problems. *Philos. Trans. R. Soc. Lond. B* 331: 303–306. [6, 12]

Schwab, M. E. and Caroni, P. 1988. Oligodendrocytes and CNS myelin are nonpermissive substrates for neurite growth and fibroblast spreading in vitro. *J. Neurosci.* 8: 2381–2393. [12]

Schwartz, A. B., Kettner, R. E. and Georgopoulos, A. P. 1988. Primate motor cortex and free arm movement to visual targets in three–dimensional space. I. Relations between cell discharge and angle of movement. *J. Neurosci.* 8: 2913–2927. [15]

Schwartz, E. A. 1987. Depolarization without calcium can release γ-aminobutyric acid from a retinal neuron. *Science* 238: 350–355. [9]

Schweizer, F. E., Schäfer, T., Taparelli, C., Grob, M., Karli, U. O., Heumann, R., Thoenen, H., Bookman, R. J. and Burger, M. M. 1989. Inhibition of exocytosis by intracellularly applied antibodies against a chromaffin granule-binding protein. *Nature* 339: 709–712. [9]

Scott, S. A. 1975. Persistence of foreign innervation on reinnervated goldfish extraocular muscles. *Science* 189: 644–646. [12]

Scott, S. A. and Muller, K. J. 1980. Synapse regeneration and signals for directed growth in the central nervous system of the leech. *Dev. Biol.* 80: 345–363. [12]

Sears, T. A. 1964a. Efferent discharges in alpha and fusimotor fibers of intercoastal nerves of the cat. *J. Physiol.* 174: 295–315. [14, 15]

Sears, T. A. 1964b. The slow potentials of thoracic respiratory motoneurones and their relation to breathing. *J. Physiol.* 175: 404–424. [15]

Seilheimer, B. and Schachner, M. 1988. Studies of adhesion molecules mediating interactions between cells of peripheral nervous system indicate a major role for L1 in mediating sensory neuron growth on Schwann cells in culture. *J. Cell Biol.* 107: 341–351. [11]

Selkoe, D. J. 1991. The molecular pathology of Alzheimer's disease. *Neuron* 6: 487–498. [10]

Sellick, P. M., Patuzzi, R. and Johnstone, B. M. 1982. Measurement of basilar membrane motion in the guinea pig using the Mossbauer technique. *J. Acoust. Soc. Am.* 72: 131–141. [14]

Shapley, R. and Perry, V. H. 1986. Cat and monkey retinal ganglion cells and their visual functional roles. *Trends Neurosci.* 9: 229–235. [16]

Shapovalov, A. I. and Shiriaev, B. I. 1980. Dual mode of junctional transmission at synapses between single primary afferent fibres and motoneurones in the amphibian. *J. Physiol.* 306: 1–15. [7]

Shatz, C. J. 1977. Anatomy of interhemispheric connections in the visual system of Boston Siamese and ordinary cats. *J. Comp. Neurol.* 173: 497–518. [17]

Shatz, C. J. 1990a. Impulse activity and the patterning of connections during CNS development. *Neuron* 5: 745–756. [11, 18]

Shatz, C. J. 1990b. Competitive interactions between retinal ganglion cells during prenatal development. *J. Neurobiol.* 21: 197–211. [18]

Shatz, C. J. and Sretavan, D. W. 1986. Interactions between retinal ganglion cells during the development of the mammalian visual system. *Annu. Rev. Neurosci.* 9: 171–207. [18]

Shatz, C. J. and Stryker, M. P. 1988. Prenatal tetrodotoxin infusion blocks segregation of retinogeniculate afferents. *Science* 242: 87–89. [18]

Sheetz, M. P., Steuer, E. R. and Schroer, T. A. 1989. The mechanism and regulation of fast axonal transport. *Trends Neurosci.* 12: 474–478. [9]

Sheetz, M. P., Wayne, D. B. and Pearlman, A. L. 1992. Extension of filopodia by motor-dependent actin assembly. *Cell Motil. Cytoskel.* 22: 160–169. [11]

Sheridan, J. D. 1978. Junctional formation and experimental modification. *In* J. Feldman, N. B. Gilula, and J. D. Pitts (eds.). *Intercellular Junctions and Synapses.* Chapman and Hall, London, pp. 37–59. [5]

Sherrington, C. S. 1906. *The Integrative Action of the Nervous System,* 1961 edition. Yale University Press, New Haven. [1, 17]

Sherrington, C. S. 1933. *The Brain and Its Mechanism.* Cambridge University Press, London. [1, 15]

Sherrington, C. S. 1951. *Man on His Nature.* Cambridge University Press, Cambridge, England. [16]

Shik, M. L. and Orlovsky, G. N. 1976. Neurophysiology of locomotor automatism. *Physiol. Rev.* 56: 465–501. [15]

Shik, M. L., Severin, F. V. and Orlovsky, G. N. 1966. Control of walking and running by means of electrical stimulation of the mid-brain. *Biofizika* 11: 756–765. [15]

Shipp, S. and Zeki, S. 1985. Segregation of pathways leading from area V2 to areas V4 and V5 of macaque monkey visual cortex. *Nature* 315: 322–325. [17]

Shkolnik, L. J. and Schwartz, J. H. 1980. Genesis and maturation of serotonergic vesicles in identified giant cerebral neuron of *Aplysia. J. Neurophysiol.* 43: 945–967. [9]

Shrager, P. 1988. Ionic channels and signal conduction in single remyelinating frog nerve fibres. *J. Physiol.* 404: 695–712. [6]

Shrager, P., Chiu, S. Y. and Ritchie, J. M. 1985. Voltage-dependent sodium and potassium channels in mammalian cultured Schwann cells. *Proc. Natl. Acad. Sci. USA* 82: 948–952. [6]

Shute, C. C. D. and Lewis, P. R. 1963. Cholinesterase-containing systems of the brain of the rat. *Nature* 199: 1160–1164. [10]

Shyng, S.-L. and Salpeter, M. M. 1990. Effect of reinnervation on the degradation rate of junctional acetylcholine receptors synthesized in denervated skeletal muscles. *J. Neurosci.* 10: 3905–3915. [12]

Shyng, S.-L., Xu, R. and Salpeter, M. M. 1991. Cyclic AMP stabilizes the degradation of original junctional acetylcholine receptors in denervated muscle. *Neuron* 6: 469–475. [12]

Siegelbaum, S. A., Belardetti, F., Camardo, J. S. and Shuster, M. J. 1986. Modulation of the serotonin-sensitive potassium channel in *Aplysia* sensory neurone cell body and growth cone. *J. Exp. Biol.* 124: 287–306. [13]

Sigel, E., Baur, R., Trube, G., Mohler, H. and Malherbe, P. 1990. The effect of subunit composition of rat brain GABA$_A$ receptors on channel function. *Neuron* 5: 703–711. [2]

Sigworth, F. J. and Neher, E. 1980. Single Na$^+$ channel currents observed in cultured rat muscle cells. *Nature* 287: 447–449. [4]

Silinsky, E. M. and Hubbard, J. I. 1973. Release of ATP from rat motor nerve terminals. *Nature* 243: 404–405. [9]

Sillito, A. M. 1979. Inhibitory mechanisms influencing complex cell orientation selectivity and their modification at high resting discharge levels. *J. Physiol.* 289: 33–53. [10]

Silver, J. and Sidman, R. S. 1980. A mechanism for the guidance and topographic patterning of retinal ganglion cell axons. *J. Comp. Neurol.* 189: 101–111. [11]

Simon, D. K. and O'Leary, D. D. M. 1989. Limited topographic specificity in the targeting and branching of mammalian retinal axons. *Dev. Biol.* 137: 125–134. [11]

Simons, D. J. and Woolsey, T. A. 1979. Functional organization in mouse barrel cortex. *Brain Res.* 165: 327–332. [14]

Sims, T. J., Waxman, S. G., Black, J. A. and Gilmore, S. A. 1985. Perinodal astrocytic processes at nodes of Ranvier in developing normal and glial cell deficient rat spinal cord. *Brain Res.* 337: 321–331. [6]

Singer, W. 1990. The formation of cooperative cell assemblies in the visual cortex. *J. Exp. Biol.* 153: 177–197. [18]

Skou, J. C. 1988. Overview: The Na,K pump. *Meth. Enzymol.* 156: 1–25. [3]

Smith, A. D., de Potter, W. P., Moerman, E. J. and Schaepdryver, A. F. 1970. Release of dopamine β-hydroxylase and chromogranin A upon stimulation of the splenic nerve. *Tissue Cell* 2: 547–568. [9]

Smith, P. J., Howes, E. A. and Treherne, J. E. 1987. Mechanisms of glial regeneration in an insect central nervous system. *J. Exp. Biol.* 132: 59–78. [6]

Smith, S. J. 1988. Neuronal cytomechanics: The actin-based motility of growth cones. *Science* 242: 708–715. [11]

Snyder, S. 1986. *Drugs and the Brain.* Scientific American Library, New York. [6]

Soejima, M. and Noma, A. 1984. Mode of regulation of the ACh-sensitive K-channel by the muscarinic receptor in rabbit atrial cells. *Pflügers Arch.* 400: 424–431. [8]

Sokoloff, L. 1977. Relation between physiological function and energy metabolism in the central nervous system. *J. Neurochem.* 29: 13–26. [17]

Sokolove, P. G. and Cooke, I. M. 1971. Inhibition of impulse activity in a sensory neuron by an electrogenic pump. *J. Gen. Physiol.* 57: 125–163. [14]

Sommer, B., Keinanen, K., Verdoorn, T. A., Wisden, W., Burnashev, N., Herb, A., Kohler, M., Takagi, T., Sakmann, B. and Seeburg, P. H. 1990. Flip and flop: A cell-specific functional switch in glutamate-operated channels of the CNS. *Science* 249: 1580–1585. [10]

Sossin, W. S., Fisher, J. M. and Scheller, R. H. 1989. Cellular and molecular biology of neuropeptide processing and packaging. *Neuron* 2: 1407–1417. [9]

Sotelo, C. and Alvarado-Mallart, R. M. 1991. The reconstruction of cerebellar circuits. *Trends Neurosci.* 14: 350–355. [12]

Sotelo, C. and Taxi, J. 1970. Ultrastructural aspects of electrotonic junctions in the spinal cord of the frog. *Brain Res.* 17: 137–141. [7]

Sotelo, C., Llinás, R. and Baker, R. 1974. Structural study of inferior olivary nucleus of the cat: Morphological correlates of electrotonic coupling. *J. Neurophysiol.* 37: 541–559. [5]

Specht, S. and Grafstein, B. 1973. Accumulation of radioactive protein in mouse cerebral cortex after injection of [^{3}H]-fucose into the eye. *Exp. Neurol.* 41: 705–722. [17]

Spector, S. A. 1985. Trophic effect on the contractile and histochemical properties of rat soleus muscle. *J. Neurosci.* 5: 2189–2196. [12]

Sperry, R. W. 1943. Effect of 180° rotation of the retinal field on visuomotor coordination. *J. Exp. Zool.* 92: 236–279. [11]

Sperry, R. W. 1963. Chemoaffinity in the orderly growth of nerve fiber patterns and connections. *Proc. Natl. Acad. Sci. USA* 50: 703–710. [11]

Sperry, R. W. 1970. Perception in the absence of neocortical commissures. *Proc. Res. Assoc. Nerv. Ment. Dis.* 48: 123–138. [17]

Spitzer, N. C. 1982. Voltage- and stage-dependent uncoupling of Rohon-Beard neurones during embryonic development of *Xenopus* tadpoles. *J. Physiol.* 330: 145–162. [5]

Squire, L. R. and Zola-Morgan, S. 1991. The medial temporal lobe memory system. *Science* 253: 1380–1386. [10]

Stadler, H. and Kiene, M.-L. 1987. Synaptic vesicles in electromotoneurones. II. Heterogeneity of populations is expressed in uptake properties, exocytosis, and insertion of a core proteoglycan in the extracellular matrix. *EMBO J.* 6: 2217–2221. [9]

Stahl, B., Müller, B., von Boxberg, Y., Cox, E. C. and Bonhoeffer, F. 1990. Biochemical characterization of a putative axonal guidance molecule of the chick visual system. *Neuron* 5: 735–743. [11]

Stanfield, B. B. and O'Leary, D. D. M. 1985. Fetal occipital cortical neurones transplanted to the rostral cortex can extend and maintain a pyramidal tract axon. *Nature* 298: 371–373. [11]

Steinbusch, H. W. M. 1984. Serotonin-immunoreactive neurons and their projections in the CNS. *In* A. Björklund, T. Hökfelt and M. Kuhar (eds.). *Handbook of Chemical Neuroanatomy, Classical Transmitters and Transmitter Receptors in the CNS,* Vol. 3, Part II. Elsevier, New York, pp. 68–125. [10]

Stent, G. S. and Kristan, W. B. 1981. Neural circuits generating rhythmic movements. *In* K. J. Muller, J. G. Nicholls and G. S. Stent (eds.). *Neurobiology of the Leech.* Cold Spring Harbor Laboratory, Cold Spring Harbor, NY, pp. 113–146. [13]

Stent, G. S. and Weisblat, D. 1982. The development of a simple nervous system. *Sci. Am.* 246(1): 136–146. [11]

Sterling, P. 1983. Microcircuitry of the cat retina. *Annu. Rev. Neurosci.* 6: 149–185. [10]

Sterling, P., Freed, M. A. and Smith, R. G. 1988. Architecture of rod and cone circuits to the on-beta ganglion cell. *J. Neurosci.* 8: 623–642. [16]

Stern-Bach, Y., Greenberg-Ofrath, N., Flechner, I. and Schuldiner, S. 1990. Identification and purification of a functional amine transporter from bovine chromaffin granules. *J. Biol. Chem.* 265: 3961–3966. [9]

Stewart, R. R., Adams, W. B. and Nicholls, J. G. 1989. Presynaptic calcium currents and facilitation of serotonin release at synapses between cultured leech neurones. *J. Exp. Biol.* 144: 1–12. [7, 13]

Stewart, R. R., Nicholls, J. G. and Adams, W. B. 1989. Na$^+$, K$^+$ and Ca^{2+} currents in identified leech neurones in culture. *J. Exp. Biol.* 141: 1–20. [13]

Stewart, R. R., Zou, D.-J., Treherne, J. M., Møllgård, K., Saunders, N. R. and Nicholls, J. G. 1991. The intact central nervous system of the newborn opossum in long-term culture: Fine structure and GABA-mediated inhibition of electrical activity. *J. Exp. Biol.* 161: 25–41. [6]

Stitt, T. N. and Hatten, M. E. 1990. Antibodies that recognize astrotactin block granule neuron binding to astroglia. *Neuron* 5: 639–649. [6]

Stone, L. S. 1944. Functional polarization in the retinal development and its reestablishment in regenerating retinae of rotated grafted eyes. *Proc. Soc. Exp. Biol. Med.* 57: 13–14. [11]

Stone, R. A., Laties, A. M., Raviola, E. and Wiesel, T. N. 1988. Increase in retinal vasoactive intestinal polypeptide after eyelid fusion in primates. *Proc. Natl. Acad. Sci. USA* 85: 257–60. [16, 18]

Streisinger, G., Walker, C., Dower, N., Knauber, D. and Singer, F. 1981. Production of clones of homozygous diploid zebrafish (*Brachydanio verio*). *Nature* 291: 293–296. [11]

Streit, W. J., Graeber, M. B. and Kreutzberg, G. W. 1988. Functional plasticity of microglia: A review. *Glia* 1: 301–307. [6]

Stroud, R. M. and Finer-Moore, J. 1985. Acetylcholine receptor structure, function and evolution. *Annu. Rev. Cell Biol.* 1: 317–351. [2]

Strumwasser, F. 1988–1989. A short history of the second messenger concept in neurons and lessons from long lasting changes in two neuronal systems producing after discharge and circadian oscillations. *J. Physiol.* (Paris) 83: 246–254. [13]

Stryer, L. 1987. The molecules of visual excitation. *Sci. Am.* 257(1): 42–50. [16]

Stryer, L. and Bourne, H. R. 1986. G proteins: A family of signal transducers. *Annu. Rev. Cell Biol.* 2: 391–419. [16]

Stryker, M. P. and Harris, W. A. 1986. Binocular impulse blockade prevents the formation of ocular dominance columns in cat visual cortex. *J. Neurosci.* 6: 2117–2133. [18]

Stryker, M. P. and Zahs, K. R. 1983. On and off sublaminae in the lateral geniculate nucleus of the ferret. *J. Neurosci.* 10: 1943–1951. [16]

Stuart, A. E. 1970. Physiological and morphological properties of motoneurones in the central nervous system of the leech. *J. Physiol.* 209: 627–646. [13]

Stuermer, C. A. O. 1988. Retinotopic organization of the developing retinotectal projection in the zebrafish embryo. *J. Neurosci.* 8: 4513–4530. [11]

Stuermer, C. A. O. and Raymond, P. A. 1989. Developing retinotectal projection in larval goldfish. *J. Comp. Neurol.* 281: 630–640. [11]

Stuermer, C. A., Rohrer, B. and Munz, H. 1990. Development of the retinotectal projection in zebrafish embryos under TTX-induced neural impulse blockade. *J. Neurosci.* 10: 3615–3626. [18]

Stühmer, W., Conti, F., Suzuki, H., Wang, X., Noda, M., Yahagi, N., Kubo, H. and Numa, S. 1989. Structural parts involved in activation and inactivation of the sodium channel. *Nature* 239: 597–603. [2, 4]

Südhof, T. C. and Jahn, R. 1991. Proteins of synaptic vesicles involved in exocytosis and membrane recycling. *Neuron* 6: 665–677. [9]

Südhof, T. C., Czernik, A. J., Kao, H.-T., Takei, K., Johnston, P. A., Horiuchi, A., Kanazir, S. D., Wagner, M. A., Perin, M. S., DeCammilli, P. and Greengard, P. 1989. Synapsins: Mosaics of shared and individual domains in a family of synaptic vesicle phosphoproteins. *Science* 245: 1474–1480. [9]

Sumikawa, K., Parker, I. and Miledi, R. 1989. Expression of neurotransmitter receptors and voltage-activated channels from brain mRNA in *Xenopus* oocytes. *Meth. Neurosci.* 1: 30–45. [2]

Sun, D., Qin, Y, Chluba, J., Epplen, J. T. and Wekerle, H. 1988. Suppression of experimentally induced autoimmune encephalomyelitis by cytolytic T–T cell interactions. *Nature* 332: 843–845. [6]

Sutherland, E. W. 1972. Studies on the mechanism of hormone action. *Science* 177: 401–408. [8]

Sutter, A., Riopelle, R. J., Harris-Warwick, R. M. and Shooter, E. M. 1979. Nerve growth factor receptors: Characterization of two distinct classes of binding sites on chick embryo sensory ganglia cells. *J. Biol. Chem.* 254: 5972–5982. [11]

Svaetichin, G. 1953. The cone action potential. *Acta Physiol. Scand.* 29: 565–600. [16]

Swanson, L. W. 1976. The locus coeruleus: A cytoarchitectonic, Golgi, and immunohistochemical study in the albino rat. *Brain Res.* 110: 39–56 [10]

Swensen, K. I., Jordan, J. R., Beyer, E. C. and Paul, D. L. 1989. Formation of gap junctions by expression of connexins in *Xenopus* oocyte pairs. *Cell* 57: 145–155. [5]

Swindale, N. V., Matsubara, J. A. and Cynader, M. S. 1987. Surface organization of orientation and direction selectivity in cat area 18. *J. Neurosci.* 7: 1414–1427. [17]

Szabo, G. and Otero, A. S. 1990. G protein mediated regulation of K$^+$ channels in heart. *Annu. Rev. Physiol.* 52: 293–305. [8]

Szatkowski, M., Barbour, B. and Attwell, D. 1990. Nonvesicular release of glutamate from glial cells by reversed electrogenic glutamate uptake. *Nature* 348: 443–446. [6]

Szentágothai, J. 1973. Neuronal and synaptic architecture of the lateral geniculate nucleus. *In* H. H. Kornhuker (ed.). *Handbook of Sensory Physiology*, Vol. 6, *Central Visual Information*. Springer-Verlag, Berlin, pp. 141–176. [16]

Takai, T., Noda, M., Mishina, M., Shimizu, S., Furutani, Y., Kayano, T., Ikeda, T., Kubo, T., Takahashi, H., Takahashi, T., Kuno, M. and Numa, S. 1985. Cloning, sequencing, and expression of cDNA for a novel subunit of acetylcholine receptor from calf muscle. *Nature* 315: 761–764. [2]

Takeuchi, A. 1977. Junctional transmission. I. Postsynaptic mechanisms. *In* E. Kandel (ed.). *Handbook of the Nervous System*, Vol. 1. American Physiological Society, Baltimore, MD, pp. 295–327. [7]

Takeuchi, A. and Takeuchi, N. 1959. Active phase of frog's end-plate potential. *J. Neurophysiol.* 22: 395–411. [7]

Takeuchi, A. and Takeuchi, N. 1960. On the permeability of the end-plate membrane during the action of transmitter. *J. Physiol.* 154: 52–67. [7]

Takeuchi, A. and Takeuchi, N. 1966. On the permeability of the presynaptic terminal of the crayfish neuromuscular junction during synaptic inhibition and the action of γ-aminobutyric acid. *J. Physiol.* 183: 433–449. [7]

Takeuchi, A. and Takeuchi, N. 1967. Anion permeability of the inhibitory post-synaptic membrane of the crayfish neuromuscular junction. *J. Physiol.* 191: 575–590. [7]

Takeuchi, N. 1963. Some properties of conductance changes at the end-plate membrane during the action of acetylcholine. *J. Physiol.* 167: 128–140. [7]

Talbot, S. A. and Marshall, W. H. 1941. Physiological studies on neural mechanisms of visual localization and discrimination. *Am. J. Ophthalmol.* 24: 1255–1264. [17]

Tamura, T., Nakatani, K., and Yau, K.-W. 1991. Calcium feedback and sensitivity regulation in primate rods. *J. Gen. Physiol.* 98: 95–130. [16]

Tanabe, T., Takeshima, H., Mikami, A., Flockerzi, V., Takahashi, H., Kangawa, K., Kojima, M., Matsuo, H., Hirose, T. and Numa, S. 1987. Primary structure of the receptor for calcium-channel blockers from skeletal muscle. *Nature* 328: 313–318. [4]

Tank, D. W., Huganir, R. L., Greengard, P. and Webb, W. W. 1983. Patch-recorded single-channel currents of the purified and reconstituted *Torpedo* acetylcholine receptor. *Proc. Natl. Acad. Sci. USA* 80: 5129–5133. [2]

Tao-Cheng, J. H., Nagy, Z. and Brightman, M. W. 1987. Tight junctions of brain endothelium in vitro are enhanced by astroglia. *J. Neurosci.* 7: 3293–3299. [6]

Tao-Cheng, J. H., Nagy, Z. and Brightman, M. W. 1990. Astrocytic orthogonal arrays of intramembranous particle assemblies are modulated by brain endothelial cells in vitro. *J. Neurocytol.* 19: 143–153. [6]

Tasaki, I. 1959. Conduction of the nerve impulse. *In* J. Field (ed.). *Handbook of Physiology*, Section 1, Vol. I. American Physiological Society, Bethesda, pp. 75–121. [5]

Tauc, L. 1982. Nonvesicular release of neurotransmitter. *Physiol. Rev.* 62: 857–893. [9]

Teschemacher, H., Ophein, K. E., Cox, B. M. and Goldstein, A. 1975. A peptide-like substance from pituitary that acts like morphine. *Life Sci.* 16: 1771–1776. [10]

Tessier-Lavigne, M. and Placzek, M. 1991. Target attraction: Are developing axons guided by chemotropism? *Trends Neurosci.* 14: 303–310. [11]

Tessier-Lavigne, M., Placzek, M., Lumsden, A. G. S., Dodd, J. and Jessell, T. M. 1988. Chemotropic guidance of developing axons in the mammalian central nervous system. *Nature* 336: 775–778. [11]

Thach, W. T. 1978. Correlation of neural discharge with pattern and force of muscular activity, joint position, and direction of next intended movement in motor cortex and cerebellum. *J. Neurophysiol.* 41: 654–676. [15]

Thach, W. T., Goodkin, H. G. and Keating, J. G. 1992. Cerebellum and the adaptive coordination of movement. *Annu. Rev. Neurosci.* 15: 403–442. [15]

Thanos, S. and Bonhoeffer, F. 1987. Axonal arborization in the developing chick retinotectal system. *J. Comp. Neurol.* 261: 155–164. [11]

Thanos, S. Bonhoeffer, F. and Rutishauser, U. 1984. Fiber–fiber interaction and tectal cues influence the development of the chick retinotectal projection. *Proc. Natl. Acad. Sci. USA* 81: 1906–1910. [11]

Thesleff, S. 1960. Supersensitivity of skeletal muscle produced by botulinum toxin. *J. Physiol.* 151: 598–607. [12]

Thoenen, H. 1991. The changing scene of neurotrophic factors. *Trends Neurosci.* 14: 165–170. [11]

Thoenen, H. and Barde, Y.-A. 1980. Physiology of nerve growth factor. *Physiol. Rev.* 60: 1284–1335. [9]

Thoenen, H., Mueller, R. A., and Axelrod, J. 1969. Increased tyrosine hydroxylase activity after drug-induced alteration of sympathetic transmission. *Nature* 221: 1264. [9]

Thoenen, H., Otten, U. and Schwab, M. 1979. Orthograde and retrograde signals for the regulation of neuronal gene expression: The peripheral sympathetic nervous system as a model. *In* F. O. Schmitt and F. G. Worden (eds.). *The Neurosciences: Fourth Study Program.* MIT Press, Cambridge, MA, pp. 911–928. [9]

Thomas, R. C. 1969. Membrane currents and intracellular sodium changes in a snail neurone during extrusion of injected sodium. *J. Physiol.* 201: 495–514. [3]

Thomas, R. C. 1972. Intracellular sodium activity and the sodium pump in snail neurones. *J. Physiol.* 220: 55–71. [3]

Thomas, R. C. 1977. The role of bicarbonate, chloride and sodium ions in the regulation of intracellular pH in snail neurones. *J. Physiol.* 273: 317–338. [3]

Thompson, S. M., Deisz, R. A. and Prince, D. A. 1988. Outward chloride/cation co-transport in mammalian cortical neurons. *Neurosci. Lett.* 89: 49–54. [3]

Thompson, W. 1983. Synapse elimination in neonatal rat muscle is sensitive to pattern of muscle use. *Nature* 302: 614–616. [11]

Thompson, W. J. and Stent, G. S. 1976. Neuronal control of heartbeat in the medicinal leech. I. Generation of the vascular constriction rhythm by heart motor neurons. *J. Comp. Physiol.* 111: 309–333. [13]

Thompson, W., Kuffler, D. P. and Jansen, J. K. S. 1979. The effect of prolonged, reversible block of nerve impulses on the elimination of polyneuronal innervation of newborn rat skeletal muscle fibers. *Neuroscience* 4: 271–281. [11]

Tibbits, T. T., Caspar, D. L. D., Phillips, W. C. and Goodenough, D. A. 1990. Diffraction diagnosis of protein folding in gap junctions. *Biophys. J.* 57: 1025–1036. [2]

Timpe, L. C., Schwarz, T. L., Tempel, B. L., Papazian, D. M., Jan, Y. N. and Jan, L. Y. 1988. Expression of functional potassium channels from *Shaker* cDNA in *Xenopus* oocytes. *Nature* 331: 143–145. [2]

Tolbert, L. P. and Oland, L. A. 1989. A role for glia in the development of organized neuropilar structures. *Trends Neurosci.* 12: 70–75. [6]

Tolbert, L. P. and Oland, L. A. 1990. Glial cells form boundaries for developing insect olfactory glomeruli. *Exp. Neurol.* 109: 19–28. [6]

Tomita, T. 1965. Electrophysiological study of the mechanisms subserving color coding in the fish retina. *Cold Spring Harbor Symp. Quant. Biol.* 30: 559–566. [16]

Tomlinson, A. and Ready, D. F. 1987. Neuronal differentiation in the *Drosophila* ommatidium. *Dev. Biol.* 120: 366–376. [11]

Tootell, R. B., Hamilton, S. L. and Switkes, E. 1988. Functional anatomy of macaque striate cortex. IV. Contrast and magnoparvo streams. *J. Neurosci.* 8: 1594–1609. [17]

Tootell, R. B., Hamilton, S. L., Silverman, M. S. and Switkes, E. 1988. Functional anatomy of macaque striate cortex. I. Ocular dominance, binocular interactions, and baseline conditions. *J. Neurosci.* 8: 1500–1530. [17]

Tootell, R. B., Switkes, E., Silverman, M. S. and Hamilton, S. L. 1988. Functional anatomy of macaque striate cortex. II. Retinotopic organization. *J. Neurosci.* 8: 1531–1568. [17]

Tootell, R. B., Silverman, M. S., Hamilton, S. L., De Valois, R. L. and Switkes, E. 1988. Functional anatomy of macaque striate cortex. III. Color. *J. Neurosci.* 8: 1569–1593. [17]

Torebjork, H. E. and Hallin, R. G. 1973. Perceptual changes accompanying controlled preferential blocking of A and C fibre responses in intact human skin nerves. *Exp. Brain Res.* 16: 321–332. [14]

Torebjork, H. E. and Ochoa, J. 1980. Specific sensations evoked by activity in single identified sensory units in man. *Acta Physiol. Scand.* 110: 445–447. [14]

Tosney, K. W. 1991. Cells and cell-interactions that guide motor axons in the developing chick embryo. *BioEssays* 13: 17–23. [11]

Toyoshima, C. and Unwin, N. 1988. Ion channel of acetylcholine receptor reconstructed from images of postsynaptic membranes. *Nature* 336: 247–250. [2]

Trautwein, W. and Hescheler, J. 1990. Regulation of cardiac L-type calcium current by phosphorylation and G proteins. *Annu. Rev. Physiol.* 52: 257–274. [8]

Treherne, J. M., Woodward, S. K. A., Varga, Z. M., Ritchie, J. M. and Nicholls, J. G. 1992. Restoration of conduction and growth of axons through injured spinal cord of neonatal opossum in culture. *Proc. Natl. Acad. Sci. USA* 89: 431–434. [6, 12]

Triller, A., Cluzeaud, F., Pfeiffer, F., Betz, H. and Korn, H. 1985. Distribution of glycine receptors at central synapses: An immunoelectron microscopy study. *J. Cell Biol.* 101: 683–688. [9]

Trimmer, J. S. and Agnew, W. S. 1989. Molecular diversity of voltage-sensitive Na channels. *Annu. Rev. Physiol.* 51: 401–418. [2]

Trimmer, J. S., Cooperman, S. S., Tomiko, S. A., Zhou, J. Y., Crean, S. M., Boyle, M. B., Kallan, R. G., Sheng, Z. H., Barchi, R. L., Sigworth, F. J., Goodman, R. H., Agnew, W. S. and Mandel, G. 1989. Primary structure and functional expression of a mammalian skeletal muscle sodium channel. *Neuron* 3: 33–46. [2]

Trisler, D. and Collins, F. 1987. Corresponding spatial gradients of TOP molecules in the developing retina and optic tectum. *Science* 237: 1208–1209. [11]

Trotier, D. 1986. A patch clamp analysis of membrane currents in salamander olfactory cells. *Pflügers Arch.* 407: 589–595. [14]

Truman, J. W. 1984. Cell death in invertebrate nervous systems. *Annu. Rev. Neurosci.* 7: 171–188. [11]

Tsien, R. W. 1987. Calcium currents in heart cells and neurons. *In* L. K. Kaczmarek and I. B. Levitan (eds.). *Neuromodulation: The Biochemical Control of Neuronal Excitability*. Oxford University Press, New York, pp. 206–242. [8]

Tsien, R. W. and Tsien, R. Y. 1990. Calcium channels, stores, and oscillations. *Annu. Rev. Cell Biol.* 6: 715–760. [8]

Tsien, R. W., Bean, B. P., Hess, P. and Nowycky, M. 1983. Calcium channels: Mechanisms of β-adrenergic modulation and ion permeation. *Cold Spring Harbor Symp. Quant. Biol.* 48: 201–212. [8]

Tsien, R. W., Lipscombe, D., Madison, D. V., Bley, K. R. and Fox, A. P. 1988. Multiple types of neuronal calcium channels and their selective modulation. *Trends Neurosci.* 11: 431–437. [4]

Tsien, R. Y. 1980. New calcium indicators and buffers with high selectivity against magnesium and protons: design, synthesis and properties of prototype structures. *Biochemistry* 19: 2396–2404. [7]

Tsien, R. Y. 1988. Fluorescent measurement and photochemical manipulation of cytosolic free calcium. *Trends Neurosci.* 11: 419–424. [3]

Tsim, K. W. K., Ruegg, M. A., Escher, G., Kroger, S. and McMahan, U. J. 1992. cDNA that encodes active agrin. *Neuron* 8: 677–689. [12]

Ts'o, D. Y. and Gilbert, C. D. 1988. The organization of chromatic and spatial interactions in the primate striate cortex. *J. Neurosci.* 8: 1712–1727. [17]

Ts'o, D. Y., Gilbert, C. D. and Wiesel, T. N. 1986. Relationships between horizontal interactions and functional architecture in cat striate cortex as revealed by cross-correlation analysis. *J. Neurosci.* 6: 1160–1170. [17]

Ts'o, D. Y., Frostig, R. D., Lieke, E. E. and Grinvald, A. 1990. Functional organization of primate visual cortex revealed by high resolution optical imaging. *Science* 249: 417–420. [17]

Tsukahara, N. 1981. Synaptic plasticity in the mammalian central nervous system. *Annu. Rev. Neurosci.* 4: 351–379. [12]

Tsuzuki, K. and Suga, N. 1988. Combination-sensitive neurons in the ventroanterior area of the auditory cortex of the mustached bat. *J. Neurophysiol.* 60: 1908–1923. [14]

Turner, A. M. and Greenough, W. T. 1985. Differential rearing effects on rat visual cortex synapses. I. Synaptic and neuronal density and synapses per neuron. *Brain Res.* 329: 195–203. [18]

Turner, D. L. and Cepko, C. L. 1987. A common progenitor for neurons and glia persists in rat retina late in development. *Nature* 328: 131–136. [11]

Tuček, S. 1978. *Acetylcholine Synthesis in Neurons.* Chapman and Hall, London. [9]

Ugolini, G. and Kuypers, H. G. J. M. 1986. Collaterals of corticospinal and pyramidal fibres to the pontine grey demonstrated by a new application of the fluorescent fibre labelling technique. *Brain Res.* 365: 211–227. [9]

Unwin, N. 1989. The structure of ion channels in membranes of excitable cells. *Neuron* 3: 665–676. [2]

Unwin, N., Toyoshima, C. and Kubalek, E. 1988. Arrangement of acetylcholine receptor subunits in the resting and desensitized states, determined by cryoelectron microscopy of crystallized *Torpedo* postsynaptic membranes. *J. Cell Biol.* 107: 1123–1138. [2]

Unwin, P. N. T. and Zampighi, G. 1980. Structure of the junction between communicating cells. *Nature* 283: 545–549. [5]

Usowicz, M. M., Porzig, H., Becker, C. and Reuter, H. 1990. Differential expression by nerve growth factor of two types of Ca^{2+} channels in rat phaeochromocytoma cell lines. *J. Physiol.* 426: 95–116. [11]

Vale, R. D. 1987. Intracellular transport using microtubule-based motors. *Annu. Rev. Cell Biol.* 3: 347–378. [9]

Vallbo, Å. B. 1990. Muscle afferent responses to isometric contractions and relaxations in humans. *J. Neurophysiol.* 63: 1307–1313. [14]

Vallee, R. B. and Bloom, G. S. 1991. Mechanisms of fast and slow axonal transport. *Annu. Rev. Neurosci.* 14: 59–92. [9]

Vallee, R. B., Shpetner, H. S. and Paschal, B. M. 1989. The role of dynein in retrograde axonal transport. *Trends Neurosci.* 12: 66–70. [9]

Valtorta, F., Jahn, R., Fesce, R., Greengard, P. and Ceccarelli, B. 1988. Synaptophysin (p38) at the frog neuromuscular junction: Its incorporation into the axolemma and recycling after intense quantal secretion. *J. Cell Biol.* 107: 2717–2727. [9]

Van der Loos, H. and Woolsey, T. A. 1973. Somatosensory cortex: Structural alterations following early injury to sense organs. *Science* 179: 395–398. [6]

Van Essen, D. C. 1979. Visual areas of the mammalian cerebral cortex. *Annu. Rev. Neurosci.* 2: 227–263. [17]

Van Essen, D. and Jansen, J. K. 1974. Reinnervation of rat diaphragm during perfusion with α-bungarotoxin. *Acta Physiol. Scand.* 91: 571–573. [12]

Vassilev, P., Scheuer, T. and Catterall, W. A. 1989. Inhibition of inactivation of single sodium channels by a site-directed antibody. *Proc. Natl. Acad. Sci. USA* 86: 8147–8151. [4]

Venable, N., Fernandez, V., Diaz, E. and Pinto-Hamuy, T. 1989. Effects of preweaning environmental enrichment on basilar dendrites of pyramidal neurons in occipital cortex: a Golgi study. *Brain Res. Dev. Brain Res.* 49: 140–144. [18]

Vertes, R. P. 1984. Brainstem control of the events of REM sleep. *Prog. Neurobiol.* 22: 241–288. [10]

Vicini, S. 1991. Pharmacologic significance of the structural heterogeneity of the $GABA_A$ receptor–chloride ion channel complex. *Neuropsychopharmacology* 4: 9–15. [10]

Villegas, J. 1981. Schwann cell relationships in the giant nerve fibre of the squid. *J. Exp. Biol.* 95: 135–151. [6]

Virchow, R. 1959. *Cellularpathologie.* (F. Chance, trans.) Hirschwald, Berlin. [6]

Vizi, S. E. and Vyskočil, F. 1979. Changes in total and quantal release of acetylcholine in the mouse diaphragm during activation and inhibition of membrane ATPase. *J. Physiol.* 286: 1–14. [7]

Vollenweider, F. X., Cuenod, M. and Do, K. Q. 1990. Effect of climbing fiber deprivation on release of endogenous aspartate, glutamate, and homocysteate in slices of rat cerebellar hemispheres and vermis. *J. Neurochem.* 54: 1533–1540. [9]

Volterra, A. and Siegelbaum, S. A. 1988. Role of two different guanine nucleotide-binding proteins in the antagonistic modulation of the S-type K^+ channel by cAMP and arachidonic acid metabolites in *Aplysia* sensory neurons. *Proc. Natl. Acad. Sci. USA* 85: 7810–7814. [8]

von Bekesy, G. 1960. *Experiments in Hearing.* McGraw-Hill, New York. [14]

von Euler, C. 1977. The functional organization of the respiratory phase-switching mechanisms. *Fed. Proc.* 36: 2375–2380. [15]

von Euler, U. S. 1956. *Noradrenaline.* Charles Thomas, Springfield, IL. [9]

Vyskočil, F., Nikosky, E. and Edwards, C. 1983. An analysis of mechanisms underlying the non-quantal release of acetylcholine at the mouse neuromuscular junction. *Neuroscience* 9: 429–435. [7]

Wachtel, H. and Kandel, E. R. 1971. Conversion of synaptic excitation to inhibition at a dual chemical synapse. 1. *Neurophysiology* 34: 56–68. [13]

Wada, H., Inagaki, N., Yamatodani, A. and Watanabe, T. 1991. Is the histaminergic neuron system a regulatory center for whole-brain activity? *Trends Neurosci.* 14: 415–418. [10]

Wagner, J. A., Carlson, S. S. and Kelly, R. B. 1978. Chemical and physical characterization of cholinergic synaptic vesicles. *Biochemistry* 17: 1199–1206. [9]

Wainer, B. H., Levey, A. I., Mufson, E. J., Mesulam, M.-M. 1984. Cholinergic systems in mammalian brain identified with antibodies against choline acetyltransferase. *Neurochem. Int.* 6: 163–182. [10]

Wall, J. T. 1988. Variable organization in cortical maps of the skin as an indication of the lifelong adaptive capacities of circuits in the mammalian brain. *Trends Neurosci.* 11: 549–557. [18]

Wallace, B. G. and Gillon, J. W. 1982. Characterization of acetylcholinesterase in individual neurons in the leech central nervous system. *J. Neurosci.* 2: 1108–1118. [10]

Wallace, B. G., Qu, Z. and Huganir, R. L. 1991. Agrin induces phosphorylation of the nicotinic acetylcholine receptor. *Neuron* 6: 869–878. [12]

Walter, J., Allsopp, T. E. and Bonhoeffer, F. 1990. A common denominator of growth cone guidance and collapse? *Trends Neurosci.* 13: 447–452. [11]

Walter, J., Henke-Fahle, S. and Bonhoeffer, F. 1987. Avoidance of posterior tectal membranes by temporal retinal axons. *Development* 101: 909–913. [11]

Walter, J., Kern-Veits, B., Huf, J., Stolze, B. and Bonhoeffer, F. 1987. Recognition of position-specific properties of tectal cell membranes by retinal axons in vitro. *Development* 101: 685–696. [11]

Wannier, T. M. J., Maier, M. A. and Hepp-Reymond, M.-C. 1991. Contrasting properties of monkey somatosensory and motor cortex neurons activated during the control of force in precision grip. *J. Neurophysiol.* 65: 572–589. [15]

Wässle, H. and Boycott, B. B. 1991. Functional architecture of the mammalian retina. *Physiol. Rev.* 71: 447–480. [16]

Wässle, H. and Chun, M. H. 1988. Dopaminergic and indoleamine-accumulating amacrine cells express GABA-like immunoreactivity in the cat retina. *J. Neurosci.* 8: 3383–3394. [10]

Wässle, H., Peichl, L. and Boycott, B. B. 1981. Morphology and topography of on- and off-alpha cells in the cat retina. *Proc. R. Soc. Lond. B* 212: 157–175. [16]

Waxman, S. G., Black, J. A., Kocsis, J. D. and Ritchie, J. M. 1989. Low density of sodium channels support conduction in axons of neonatal rat optic nerve. *Proc. Natl. Acad. Sci. USA* 86: 1406–1410. [4]

Wehner, R. 1989. Neurobiology of polarization vision. *Trends Neurosci.* 12: 353–359. [13]

Wehner, R. and Menzel, R. 1990. Do insects have cognitive maps? *Annu. Rev. Neurosci.* 13: 403–414. [13]

Weiner, N. and Rabadjija, M. 1968. The effect of nerve stimulation on the synthesis and metabolism of norepinephrine from cat spleen during sympathetic nerve stimulation. *J. Pharmacol. Exp. Therap.* 160: 61–71. [9]

Weinrich, D. 1970. Ionic mechanisms of post-tetanic potentiation at the neuromuscular junction of the frog. *J. Physiol.* 212: 431–446. [7]

Weisblat, D. A. 1981. Development of the nervous system. In K. J. Muller, J. G. Nicholls and G. S. Stent (eds.). *Neurobiology of the Leech.* Cold Spring Harbor Laboratory, Cold Spring Harbor, NY, pp. 173–195. [11, 13]

Weisblat, D. A. and Shankland, M. 1985. Cell lineage and segmentation in the leech. *Philos. Trans. R. Soc. Lond. B* 312: 39–56. [13]

Weisblat, D. A., Kim, S.-Y. and Stent, G. S. 1984. Embryonic origins of cells in the leech *Heliobdella triserialis*. *Dev. Biol.* 104: 65–85. [6]

Weiss, E. R., Kelleher, D. J., Woon, C. W., Soparkar, S., Osawa, S., Heasley, L. E. and Johnson, G. L. 1988. Receptor activation of G proteins. *FASEB J.* 2: 2841–2848. [2]

Weiss, P. 1936. Selectivity controlling the central-peripheral relations in the nervous system. *Biol. Rev.* 11: 494–531. [11]

Weiss, P. and Hiscoe, H. B. 1948. Experiments on the mechanism of nerve growth. *J. Exp. Zool.* 107: 315–395. [9]

Wekerle, H., Linington, C., Lassmann, H. and Meyermann, R. 1986. Cellular immune reactivity within the CNS. *Trends Neurosci.* 9: 271–277. [6]

Wekerle, H., Sun, D., Oropeza-Wekerle, R. L. and Meyermann, R. 1987. Immune reactivity in the nervous system: Modulation of T-lymphocyte activation by glial cells. *J. Exp. Biol.* 132: 43–57. [6]

Welcher, A. A., Suter, U., De Leon, M., Bitler, C. M. and Shooter, E. M. 1991. Molecular approaches to nerve regeneration. *Philos. Trans. R. Soc. Lond. B* 331: 295–301. [11]

Welker, C., 1976. Microelectrode delineation of fine grain somatotopic organization of SM1 cerebral neocortex in albino rat. *J. Comp. Neurol.* 166: 173–190. [14]

Welker, E. and Van der Loos, H. 1986. Quantitative correlation between barrel-field size and the sensory innervation of the whiskerpad: A comparative study in six strains of mice bred for different patterns of mystacial vibrissae. *J. Neurosci.* 6: 3355–3373. [14, 18]

Werner, R., Miller, T., Azarnia, R. and Dahl, G. 1985. Translation and expression of cell–cell channel mRNA in *Xenopus* oocytes. *J. Memb. Biol.* 87: 253–268. [5]

Wernig, A. 1972. Changes in statistical parameters during facilitation at the crayfish neuromuscular junction. *J. Physiol.* 226: 751–759. [7]

Werz, M. A. and Macdonald, R. L. 1983. Opioid peptides selective for mu- and delta-opiate receptors reduce calcium-dependent action potential duration by increasing potassium conductance. *Neurosci. Lett.* 42: 173–178 [10]

Westerfield, M., Liu, D. W., Kimmel, C. B. and Walker, C. 1990. Pathfinding and synapse formation in a zebrafish mutant lacking functional acetylcholine receptors. *Neuron* 4: 867–874. [11]

Weston, J. 1970. The migration and differentiation of neural crest cells. *Adv. Morphogen.* 8: 41–114. [11]

Wickelgren, W. O., Leonard, J. P., Grimes, M. J. and Clark, R. D. 1985. Ultrastructural correlates of transmitter release in presynaptic areas of lamprey reticulospinal axons. *J. Neurosci.* 5: 1188–1201. [9]

Wiesel, T. N. 1982. The postnatal development of the visual cortex and the influence of environment. *Nature* 299: 583–591. [18]

Wiesel, T. N. and Hubel, D. H. 1963a. Effects of visual deprivation on morphology and physiology of cells in the cat's lateral geniculate body. *J. Neurophysiol.* 26: 978–993. [18]

Wiesel, T. N. and Hubel, D. H. 1963b. Single-cell responses in striate cortex of kittens deprived of vision in one eye. *J. Neurophysiol.* 26:1003–1017. [18]

Wiesel, T. N. and Hubel, D. H. 1965a. Comparison of the effects of unilateral and bilateral eye closure on cortical unit responses in kittens. *J. Neurophysiol.* 28: 1029–1040. [18]

Wiesel, T. N. and Hubel, D. H. 1965b. Extent of recovery from the effects of visual deprivation in kittens. *J. Neurophysiol.* 28: 1060–1072. [18]

Wiesel, T. N. and Hubel, D. H. 1966. Spatial and chromatic interactions in the lateral geniculate body of the rhesus monkey. *J. Neurophysiol.* 29: 1115–1156. [17]

Wiesel, T. N. and Hubel, D. H. 1974. Ordered arrangement of orientation columns in monkeys lacking visual experience. *J. Comp. Neurol.* 158: 307–318. [18]

Wigston, D. J. and Sanes, J. R. 1985. Selective reinnervation of intercostal muscles transplanted from different segmental levels to a common site. *J. Neurosci.* 5: 1208–1221. [12]

Wilkinson, D. G. and Krumlauf, R. 1990. Molecular approaches to the segmentation of the hindbrain. *Trends Neurosci.* 13: 335–339. [11]

Willard, A. 1981. Effects of serotonin on the generation of the motor program for swimming by the medicinal leech. *J. Neurosci.* 1: 936–944. [13]

Williams, J. H., Errington, M. L., Lynch, M. A. and Bliss, T. V. P. 1989. Arachidonic acid induces a long-term activity-dependent enhancement of synaptic transmission in the hippocampus. *Nature* 341: 739–742. [10]

Williams, P. L. and Warwick, R. 1975. *Functional Neuroanatomy of Man.* Saunders, Philadelphia. [1]

Wise, D. S., Schoenborn, B. P. and Karlin, A. 1981. Structure of acetylcholine receptor dimer determined by neutron scattering and electron microscopy. *J. Biol. Chem.* 256: 4124–4126. [2]

Witzemann, V., Brenner, H.-R. and Sakmann, B. 1991. Neural factors regulate AChR subunit mRNAs at rat neuromuscular synapses. *J. Cell Biol.* 114: 125–141. [12]

Wolfe, L. S. 1989. Eicosanoids. *In* G. Siegel, B. Agranoff, R. W. Albers and P. Molinoff (eds.). *Basic Neurochemistry.* Raven Press, New York, pp. 399–414. [8]

Wong, L. A. and Gallagher J. P. 1991. Pharmacology of nicotinic receptor-mediated inhibition in rat dorsolateral septal neurones. *J. Physiol.* 436: 325–346. [7]

Wong-Riley, M. T. 1979. Changes in the visual system of monocularly sutured or enucleated cats demonstrable with cytochrome oxidase histochemistry. *Brain Res.* 171: 11–28. [17]

Wong-Riley, M. T. 1989. Cytochrome oxidase: An endogenous metabolic marker for neuronal activity. *Trends Neurosci.* 12: 94–101. [17]

Wood, P. M., Schachner, M. and Bunge, R. P. 1990. Inhibition of Schwann cell myelination in vitro by antibody to the L1 adhesion molecule. *J. Neurosci.* 10: 3635–3645. [11]

Wood, W. B. (ed.) 1988. *The Nematode Caenorhabditis elegans.* Cold Spring Harbor Laboratory, Cold Spring Harbor, NY. [13]

Woolsey, T. A. and Van der Loos, H. 1970. The structural organization of layer IV in the somatosensory region (SI) of mouse cerebral cortex: The description of a cortical field composed of discrete cytoarchitectonic units. *Brain Res.* 17: 205–242. [14]

Yahr, M. D. and Bergmann, K. J. 1987. *Parkinson's Disease.* (*Advances in Neurology* Vol. 45). Raven Press, New York. [10]

Yamamori, T., Fukada, K., Aebersold, R., Korsching, S., Fann, M. J. and Patterson, P. H. 1989. The cholinergic neuronal differentiation factor from heart cells is identical to leukemia inhibitory factor. *Science* 246: 1412–1416. [11]

Yamashita, M. and Wässle, H. 1991. Responses of rod bipolar cells isolated from the rat retina to the glutamate agonist 2-amino-4-phosphonobutyric acid (APB). *J. Neurosci.* 11: 2372–2382. [16]

Yancey, S. B., John, S. A., Lal, R., Austin, B. J., and Revel, J.-P. 1989. The 43-kD polypeptide of heart gap junctions: Immunolocalization, topology, and functional domains. *J. Cell Biol.* 108: 2241–2254. [5]

Yawo, H. 1990. Voltage-activated calcium currents in presynaptic nerve terminals of the chicken ciliary ganglion. *J. Physiol.* 428: 199–213. [7]

Yau, K. W. 1976. Receptive fields, geometry and conduction block of sensory neurones in the central nervous system of the leech. *J. Physiol.* 263: 513–538. [5, 13]

Yau, K. W. and Baylor, D. A. 1989. Cyclic GMP-activated conductance of retinal photoreceptor cells. *Annu. Rev. Neurosci.* 12: 289–327. [16]

Yau, K. W. and Nakatani, K. 1985. Light-suppressible, cyclic GMP-sensitive conductance in the plasma membrane of the truncated rod outer segment. *Nature* 317: 252–255. [16]

Yellen, G. 1982. Single Ca^+-activated nonselective cation channels in neuroblastoma. *Nature* 296: 357–359. [3]

Young, J. Z. 1936. The giant nerve fibres and epistellar body of cephalopods. *Q. J. Microsc. Sci.* 78: 367–386. [3]

Young, R. R. and Delwaide, P. J. (eds.). 1992. *The Principles and Practice of Restorative Neurosurgery.* Butterworth-Heinemann, Guilford, England. [19]

Yurek, D. M. and Sladek, J. R., Jr. 1990. Dopamine cell replacement: Parkinson's disease. *Annu. Rev. Neurosci.* 13: 415–440. [10]

Zagotta, W. N., Hoshi, T. and Aldrich, R. W. 1990. Restoration of inactivation in mutants of *Shaker* potassium channels by a peptide derived from ShB. *Science* 250: 568–571. [4]

Zecevic, D., Wu, J. Y., Cohen, L. B., London, J. A., Hopp, H. P. and Falk, C. X. 1989. Hundreds of neurons in the *Aplysia* abdominal ganglion are active during the gill-withdrawal reflex. *J. Neurosci.* 9: 3681–3689. [13]

Zeki, S. M. 1978. The third visual complex of rhesus monkey prestriate cortex. *J. Physiol.* 277: 245–272. [17]

Zeki, S. 1990. Colour vision and functional specialisation in the visual cortex. *Disc. in Neurosci.* 6: 1–64. [17]

Zeki, S. and Shipp, S. 1988. The functional logic of cortical connections. *Nature* 335: 311–317. [17]

Zeki, S., Watson, J. D., Lueck, C. J., Friston, K. J., Kennard, C. and Frackowiak, R. S. 1991. A direct demonstration of functional specialization in human visual cortex. *J. Neurosci.* 11: 641–649. [17]

Zetterstrom, T. and Fillenz, M. 1990. Adenosine agonists can both inhibit and enhance in vivo striatal dopamine release. *Eur. J. Pharmacol.* 180: 137–143. [9]

Zigmond, R. E., Schwarzchild, M. A. and Rittenhouse, A. R. 1989. Acute regulation of tyrosine hydroxylase by nerve activity and by neurotransmitters via phosphorylation. *Annu. Rev. Neurosci.* 12: 415–461. [9]

Zimmerman, H. and Denston, C. R. 1977a. Recycling of synaptic vesicles in the cholinergic synapses of the *Torpedo* electric organ during induced transmitter release. *Neuroscience* 2: 695–714. [9]

Zimmerman, H. and Denston, C. R. 1977b. Separation of synaptic vesicles of different functional states from the cholinergic synapses of the *Torpedo* electric organ. *Neuroscience* 2: 715–730. [9]

Zipser, B. and McKay, R. 1981. Monoclonal antibodies distinguish identifiable neurones in the leech. *Nature* 289: 549–554. [13]

INDEX

ABOUT THE BOOK

This book is set in Palatino, a face designed by the contemporary German typographer Hermann Zapf. Inspired by the typography of the Italian Renaissance, Palatino letters derive their elegance from the natural motion of the edged pen. Palatino is a highly readable face especially suited for extended use in books.

Many of the drawings for the book were prepared by Laszlo Meszoly, who worked in collaboration with the authors. J/B Woolsey Associates of Philadelphia rendered additional artwork for this edition. Gretchen Becker was copyeditor.

The type was set by DEKR Corporation, and the book was manufactured by R. R. Donnelley & Sons. Joseph J. Vesely supervised production.